Radiative Transfer in the Atmosphere and Ocean

Radiative transfer is important to a range of disciplines, from the study of the greenhouse warming to stellar atmospheres and ocean optics. This text provides a foundation of the theoretical and practical aspects of radiative transfer for senior undergraduate and graduate students of atmospheric, oceanic, and environmental sciences. With an emphasis on formulation, judicial approximations and numerical solutions of the radiative transfer equation, *Radiative Transfer in the Atmosphere and Ocean* fills a gap between descriptive texts covering the physical processes and the practical numerical approaches needed in research. Designed to convey physical insight into the transfer process, it can also be used as a self-contained manual for practitioners who require accurate modeling of the effects of solar and infrared radiation on natural systems.

Radiative Transfer in the Atmosphere and Ocean includes a unified treatment of radiation within both the atmosphere and ocean, boundary properties (such as reflection and absorptance of solid surfaces), heuristic models (Lorentz atom, two-level atom, rotating vibrator), and extensive use of two-stream and approximate methods. State-of-the-art computational methods are illustrated by a thorough treatment of the discrete-ordinates technique and the correlated-k band absorption method. The former method provides a theoretical foundation for users of the well-known public domain computer code DISORT. Exercises and problem sets provide practice in both formulation and solution techniques. Applications to the subjects of solar UV penetration of the atmosphere/ocean system, and the greenhouse effect serve to illustrate the use of such techniques in modern research.

This self-contained, systematic treatment will prepare the student in solving radiative transfer problems across a broad range of subjects.

Gary E. Thomas is a Professor in the Department of Astrophysics and Planetary Sciences and the Department of Atmospheric and Oceanic Sciences at the University of Colorado, Boulder.

Knut Stamnes is a Professor in the Geophysical Institute and the Department of Physics at the University of Alaska, Fairbanks.

Cambridge Atmospheric and Space Science Series

Editors: A. J. Dessler, J. T. Houghton, and M. J. Rycroft

This series of upper-level texts and research monographs covers the physics and chemistry of different regions of the Earth's atmosphere, from the troposphere and stratosphere, up through the ionosphere and magnetosphere, and out to the interplanetary medium.

Cambridge Atmospheric and Space Science Series

TITLES IN PRINT IN THIS SERIES

M. H. Rees
Physics and Chemistry of the Upper Atmosphere

Roger Daley
Atmospheric Data Analysis

Ya. L. Al'pert
Space Plasma, Volumes 1 and 2

J. R. Garratt
The Atmospheric Boundary Layer

J. K. Hargreaves
The Solar–Terrestrial Environment

Sergei Sazhin
Whistler-Mode Waves in a Hot Plasma

S. Peter Gary
Theory of Space Plasma Microinstabilities

Martin Walt
Introduction to Geomagnetically Trapped Radiation

Tamas I. Gombosi
Gaskinetic Theory

Boris A. Kagan
Ocean–Atmosphere Interaction and Climate Modelling

Ian N. James
Introduction to Circulating Atmospheres

J. C. King and J. Turner
Antarctic Meteorology and Climatology

J. F. Lemaire and K. I. Gringauz
The Earth's Plasmasphere

Daniel Hastings and Henry Garrett
Spacecraft–Environment Interactions

Thomas E. Cravens
Physics of Solar System Plasmas

John Green
Atmospheric Dynamics

Tamas I. Gombosi
Physics of the Space Environment

Gary E. Thomas and Knut Stamnes
Radiative Transfer in the Atmosphere and Ocean

Radiative Transfer in the Atmosphere and Ocean

Gary E. Thomas
University of Colorado, Boulder

and

Knut Stamnes
University of Alaska, Fairbanks

PUBLISHED BY THE PRESS SYNDICATE OF THE UNIVERSITY OF CAMBRIDGE
The Pitt Building, Trumpington Street, Cambridge, United Kingdom

CAMBRIDGE UNIVERSITY PRESS
The Edinburgh Building, Cambridge CB2 2RU, UK http://www.cup.cam.ac.uk
40 West 20th Street, New York, NY 10011-4211, USA http://www.cup.org
10 Stamford Road, Oakleigh, Melbourne 3166, Australia

First published 1999

Printed in the United States of America

Typeset in Times $10\frac{1}{4}/13\frac{1}{2}$ pt. and Joanna in LaTeX 2_ε [TB]

A catalog record for this book is available from the British Library

Library of Congress Cataloging in Publication Data

Thomas, Gary E.

Radiative transfer in the atmosphere and ocean / Gary E. Thomas,

Knut Stamnes.

p. cm.

ISBN 0-521-40124-0 (hc)

1. Radiative transfer. 2. Atmospheric radiation.

3. Oceanography. 4. Environmental sciences. I. Stamnes, Knut.

II. Title.

QC175.25.R3T48 1999

551.46′01 – dc20 95-47502

CIP

ISBN 0 521 40124 0 hardback

Dedication by G. E. Thomas

My Boomerang Won't Come Back

Before the book was completed, my beloved son, Curt, age 32, was killed in an auto accident. Curt was passionately dedicated to teaching and to the pursuit of truth wherever it led him. Despite my grief, I felt that I must continue to carry out in my own way his ardent desire to share his wisdom and knowledge. I truly believe that Curt's spirit has sustained me to the end of an arduous, but ultimately rewarding, process. This book is dedicated to Curt and to teachers like him everywhere.

Dedication by K. Stamnes

I dedicate this book to the memory of my father, Alfred, whose keen interest in basic education encouraged me to pursue an academic path in spite of my early desire to follow in his footsteps and become a commercial fisherman. My father's honesty and integrity, intellectual courage, and independence have been a constant source of inspiration throughout my early adventures in commercial fishing and subsequent endeavors in research and education.

Contents

Illustrations

Preface

The subject of radiative transfer has matured to the point of being a well-developed tool, which has been adapted over the years to a host of disciplines, ranging from atmospheric and ocean optics to stellar atmospheres. It has also become a part of many engineering curricula, since its industrial applications (particularly for the infrared) are wide ranging. As a result of this broadness, developments of radiative transfer theory in many separate fields have grown up in isolation. In comparing the literature in these various disciplines, one finds a bewildering multiplicity of approaches, which often obscures the fact that the same fundamental core is present, namely the radiative transfer equation. The same can be said for the two fields of atmospheric radiation and ocean optics. These have evolved along largely separate paths, with their own sets of jargon and nomenclature. However, in view of the fact that there is a growing need for interdisciplinary research involving the coupled atmosphere–ocean system, we feel that the time has come to write a textbook that acknowledges the following basic fact: *The radiation that enters, or is emitted by, the ocean encounters the same basic processes of scattering and absorption as those involved in atmospheric radiation.* There are no *inherently different* optical properties between atmospheric and aqueous media. Because the two media share a common interface that readily passes radiative energy, there is even more need for a unified approach.

Coming from an atmospheric background, we must confess to having built-in biases toward the nomenclature and jargon of atmospheric radiation. To counteract this, we inform the reader of its oceanic counterpart whenever an atmospheric radiation variable is defined. We also present a table of nomenclature comparing the two usages. A glossary of terms should also be helpful in bridging the gap between fields.

This book is an outgrowth of two graduate courses that have been taught more or less regularly by G. E. Thomas at the University of Colorado for the past thirty years. Its content has also greatly benefitted from a course taught by K. Stamnes at the University of Alaska for the past fifteen years. A more or less common body of notes was used in the teaching of those courses, and a common philosophy has evolved. This philosophy is based on the notion that students benefit most from a single-minded systematic approach, that of derivation, formulation, and solution of the radiative transfer equation. Along the way, numerous examples and problems are worked out, and physical interpretation is constantly stressed.

The book starts with an introduction of the basic concepts of absorption, emission, radiative equilibrium, radiative forcing, and feedbacks. A brief review is provided of atmospheric and oceanic vertical structures and their basic optical properties. In Chapter 2 we define the basic state variables of the theory, such as intensity (radiance) and flux (irradiance), as well as prove several useful theorems. Here we introduce the fundamental law of extinction and a general form of the differential equation of radiative transfer. In Chapter 3 we introduce the first of the two types of light–matter interactions, that of *scattering*. A discussion of scattering in its various guises (Rayleigh, resonance, Mie–Debye, etc.) is followed by a simple-harmonic oscillator treatment of the line-broadening process, from which follows the important Lorentz line-broadening formula. Then Chapter 4 considers the second basic type of interaction, that of *absorption*. The topics of thermal emission, the Planck Law, and local thermodynamic equilibrium (LTE) are illustrated through the use of the two-level atom concept. This approach also illuminates the importance of collisions (and the radiation field itself) in maintaining LTE. Chapter 5 concerns itself with the radiative properties of solid, aqueous, and atmospheric media, particularly with respect to their boundary surfaces. The solution of the radiative transfer equation in the limit of zero scattering is then generalized to obtaining a formal solution in the general case (including both scattering and absorption). Half-range quantities in a slab medium are defined. Chapter 6 sets the stage for solving the radiative transfer equation in emphasizing the *formulation* of problems as an essential step in their ultimate solution. Chapter 7 is almost entirely devoted to the two-stream (or two-flow analysis in ocean optics) approximate solution of various prototype problems. In our opinion, this is the simplest and most intuitive solution technique available. Even though its pedagogic value often exceeds its computational accuracy, there is usually something in this method to learn by even the most experienced of workers. Chapter 8 describes a solution technique that is a natural outgrowth of the two-stream method wherein the number of streams is allowed to increase until the solution is as accurate as desired.

The last three chapters deal with various applications of the theory. Chapter 9 contains a detailed accounting of shortwave transfer of solar radiation through various types of media (clear air, clouds, water, and snow), with emphasis on UV transmission as it is affected by atmospheric ozone. Transfer of longwave transfer of planetary infrared radiation is the subject of Chapters 10 and 11. Finally, Chapter 12 is concerned with radiation and climate, in particular the greenhouse problem. Concepts employed by general circulation climate models are illustrated by the use of simple two-stream ideas, including the joint effects of radiation and convection; radiative forcing from greenhouse gases, clouds, and aerosols; climate gain and feedback; the runaway and "anti-greenhouse effect"; etc.

The book contains five appendices (A–E), but fourteen additional appendices (F–S) are referred to in the text. These (including other supplementary material) are available through the following web page: `http://lasp.colorado.edu/~rttext`

This book assumes that the reader is familiar with the fundamentals of electromagnetic theory, optics, thermodynamics, and the kinetic theory of gases, with some basic knowledge of the quantum structure of atoms and molecules. Familiarity with the fundamentals of integral calculus and ordinary and partial differential equations is required. An elementary knowledge of linear algebra and numerical analysis will make it easier to follow Chapter 8. The book is most suitable as a text for advanced seniors and first-year graduate students at U.S. universities. It is intended as a first text for students who need not only a basic theoretical knowledge but an exposure to modern methods of solution of atmospheric and oceanic radiation problems. It could be preceded by, or supplemented with, a course at the same level describing the physical processes of light–matter interactions. It could be followed with courses on more advanced topics, such as remote sensing, inversion techniques, climate modeling of the air–sea system, or computer-intensive methods involving problems coupling radiation with dynamics, cloud microphysics, photochemistry, or photobiology.

There are an enormous number of research papers and books on this subject already. So why do we add another book? Our motivation was to write a tutorial that had the single theme of the *practical* mathematical solution of the radiative transfer equation. Although this would seem to be an obviously desirable quality, we felt that its uniform application was lacking in all the texts we used over the years. However, we would be remiss if we did not acknowledge an enormous debt to the books by the late S. Chandrasekhar (*Radiative Transfer*, 1949) and R. M. Goody (*Atmospheric Radiation, I, Theoretical Basis*, 1964). These two masterworks have inspired us throughout our academic careers, and they still contain much to teach us even today. Of course, there are many other wonderful sources of information. We have acknowledged only a few original contributors to the ideas and concepts we so liberally have borrowed from scores of others. We include a few key references at the end of each chapter, from which more detail can be obtained. The reader must dip into the vast literature to acquire a historical view of who actually thought of the concept first, and in what context.

Acknowledgments

Authors traditionally leave the acknowledgments of the contributions of their wives and family to the last. We hereby break with this tradition by thanking them upfront. We express our most heartfelt thanks to Susan and Jennifer Thomas, Anja Moen, and Kaja and Snorre Stamnes for their loving support and encouragement during the past half-dozen years. They cheerfully relinquished precious time we could have spent together, to an unspeaking, impersonal, yet demanding competitor.

We are also indebted to the scores of graduate students who have labored mightily over earlier versions of our class notes, both at the University of Colorado and the University of Alaska. We are indebted to D. Anderson, P. Flatau, J. London, J. J. Olivero, J. J. Stamnes, M. Niska, B. L. Lindner, and S. Warren for taking the time to read various chapters of the manuscript and offering many helpful suggestions for improvements. Although not our only editor, Catherine Flack was our most enduring and supportive. She never flagged in her enthusiasm for the book, even in the darkest period when it appeared least likely to ever see the light of day. During the time of the most intensive writing, G. E. Thomas was supported by the Atmospheric Research Section of the National Science Foundation. K. Stamnes was supported by the Quantitative Links and ARM Program of the Department of Energy, the Atmospheric Chemistry Program of NASA, and the Office of Polar Programs and the Atmospheric Chemistry Section of the National Science Foundation. We would also like to acknowledge the considerable assistance of Dr. Arve Kylling who proofread a large part of the book and worked out a large number of the exercises. The assistance of Eric Olson and Jason Parsons on the typesetting of the book in LaTeX is also greatly appreciated. Sharon Kessey provided much help with the improvement of the figures that were scanned electronically from a variety of sources. Finally, we acknowledge the substantial assistance by Hans Eide in improving the quality of the figures made in IDL (Interactive Data Language), in digitizing several of the figures, and for managing the assemblage of the text and figures of the various chapters into an organized LaTeX typescript ("the book") on a workstation in a UNIX environment.

For those interested in the technical aspects, the manuscript was typeset first in TeX and later in LaTeX. Many of the figures were produced by IDL, Aldus SuperPaint, and Jandel SigmaPlot graphics software on Apple MacIntosh desktop computers.

Chapter 1

Basic Properties of Radiation, Atmospheres, and Oceans

1.1 Introduction

This chapter presents a brief overview of the spectra of the shortwave solar and longwave terrestrial radiation fields and the basic structure of atmospheres and oceans. Some general properties of the emission spectra of the Sun and the Earth are described. Their broad features are shown to be understandable from a few basic radiative transfer principles. We introduce the four basic types of matter that interact with radiation: gaseous matter, aqueous matter, particles, and surfaces. The stratified vertical structure of the bulk properties of an atmosphere or ocean is shown to be a consequence of hydrostatic balance. The vertical temperature structure of the Earth's atmosphere is shown to result mainly from radiative processes. Optical paths in stratified media are described for a general line-of-sight direction. Radiative equilibrium, the greenhouse effect, feedbacks, and radiative forcing are introduced as examples of concepts to be dealt with in greater detail in Chapter 12.

The ocean's vertical temperature structure, and its variations with season, are discussed as resulting from solar heating, radiative cooling, latent heat exchange, and vertical mixing of water masses of different temperature and salinity. Its optical properties are briefly described, along with ocean color. Section 1.7 prepares the reader for the notation and units used consistently throughout the book.

1.2 Parts of the Spectrum

In Table 1.1, we summarize the nomenclature attached to the various parts of the visible and infrared spectrum. The spectral variable is the *wavelength* λ. Here $\lambda = c/\nu$, where

Table 1.1. *Subregions of the spectrum.*

Subregion	Range	Solar variability	Comments
X rays	$\lambda < 10$ nm	10–100%	Photoionizes all thermosphere species.
Extreme UV	$10 < \lambda < 100$ nm	50%	Photoionizes O_2 and N_2. Photodissociates O_2.
Far UV	$100 < \lambda < 200$ nm	7–80%	Dissociates O_2. Discrete electronic excitation of atomic resonance lines.
Middle UV, or UV-C	$200 < \lambda < 280$ nm	1–2%	Dissociates O_3 in intense Hartley bands. Potentially lethal to biosphere.
UV-B	$280 < \lambda < 320$ nm	<1%	Some radiation reaches surface, depending on O_3 optical depth. Responsible for skin erythema.
UV-A	$320 < \lambda < 400$ nm	<1%	Reaches surface. Benign to humans. Scattered by clouds, aerosols, and molecules.
Visible, or PAR[a]	$400 < \lambda < 700$ nm	≤0.1%	Absorbed by ocean, land. Scattered by clouds, aerosols, and molecules. Primary energy source for biosphere and climate system.
Near IR	$0.7 < \lambda < 3.5$ μm		Absorbed by O_2, H_2O, CO_2 in discrete vibrational bands.
Thermal IR	$3.5 < \lambda < 100$ μm		Emitted and absorbed by surfaces and IR-active gases.

Note: [a]PAR stands for "photosynthetically active radiation."

c is the speed of light and ν is the *frequency* ([s^{-1}] or [Hz]). In the IR λ is usually expressed in *micrometers* (or more commonly microns, where 1 μm $= 10^{-6}$ m). In the UV and visible spectral range, λ is expressed in *nanometers* (1 nm $= 10^{-9}$ m). A wavelength unit widely used in astrophysics and laboratory spectroscopy is the *Ångström* (1 Å $= 10^{-10}$ m). For completeness we list both X rays and the shorter-wavelength UV regions, even though they are not discussed in this book. The third column lists the known solar variability (in percent), defined as the maximum minus minimum divided by the minimum. We also provide brief comments on how radiation in each spectral subregion interacts with the Earth's atmosphere. A common usage is to denote the solar part of the spectrum as *shortwave* radiation and the thermal infrared (IR) as *longwave* radiation. The latter is sometimes referred to as *terrestrial* radiation.

1.2.1 Extraterrestrial Solar Flux

In this section we consider some elementary aspects of solar radiation and the origin of its deviations from blackbody behavior. We will assume the reader is familiar with the concept of *absorption opacity*, or *optical depth*, $\tau(\nu)$ at frequency ν. The basic ideas are reviewed in Appendix G and covered more thoroughly in Chapter 2.

In Fig. 1.1 we show the measured *spectral flux*, or *irradiance*, of the Sun's radiative energy at a distance of one astronomical unit $r_\oplus$ ($r_\oplus = 1.5 \times 10^6$ km).[1] Integrated over all frequencies, this quantity is called the *solar constant*, S [$\mathrm{W \cdot m^{-2}}$]. These data were taken by a spectrometer on board an Earth-orbiting satellite, beyond the influences of the atmosphere.[2] The solar constant is not actually a constant but is slightly variable. For this reason, the modern term is the *total solar irradiance*, whose value[3] is about 1368 $\mathrm{W \cdot m^{-2}}$. The total solar irradiance S represents the total instantaneous radiant energy falling normally on a unit surface located at the distance $r_\oplus$ from the Sun. It is the basic forcing of the Earth's heat engine, and indeed for all planetary bodies that derive their energy primarily from the Sun. The quantity $S(r_\oplus^2/r^2)$ is the total instantaneous radiant energy falling normally on a unit surface at the solar distance r.

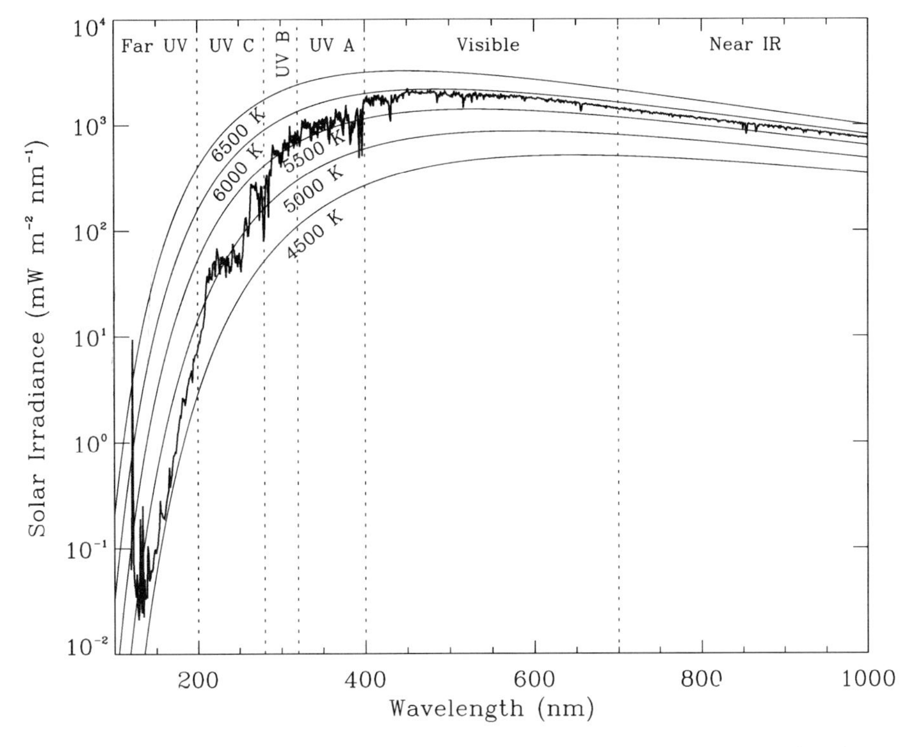

Figure 1.1 Extraterrestrial solar flux, or irradiance, measured by a spectrometer on board an Earth-orbiting satellite. The UV spectrum ($119 < \lambda < 420$ nm) was measured by the SOLSTICE instrument on the *UARS* satellite (modified from a diagram provided by G. J. Rottmann, private communication, 1995). The vertical lines divide the various spectral subranges defined in Table 1.1. The smooth curves are calculated blackbody spectra for a number of emission temperatures.

Also shown in Fig. 1.1 are spectra of an ideal blackbody at several temperatures. Requiring that the total energy emitted be the same as a blackbody, one finds that the Sun's effective temperature is 5,778 K. If the radiating layers of the Sun had a uniform temperature at all depths, its spectrum would indeed match one of the theoretical blackbody curves exactly. The interesting deviations seen in the solar spectrum can be said to be a result of emission from a *nonisothermal atmosphere*. Radiative transfer lies at the heart of the explanation for this behavior.

We can explain the visible solar spectrum qualitatively by considering two characteristics of atmospheres – 1. their absorption opacity $\tau(\nu)$ depends upon frequency and 2. their temperature varies with atmospheric depth – and one basic rule – that a radiating body emits its energy to space most efficiently at wavelengths where the opacity is approximately unity. This rule is explained in terms of the competing effects of absorption and emission. In spectral regions where the atmosphere is transparent ($\tau(\nu) \ll 1$), it neither emits nor absorbs efficiently. In contrast, where it is opaque ($\tau(\nu) \gg 1$), its radiative energy is prevented from exiting the medium, that is, it is reabsorbed by surrounding regions. At $\tau(\nu) \approx 1$, a balance is struck between these opposing influences.

At visible wavelengths, the Sun's opacity is unity deep within the solar atmosphere in the relatively cool *photosphere*, where the temperature is $\cong$5,780 K. Regions as cool as 4,500 K are apparent at 140–180 nm (see Fig. 1.1). At shorter wavelengths the opacity increases, thereby raising the effective emission height into the higher-temperature *chromosphere*. The solar spectrum can be thought of as a "map" of the vertical temperature structure of the Sun. The map can be read provided one has knowledge of the dependence of opacity of the solar atmosphere on wavelength.

1.2.2 Terrestrial Infrared Flux

An understanding of radiative transfer is also essential for understanding the energy output of the Earth, defined to be the spectral region $\lambda > 3.5$ μm. Figure 1.2 shows the IR emission spectrum measured from a down-looking orbiting spacecraft, taken at three different geographic locations.[4] Also shown are blackbody curves for typical terrestrial temperatures. The spectral variable in this case is *wavenumber* $\tilde{\nu} = 1/\lambda$, commonly expressed in units of [cm^{-1}]. Again, as for the solar spectrum, the deviations are attributed to the nonisothermal character of the Earth's atmosphere. The spectral regions of minimum emission arise from the upper cold regions of the Earth's troposphere where the opacity of the overlying regions is $\sim$1. Those of highest emission originate from the warm surface in transparent spectral regions ("windows"), with the exception of the Antarctic spectrum, where the surface is actually colder than the overlying atmosphere (see Fig. 1.2). In this somewhat anomalous situation, the lower-opacity region is one of higher radiative emission because of the greater rate of emission of the warm air. Again, the deviations from blackbody behavior can be understood qualitatively in terms of the temperature structure of the Earth's atmosphere and the variation with frequency of the IR absorption opacity.

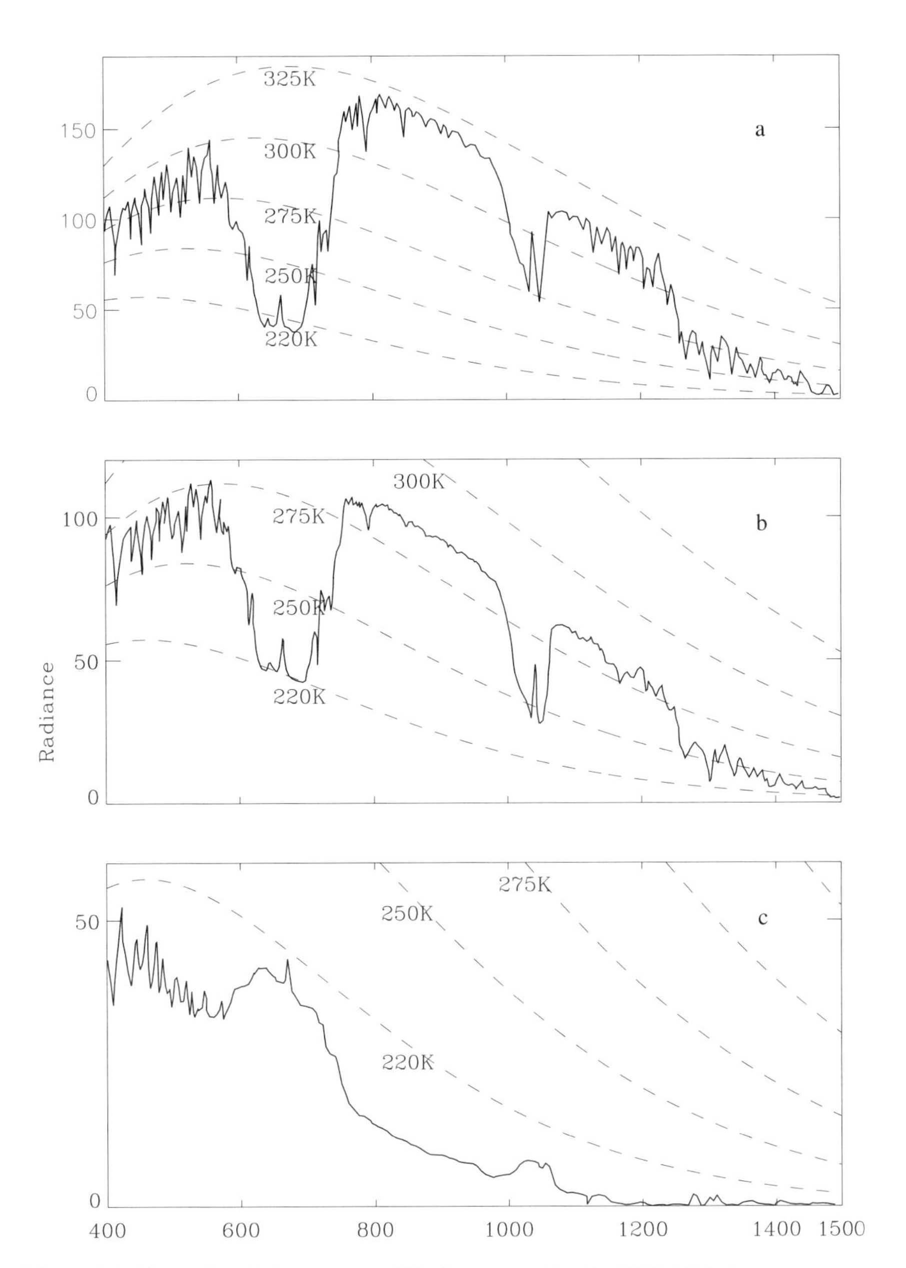

Figure 1.2 Thermal emission spectra of Earth measured by the IRIS Michelson interferometer instrument on the *Nimbus 4* spacecraft (see Endnote 4). Shown also are the radiances of blackbodies at several temperatures. (a) Sahara region; (b) Mediterranean; (c) Antarctic.

1.3 Radiative Interaction with Planetary Media

1.3.1 Feedback Processes

The properties of planetary media (chemical and dynamical) may themselves be affected by radiation, on all spatial scales. These changes may then further influence the way the media interact with radiation, etc. On the macrophysical (much greater than molecular) scale, we will mention two examples: 1. During daytime, solar radiation heats the Earth's surface and atmosphere. Often there results a fluid instability, which causes air to be set into convective motion, some air parcels moving upward, others downward. Upward air motion causes adiabatic cooling and, if the atmosphere is sufficiently moist, will lead to condensation and cloud formation. Clouds will alter the distribution of incoming sunlight and absorb and emit IR radiation, and thus affect the heating, etc. 2. A second example is that of ocean photosynthesis. The concentrations of light-absorbing phytoplankton determine the depth dependence of the radiation field, which itself governs the viability of such organisms.

If we had to concern ourselves with these "chicken-and-egg" problems of simultaneous mutual interactions of the medium and the radiation, this book would be very different and the subject much more difficult. Throughout, we will deal with the optical properties of planetary media as if they were specified *a priori*. This is a useful fiction for our purposes. Our neglect of these so-called feedback processes necessarily rules out a treatment of many interesting climate phenomena (cloud–radiation interactions, ocean photo-biologic feedback on climate change, etc.). Fortunately these feedback processes do not alter the fundamental description of the light–matter interaction. The speed of light is so high that the radiation field adjusts itself instantaneously to its environment. As a result radiative transfer is essentially a quasi-static phenomenon, and consequently its interaction with matter can usually be treated separately from other physical influences.

On the microphysical (molecular) scale, the presence of radiation can alter the basic optical properties of matter itself. Radiative heating leads to a redistribution of quantized states of excitation (for example the vibratory motion of molecules), which in turn alters the light interaction properties of the gas. In other words, the absorptive and emissive properties of a gas depend upon its temperature, which is itself affected by radiative heating. Again a fortunate circumstance usually allows us to decouple these two situations, so that the gas temperature may be considered to be an externally specified quantity, independent of the radiation field. This is contingent on the gas density being sufficiently high, so that *Kirchhoff's Law* is obeyed (§5.2.1). This condition is easily met for the lower portions of most planetary atmospheres and for the ocean.

1.3.2 Types of Matter that Affect Radiation

Pretending that they are independent of the radiation, we now focus on those aspects of oceans and atmospheres that are important in modifying the radiation field. For our purposes, there are four forms of matter that can affect radiation:

Gaseous matter: Under local thermodynamic equilibrium conditions (§5.2.1), the density ρ, temperature T, and chemical composition are normally all that is required to determine the optical properties. Gas pressure p should also be included in this list, although it is not independent of ρ and T. Gas pressure, through its collisional effects on the quantized excited states of the molecules, affects absorption of light by altering the line strengths, as well as the line positions in frequency and their spectral width (§3.3.3). The quantities ρ, T, and p are related to one another by an empirical "real-gas" formula, although it is almost always an adequate approximation to use the ideal gas law (see the following section).

Aqueous matter: As in gaseous media, density largely determines the optical properties of pure ocean water. Salinity, which is important for ocean dynamics, is unimportant for the optical properties. However, "pure sea water" hardly exists outside the laboratory. "Impurities" usually dominate the optical properties of natural bodies of water.

Particles: The atmospheric particulate population consists of suspended particles (aerosols) and condensed water (*hygrosols*). The latter is the generic term for water droplets and ice crystals, or combinations with dust. Airborne particles may be of biological origin or originate from pulverization of solid surfaces. Particles are frequently chemically or physically altered by the ambient medium, and these alterations can affect their optical properties. Particles with sizes comparable to the wavelength take on optical characteristics that can be quite different from their parent-solid bulk optical properties (§4.2). Oceanic particles consist of a large variety of dissolved and suspended organic and inorganic substances, such as the variously pigmented phytoplankton and organic yellow substances.[5]

Solid and ocean surfaces: The atmospheres of the terrestrial planets are all in contact with surfaces, which vary greatly in their visible-light reflectance and absorptance properties (§5.2). In many applications their strong continous absorption in the IR allows them to be treated as thermally emitting blackbodies, an enormous theoretical simplification. Knowledge of the visible reflectance of underlying land and ocean surfaces is necessary for calculating the diffuse radiation field emergent from the atmosphere. In addition, the reflectance of the ocean bottom in shallow seas has an important influence on the diffuse radiation field in the ocean and on the radiation leaving the ocean surface.

1.4 Vertical Structure of Planetary Atmospheres

It is useful to describe those general aspects of similarity and dissimilarity of oceans and atmospheres. First, they are similar in that they are both *fluids*, that is, they readily flow under the influences of gravity and pressure differences. Also, they both obey the basic equation of *hydrostatic equilibrium*. A fundamental difference is that atmospheres are highly *compressible*, whereas oceans are nearly *incompressible*. A quantitative difference arises from the fact that the average density of water (1×10^3 [$\text{kg} \cdot \text{m}^{-3}$]) is much higher than that of most planetary atmospheres. For the Earth's atmosphere on a clear day at sea level, a visible light photon can traverse unattenuated a horizontal path many hundreds of kilometers long. In the ocean, it penetrates at most a few hundred meters before being attenuated. Of course for sufficient depths in the atmospheres of Venus and of the giant outer planets, the atmospheric density can approach or even exceed that of water.

1.4.1 Hydrostatic and Ideal Gas Laws

In this section, we describe some important *bulk properties* of the atmosphere and ocean, described by its density, pressure, temperature, and index of refraction. As a result of gas being highly compressible, the *atmospheric density*, ρ [$\text{kg} \cdot \text{m}^{-3}$], the mass per unit volume, varies strongly with height, z. For both atmospheres and oceans in a state of rest, the pressure, p, must support the weight of the fluid above it. This is called a state of *hydrostatic equilibrium*. With increasing height in the atmosphere, the density decreases as the pressure decreases (*Boyle's Law*). With increasing depth in the ocean, this also holds true but the density change is slight.

Consider the atmospheric case. In differential form, the weight of the air (mass times the acceleration of gravity g) in a small volume element dV is gdM, where dM is the mass of the air inside the volume. Now $dM = \rho dV = \rho dAdz$, and the net force exerted by the surrounding gas on the parcel is $-dpdA$. The differential dp is the change in pressure over the small height change dz. The minus sign comes from the fact that the pressure at $z + dz$ is smaller than at z, and the upward buoyancy force must be positive. Equating the two forces, $-dpdA = g\rho dAdz$, we find upon cancellation of the dA term

$$dp = -g\rho dz. \tag{1.1}$$

For planetary atmospheres of moderate and low density, and specifically the Earth's atmosphere, the equation of state is closely approximated by the *ideal gas law*,

$$\rho = \frac{\bar{M}p}{RT} = \bar{M}n, \tag{1.2}$$

where $\bar{M}$ is the *mean molecular mass*, R is the *gas constant per mole*, and n is the total *concentration* of molecules (number of molecules per unit volume). More detailed descriptions of each of the above quantities are given below. Substituting Eq. 1.2 into Eq. 1.1, we find

$$\frac{dp}{p} = -\frac{dz}{H}, \quad \text{where } H \equiv \frac{RT}{\bar{M}g}. \tag{1.3}$$

H is called the *atmospheric scale height*.

Earth's atmosphere consists of a mixture of long-lived, permanent species, together with many other minor species molecules. The ith species contributes to the overall bulk density through its molecular mass $M(i)$ and its concentration n_i. Air contains a "standard" mixture of 78% nitrogen ($M(N_2) = 28$), 21% oxygen ($M(O_2) = 32$), and about 1% argon ($M(Ar) = 40$). The proportions of these constituents are essentially constant up to about 95 km. Above this height, the *homopause*, $\bar{M}$, decreases as the composition changes to lower-mass species, such as O, N, and He. This results from photodissociation of the heavier molecules, combined with gravitational separation tending to place the lighter elements at the highest levels.

Integrating Eq. 1.3 from $z' = z_o$ to $z' = z$, and using the ideal gas law, Eq. 1.2, we obtain

Three forms of the hydrostatic equation

$$p(z) = p(z_0) \exp\left[-\int_{z_0}^{z} dz'/H(z')\right], \tag{1.4a}$$

$$n(z) = n(z_0)\frac{T(z_0)}{T(z)} \exp\left[-\int_{z_0}^{z} dz'/H(z')\right], \tag{1.4b}$$

$$\rho(z) = \rho(z_0)\frac{T(z_0)}{T(z)} \exp\left[-\int_{z_0}^{z} dz'/H(z')\right]. \tag{1.4c}$$

It is clear that from a knowledge of surface pressure $p(z_0)$ and the variation of scale height $H(z)$ from z_0 to z, Eqs. 1.4 allow us to determine the bulk gas properties at any height z.

The various quantities and their units [ ] are defined below:

z, z_0: Height [m] above a reference level z_0. z_0 is usually the surface, $z_0 = 0$. The more common atmospheric unit is [km].

ρ: Gas density or mass per unit volume. $\rho = 1.293$ [kg · m^{-3}] at 0°C and 1 bar, which is the condition of standard temperature and pressure (STP).

p: Gas pressure [N · m^{-2}] at height z. p is the sum of the *partial pressures* $\sum p_i$. Pressure is expressed in bars, 1 [bar] = 101,325 [N · m^{-2}], or Pascals [Pa]. Another common unit is the *millibar* ([mbar]) or *hectoPascal*. 1 [mbar] = 1 [hPa] = 10^{-3} [bar] = 10^2 [Pa].

T: Atmospheric temperature in Kelvins [K]. In everyday use, degrees Celsius [C] is more frequent. $T(C) = 273.2 + T(K)$.

H: Atmospheric scale height [m], more commonly [km], equal to $RT/\bar{M}g \equiv R_aT/g$. Up to the homopause, $H \approx 29.3\ T$ (T in [K]) [m].

R, R_a: Molar gas constant, 8.3143×10^3 [J · K^{-1} · kmol^{-1}], and specific gas constant = 2.87×10^2 [J · K^{-1} · kg^{-1}], respectively. The latter applies to air below the homopause. One kilogram-mole [kmol] is the quantity of matter in a volume having a mass equal to its molecular weight. One kmol contains the number N_a (Avogadro's number) $= 6.022 \times 10^{26}$ molecules.

n: Total gas concentration [m^{-3}], equal to the sum of the individual gas concentrations $\sum n_i$.

$\bar{M}$: Mean molecular mass per kilogram-mole [$kg \cdot kmol^{-1}$]. For dry air, $\bar{M} = 28.964$ $kg \cdot kmol^{-1}$ up to the homopause. $\bar{M} = \sum n_i M_i / n$, where M_i are the molecular masses of the individual species.

g: Acceleration due to gravity [$m \cdot s^{-2}$]. $g(z) = g(z_0)(z_0 + R_\oplus)^2/(z + R_\oplus)^2$, where $g(z_0) = 9.807$ [$m \cdot s^{-2}$] and $R_\oplus$ is the Earth's mean radius, equal to 6,371 [km]. Strictly speaking, g also should include the small ($\leq 1\%$) outward centripetal force due to the Earth's rotation, which varies with latitude.

The following approximation is often made to simplify the form of the hydrostatic equation. Assuming that g, T, and $\bar{M}$ are constant with height, we can integrate the argument of the exponential in Eqs. 1.4 to obtain

$$\frac{p(z)}{p(z_0)} \approx e^{-(z-z_0)/H}, \tag{1.5a}$$

$$\frac{n(z)}{n(z_0)} \approx e^{-(z-z_0)/H}, \tag{1.5b}$$

$$\frac{\rho(z)}{\rho(z_0)} \approx e^{-(z-z_0)/H}. \tag{1.5c}$$

The above equations show that H is an e-fold height for density.

Another important property of ideal gases is given by *Dalton's Law*, which states that the total pressure equals the sum of the *partial pressures* p_i, where i denotes the ith gas species. As long as the gases are well mixed, each separate species concentration n_i obeys the same hydrostatic equation. However, this does not apply to short-lived species (time scales short compared to a mixing time scale, typically ~hours to a few days in the troposphere). Examples are ozone (O_3), which is chemically destroyed or created, and water (H_2O), which undergoes phase changes on short time scales. The hydrostatic equation does not apply to these species, and empirical or theoretically modeled determinations of species concentrations are then needed. This is one of the principal tasks of the subject of *aeronomy*.

The hydrostatic equation shows that atmospheric bulk density and pressure change rapidly with height with an e-folding height of 6 to 8 km. An important property of an atmosphere or ocean is its tendency to arrange itself into a vertically stratified and horizontally homogeneous medium. Quasi-horizontal motions tend to homogenize properties along constant-pressure (more correctly constant-entropy) surfaces. Horizontal variations do occur (after all, this is the origin of weather), particularly in temperature, but usually on spatial scales much greater than a scale height. However, all over the Earth, the pressure and density at sea level are nearly the same on a horizontal plane. Even during severe weather disturbances, the horizontal surface pressure difference is no greater than about 15%. For our purposes, the atmosphere (or ocean) may be regarded as *locally stratified*.

The local stratification property does not usually apply to particles. They tend to have highly localized sources and/or sinks; for example, cloud droplets usually occur only in local pockets of rising air as a result of water condensation. Clouds, particularly of convective origin, may have horizontal scale sizes considerably less than a scale height. In contrast, stratiform clouds get their name from being horizontally homogeneous. Horizontal air motions tend to stratify noncondensible particulates, such as dust or pollen. Ocean currents tend to equalize particle concentrations more rapidly in the horizontal plane than in the vertical. However, inhomogeneities such as phytoplankton "blooms" tend to occur in regions of local upwelling, in analogy to atmospheric clouds. The assumption of local stratification is also called that of a *plane-parallel* or *slab* medium, when referring to its radiation field.

Since a knowledge of temperature is essential to understanding the bulk structure of an atmosphere, we will consider temperature as a key atmospheric variable. (The other key variable is composition, which largely determines the optical properties.) What physical processes give rise to the observed temperature structure? This is a vast subject, and we can only give a few examples. First of all, consider that part of an atmosphere in thermal contact with the warm surface of the land or ocean. During daytime, the upward transport of heat tends to expand the air near the surface. Due to the tendency for pressure to remain constant, the air density decreases, according to Eq. 1.2. Lower density means that the air is more buoyant than the overlying cooler air, and a convective instability may occur, in which air is set into small-scale turbulent motion. The rising air cools by expansion, displacing cooler air. Because of its higher density, the upper cooler air sinks and compressively heats. If allowed to approach equilibrium, a temperature gradient is set up that, in the absence of any other heating/cooling processes (i.e., it is an *adiabatic process*), is given by

$$\left(\frac{\partial T}{\partial z}\right)_{\text{ad}} = -\frac{g}{c_p} \quad \text{(dry atmosphere)}, \tag{1.6}$$

where c_p is the specific heat of (dry) air $= 1{,}006$ $[\text{J} \cdot \text{kg}^{-1} \cdot \text{K}^{-1}]$. This is called the *dry adiabatic lapse rate*, whose value is approximately -9.8 K $\cdot$ km^{-1} for Earth. For moist air, condensation and release of latent heat cause the gradient to be considerably smaller (in magnitude) than $-g/c_p$. $\partial T/\partial z$ can be as small as -3 K $\cdot$ km^{-1}, depending upon the moisture content of the air.[6]

The tendency toward an adiabatic lapse rate for an atmosphere heated from below is a fundamental property of planetary atmospheres. The observed mean temperature profile for Earth displayed in Fig. 1.3 shows a nearly linear temperature lapse rate of -6.5 K$\cdot$km^{-1} in the lowest part of the troposphere.[7] The region of declining temperature is called the *troposphere*, and its upper boundary the *tropopause*. The explanation of this temperature minimum, and the increasing temperature in the upper region, the *stratosphere*, lies in radiative processes, namely in the in situ solar absorption by the ozone layer. The thermal transfer process will be described in more detail in Chapter 9.

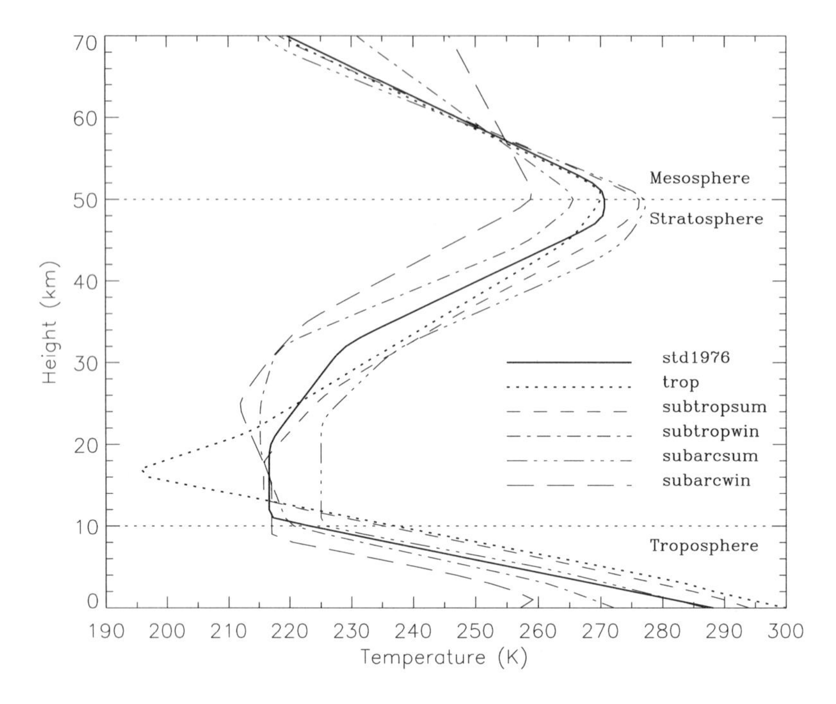

Figure 1.3 Standard empirical model temperature profiles for various locations and seasons. The 1978 standard atmosphere is appropriate for the global mean. The tropical atmosphere is valid for latitudes less than 30°; the subtropical model for 30–45°; the subarctic atmosphere for 45–60°; and the arctic one for 60–90°.

The upper region of declining atmospheric temperature in the 50–95 km height range is called the *mesosphere*. At the upper boundary of the mesosphere, a second temperature minimum occurs, called the *mesopause*. Again, the declining mesospheric temperature is a consequence of radiative effects, in this case, a decreasing ozone concentration and an increased efficiency of IR cooling. Low ozone concentrations at these heights imply reduced solar heating, and IR radiative cooling causes this atmospheric region to be significantly colder than the stratosphere, particularly at the mesopause at about 85 km. However, this is not the whole story, as dynamical processes are also important in this region.

The uppermost atmospheric region above 90 km, the *thermosphere*, is a very hot, tenuous region forced by photoionization heating due to energetic UV and X-ray absorption. In the polar regions, energetic auroral electrons are also important for the heat budget. The thermal structure is a result of imbedded solar heating and cooling by airglow emission, combined with transport by thermal conduction and dynamical advection. More details about the wavelength dependence of the solar radiation deposited in this region are provided in Table 1.1.

1.4.2 Minor Species in the Atmosphere

There are dozens of chemical species in the Earth's atmosphere, some naturally occurring, others of anthropogenic origin. The ones of greatest interest are those that interact with UV and IR radiation, the polyatomic *radiatively active* gases. For reasons discussed in Chapter 4, the principal gases N_2, O_2, and Ar are generally unimportant for visible and IR absorption. The study of atmospheric chemistry has importance in problems of air pollution, climate, and the health of the biosphere, to mention a few applications. Shortwave radiation plays a crucial role in determining the concentrations of *photochemically active* species through the process of *photolysis*, in which molecules are split up into smaller fragments (atoms and molecules) as illustrated in Fig. 1.4a. One of these species, the ozone (O_3) molecule, is of paramount importance in providing life a protective shield to biologically damaging ultraviolet radiation. The response of biological systems to ultraviolet radiation is illustrated in Fig. 1.4b. These topics are further discussed in Chapter 9.

A species of key importance to terrestrial radiation, and thus to its heat balance, is the water vapor molecule (H_2O). Although present in fractions of 1% by volume, it dominates the absorption of infrared radiation. Water vapor (along with several other minor species) has the dual properties of allowing relatively free passage of shortwave solar radiation while impeding the release to space of thermal infrared radiation. This "trapping" of radiative energy causes the Earth's surface temperature to be significantly higher than would be the case in the absence of those gases. The resemblance of this mechanism to what happens in a glass hot house, or greenhouse, is why they are called *greenhouse gases*. Additional species that trap infrared radiation are carbon dioxide (CO_2), methane (CH_4), nitrous oxide (N_2O), and the chlorofluorocarbons (CFCs). Since these chemicals have a component of anthropogenic origin (so-called biogenic gases), the effects of their increasing concentration on the heat budget is of great societal concern.

1.4.3 Optical Line-of-Sight Paths

In atmospheric radiation problems, we frequently require a line-of-sight integration over the ith species density, given by $\rho_i(z)$. In what follows we ignore atmospheric refraction. Thus, light rays follow straight lines, and we need to evaluate the line-of-sight *slant column mass*, $\mathcal{M}_i$, between points P_1 and P_2, along the direction of propagation $\hat{\Omega}$. We will deal with a well-mixed species, such as CO_2. Assuming constant temperature and gravity, we integrate along the line-of-sight distance variable s. Using Eq. 1.5c, we find

$$\mathcal{M}_i(1,2) = \int_1^2 ds\rho_i(s) \approx \rho_i(z_0) \int_1^2 ds e^{-[z_2(s)-z_1(s)]/H_i}. \tag{1.7}$$

Referring to Fig. 1.5, we transform from the variable ds to dz; $ds = dz \sec\theta$, where θ is the polar angle ($0 \le \theta < \pi/2$) made by the vector $\hat{\Omega}$ with the vertical, $\cos\theta = |\hat{z} \cdot \hat{\Omega}|$. $\hat{z}$ is a unit vector in the positive z direction. Integration of the above equation yields

$$\begin{aligned}\mathcal{M}_i(1,2) &= \rho_i(z_0) H_i \sec\theta \left(e^{-z_1/H_i} - e^{-z_2/H_i}\right) \\ &= [\rho_i(z_1) - \rho_i(z_2)] H_i \sec\theta.\end{aligned} \tag{1.8}$$

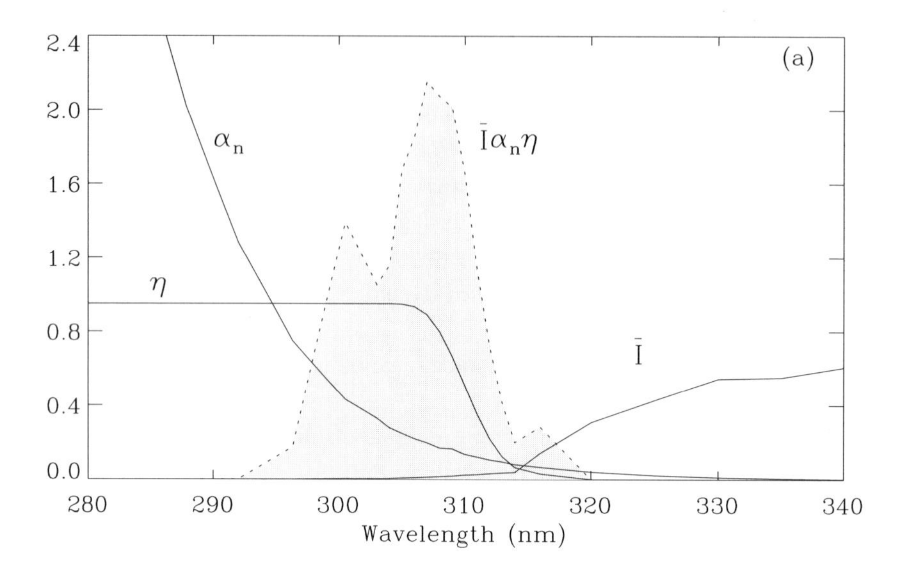

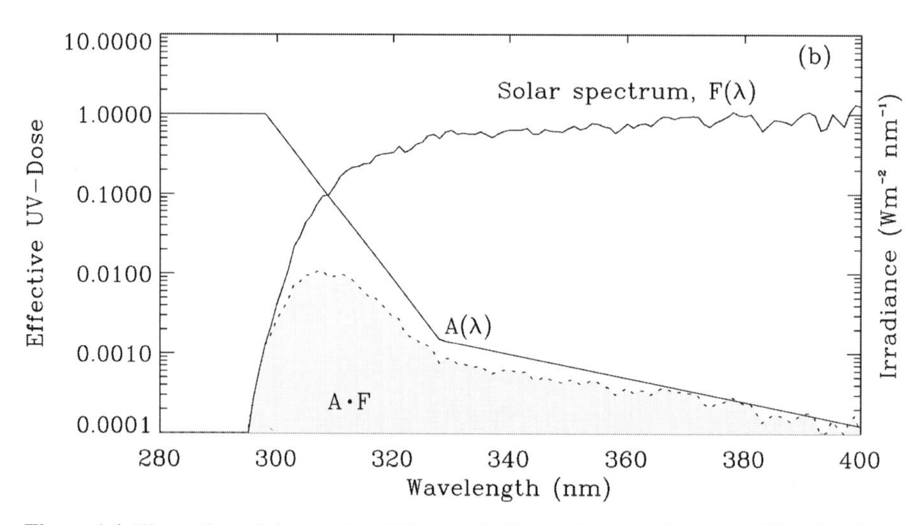

Figure 1.4 Illustration of chemical and biological effects of solar radiation. (a) The photolysis rate is determined by the integral of the product of the "actinic flux" of radiation, $\bar{I}_\nu$ ($\times 10^{-15}$ s^{-1} cm^{-1}), the photoabsorption cross section, $\alpha_n(\nu)$ ($\times 10^{-17}$ cm^2), and the quantum yield $\eta(\nu)$. The net effect $\bar{I}_\nu \alpha_n(\nu)\eta(\nu)$ ($\times 10^{-7}$ nm^{-1}) is determined by the area under the product curve as indicated. (b) The UV exposure is determined by the product of a biological "action spectrum", $A(\nu)$, and the solar irradiance, F_ν. The instantaneous dose rate is the area under the product curve as indicated.

We find that the species slant column mass between two atmospheric points separated by the heights z_1 and z_2, and whose direction is θ from the normal, is the difference of two quantities $\mathcal{M}_i(1,2) = \mathcal{M}_i(z_1,\theta) - \mathcal{M}_i(z_2,\theta)$, where

$$\mathcal{M}_i(z,\theta) = \rho_i(z) H_i \sec\theta = \frac{p_i(z)}{g} \sec\theta \tag{1.9}$$

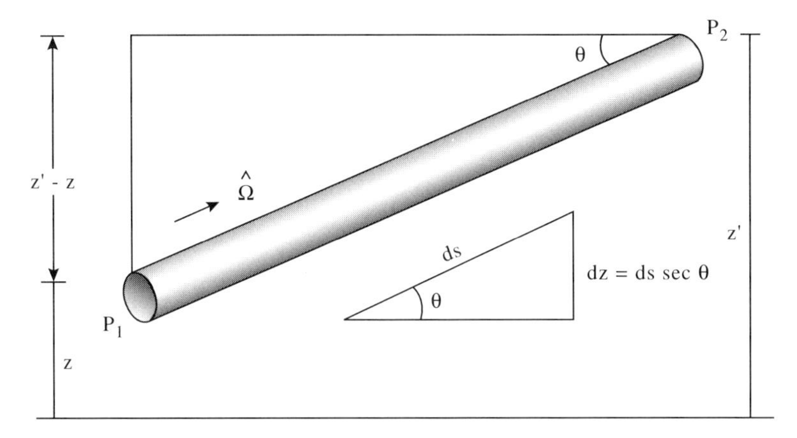

Figure 1.5 Geometry of the slant-column number. The right-circular cylinder of unit geometrical cross section contains $\mathcal{N}$ molecules (number per m^2).

is the slant column mass between the height z and $z \to \infty$ along the direction $\hat{\Omega}(\theta)$. In the second result we have used the ideal gas law and the definition of H_i. Another common notation for *vertical column mass* is u.

As discussed in Appendix G and in Chapter 2 the optical path is obtained by multiplying $\mathcal{M}_i$ by a mass absorption coefficient α_m. This assumes α_m does not vary along the path – otherwise it must be inside the integration over path length. We can visualize $\mathcal{M}_i(1,2)$ as being the species atmospheric mass contained within a geometric cylinder of unit cross section extending from points P_1 to P_2. For P_2 outside the atmosphere, $\mathcal{M}_i(z,\theta)$ is the species atmospheric mass in an infinitely long slant column above z in the direction θ. The above equation also shows that the quantity H_i may be interpreted as the equivalent vertical thickness, if the species were extracted from the air and compressed everywhere to a density equal to its surface density.

The above equations are invalid for nearly horizontal paths, for which θ is near 90°. In the range $\theta > 82^\circ$, it is necessary to consider the spherical nature of the atmosphere. In fact, $\mathcal{M}_i$ does not diverge as $\theta \to \pi/2$, as predicted by Eq. 1.9, but reaches a finite value, depending upon the radius of the planet (§6.4). Curvature effects are considered further in §6.4.

Analogous to column mass we define the *column number* $\mathcal{N}_i(1,2)$ as the *number* of molecules (of composition i, each with molecular mass M_i) in a cylinder of unit cross section:

$$\mathcal{N}_i(1,2) = \int_1^2 ds\, n_i = \mathcal{M}_i(1,2)/M_i = \mathcal{N}_i(z_2,\theta) - \mathcal{N}_i(z_1,\theta). \qquad (1.10)$$

Although $\mathcal{N}_i$ has the mks units of [molecules $\cdot$ m^{-2}], it is more commonly expressed in units of atm $\cdot$ cm. This unit is the length of a column filled with the species of interest and compressed to standard temperature and pressure (STP), which are 273.16 K

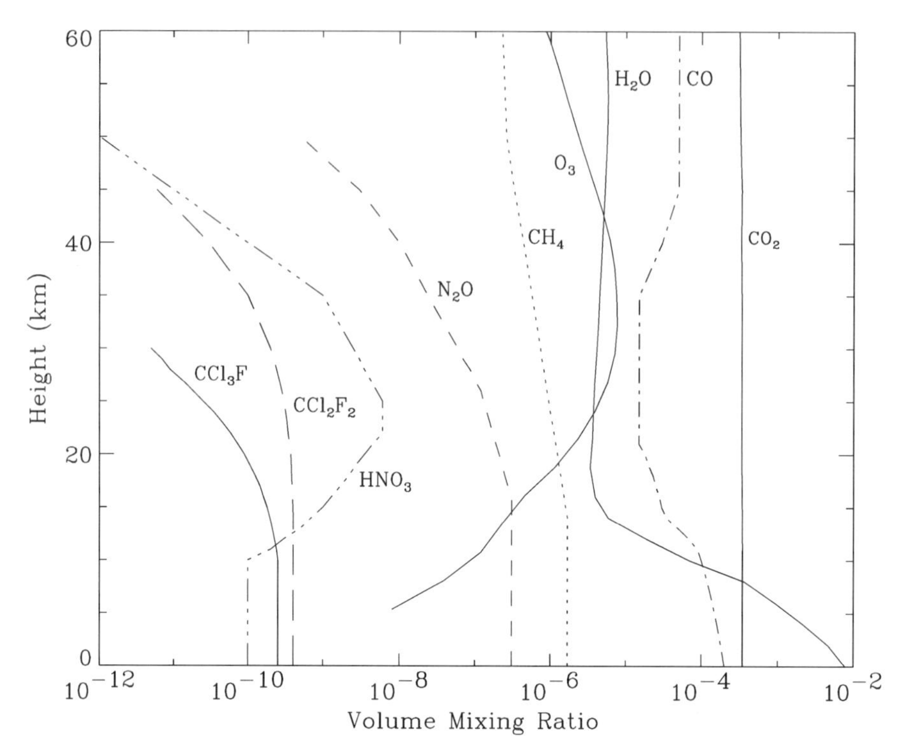

Figure 1.6 Radiatively active atmospheric constituent height profiles of volume mixing ratio. The ozone and water (above 25 km) profiles are taken from globally averaged data from the HALOE experiment on board the *UARS* satellite. The remaining profiles are taken from standard atmosphere compilations and from balloon measurements.

and 1013.25 mb, respectively. To convert from [molecules $\cdot$ cm^{-2}] to [atm $\cdot$ cm], one divides $\mathcal{N}_i$ by *Lochmidt's number*, the air density at STP, $n_{\rm L} = 2.687 \times 10^{19}$ cm^{-3}.

Well-mixed species (those with long photochemical lifetimes) such as N_2, O_2, and CO_2 may be described by the above equations up to the homopause.[8] However, most chemical species are photochemically produced and lost over time scales short compared with a mixing time scale (see Fig. 1.6), and we cannot use Eqs. 1.7–1.9 for the column number or column mass. Nevertheless, the dominant height variation of many species, for example N_2O and CH_4, are at least quasi-exponential. A convenient variable to describe deviations of the species height distribution from the hydrostatic case is the *mixing ratio*. This is defined either in terms of the *mass mixing ratio*, $w^i_m(z) = \rho_i(z)/\rho(z)$, or the *volume mixing ratio*, $w^i(z) = n_i(z)/n(z)$, where $\rho(z)$ is the total density and $n(z)$ is the total concentration of the ambient atmosphere. Quite often we are dealing with constituents with very small mixing ratios. Typically w^i is specified in parts per million by volume [ppmv] or parts per billion [ppbv]. w^i is also sometimes called the *molar fraction* of species i. w^i_m is usually specified in [g $\cdot$ g^{-1}] (grams per gram). Figure 1.6 displays the volume mixing ratios of various species versus height, derived from measurements and modeling studies. This diagram is intended to be illustrative, rather than an accurate representation, since individual species can vary by large factors with time, season, and with the diurnal cycle.

1.4.4 Radiative Equilibrium and the Thermal Structure of Atmospheres

Atmospheric temperature plays a key role in the energy balance of any planet. A planet is heated by absorption of solar radiation and cooled by emission of thermal infrared radiation. *Kirchhoff's Law* (§5.2.1) states that matter at a finite temperature T will emit at a rate that depends upon both the absorptive properties of the matter and the absolute temperature. Thus a planet sheds its energy by radiating to space solely by this thermal emission process. In general, these two rates do not balance one another at any one location or time. As a result of this instantaneous imbalance of heating/cooling, the temperature of a rotating planet with an axial tilt will undergo both diurnal and seasonal cycles of temperature. However, the highs and lows tend to remain the same, and when averaged over the entire surface area and over an orbital period, the mean temperature is expected to remain constant, in the absence of internal or external changes. This *negative feedback* of the planetary temperature (through Kirchhoff's Law) acts like a giant thermostat, continually adjusting the rate of energy loss to compensate for excess heating or cooling.

Assuming that the local thermal emission is balanced by the local rate of heating due to absorption of radiation at all wavelengths is called *local radiative equilibrium* (§5.2.1). This state of affairs is only met in special situations in the Earth's atmosphere, for example in the upper stratosphere. However, analysis of outgoing and incoming radiation shows that averaging the energy gains and losses over the entire planet and over a suitably long time interval results in a very close balance, within the accuracy of the measurements. This is called a state of *planetary radiative equilibrium*. Satellite measurements verify this condition to a high accuracy, when averaged over the course of a year. Climatic records show that the Earth's climate has been remarkably stable over geologic time. The geologic record shows that climatic change has indeed occurred, but life's continuing existence for 3.8 billion years attests to a basic climate stability. Of current concern are fluctuations in climate that could be so extreme, in both their magnitude and suddeness, that they could place great stress on the biosphere. For example, changes in rainfall patterns could influence agricultural productivity and the world's food supply. Fluctuations due to natural causes (for example, due to variability of the solar "constant") and to inadvertent human influence (such as carbon dioxide forcing) need to be clearly understood. We will consider only the most basic ideas of energy balance in this chapter, and we will take up this subject again in the final chapter.

Other planetary atmospheres have temperature structures that pose their own particular challenges. For example, the high surface temperature of the planet Venus ($T_s = 750$ K) requires a very powerful greenhouse mechanism.[9] A second example is the Martian atmospheric heat budget. Mars undergoes substantial global change in its

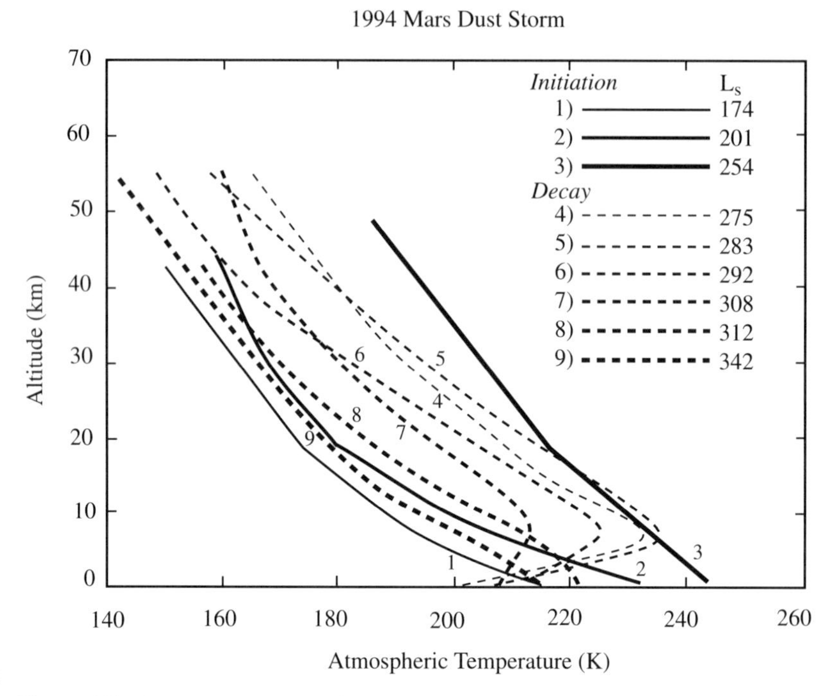

Figure 1.7 Temperature profiles of the atmosphere of Mars, derived from emission/absorption spectra of the carbon monoxide microwave lines at 1.3 mm and 2.6 mm (see Endnote 10). They are derived from full-disk averaged spectra obtained at the Kitt Peak NRAO radio telescope during the period January–June, 1994. The spectral dependence of the pressure-broadened line is used to retrieve atmospheric temperature versus pressure with roughly 10–15 km vertical resolution. The various profiles reflect the solar heating rates due to varying degrees of dust loading during the global dust storm in 1994. L_s denotes the longitude of Mars relative to a fixed point in its orbit, the spring equinox.

temperature structure, as shown in Fig. 1.7.[10] This short-term variability in temperature is explained by the presence of differing heating distributions over the planet due to variable dust loading. Solar absorption by suspended dust particles is responsible for the temperature inversion.

Consider the total thermal energy of the entire climate system (excluding geothermal energy), E. Dividing this by the total surface area of the Earth, $4\pi R^2$, yields the average column-integrated energy $\bar{E} \equiv E/4\pi R^2$. According to the *First Law of Thermodynamics*, the time rate of change of the globally averaged column energy is given by

$$\frac{\partial \bar{E}}{\partial t} \equiv \bar{N} = (1 - \bar{\rho})\bar{F}^{\mathrm{s}} - \bar{F}_{\mathrm{TOA}}. \qquad (1.11)$$

Here $\bar{N}$ is called the mean *radiative forcing* and is equal to the net flux at the top of the atmosphere. The incoming flow is the absorbed solar radiation $(1 - \bar{\rho})\bar{F}^{\mathrm{s}}$, where $\bar{\rho}$ is the spherical albedo and $\bar{F}^{\mathrm{s}}$ is the average flux falling on the planet. The second term is the outgoing thermal emission $\bar{F}_{\text{TOA}}$ at the top of the atmosphere (TOA). Characterizing the mean outgoing IR radiation in terms of an *effective temperature*, T_e, we get $\bar{F}_{\text{TOA}} = \sigma_B T_e^4$, where $\sigma_B = 5.67 \times 10^{-8}$ [W · m^{-2} · K^{-4}] is the Stefan–Boltzmann constant. Further averaging over one year (or several years) yields a time- and space-averaged radiative forcing $\langle \bar{N} \rangle$, which is very close to zero. Assuming $\langle \bar{N} \rangle = 0$ implies planetary radiative equilibrium. In this situation, the effective temperature is thus

$$T_e = \left[\frac{(1 - \bar{\rho})\bar{F}^S}{\sigma_B} \right]^{1/4}. \tag{1.12}$$

The spherical albedo of Earth is about 30%, and $\bar{F}^S$ is one quarter of the solar constant, $\bar{F}^S = S/4 = 342$ [W · m^{-2}]. (The factor of 1/4 is the ratio of the total surface area to the disk area of the Earth intercepting the radiation.) We find $T_e = 255$ K or -18°C. This value is considerably lower than Earth's mean surface temperature T_s, which is 288 K or 15°C. In fact, T_e is the temperature of a fictitious airless body at 1 AU (e g., the Moon) that has the same albedo as the Earth. Actually the Moon's albedo is very low, $\bar{\rho} = 0.07$, and thus its effective temperature is higher than Earth's, $T_e = 367$ K. With its dark oceans, Earth would have an albedo similar to that of the Moon, were it not for its extensive cloud cover. The effective temperature is a useful property for all planets, even the gas giants (Jupiter and Saturn), which possess their own internal energy sources. For the Earth the internal (geothermal) energy production is only ~0.01% of the solar input and can therefore be disregarded.

The fact that $T_s > T_e$ on terrestrial-like planets with substantial atmospheres is a result of their being relatively transparent in the visible part of the spectrum and relatively opaque in the infrared. A portion of the IR radiation emitted by the surface and atmosphere is absorbed by polyatomic molecules in the atmosphere (the greenhouse gases). Emission of this absorbed energy takes place in all directions. Some of this energy is subsequently released to space, but a significant part is emitted downward, where it may warm the surface. The greenhouse effect is thus the additional warming of the surface by the downwelling IR "sky emission." In fact the warming of the surface by absorption of diffuse sky and cloud radiation is nearly twice the direct solar heating. It is sometimes loosely stated that the IR radiation is "trapped" between the surface and the atmosphere, as if somehow it never escapes. In reality, the speed of the energy flow is limited only by the radiative lifetimes of the molecular excited states and the speed of light. In the opaque parts of the spectrum there is indeed a radiative energy buildup (the energy density is raised), but in other parts of the spectrum, the radiation readily escapes to space. This is particularly apparent in the 8–12 μm spectral "window," where the atmosphere is relatively transparent. According to the principle mentioned in §1.2, this radiating level (at a height of about 5 km) is located where the IR opacity is of order unity.

In Chapter 12, we examine in greater detail the temperature structure of the Earth's atmosphere. For this brief introduction, we will illustrate the basic idea by using a zero-dimensional model of the energy balance.[11]

1.4.5 Climate Change: Radiative Forcing and Feedbacks

The notion of radiative forcing of the temperature structure of the atmosphere is in widespread use in modern climate studies. Since the ocean is intimately involved in the exchange of energy of the planet with the Sun, we will refer to the combined ocean–atmosphere as the climate "system". Although the specification of the forcing is straightforward, the job of describing its *response* to this forcing is a complex task, a problem that is far from solution even with today's elaborate general circulation models (GCMs). Nevertheless it is useful to consider a simplified model in which the response is described in terms of feedback mechanisms already mentioned briefly. As a way of introducing some of these ideas, we will begin with the definition of radiative forcing and illustrate its usefulness in quantifying the relative effects of various perturbations (external or internal) to the climate. We will then discuss briefly the ideas of how feedbacks influence the response of the climate system. These concepts are described in greater detail in Chapter 12.

To relate the radiative forcing to the surface temperature, we first consider the outgoing flux to consist of the sum of two terms: the dominant term due to the surface contribution and the second due to emission from the atmosphere. Neglecting the smaller atmospheric contribution, and approximating the surface as black, we can approximate the outgoing flux by $F_{\mathrm{TOA}} \approx \mathcal{T}_{\mathrm{F}}\sigma_{\mathrm{B}}T_{\mathrm{s}}^4$. $\mathcal{T}_{\mathrm{F}}$ is the flux transmittance, the fraction of radiation that survives passage through the atmosphere. If we now correct this result for the neglected term, we can rewrite the above expression in terms of an effective flux transmittance, $\mathcal{T}_{\mathrm{eff}}$ $(0 < \mathcal{T}_{\mathrm{F}} < \mathcal{T}_{\mathrm{eff}} < 1)$, so that

$$F_{\mathrm{TOA}}(T_{\mathrm{s}}) = \mathcal{T}_{\mathrm{eff}}\sigma_{\mathrm{B}}T_{\mathrm{s}}^4. \tag{1.13}$$

The above equation states that the radiative forcing is a function of surface temperature.

We now consider a perturbation to the radiative forcing, such that $N(T_{\mathrm{s}})$ is changed to $N(T_{\mathrm{s}}) + \Delta N$. (We drop the average ($\langle \bar{N} \rangle$) notation, since by now there should be no confusion that we are dealing with globally and time-averaged quantities.) Assuming that the perturbed atmosphere relaxes to a new equilibrium state, and assuming small changes, we can write the total derivative of N as the sum of the radiative forcing (the "cause") and the atmospheric response (the "effect"):

$$\overbrace{\Delta N}^{\textit{forcing}} + \overbrace{\frac{\partial N}{\partial T_{\mathrm{s}}}\Delta T_{\mathrm{s}}}^{\textit{response}} = 0. \tag{1.14}$$

Solving for the surface temperature response, we have

$$\Delta T_{\mathrm{s}}^{\mathrm{d}} = \alpha \Delta N, \quad \text{where } \alpha \equiv -(\partial N/\partial T_{\mathrm{s}})^{-1} = \left[\frac{\partial F_{\mathrm{TOA}}}{\partial T_{\mathrm{s}}} - \frac{\partial(1-\bar{\rho})\bar{F}^{\mathrm{s}})}{\partial T_{\mathrm{s}}}\right]^{-1}. \tag{1.15}$$

The factor α is called the *climate sensitivity*. We have used the notation $\Delta T_{\mathrm{s}}^{\mathrm{d}}$ to denote the *direct* temperature response.

Example 1.1 Climate Response to a CO_2 Doubling

As will be discussed in Chapter 12, the radiative forcing from a doubling of CO_2 may be calculated from a detailed spectral radiative transfer model. Its value is $\Delta N \sim 4\ \mathrm{W \cdot m^{-2}}$. We can then find the temperature response due solely to the (negative) feedback of the change of surface flux with temperature, or in other words, due to the change in the F_{TOA} term. $\alpha = [\partial(\sigma_B T_s^4 \mathcal{T}_{\mathrm{eff}})/\partial T_S]^{-1} = [4\sigma_B T_s^3 \mathcal{T}_{\mathrm{eff}}]^{-1} = T_s/4F_{\mathrm{TOA}} = 288/(4 \times 240) = 0.3\ \mathrm{W \cdot m^{-2} \cdot K^{-1}}$. Then the direct response of the surface temperature, in the absence of feedbacks, ΔT_s^{d}, is $0.3 \times 4 = 1.2$ K.

Example 1.2 Climate Response to a Change in the Solar Constant

Suppose the solar constant were to decrease by 1%. What is the change in the surface temperature, assuming no feedbacks except for the negative feedback of the reduced thermal emission? Since we ignore changes in the albedo, the climate sensitivity is simply $\alpha = [\partial F_{\mathrm{TOA}}/\partial T_s]^{-1}$. From Eq. 1.11, $\Delta N = (1 - \bar{\rho})\Delta F^{\mathrm{s}}$. The change in surface temperature is thus $\Delta T_s = \alpha(1 - \bar{\rho})\Delta F^{\mathrm{s}}$. From Eq. 1.13, we have

$$\alpha = [\partial F_{\mathrm{TOA}}/\partial T_s]^{-1} = \left[\partial\left(\mathcal{T}_{\mathrm{eff}}\sigma_B T_s^4\right)/\partial T_s\right]^{-1} = [(4/T_s)F_{\mathrm{TOA}}]^{-1}.$$

Since the unperturbed system is in radiative equilibrium, $F_{\mathrm{TOA}} = (1 - \bar{\rho})F^{\mathrm{s}}$, and therefore

$$\Delta T_s^{\mathrm{d}} = \left(\frac{T_s}{4}\right)\frac{\Delta F^{\mathrm{s}}}{F^{\mathrm{s}}}.$$

Setting $\Delta F^{\mathrm{s}}/F^{\mathrm{s}} = -0.01$ and $T_s = 288$ K, we find $\Delta T_s^{\mathrm{d}} = -0.72$ K. Note that this would partially offset the warming (+1.2 K) due to a simultaneous doubling of CO_2.

We now consider the indirect effects on the surface temperature, brought about by temperature-dependent processes. An effect is called a *positive feedback* if the changes tend to amplify the temperature response. For example, increased temperature tends to increase evaporation and thus raise the humidity. Since water vapor is a greenhouse gas, increased IR opacity causes a further increase in temperature. In contrast, a *negative feedback* tends to dampen the temperature response. For example, increased low-level cloudiness could result from a warmer climate, and more clouds cause a higher albedo. This in turn tends to cool the Earth. We can formalize this behavior by considering the parameter Q (such as albedo) that depends upon the surface temperature. Then the direct forcing ΔN is augmented by an additional term $(\partial N/\partial Q)(\partial Q/\partial T_s)\Delta T_s$. Thus $\Delta N + (\partial N/\partial Q)(\partial Q/\partial T_s)\Delta T_s = -(\partial F_{\mathrm{TOA}}/\partial T_s)\Delta T_s$. Solving for the climate sensitivity we find

$$\alpha = [4F_{\mathrm{TOA}}/T_s - (\partial N/\partial Q)(\partial Q/\partial T_s)]^{-1}. \tag{1.16}$$

For an arbitrary number of forcings Q_i, we can use the chain rule of differentiation to obtain a general expression for the climate sensitivity:

$$\alpha = \left[4F_{\mathrm{TOA}}/T_s - \sum_i (\partial N/\partial Q_i)(\partial Q_i/\partial T_s)\right]^{-1}. \tag{1.17}$$

The temperature response can now be written as

$$\Delta T_{\mathrm{s}} = \alpha \Delta N = f \Delta T_{\mathrm{s}}^{\mathrm{d}}, \tag{1.18}$$

where the *gain* of the climate system is

$$f = \left[1 - \sum_i \lambda_i\right]^{-1}, \quad \text{where } \lambda_i \equiv \frac{(\partial N/\partial Q_i)(\partial Q_i/\partial T_{\mathrm{s}})}{(4F_{\mathrm{TOA}}/T_{\mathrm{s}})}. \tag{1.19}$$

The gain $f = 1$ in the absence of feedbacks, $f > 1$ when the net feedbacks are positive ($\sum \lambda_i > 0$), and $f < 1$ if the net feedbacks are negative ($\sum \lambda_i < 0$).

Equation 1.18 would appear to tell us that the temperature response is independent of the nature of the forcing mechanism, as long as it has the same numerical value. In particular, it suggests that increases in two different greenhouse gas (such as CO_2 and CH_4) have the same effect on climate if their individual forcings, ΔN_i, are the same. Detailed model calculations verify that this is indeed a good approximation, with two notable exceptions: 1. stratospheric ozone forcing, which has a very different altitude dependence than that of a well-mixed tropospheric gas, and 2. regionally concentrated sulfate aerosols, which act in the global average in a different way than the well-mixed, long-lived greenhouse gases.

The radiative forcings contributed by individual greenhouse gases over the past several centuries are shown in Fig. 1.8a for different epochs.[12] These results were computed from a detailed radiative transfer model. The radiative forcing concept can be extended to include clouds, where the forcing consists of both shortwave (due to albedo changes) and longwave contributions (due to greenhouse trapping by cloud particles). The detailed response of the climate system will depend upon the specifics of the perturbation, and the quantities λ_i and α must be determined from a detailed climate model. Current estimates for the epoch 1850 to present are shown in Fig. 1.8b, including aerosol and solar forcing. The major indirect effects are a depletion of stratospheric ozone caused by the chlorofluorocarbons and other halocarbons and an increase in the concentration of tropospheric ozone.

The change in radiative forcing due to a CO_2 increase occurring over the period 1900–1990 is estimated to be $\Delta N = 1.92\ \mathrm{W \cdot m^{-2}}$. Since this corresponds to a direct temperature change $\Delta T_{\mathrm{s}}^{\mathrm{d}}$ of 0.356°C, and the actual temperature response was $\Delta T_{\mathrm{s}} =$ 0.5°C, the gain for the current climate is estimated to be 1.4. This does not take into account the time lag of the response. A radiative forcing from CO_2 doubling (see Example 1.1) of $4\ \mathrm{W \cdot m^{-2}}$ yields accurately a computed steady-state temperature change of about $3^\circ \pm 1.5^\circ$C. Again, from a direct response of 1.2°C, this yields a gain of 2.5 ± 1.25, which includes the effects of time lag.

The temperature is just one response among many. The dynamics and hydrologic cycle will also be affected by radiative forcing, and these will have additional influences on the biosphere and oceans. One might ask: What role do oceans play in climate change? The simple system discussed above does not seem to require the ocean at all!

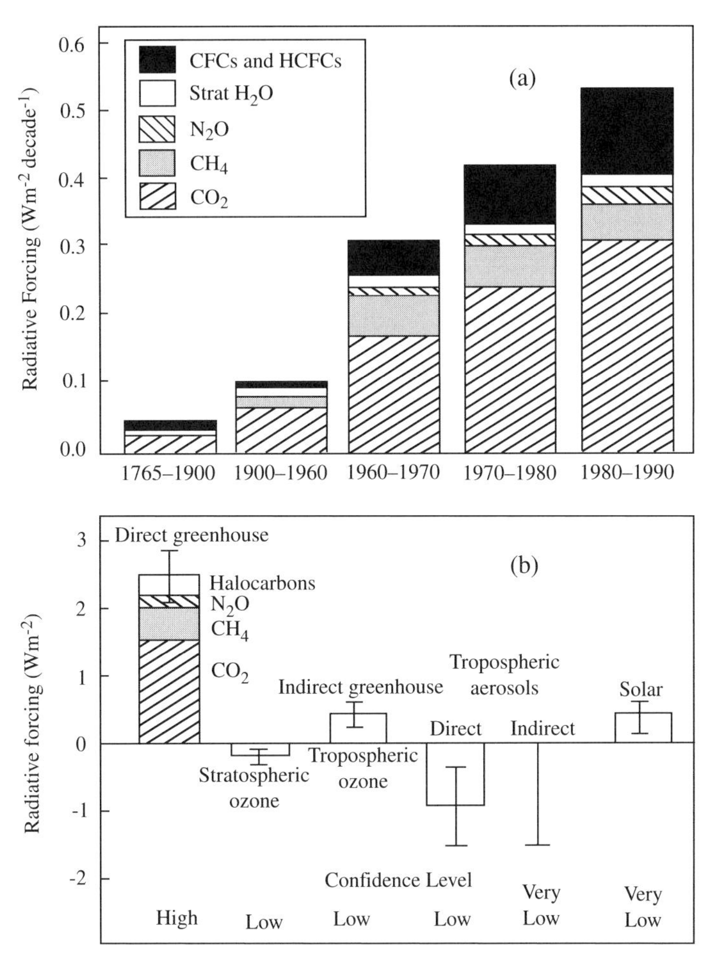

Figure 1.8 (a) The degree of radiative forcing produced by selected greenhouse gases in five different epochs (see Endnote 3). Until about 1960, nearly all the forcing was due to CO_2; today the other greenhouse gases combined nearly equal the CO_2 forcing. (b) Estimates of the globally averaged radiative forcing due to changes in greenhouse gases and aerosols from pre-industrial times to the present day and changes in solar variability from 1850 to the present day. The height of the bar indicates a mid-range estimate of the forcing. The lines show the possible range of values. An indication of relative confidence levels in the estimates is given below each bar.

In fact, oceans play a crucial role in the *time-dependent* response. The oceans absorb and give up heat to the atmosphere, over time scales that are very long compared to atmospheric time scales. The time scale for atmospheric changes is of the order of one year. For the ocean, several hundred years may ensue before the ocean has fully responded to a climate perturbation. This causes a time lag in the response of the climate system, and it gives rise to the notion of an "unrealized warming." We now turn to some basic properties of the ocean.

1.5 Density Structure of the Ocean

Some basic state variables of the ocean (such as density and temperature) are the same as those of the atmosphere. The hydrostatic equation (Eq. 1.1) is unchanged. However, the equation of state is quite different from the ideal gas formula. In particular, the density of water is *not* directly proportional to p/T but is a much weaker nonlinear function of these variables. It also depends upon the salinity S. This dependence may be expressed in terms of the empirical formula $\rho = \rho(S, T, p = 0)/[1 - p/K(S, T, p)]$, where K is the bulk modulus given by low-order polynomial fits in each variable S, T, and p. As a numerical example, the density of pure water ($S = 0$) at 5°C increases from 1,000 to 1,044 kg · m^{-3} from the surface to a depth where the pressure is 1,000 bars (about 10 km depth). Even though the changes of ρ are small, they nevertheless have important implications for the stability and circulation of ocean water. However, such effects can usually be ignored from the optical point of view.

The variation of pressure with depth is obtained by integrating Eq. 1.1 from the sea surface downward to the depth h:

$$p(h) = p_{\mathrm{a}} + \int_0^h dh'\rho(h')g \approx p_{\mathrm{a}} + \bar{\rho}gh, \tag{1.20}$$

where p_{a} is the atmospheric pressure at the sea surface, and $\bar{\rho}$ is the depth-averaged value of the density. This result shows that the total hydrostatic pressure is equal to the sea-level atmospheric pressure plus the weight of the overlying column of water ($\bar{\rho}gh$), which varies linearly with depth.

As in atmospheric radiation problems, we are interested in the line-of-sight slant-path mass of water between two points, P_1 and P_2. If the line of sight makes an angle θ with the upward vertical direction (see Fig. 1.5), we have

$$\mathcal{M}(1, 2) = \bar{\rho}h_2 \sec\theta - \bar{\rho}h_1 \sec\theta, \tag{1.21}$$

where we assumed that $h_2 > h_1$. Comparing Eq. 1.9 to Eq. 1.21, we see that the result for the ocean is the same as for the atmosphere, provided we replace the atmospheric scale height H with the ocean depth h.

Knowledge of the slant-path water mass is usually sufficient for shortwave radiative transfer studies in very pure waters, in which only *Rayleigh scattering* (§3.2, §9.3.1) needs to be considered. In this case all water molecules take part in the light scattering process in a similar way to the light scattering by air molecules that causes the blue sky. However, in most practical radiation problems, water impurities are of crucial importance. These are dissolved organic substances and suspended minerals and organic matter. Their concentrations can vary in almost arbitrary ways in the vertical. Fortunately, the vertical stratification approximation is often valid.

1.6 Vertical Structure of the Ocean

The energy budget of the ocean is a subject of considerable interest for modern studies of climate, because of the intimate coupling of the ocean (covering nearly three fourths of the total world surface area) with the atmosphere. In the following we describe briefly a few features of the ocean of relevance to its interaction with radiation.

1.6.1 The Mixed Layer and the Deep Ocean

The ocean's vertical structure can be crudely subdivided into two regions: an upper *mixed layer*, typically 50 to 200 m in depth, and beneath it, the deep ocean having an average depth of 4 km. The mixed layer is characterized by a nearly uniform temperature and salinity. Its homogeneous nature is maintained by turbulent transport caused by mechanical stirring by wind stress, and less frequently, by spontaneous overturning when more dense water overlays less dense regions. The latter situation can occur at night and during high latitude in winter when the cooling surface water becomes denser than the underlying layers. The transition to colder, denser bottom water is characterized by an abrupt decrease in temperature, the *thermocline*, and an abrupt increase in density, the *pycnocline*. Accompanying these changes is a large increase in the fluid stability, so that small-scale mixing ceases to become important in transporting heat and salinity downward. Due to precipitation, the ocean receives fresh (less dense) water. Melting of sea ice also freshens the sea surface. Counteracting this tendency is evaporation, which tends to leave more saline water behind. Figure 1.9 shows typical depth profiles of temperature at different latitudes.[13] Water density (not shown) tends to follow profiles that are inverse to temperature, so that density tends

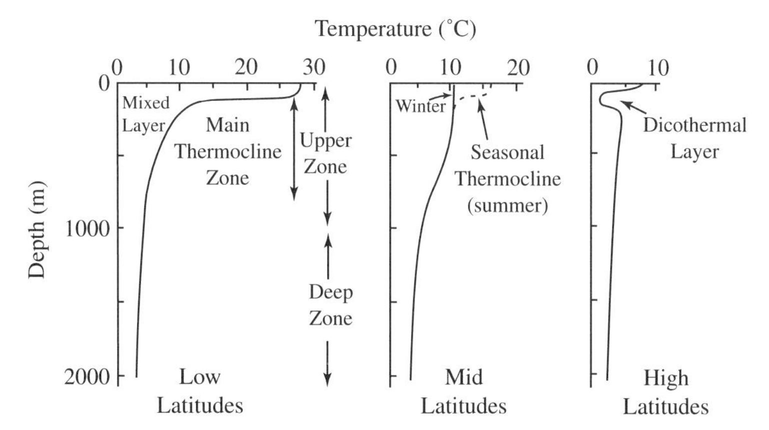

Figure 1.9 Typical mean temperature/depth profiles for the open ocean (see Endnote 13). The *dicothermal layer* refers to a layer of cold water that often occurs in northern high latitudes between 50 and 100 m. Stability in this layer is maintained by an increase in salinity with depth.

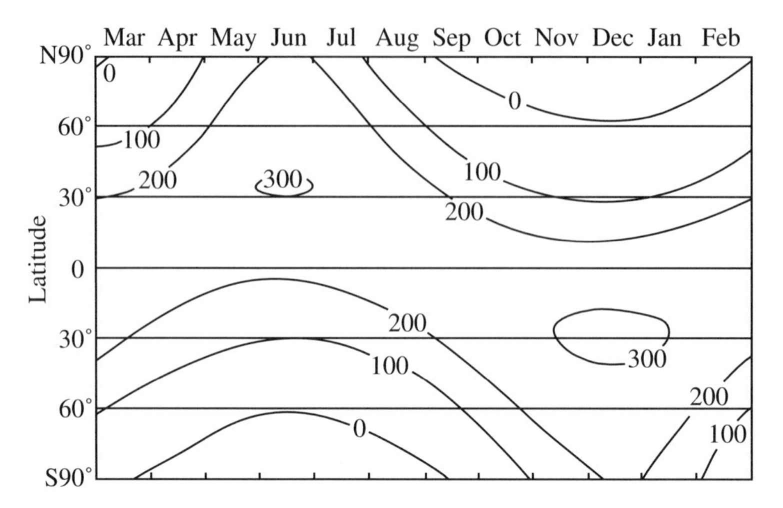

Figure 1.10 Daily shortwave radiation received at the sea surface in the absence of clouds (see Endnote 13). Units are [W · m^{-2}].

to increase with depth. In the polar oceans, and at lower latitudes during winter, the thermocline and pycnocline may cease to exist, and the water temperature and salinity are nearly constant with depth.

The absorption of sunlight (direct plus diffuse) within a shallow layer generally 1 to 10 m thick is the ocean's ultimate source of energy. Periodic heating over the year occurs as a result of the varying solar incidence angle and the length of the day. The global distribution of the daily shortwave radiation received at the sea surface under clear conditions is shown in Fig. 1.10.[14] The annual variation of the diurnally averaged mixed-layer temperature at a northern hemisphere site is shown in Fig. 1.11.[15] Interpretation of this behavior requires an understanding of the physical properties of seawater. Generally speaking, when the temperature increases, the density decreases slightly. Thus surface heating creates a lighter and hence more buoyant upper layer, which is resistant to downward mixing. Vertical mixing would otherwise distribute the absorbed energy uniformly throughout the layer and entrain colder water from below, thus thickening the layer. Thus the shallower the mixed layer, the higher its temperature. Conversely, the deeper the mixed layer the lower the temperature.

1.6.2 Seasonal Variations of Ocean Properties

At a given location, the depth of the mixed layer varies with season, and with the degree of mechanical stirring induced by winds. For example, a storm of duration of several days has been observed to lower the thermocline by 30–50 m, while cooling the surface by 1–2 K. In Fig. 1.11 the mixed layer is seen to be deepest when the solar heating is a minimum (during winter). During winter, the ocean emits more thermal infrared energy than it receives from the Sun, and a net cooling occurs. The deepening of the mixed layer occurs because the colder, denser (and thus less stable) water can be mixed

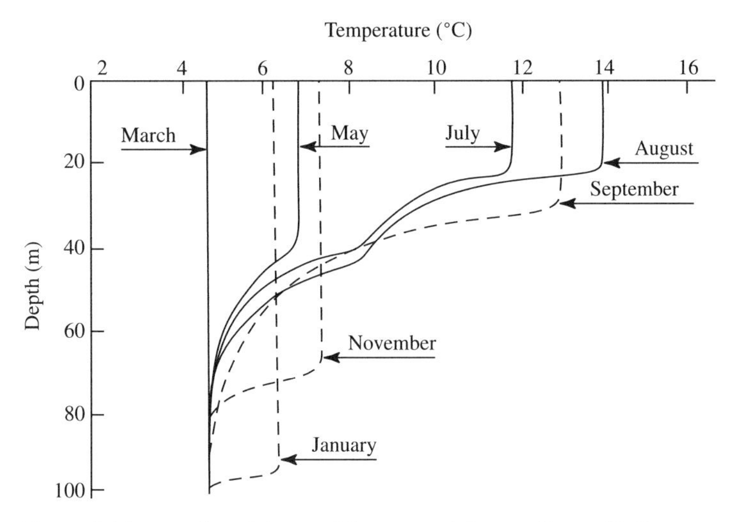

Figure 1.11 Growth and decay of the seasonal thermocline at 50°N, 145°W in the eastern North Pacific (see Endnote 13).

to greater depths than during summer. During the summer, the ocean receives more energy than it radiates, and the ocean surface warms. The upper levels become more buoyant and less susceptible to downward mixing. The mixed layer becomes shallower and warmer as the summer season progresses. As autumn approaches, the warming lessens and the water becomes less stable. Downward mixing causes the thermocline to deepen, and energy is shared by a larger volume. Cooling occurs during autumn more as a result of this downward mixing than from increased net radiative energy loss. Thus the mixed layer's annual temperature cycle, although driven by solar heating at the surface, is controlled more by turbulent heat transport than by radiative transfer.

The deep ocean shows virtually no seasonal variation. Its temperature decreases slowly with depth, approaching 0°C at abyssal depths. Its temperature is governed by larger-scale processes of overturning. It is interesting that downwelling occurs primarily in the polar regions. The return of colder water to the surface by upwelling occurs over a much greater fraction of the world ocean than that of downwelling. The time scale of this overturning process is of the order of several centuries. If the Sun were to suddenly go out, the Earth's oceans would not freeze solid for several hundred years.

1.6.3 Sea-Surface Temperature

Another important aspect of water is its very high IR opacity. Unit optical depth occurs within an extremely small distance near the surface. If the water is calm, the daytime "skin temperature" may be appreciably different than that of water just below the surface, due to the fact that even a thin layer of water takes up a considerable amount of energy. Due to its high opacity (see Fig. 1.12), which means a low "radiative conductivity," the downward transport of IR radiation through water is negligible. The surface sheds its energy upward by radiation and by exchange of latent heat with the overlying atmosphere via evaporation and precipitation. The sea-surface temperature measured by infrared sensors always refers to this skin layer. Interpreting this temperature as that of the upper mixed layer requires the assumption that the water is well stirred. Under calm conditions, this can lead to an error of several degrees Celsius, a substantial error in remote sensing applications.

Because the ocean has a high IR absorptance, then by *Kirchhoff's Law* (Chapter 5) it is also an efficient emitter of thermal radiation. The more physically relevant quantity is not the IR emission rate, but the *net* cooling, that is, the difference between the energy emitted and that received. Measurements show that as the ocean temperature increases, the net cooling actually decreases. This is due to the strong dependence of the atmospheric radiation upon the atmospheric water vapor content. As the ocean temperature increases, the excess of evaporation over condensation causes the absolute humidity to rise. As we will find in Chapter 12, the longwave opacity of the atmosphere is dominated by water vapor, particularly over the ocean. The increased warming of the ocean from downwelling atmospheric radiation more than compensates for the T^4 increase of the emitted surface flux.

1.6.4 Ocean Spectral Reflectance and Opacity

Due to its low visible reflectance (of the order of 7%) the ocean absorbs nearly all incident shortwave radiation. Despite the fact that the oceans receive most of their energy directly from the Sun (the remainder comes from sensible heat transport from the atmosphere) only this initial radiative "forcing" is of concern to us, since the issue of its subsequent transport is almost entirely one of dynamical transport by turbulence and large-scale overturning.

As shown in Fig. 1.12, the ocean's shortwave opacity exhibits a deep minimum in the visible spectrum, where incoming radiative energy can be deposited up to 100 m in depth.[16] This transparency "window" for visible radiation for liquid water is analogous to a similar atmospheric transparency window (see Fig. 9.10). That these two windows are both situated at wavelengths 0.4–0.6 μm, around the peak of the solar spectrum, is one of the most remarkable coincidences of nature, for it makes it possible for life to exist – indeed the water window was probably necessary for the rise of life itself, which is believed to have developed, if not originated, in the oceans. Just as remarkable is the very high opacity of water to the UV. For example, Fig. 1.12 shows that the *penetration depth* (the inverse of the absorption coefficient) for $\lambda < 0.35$ μm is less than 10 cm. Long before the primitive biosphere was able to generate a UV-protective ozone layer (estimated to have formed about one billion years ago) oceanic life and in particular phytoplankton could exist in a safe haven between depths of 10 cm and tens of meters, shielded from harmful UV, but bathed in visible radiation necessary for photosynthesis.

Finally, we mention the ocean color, which ranges from deep blue to green or greenish-yellow. A deep indigo color is characteristic of tropical waters, where comparitively little biology exists (oceanic "deserts"). At higher latitudes, the color changes from green-blue to green in polar regions. Consider first the reason for the blue color of

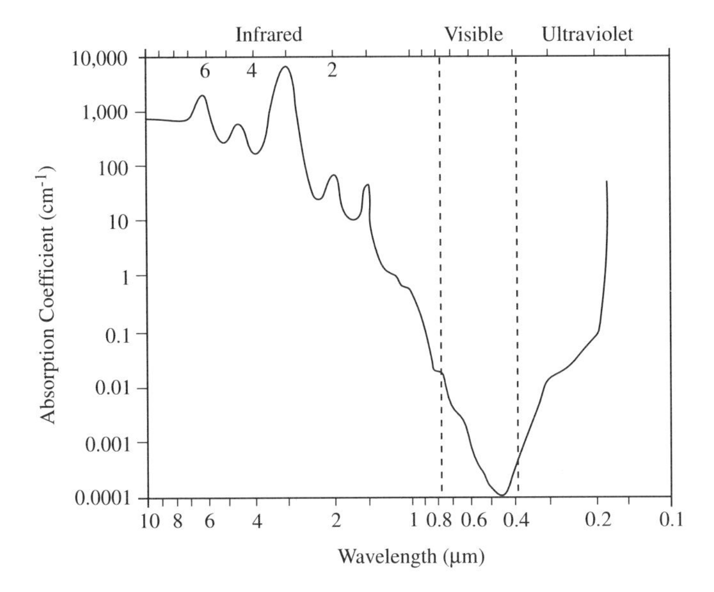

Figure 1.12 Typical absorption coefficient of the ocean (see Endnote 16).

open ocean waters containing very little particulate matter. Both absorption and scattering processes are involved. At a few meters' depth, the red and yellow wavelengths will be selectively removed, as shown in Fig. 1.12. Thus only blue light remains, and since the water molecules themselves are responsible for the scattering, these shorter (blue) wavelengths are favored by the Rayleigh scattering process (see §3.3.7). The upward scattered light will also selectively leave the surface, giving a deep-blue color to the water. In more fertile waters, green phytoplankton and yellow substances will absorb the blue light, and the color will shift toward the green end of the spectrum. Certain colors (such as the "red tide") are characteristic of particular phytoplankton blooms. The coloration of the ocean has in recent times become a subject of commercial interest. Satellite remote sensing of ocean color is a valuable enterprise to fisheries and for making surveys of ocean fertility in global change studies.

1.7 Remarks on Nomenclature, Notation, and Units

By *nomenclature* we mean the names and symbols of the various quantities in the theory, a subject covered in detail in Appendix A. By *units* we mean the particular standard: English, cgs, or mksa; the latter stands for meter-kilogram-second-ampere. The older cgs system (centimeter-gram-seconds) of units has been consistently used in the astrophysical literature. This system was also used in the atmospheric and oceanic sciences until the late 1950s. At that time the Système Internationale des Unités (SI),[17] already in wide use in the engineering community, was adopted by the various international and national scientific bodies. The SI system employs mksa units and encompasses a large variety of physical measurements. The situation is more unfortunate with regard to nomenclature. The beginning reader will encounter a bewildering variety of nomenclature particularly in the older literature. There has been some convergence toward a uniform system in the atmospheric literature. However, ocean optics has followed its own system of nomenclature and units. See Appendix D for more details.

In the following we present our notation scheme for future reference. We indicate units by enclosing the abbreviation of the quantity in brackets [ ]. We use the SI units in this book (discussed more fully in Appendix A). The abbreviations for the various mksa units are length in meters [m], mass in kilograms [kg], and time in seconds [s]. Derivative units are as follows: for wave frequency, ν, cycles per second or hertz [Hz]; wavelength λ, nanometers (1×10^{-9} m), [nm] or micrometers (1×10^{-6} m), [μm]; wavenumber [cm^{-1}]; radians [rad]; steradians [sr]; watts (joules per second), [W]; absolute temperature in kelvins [K] (*not* °K!); force in newtons [N]; hydrostatic pressure in newtons per unit area [$\mathrm{N} \cdot \mathrm{m}^{-2}$] (or pascals [Pa]); k- as a prefix means kilo-, as in [km] $= 1 \times 10^3$ [m], etc.

The reader will quickly note the mixing of usages of frequency ν, wavelength λ, and wavenumber $\tilde{\nu}$. In fact, this is deliberate. When referring to a spectral function, say $\tau(\nu)$, we prefer frequency as the variable, for the somewhat pedantic reason that frequency is a more fundamental property of the wave. Unlike wavelength and wavenumber its value is independent of the medium. However, when we want to identify specific parts of the spectrum, we will use units of wavelength (nm or μm in the visible and near-IR), and wavenumber (cm^{-1} in the thermal IR). We will not distinguish between the small differences of λ in vacuum and air.

We use boldface for vectors (**r**), a circumflex for unit vectors ($\hat{n}$), an overbar for an angular average ($\bar{I}_\nu$), and angle brackets for time, spatial, or frequency averages, for example, $\langle A \rangle$. A dot ($\cdot$) separating two vectors (such as $\mathbf{A} \cdot \mathbf{B}$) denotes the scalar product. Integrals are written with the differential closest to its integral sign. Thus we write

$$\int_0^{2\pi} d\phi \int_0^{\pi} d\theta F(\theta, \phi)$$

rather than

$$\int_0^{2\pi} \int_0^{\pi} F(\theta, \phi) d\theta d\phi.$$

We use the symbol $\equiv$ to denote "is defined by." We deviate from the practice of many texts in the ordering of the arguments of such quantities as the scattering phase function p (§5.3), which contains both the incident ($\hat{\Omega}'$) and outgoing ($\hat{\Omega}$) unit vectors. For example, many authors usually write $p(\hat{\Omega}, \hat{\Omega}')$ so that the second variable is the "input" and the first variable is the "output." We prefer to write it in the order from left to right, that is, from incident to outgoing directions, $p(\hat{\Omega}' \rightarrow \hat{\Omega})$, or we leave out the right arrow and write simply $p(\hat{\Omega}', \hat{\Omega})$. Another example is the hemispherical-directional reflectance (§5.2), which we write as $\rho(-2\pi, +\hat{\Omega})$, which tells us that uniform illumination (-2π) is incident from above on a surface, and we are interested in a specific upward (+) direction $\hat{\Omega}$ for the outgoing beam of radiation.

1.8 Summary

In this chapter we set the stage by introducing the basic ideas of absorption and emission and showing how they can explain the deviations of both the solar emission spectrum and the terrestrial infrared emission spectrum from that of a classic blackbody. We provided a brief summary of how hydrostatic balance affects the vertical structure of the bulk properties of the atmosphere and ocean. Temperature is affected by solar heating, IR opacity (the greenhouse effect), and by convective transport of heat. A discussion of line-of-sight optical paths was followed by an introduction of the concept of radiative equilibrium. A simple model of climate change illustrated the concept of radiative forcing and feedbacks. The temperature and density structure of the oceanic mixed layer was discussed in terms of solar heating and conductive mixing. We discussed optical properties of the ocean such as the transparent spectral window in the visible, the highly opaque spectral regions in the UV and IR, and a low visible reflectance.

Problems

1.1 An early study of the ice-albedo feedback problem was made by M. Budyko.[18] He showed for a 1% reduction in the solar constant that $\Delta T_s = -5°C$; and for a 1.5% reduction, $\Delta T_s = -9°C$. This sensitivity is much greater than the direct effect discussed in Example 1.2 and is due to the equatorward advance in the permanent ice line as the Earth cools. In the 1.5% reduction case, the ice line is at 50°N latitude, corresponding to the southward advance of Quaternary glaciation. For a slightly greater (1.6%) reduction, $\Delta T_s = -33°C$. Here the Earth is completely glaciated.

(a) Find the values of albedo $\bar{\rho}$ corresponding to these three solar flux reduction scenarios. Assume that $\mathcal{T}_{\text{eff}}$ remains fixed in Eq. 1.13.

(b) Plot $\bar{\rho}$ versus T_s and estimate the value of $\partial\bar{\rho}/\partial T_s$ for each scenario. Take into account the variation of F_{TOA} and T_s.

(c) Plot the climate sensitivity α and the gain f versus $\Delta F^s/F^s$. Comment on the steep increases of α and f as $\Delta F^s/F^s$ reaches the critical value of 0.016. How does this illustrate the "runaway glaciation" effect?

1.2 The diurnally averaged solar radiation at the top of the atmosphere available for heating may be parameterized as follows:[19]

$$F_a^s(\mu, t) = 342 \times \max\left\{0, 1.0 - 0.796\mu\cos[2\pi(t-0.75)] \right.$$
$$\left. + \{0.147\cos[4\pi(t-0.75)] - 0.477\}\frac{(3\mu^2-1)}{2}\right\},$$

where $\mu = \sin\lambda$ and λ is the latitude. Time t is measured in years, starting from the vernal equinox.

Plot isolines of F_a^s versus λ and t and compare with Fig. 1.10. What accounts for the differences between these two results?

1.3 Assume that the albedo of the Earth's system responds to a change in solar flux to *exactly* compensate, so as to leave the surface temperature unaffected. This is called *homeostasis*, and it is the hypothesized response envisioned in some "strong versions" of the *Gaia hypothesis*.[20] Show that if the solar flux were to increase by 1%, the albedo in this hypothetical system would respond by increasing by 0.7%.

1.4 The effect of the various greenhouse gases on the radiative forcing may be crudely estimated by setting $\mathcal{T}_{\text{eff}} = \exp[-\sum_i \tau_i]$, where τ_i is the opacity due to the ith species. Suppose that the concentration of a single species (say CO_2) were increased from τ_i to $\tau_i + \Delta\tau_i$.

(a) Show that the radiative forcing is given by $N_i = \Delta N_i = +F_{\text{TOA}}\Delta\tau_i$.

(b) Show that for a CO_2 doubling (for which $N_i = 4\ \text{W}\cdot\text{m}^{-2}$), the required change in opacity is $\Delta\tau_i/\sum_i \tau_i = 0.034$, that is, only a 3% change in effective opacity is implied. What factors tend to reduce the radiative effectiveness of a CO_2 doubling?

Notes

1 The visible spectrum is taken from a variety of sources. See *Atmospheric Ozone* , 1985, World Meteorological Organization Global Ozone Research and Monitoring Project Report No. 16, 355–62, 1985; Nicolet, M., "Solar spectral irradiances with their diversity between 120 nm and 900 nm," *Planet. Space Sci.*, **37**, 1249–89, 1989.

2 Rottman, G. J., T. N. Woods, and T. P. Sparn, "Solar stellar irradiance comparison experiment I. Instrument design and operation," *J. Geophys. Res.*, **98**, 10,667–78, 1993.

3 *Climate Change 1995: The Science of Climate Change, Contribution to Working Group 1 to the Second Assessment Report of the Intergovernmental Panel on Climate Change*, ed. by J. Houghton et al., Cambridge Univ. Press, Cambridge, p. 57, 1996.

4 Hanel, R. A. and B. J. Conrath, "Thermal emission spectra of the Earth and atmosphere obtained from the *Nimbus 4* Michelson Interferometer Experiment," NASA Report X-620-70-244, 1970.

5 Yellow substances are a large class of organic compounds derived mainly from the remains and metabolic products of marine plants and animals. (See Jerlov, N. G., *Optical Oceanography*, Elsevier, Amsterdam, 1968.)

6 The derivations of the dry and moist adiabatic lapse rates are found in, e.g., Chamberlain, J. W. and D. M. Hunten, *Theory of Planetary Atmospheres*, Academic Press, Inc., Orlando, 1987.

7 McClatchey, R. A. et al., *Optical Properties of the Atmosphere*, AFCRL-71-0279, Air Force Cambridge Research Laboratories, 85 pp., 1971.

8 Russell, J. M. III, L. L. Gordley, J. H. Park, S. R. Drayson, W. D. Hesketh, R. J. Cicerone, A. F. Tuck, J. E. Frederick, J. E. Harries, and P. J. Crutzen, "The halogen occultation

experiment, *J. Geophys. Res.*, **98**, 10,777–97, 1993; Anderson, G. P., S. A. Clough, F. X. Kneizys, J. H. Chetwynd, and E. P. Shettle, 1987: *AFGL Atmospheric Constituent Profiles (0–120 km)*, AFGL-TR-86-0110, AFGL (OPI), Hanscom AFB, MA 01736. The water vapor values below 25 km are taken from the U.S. 1976 Standard Atmosphere (Appendix C).

9 Tomasko, M., "The thermal balance of the lower atmosphere of Venus," in *Venus*, ed. D. M. Hunten et al., University of Arizona Press, Tucson, 1983.

10 Clancy, R. T., R. M. Haberle, E. Lellouch, Y. N. Billawala, B. J. Sandor and D. J. Rudy, "Microwave observations of a 1994 Mars global dust storm," *Bull. Amer. Astron. Soc.*, 26, 1130, 1994.

11 For example, see D. L. Hartmann, "Modeling climate change", in *Global Climate and Ecosystems Change*, ed. G. MacDonald and I. Settorio, Plenum Press, New York, pp. 97–140, 1990.

12 Graedel, T. and P. C. Crutzen, *Atmospheric Change: An Earth System Perspective*, W. H. Freeman, New York, 1993; *Climate Change 1994: Radiative Forcing of Climate Change and an Evaluation of the IPCC 1S92 Emission Scenarios*, J. T. Houghton, L. G. Meira Filho, J. Bruce, Hoesung Lee, B. A. Callandar, E. Haites, N. Harris and K. Maskell (eds.), Cambridge University Press, Cambridge, UK, 339 pp., 1994.

13 Pickard, G. L. and W. J. Emery, *Descriptive Physical Oceanography*, 4th ed., Pergamon Press, New York, 1982.

14 Adapted from Pickard and Emery; see Endnote 13.

15 Adapted from Pickard and Emery; see Endnote 13.

16 Figure 1.12 is adapted from N. Wells, *The Atmosphere and Ocean*, Taylor and Francis, London, Philadalphia, 1986.

17 The SI system of units is described in Page, C. H. and P. Vigoureux (eds.), *The International System of Units (SI)*, National Bureau of Standards Special Publication 330, Washington, DC, 42 pp., 1972. The Radiation Commission of the International Association of Meteorology and Atmospheric Physics has adopted a standardized system of terminology for atmospheric radiation: Raschke, E. (ed.), *Terminology and Units of Radiation Quantities and Measurements*, Radiation Commission (IAMAP), Boulder, Colorado, 17 pp., 1980. However, this system has not yet affected the ocean optics and astrophysical literature.

18 Budyko, M., "The effects of solar radiation variations on the climate of the Earth", *Tellus*, **21**, 611–19, 1969. See also §8.5.3 in K-N. Liou, *An Introduction to Atmospheric Radiation*, Academic Press, New York, 1980.

19 Saravanan, R. and J. C. McWilliams, "Multiple equilibria, natural variability and climate transitions in an idealized ocean–atmosphere model," *J. Climate*, **8**, 2296–323, 1995.

20 Margulis, L. and J. E. Lovelock, "Gaia and geognosy," in *Global Ecology*, Academic Press, New York, 1989.

Chapter 2

Basic State Variables and the Radiative Transfer Equation

2.1 Introduction

In this book we are mostly concerned with the flow of radiative energy through atmospheres and oceans.[1] We will ignore polarization effects, which means that we disregard the Q, U, and V components of the Stokes vector[2] and consider only the first (intensity) component I. This approach is known as the *scalar approximation*, in contrast to the more accurate *vector* description. In general, this approximation is valid for longwave radiation where thermal emission and absorption dominate scattering processes. However, at shorter wavelengths where scattering is important, the radiation is generally partially polarized. For example, polarization is a basic part of a description of scattering of sunlight in a clear atmosphere or in pure water (so-called Rayleigh scattering). Generally, a coupling occurs between the various Stokes components, and an accurate description requires the full Stokes vector representation.

Of central importance in the theory is the *scalar intensity*, which plays as central a role in radiative transfer theory[3] as the *wave function* plays in quantum theory. Its full specification as a function of position, direction, and frequency variables conveys all of the desired information about the radiation field (except for polarization).

In this chapter, we define the basic state variable of the theory, the scalar intensity. We first review the most basic concepts of geometrical optics, those of pencils and beams of light. We then define the various *state variables* in a transparent medium, involving flow of radiative energy in beams traveling in specific directions and over a hemisphere. Several theorems governing the propagation of intensity are described. The Extinction Law is stated in both differential and integral forms. The radiative transfer equation is shown to be a consequence of extinction and the existence of radiation sources.

A brief description of our notation is in order. Radiation state variables are described in terms of both *spectral* (or *monochromatic*) quantities and frequency-integrated quantities. Frequency is measured in cycles per second or *hertz*, abbreviated as [Hz]. For spectral quantities, we may visualize a small frequency interval over which all properties of the radiation and its interaction with matter are constant. A general quantity, f_ν, is identified with a frequency (ν) subscript if it is defined on a per-frequency basis. If f is a *function* of frequency, it is written $f(\nu)$. Of course, spectral quantities can also be expressed[4] as a function of wavelength, λ [nm] or [μm], or wavenumber, $\tilde{\nu}$ [cm^{-1}]. Frequency-integrated quantities are of the form $\int d\nu f_\nu$.

2.2 Geometrical Optics

It is important to point out that the basic assumptions of the radiative transfer theory are the same as those of geometrical optics. Indeed if there is no scattering or thermal emission, the radiative transfer equation reduces to the intensity law of geometrical optics. The concept of a sharply defined *pencil* of radiation was first defined in geometrical optics. A radiation pencil is realized physically by allowing light emanating from a point source to pass through a small opening in an opaque screen. This light may be viewed by allowing it to fall on a second screen. If we were to examine this spot of light near its boundary, we would notice that the edge would not be geometrically sharp. Instead we would find a series of bright and dark bands, called *diffraction fringes*. The size of the region over which these bands occur is of the order of the wavelength of light, λ. If the diameter of the cross-sectional area of the pencil is very much larger than λ, diffraction effects are small, and we may speak of a sharply bounded pencil of rays. The propagation of light may then be described in purely geometrical terms, and energy transport will occur along the direction of the light rays. These rays are not necessarily straight lines. In general they are curves whose directions are determined by the gradient of the *index of refraction*, m. The real part of m is the ratio of the speed of energy propagation in a vacuum to that in the medium. It is the most important light–matter interaction parameter in geometrical optics. Absorption along the ray may be shown to depend upon the imaginary part of the complex index of refraction. The fact that m varies with frequency is known as the phenomenon of *dispersion*.

In geometrical optics theory, interference and diffraction of light are unimportant. The same is true of the radiative transfer theory. In this book we set the index of refraction equal to a constant value pertaining to either air or water. Thus, we ignore both dispersion and ray bending. However, variation of m with position giving rise to refraction or ray bending cannot be ignored in remote sensing applications in which the rays traverse a significant atmospheric path length. Refraction is also important in radiative transfer through the ocean–atmosphere interface, at which the index of refraction changes abruptly. In this case it is usually sufficient to assume that the index of refraction is unity for air and equal to 1.33 for water.

For our purposes the concept of incoherent (noninterfering) beams of radiation is more convenient than the concept of ray pencils. We define a *beam* in analogy with a plane wave: It carries energy in a specific propagation direction (the ray direction) and has infinite extent in the transverse direction.[5] We will use *ray* and *beam direction* synonymously throughout this book. When a beam of sunlight is incident on a scattering medium (e.g., the Earth's atmosphere), it splits into an infinite number of incoherent beams propagating in different directions. Similarly, when a beam is incident on a diffusely reflecting surface (e.g., a "rough" ocean surface or a plant canopy), the reflected radiation splits into an infinite number of incoherent beams traveling in different directions. However, if a beam is incident on a perfectly smooth, plane interface, it will give rise to one reflected and one transmitted beam. The directions of these two beams follow from the geometrical optics laws of reflection and refraction (*Snell's Law*), while their states of polarization follow from *Fresnel's equations* (Appendix E).

It will also be convenient to define an *angular beam* as an incoherent sum of beams propagating in various directions inside a small cone of solid angle $d\omega$ centered around the direction $\hat{\Omega}$.

2.3 Radiative Flux or Irradiance

We consider the flow of radiative energy across a surface element dA, located at a specific position, and having a unit normal $\hat{n}$ (see Fig. 2.1). As mentioned above, the energy flow is visualized as being carried by incoherent (noninteracting) angular beams of radiation moving in all directions. Because the beams traveling in different directions do not interact, we may treat them separately. The net rate of radiative energy

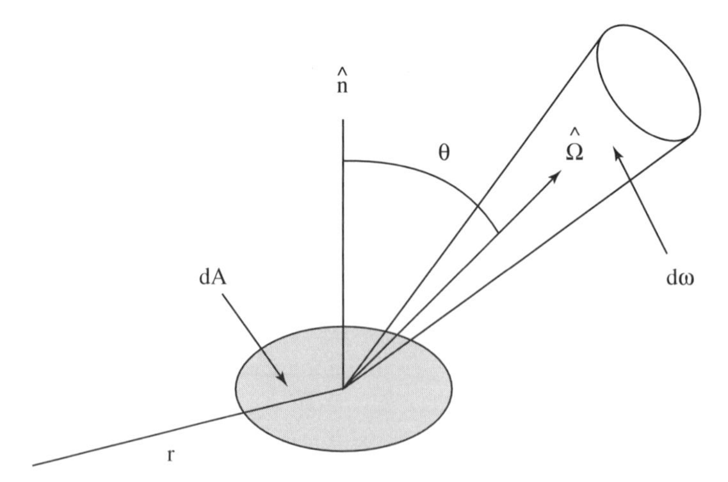

Figure 2.1 The flow of radiative energy carried by a beam in the direction $\hat{\Omega}$ through the transparent surface element dA. The flow direction $\hat{\Omega}$ is at an angle θ with respect to the surface normal $\hat{n}$ ($\cos\theta = \hat{n} \cdot \hat{\Omega}$).

flow, or power, per unit area within the small spectral range ν to $\nu + d\nu$ is called the spectral *net flux*. The spectral net flux F_ν is expressed in terms of the net energy d^3E that crosses the area dA, in the time interval $t, t + dt$, within the frequency interval $\nu, \nu + d\nu$, as

$$F_\nu = \frac{d^3E}{dAdtd\nu} \quad [\mathrm{W \cdot m^{-2} \cdot Hz^{-1}}].$$

The quantities d^3E and $dAdtd\nu$ are third-order differential quantities[6] that are considered positive if the flow is into the hemisphere centered on the direction $\hat{n}$ and negative if the flow is into the opposite hemisphere centered on $-\hat{n}$. A general radiation field consists of angular beams traveling in all directions. It is convenient to separate the energy flow into oppositely directed positive energy flows in two separate hemispheres d^3E^+ and d^3E^-. Each of these partial flows carries a positive amount of energy. We define the spectral *hemispherical fluxes* F_ν^+ and F_ν^- as

$$F_\nu^+ = \frac{d^3E^+}{dAdtd\nu}, \qquad F_\nu^- = \frac{d^3E^-}{dAdtd\nu}. \tag{2.1}$$

The net energy flow in the positive direction is $d^3E = d^3E^+ - d^3E^-$. In the same way the spectral *net flux* is written as the difference of two positive quantities $F_\nu = F_\nu^+ - F_\nu^-$. Summing over all frequencies, we obtain the *net flux*, or *net irradiance*:

$$F = \int_0^\infty d\nu F_\nu \quad [\mathrm{W \cdot m^{-2}}].$$

The spectral net flux F_λ within a small wavelength range λ to $\lambda + d\lambda$ is defined within the wavelength interval $d\lambda$, related to the frequency interval $d\nu$. Thus, $F_\nu\,|d\nu| = F_\lambda|d\lambda|$ and

$$F_\lambda = F_\nu\,|d\nu/d\lambda| = F_\nu(c/\lambda^2) \quad [\mathrm{W \cdot m^{-2} \cdot nm^{-1}}].$$

Similarly, the spectral net flux per wavenumber ($\tilde{\nu} = \nu/c$) is

$$F_{\tilde{\nu}} = F_\nu|d\nu/d\tilde{\nu}| = F_\nu c \quad [\mathrm{W \cdot m^{-2} \cdot cm}].$$

The spectral net flux is also the component of a spectral *radiative flux vector* $\mathbf{F}_\nu(\mathbf{r})$, which points in the reference direction $\hat{\Omega}$,

$$\mathbf{F}_\nu(\mathbf{r}) = F_\nu(\mathbf{r})\hat{\Omega}.$$

(This should not be confused with the four-dimensional *Stokes vector*.) The vector $\mathbf{F}_\nu(\mathbf{r})$ is the generalization of the *Poynting vector* used to describe streams traveling in arbitrary directions. Clearly, the scalar spectral net flux is the component of $\mathbf{F}_\nu(\mathbf{r})$ in the direction $\hat{\Omega}$, that is,

$$F_\nu(\mathbf{r}) = \hat{\Omega} \cdot \mathbf{F}_\nu(\mathbf{r}).$$

That part of the net flux originating in reflection or emission from a surface is sometimes referred to as the *radiant exitance*.

2.4 Spectral Intensity and Its Angular Moments

The net flux through a surface element dA depends upon the cumulative effect of all the angular beams crossing it in different directions. This quantity conveys little information about the *directional dependence* of the energy flow. A more precise description of the energy flow is desirable, especially in remote sensing applications. In principle, what is needed is the specification of the vector flux for every direction. Alternatively, the directional dependence can be visualized in terms of a *distribution* of energy flow in all 4π directions. In any particular direction the energy flow is associated with the angular beam traveling in that direction.

We denote by d^4E any small subset of the total energy flow within a solid angle $d\omega$ around a certain direction $\hat{\Omega}$ in the time interval dt and over a small increment of frequency or energy. We require that this subset of radiation has passed through a surface element dA whose orientation is defined by its unit normal $\hat{n}$. The geometry is illustrated in Fig. 2.1. The angle between $\hat{n}$ and the direction of propagation $\hat{\Omega}$ is denoted by θ. The energy per unit area, per unit solid angle, per unit frequency, and per unit time defines the *spectral intensity*.[7]

The *spectral intensity* or *radiance* I_ν is defined as the ratio

$$I_\nu = \frac{d^4E}{\cos\theta dAdtd\omega d\nu} \quad [\mathrm{W \cdot m^{-2} \cdot sr^{-1} \cdot Hz^{-1}}]. \tag{2.2}$$

Note that, in addition to dividing by $d\omega d\nu dt$, we have divided by the factor $\cos\theta = \hat{n}\cdot\hat{\Omega}$. This factor multiplied by dA is the *projection* of the surface element onto the plane normal to $\hat{\Omega}$. Note also that if $\hat{n}$ and $\hat{\Omega}$ are directed into opposite hemispheres, then $\hat{n}\cdot\hat{\Omega}$ is negative. The energy flow is also negative in this case, by definition, so that the ratio $d^4E/\cos\theta$ remains positive. *Intensity is always positive.* The infinitesimal cone of solid angle $d\omega$ is visualized as a truncated cone with base $dA\cos\theta$.

Intensity is seen to be a *scalar quantity*, describing an angular variation of energy flow and how this angular variation itself depends upon position. In addition to the dependence on the coordinate variable $\mathbf{r}$, the angular variable $\hat{\Omega}$, and the frequency variable ν, it may also be time dependent, bringing to seven the number of independent variables: three in space, two in angle, one in frequency, and one in time.[8] Time dependence must be considered when the incident radiation field or the properties of the medium change over the very short time scale of light transport within the medium. Except in active remote sensing applications, such as lidar and radar, this complication may be confidently ignored. Also, planetary media have approximate planar uniformity. Therefore, it often suffices to use only one position variable (height in the atmosphere and depth in the ocean). This reduces the number of variables to four, which still poses a formidable task. Fortunately, such a detailed specification is seldom necessary, particularly in calculations of the radiative energy balance. It often happens that only a few angular moments are needed. Our usual situation is such that only two variables (frequency and a height variable) remain and the mathematics becomes tractable. This is particularly true in problems where the frequency enters as a parameter, so that the problem can be solved for a single frequency and then repeated for all frequencies contributing to the radiation field. In the following sections, we examine two angular moments[9] of the intensity: the flux and the energy density.

2.4.1 Relationship between Flux and Intensity

This relationship follows from Eq. 2.2, which we rewrite as

$$d^4E = I_\nu \cos\theta \, dA \, dt \, d\omega \, d\nu. \tag{2.3}$$

The rate at which energy flows into each hemisphere is obtained by integration of the separate energy flows

$$d^3E^+ = \int_+ d^4E^+, \qquad d^3E^- = \int_- d^4E^-. \tag{2.4}$$

In these expressions, the subscript (+) on the integral signs denotes integration over the hemisphere defined by $+\hat{n}$. Similarly, we use a (−) subscript to denote integration over the hemisphere defined by $-\hat{n}$. Use of Eqs. 2.1, 2.3, and 2.4 yields expressions for the half-range fluxes defined as positive quantities:

$$F_\nu^+ = \frac{d^3E^+}{dAdtd\nu} = \int_+ d\omega \cos\theta \, I_\nu, \qquad F_\nu^- = \frac{d^3E^-}{dAdtd\nu} = -\int_- d\omega \cos\theta \, I_\nu. \tag{2.5}$$

Combination of the half-range fluxes yields the *net flux*.

The spectral net flux is the integration of $I_\nu \cos\theta$ over all solid angles,

$$F_\nu = F_\nu^+ - F_\nu^- = \int_{4\pi} d\omega \cos\theta \, I_\nu \quad [\mathrm{W \cdot m^{-2} \cdot Hz^{-1}}]. \tag{2.6}$$

Note that in the latter of Eqs. 2.5, $\cos\theta$ is negative so that $F_\nu^- > 0$. Similarly, in Eq. 2.6, $\cos\theta$ is negative over half the angular integration domain. The negative contribution to F_ν from the component F_ν^- tends to offset the positive contribution from the oppositely directed component F_ν^+. If the spectral intensity $I_\nu(\mathbf{r}, \hat{\Omega})$ at a point is independent of direction $\hat{\Omega}$, it is said to be *isotropic*. If it is independent of position, it is called *homogeneous*. The spectral intensity is both isotropic and homogeneous in the special case of *thermodynamic equilibrium*, where the net flux is zero everywhere in the medium. This follows from the fact that even though the hemispherical fluxes are finite, they are of equal magnitude and opposite direction. Therefore no *net* energy flow can occur in this equilibrium case (see §4.3.1 for more details).

2.4.2 Average Intensity and Energy Density

Averaging the directionally dependent intensity over all directions at a given point **r** yields a scalar quantity dependent upon the position variable **r**. This quantity is called the *spectral average intensity*:

$$\bar{I}_\nu(\mathbf{r}) = (1/4\pi) \int_{4\pi} d\omega I_\nu(\mathbf{r}, \hat{\Omega}) \quad [\mathrm{W \cdot m^{-2} \cdot Hz^{-1}}]. \tag{2.7}$$

In general, the overbar ($\bar{\ }$) indicates an average over the sphere. $\bar{I}_\nu$ is proportional to the spectral *energy density* of the radiation field $\mathcal{U}_\nu$, the radiative energy that resides within a unit volume. The *actinic flux* used by photochemists is simply $4\pi \bar{I}_\nu$. The spectral energy density $\mathcal{U}_\nu$ can be related to the intensity in the following way. Consider a small cylindrical volume element of area dA having a unit normal $\hat{n}$, whose length, cdt, is the distance light travels in time dt (c is the light speed). Let the direction of propagation $\hat{\Omega}$ be at an angle θ with respect to $\hat{n}$. The volume of this element is $dV = dA \cos\theta cdt$, where $dA \cos\theta$ is the projection of the surface element dA onto the plane normal to $\hat{\Omega}$. The energy per unit volume per unit frequency residing in the volume between t and $t + dt$ is therefore

$$d\mathcal{U}_\nu = \frac{d^4E}{dVd\nu} = \frac{I_\nu \cos\theta dA dt d\nu d\omega}{dA \cos\theta c dt d\nu} = \frac{I_\nu}{c} d\omega. \tag{2.8}$$

If we consider the energy density in the vicinity of a collection of incoherent beams traveling in all 4π directions, we must integrate this expression over all solid angles $d\omega$ to obtain

$$\mathcal{U}_\nu = \int_{4\pi} d\mathcal{U}_\nu = \frac{1}{c} \int_{4\pi} d\omega I_\nu = \frac{4\pi}{c} \bar{I}_\nu \quad [\mathrm{J \cdot m^{-3} \cdot Hz^{-1}}]. \tag{2.9}$$

The *total energy density* is the sum over all frequencies:

$$\mathcal{U} = \int_0^\infty d\nu \mathcal{U}_\nu \quad [\mathrm{J \cdot m^{-3}}]. \tag{2.10}$$

In the following examples we will examine some special cases of particularly simple symmetry, which have proven to be useful mathematical idealizations.

Example 2.1 Isotropic Distribution

Let us assume that the spectral intensity is independent of direction, that is, $I_\nu(\hat{\Omega}) = I_\nu =$ constant. This assumption applies to a medium in thermodynamic equilibrium and is approximately valid deep inside a dense medium. The flux and energy density are easily evaluated as

$$F_\nu^+ = F_\nu^- = \pi I_\nu, \tag{2.11}$$

$$F_\nu = F_\nu^+ - F_\nu^- = 0, \tag{2.12}$$

$$\mathcal{U}_\nu = \frac{4\pi I_\nu}{c}. \tag{2.13}$$

Note that the two contributions to the flux from the opposite hemispheres cancel. Mathematically, this occurs because $\cos\theta$ is an odd function of θ in the interval $[0, \pi]$, and physically because of a balance between upward- and downward-flowing beams.

Example 2.2 Hemispherically Isotropic Distribution

This situation is the essence of the *two-stream approximation* (Chapter 7). It yields the simplest nontrivial description of diffuse radiation having a nonzero net energy transport. Let I_ν^+ denote the (constant) value of the intensity everywhere in the positive hemisphere, and let I_ν^- denote the (constant) value of the intensity everywhere in the opposite hemisphere. $I_\nu^+ \neq I_\nu^-$ by assumption. For a slab medium, an angular distribution of intensity has been replaced by a pair of numbers at each point in the medium.[10] In some situations, this is too inaccurate. However, in situations where the spectral intensity is close to being isotropic, it leads to results of surprisingly high accuracy, as discussed in Chapter 7. As in the isotropic case, we solve for the various moments:

$$F_\nu = I_\nu^+ \int_0^{2\pi} d\phi \int_0^{\pi/2} d\theta \sin\theta\cos\theta + I_\nu^- \int_0^{2\pi} d\phi \int_{\pi/2}^{\pi} d\theta \sin\theta\cos\theta = \pi(I_\nu^+ - I_\nu^-),$$

$$\begin{aligned} \mathcal{U}_\nu &= \frac{I_\nu^+}{c} \int_0^{2\pi} d\phi \int_0^{\pi/2} d\theta \sin\theta + \frac{I_\nu^-}{c} \int_0^{2\pi} d\phi \int_{\pi/2}^{\pi} d\theta \sin\theta \\ &= \frac{2\pi}{c}(I_\nu^+ + I_\nu^-). \end{aligned} \tag{2.14}$$

Example 2.3 Collimated Distribution

This is a commonly used approximation for the intensity of an incoming solar beam, in which the finite size of the Sun is ignored.[11] We write the solar intensity in the general direction $\hat{\Omega}$ as

$$I_\nu^{\rm s}(\hat{\Omega}) = F_\nu^{\rm s}\delta(\hat{\Omega} - \hat{\Omega}_0). \tag{2.15}$$

$F_\nu^{\rm s}$ is the flux carried by the beam across a plane normal to the direction of incidence $\hat{\Omega}_0$. $\hat{\Omega}_0$ has the polar angle θ_0 and the azimuthal angle ϕ_0. $\delta(\hat{\Omega} - \hat{\Omega}_0) = \delta(\phi - \phi_0)\delta(\cos\theta - \cos\theta_0)$ is the two-dimensional *Dirac δ-function*. According to the mathematical properties of the δ-function,[12] the intensity is zero in all but a single direction $\hat{\Omega}_0$ (where it is infinite). Equation 2.15 is physically meaningful only when it is integrated over some finite solid angle. Note that since the δ-function has the units of inverse solid angle, $I_\nu^{\rm s}$ has the correct units (energy per unit area, per unit frequency, per unit time, per unit solid angle). The moments for a collimated beam are

$$F_\nu = F_\nu^{\rm s} \int_0^{2\pi} d\phi\delta(\phi - \phi_0) \int_{-1}^{1} du u\delta(u - \mu_0) = F_\nu^{\rm s}\cos\theta_0,$$

$$\mathcal{U}_\nu = \frac{F_\nu^{\rm s}}{c} \int_0^{2\pi} d\phi\delta(\phi - \phi_0) \int_{-1}^{+1} du\delta(u - \mu_0) = \frac{F_\nu^{\rm s}}{c}.$$

Here we have set $u \equiv \cos\theta$ and $\mu_0 \equiv \cos\theta_0$.

Example 2.4 Azimuthally Symmetric Distribution

In this special case, the intensity distribution is constant with azimuthal angle ϕ. Of course, all other variables are held constant. This describes the singly scattered part of the radiation field in a coordinate system in which the z axis is along the incoming direction of a collimated beam. It is also valid for the multiply scattered radiation in a *slab* or *plane-parallel* geometry when the scattering phase function (§5.3) is *isotropic*, a frequently used idealization in the theory of radiative transfer (see §6.2). Further, it applies to a radiation field produced by thermal emission and, of course, to an isotropic or hemispherically isotropic distribution. The angular moments are

$$F_\nu = \int_0^{2\pi} d\phi \int_{-1}^{+1} du u I_\nu(u) = 2\pi \int_{-1}^{+1} du u I_\nu(u),$$

$$\mathcal{U}_\nu = \frac{1}{c} \int_0^{2\pi} d\phi \int_{-1}^{+1} du I_\nu(u) = (4\pi/c)\bar{I}_\nu.$$

2.5 Some Theorems on Intensity

Perhaps the most important property of the intensity is expressed in the following theorem:

Theorem I *In a transparent medium, the intensity is constant along a ray.*

To prove this we begin by imagining two arbitrarily oriented surface elements separated by a distance r (see Fig. 2.2). A ray PP' connects the two areas and defines the direction $\hat{\Omega}$. From Eq. 2.2 the amount of energy crossing dA in time dt and entering the solid angle $d\omega$ is given by

$$d^4E = I_\nu(P, \hat{\Omega}) dA \cos\theta \; d\nu \; dt \; d\omega.$$

Similarly, the energy passing through dA' in time dt into solid angle $d\omega'$ (subtended by the surface element dA) is the same amount of energy

$$d^4E = I_\nu(P', \hat{\Omega}) dA' \cos\theta' \; d\nu \; dt \; d\omega'.$$

Our objective is to prove that $I_\nu(P, \hat{\Omega}) = I_\nu(P', \hat{\Omega})$. Solving for $I_\nu(P, \hat{\Omega})$ from the first of the above equations and substituting the expression for d^4E from the second equation, we have

$$I_\nu(P, \hat{\Omega}) = \frac{d^4E}{dA \cos\theta \; d\nu \; dt \; d\omega} = \frac{I_\nu(P', \hat{\Omega}) dA' \cos\theta' d\omega'}{dA \cos\theta \; d\omega}.$$

But since $d\omega = dA' \cos\theta'/r^2$ and $d\omega' = dA \cos\theta/r^2$, we see that $I_\nu(P, \hat{\Omega}) = I_\nu(P', \hat{\Omega})$. Theorem I may be generalized to apply to a beam that is reflected any number of times by perfectly reflecting mirrors:

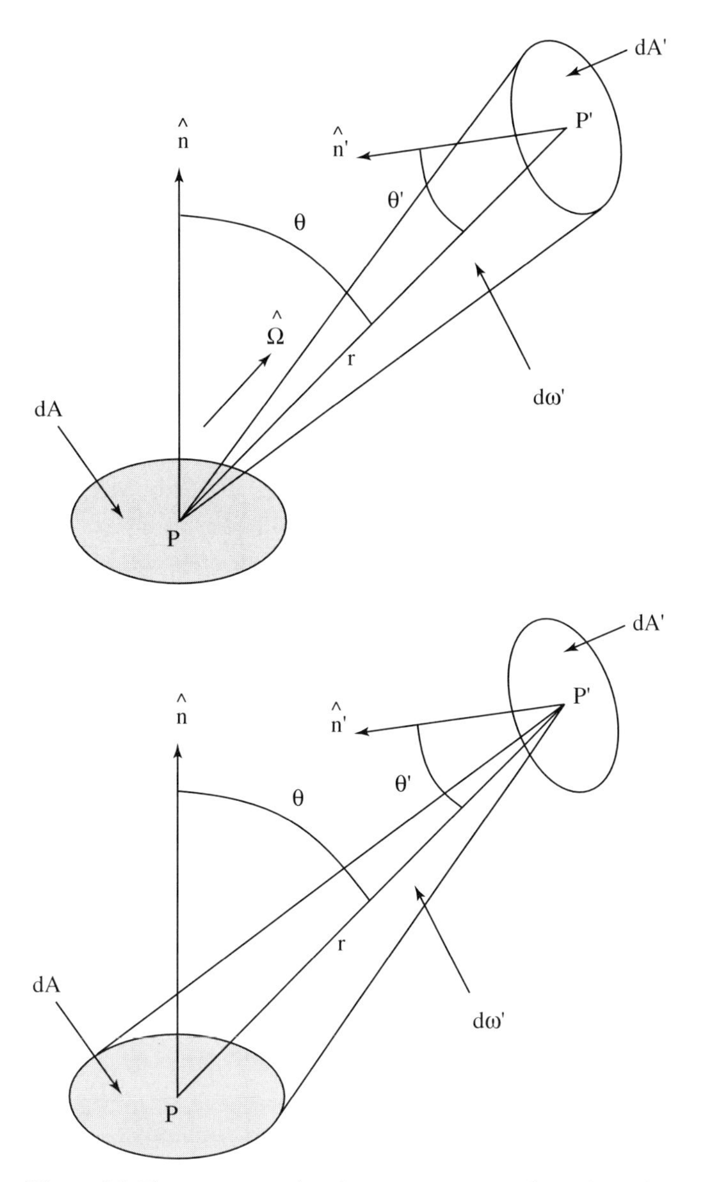

Figure 2.2 The energy crossing the transparent area dA and entering into the solid angle $d\omega$ is the same as that which crosses the area dA' and is contained within the solid angle $d\omega'$. r is the distance between P and P'.

Theorem II *The intensity remains constant along a ray upon perfect reflection by any mirror or combination of mirrors.*

This theorem may be proven by a method similar to that of Theorem I (see Problem 2.1). A third property of the intensity applies to refraction in a transparent medium of variable index of refraction. The theorem also applies for discrete changes in $m(\nu)$ as long as reflection at the interfaces can be neglected.

Theorem III *The quantity $I_\nu/m^2(\nu)$ remains constant along a ray in a transparent medium, provided that the reflectance at each interface can be neglected.*

Again, a similar method is used to prove this theorem.[13] The law of refraction (*Snell's Law*) is used to relate the incident and refracted angles (see Problem 2.3). The quantity $I_\nu/m^2(\nu)$ is called the *basic radiance*. Clearly, Theorem I is a special case of Theorem III.

2.5.1 Intensity and Flux from an Extended Source

At first sight Theorems I–III may appear quite strange – we might expect the intensity to weaken as the beam diverges. Let us consider a specific example of an interplanetary traveler moving outward from the Sun. We will assume that the Sun emits like an isothermal blackbody (see §4.3.1). This means that the intensity emitted is isotropic and constant over the entire surface. We assume also that there is negligible interplanetary attenuation. To determine the intensity at a distance r from the Sun's center we need to determine the energy ΔE per unit time and per unit area arriving within a finite solid angle $\Delta\omega$ centered on the Sun and just barely containing the Sun. We now move away from the Sun to a distance $r' > r$ and make another measurement $\Delta E'$ within the solid angle $\Delta\omega'$, again just barely containing the Sun. If the measurements are made along the same ray then from conservation of energy the ratios

$$\frac{\Delta E}{\Delta\omega} = \frac{\Delta E'}{\Delta\omega'} = \text{constant.}$$

Although the amount of energy $\Delta E'$ contained within the smaller solid angle $\Delta\omega'$ is smaller (by the ratio of the squares of the distances (r^2/r'^2)) the ratio of the solid angles is also smaller by the same factor. However, the *fluxes* at the two points r and r' are different, because the energy flow received at the two distances *per unit area* are clearly different. We can relate flux and intensity from Eq. 2.6. If I_ν is the (hemispherically isotropic) value of the intensity leaving the Sun's surface, then since the intensity for all beams not intersecting the Sun is zero, the net flux at a distance r from the center of the Sun is

$$F_\nu(r) = \int_{\Delta\omega} d\omega I_\nu \cos\theta = I_\nu \int_0^{2\pi} d\phi \int_0^{\alpha} d\theta \sin\theta\cos\theta,$$

where α is the angle subtended by the Sun's radius a at the distance r from the center of the Sun.

The integration is easily carried out, using $\sin\alpha = a/r$, to yield

$$F_\nu = 2\pi I_\nu \int_0^{\alpha} d\theta \sin\theta\cos\theta = \pi I_\nu \sin^2\alpha = \frac{\pi I_\nu a^2}{r^2}.$$

From Eq. 2.14, πI_ν is the outward spectral flux at the surface, $F_\nu(r = a)$. $F_\nu(r)$ varies as the *inverse square* of the distance r *as a result of the constancy of the intensity along a ray*. Of course the inverse-square law is demanded by energy conservation.

2.6 Perception of Brightness: Analogy with Radiance

A quantity analogous to intensity is *brightness*, which is actually the perception of the human eye of differing *luminances* of two objects in a scene. *Luminance* is the frequency integral of the intensity weighted by the spectral response function of the human eye.[14] *Illumination* is the photometric counterpart of flux. A difficulty with the concept of brightness is that the same object viewed against a dark background may appear relatively "bright," whereas when seen against a bright background it may appear relatively "dark." For example, falling snowflakes seen against the bright sky appear to be black, but seen against a dark woods they appear to be white. The analogy of brightness with luminance is more satisfactory if we consider the appearance of an extended object against a background of fixed luminance.

Consider the extended object to be a lady's white dress, which we assume reflects visible light like a *Lambert reflector* (§5.2). This amounts to assuming that the reflected intensity is quasi-isotropic. In Fig. 2.3 the lady's dress does *not* appear to "dim" as she walks away from an observer. Although she recedes in the distance, and the solid angle subtended by the dress diminishes as the square of the distance, the dress appears to the

Figure 2.3 Successive images of lady in white dress.

eye to remain at a constant brightness. Thus although the total energy (illumination) received by the eye decreases, the energy falling on an individual pixel (analogous to luminance or brightness) remains constant. For our purpose, we can directly relate intensity and luminance. Thus, from Theorem I her image does not appear to become dimmer with distance. However, as she recedes further, her image will eventually shrink to the size of less than one pixel. At this point (not shown in the figure), any further increase in distance between object and observer means that the energy falling on a pixel also begins to decrease. Thus, the eye switches over from perceiving luminance to illumination beyond this distance (about 2.5 km). Therefore the lady's image becomes a point source, which dims inversely as the distance squared. The angle subtended by the lady has become *smaller* than the angular resolution of the eye and the eye senses (luminous) flux.

If one were equipped with a telescope, the magnified image would of course cover a larger number of pixels. However, according to Theorem II, the amount of light falling on a single pixel (and therefore the brightness) is unchanged. The magnified image will *never* appear to be brighter, only bigger, even in the largest telescope. The only difference is that the lady's dress would be visible as an extended object to a greater distance. As before, a constant intensity (brightness) is perceived until her image subtends an angle smaller than the combined resolving power of the telescope and observer. The above statements follow directly from Theorems II and III (see also Problems 2.1 and 2.2).

2.7 The Extinction Law

We now introduce the specific interaction properties that comprise the essential elements in the radiative transfer theory. They are defined in terms of the most important principle in the theory, the *Extinction Law*, more commonly known as *Beer's Law*,[15] the *Beer–Lambert Law*, or the *Beer–Lambert–Bouguer Law*. Consider a small volume dV containing matter described by n [m^{-3}], the *concentration*, defined as the ratio of the number of particles dN divided by the volume dV. The particles are assumed to be *optically active*, that is, they have a nonnegligible effect on radiation that passes through the volume. Other (optically inactive) particles may be present, but these may be ignored for the present purposes. For convenience, the volume dV is considered to be a slab of infinitesimal thickness ds and area dA (Fig. 2.4). Suppose a beam of radiation is incident normally on the slab. From Eq. 2.3 the differential of energy falling on the front surface is

$$d^4E = I_\nu \, dA \, dt \, d\nu \, d\omega.$$

As the beam of radiation passes through the slab, it interacts with the particles through either absorption or scattering and a reduced amount of energy emerges at the opposite side. The beam of radiation is said to have suffered *extinction*. The energy

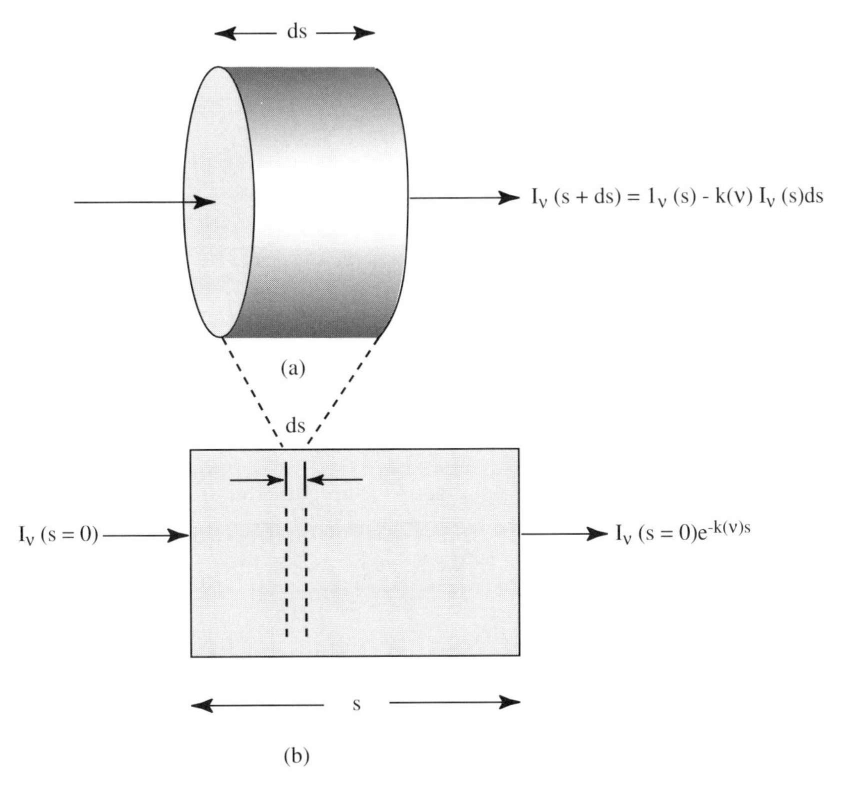

Figure 2.4 (a) Intensity passing through a thin slab suffers extinction proportional to the path length ds. (b) Intensity passing through a finite path length s suffers exponential extinction.

emerging from the back face of the slab is given by the original energy, less an amount given by

$$\delta(d^4E) = dI_\nu \, dA \, dt \, d\nu \, d\omega, \tag{2.16}$$

where dI_ν is the differential loss in intensity along the length ds.

It is found experimentally that the degree of weakening depends linearly upon both the incident intensity and the amount of optically active matter along the beam direction (proportional to the length ds):

The Extinction Law (differential form)

$$dI_\nu \propto -I_\nu ds. \tag{2.17}$$

We define the constant of proportionality in Eq. 2.17 to be the *extinction coefficient* k. There are three different ways of defining extinction: in terms of the path length itself ds, the *mass path* $d\mathcal{M} = \rho ds$, or the *column number* $d\mathcal{N} = nds$ of the absorbing/scattering species concentration. Here ρ and n are the *mass density* [kg $\cdot$ m^{-3}] and *particle density* [m^{-3}] of the optically active gas. This leads to the following three definitions of extinction:

$$k(\nu) \equiv -\frac{dI_\nu}{I_\nu ds} \quad \textit{extinction coefficient } [\mathrm{m}^{-1}], \tag{2.18}$$

$$k_m(\nu) \equiv -\frac{dI_\nu}{I_\nu \rho ds}$$

$$= -\frac{dI_\nu}{I_\nu d\mathcal{M}} \quad \textit{mass extinction coefficient } [\mathrm{m}^2 \cdot \mathrm{kg}^{-1}], \tag{2.19}$$

$$k_n(\nu) \equiv -\frac{dI_\nu}{I_\nu n ds}$$

$$= -\frac{dI_\nu}{I_\nu d\mathcal{N}} \quad \textit{extinction cross section } [\mathrm{m}^2]. \tag{2.20}$$

The extinction cross section k_n is analogous to the collision cross section in atomic physics. It is interpreted as the *effective area* of the molecule or particle presented to the incident beam, resulting in some of the beam being absorbed or deflected into other directions.

A more common form of the extinction law is obtained by integration over a finite distance along the beam. Let the total length of the straight-line path through an extended section of the medium be s, and let an intermediate path length be $s' \leq s$. Denoting the intensity entering the medium at $s' = 0$ as $I_\nu(s' = 0, \hat{\Omega})$, we need to find the intensity $I_\nu(s' = s, \hat{\Omega})$, where $\hat{\Omega}$ denotes the propagation direction of the beam. Integrating Eqs. 2.18, 2.19, and 2.20 from $s = 0$ to $s' = s$, we obtain

$$\tau_s(\nu) \equiv -\ln \left[\frac{I_\nu(s' = s, \hat{\Omega})}{I_\nu(s' = 0, \hat{\Omega})} \right], \tag{2.21}$$

where

$$\tau_s(\nu) \equiv \int_0^s ds' \, k(\nu) \equiv \int_0^s ds' \, k_m(\nu)\rho \equiv \int_0^s ds' \, k_n(\nu) n. \tag{2.22}$$

Here τ_s is the *extinction optical path* or *opacity* along the path s. The dimensionless quantity τ_s is a measure of the strength and number of optically active particles along a beam. Solving for the intensity at $s' = s$ by taking the antilogarithm of Eq. 2.21, we obtain

The Extinction Law (integrated form)

$$I_\nu(s, \hat{\Omega}) = I_\nu(0, \hat{\Omega}) \exp[-\tau_s(\nu)]. \tag{2.23}$$

The intensity is seen to decay exponentially with optical path along the beam direction.[16] Equation 2.23 reduces to Theorem I when the optical path is zero, resulting in the statement that the intensity remains constant along the beam direction in the absence of extinction.

2.7.1 Extinction = Scattering + Absorption

We have dealt with extinction as if it were one process, whereas it is actually caused by two distinctly different phenomena. It is clear that the attenuation of a light beam in a specific direction can be obtained by either absorption or scattering. This is obvious for absorption but some care needs to be given as to how scattering also weakens the beam. Since this process diverts the radiation into beams propagating in other directions, it must necessarily result in a loss of energy in the initial beam along $\hat{\Omega}$. However, suppose a photon in the beam is deflected only a very small amount. A detector of finite angular resolution may then measure the presence of this scattered photon along with the original (unscattered) beam photons. This can be a difficult experimental problem, since small-angle scattering from particulate matter (the so-called diffraction peak) can be orders of magnitude more efficient than large-angle scattering. It is also possible that photons in a different beam, propagating in direction $\hat{\Omega}''$, might be deflected *into* the direction $\hat{\Omega}$, and thus they could become confused with the original beam. This deflection is a consequence of *multiple scattering* mentioned earlier. The solution of the multiple scattering problem is a major concern of this book. The deflection problem is solved in principle provided we determine the multiple scattering contribution to the radiation in the direction of the original beam (see §5.6).

The extinction optical path τ_s of a mixture of scattering/absorbing molecules and particles is defined as the sum of the individual scattering optical path, $\tau_{sc}(\nu)$, and the absorption optical path, $\tau_a(\nu)$, that is,

$$\tau_s(\nu) = \tau_{sc}(\nu) + \tau_a(\nu),$$

where

$$\tau_{sc}(\nu) = \sum_i \int_0^s ds' \sigma^i(\nu, s') = \sum_i \int_0^s ds' \sigma^i_m(\nu) \rho_i(s') = \sum_i \int_0^s ds' \sigma^i_n(\nu) n_i(s'), \tag{2.24}$$

and

$$\tau_a(\nu) = \sum_i \int_0^s ds' \alpha^i(\nu, s') = \sum_i \int_0^s ds' \alpha^i_m(\nu) \rho_i(s') = \sum_i \int_0^s ds' \alpha^i_n(\nu) n_i(s'). \tag{2.25}$$

The sum is over all optically active species. Here ρ_i and n_i are the mass densities and concentrations of the ith optically active species (either molecule or particle). The quantities α^i, α^i_m, and α^i_n are called the *absorption coefficient*, the *mass absorption coefficient*, and the *absorption cross section* of the ith constituent (molecule or

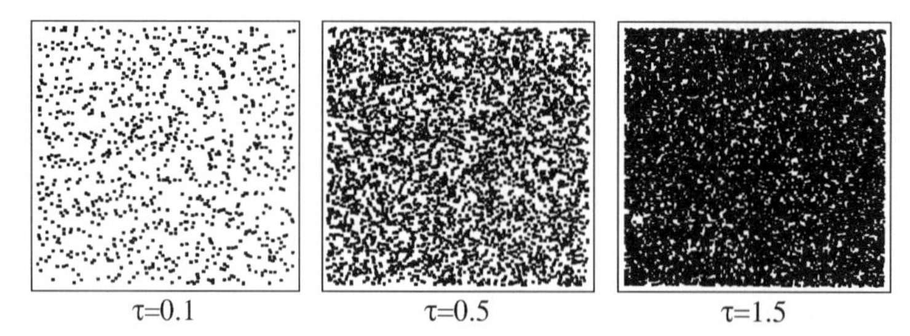

Figure 2.5 Leaf shadows cast by the overhead Sun on level ground due to three idealized leaf canopies. Each randomly distributed leaf shadow has an area $a = 0.01\%$ of the total ground area A. The optical path is given by the total leaf area per unit projected ground area, $\tau = Na/A$, where N is the number of leaf shadows. For the cases from left to right, $N = 10^3, 5 \times 10^3$, and 1.5×10^4, corresponding to $\tau = 0.1, 0.5$, and 1.5, respectively. Note that for $\tau = 1.5$, overlapping of leaf shadows is important for determining the total shaded area. The canopy beam transmittance (see §5.2) $e^{-\tau}$ is interpreted as the fraction of sunlit area, which for small τ is $\approx 1 - \tau$.

particle), respectively. Similarly, the quantities σ^i are the corresponding *scattering coefficient*, etc.

The fact that the various extinctions are independent may be easily justified by noting that within a small volume element, the particles do not "shadow," or overlap, one another. If this were the case, one could always decrease the size of the volume element until the shadowing becomes negligible. Of course, over a longer path the effects of both scattering and absorption accumulate, and overlapping can become a dominant effect. For example, in a leaf canopy with a large leaf density (Fig. 2.5) the shadows tend to overlap, a tendency that allows a nonzero transmission even for a leaf density area per unit area (optical path) that can be much greater than unity.

The extinction coefficient will generally vary with the physical conditions of the medium and will therefore be a function of space and possibly time. For example, the rate of collisional excitation controls the populations of excited molecular levels, and therefore it affects the variation of the absorption coefficient with frequency. Also, absorption within the rotational structure of a single molecular band varies as the gas temperature changes.

In most atmospheric and oceanic applications, the extinction coefficient does not vary with the direction of propagation, $\hat{\Omega}$. A medium in which this is valid is called *isotropic*. If this is not the case, the medium is said to be *anisotropic*. For example, falling raindrops or snowflakes tend to be aerodynamically oriented, causing the extinction coefficient to be largest in the vertical direction. Also, leaf canopies generally have extinction cross sections that vary with incidence angle.

Example 2.5 Photon Mean Free Path

The quantity $e^{-\tau_s(\nu)}$, called the *beam transmittance* $\mathcal{T}_b$, is proportional to the probability $P(s)$ that a photon, beginning its path at $s' = 0$, will travel unhindered through the medium to the point

$s' = s$. The (normalized) probability is

$$P(s) = \frac{e^{-\tau_s}}{\int_0^\infty ds' e^{-\tau_{s'}}} \quad (0 \leq P(s) \leq 1).$$

Since radiative transfer processes are statistical in nature, most quantities in transfer theory may be interpreted as probabilities, or probability distributions. To illustrate how extinction may be interpreted in this way, consider a situation in which all photons ultimately suffer an extinction event along a particular line of sight. For example, the photons from a streetlight on a foggy night will eventually be scattered out of the direct beam. (For a dense water fog, visible light absorption is weak so that shortwave extinction is identical to scattering.) To a distant observer, the streetlight itself will be indistinguishable from the surrounding scattered "halo." Such a medium in which $\tau_s \gg 1$ is called an *optically thick medium*. As the observer approaches the lamp, the light bulb becomes more and more distinguishable, that is, more and more photons have survived the direct path. Finally, at the lamp itself, none of the emerging photons has undergone scattering. We see that the probability that a photon has escaped scattering is just $P(s)$. More precisely, $P(s)ds$ is the probability of finding a photon with free path length lying between s and $s + ds$. $P(s)$ versus s may also be thought of as a distribution of photon free paths along a given line. The average straight-line distance traveled by a photon in an optically thick medium is the *photon mean free path* (mfp) given by

$$mfp = \langle s \rangle = \int_0^\infty ds s P(s) = \frac{1}{k(\nu)},$$

which follows from the relation $d\tau_s(\nu)/ds = k(\nu)$. *The extinction coefficient is seen to be the reciprocal of the mean free path.* Note that this quantity is usually a function of frequency. It is possible to define separately mean free paths for scattering, absorption, and extinction.

2.8 The Differential Equation of Radiative Transfer

We define formally the *emission* of radiative energy by a differential volume element within the medium. We ignore any time dependence of the radiation field.[17] Consider again a slab of thickness ds and cross-sectional area dA, filled with an optically active material giving rise to radiative energy of frequency ν in time dt. This energy emerges from the slab as an angular beam within the solid angle $d\omega$ around $\hat{\Omega}$. The *emission coefficient* is defined as the ratio

$$j_\nu(\mathbf{r}, \hat{\Omega}) \equiv \frac{d^4E}{dA\, ds\, dt\, d\nu\, d\omega} = \frac{d^4E}{dV\, dt\, d\nu\, d\omega} \quad [\mathrm{W \cdot m^{-3} \cdot Hz^{-1} \cdot sr^{-1}}].$$

We now have general definitions for both the loss and the gain of radiative energy of a beam, and we may therefore write an equation for the net rate of change of the intensity along the beam direction. Combining the extinction law with the definition of emission, we have

$$dI_\nu = -k(\nu) I_\nu ds + j_\nu ds, \tag{2.26}$$

where $k(\nu) = \sigma(\nu) + \alpha(\nu)$. Dividing Eq. 2.26 by $k(\nu)ds$, the differential optical path $d\tau_s$, we find

$$\frac{dI_\nu}{d\tau_s} = -I_\nu + \frac{j_\nu}{k(\nu)}.$$

The ratio $j_\nu/k(\nu)$ is given the special name of the *source function*:

$$S_\nu \equiv \frac{j_\nu}{k(\nu)} \quad [\mathrm{W \cdot m^{-2} \cdot Hz^{-1} \cdot sr^{-1}}]. \tag{2.27}$$

Our fundamental equation may be written in three mathematically equivalent ways.

The differential equation of radiative transfer:

$$\frac{dI_\nu}{d\tau_s} = -I_\nu + S_\nu, \qquad \frac{dI_\nu}{ds} = -k(\nu)I_\nu + k(\nu)S_\nu,$$
$$\hat{\Omega} \cdot \nabla I_\nu = -k(\nu)I_\nu + k(\nu)S_\nu. \tag{2.28}$$

In the third form we have used an alternate form for the directional derivative involving the gradient operator ∇, which emphasizes that we are describing a rate of change of intensity along the beam in the direction $\hat{\Omega}$. As described in §6.4, $\hat{\Omega} \cdot \nabla I_\nu$ is sometimes called the *streaming term*. A derivation of the extra term $(1/c)\partial I_\nu/\partial t$ that must be added to the left-hand side of Eq. 2.29 when time dependence cannot be ignored is provided in Example 2.6 below. Equation 2.29 is an expression of *continuity of the photon distribution function* (see Example 2.6). Photons are lost from the beam by extinction and added to it by emission. Both I_ν and S_ν are generally functions of both position (**r**) and direction ($\hat{\Omega}$). They also depend parametrically on frequency.

Example 2.6 Alternative Derivation of the Equation of Radiative Transfer

To arrive at the radiative transfer equation describing the spatial and temporal evolution of the angular intensity (or radiance), we may begin with the Boltzmann equation for the photon gas. We define the photon distribution function, $f_\nu(\mathbf{r}, \hat{\Omega}, t)$ $[\mathrm{m^{-3} \cdot sr^{-1} \cdot Hz^{-1}}]$ such that $f_\nu(\mathbf{r}, \hat{\Omega}, t)cdtdA \cos\theta d\nu d\omega$ is the number of photons with frequencies between ν and $\nu + d\nu$ crossing a surface element dA in direction $\hat{\Omega}$ into solid angle $d\omega$ in time dt. Since each photon carries energy $h\nu$, the amount of energy associated with these photons is

$$d^4E = ch\nu f_\nu \, dA \, \cos\theta \, d\nu \, d\omega \, dt.$$

The intensity is defined such that the same amount of energy is given by

$$d^4E = I_\nu(\mathbf{r}, \hat{\Omega}, t) dA \, \cos\theta \, d\nu \, d\omega \, dt.$$

Comparing the two expressions we find that the intensity and the photon distribution function are related by

$$I_\nu(\mathbf{r}, \hat{\Omega}, t) = ch\nu f_\nu(\mathbf{r}, \hat{\Omega}, t). \tag{2.29}$$

The temporal and spatial evolution of a general distribution function is described by the Boltzmann equation

$$\frac{\partial f_\nu}{\partial t} + \nabla_{\mathbf{r}} \cdot (\mathbf{v} f_\nu) + \nabla_{\mathbf{v}} \cdot (\mathbf{a} f) = Q_\nu(\mathbf{r}, \hat{\Omega}, t), \tag{2.30}$$

where $\mathbf{v}$ stands for velocity, $\mathbf{a}$ stands for acceleration, and $\nabla_{\mathbf{r}}$ and $\nabla_{\mathbf{v}}$ stand for the divergence operator in configuration and velocity space, respectively. $Q_\nu(\mathbf{r}, \hat{\Omega}, \mathbf{t})$ represents the source and loss terms due to scattering, extinction, and emission. For $Q_\nu = 0$, Eq. 2.30 is the well-known *Liouville's Theorem*, which is just Theorem I. If we ignore general relativistic effects there are no forces acting on the photons ($\mathbf{a} = \mathbf{0}$), and they travel in straight lines with velocity c between collisions. Therefore, Eq. 2.30 becomes

$$\frac{\partial f_\nu}{\partial t} + c(\hat{\Omega} \cdot \nabla) f_\nu = Q_\nu(\mathbf{r}, \Omega, t),$$

which we recognize as a continuity equation for the photon gas. Using Eq. 2.29, we obtain

$$\frac{1}{c}\frac{\partial I_\nu}{\partial t} + (\hat{\Omega} \cdot \nabla) I_\nu(\mathbf{r}, \hat{\Omega}) = h\nu Q(\mathbf{r}, \hat{\Omega}).$$

The left side of this equation is identical to that of Eq. 2.29 if we ignore the time-dependent term, $(1/c)(\partial I_\nu/\partial t)$, and the right side will also become the same as that of Eq. 2.29 if we assume that the photons interact with a medium that scatters, absorbs, and emits radiation.

2.9 Summary

We have defined the basic state variables of the radiative transfer theory, the scalar intensity I_ν and the net flux F_ν. Since the flux is derivable from the intensity (but not vice versa), the intensity is the more basic variable. Although I_ν contains all the information concerning the radiative energy flow in the gas, it does not convey any knowledge of the state of polarization. Angular moments of the spectral intensity were defined, and these were related to the spectral net flux and the spectral energy density $\mathcal{U}_\nu$. Four idealized angular distributions of I_ν were shown to give simple expressions for the angular moments. Theorems describing the behavior of I_ν in a homogeneous, transparent medium were given. The notion of intensity as *visual brightness* was discussed as a good analogy, but not as a one-to-one correspondence. The Extinction Law was considered to be a given result, ultimately based on experiment. The important concepts of extinction, absorption, and scattering optical depths were introduced. The differential radiative transfer equation was derived from consideration of the sources and sinks of radiation along a ray.

Problems

2.1 Let a small object be located at P' on the optical axis of a concave mirror. The object has an area dA' and is oriented normally to the axis. This object forms a real image at the distance P. Show that the intensity leaving the object at P' and the (unattenuated) intensity arriving at P are equal. Thus prove Theorem II for the case of singly reflected light from a concave mirror. Argue that the principle can be generalized to show that the image has the same distribution of intensity as the original object.

2.2 Consider two transparent media of indices of refraction m_1 and m_2, separated by an interface of arbitrary shape. Prove Theorem III by using *Snell's Law* to show that for a bundle of beams passing through the interface, the intensity I_1 within the bundle in medium 1 is related to the intensity I_2 in medium 2 within the refracted beam through $I_1/m_1^2 = I_2/m_2^2$.

2.3 A Lambertian disk of radius a emits a quasi-isotropic intensity $\mathcal{I}$. Show that the outward flux at a point lying on the axis of the disk a distance z from the center of the disk is given by

$$F(z) = \pi\mathcal{I}\frac{a^2}{(a^2 + z^2)}.$$

Notes

1 Radiative transfer theory originated in astrophysics. Even though it is a bit out of date, the basic reference for radiative transfer in planetary atmospheres is still Chandrasekhar, S., *Radiative Transfer*, Clarendon Press, Oxford, reprinted by Dover Publications, New York, 1960. A well-written history of the field of atmospheric radiation up to the space age is by Möller, F., Radiation in the atmosphere, pp. 43–71, in Meteorological Challenges: A History, McIntyre, D. P. (ed.), Information Canada, Ottawa, CA, 1972. The classic reference for transfer of radiation in oceans is Jerlov, N. G., *Optical Oceanography*, Elsevier, Amsterdam, 1968. A more modern treatment on marine bio-optics is Kirk, J. T. O., *Light and Photosynthesis in Aquatic Ecosystems*, 2nd ed., Cambridge University Press, New York, 1994.

2 To include the effects of polarization, a partially incoherent light wave is written as a four-vector, with the components commonly known as I, Q, U, and V. The first component is the ordinary intensity.

3 Basic radiative transfer field variables are clearly discussed in the classic astrophysical reference Milne, E. A., *Thermodynamics of the Stars, Handbuch der Astrophysik*, 3, Part I, Chapter 2, 1930, pp. 65–255, reprinted in *Selected Papers on the Transfer of Radiation*, ed. D. H. Menzel, Dover, New York, 1966. Modern treatments of radiative transfer are found in many books, e.g., Mihalas, D., *Stellar Atmospheres*, W. H. Freeman, San Francisco, 1978, and Liou, K.-N., *An Introduction to Atmospheric Radiation*, Academic Press, New York, 1980.

4 The wavelength should be reported *in vacuum*. The small differences are very important in precise spectroscopy, and in line-by-line calculations.

5 Note that this beam concept is different from that used in optics and antenna theory. A laser beam or an antenna beam usually has a finite spatial extent in the transverse direction.

6 Note that finite quantities, such as F_ν, are ratios of two infinitesimal quantities of the same order.

7 The reader should not confuse our definition of intensity with a quantity with the same name, but with different units and interpretation. See, for example, *American National Standard: Nomenclature and Definitions for Illuminating Engineering*, Z7.1, Illuminating Engineering Society, New York, 1967. Its definition is $d^3E/d\omega d\nu dt$ [$\mathrm{W \cdot sr^{-1} \cdot Hz^{-1}}$]. It is a measure of the radiative power carried per solid angle and is applicable to point sources of radiation. It rarely appears in problems dealing with extended sources, such as occurs in natural radiation, and thus plays no role in our scheme.

8 The intensity distribution is analogous to the distribution in six-dimensional phase space of molecular velocities and positions in the kinetic theory of gases as a function of time (i.e., three spatial cordinates, three components of the velocity, and time).

9 The second angular moment of the intensity is proportional to the *radiation pressure*, which is of interest in stellar atmospheres but of no importance in atmospheric/oceanic radiative transfer.

10 In a three-dimensional medium with no symmetry, we require three pairs of numbers, for the x, y, and z directions.

11 At the Earth's mean distance the Sun's disk subtends 32 minutes of arc, or 6.8×10^{-5} sr.

12 For more information on the properties of the Dirac δ-function, see Appendix F.

13 See Appendix E for a derivation of Theorem III.

14 When a radiative quantity, such as I_ν, F_ν, etc., is integrated in this fashion, it becomes a *photometric* quantity and is usually accompanied by the term *luminous* (luminous intensity, luminous flux, etc.). Excellent discussions of the relationships among irradiance, radiance, luminance, and brightness are given in Bohren, C. F., "All that's best of dark and bright," *Weatherwise*, **43**, 160–3, June, 1990. (See also Bohren, C. F., Chapter 15 of the same title, in *What Light Through Yonder Window Breaks?*, Wiley, New York, 1991.) An alternate definition of the intensity, as well as an up-to-date treatment of modern radiative transfer theory and practice, is given in Lenoble, J., *Atmospheric Radiative Transfer*, Deepak Publishing, Hampton, VA, 1993.

15 August Beer (1825–63) was not the first to expound this principle, nor was Johann Lambert (1728–77). The honor actually belongs to Pierre Bouguer (1698–1758), considered to be the father of photometry. We have settled on the historically neutral term "Extinction Law," which at least is descriptive. This is one of many examples in our science in which prematurely assigning the "discoverer's" name to a physical principle can lead to later difficulties. Another example is the term "Mie-scattering," which gives the false impression that Mie was the originator.

16 If the medium consists of closely spaced (but randomly distributed) particles, such as in nonporous soil or a dense leaf canopy, the extinction law must be modified to include the *filling factor* F $(0 \leq F \leq 1)$, which is the total particle volume per unit volume $= n\langle V\rangle$, where $\langle V\rangle$ is the average particle volume. The modification consists of replacing n in Eq. 3.6 with an effective particle density $n_e = n[-\log(1 - F)/F]$. When $F \ll 1$, the bracketed quantity $[\ ] \to 1$, which is valid for a *dilute* medium such as an atmosphere. For $F \to 1$, $n_e \to \infty$ and the medium becomes *opaque*.

17 There are few natural processes (with the possible exception of a solar eclipse) that proceed rapidly enough to compete with the time scale of light travel. Most time-dependent problems of atmospheric radiative transfer are governed by the slower (rate-determining) processes of heating and cooling, or by the changing illumination conditions, for example, at the shadow line.

Chapter 3

Basic Scattering Processes

3.1 Introduction

In the next two chapters we will study the physical basis for the three types of light–matter interactions that are important in planetary media – scattering, absorption, and emission. In this chapter we concentrate on scattering, which may be thought of as the first step in both the emission and absorption processes. The classical concept of the Lorentz atom is first used to visualize the process of scattering, which encompasses both coherent processes, such as refraction and reflection, as well as the many incoherent processes that comprise the main topics of this chapter.

Consideration of the classical interaction of a plane wave[1] with an isolated damped, simple harmonic oscillator helps to introduce the concept of the cross section. The scattering cross section is expressed in terms of the frequency of the incident light, the natural frequency of the oscillator, and the damping rate. A simple extension of the concept is then made to scattering involving excited quantum states. This approach also helps to understand three different scattering processes (Rayleigh, resonance, and Thomson scattering) using one unified description. It also gives the Lorentz profile for absorption in terms of the classical damping rate, which apart from a numerical constant agrees with the quantum mechanical result. This approach also allows for a description of the two principal mechanisms responsible for broadening of absorption lines in realistic molecular media: pressure broadening and Doppler broadening.

Radiation interacts with matter in three different ways: through emission, absorption, and scattering. We first contrast these three interactions in terms of their energy conversions between internal energy states of matter, E_{I} (which includes kinetic energy of motion), and radiative energy, E_{R}. It is convenient to consider monochromatic radiation. *Emission* converts internal energy to radiative energy ($E_{\mathrm{I}} \rightarrow E_{\mathrm{R}}$). *Absorption* converts radiative energy to internal energy ($E_R \rightarrow E_I$). *Scattering* is a double conversion ($E_{R'} \rightarrow E_I \rightarrow E_R$) where the radiative energy $E_{R'}$ is first absorbed by matter ($E_{R'} \rightarrow E_I$) and then radiated ($E_I \rightarrow E_R$). Thus, the radiated field, denoted by R, is generally modified in frequency, direction of propagation, and polarization relative to the absorbed field.

Some general relationships among these interactions follow from the energy-conversion viewpoint. For example, emission and absorption appear to be inverse processes. Indeed, in the special case of thermodynamic equilibrium (§5.3) the rates of emission and absorption are identical. Similarly, we might think of scattering as being simply a combination of absorption followed by emission, although the scattering process actually proceeds via *virtual states* and does not require absorption and then reemission. It is nonetheless an accurate description in certain situations,[2] and it serves as a useful mental picture of scattering.

In the continuum view, matter can be divided into finer and finer elements with no limits on the smallness of the values of the charge or matter within the elementary volumes. In contrast, the atomic theory is based on the notion of a fundamental *discreteness* of matter, thus placing a limit on the size of these basic volume elements. Our microscopic description is based on the interactions of light with these basic "building blocks," assumed to consist only of mass and positive and negative electric charges bound together by elastic forces. If we imagine these elements to be endowed with certain internal energy modes ("excitation"), which coincide with those derived from quantum theory (or determined from experiment), then Maxwell's theory provides us with all the tools we need to understand the interactions of these elements with electromagnetic radiation. Although individual atoms are the actual agents of absorption, emission, and scattering, the mathematics of the classical theory often requires us to consider the matter to consist of an infinitely divisible *continuous* distribution of charge. Fortunately, the dimensions of atoms are so small that these two contradictory views never pose any practical problems. For example, we will use mathematically smooth functions to describe how the density of matter varies with space or how the speeds of molecules vary within a small volume of space without concern with the basic granularity of matter. We will choose volume elements in either real space, velocity space, or energy space that are sufficiently small that we can consider the properties of matter to be uniform within that element. Moreover, these volumes will be large enough to contain a sufficiently large number of atoms so that the granularity can be ignored.

With regard to the radiation field, we again take two apparently contradictory views. The classical point of view is that the electromagnetic field is a continous function of space and time, with a well-defined electric and magnetic field at every location and instant of time. In this classical picture the radiative energy within the small frequency range ν and $\nu + d\nu$ is a continuous function of ν and there is no limit to how small the energy differences can be. The quantum view of the radiation field is that of concentrations at discrete values of energy that are separated in increments of the minimum energy $h\nu$ at a given frequency. Here h is Planck's constant. The total energy density is determined by the total number of radiation quanta (*photons*) times the energy $h\nu$ per quantum. As in the case of atoms, we can define the interval $d\nu$ to be sufficiently small so that the energy may be considered to be constant over $d\nu$, but large enough to contain enough photons so that the discretization in energy is unobservable.

We will sometimes refer to radiation in terms of a "field" and other times in terms of "photons." This deliberate looseness allows us a flexibility in visualizing light–matter interactions, sometimes in terms of light particles, whereas other times in terms of a continuous distribution of an electromagnetic field. However, our actual mathematical description will usually be based on the classical theory. The classical approach can be extended to the description of discrete spectral line absorption (a distinct quantum process). In this "semiclassical" theory, the atom is described in quantum terms but the radiation is treated as a classical entity. Although this theory has been very successful, some phenomena (such as spontaneous emission) require for their complete explanation that both the matter and the radiation be quantized (called "quantum electrodynamics"). Fortunately, quantum effects can fairly easily be incorporated in the classical approach. Although not rigorous in any sense, this artifice usually leads to results consistent with observation. For example, the Planck formula for the frequency distribution within a blackbody cavity stems from quantum theory. Its adoption in the classical theory is straightforward. We will unapologetically mix classical and quantum concepts throughout the book.

3.2 Lorentz Theory for Radiation–Matter Interactions

In 1910, Lorentz[3] put forth a very successful microscopic theory of matter, picturing the electrically neutral atoms of a substance to consist of negative charges (electrons) and equal positive charges (the nucleus) bound together by elastic forces. These elastic forces are proportional to the distance of the charges from the center of charge (*Hooke's Law*). The question of the nature of these forces is not our concern here, as this requires quantum theory for a satisfactory answer. However, the Lorentz theory combined with the familiar Coulomb forces between electrical charges, and the Maxwell theory of the electromagnetic field, provided the prequantum world with a satisfactory explanation of a vast number of phenomena. Some of the constants resulting from the Lorentz theory needed to be adjusted to agree with experiments, which later on were explained in a more basic way with quantum theory. In addition, the field equations of Maxwell served to explain nearly all properties of radiation as an electromagnetic phenomenon.

A dramatic failure of the classical theory was its inability to predict the blackbody frequency distribution law. This failure eventually led Planck in 1900 to his paradigm-shattering notion of quantized energy states of matter. This advance, plus the failure of the classical theory to explain the photoelectric effect, led Einstein in 1914 to postulate that light itself is quantized.[4] The new quantum theory eventually replaced the old classical theory because of its successful application to a very broad range of phenomena. However, the Lorentz theory has survived to the present day, not because it in any way competes with the newer theory, but because it has important advantages of concrete visualization that the quantum theory often lacks. For this pedagogical reason (and because it usually gives the correct answers when the unknown constants are provided by the accurate quantum theory) we will use the Lorentz theory throughout the book. Note, however, that the interpretation of many modern optical phenomena involving so-called coherent radiation requires quantum theory. Fortunately, to explain the propagation of natural (incoherent) radiation we seldom need to resort to these sophisticated descriptions. On the one hand, accurate numerical values of many of the interaction parameters required in the radiative transfer theory (for example molecular absorption cross sections) cannot be provided by the classical theory. However, cross-section calculations for spherical particles[5] are accurately described by the classical *Mie–Debye* theory. Except in certain simple situations, we will consider the interaction parameters to be given, either from quantum-theoretical calculations, from the Mie–Debye theory, or from laboratory measurements.

3.2.1 Scattering and Collective Effects in a Uniform Medium

We now use the Lorentz model and the classical radiation theory to visualize how light is affected in its passage through matter. We consider atoms and molecules to behave in basically similar ways. Consider a monochromatic plane wave incident upon a dielectric medium consisting of a uniform distribution of nonabsorbing Lorentz atoms. The plane wave has a fixed frequency, phase, and polarization (orientation of the electric field direction). The imposed electric field creates within each atom an oscillating charge separation, which varies in time with the same period as that of the incident field. The strength of the interaction is measured by the *induced dipole moment* $\mathbf{p}$, which is proportional to two quantities: (a) the polarizability $\alpha_{\rm p}$, which depends upon the bonding forces between the constituent positive and negative charges, and (b) the imposed electric field intensity $\mathbf{E}'$. For simplicity we consider an isotropic medium, for which $\alpha_{\rm p}$ is a scalar.[6] The induced dipole (the product of the electronic charge and its displacement from the equilibrium arrangement within the atom) is mostly due to the oscillatory motion of the bound electrons because they are much lighter than the positive nucleus.

Consider the effect of the incident wave on an isolated atom. Electromagnetic theory predicts that an oscillating charge will radiate an outgoing electromagnetic wave of the same frequency as the oscillation frequency. In general, this radiated or scattered wave will have a definite phase shift with respect to the incoming plane wave. Thus, the scattered wave is *coherent* with the incoming wave. In this simplest of situations, it propagates outward as a spherical wave with the typical dipole radiation pattern.

The effect of a single nonabsorbing atom is thus to divert the flow of radiative energy, but not to destroy it. However, the *collective* action of a uniform, optically dense medium is quite different. In fact, no overall scattering will occur! This is despite the fact that each spherical wave may interact strongly with other waves in the manner just described. To understand this paradox, consider the fact that pure glass, water, or air transmits light freely with (at most) some bending of the rays at interfaces where there is a change in the index of refraction. Moreover, away from interfaces, the basic radiance (I/m_r^2) is not attenuated along the ray (see Theorem III of §2.5). To understand how the induced dipole picture is compatible with the notion of an unattenuated refracted ray, we must consider the interactions of the scattered waves with the incident field and with one another. The medium is assumed to be perfectly uniform. As discussed earlier, each atom is forced to radiate spherically outgoing waves that are coherent with the incident wave. The net radiation field therefore is a coherent superposition of these scattered waves and the incident wave. Because of the coherence, the separate electric fields must be added with due regard to their relative phases.

Figure 3.1 illustrates light incident on a smooth plane boundary separating a vacuum (to the left) from a semi-infinite medium (to the right). We assume that the plane wave falls on this boundary at normal incidence. If the medium is perfectly uniform, then for every point P' on the boundary we can locate a second point P'' such that for a given direction of observation the path length difference is $\lambda/2$, where λ is the wavelength. The two scattered waves from P' and P'' cancel in this direction through destructive interference. For all other directions of observation (other than the forward direction) we can always find other pairs of points for which perfect cancellation occurs. As a result of destructive interference the incident wave is completely extinguished inside the medium (the *Ewald–Oseen Extinction Theorem*). If the medium to the right is a slab of finite extension, then all that remains of the incident light is the transmitted and reflected rays, in the forward and backward directions, respectively. We might expect the net result for the transmitted ray to be the same as if there were no medium at all. This is not the case – as a result of the repeated scatterings and reemissions along the ray, its forward progress turns out to be slowed down by a factor m_r, the index of refraction of the medium.

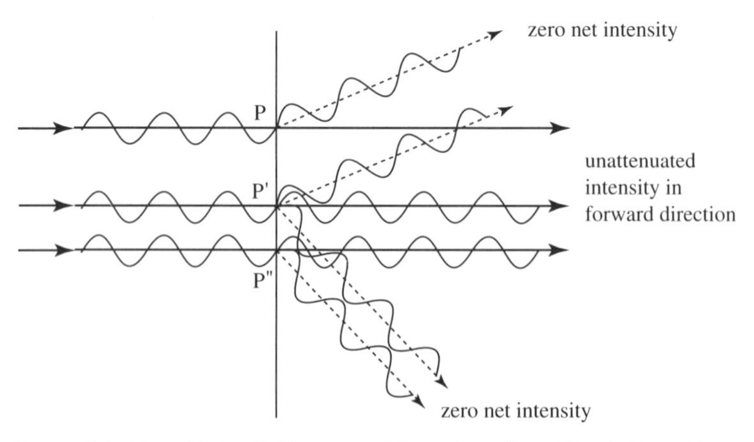

Figure 3.1 The radiation fields scattered from the points P'' and P' are 90° different in phase and therefore interfere destructively.

If the incident light falls obliquely on the smooth planar interface of Fig. 3.1 and the medium is assumed to be a finite slab, then the scattered waves will be found to survive in two directions *outside* the medium corresponding to those of the familiar specularly reflected and transmitted rays in geometric optics. The direction of propagation of the refracted ray within the medium is given by *Snell's Law* (see Appendix E). Thus, both refraction and reflection are manifestations of a coherent superposition of waves, each caused by a single, more fundamental scattering process. In fact, any type of "reflection" when analyzed in detail will be found to be the result of the same basic type of coherent scattering described above. As discussed in the following section, the myriad of collective processes occurring at the surfaces and interiors of most natural media modify beyond recognition the basic underlying scattering pattern of each individual atom or molecule.

3.2.2 Scattering from Density Irregularities

Since no real substance is perfectly uniform, we now consider the case in which various irregularities, or inhomogeneities, are distributed throughout the medium. These irregularities can take various forms. For example, even in pure solids, crystal defects may be present, or there may be irregularities in the orientation of the atoms or in the density (and thus in the index of refraction) from place to place. Fluctuations occur in the number density n (or in other thermodynamic properties depending upon density) due to the fact that no actual substance is perfectly uniform. The atomic nature of the substance causes it to have statistical variations in density that are present in all phases, but these are most apparent in gases and liquids. These density inhomogeneities, Δn (whose spatial scales are small compared to λ), give rise to corresponding changes in the number of induced dipoles per unit volume, $\mathcal{P} = n\alpha_{\rm p}\mathbf{E}'$, where $\mathcal{P}$ is the *bulk polarization* of the medium, $\alpha_{\rm p}$ is the *polarizability*, and $\mathbf{E}'$ is the imposed electric field. The electric and magnetic fields at some distance away from two source points P' and P'' will interfere as in a uniform medium (see Fig. 3.1), but in this case the result is an incomplete cancellation. The "excess" (surviving) electric field $\mathbf{E}$ in the scattered wave is proportional[7] to the change in the bulk polarization, $\Delta\mathcal{P}$, so that $\mathbf{E} \propto \Delta\mathcal{P} = \alpha_{\rm p}\Delta n$. Since the intensity of an electromagnetic wave is proportional to the square of the electric field amplitude (averaged over a wave period), $\langle \Delta I \rangle \propto \langle (\Delta E)^2 \rangle \propto \langle \Delta\mathcal{P} \rangle^2 \propto \langle (\Delta n)^2 \rangle$. Now the statistical theory for fluctuations in an ideal gas predicts that $\langle (\Delta n)^2 \rangle \propto n$. Thus, we are led to the conclusion that for an ideal gas the scattered intensity is proportional to the number of atoms per unit volume; that is, *the scattering behaves as if the atoms scatter independently of one another*. This remarkable result[8] was understood and used by Lord Rayleigh in his classic explanation of the blue sky in a series of papers between 1899 and 1903. Mathematical proofs were provided by R. Smoluchowski in 1908 and by A. Einstein in 1910. A consequence of this result is that light can be simultaneously refracted and scattered by air molecules. This foremost example of scattering is called *Rayleigh scattering*.

3.2.3 Scattering in Random Media

We have seen that, although a uniform medium transmits light in a collective transparent manner, the randomly distributed inhomogeneities or imperfections (in an otherwise uniform medium) scatter light as if the individual atoms were unaffected by the radiation from their neighbors. With the exception of the forward direction, it can be shown that for most planetary media, the individual scattered spherical wavelets have no permanent phase relationships.[9] This randomness of the phases of the superposed scattered wavelets implies that the net intensity due to all the scattering centers is simply the sum of the individual intensities. This property of *incoherent scattering* is shared by a very large variety of planetary materials, consisting as they usually do of mixtures of substances and of several types of inhomogeneities. For example, the presence of small bubbles of air trapped in water or ice will give rise to a milky appearance owing to the spatial inhomogeneities in the index of refraction. We shall generally refer to the scattering centers in such substances (so-called random media) as *particles*. They are generally different in composition from the atoms of the ambient media and are distributed randomly within the ambient media. The assumption of *independent scatterers* is violated if the particles are too closely packed. The average spacing between particles should be several times their diameters to prevent intermolecular forces from causing correlations between neighboring scattering centers.[10] This requirement would appear to rule out independent scattering in aqueous media where such short-range correlations are the very essence of the liquid state. In particular, pure water is composed of transitory water clusters of random size held together by hydrogen bond forces. Seawater[11] contains a diversity of ion clusters, depending upon the various types of dissolved salts. Scattering therefore occurs from the clusters rather than from individual water molecules. An important point is that these clusters are much smaller than the wavelength of visible light. With regard to their spatial correlations, the long-range forces between adjacent clusters are of critical importance to the thermodynamic properties (such as viscosity). The wavelength dependence of scattering in pure water is $\sim\lambda^{-4.3}$, suggesting that there is indeed incoherency between the scattered wavelets. From the optical point of view the clusters are uncorrelated.

Aerosols (solid or liquid particles suspended in the atmosphere or ocean) may have important radiative effects. Even though the concentrations of scattering "particles" may be present only in trace amounts, they usually have much larger scattering cross sections than the molecules. Thus they often have an important influence on the transfer of radiation.[12] The distribution of sky brightness may be severely altered by atmospheric aerosols (dust, soot, smoke particles, cloud water droplets, raindrops, ice crystals, etc.), which may be present only in the parts per million by volume. Similarly, suspended organic and inorganic particles in seawater, which may be of minor importance to ocean chemistry, may nevertheless be of dominant importance to the radiation field in ocean water. Most natural surface materials (soil, snow, vegetation canopies) are classified as random media. These materials are composed of randomly distributed collections of diverse scattering elements, which scatter light incoherently.

A counterexample is the surface of calm water, for which cooperative effects account for the specularly reflected and transmitted light. In such cases, the radiative transfer theory cannot be used, and it is necessary to use Maxwell's equations, as, for example, in the Fresnel theory of reflection (§6.6).

In most media of interest to us, the dimensions of scattering particles are comparable to, or exceed, the wavelength of light. In such cases, their "radiation patterns" are often very complicated (see §6.7 and §6.8). As in the case of small density irregularities, the intensity of the direct beam is largely a result of collective, coherent effects, but it is also weakened by the fact that the secondary radiation diverts energy into other directions. We will refer to this secondary scattered radiation as *diffuse radiation*, because in contrast to direct (collimated or unidirectional) radiation it is distributed over many directions, in general through 4π steradians.

Scattering in random media occurs over the very small spatial scales of the particles themselves. We may contrast this with light interactions with irregularities having larger spatial scales. For example, convection or overturning of air parcels causes a mixing of irregular warm and cool air masses over scales of the order of centimeters to meters. Variations of air density and temperature lead to variations over the same scale in the index of refraction of the air. This in turn alters the direction of light rays in a chaotic manner and explains the twinkling of stars (scintillation). On a still larger scale, the air density of a planetary atmosphere declines exponentially with height (§1.4) over characteristic scales of the order of 5 to 8 km. The distortion of the image of the setting Sun is a result of the vertical gradient of the index of refraction of air. Refraction must be dealt with by considering the coherent wave nature of the radiation. We will not consider such phenomena in this book, although it may be introduced into the incoherent scattering theory in a straightforward manner.

3.2.4 First-Order and Multiple Scattering

In an earlier section we discussed the fact that, in uniform media, mutual interactions between the various scattered waves and between each scattered wave and the incident wave are of utmost importance. In random media where the particles scatter independently of one another, their individual contributions add together as if there were no mutual interactions. Consider the illumination of the atmosphere by the Sun. Assume that the particles are well separated, so that each is subjected to direct solar radiation. A small portion of the direct radiation incident on the particle will be scattered and thereby gives rise to scattered or diffuse radiation. If the diffuse radiation arriving from all parts of the medium is negligible compared with the direct radiation, the medium is said to be *optically thin* and diffuse radiation is unimportant. If we were to double the number of scatterers in an optically thin medium, the scattered or diffuse radiation would also be doubled. However, it often happens that the diffuse radiation itself is an important additional source of radiation, becoming a source for still more scattering, etc. The diffuse radiation arising from scattering of the direct solar beam is called *first-order* or *primary* scattering. If additional scattering events need to be included, the radiation is said to be *multiply scattered*, and the medium in which this is important is said to be *optically thick*. Thus, in many situations of interest in planetary media, the radiation field is determined not only by the transmitted incident radiation field but also by the "self-illumination" from the medium itself. This incoherent multiple scattering could be regarded as a collective effect, but it should not be confused with the coherent, collective effects already discussed. We will describe multiple scattering in greater detail in §5.3.

In the following two sections we derive some basic equations for the interaction cross sections of matter with radiation. Our treatment only scratches the surface of a vast and complex subject. We present below a simple example of how the interaction works in a very specific, simplified situation. It also provides useful quantitative results for cross sections and line profiles in real media. For an isolated molecule and including only the natural damping interaction, the analysis yields: (1) a strong resonant interaction, which occurs when the light frequency is very near one of the natural oscillation frequencies of the molecule, and (2) a much weaker interaction, which affects all light frequencies, and which provides a very good model for Rayleigh scattering. In either case, the interaction is that of elastic scattering (see §3.3.3). Our treatment does not explicitly consider the coherence of the incoming and outgoing waves, although this is necessary in order to derive the corresponding extinction of the incoming beam. Although the combination of both the direct and scattered fields is important in dense media, where the oscillators themselves affect the local electric field, here we ignore this effect.

A simple generalization of the meaning of the damping constant to include collisional effects provides a first-order description of pressure broadening. The Lorentz line profile predicted by this simple model is in very good agreement with high-spectral-resolution measurements. We also include the Doppler-broadening effects of thermal motions on the line profile. We then describe the net result of pressure and Doppler broadening, the so-called Voigt broadening. The Rayleigh angular scattering pattern is then derived from the same simple model.

3.3 Scattering from a Damped Simple Harmonic Oscillator

In certain applications, it is permissible to treat a molecule as a simple harmonic oscillator with a single natural oscillation frequency ω_0. To avoid continual appearances of the factor 2π, it is convenient to deal with the angular frequency ω, rather than the frequency ν. The molecule is assumed to consist of an electron bound to a positively charged nucleus with a certain "spring-constant," related to the natural oscillator frequency. Later we will remove some of the restrictions of the simple model.

When this simple system is irradiated by a linearly polarized, monochromatic, plane electromagnetic wave of angular frequency ω, the electron undergoes a harmonic acceleration in response to the oscillating electric field. The nucleus, being much more massive than an electron, is considered to be a rigid support, and its motion may be neglected. The relative displacement of positive and negative electrical charges causes the formation of an induced electric dipole. According to classical theory, acceleration of an electric charge gives rise to the emission of electromagnetic radiation. A large-scale example is a dipole[13] antenna, which emits radio waves. Without energy loss, absorption of light by the oscillator increases its motion indefinitely. Loss of energy in a mechanical oscillator, such as a spring, occurs as the result of a frictional *damping force*, which is approximated as being proportional to the velocity. To account for the energy loss due to the emitted wave, a damping force must exist. For an isolated molecule, this damping force may be thought of as a *radiation resistance*.[14] The classical radiative damping force (assumed to be suitably small) is given by $\mathbf{F} = -m_e \gamma \mathbf{v}$, where

$$\gamma = e^2 \omega_0^2 / 6\pi \epsilon_o m_e c^3. \tag{3.1}$$

Here m_e is the electron mass, $\mathbf{v}$ is its velocity, e is its charge, ω_0 is the natural angular frequency, ϵ_o is the vacuum permittivity, and c is the speed of light in a vacuum. In an insulating solid, or in a gas where the electron is subjected to an additional force from collisions with the lattice or with other molecules, the *damping rate* γ takes the form of a collisional frequency, given by the inverse of the mean time between collisions.

The power emitted by an accelerated charge may be found by considering the equation of motion of a damped, simple harmonic oscillator, subject to a forcing electric field[15] of amplitude $\mathbf{E}'$ and angular frequency ω. According to classical theory,[16] a charge set into accelerated motion radiates an electromagnetic wave with the time-averaged power given by (see Problem 3.1)

$$P(\omega) = \frac{e^4 E'^2}{12\pi m_e^2 \epsilon_o c^3} \left[\frac{\omega^4}{\left(\omega_0^2 - \omega^2\right)^2 + \gamma^2 \omega^2} \right] \quad [\mathrm{W}]. \tag{3.2}$$

We now take the ratio of the above expression for the scattered power to the power carried in the incident field per unit area. The latter is simply $\epsilon_o c E'^2/2$. This ratio is just the total *scattering cross section*,

$$\sigma_n(\omega) = \frac{P(\omega)}{\epsilon_o c E'^2/2} = \frac{e^4}{6\pi m_e^2 \epsilon_o^2 c^4} \left[\frac{\omega^4}{\left(\omega_0^2 - \omega^2\right)^2 + \gamma^2 \omega^2} \right] \quad [\mathrm{m}^2]. \tag{3.3}$$

Below we consider two special cases of this general result.

3.3.1 Case (1): Resonance Scattering and the Lorentz Profile

Here we allow the frequency of the incident light to be "tuned" to a discrete energy level of a molecule (*resonance scattering*). In this case the strength of the interaction is typically many orders of magnitude greater than the nonresonant interaction. Let the driving frequency ω be very close to resonance with the natural oscillation frequency ω_0. We simplify Eq. 3.3 with the provision that $\delta\omega \equiv \omega_0 - \omega \ll \omega$. Then $\omega_0^2 - \omega^2 = (\omega + \delta\omega)^2 - \omega^2 = 2\omega\delta\omega + (\delta\omega)^2 \approx 2\omega(\omega_0 - \omega)$. Substituting into Eq. 3.3, using the definition of γ from Eq. 3.1, we find upon simplication

$$\sigma_n^{\rm res}(\omega) = \frac{e^2}{m_e \epsilon_o c}\left[\frac{(\gamma/4)}{(\omega_0-\omega)^2+(\gamma/2)^2}\right]. \tag{3.4}$$

Returning to ordinary frequency using $\omega = 2\pi\nu$, and $\omega_0 = 2\pi\nu_0$, we have

$$\sigma_n^{\rm res}(\nu) = \frac{e^2}{4m_e \epsilon_o c}\frac{1}{\pi}\left[\frac{(\gamma/4\pi)}{(\nu_0-\nu)^2+(\gamma/4\pi)^2}\right]. \tag{3.5}$$

The frequency-dependent part of this result is called

the Lorentz profile:

$$\Phi_{\rm L}(\nu) = \frac{\gamma/4\pi}{\pi[(\nu_0-\nu)^2+(\gamma/4\pi)^2]}. \tag{3.6}$$

The frequency *line width* $[\mathrm{s}^{-1}]$ (full width at half-maximum) of the Lorentz profile is just the damping parameter γ. The profile $\Phi_{\rm L}(\nu)$ is normalized. This can be seen by changing variables to $x = 4\pi(\nu-\nu_0)/\gamma$. Integrating over x, and noting that $4\pi\nu_0/\gamma \gg 1$, one finds

$$\int_0^\infty d\nu \Phi_{\rm L}(\nu) = \frac{1}{\pi}\int_{-4\pi\nu_0/\gamma}^{\infty}\frac{dx}{1+x^2} \to \frac{1}{\pi}\int_{-\infty}^{+\infty}\frac{dx}{1+x^2} = 1. \tag{3.7}$$

Since the Lorentz profile is normalized, we find by integrating over all frequencies (we have assumed only one natural frequency)

$$\int_0^\infty d\nu \sigma_n^{\rm res}(\nu) = \frac{e^2}{4m_e\epsilon_o c}. \tag{3.8}$$

Thus, the integrated, or total, classical cross section is constant, depending only upon fundamental atomic constants.[17] This simple result is independent of the damping constant and originates from the area-preserving frequency dependence of Eq. 3.6. This property of the Lorentz profile is illustrated in Fig. 3.2. Increasing the value of γ decreases the strength of the spectral line in the *line core*, the region $|\nu-\nu_0| \le \gamma/4\pi$. However, increasing γ strengthens the *line wings*, the region where $|\nu-\nu_0| > \gamma/4\pi$. In the distant parts of the line, $\Phi_{\rm L}(\nu)$ varies as ν^{-2}.

The above expression was derived from strictly classical considerations of a single electron forced by the oscillating electric field of the incident wave. In actuality, there is more than one resonant frequency, so that we refer to the ith frequency or quantum transition. In addition no consideration has been given to the quantum-mechanical character of the process, which involves the notion of a transition from a ground state to a quantized excited state. It is found that the more correct derivation[18] yields a nearly identical expression, the difference being that it contains an extra multiplicative factor, called the *oscillator strength*, f_i. Thus we may write our expression for the

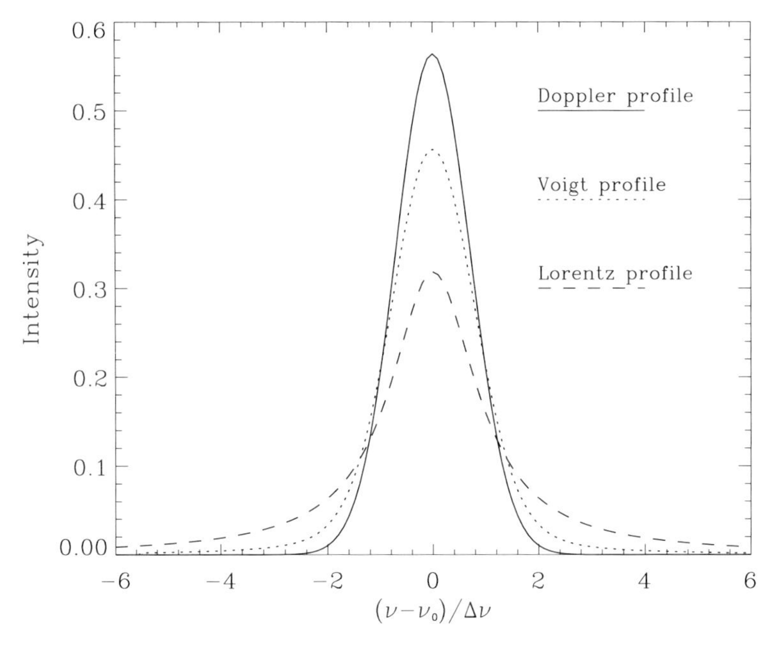

Figure 3.2 Comparison of normalized Lorentz, Voigt, and Doppler profiles versus $x = (\nu - \nu_0)/\Delta\nu$. $\Delta\nu$ is the Doppler width $\alpha_{\rm D}$ for both Doppler broadening and Voigt broadening and is the Lorentz width $\alpha_{\rm L}$ for Lorentz broadening. $a = \alpha_{\rm L}/\alpha_{\rm D} = 1$ was used for the Voigt profile.

resonance cross section in the following form, showing the relationships between the line profile, the oscillator strength, and the *line strength* $\mathcal{S}_i$ (see Eq. 4.59):

$$\sigma_{\rm ni}^{\rm res}(\nu) = \frac{e^2 f_i}{4m_e\epsilon_0 c}\Phi_{\rm L}(\nu) \equiv \mathcal{S}_i\Phi_{\rm L}(\nu). \tag{3.9}$$

3.3.2 Conservative and Nonconservative Scattering

It is convenient to classify scattering processes, depending upon whether or not the photon changes frequency upon scattering. In analogy with elastic and inelastic collisions between two material particles, the issue is whether or not there is a net change of energy and momentum following the "collision." Light carries momentum (of order $h\nu/c$), and one might expect the absorption and subsequent reemission of light would impart both a momentum impulse[19] and a change of energy (frequency) of the photon. For the UV/visible/IR radiation fields of interest to us, this process is negligible compared with the damping (or more correctly, broadening) effects resulting from the uncertainty of the upper-state lifetime. For resonance scattering, Rayleigh scattering, and Mie–Debye scattering the change in emitted frequency is very small, compared with the incident frequency. The term *coherent scattering* is sometimes used to describe this case, since the process involves interference between the incident and scattered waves. However, this term could lead to confusion, since it might lead one to believe that it does not apply to scattering from random media. It also implies that if absorption occurs, the process is incoherent, which is untrue.

Inelastic scattering involves an exchange of internal energy of the medium with that of the radiation field. Typically, the exchange results in a net loss of radiative energy and a gain of internal energy of the medium in the form of heat or chemical energy. Such processes are possible when the matter contains a number of accessible energy states. For example, *rotational Raman scattering* is the process by which the excited "virtual" state of a molecule radiatively decays, not to the original state (Rayleigh scattering), but to a state higher in rotational energy (the *Stokes component*) or lower in rotational energy (the *anti-Stokes component*). If molecules are illuminated by monochromatic light from a laser, the scattered spectrum will be a series of closely spaced lines on either side of the central *Cabannes line*.[20] An understanding of the basic aspects of rotational Raman scattering can be obtained from classical considerations. We replace our notion of an induced dipole fixed in space by one that is also rotating. The dipole will then emit not only its fundamental ("carrier") frequency, but "sideband" frequencies consisting of sums and differences of the basic frequency and the frequencies corresponding to the energies of rotation. Because rotational energy states are quantized, this results in a number of discrete "beat frequencies" (see Problem 3.3). A closely related process is *fluorescence*, which applies to two bound states, either vibrational or electronic. Here the molecule is originally in its ground (lowest energy) state and returns after scattering to a higher energy state. The scattered photon will, of course, have less energy than the incident photon.

Raman scattering and fluorescence may be classified as "cross-wavelength" processes, in that scattered photons of more than one discrete frequency are involved. Raman scattering is important for some ocean and lidar applications. We will not concern ourselves with such processes, since they occur in rather specialized applications (such as in so-called Raman lidars).

Another classification is whether the scattering process results in partial absorption of the light energy. If there is negligible absorption, it is said to be *conservative*, referring to the conservation of radiative energy. If there is some absorption, it is *nonconservative*. These terms may be considered to be synonomous with elastic and inelastic scattering. To ensure that we do not confuse the terms "elastic" and "inelastic" with processes involving collisions of molecules (see Chapter 4), we prefer to use the terms "conservative" and "nonconservative" when referring to light scattering.

3.3.3 Natural Broadening

For an isolated molecule, the notion of the damping parameter γ may be extended to a more realistic interpretation as the inverse lifetime of an excited quantum state t_r, as discussed further in Chapter 4. As seen from Eq. 3.6 the shorter the upper-state lifetime, the broader the frequency line width of the profile, which we temporarily denote as $\Delta\nu$. This inverse relationship is consistent with *Heisenberg's Uncertainty Principle*, relating the uncertainty in knowledge of the energy of a quantum system $\Delta E = h\Delta\nu$ to the uncertainty in knowledge of the lifetime Δt of the energy state. In this context, Heisenberg's relationship is written $\Delta E \Delta t \approx (h/2\pi)$. It follows immediately that $\Delta t \equiv t_r \approx (2\pi\,\Delta\nu)^{-1}$. The situation where the absorption line is dampened solely by the natural lifetime of the upper level is called *natural broadening*.

3.3.4 Pressure Broadening

Natural dampening applies to an isolated molecule, that is, one unperturbed by collisions with its neighboring molecules within a radiative lifetime, t_r. For a strong (*allowed*) transition in the shortwave spectrum, a typical value is $t_r \sim 10^{-8}$ s. Thus the natural line width α_N is $\sim 1/2\pi t_r \sim 1 \times 10^7$ s^{-1}. In wavenumber units $\alpha_N = 1/2\pi t_r c = 5 \times 10^{-4}$ cm^{-1}, which is very much smaller than that observed in atmospheric spectra. For vibrational and rotational transitions in the IR, t_r is much longer, of the order of $t_r \sim 10^{-1}$ to 10^1 s, with even smaller values of α_N. Thus natural broadening is completely negligible in atmospheric applications.

Collisions between molecules result in collisionally induced transitions, which occur temporarily in the joint system of two molecules in close vicinity. The net effect of these nearly resonant transitions on the emitted energy is very small, since roughly half of the transitions are excitations, and the other half deexcitations of energy states. These processes effectively reduce the lifetime of the upper state, and thus they broaden the line. The reduced lifetime is called the optical lifetime t_{opt}. It may be shown that under rather general conditions (see Problem 3.6) the collision process leads to a Lorentz profile, with a line width $\alpha_L \approx 1/2\pi t_{opt}$, of

$$\Phi_L(\nu) = \frac{\alpha_L}{\pi\left[(\nu - \nu_0)^2 + \alpha_L^2\right]}. \tag{3.10}$$

The theory of collisional line broadening is quite complicated. A number of methods have been devised, but all make simplifying assumptions. However, common to all the theories is the predicted linear dependence of α_L on the number density of the perturbing molecules, and upon the relative speed of the collision partners, v_{rel}. Thus for an arbitrary number density n and temperature T, since $v_{rel} \sim \sqrt{T}$, we may scale the pressure-broadened Lorentz width

$$\alpha_L \approx \alpha_L(\mathrm{STP}) \frac{n v_{rel}}{n_L v_{rel}(\mathrm{STP})} = \alpha_L(\mathrm{STP}) \frac{n\sqrt{T}}{n_L\sqrt{T_o}}, \tag{3.11}$$

where n_L is Loschmidt's number (the number density of air at STP) and T_o is standard temperature (272.2 K).

A typical value for t_{opt} is 10^{-10} s at STP, yielding a line width of 2×10^9 s^{-1}, or in wavenumber units $\alpha_L = 0.05$ cm^{-1}. This value greatly exceeds both the natural line width and the Doppler line width (see the next section), so that pressure broadening is dominant at STP for all wavelengths.

3.3.5 Doppler Broadening

The second major source of line broadening is that due to small *Doppler shifting* of the emitted and absorbed frequencies. As shown later, the shift is of order $v_{\rm rel}\nu/c = v_{\rm rel}\tilde{\nu}$, where $v_{\rm rel}$ is the relative speed and c is the light speed. For CO_2 at STP, at the wavelength of the strong 15-μm band, the Doppler shift is $\sim 8 \times 10^{-4}$ cm^{-1}. Thus Doppler broadening is negligible near the surface but grows in importance with height, since it varies as $\sqrt{T}$ and $\alpha_{\rm L}(z)$ falls off exponentially with z. They are equal in importance where $n/n_{\rm o} \approx 0.016$, which occurs at a height of about 30 km (see Appendix C). Below we examine the Doppler broadening in greater detail.

Doppler broadening stems from the simple fact that molecules are in motion when they absorb and when they emit. Consider a photon scattered in a resonance line. Since this process involves a photon arriving and leaving in different directions, there is a relative Doppler shift between the two events. Given that one always observes the net effect of an ensemble of scattering molecules, the result will be a spreading of the frequency of an initially monochromatic photon. Absorbing molecules will be distributed along the line of sight according to the *Maxwell–Boltzmann Law*, and therefore they will absorb in proportion to the number having a certain velocity component. Finally, suppose the molecules are excited by collisions and undergo many elastic collisions so that the "memory" of the direction of the colliding molecule is lost. Then the excited molecules will emit according to their velocity distribution, which is Maxwellian by assumption. In all three cases (scattering, absorption, and emission), we find that the line profile is dominated by thermal Doppler shifts, if the spread in frequency is larger than that caused by natural or pressure broadening.

We will describe the effects of thermal broadening on absorption. Consider the frequency of the photon in two reference frames (ν and ν'). Here ν is the frequency in the *laboratory frame*, which is the normal frame of an observer, and ν' is the frequency in the atom's (rest) frame. The relationship between these two frequencies is obtained by considering an atom with speed v moving toward the observer with a line-of-sight velocity $v\cos\theta$, where θ is the angle between the direction of motion and the line of sight. Suppose the molecule receives a photon of frequency ν in the lab frame. If the molecule were at rest with respect to the lab frame, in one second it would "see" exactly c/λ oscillations, where λ is the irradiating monochromatic wavelength. However, because of the fact that the molecule is moving toward the emitter, it "sees" an additional number of oscillations equal to the distance it travels in one second, divided by the wavelength. Thus, the number of oscillations the atom encounters is $c/\lambda + v\cos\theta/\lambda$. This quantity is just the frequency seen in the atom's frame, $\nu' = c/\lambda'$. Hence

$$\nu' = \nu + \frac{v\cos\theta}{\lambda} = \nu + \nu(v/c)\cos\theta = \nu[1 + (v/c)\cos\theta]. \qquad (3.12)$$

The molecule will therefore absorb according to its absorption cross section at the shifted frequency ν'. Suppose we align the rectangular coordinate system so that the x direction coincides with the line of sight. The absorption cross section appropriate to the molecule moving with the velocity component v_x is $\sigma_n(\nu + \nu v_x/c)$. The number of molecules moving with this velocity component is given by the Maxwell–Boltzmann distribution

$$f(v_x)dv_x = \left(\frac{m}{2\pi k_B T}\right)^{1/2} e^{-v_x^2/v_o^2} dv_x, \tag{3.13}$$

where $v_o = \sqrt{2k_B T/m}$ is the mean speed of the molecules. The cross section at the frequency ν due to all line-of-sight components is given by

$$\sigma_n(\nu) = \int_{-\infty}^{+\infty} dv_x f(v_x)\sigma_n[\nu(1 + (v_x/c)]$$
$$= \left(\frac{m}{2\pi k_B T}\right)^{1/2} \int_{-\infty}^{+\infty} dv_x e^{-v_x^2/v_o^2} \sigma_n(\nu + \nu v_x/c). \tag{3.14}$$

We now need to assume the functional form for the cross section in the molecule's frame. For simplicity we first assume that the broadening in the molecule's frame is much less than the thermal broadening. In effect, we are assuming that the molecule absorbs according to an infinitely narrow peak, that is, $\sigma_n(\nu) \approx \mathcal{S}\delta(\nu' - \nu_0) = \mathcal{S}\delta(\nu - \nu_0 + \nu v_x/c)$. Substituting in the above integral and integrating, we obtain

$$\sigma_n(\nu) = \mathcal{S}\left(\frac{m}{2\pi k_B T}\right)^{1/2} \exp\left[-c^2(\nu - \nu_0)^2/\nu_0^2 v_o^2\right]. \tag{3.15}$$

Letting $\alpha_D \equiv \nu_0 v_o/c$, which is called the *Doppler width*, we find

$$\sigma_n(\nu) = \mathcal{S}\Phi_D(\nu) = \frac{\mathcal{S}}{\sqrt{\pi}\alpha_D} \exp\left[-(\nu - \nu_0)^2/\alpha_D^2\right]. \tag{3.16}$$

It can be verified that the Doppler line profile $\Phi_D(\nu)$ is properly normalized. The mathematical form of Eq. 3.16 is recognized as a *Gaussian* function, of $(1/e)$-width α_D, and line width $\alpha_D\sqrt{\ln 2}$.

We now consider the more general case in which the broadening in the rest frame cannot be ignored compared with the thermal broadening. Suppose this is given by Lorentz broadening, with a total (natural- and pressure-broadened) line width α_L. Then substituting Eq. 3.10 into Eq. 3.14, and simplifying the notation, we find

$$\sigma_n(\nu) = \mathcal{S}\frac{a}{\pi^{3/2}\alpha_D} \int_{-\infty}^{+\infty} \frac{dy e^{-y^2}}{(v - y)^2 + a^2} \equiv \mathcal{S}\Phi_V(\nu), \tag{3.17}$$

where the *damping ratio* is $a \equiv \alpha_L/\alpha_D$ and $v \equiv (\nu - \nu_0)/\alpha_D$. The function $\Phi_V(\nu)$ is called the *Voigt profile*, which can be shown to be properly normalized. It represents the combined effects of both Lorentz and Doppler broadening. We note that for small damping ratios ($a \to 0$), we retrieve the Doppler result. The Voigt profile shows a Doppler-like behavior in the line core and Lorentz-like (ν^{-2}) behavior in the line wings. For $a > 1$, the Voigt profile resembles the Lorentz profile for all frequencies. The Voigt profile must be evaluated by numerical integration, and to this end, many authors have published algorithms suitable for efficient computation.[21]

3.3.6 Realistic Line-Broadening Processes

The collisional processes considered up to now are *adiabatic interactions*, implying only very gentle interactions with other molecules. Unfortunately, the quantum-mechanical line-broadening theory describing more realistic collisional perturbations of the upper state is extremely difficult. *Collisional narrowing* of lines can occur, as well as asymmetry and shifting of the line profile away from its unperturbed location. The far wings of lines are most affected by such complications. Empirical corrections are sometimes applied, in which the power of the exponent b in the $\nu^{-\mathrm{b}}$ formula is altered from its canonical value of $b = 2$ to obtain *super-Lorentzian* ($b < 2$) and *sub-Lorentzian* ($b > 2$) wing behavior. Fortunately, the line core and near wings remain Lorentzian even under severe collisional interactions. In practice, neighboring lines often begin to encroach on one another with increasing gas pressure, and line overlapping usually is more important than far-wing effects. The net effect of increasing pressure is greater line broadening, in agreement with Eq. 3.11.

In liquids and solids, the effects of nearby molecules are of course even more important than in the densest gases. Their absorption spectra may be extremely complex resulting from the myriad of energy states created by the mutual interactions. In most situations, first-principles analysis is impossible. Fortunately, the spectra frequently overlap to the point where the absorption spectra appear to be nearly continuous and slowly varying with frequency. Then the situation is actually simplified from the point of view of the radiative transfer. It is then sufficient to use low-spectral-resolution measurements and tabulation of the optical properties, provided one can collect samples of the material for transmission experiments in the laboratory. In practice, some materials are so opaque that standard transmission experiments are not possible. Reflection and absorption experiments are required for such substances, combined with the use of *Fresnel's equations* (see Appendix E) and various theoretical relationships (for example, the *Kramers–Krönig relations*). Standard tabulations of optical properties for continuous media are usually either in the form of real and imaginary indices of refraction or involve the complex dielectric constant, from which the absorption and scattering coefficients may be derived.

3.3.7 Case (2): Rayleigh Scattering

Suppose the driving frequency is much less than the natural frequency, $\omega \ll \omega_0$. Then Eq. 3.3 becomes

$$\sigma_n^{\mathrm{RAY}} = \frac{e^4\omega^4}{6\pi m_e^2\epsilon_0^2 c^4\omega_0^4} = \frac{1}{6\pi}\left(\frac{\omega}{c}\right)^4\left(\frac{e^2}{m_e\epsilon_0\omega_0^2}\right)^2. \tag{3.18}$$

This result displays the well-known ω^4 dependence of Rayleigh scattering.

The Rayleigh cross section may be related to the molecular polarizability, α_p. This quantity was previously defined (see §3.2) as a relationship between the induced dipole moment and the imposed electric field. From the harmonic oscillator solution, we find for $\omega \ll \omega_0$ the result $\alpha_p = e^2/4\pi m_e \epsilon_o \omega_0^2$ (Problem 3.1). Using this result in Eq. 3.18, and transforming to wavelength $\lambda = c/\nu$, we find

$$\sigma_n^{\mathrm{RAY}}(\lambda) = \frac{8\pi}{3}\left(\frac{2\pi}{\lambda}\right)^4 \alpha_p^2. \tag{3.19}$$

As discussed in §3.2.3 it is permissible to assume that we can add together the separate molecular contributions for a gaseous medium consisting of scatterers with random orientations and positions. A dilute mixture of gases, such as air, can be described in terms of a weighted average of the real refractive indices. This is denoted by m_r and may be related to the mean polarizability through

The Lorentz–Lorenz equation:

$$\alpha_p = (m_r - 1)/2\pi n.$$

Our final form for the macroscopic Rayleigh scattering coefficient is thus

$$\sigma^{\mathrm{RAY}}(\lambda) \equiv \sigma_n^{\mathrm{RAY}} n = \frac{32\pi^3(m_r - 1)^2}{3\lambda^4 n} \quad [\mathrm{m}^{-1}]. \tag{3.20}$$

For purely scattering dielectric spheres of radius $a \ll \lambda$, the polarizability was shown by Lorentz to be given by

$$\alpha_p = \frac{m_r^2 - 1}{m_r^2 + 2} a^3. \tag{3.21}$$

The cross section then follows immediately from Eq. 3.19. Other forms for the Rayleigh scattering cross section follow if the particles are anisotropic, nonspherical, partially absorbing, inhomogeneous, etc. A third limit, $\omega \gg \omega_0$, is considered in Problem 3.2.

Since m_r varies with wavelength, the actual cross section departs somewhat from the λ^{-4} behavior. A convenient numerical formula (accurate to 0.3%) for the Rayleigh scattering cross section for air is[22]

$$\sigma_n^{\mathrm{RAY}} = \lambda^{-4} \sum_{i=0}^{3} a_i \lambda^{-2i} \times 10^{-28} \quad [\mathrm{cm}^2] \quad (0.205 < \lambda < 1.05\ \mu\mathrm{m}),$$

where the coefficients are $a_0 = 3.9729066$, $a_1 = 4.6547659 \times 10^{-2}$, $a_2 = 4.5055995 \times 10^{-4}$, and $a_3 = 2.3229848 \times 10^{-5}$.

Although the classical Lorentz dispersion theory (§3.2) is still quite useful in understanding some phenomena in liquids and solids, it has been more profitable to treat them as continuous media. The scattering is considered to occur as a result of optical inhomogeneity arising from impurities and imperfections, as well as statistical fluctuations of density and concentration. These fluctuations are due to various types of collective oscillations, set up by thermal motions. This approach was pioneered by Smoluchowski and Einstein in the early twentieth century. As discussed in §3.2, it has been long understood that even scattering from gases requires such a description, because the coherent interference of the scattered waves would predict zero scattering in a homogeneous medium. Fortunately, gases in planetary atmospheres generally scatter as if there were no mutual interactions (except in the forward direction).

3.4 The Scattering Phase Function

So far we have ignored the directional dependence of the scattered radiation. We now consider the determination of the scattering phase function. We wish to describe the amount of radiation emanating from a small volume element as a result of scattering due to radiation coming from the Sun or from other parts of the medium. The angle Θ between the directions of incidence $\hat{\Omega}'$ and observation $\hat{\Omega}$ is given by $\cos\Theta = \hat{\Omega}' \cdot \hat{\Omega}$. This angle is called the *scattering angle*. The term *forward scattering* refers to observation directions for which $\Theta < \pi/2$, and we use *backward scattering* for $\Theta > \pi/2$. The *total* scattering cross section was defined in Eq. 3.3 as the total power per unit area scattered *in all directions* divided by the incident power per unit area of the incident plane wave. Similarly, the scattered power per unit area per steradian *in a particular direction of observation* divided by the power per unit area of the incident plane wave is called the *angular* scattering cross section, $\sigma_n(\Theta)$ [$\mathrm{m^2 \cdot sr^{-1}}$]. Azimuthal asymmetry (dependence on the ϕ coordinate) of the scattering phase function will usually disappear when averaging over all orientations of scatterers. Thus it is almost always permissible to assume that the scattering cross section is the same everywhere along a cone of half-angle Θ. To determine the scattering cross section as a function of Θ, we form the following scalar product:

$$\hat{\Omega}' \cdot \hat{\Omega} = \cos\Theta = \Omega_{x'}\Omega_x + \Omega_{y'}\Omega_y + \Omega_{z'}\Omega_z.$$

The rectangular components Ω_x, Ω_y, and Ω_z are illustrated in Fig. 3.3. Carrying out the multiplications and noting that $\cos(\phi' - \phi) = \cos\phi' \cos\phi + \sin\phi' \sin\phi$, we find (Problem 3.4)

$$\cos\Theta = \cos\theta' \cos\theta + \sin\theta' \sin\theta \cos(\phi' - \phi). \tag{3.22}$$

This result is recognized as the familiar *cosine law* of spherical geometry.

Consider a medium consisting of just one type of particle of number density n [$\mathrm{m^{-3}}$]. $\sigma_n(\cos\Theta)$ is the angular cross section per particle and $\sigma(\cos\Theta) = n\sigma_n(\cos\Theta)$ is the *angular scattering coefficient* [$\mathrm{m^{-1} \cdot sr^{-1}}$]. It is convenient to introduce a dimensionless quantity that characterizes the scattering process. We define the *phase function*[23] as the normalized angular scattering cross section as follows:

$$p(\cos\Theta) \equiv \frac{n\sigma_n(\cos\Theta)}{n\int_{4\pi} d\omega\sigma_n(\cos\Theta)} \quad [\mathrm{sr^{-1}}]. \tag{3.23}$$

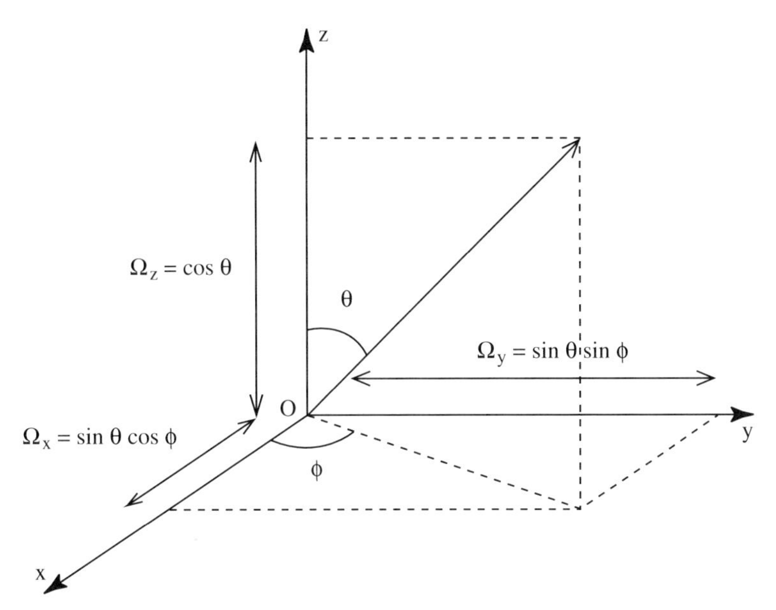

Figure 3.3 Illustration of the relationship between Cartesian and spherical coordinates. The rectangular components of the unit vector $\hat{\Omega}$ are $\Omega_x = \sin\theta\cos\phi$, $\Omega_y = \sin\theta\sin\phi$, and $\Omega_z = \cos\theta$.

The normalization is

$$\int_{4\pi} d\omega \, \frac{p(\cos\Theta)}{4\pi} = \int_0^{2\pi} d\phi \int_0^{\pi} d\theta \sin\theta \, \frac{p(\theta', \phi'; \theta, \phi)}{4\pi} = 1. \tag{3.24}$$

Since $p/4\pi$ varies between 0 and 1 this suggests a probabilistic interpretation. Given that a scattering event has occurred, the probability of scattering in the direction $\hat{\Omega}$ into the solid angle $d\omega$ centered around $\hat{\Omega}$ is $p(\hat{\Omega}', \hat{\Omega})d\omega/4\pi = p(\cos\Theta)d\omega/4\pi$.

3.4.1 Rayleigh-Scattering Phase Function

The radiation pattern for the far field of a classical dipole is proportional to $\Pi \sin^2\theta$, where θ is the polar angle as measured from the axis defined by the induced field, and Π is the induced dipole moment along that axis. The scattered radiation therefore maximizes in the plane normal to the dipole and vanishes on the axis of the dipole itself. We now consider how this translates into a normalized angular scattering cross section or phase function. This is the probability $p(\Theta)$ of scattering per unit solid angle, defined above, which depends upon the projection of the induced dipole moment in the direction Θ of the scattered radiation.

As usual we denote the direction of propagation of the incident and scattered waves to be $\hat{\Omega}'$ and $\hat{\Omega}$, respectively. It is convenient to use as a reference the *scattering plane*, defined as the plane containing $\hat{\Omega}'$ and $\hat{\Omega}$. For the present purpose it is sufficient to consider two linearly polarized incident waves, one with its electric field parallel to (or

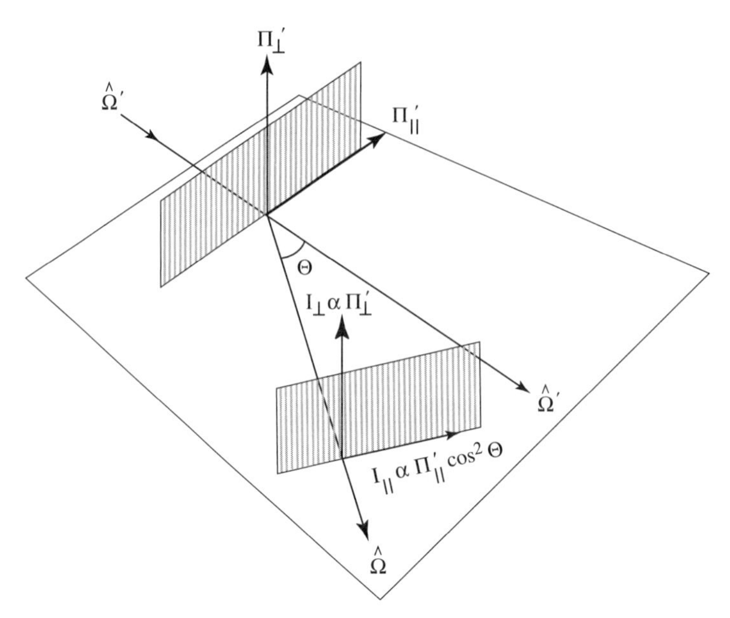

Figure 3.4 Illustration of the two transverse components of Rayleigh-scattered light. $\hat{\Omega}'$ and $\hat{\Omega}$ are the incident and scattered propagation vectors, respectively. $\Pi_\perp$ and $\Pi_\parallel$ are the induced dipole moments for incident electric fields that are linearly polarized in the directions perpendicular to, and parallel with, the scattering plane (shown as the white rectangle), respectively. $I_\perp$ and $I_\parallel$ are the corresponding scattered intensities in direction $\hat{\Omega}$ associated with the induced dipoles. The plane defined by $\Pi_\perp$ and $\Pi_\parallel$ as well as the plane defined by $I_\perp$ and $I_\parallel$ (both shown as shaded) are normal to the scattering plane.

in) the scattering plane, and the other with its electric field orthogonal to the scattering plane. As indicated in Fig. 3.4 these incident waves give rise to induced dipoles ($\Pi_\parallel$ and $\Pi_\perp$) along the respective incident fields. Referring to Fig. 3.4, we see that if the incident electric field lies in the scattering plane, then the angle θ between the induced dipole and the direction of scattering is $(\pi/2 + \Theta)$, where Θ is the scattering angle. Thus, the scattered light intensity is $I = I_\parallel \propto \Pi_\parallel \sin^2(\pi/2 + \Theta) = \Pi_\parallel \cos^2 \Theta$. If, however, the incident plane wave is linearly polarized perpendicular to the scattering plane, then the angle θ between the induced dipole and the direction of scattering is $\pi/2$, and the scattered intensity is simply proportional to the strength of the induced dipole, that is, $I = I_\perp \propto \Pi_\perp$.

We are concerned in this book with natural, unpolarized incident light, which can be treated as a sum of two orthogonal, linearly polarized waves having no coherent relationship. Furthermore they are of equal intensity ($I_\perp = I_\parallel = I/2$). Thus we find that for incident unpolarized light the scattered intensity and the *linear polarization* are

$$
\begin{aligned}
I_{\text{RAY}}(\Theta) &\propto (I_\perp + I_\parallel) \propto I(1 + \cos^2 \Theta), \\
P_{\text{RAY}}(\Theta) &\equiv \frac{I_\perp - I_\parallel}{I_\perp + I_\parallel} = \frac{1 - \cos^2 \Theta}{1 + \cos^2 \Theta}.
\end{aligned}
\tag{3.25}
$$

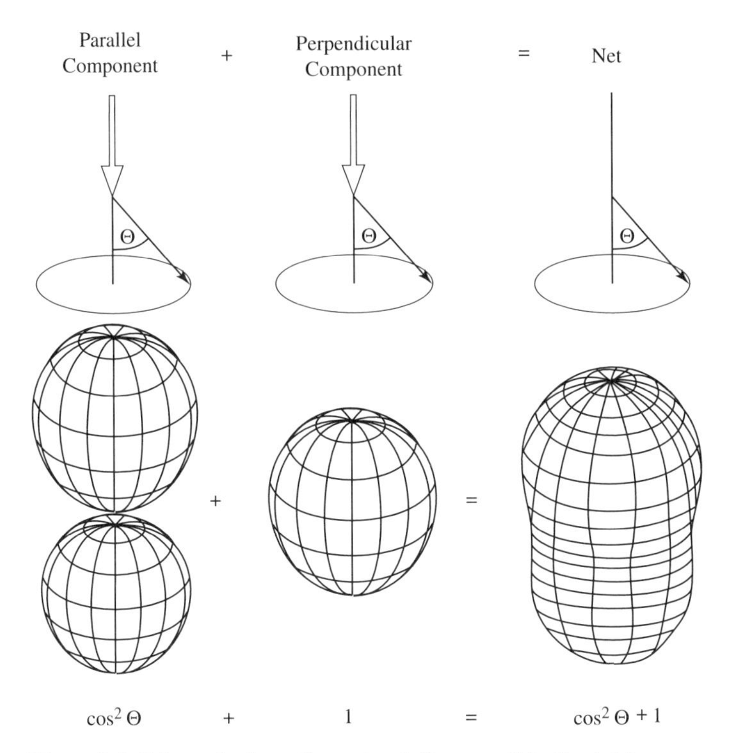

Figure 3.5 Schematic three-dimensional diagram of the Rayleigh-scattering phase function, or scattering pattern, for the case of incident unpolarized light. The intensity of the parallel component of the incident light is scattered according to the pure dipole ($\cos^2 \Theta$) law, as a result of the projection of the induced dipole along the scattered light direction Θ (see Fig. 3.4). In contrast, the intensity of the scattered component perpendicular to the scattering plane is the same for all Θ and thus is isotropic. The sum of the two patterns yields the scattering pattern for unpolarized light.

The proportionality constant follows from noting that $d\omega = d\phi \sin\theta d\theta$ in a polar coordinate system where $\hat{\Omega}'$ is along the z axis. Therefore

$$\frac{1}{4\pi}\int_{4\pi} d\omega(1+\cos^2\theta) = \frac{1}{4\pi}\int_0^{2\pi} d\phi \int_0^{\pi} d\theta \sin\theta(1+\cos^2\theta) = \frac{4}{3},$$

which implies that

$$p_{\mathrm{RAY}}(\Theta) = \frac{3}{4}(1+\cos^2\Theta). \tag{3.26}$$

Figure 3.5 illustrates how the two components, one isotropic, the other a dipole, combine to yield the Rayleigh phase function.

It should be kept in mind that even though we assumed that the incident light beam is unpolarized, the scattered light from a Rayleigh-scattering medium is polarized. This follows from Eq. 3.25 and the fact that $I_{\|}$ is not usually equal to $I_{\perp}$. In fact for

$\Theta = 90^\circ$, Eq. 3.25 shows that the scattered light is 100% polarized, in the $\perp$ direction. However, its importance is diminished in the presence of multiple scattering, which mixes light beams of many polarization states. In practice, if one is interested only in the flow of energy through a Rayleigh-scattering medium, errors in methods that ignore polarization and that use Eq. 3.26 for all orders of scattering are small, typically less than a few percent. However, angular distributions are subject to more severe errors, depending upon the optical thickness, among other factors. Another consideration for air molecules is the nonspherical shapes of N_2 and O_2. Their effect is to introduce a small anisotropic correction to the above formulas, since the two induced moments are slightly unequal ($\Pi_\parallel \neq \Pi_\perp$). One consequence of anisotropy is that the light scattered from air molecules through 90° is slightly depolarized ($P(90^\circ) \sim 96\%$).

3.5 Mie–Debye Scattering

For particles not small compared with the wavelength, the interaction properties must be found by solving a complicated boundary-value problem for the electric and magnetic fields. For spherical particles this theory is well established, and the detailed derivations and applications are found in many references and texts. Here we gather together the most important formulas, which convey the flavor of the input and output quantities, and how the calculation is carried out. In the limit of large particles ($r \gg \lambda$, where r is the radius of the spherical particle), the Mie–Debye results may be derived from geometrical optics.

The scattering cross section σ_n and *scattering efficiency* ($Q_s \equiv \sigma_n/\pi r^2$) are calculated from the multipole expansion expression

$$\sigma_n = \frac{2\pi}{k^2}\sum_{n=1}^{\infty}(2n+1)(|a_n|^2 + |b_n|^2), \tag{3.27}$$

where $k \equiv \omega/c = 2\pi/\lambda$ is the *angular wavenumber*. The extinction cross section ($k_n \equiv \sigma_n + \alpha_n$) is given by

$$k_n = \frac{2\pi}{k^2}\sum_{n=1}^{\infty}(2n+1)\Re(a_n + b_n), \tag{3.28}$$

where $\Re$ denotes the real part of the expression. The coefficients a_n and b_n, which may be complex, are derived from the formulas

$$a_n = \frac{m\psi_n(mx)\psi_n'(x) - \psi_n(x)\psi_n'(mx)}{m\psi_n(mx)\xi_n'(x) - \xi_n(x)\psi_n'(mx)}, \tag{3.29}$$

$$b_n = \frac{\psi_n(mx)\psi_n'(x) - m\psi_n(x)\psi_n'(mx)}{\psi_n(mx)\xi_n'(x) - m\xi_n(x)\psi_n'(mx)}, \tag{3.30}$$

where $m \equiv m_p/m_{air}$, and m_p and m_{air} are the refractive indices of the particle and the surrounding medium (in this case air), respectively.

The functions ψ_n and ξ_n are the *Ricatti–Bessel functions*, which are related to the spherical Bessel functions. Primes (′) denote differentiation with respect to the argument. The quantity $x \equiv 2\pi r/\lambda$, where λ is the wavelength inside the medium ($\lambda = \lambda_o/m_p$), is called the *size parameter*.

The phase function for scattering is given by

$$p(\Theta) = \frac{1}{2}(|S_1|^2 + |S_2|^2), \tag{3.31}$$

where S_1 and S_2 are the *scattering amplitudes*,

$$S_1 = \sum_{n=1}^{\infty} \frac{(2n+1)}{n(n+1)}(a_n\pi_n + b_n\tau_n), \qquad S_2 = \sum_{n=1}^{\infty} \frac{(2n+1)}{n(n+1)}(a_n\tau_n + b_n\pi_n), \tag{3.32}$$

and where

$$\pi_n(\Theta) = \frac{P_n^1(\Theta)}{\sin\Theta}, \qquad \tau_n(\Theta) = \frac{dP_n^1(\Theta)}{d\Theta}. \tag{3.33}$$

$P_n^1(\Theta)$ is the *associated Legendre polynomial*. (For example, $P_1^1 = \sin\Theta$, $\pi_1 = 1$, and $\tau_1 = \sin\Theta$.)

The Rayleigh-scattering formula (see Eqs. 3.20 and 3.21) is retrieved from Eq. 3.27 by expanding the functions ψ_n and ξ_n in powers of x and retaining only the lowest-order terms.[24]

3.6 Summary

Scattering processes occur in both homogeneous and inhomogeneous media. The basic difference is that in the former, the scattered wavelets cancel and only the direct beam is preserved. In the latter, because the scatterers are subject to statistical variations (in time and space) the scattered wavelets do not cancel. In both gaseous and aqueous media of interest to us, the squares of the electric fields in the scattered waves are added as if all the particles were independent scatterers. In both cases the direct beam experiences the collective effects of refraction and reflection. In the classical picture, the scattering process occurs as a result of the incident electric field inducing temporary electric dipoles, which then act as sources of secondary radiation. The radiation issuing from these centers is called first-order scattering. Radiation produced by further interactions of the first-order radiation with the medium is called multiple scattering. Scattering processes in which there is no net energy exchange between the gas and the photons are classified as *conservative*. If some absorption occurs, the process is called *nonconservative*.

The frequency dependence of the absorption line profile was illustrated by solution of the forced classical damped harmonic oscillator. The concepts of resonance and Rayleigh scattering follow from various limiting forms of the classical oscillator solution. The Lorentz line-broadening profile as well as the Rayleigh phase function also follow from this formulation. The classical picture of an induced dipole was extended to describe scattering as an excitation of a quantized state. The concept of pressure broadening is understood as the extension of the harmonic oscillator damping to include perturbations of the excited state by elastic collisions. Doppler and mixed (Voigt) line broadening were shown to result from the spectral line shifts from a gaseous ensemble in which there is a dispersion of velocities, described by the Maxwell–Boltzmann distribution. The Rayleigh phase function applying to unpolarized incident light was derived from consideration of the components of the induced dipole in directions normal to, and perpendicular to, the scattering plane. Mie–Debye scattering theory was briefly described for spherical particles.

Problems

3.1 The equation of motion of a damped, simple harmonic oscillator, subject to a forcing electric field of amplitude $\mathbf{E}'$ and angular frequency ω, may be written

$$m_{\mathrm{e}}\left(\frac{d^2\mathbf{z}}{dt^2}+\omega_0^2\mathbf{z}\right)=-e\Re(\mathbf{E}'\mathbf{e}^{i\omega t})-m_{\mathrm{e}}\gamma\frac{d\mathbf{z}}{dt}. \tag{3.34}$$

Here $\mathbf{z}$ is the vector displacement which is in the same direction as the imposed electric field vector $\mathbf{E}'$. $\Re(\,)$ denotes the real part of a complex quantity.

(a) Derive Eq. 3.2.

(b) Show that for $\omega \ll \omega_0$ the polarizability is given by

$$\alpha_{\mathrm{p}} = e^2/4\pi m_{\mathrm{e}}\epsilon_{\mathrm{o}}\omega_0^2.$$

3.2

(a) Taking the high-frequency limit of Eq. 3.3 ($\omega \gg \omega_0$), show that the total scattering cross section is given by

$$\sigma_n = \frac{8\pi}{3}r_0^3,$$

where $r_0 = e^2 m_e c^2/4\pi\epsilon_{\mathrm{o}}$ is the classical electron radius ($r_0 = 2.82 \times 10^{-15}$ [m]). σ_n, the *Thomson-scattering* cross section, gives the scattering cross section for a free electron.

(b) Derive the above equation for r_0 by equating the electrostatic self-energy of an electron to its relativistic equivalent $m_e c^2$.

3.3 A qualitative understanding of Raman scattering is gotten by considering an analogy with Rayleigh scattering. Consider the polarizability to oscillate with a frequency corresponding to the vibrational (or rotational) energy,

$$\alpha = \alpha_0 + \alpha_k \cos(\omega_k t + \delta_k).$$

Show that the scattering consists not only of the principal (Rayleigh) component, but also two other frequency components, the *Stokes line* having a frequency $(\omega_0 - \omega_k)$, and an *anti-Stokes line* with frequency $(\omega_0 + \omega_k)$.

3.4 Derive Eq. 3.22.

3.5 Show that a line that is broadened both naturally (line width α_N) and by pressure effects (line width α_P) is a Lorentzian with a line width given by $\alpha = \alpha_N + \alpha_P$.

3.6 A classical view of collisional broadening results from imagining the molecules to be radiating simple harmonic oscillators, perturbed by elastic collisions that change the phase of the oscillators in a random manner. Assume that collisions occur with a probability $p(t) = e^{-t/t_{opt}}$, where t and t_{opt} are the time, and mean time, between collisions, respectively.

First find the Fourier cosine transform $g(\nu)$ of the ensemble of wave trains of varying length ct by carrying out the integration

$$g(\nu, t) = \sqrt{\frac{2}{\pi}} \int_{-t/2}^{t/2} dt' \cos 2\pi \nu_0 t' \cos 2\pi \nu t'.$$

Ignore the term involving $(\nu + \nu_o)^{-1}$. Then evaluate the power spectrum by convoluting the Poisson distribution of collision times $p(t)$ with the square of the cosine transform (since the intensity is proportional to the square of the electric field amplitude),

$$E(\nu) = \int_0^{\infty} dt p(t) g^2(\nu, t).$$

Show that the result is that $E(\nu) \propto \phi_L(\nu)$, the Lorentz profile, where the line width is $\alpha_L = 1/(2\pi t_{opt})$.

3.7 Generalize Eq. 3.23 to describe the phase function for a mixture of particles consisting of different scattering cross sections and phase functions, having densities n_i $(i = 1, \ldots, n)$.

Notes

1 The concept of a plane wave is discussed in detail in Appendix H.

2 In *fluorescence scattering*, the accurate quantum-mechanical probability of scattering is given by the product of the separate absorption and emission probabilities. In this case the intermediate (excited) state of matter is an observable property.

3 The classical Lorentz theory of the interaction of radiation and matter is given a consistent treatment in the well-written textbook, J. M. Stone, *Radiation and Optics, An Introduction to the Classical Theory*, McGraw-Hill, New York 1963.

4 Only later in 1925 did these light particles become known as "photons."

5 This type of scattering is often called *Mie scattering*. This term should be reserved strictly for scattering from *homogeneous spherical* particles for which Gustav Mie in 1908 brought together the contributions of others and added his own improvements: "Beiträge zur Optik trüber Medien, Speziell Kolloidaler Metallösungen", *Annalen der Physik*, **25**, 377–445, 1908.

6 α_p is written as a scalar here, but it can be a tensor quantity in so-called anisotropic media. In other words, it is possible in such media for the induced dipole to have components normal to the direction of the imposed electric field.

7 The proportionality of the **E** field to the induced dipole moment per unit volume $\mathcal{P}$ is a consequence of Maxwell's equations. The electric fields in the "far field" can be shown to be given by $\mathbf{E} \propto \partial^2\mathcal{P}/\partial^2 t$. Since both **E** and $\mathcal{P}$ vary in time as $\exp(i\omega t)$, then $\mathbf{E} \propto \omega^2\mathcal{P} \propto \lambda^{-2}\mathcal{P}$. Since the scattered intensity varies as E^2, then $I \propto \lambda^{-4}\mathcal{P}^2$.

8 This result applies only to ideal gases, for which there are no mutual interactions of the molecules. A further restriction is that the index of refraction must be close to unity, which is valid for all transparent gases. Condensed media, for example liquid water, do not share this property, although they usually still maintain an approximate λ^{-4} behavior. Scattering from inhomogeneities in such materials is a valuable tool for studying both the atomic and thermodynamic properties of the substance.

9 Lack of permanent phase relationships is a consequence not only of the random spatial distribution of scatterers, but also because their orientations and thermal velocities are uncorrelated. There are many synthetic substances for which this is not the case. For example, a liquid crystal owes much of its peculiar optical properties to the fact that the orientations of the various scattering centers are highly correlated. Another example is an ionized medium that will affect the passage of a radio wave in a collective fashion as a result of the spatial coherence brought on by the mutual electrical forces between charges.

10 The standard reference on single-scattering theory is Van de Hulst, H. C., *Light Scattering by Small Particles*, Dover, New York, 1981.

11 The subject of absorption and scattering in the ocean is covered in N. G. Jerlov, *Optical Oceanography*, Elsevier, Amsterdam, 1968 and in Chapters 4 and 5 of J. Dera, *Marine Physics*, Elsevier, Amsterdam, 1992. A modern treatise is that of C. D. Mobley, *Light and Water: Radiative Transfer in Natural Waters*, Academic Press, San Diego, 1994.

12 On the average, one out of ten solar photons encounter an aerosol particle before being absorbed or scattered back into space.

13 Even though the induced magnetic dipole arising from the oscillating current is vital in the emission process, for the present purposes it is possible to ignore it altogether.

14 Although this classical notion has been fraught with conceptual difficulties, it has nevertheless remained a useful concept since its predictions are consistent with experiment.

15 If the oscillator is within an insulating solid, and therefore subject to the induced electric fields from neighboring molecules in the lattice, $\mathbf{E}'$ should be the *local field*, $\mathbf{E}'_{\mathrm{loc}}$, and not the externally applied electric field intensity of the incident light wave. It is a complicated and not completely solved problem to relate these two fields or to relate

the molecular properties, such as the polarizability, to the bulk optical constants, such as the index of refraction. One commonly used relationship is the *Clausius–Mosetti equation*. A nice derivation is found in Hapke, B., *Theory of Reflectance and Emittance Spectroscopy*, Cambridge University Press, Cambridge, 1993.

16 A good treatment is found in Zahn, M., *Electromagnetic Theory, A Problem Solving Approach*, Wiley, New York, 1979. Another is in the recent book, Heald, M. A. and J. B. Marion, *Classical Electromagnetic Radiation*, 3rd ed., Saunders College Publ. Co., Ft. Worth, 1995.

17 The reader may find this formula given in other references in cgs units, in which case the result for the integrated cross section is $\pi e^2/m_e c$. The difference is the substitution of the term e^2 in cgs units for $e^2/4\pi\epsilon_0$ in SI units and stems from the difference in the Couloumb-Law expression in the two systems of units for the force between two point charges separated by a distance r, e^2/r^2 (cgs) and $e^2/4\pi\epsilon_0 r^2$ (SI).

18 Many references exist. One useful one is Shu, F. H., *The Physics of Astrophysics, Vol. 1: Radiation*, University Science Books, Mill Valley, CA, 1991.

19 The so-called radiation reaction, the small impulse delivered to the atom as a result of the emission process, actually accounts for the phase difference of the scattered and incident waves.

20 The accepted terminology is that the sum of the *Cabannes line* and the rotational lines is the *Rayleigh line*. See Young, A. T., "Rayleigh scattering," *Physics Today*, **35**, 42–8, 1982.

21 A thorough discussion of computational algorithms for the Voigt function is that of Armstrong, B. H., "Spectrum line profiles: The Voigt function," *J. Quant. Spectrosc. Radiative Transfer*, **7**, 61–88, 1967. The algorithm used in the Voigt subroutine in the IDL graphics package is from this reference. A discussion of the computational strategies for computing the Voigt function is given by Uchiyama, A., "Line-by-line computation of the atmospheric absorption spectrum using the decomposed Voigt line shape," *J. Quant. Spectrosc. Radiative Transfer*, **47**, 521–32, 1992.

22 This formula was provided by M. Callan, University of Colorado, who fitted the numerical results of D. R. Bates, "Rayleigh scattering by air," *Planet. Space Sci.*, **32**, 785–90, 1984.

23 The term *phase* comes from the original astronomical usage, which refers to the variation of the planetary brightness versus *phase angle* α, the angle between the two vectors from the Earth to the planet and from the Sun to the planet $(\pi - \Theta)$. It should not be confused with the phase of an electromagnetic wave.

24 See Chapter 5 of Bohren, C. F. and D. R. Huffman, *Absorption and Scattering of Light by Small Particles*, Wiley, New York, 1983.

Chapter 4

Absorption by Solid, Aqueous, and Gaseous Media

4.1 Introduction

Most particles in the atmosphere and the ocean (except those responsible for the density irregularities leading to Rayleigh scattering) are also absorbers of radiation. Absorption causes the incident radiation to be further weakened (in addition to scattering) by losses within the particles themselves. As explained in Chapter 3 the net effect of scattering and absorption is called attenuation or extinction. Absorption is inherently a quantum process resulting from the fact that matter contains energy levels that can be excited by the absorption of radiation. A transition from an initial quantum state to a higher-energy state is highly dependent on the frequency or energy of the incident light. When the photon energy is close to the energy difference between the initial and final state, the atoms and light may be said to be in resonance, and the absorption is comparatively high. Conversely, when the photon energy is not close to the transition energy, the absorption is often much weaker than the scattering and is not easily measurable. This energy selectivity is the outstanding characteristic of absorption. (In contrast, scattering is generally much less selective and usually has a smoothly varying efficiency with wavelength.) Selective absorption causes individual molecular absorption spectra to be very complex. The resonances are usually very sharp, and because of the many modes of excitation of molecules (particularly the polyatomic molecules of greatest interest to us) there may exist tens to hundreds of thousands of discrete absorption lines in molecular spectra. The dominant characteristic of such spectra is the presence of dark regions in the absorption spectrum. These broad spectral features are called molecular absorption bands, in which lines are clustered closely in frequency groups. Under low resolution these bands appear to be continuous functions of frequency. Figure 4.1 shows the calculated transmittance spectrum of air, including the separate contributions from molecules (Rayleigh scattering and near-IR absorption by water vapor and molecular oxygen) and aerosol scattering.[1]

In this chapter we mainly consider radiative processes that occur in the solar near infrared ($1 \leq \lambda \leq 3$ μm) and the thermal infrared ($\lambda \geq 3$ μm). We will frequently refer to the former as the near-IR and to the latter simply as the IR. Here, in contrast to the visible spectrum, extinction is dominated by absorption. This state of affairs is due to the multitude of quantum states that become accessible to low-energy photons. IR radiative transfer occurs as a series of emissions and absorptions. Unlike scattering, the temperature of the medium plays a vital role in the IR, since in the high-density media of interest to us in this book, it controls the rate of emission through Kirchhoff's Law (Chapter 5). We briefly touch on the topic of absorption in the UV and visible, which involves not only rotation and vibration but also electronic excitation. Electronic band

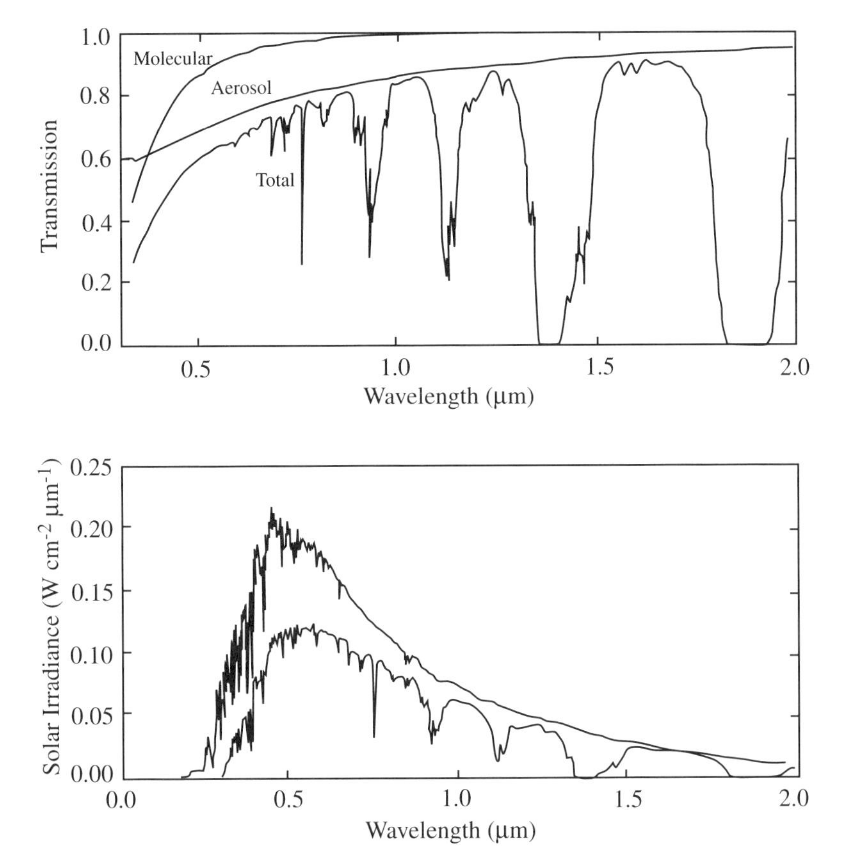

Figure 4.1 Upper panel: Low-resolution transmittance spectrum for three atmospheric components, molecular scattering, aerosol scattering, and molecular absorption.[3] The atmospheric absorption bands are mainly due to IR absorption by H_2O. Lower panel: The upper curve is the Lowtran solar irradiance outside the atmosphere. The lower curve is the solar irradiance observed vertically through the (clear) atmosphere.

spectra for simple diatomic molecules illustrate the more complicated band spectra of polyatomic molecules. Although electronic transitions play a minor role in the overall radiative energy balance of planetary atmospheres, they are of central importance for determining the composition and thermal structure of the upper atmosphere.

4.2 Absorption on Surfaces, on Aerosols, and within Aqueous Media

The general physical consequence of light absorption is the deposition of energy in the medium. If light absorption leads to excitation of a *bound* excited state, the energy of the excited state is usually promptly converted into thermal energy, by means of a collision with a neighboring gas molecule, or for a solid or liquid, by a dissipation of the energy through vibrations of the surrounding lattice or fluid cluster. Alternatively, the chemical energy of the medium may be altered. Because of its increased reactivity, an individual excited molecule may participate in a photo-induced or *photochemical reaction* with its neighbors. At higher photon energies (in the UV and at shorter wavelengths) the excited molecular state may be *unbound*. In this case the kinetic energy of the resultant atomic (or molecular) fragment is not quantized but is a continuous function of the incident photon energy. This is the process of *photodissociation*, which is important for the photochemistry and heating of the Earth's middle atmosphere (e.g., for the formation of ozone). At still higher photon energies, the absorption into an unbound electronic state may cause an electron to be removed from its parent molecule, leaving behind a positively charged *ion*. This is the basic mechanism giving rise to the ionosphere, and it accounts for the existence of free electrons whenever high-energy radiation is present. The rate of these photoabsorption processes depends upon the spectrum and strength of the radiation field, and therefore its accurate determination requires the application of the principles of radiative transfer.

4.2.1 Solids

In general, the absorption, reflection, and transmission properties of solids vary with frequency in a complex fashion. They may vary smoothly with frequency, or they sharply change in the neighborhood of *resonances*, where the absorbed energy coincides with energy differences between various types of quantized states of the solid. Consider first the high-energy process whereby an electron is transferred from a lower-energy state into an unoccupied higher-energy state. Many such states cluster together in "bands," so that we speak of a transition involving two bands. A fundamental difference between insulators (such as water and most soil minerals) and conductors (metals) is the disposition of these bands in energy. Conductors have incompletely filled bands or bands that overlap in energy with adjacent unoccupied bands. The availability of nearby unoccupied energy levels makes it possible for low-energy photons to be absorbed. However, the bands in insulators are well separated, such that the low-energy bands are filled, and the upper-energy bands are unfilled. Thus, only photons with energy greater than the "band gap" may be absorbed. Generally speaking, conductors are highly absorbing and reflecting[4] in the visible and IR, whereas insulators are more or less transparent over this spectral range, becoming absorbing in the UV.

Example 4.1 Color and Brightness of an Object

The overlapping of numerous absorption lines in solids does not mean that broad spectral absorption features, absorption "edges," etc. are absent. In fact, it is this selectivity that is responsible for nearly all *color* of objects in our natural environment. Objects are red because of their selective absorption of blue light. The color of the yellow substance in oceans is a result of its strong absorption of blue and UV light, etc. An exception to this general rule is Rayleigh scattering, which is responsible for the blue color of clear skies, oceans, blue-jay feathers, and the eyes of newborn infants. Most other blue colors (certainly one's blue sweater) are due to selective absorption of red and yellow light by various absorbing pigments. Of course, scattering contributes to the perception of the texture, sheen, etc. of objects, but it is the selective removal of various wavelengths that gives objects their characteristic hue.

In the thermal IR, absorption causes excitation of lattice vibrations (*phonons*), molecular vibrational states, and so-called intermolecular vibrations. The latter is associated with collective interactions between molecules and naturally depends more sensitively on the density and phase of the material. For *polar substances*, materials consisting of molecules that have a permanent electric dipole moment (such as water), the oscillating electric field tends to align the dipoles. This happens in the microwave spectrum where the collisional relaxation time (defined in §3.3.3) of the water molecules is $\sim 3 \times 10^{-11}$ s, which is very short compared with the wave period. Thermal motions tend to convert this energy of alignment into heat. This process is called *Debye relaxation*, and it is the mechanism responsible for intense absorption of microwave energy by liquid water. Because the damping effects are much less significant in water ice (damping times are about 1×10^{-3} s), radar backscatter cross sections depend sensitively upon whether the particles are raindrops or hailstones.

Most laboratory results for the reflecting, transmitting, and absorbing properties of solids apply only to "smooth" (polished) surfaces. However, most natural surfaces are irregular over many size scales, and laboratory results are not immediately applicable. Determination of the boundary properties of a rough surface, or of small suspended particles, requires a more fundamental knowledge of the bulk properties of a substance. This knowledge is embodied in the *optical constants*, the real (m_r) and imaginary (m_i) indices of refraction. The phase velocity of the wave is c/m_r. For absorption of a plane wave in an infinite dielectric medium, m_i determines the absorption coefficient through the dispersion relation $\alpha = 2\pi m_i/\lambda$. How are the optical constants (they are not actually constants, since they vary with frequency) determined? This might typically involve working backward from measurements of transmittance and reflectance of a thin sample, via *Fresnel's equations* (Appendix E). Given the optical constants, it is possible to determine through theory the scattering and absorptive properties of polished pure solids. Since most natural substance are irregular and of mixed composition, it is usually necessary to perform experiments on the bulk samples.

4.2.2 Aerosols

What happens to the optical properties when the material is finely divided into small particles? If the particles are large compared to the wavelength of light, then the principles of geometrical optics will apply, and ray-tracing techniques may be used to derive the appropriate interaction properties. If the particle dimensions are smaller than several hundred wavelengths, the concepts of transmittance, reflectance, and absorptance are not useful. At this point we must deal with the properties applicable to dispersed matter, that is, absorption and scattering coefficients. Radiation that penetrates the particles undergoes interference effects, which depend sensitively upon the size and shape of the particles. Interference can affect both the absorption and scattering in very different ways than in the bulk state. But are the bulk optical constants still relevant to small particles? Experience shows that *the same bulk optical constants (m_r and m_i) apply down to the smallest (0.1-μm radius) particles of practical interest to us.*

Provided aerosol particles are homogeneous, spherical, and of known composition, their absorption and scattering coefficients can be determined by solving a classical boundary-value problem, as discussed in §3.5. With the advent of fast computers, numerical solutions for other idealized shapes, such as spheroids, have become possible. Also, approximate techniques, such as the *discrete-dipole method*, have been developed to handle arbitrarily shaped particles, but with rather severe demands on computer time. The point is that if we are given an ensemble of independently scattering and absorbing particles, of known shapes, sizes, and composition for which the optical constants are known, it is possible, at least in principle, to compute the scattering and absorption coefficients. Because uncertainties in the knowledge of the aerosols are often thought to be more serious than shape effects (such as in the size distribution), equivalent spherical-particle assumptions are commonly resorted to, even for extreme nonspherical particles, such as snow crystals. The effects of particle shapes on radiative transfer remain to be established.

4.2.3 Liquids

Absorption in pure liquids results from the mutual interactions between the intermolecular forces, which may be thought of as collective excitation modes. As a consequence of this added complexity:

i. It is very difficult to calculate from first principles the quantitative details of the transitions (such as absorption line strengths and band frequency positions). Laboratory and/or in situ measurements of absorption spectra are therefore essential.

ii. The number of transitions is so large that overlapping of adjacent spectral absorption lines (or bands) yields an almost continuous absorption spectrum. Paradoxically, their complexity causes condensed media to have a much simpler absorption spectrum than that of its constituent molecules. Radiative transfer in aquatic media has a significant practical advantage over that in the

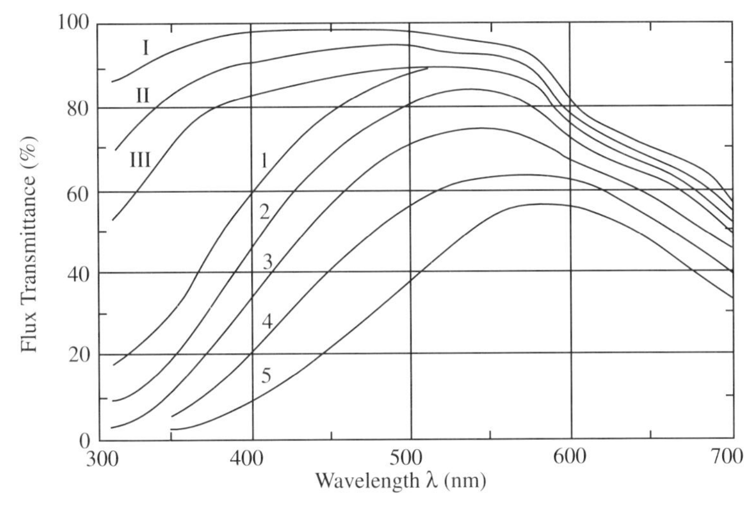

Figure 4.2 Spectral variation of the flux transmittance through a 1-m layer of water in the euphotic zone. Types I, II, and III are different ocean water types, and types 1–5 describe different coastal water types. (See Endnote 5.)

atmosphere: The spectral sampling interval for the oceanic radiation field can be orders of magnitude larger than that required in atmospheric radiation problems to achieve the same accuracy. (See Chapter 10 for more details.)

Unfortunately the above advantage is offset by the fact that, except for the purest waters, the optical properties of the ocean are largely governed by dissolved and suspended impurities, of both inorganic and organic origin. The compositional variability from location to location makes it difficult to create "standard" optical models, such as those used widely in atmospheric studies. An attempt to optically classify various water types[5] is shown in Fig. 4.2.3. Generally speaking, seawater is most transparent in the 400–600 nm region.

4.3 Molecular Absorption in Gases

Atmospheric molecules are highly selective in their ability to absorb radiation. This is particularly true in the thermal infrared part of the spectrum, where a large number of spectral absorption features occur. Figure 4.3 shows synthetic IR atmospheric radiance along a vertical path looking down at the surface from several altitudes. The surface emits thermal radiation (dashed curve), which is attenuated upon passage through the atmosphere (solid line). To delineate the effect of absorption on the transmitted radiation field, the atmospheric emission was not included in this computation, which was made at moderate spectral resolution (0.1 μm, or $\approx$2 cm^{-1}).[6] This figure shows that highly opaque regions exist side by side with transparent regions. These *molecular bands* exist where there are bunched coincidences of photon energies and excitation

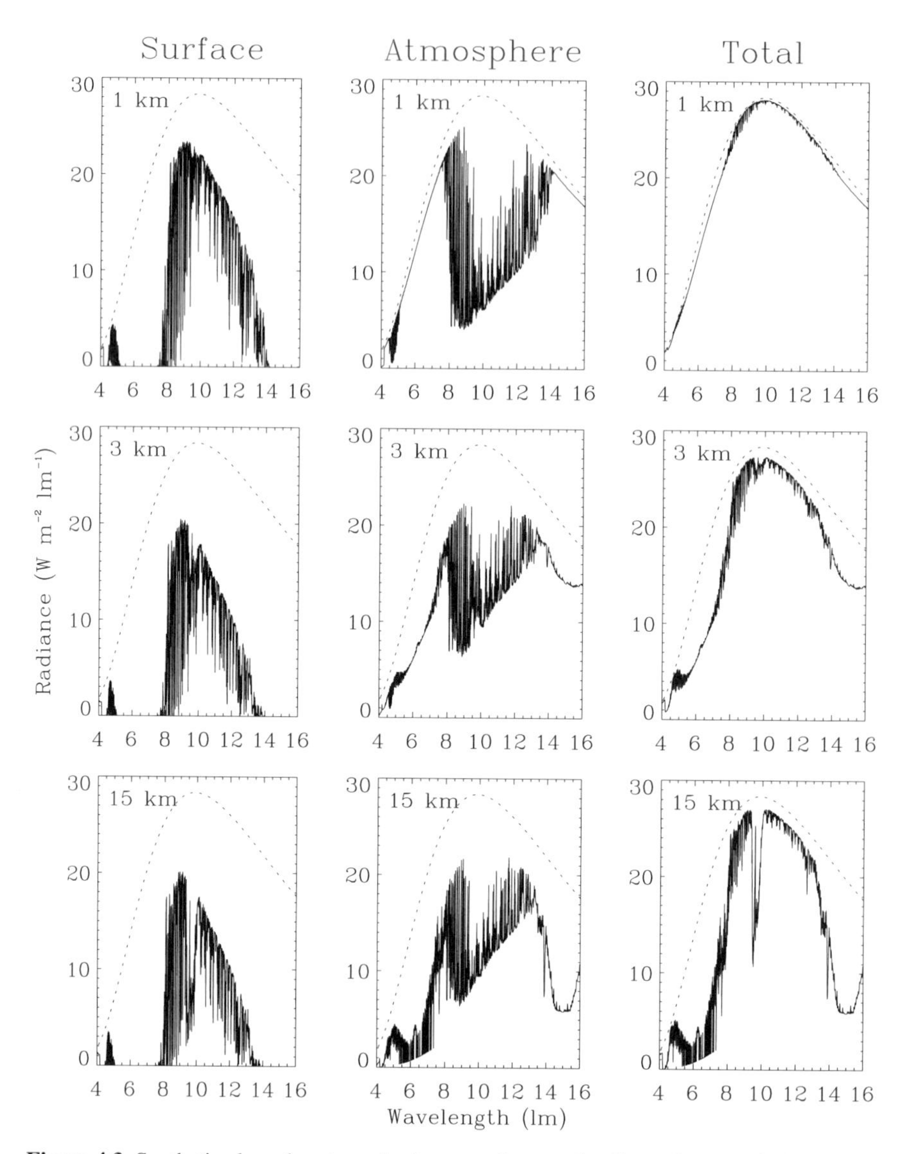

Figure 4.3 Synthetic clear-sky atmospheric upward spectral radiance for a vertical path from sea level to 1, 3, and 15 km. The computation was made using the MODTRAN code. The dashed line shows the blackbody curve evaluated at the surface temperature of 294 K. Left panels: Atmospheric emission was ignored. Middle panels: Same as left panel, but for atmospheric emission (no surface emission). Right panels: Same as left panel, but for emission from both surface and atmosphere.

energies of the various quantum levels. The excitation energies coincide with those of various normal modes of molecular vibration.

In the middle panels of Fig. 4.3 we show the synthetic spectrum computed in the same way, except that here we ignored the surface emission but included the atmospheric emission. Note that the absorption features in the previous figure appear as emission features, as a consequence of Kirchhoff's Law. Also, in the opaque region

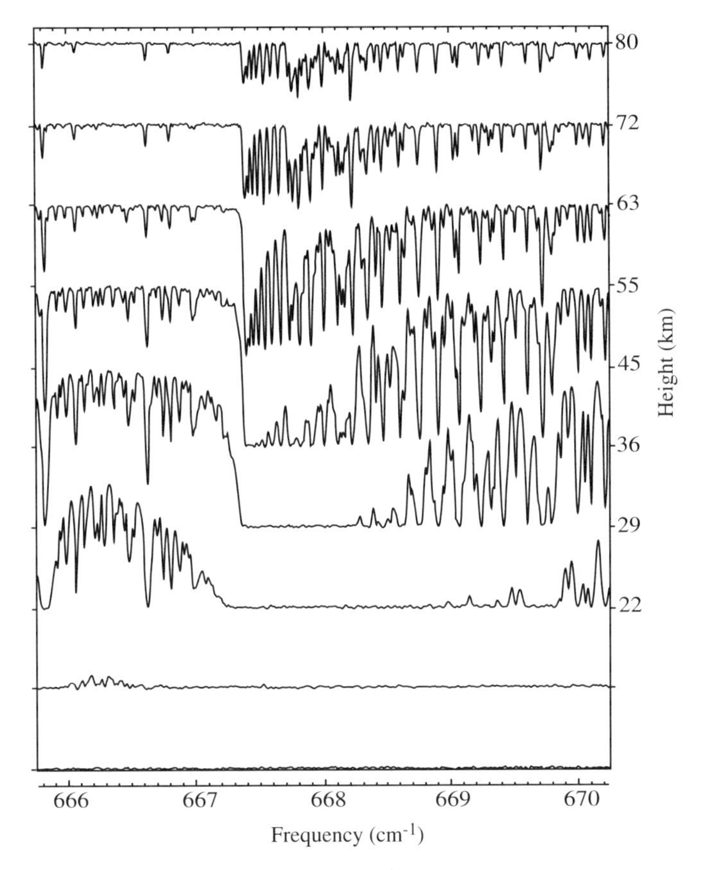

Figure 4.4 High-resolution (0.01 cm^{-1}) transmittance spectrum of the Earth's stratosphere and mesosphere measured by the ATMOS Michelson interferometer experiment from the NASA Space Shuttle. (See Endnote 7.) Shown is a section of the 667-cm^{-1} (15 μm) CO_2 band. Each curve applies to a horizontal line of sight identified by the minimum height of the tangent ray of solar radiation through the atmosphere. Note the intense absorption in the Q-branch near 667–668 cm^{-1}. Each curve is shifted vertically for clarity. A sliding transmittance scale (from 0 to 1) extends over a vertical distance of two tick marks.

between 5 and 7 μm, and between 14 and 16 μm, the intensity in the lower atmosphere (1 km level) is close to the Planck curve as expected. The total contribution to the upward intensity from both surface and atmospheric emission is shown in the right panel of Fig. 4.3 at the same atmospheric levels. At high altitude, the spectrum resembles what a downward-looking sensor above the atmosphere would observe.

At higher spectral resolution, molecular bands reveal their underlying structure – that of closely spaced lines. Figure 4.4 shows a small portion of the measured transmittance in the strong 15-μm band of CO_2. These data were taken by a Michelson interferometer[8] on board the Space Shuttle *Challenger* in 1985. The transmittance is the ratio of the irradiance, measured along a line of sight through the atmosphere, to the extraterrestrial solar irradiance. Each absorption line corresponds to a transition

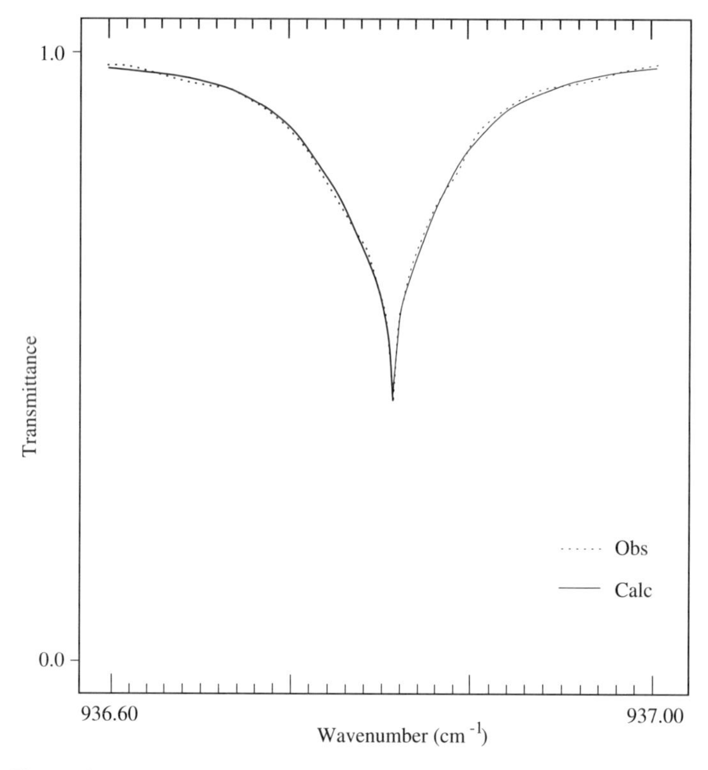

Figure 4.5 Dots: Ultra-high-resolution measurement of an individual molecular absorption line of CO_2 in the 936-cm^{-1} region from solar spectra obtained from Denver, Colorado during sunset on January 12, 1989, with a BOMEM model DA3.002 interferometer (see Endnote 9). Solid curve: theoretical line profile (see Eq. 10.50).

between two quantum states of a specific molecule. The frequency of each transition is a unique "fingerprint" of the particular species. Under still higher resolution (see Fig. 4.5), an individual line has a distinctive spectral width.[9] At altitudes below about 50 km in the Earth's atmosphere, both the strength of the line and its spectral width depend upon atmospheric pressure and temperature.

Regions of high transparency are called *spectral windows*. These are very important for atmospheric remote sensing of the planetary surface. The 10–12 μm window is particularly significant for cooling the Earth's surface, which emits approximately like a blackbody with a maximum near 10 μm (see Fig. 4.3). The effect of windows was illustrated in Fig. 1.2, which shows the emission spectrum of the Earth measured by a high-resolution interferometer from an orbiting spacecraft. In the high transparency regions, the Earth's surface emission is evident. The contribution of the upwelling atmospheric radiation occurs within the opaque bands, at an effective temperature lower than that of the surface. The emitted radiance is reduced in the regions of high opacity, because the radiation received by the satellite instrument is emitted from the upper colder atmospheric regions, where the lines are optically thin. Notice in the case of the Antarctic, where the surface is colder than the atmosphere, more radiation is emitted from the warmer atmosphere in the vicinity of the bands than from the surface (in the windows).

4.3.1 Thermal Emission and Radiation Laws

Thermal emission is the inverse of absorption. Every particle of matter at a temperature greater than absolute zero contains excited quantum states. The spontaneous decay of these states is accompanied by the creation of radiative energy. If only thermal emission were acting, all the molecules would eventually revert to their ground-state levels, and all the energy would reside in the radiation field. Of course, in reality the medium and its radiation field are continually exchanging energy by absorption and emission. For example, solar radiation is absorbed by a planet, with some of the energy going into thermal energy and some into mechanical energy (fluid motion) and chemical energy (change of state). The remainder of the absorbed energy goes into emission of thermal radiation, which is reabsorbed or lost to space. When the incoming and outgoing radiative powers are in balance, the planet is said to be in *planetary radiative equilibrium*.[10] A more restricted equilibrium occurs when the amount of energy absorbed locally is equal to that emitted locally. We call this simply *radiative equilibrium*.

Other kinds of equilibria refer to the temperature and motion fields and to the chemical composition. *Thermal equilibrium* occurs in a constant-temperature medium. No heat flows in the absence of a temperature gradient. *Mechanical equilibrium* occurs when there are no net forces or stresses anywhere in the medium. Consequently there is no bulk motion of the fluid. *Chemical equilibrium* occurs when the rates of all chemical reactions are balanced by their inverse reactions, so that the chemical composition is fixed throughout the medium.

When all these equilibria occur, we have the most general state of *thermodynamic equilibrium*. To attain this situation requires a closed system, called a *blackbody cavity*, or *hohlraum*,[11] with insulating walls completely isolating it from external influences. Planetary media, being "open" systems in the thermodynamic sense, would at first glance appear to be far from such an artificial condition. However, as we shall see, atmospheres and oceans do share certain properties with a medium in thermodynamic equilibrium. For this reason we are particularly interested in the properties of the equilibrium radiation field and its interaction with matter within a hohlraum. These properties are found to depend only upon the temperature of the medium and are totally independent of the nature of the matter occupying the hohlraum. They are expressed in the famous Radiation Laws, first laid out by Kirchhoff in 1860.[12]

4.3.2 Planck's Spectral Distribution Law

Suppose that a tiny opening is made in a containing wall of a hohlraum. Consider first the effect this opening has on incident radiation. It is clear that it will completely absorb the incident radiation, for its likelihood of being reflected inside the container and making it back out is negligibly small. For this reason the opening is perfectly absorbing, or "black." The radiation that escapes the enclosure through the opening will have reached thermal equilibrium with the matter within the enclosure. The loss of energy due to this leakage is assumed to be very small compared with the total energy. The radiation emanating from this "black" surface is called *blackbody radiation.* (This can be a confusing term for beginners, since it actually refers to the *absorbing* properties of the surface giving rise to the emission.)

By introducing his hypothesis of quantized oscillators in a radiating body, Planck in 1901 derived the expression for the hemispherical blackbody spectral radiative flux,

$$F_\nu^{\mathrm{BB}} = \frac{m_\mathrm{r}^2}{c^2}\frac{2\pi h\nu^3}{\left(e^{h\nu/k_\mathrm{B}T} - 1\right)}, \tag{4.1}$$

where h is Planck's constant, m_r is the real index of refraction, and k_B is Boltzmann's constant. (See Example 4.2 for a derivation of the blackbody formula.) Throughout the hohlraum the radiation field is isotropic and unpolarized. An equal and opposite hemispherical flux opposes F_ν^{BB}, so that the net flux is everywhere zero (see Eq. 2.10).

Approximations for F_ν^{BB} are *Wien's limit*, for high energies:

$$F_\nu^{\mathrm{BB}} \approx \frac{m_\mathrm{r}^2}{c^2} 2\pi h\nu^3 e^{-h\nu/k_\mathrm{B}T} \quad (h\nu/k_\mathrm{B}T \gg 1) \tag{4.2}$$

and the *Rayleigh–Jeans limit*, for very low energies:

$$F_\nu^{\mathrm{BB}} \approx \frac{2\pi\nu^2 m_\mathrm{r}^2 k_\mathrm{B}T}{c^2} \quad (h\nu/k_\mathrm{B}T \ll 1). \tag{4.3}$$

The latter is useful in the microwave spectral region for $\lambda > 1$ mm.

As a result of the isotropy, the blackbody intensity is related to the hemispherical flux through $F_\nu^{\mathrm{BB}} = \pi I_\nu^{\mathrm{BB}} = \pi B_\nu$ (see Eq. 2.11). Here B_ν is

The Planck function

$$I_\nu^{\mathrm{BB}} = B_\nu(T) \equiv \frac{m_\mathrm{r}^2}{c^2}\frac{2h\nu^3}{\left(e^{h\nu/k_\mathrm{B}T} - 1\right)}. \tag{4.4}$$

$B_\nu(T)$ has the same units as intensity. The closely related function B_λ is illustrated in Fig. 4.6 for a range of terrestrial temperatures. It should be mentioned that the preceding equations apply separately to each polarization component of the electric field. In this book we usually deal with unpolarized light, which is the sum of these (equal) components. As indicated earlier, when we deal with gases we will set[13] $m_\mathrm{r} = 1$.

By differentiating the Planck function and equating the result to zero, one may easily show that the spectral distribution of blackbody radiation has its maximum value at the frequency ν_m, or wavelength λ_m, where

$$\nu_m T = 5.88 \times 10^{10} \quad [\mathrm{Hz}\cdot\mathrm{K}] \quad \text{or} \quad \lambda_m T = 2{,}897.8 \quad [\mu\mathrm{m}\cdot\mathrm{K}]. \tag{4.5}$$

This important result, known as the *Wien Displacement Law*, states that *the wavelength of peak blackbody emission is inversely proportional to temperature.*

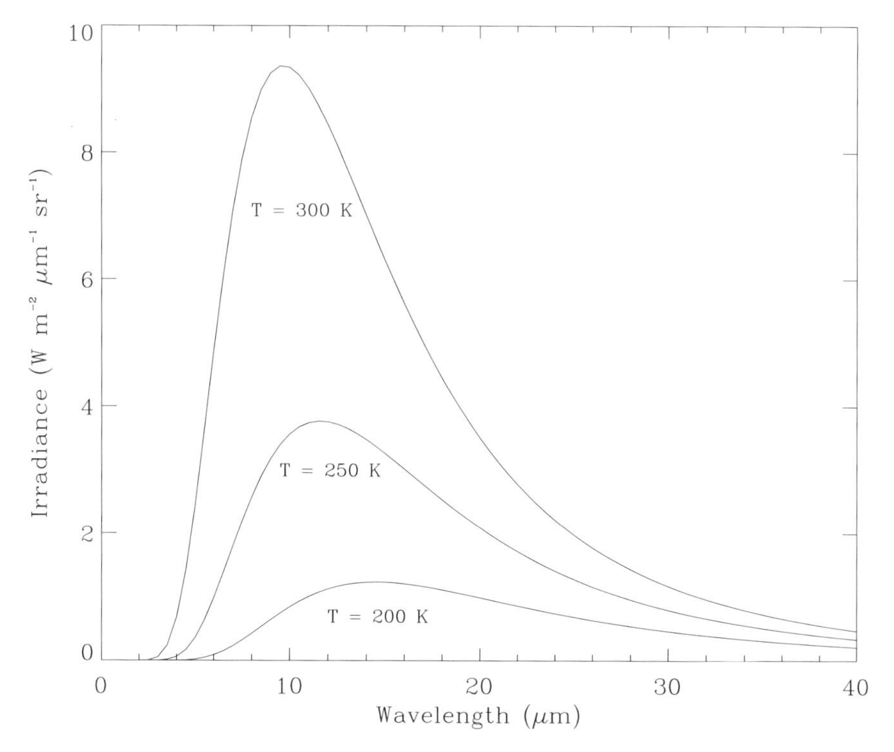

Figure 4.6 The blackbody intensity B_λ versus wavelength λ. The relationship between B_ν and B_λ is $B_\lambda d|\lambda| = B_\nu |d\nu|$. Since $\lambda = c/\nu$ then $|d\nu| = (c/\lambda^2)|d\lambda|$.

The Stefan–Boltzmann Law

From Eqs. 2.1 and 2.5, the frequency-integrated hemispherical flux leaving the hohlraum is

$$F^{\mathrm{BB}} = \int_0^\infty d\nu \int_{2\pi} d\omega \cos\theta \, I_\nu^{\mathrm{BB}} = \pi \int_0^\infty d\nu B_\nu(T). \tag{4.6}$$

Substituting Eq. 4.4 in Eq. 4.6 and setting $x = h\nu/k_{\mathrm{B}}T$ and $m_{\mathrm{r}} = 1$, we obtain the expression for the frequency-integrated emergent blackbody flux:

$$F^{\mathrm{BB}} = \frac{2\pi (k_{\mathrm{B}}T)^4}{h^3c^2} \int_0^\infty \frac{dx \, x^3}{(e^x - 1)} = \left(\frac{2\pi^5 k_{\mathrm{B}}^4}{15h^3c^2}\right) T^4,$$

where we have used the result that the value of the definite integral is $\pi^4/15$. We have derived the important

Stefan–Boltzmann Law:

$$F^{\mathrm{BB}} = \sigma_{\mathrm{B}} T^4, \tag{4.7}$$

where $\sigma_B = 2\pi^5 k_B^4/15h^3c^2 = 5.6703 \times 10^{-8}$ [$W \cdot m^{-2} \cdot K^{-4}$] is the *Stefan–Boltzmann constant*.

Example 4.2 Derivation of the Planck Radiation Law

The derivation consists of two parts: determining the average energy per photon state and finding the density of states within the hohlraum. We begin with a general statistical law describing the distribution of states in a system in thermodynamic equilibrium. If there are N total states, with individual discrete energies $\mathcal{E}_i$ $(i = 1, 2, \ldots, N)$, each having a degeneracy g_i, the *canonical*, or *Gibbs distribution* describes the probability of a particular energy state occurring for a system at temperature T. It is proven in statistical mechanics treatises that

$$p(\mathcal{E}_i) \propto g_i e^{-\mathcal{E}_i/k_B T}. \tag{4.8}$$

Since the probability summed over all states is unity, this distribution may be written as

$$p(\mathcal{E}_i) = \frac{g_i e^{-\mathcal{E}_i/k_B T}}{\sum_{j=0}^{N} g_j e^{-\mathcal{E}_j/k_B T}} \equiv \frac{g_i e^{-\mathcal{E}_i/k_B T}}{Q_p(T)}. \tag{4.9}$$

Here Q_p is called the photon *partition function* (see §4.5.3 for more detailed examples of partition functions).

Equation 4.9 describes the *Boltzmann distribution* of discrete energy states in thermodynamic equilibrium (see §4.3.4 and Eq. 4.19). It may also be transformed to yield the equilibrium distribution of molecular velocities in a gas. It is necessary to allow the range of energies to be continuous and to convert the sums to integrals (see Eq. 4.12). Here we are interested in deriving the equilibrium distribution of photon energies, B_ν.

We next ask, "What is the mean energy of the quantum states having the frequency ν?" If each state contains n photons of energy $h\nu$, then the energy of the nth state is $E_n = nh\nu$. The mean energy is thus the sum of E_n weighted by the probability of that energy occurring, $p(E_n) = g_n e^{-\beta E_n}/Q_p(T)$, where $\beta \equiv (k_B T)^{-1}$. Thus

$$\langle E(\nu)\rangle = \frac{\sum_{n=0}^{\infty} nh\nu e^{-\beta nh\nu}}{\sum_{n=0}^{\infty} e^{-\beta nh\nu}}.$$

We have factored out the common factor g_n, which is the same for all states. To evaluate this expression explicitly, we note that

$$\langle E(\nu)\rangle = -\sum \frac{\partial}{\partial\beta}[e^{-\beta nh\nu}] \Big/ \sum e^{-\beta nh\nu} = -\frac{d}{d\beta} \ln\left[\sum e^{-\beta nh\nu}\right],$$

where we have interchanged differentiation and summation. The sum can be evaluated as a geometric series $\sum x^n = (1-x)^{-1}$ where $x \equiv e^{-\beta h\nu}$. Thus

$$\langle E(\nu)\rangle = -\frac{d}{d\beta} \ln \frac{1}{(1 - e^{-\beta h\nu})} = \frac{d}{d\beta} \ln(1 - e^{-\beta h\nu}) = \frac{h\nu}{e^{h\nu/k_B T} - 1}.$$

For vanishing temperature, $\langle E(\nu)\rangle \approx h\nu \exp(-h\nu/k_B T) \to 0$. For high temperature, $\langle E(\nu)\rangle \approx k_B T$. The latter result tells us that the energy per photon is that predicted by the classical *Equipartition Theorem*, which states that the particle has $k_B T/2$ of energy for every degree of freedom. The photon has two degrees of freedom, corresponding to the two polarization components. In quantum terminology, it has two spin directions ("up" and "down").

We now make the correspondence of an oscillator with a standing electromagnetic wave in the cavity (see Appendix H). For a given frequency ν a standing wave can have a host of discrete energies, given by $\mathcal{E}_{\rm p} = (h/2\pi)c|\mathbf{k}| = h\nu$. Here $\mathbf{k}$ is the quantized wavenumber vector, having components $k_x = n_x(\pi/L)$, $k_y = n_y(\pi/L)$, and $k_z = n_z(\pi/L)$, where L is the length of one side of the cubical cavity. We first need to know the number of standing waves $\Phi(k)dk$ having wavenumber lying between k and $k+dk$. Multiplying this number by the mean energy $\langle E(\nu)\rangle$, we obtain the expression for the energy density per unit volume. We can find $\Phi(k)$ by appealing to a simple geometrical construction, where the vector $\mathbf{k}$ is drawn in the pseudo-space of k_x, k_y, and k_z. Not all values of k_x, etc. are allowed – only those satisfying $k_x = \pi/L, 2\pi/L, 3\pi/L, \ldots$. We now visualize a volume element in this space, which is defined by incrementing each of the values of the k components by one. Then for k large, the element is approximately a cube of side π/L whose volume is $(\pi/L)^3$. A given energy state defined by the coordinates (k_x, k_y, k_z) "fills" the above volume element. A one-to-one correspondence exists between the volume $(\pi/L)^3$ and a specific energy state. Thus to count the total number of states out to some radial distance k in this pseudo-space, all we need to do is to evaluate the total volume and divide by $(\pi/L)^3$. There is some bookkeeping we must do before we get the correct answer. Since k is positive, we should evaluate only one quadrant of the total sphere in k-space; that is, the volume should be $4\pi k^3/3$ divided by 8. As mentioned earlier, for every specific value of $\mathbf{k}$, there are two independent states, corresponding to the two polarization states. Therefore the total number of states $N(k)$ for which the k-values are less than or equal to k is

$$N(k) = \frac{2\times(4\pi k^3/3)}{8(\pi/L)^3} = \frac{Vk^3}{3\pi^2},$$

where $V = L^3$ is the cavity volume. The density of states (number per unit volume) is therefore

$$\Phi(k) = \frac{d}{dk}\frac{N(k)}{V} = \frac{d}{dk}\left(\frac{k^3}{3\pi^2}\right) = \frac{k^2}{\pi^2}.$$

We now return to frequency space and find the number of states between ν and $\nu + d\nu$. Since $k = (2\pi\nu/c)$, then $k^2 = 4\pi^2\nu^2/c^2$ and $dk = 2\pi d\nu/c$. Therefore

$$\Phi(\nu)d\nu = \frac{4\nu^2}{c^2}\left(\frac{2\pi d\nu}{c}\right) = \frac{8\pi\nu^2 d\nu}{c^3}.$$

Finally, the energy density is the number of oscillators per unit volume multiplied by the average energy

$$\mathcal{U}_\nu = \frac{8\pi\nu^2\langle E(\nu)\rangle}{c^3} = \frac{8\pi\nu^2 h\nu}{c^3\left(e^{h\nu/k_{\rm B}T}-1\right)}.$$

Since the radiation is isotropic, $\mathcal{U}_\nu = 4\pi\bar{I}_\nu/c = 4\pi I_\nu/c$ (see Eq. 2.13), we find the blackbody intensity formula

$$I_\nu^{\rm BB} = B_\nu(T) = \frac{c}{4\pi}\mathcal{U}_\nu = \frac{2h\nu^3}{c^2}\frac{1}{e^{h\nu/k_{\rm B}T}-1}.$$

4.3.3 Radiative Excitation Processes in Molecules

A rigorous treatment of the interaction of matter and radiation requires both the matter and the radiation to be a fully coupled, quantized assembly. In fact, the phenomenon of spontaneous emission requires such a description, a discipline known as "quantum electrodynamics," which is certainly beyond the scope of this text. For the present purpose we use a hybrid approach, the *semiclassical theory*, in which the radiation is described by the classical electromagnetic theory, whereas the structure of matter is specified by the quantum theory. In the older quantum theory, the connection between radiation and matter was specified by the Einstein equations, to be discussed in the next section. In the modern theory, the coefficients of interaction are calculated by quantum mechanical *Perturbation Theory*. Mixing of classical and quantum physics, although not rigorous, has proven to be immensely successful and is also conceptually and mathematically simpler. We will adopt this approach here.

In the semiclassical theory we will continue to describe the radiation field in terms of its intensity (radiance), flux (irradiance), and energy density. The "matter field" we will describe in terms of the populations of the excited states. These states may be either discrete or continuous. Absorption of a photon of energy $E = h\nu$ results in excitation of a state (either molecular or atomic), thus reducing the population of the initial state population n_0 (normally the ground state) by one, and increasing the excited state population n_i by one. Here $n_i (i = 1, 2, \ldots)$ is the number of molecules (atoms) per unit volume in the state having energy E_i. However, an excited state can *decay*, either by spontaneous or induced emission (to be described in detail later), and the reverse will happen. The continual exchange of energy between the matter and radiation is described in terms of rate equations for the various processes.

Photon–matter processes are classified in terms of three basic kinds of interactions: (a) *bound–bound processes* describe the exchange of energy when the initial and excited states are both discrete states, schematically $(E_i \to E_j)$; (b) *bound–free processes* describe transitions between discrete and continuous states, schematically $(E_i \to E_j, E_j + dE_j)$; and (c) *free–free processes* describe transitions between two continuous states, schematically $(E_i, E_i + dE_i \to E_j, E_j + dE_j)$. Only the bound–bound processes are important in infrared radiative transfer in planetary atmospheres. Process (b) is important in some ultraviolet absorption processes, such as the absorption by ozone in the *Hartley bands* (200–300 nm), in which an oxygen atom is removed from the ozone molecule in the process of photoabsorption.

4.3.4 Inelastic Collisional Processes

We will use chemical notation for shorthand purposes. Let AB be a molecule with atoms A and B bound together, and let M be a second molecule of unspecified nature, which we will designate as a "third body." Denoting $(KE)'$ and (KE) as the sum of the kinetic energies of the reactants and products respectively, we then consider the following collisional "reaction":

$$AB + M + (KE)' \to (AB)^* + M + (KE),$$

where the notation $(AB)^*$ indicates internal excitation of the AB molecule (electronic, vibrational, rotational, or some combination). The above reaction describes a *collisional excitation* of the molecule AB. In this case, $(KE) < (KE)'$ of course, and energy is extracted from the thermal "pool" and placed into energy of excitation. The inverse reaction is

$$(AB)^* + M + (KE) \rightarrow AB + M + (KE)',$$

which is called *collisional deexcitation*, or *collisional quenching*. The above are examples of *inelastic collisions*, in which energy is transferred from kinetic to internal excitation energy, or vice versa.[14] They are of great importance in the theory of absorption. An *elastic collision* is one in which there is no net transfer of energy from kinetic to excitation energy, although the collision partners will end up with different shares of the total kinetic energy. In other words, exchanges of momentum and kinetic energy occur but the totals of each remain the same.

To quantify the collisional and radiative processes, we need to describe the *rates* at which the various reactions occur. We will use the same general framework as we used in describing radiative processes, namely that of *cross sections*. The *collision cross section* σ is defined analogously to the radiative cross section, provided the differential flux of incoming particles is used, rather than the photon flux. This differential flux is the analog of the intensity, being defined as the flux per unit solid angle. The change of this flux over a distance (say ds) is proportional to the product of the flux and the number of target molecules within a cylinder of unit cross section having a length ds. The constant of proportionality is σ, which depends upon the properties of incident and target molecules, as well as their relative speeds. There are various ways we can define the change of the flux over ds. We can specify that the incident molecules change their directions but not their speeds. In this case, the cross section is said to be the *elastic cross section*. We can be more specific and ask that the molecules be deflected in a particular direction Θ (per unit solid angle). This gives rise to the concept of a *differential* elastic collision cross section, $d\sigma_{\text{el}}(\Theta)/d\omega$, which is analogous to the product of the photon cross section and the phase function (see Chapter 3). The total elastic cross section is given by

$$\sigma_{\text{el}} = \int_{4\pi} d\omega \frac{d\sigma_{\text{el}}(\Theta)}{d\omega}. \tag{4.10}$$

Alternatively, the *inelastic cross section* σ_{in} involves a change in the molecules' internal excitation. It is important to indicate the particular excited state of interest, in which case we have a smaller *partial* cross section. The total inelastic cross section is the sum of all the partial cross sections.

We now consider typical orders of magnitude of collisional cross sections. Elastic cross sections are of order 10^{-19} m^2. Inelastic cross sections are much smaller,[15] of order 10^{-23}–10^{-25} m^2. This large difference will be important later on when we consider the maintenance of various thermal equilibrium distributions.

The rate of increase of the population of excited states, $d[AB]^*/dt$, due to a particular inelastic process is defined in terms of the product of the reactants and, because of the dependence on relative speeds, the sum (integral) over all possible relative speeds:

$$\frac{d[AB]^*}{dt} = [AB][M] \int d^3\, |\mathbf{v}_{AB} - \mathbf{v}_M| f_{AB}(\mathbf{v}_{AB}) f_M(\mathbf{v}_M) \frac{d\sigma_{\text{in}}}{d\omega}. \tag{4.11}$$

Here $[AB]$ denotes the concentration of species AB, and f_{AB} and v_{AB} are the velocity distribution and velocity of species AB, etc. The integral is over all possible relative velocities $|\mathbf{v}_{AB} - \mathbf{v}_M|$ of AB and M. This amounts to an integration over all relative energies and directions of the colliding particles. The velocity distributions of a species of mass m in thermal equilibrium is given by the familiar *Maxwell–Boltzmann distribution* of velocities:

$$f_{\text{MB}}(\mathbf{v}) = \left[\frac{m}{2\pi k_{\text{B}} T}\right]^{3/2} \exp\left[-\frac{v_x^2 + v_y^2 + v_z^2}{(2k_{\text{B}}T/m)}\right], \tag{4.12}$$

where v_x, v_y, and v_z are the Cartesian components of the velocity, and k_{B} is Boltzmann's constant. The Maxwell–Boltzmann distribution, given by Eq. 4.12, is maintained by *elastic* collisions between molecules of the gas.

Equation 4.11 can be written more compactly as

$$\frac{d[AB]^*}{dt} = k_{\text{in}}[AB][M] \quad [\text{cm}^3 \cdot \text{s}^{-1}], \tag{4.13}$$

where k_{in} is called the *collisional excitation coefficient* for the particular inelastic collision of interest. Rate constants are commonly reported in cgs units.

Numerical evaluation of Eq. 4.11 using actual values of $d\sigma_{\text{in}}/d\omega$ show that the reaction rate coefficient can usually be approximated as

$$k_{\text{in}} \approx a \left(\frac{T}{300}\right)^b e^{-c/T} \quad [\text{cm}^3 \cdot \text{s}^{-1}], \tag{4.14}$$

where a is a combination of molecular constants, b is a dimensionless constant of order unity, and c is the *activation temperature* in kelvins; $c \equiv \Delta E/k_{\text{B}}$, where ΔE is the energy of excitation; and $e^{-c/T}$ is called the *Boltzmann factor*. Equation 4.14 is in a form convenient for comparison with data, in which case a, b, and c are empirical constants to be determined by fitting Eq. 4.14 to laboratory measurements. A similar expression is obtained for the decrease of the excited-state populations due to collisions (collisional quenching):

$$\frac{d[AB]^*}{dt} = -k'_{\text{in}}[AB^*][M], \tag{4.15}$$

where k'_{in} is the appropriate *quenching coefficient*, which is related to the excitation coefficient k_{in} through the *Principle of Detailed Balance* (see Endnote 19 of Chapter 6).

The coefficients for either elastic or inelastic collisions are written in terms of the *mean molecular speed*

$$\langle v \rangle = \int dv v f_{\mathrm{MB}}(v) = (8k_{\mathrm{B}}T/\pi m)^{1/2}, \tag{4.16}$$

where $f_{\mathrm{MB}}(v)$ is given by

$$f_{\mathrm{MB}}(v) = 4\pi \left(\frac{m}{2\pi k_{\mathrm{B}} T} \right)^{3/2} v^2 e^{-mv^2/k_{\mathrm{B}}T}. \tag{4.17}$$

From the units [$\mathrm{cm}^3 \cdot \mathrm{s}^{-1}$] of the collisional coefficient (either k_{in} or the corresponding rate of elastic collisions, k_{el}), it is clear that k is just the effective volume swept out by a moving molecule per unit time. If we imagine the other molecules to be stationary, this volume is the product of the appropriate cross section and the relative speed ($\sim\langle v \rangle$). Then $k_{\mathrm{el}} \approx \sigma_{\mathrm{el}}\langle v \rangle$, and $k_{\mathrm{in}} \approx \sigma_{in}\langle v \rangle$. For $T = 300$ K, and assuming N_2 molecules as third bodies, then $\langle v \rangle \approx 480$ [$\mathrm{m} \cdot \mathrm{s}^{-1}$]. For elastic collisions, $\sigma_{\mathrm{el}} \approx 1 \times 10^{-15}$ [cm^2], and the values for the reaction rate coefficients are

$$k_{\mathrm{el}} \approx 5 \times 10^{-10}\,(T/300)^{1/2}, \qquad k_{\mathrm{in}} \approx 5 \times 10^{-14}\text{–}10^{-16}\ [\mathrm{cm}^3 \cdot \mathrm{s}^{-1}]. \tag{4.18}$$

Thus inelastic collisions proceed at about 10^{-4} to 10^{-6} the rate of elastic collisions.

4.3.5 Maintenance of Thermal Equilibrium Distributions

A second important statistical distribution valid in thermodynamic equilibrium is that of the *Boltzmann distribution* of excited states,

$$\frac{n_j}{n_i} = \frac{g_j}{g_i} e^{-(E_j - E_i)/kT}, \tag{4.19}$$

where n_j, E_j, and g_j denote the volume density, energy, and the statistical weight of the jth excited state, respectively. Since the excited-state populations are established by inelastic collisions, it is clear that these processes maintain the Boltzmann distribution. Since these processes occur much less frequently than elastic collisions (which maintain the Maxwell–Boltzmann distribution of velocities), we therefore expect that for low gas density, the Boltzmann distribution will become invalid.

The third distribution of interest is the Planck distribution of photon energies,

$$B_\nu(T) = \frac{2h\nu^3}{c^2} \frac{1}{e^{h\nu/kT} - 1}. \tag{4.20}$$

As discussed further below, this is a distribution that is maintained by emission and absorption of photons – these processes in turn are determined by the populations of the various excited states in the medium, or in other words, are also maintained by inelastic collision processes. In strict thermodynamic equilibrium (TE), $B_\nu(T)$ not only describes the blackbody radiation field I_ν^{BB}, but it also describes the *source function* S_ν (§2.8). From Kirchhoff's Law (§5.3) the emission coefficient (rate of thermal emission per unit volume) is given by $j_\nu = \alpha(\nu)B_\nu(T)$. Here $\alpha(\nu)$ is the absorption coefficient. (We ignore scattering for the present.) Since $S_\nu \equiv j_\nu/\alpha(\nu)$, then in TE, $S_\nu = B_\nu(T)$. This result also follows from the radiative transfer equation (Eq. 2.28) $dI_\nu/d\tau_\nu = -I_\nu + S_\nu$. In TE the intensity is uniform and isotropic, and $dI_\nu/d\tau_\nu = 0$. Thus $S_\nu = I_\nu^{BB} = B_\nu$, which applies to a closed system in equilibrium.

What about an "open system," such as an atmosphere or ocean, that receives energy from the Sun and radiates energy to space? Under rather general conditions, such systems also share certain properties of a system in TE. As will be shown in greater detail for a two-level atom, the source function is Planckian (but $I_\nu \neq B_\nu$) in *local thermodynamic equilibrium*, or LTE. This applies if the photon frequency is sufficiently low ($\nu < k_B T/h$) and the density is sufficiently high for the rate of collisional excitation/deexcitation processes to greatly exceed the corresponding radiative processes. This is the same condition for which the Boltzmann distribution is valid. In fact if Eq. 4.19 is valid, this is a necessary and sufficient condition for LTE.

Local thermodynamic equilibrium applies if the gas density is sufficiently high to ensure that the collisional lifetime of an excited state t_{coll} is much smaller than the radiative lifetime t_{rad}. This condition is fulfilled in the thermal IR ($\lambda > 3$ μm). For vibrational energy states, t_{rad} is typically 0.1–1 s. For the 15-μm band of CO_2 under STP conditions (see §1.4.3), $t_{coll} \sim [AB]^*/d[AB]^*/dt = 1/k'_{in}[M] \sim 10^{-5}$ s. Therefore we would expect that LTE would exist in this band down to pressures of 0.01–0.1 mb, which occurs near 75 km. This is called the *level of vibrational relaxation*, since above this height, the upper state "relaxes" to a population different than that given by the Boltzmann distribution. This situation is called *nonlocal thermodynamic equilibrium* (NLTE). Finding the source function in the NLTE case is considerably more complicated than in the LTE case. The interplay of collisional and radiative processes in the general NLTE case is illustrated by considering the example of a two-level atom.

4.4 The Two-Level Atom

We now consider the more realistic situation in which all the radiative and collisional processes act together. We will assume that the values of the rate processes are given, and we will derive equations describing the transfer of radiation through the system. Many of the properties of a complex system are embodied in the *two-level atom* concept, which envisions an atom with only two discrete energy levels. Altogether, we must consider five separate processes (see Fig. 4.7) connecting the two energy levels of the atom. We begin by considering the radiative processes.

4.4.1 Microscopic Radiative Transfer Equation

The radiation field is assumed to be a result of transitions from the single excited level (state 2) to the ground level (state 1) of a radiatively active species. The gas will be a

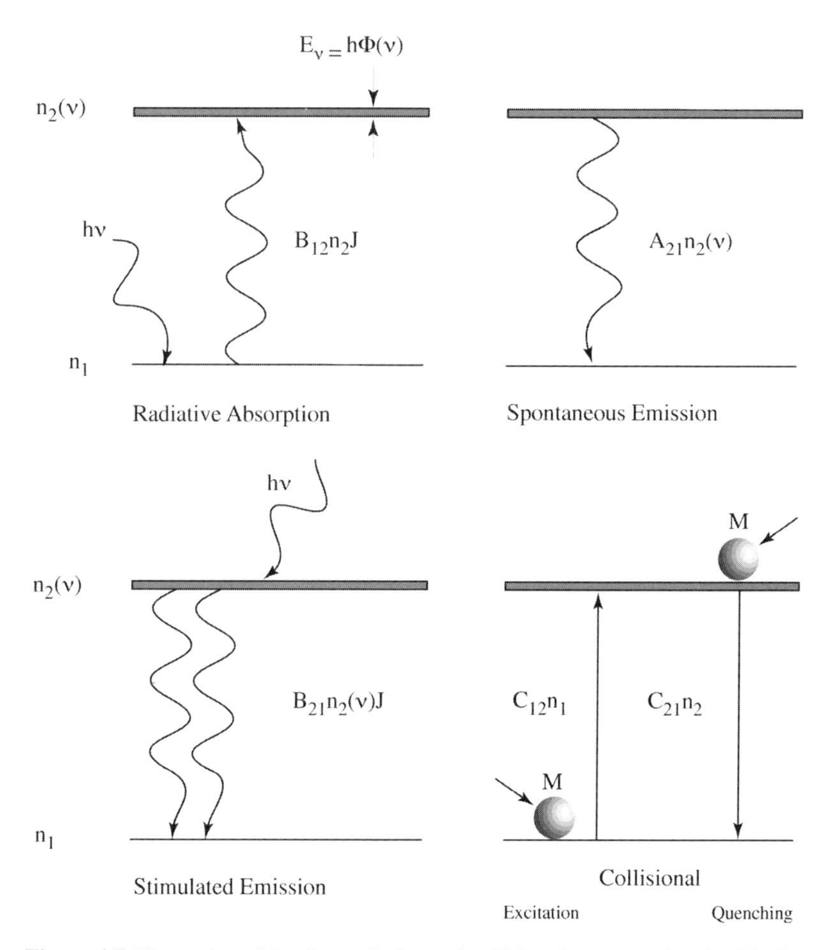

Figure 4.7 Illustration of the five radiative and collisional processes involved in the rate of population of energy levels in a two-level atom.

two-component mixture consisting of the radiatively active species and a radiatively inert "buffer" gas. The latter plays the role of collisionally transforming the excited level to the ground state and vice versa. The populations in the two levels are denoted n_1 and n_2. The sum of the two populations is a constant, equal to the density of the radiatively active species, n. The average energy difference between the states is $E_{21} = h\nu_0$, but there is assumed to be a small spread in frequencies, due to spectral broadening. The radiative processes (see Fig. 4.7) are:

1. *absorption*: $h\nu + n_1 \rightarrow n_2$,
2. *spontaneous emission*: $n_2 \rightarrow n_1 + h\nu$, and
3. *stimulated emission*: $n_2 + h\nu \rightarrow n_1 + h\nu + h\nu$.

Process 3 is one in which the emitted radiation is exactly coherent with the incident radiation, in both direction and phase. Processes 1 and 3 may be understood from classical physics, but 2 requires quantum theory for a fundamental description. In the semiclassical theory, we assign a rate to this process that is independent of the surroundings of the atom.[16]

The rates at which the three radiative processes occur were first derived by Einstein in 1905. The rate efficiencies are described by the *Einstein coefficients* B_{12}, A_{21}, and B_{21}. We now consider the rate equations for each individual process: Process 1 describes the rate at which absorption depletes the lower state. It is proportional to the number of atoms in the ground state n_1, to the absorption cross section $\alpha_n(\nu)$, and to the number of photons in solid angle $d\omega$, $(I_\nu/h\nu)d\omega$. Integrating over all frequencies[17] and photon directions, we find

$$(dn_2/dt)_{\text{abs}} = n_1 \int_0^\infty d\nu \int_{4\pi} d\omega \alpha_n(\nu)(I_\nu/h\nu) = 4\pi n_1 \int_0^\infty d\nu \alpha_n(\nu)(\bar{I}_\nu/h\nu). \tag{4.21}$$

Process 2 is the rate at which photons depopulate the upper state. Einstein asserted that its rate may be written

$$(dn_2/dt)_{\text{spon}} = -A_{21}n_2. \tag{4.22}$$

The above equation shows that the excited states decay via this process *independently of their surroundings*. Stimulated emission, Process 3, is given by an expression similar to Eq. 4.21, since the rate is also proportional to the number of photons available:

$$\begin{aligned}(dn_2/dt)_{\text{stim}} &= -n_2 \int_0^\infty d\nu \int_{4\pi} d\omega \alpha_n(stim;\, \nu)(I_\nu/h\nu) \\ &= -4\pi n_2 \int_0^\infty d\nu \alpha_n(stim;\, \nu)(\bar{I}_\nu/h\nu),\end{aligned} \tag{4.23}$$

where $\alpha_n(stim;\, \nu)$ is the absorption cross section for stimulated emission.

We now write the above absorption cross sections in terms of the Einstein coefficients, $\alpha_n(\nu) \equiv h\nu B_{12}\Phi(\nu)/4\pi$ and $\alpha_n(stim;\, \nu) \equiv h\nu B_{21}\Phi(\nu)/4\pi$. The function $\Phi(\nu)$ is the normalized *line-profile function*, defined so that

$$\int_0^\infty d\nu \Phi(\nu) = 1. \tag{4.24}$$

There is no a priori reason why the line profiles for stimulated emission and absorption should generally be the same. We made the assumption because there are many situations where it is an excellent approximation.[18] The properties of the atom (molecule) and its surroundings determine the line shape. We assume for now that $\Phi(\nu)$ is known and is independent of position.

As we might suspect, the rates of the above three processes are related. In fact, we will show that it is sufficient to know the value of one Einstein coefficient to determine the other two. For this purpose we use the common approach of assuming a special case (that of TE) and then arguing that the result so obtained has more general validity. In TE, $I_\nu = B_\nu$, and the populations n_1 and n_2 are related through the Boltzmann equation, Eq. 4.19. We denote the ratio of the two populations in TE as n_2^*/n_1^* to distinguish it from the more general ratio n_2/n_1. From Eq. 4.19, we have

$$n_2^*/n_1^* = (g_2/g_1)\exp(-h\nu_0/k_B T), \tag{4.25}$$

where the g_i are the statistical weights. (Note that we have used the average energy difference between the two states $E_{21} = h\nu_0$.) Assuming time-independent conditions, we have $dn_1^*/dt = -dn_2^*/dt = 0$. The radiative rates must all balance[19] and

$$dn_1^*/dt = 0 = n_2^* A_{21} + n_2^* B_{21} \int_0^\infty d\nu B_\nu \Phi(\nu) - n_1^* B_{12} \int_0^\infty d\nu B_\nu \Phi(\nu). \tag{4.26}$$

We will simplify the above expression by using the fact that the Planck function B_ν is slowly varying over the line profile. Removing this from the frequency integration, and using the normalization property of $\Phi(\nu)$, Eq. 4.24, we find

$$n_2^* A_{21} + n_2^* B_{21} B_{\nu_0} = n_1^* B_{12} B_{\nu_0}.$$

Solving for B_{ν_0} we find

$$B_{\nu_0} = \frac{(A_{21}/B_{21})}{(g_1 B_{12}/g_2 B_{21})\, e^{h\nu_0/k_B T} - 1}. \tag{4.27}$$

But we already know the functional form of the Planck function:

$$B_{\nu_0} = \frac{2h\nu_0^3/c^2}{e^{h\nu_0/k_B T} - 1}. \tag{4.28}$$

Making the correspondence of the above two equations, we obtain the following two expressions:

The Einstein Relations

$$A_{21} = \left(2h\nu_0^3/c^2\right) B_{21}, \tag{4.29a}$$

$$g_1 B_{12} = g_2 B_{21}. \tag{4.29b}$$

These relationships are independent of the state of the gas, in particular of the temperature or density, and therefore they must involve only the basic properties of the atom itself. We therefore assert that *the Einstein relations are quite general and are independent of the situation assumed in their derivation.* Thus, they should apply to the more general situation of NLTE.

We now use the above relationships to write down the continuity equation for photons, which is just our familiar radiative transfer equation. We first note that dI_ν/ds is the rate at which radiative energy is lost, or gained, along a beam. Then we can write this quantity as being equal to the gains less the losses due to the three radiative processes. The result is

The Microscopic Radiative Transfer Equation,

$$\frac{dI_\nu}{ds} = -\frac{h\nu_0}{4\pi} n_1 B_{12} I_\nu \Phi(\nu) + \frac{h\nu_0}{4\pi} n_2 B_{21} I_\nu \Phi(\nu) + \frac{h\nu_0}{4\pi} n_2 A_{21} \Phi(\nu). \tag{4.30}$$

We have introduced the additional assumption that the line profile for spontaneous emission is also given by $\Phi(\nu)$. Equation 4.30 may now be related to our conventional radiative transfer equation, Eq. 2.28, which we can now call the *macroscopic* radiative transfer equation:

$$\frac{dI_\nu}{ds} = -k(\nu)\,(I_\nu - S_\nu). \tag{4.31}$$

Equating the factors multiplying I_ν in Eqs. 4.30 and 4.31, we find

$$k(\nu) = \frac{h\nu_0}{4\pi} \Phi(\nu)\,(n_1 B_{12} - n_2 B_{21}). \tag{4.32}$$

This relationship allows us to relate microscopic quantities to macroscopic quantities. Consider the above equation in the case of LTE. Replacing the quantities n_1 and n_2 by n_1^* and n_2^*, we have

$$k^*(\nu) = \frac{h\nu_0}{4\pi} \Phi(\nu) n_1^* B_{12} \left(1 - \frac{n_2^* B_{21}}{n_1^* B_{12}}\right), \tag{4.33}$$

where we denote by $k^*(\nu)$ the LTE value of the effective extinction coefficient. From the Boltzmann relation for the ratio n_2^*/n_1^* (Eq. 4.25) and the Einstein relation $g_2 B_{21} = g_1 B_{12}$, we find

$$k^*(\nu) = \frac{h\nu_0}{4\pi} \Phi(\nu) n_1^* B_{12} \left(1 - e^{-h\nu_0/k_B T}\right). \tag{4.34}$$

The above equation is the extinction coefficient in LTE, *corrected for stimulated emission*. It is clear that stimulated emission is simply *negative absorption*, since the emitted photon is coherent with, and in the same direction as, the incident photon. Thus, our macroscopic equation needs a slight adjustment for the LTE situation, such that

$$\frac{dI_\nu}{ds} = -k^*(\nu)\,(I_\nu - S_\nu), \tag{4.35}$$

where $k^*(\nu)$ is given by Eq. 4.34. The more general NLTE equation would use the expression in Eq. 4.32 for $k(\nu)$. In many atmospheric problems, the factor $e^{-h\nu_0/k_B T} \ll 1$ and the effect of stimulated emission is negligible.

We now equate the source terms in Eqs. 4.30 and 4.31. Using Eq. 4.32 for $k(\nu)$, we find

$$S_\nu = S_{\nu_0} = \frac{n_2 A_{21}}{n_1 B_{12} - n_2 B_{21}} = \frac{2h\nu_0^3/c^2}{(n_1 g_2/n_2 g_1) - 1}, \tag{4.36}$$

where we used the Einstein relationships, Eqs. 4.29. An important aspect of Eq. 4.36 is that the frequency dependence of the source function has vanished. This stems directly from the assumptions that the line profiles for stimulated emission, spontaneous emission, and absorption are identical. This approximation is called *complete frequency redistribution*. Note that if we assume that $n_2/n_1 = n_2^*/n_1^*$, that is, make the LTE assumption, then the source function becomes the Planck function, Eq. 4.28, as can be seen from Eq. 4.25. Therefore, Eq. 4.36 is the expression for the NLTE source function. However, it is not very useful to express it in terms of another unknown, the ratio of the two populations. The equation for determining this unknown ratio comes from considering inelastic collisions.

4.4.2 Effects of Collisions on State Populations

So far we have considered only the effects of radiation on the excited states. We now take into account the additional effects of collisional excitation and quenching. We previously defined these rates in terms of the product of reactant concentrations and a reaction rate coefficient. For purposes of simplifying the notation, we define the collisional excitation rate per atom as $k_{\text{in}}[M] \equiv C_{12}$ and the collisional quenching rate per atom as $k'_{\text{in}}[M] \equiv C_{21}$. We may now write down the rate at which both collisions and radiation populate the excited state. In a steady state we set this rate equal to zero:

$$\begin{aligned} \frac{dn_1}{dt} &= -n_1 C_{12} - n_1 B_{12} \int_0^\infty d\nu \Phi(\nu) \bar{I}_\nu \\ &\quad + n_2 C_{21} + n_2 B_{21} \int_0^\infty d\nu \Phi(\nu) \bar{I}_\nu + n_2 A_{21} = 0. \end{aligned} \tag{4.37}$$

This equation is called the *statistical equilibrium equation*. It provides a second equation, which, in addition to Eq. 4.36, allows us to solve for both unknowns, n_2/n_1 and the source function. But first we will consider some relationships between the collisional rates by once again invoking the principle of detailed balance.

In deriving the Einstein relationships, we considered the state of TE, in which the radiative processes are in balance with one another, without regard to collisional processes. We use the same idea with collisions and ignore radiative processes. Assuming TE, we set the two rates equal:

$$n_2^* C_{21} = n_1^* C_{12}. \tag{4.38}$$

Using the definitions of the coefficients, and invoking the Boltzmann distribution of excited states, Eq. 4.19, we find

$$C_{21} = C_{12}\frac{g_1}{g_2}e^{h\nu_0/k_BT}. \tag{4.39}$$

As in the case of the Einstein relationships we will argue that the above relationship is more general than the assumption used in deriving it. We cannot argue that Eq. 4.39 describes an inherent atomic property, because of its dependence on the temperature. However, we observe that the collisional excitation rate (Eq. 4.11) is determined by an integration over the product of the Maxwell–Boltzmann velocity distributions of the reactants. We also recall that these distributions are maintained by elastic collisions, which are millions of times more efficient than inelastic collisions. Thus, we would expect that Eq. 4.39 would be valid in nonequilibrium situations, as long as the velocity distribution is Maxwellian. To emphasize that there may be several different temperatures in a NLTE situation, the quantity entering the Maxwell–Boltzmann distribution is often referred to as the *kinetic* or *translational* temperature.

We now return to the statistical equilibrium equation. Solving for n_1/n_2 from Eq. 4.37 we obtain

$$\frac{n_2}{n_1} = \frac{C_{12} + B_{12}J}{C_{21} + B_{21}J + A_{21}}, \quad \text{where } J \equiv \int_0^\infty d\nu \Phi(\nu)\bar{I}_\nu. \tag{4.40}$$

Equation 4.40 is in the form of the ratio of the net rate of excitation to the net rate of quenching, or the "source" divided by the "sink" of excited states.

We consider Eqs. 4.36 and 4.40 to be two equations in the two unknowns, n_1/n_2 and J. The quantity J depends upon the radiation field $\bar{I}_\nu$, which can be determined from the source function equation for isotropic scattering, in the usual way. Rewriting the statistical equilibrium equation, using Eq. 4.39 to eliminate the collisional rate C_{12}, and Eqs. 4.29 to eliminate the Einstein coefficient B_{12}, we find

$$\frac{n_1}{n_2} = \frac{A_{21} + B_{21}J + C_{21}}{(g_2/g_1)C_{21}e^{-h\nu_0/k_BT} + (g_2/g_1)B_{21}J}. \tag{4.41}$$

Note carefully that T is understood to be the kinetic temperature of the gas. Thus the velocity distribution of the atoms in the gas is in LTE, while the populations of the energy states may be far from an LTE distribution. We rewrite the above equation as

$$\frac{n_1 g_2}{n_2 g_1} = \frac{A_{21} + B_{21}J + C_{21}}{B_{21}J + C_{21}e^{-h\nu_0/k_BT}}. \tag{4.42}$$

We now have the combination that appears in the denominator of Eq. 4.36. Substitution of Eq. 4.42 into 4.36 yields

$$S_{\nu_0} = \frac{2h\nu_0^3}{c^2}\left[\frac{A_{21} + B_{21}J + C_{21}}{B_{21}J + C_{21}e^{-h\nu_0/k_BT}} - 1\right]^{-1}.$$

Clearing fractions, we find

$$S_{\nu_0} = \frac{\left(2h\nu_0^3/c^2\right)\left(B_{21}J + C_{21}e^{-h\nu_0/k_BT}\right)}{A_{21} + B_{21}J + C_{21} - B_{21}J - C_{21}e^{-h\nu_0/k_BT}},$$

and using the first of the Einstein relations, Eq. 4.29, we obtain

$$S_{\nu_0} = \frac{J + \left(2h\nu_0^3/c^2\right)(C_{21}/A_{21})e^{-h\nu_0/k_BT}}{1 + (C_{21}/A_{21})\left(1 - e^{-h\nu_0/k_BT}\right)}.$$

Defining a new parameter, ϵ_ν,

$$\epsilon_\nu \equiv \frac{C_{21}}{C_{21} + A_{21}\left(1 - e^{-h\nu_0/k_BT}\right)^{-1}}, \tag{4.43}$$

we find with some additional manipulation and using Eq. 4.28

The NLTE source function

$$S_{\nu_0} = \epsilon_\nu B_{\nu_0} + (1 - \epsilon_\nu)J. \tag{4.44}$$

We have shown that the NLTE source function is the sum of two terms: a thermal emission term plus a term that represents the scattering contribution to the source function. $\epsilon_\nu \Phi(\nu)$ is interpreted as the emittance per unit volume, that is, its efficiency as a blackbody emitter as a function of frequency within the spectral line. In terms of the macroscopic absorption ($\alpha(\nu)$) and extinction ($k(\nu)$) coefficients, $\epsilon_\nu = \alpha(\nu)/k(\nu)$ (note that the frequency dependence cancels in the ratio). The emission coefficient is obtained from its definition ($j_\nu = S_\nu/k(\nu)$), yielding

$$j_\nu = \alpha(\nu)B_\nu(T) + \int_0^\infty d\nu\sigma(\nu)\bar{I}_\nu, \tag{4.45}$$

where we used the relationship $\sigma(\nu) = k(\nu) - \alpha(\nu)$. The first term in the above equation is the expression of Kirchhoff's Law for a volume element, which states that the thermal emission is the product of the absorption coefficient and the Planck function (§5.3.1). The second term is the contribution to the volume emission from scattering within the volume (see §5.3.2).

The quantity ϵ_ν is a measure of the *coupling* between the gas and the radiation field. When it is large ($\epsilon_\nu \to 1$), the coupling is strong, and there is a rapid exchange between kinetic and internal energy. In this limit, $S_{\nu_0} \to B_{\nu_0}$, which is just the LTE limit. In the opposite case of weak coupling ($\epsilon_\nu \to 0$), the source function approaches the pure-scattering limit

$$S_{\nu_0} \to J = \int_0^\infty d\nu\Phi(\nu)\bar{I}_\nu, \tag{4.46}$$

which might be called an extreme condition of NLTE, in which the excited states are populated exclusively by radiation and collisions no longer play a role. In a planetary or stellar atmosphere, as one moves upward into lower densities and pressures, there will be a transition from LTE to NLTE as the coupling between the gas and the radiation field disappears.

4.5 Absorption in Molecular Lines and Bands

We are concerned with molecular absorption by solar near-infrared (1 μm to 3 μm) and thermal infrared radiation, which occupies the spectrum from about 3 μm to 100 μm. We shall refer to this entire range, 1 μm to 100 μm (100–10,000 cm^{-1}), generically as the IR spectral range. The molecular excited states of interest are those of vibration (500–10,000 cm^{-1}) and rotation (1–500 cm^{-1}). This range of energies contrasts with that of the higher-lying electronic states (10,000–100,000 cm^{-1}), which interact primarily with visible and ultraviolet radiation. To a first approximation the internal excitation energy is the sum of these three types of energies, electronic (E_e), vibrational (E_v), and rotational (E_r). We shall also be concerned with the kinetic energy of the molecules, since it plays an important indirect role in determining the populations of the various absorbing states.

Mastery of the subject of IR spectroscopy demands a thorough familiarity with quantum mechanics, a subject beyond the scope of this book. Our approach is to consider only a few of the simpler ideas underlying the physics of vibrational and rotational spectra. Fortunately an understanding of the radiative transfer process itself does not require detailed spectroscopic knowledge. This situation has been made possible in recent years by the availability of accurate compilations of line strengths and frequencies for all the major terrestrial molecular species. We will follow this empirical approach, as opposed to the more traditional and perhaps more intellectually satisfying spectroscopic approach.

We first consider some elementary physics of the absorption process. It is important to note that the major molecular species (O_2, N_2) of the Earth's atmosphere have essentially no importance for IR absorption. This follows from the symmetrical structure of homonuclear, diatomic molecules, as will be explained shortly. Four of the most important IR-absorbing molecular species are the minor constituent polyatomic molecules, water vapor (H_2O), carbon dioxide (CO_2), ozone (O_3), and methane (CH_4). Dozens of other species have a small effect on the heat budget, when considered collectively; also, these minor species (and their isotopic variants) are important in remote sensing. We will discuss only these representative species.

The absorption of light gives rise to excited states, which may be a combination of electronic, vibrational, and rotational motion (we will ignore the small effects associated with nuclear spin). We will begin with a consideration of molecular vibration, ignoring for the time being electronic or rotational energy. Separating the three is a useful abstraction, because the total internal energy of a molecule is given approximately by the sum of the three kinds of energy. Before discussing specifics, we will attempt to provide a visualization of the physics of the absorption process, in the same spirit as our earlier discussion of the Lorentz atom with regard to the scattering process (§3.2). The constituent atoms are held together in a semirigid structure by attractive forces provided by the electron "cloud," which is more or less shared by all the atoms. The bonding forces can be either electrostatic (ionic bonding) or quantum mechanical (exchange or covalent bonding). The nature of these forces does not concern us here. We need only consider their behavior as "springs" binding the various positively charged nuclei together. The simplest example is a diatomic molecule, which acts in many ways like a classical oscillator. Upon being "struck," either by a collision with another molecule or by absorption of a photon of the proper frequency, the constituent atoms are set into internal motion, alternately stretching and compressing the molecule. In polyatomic molecules the bonds may also "bend" so that the angles between the various axes may also oscillate. Classically, the energy of oscillation of a molecule can vary continuously, but in reality, the number of energy states is a discrete set, due to the quantum nature of energy states.

According to classical-mechanical analyses, the internal motion of a semirigid system, no matter how complicated, can be decomposed into a sum of elementary motions, called *normal modes*. A diatomic molecule, modeled by a simple harmonic oscillator, has only one normal mode of oscillation, along the internuclear axis. However, with increasing complexity of the molecule, more normal modes are possible. The general rule is that if a molecule has N atoms, the number of independent modes (or degrees of freedom) is $3N - 6$ for a nonlinear molecule ($N > 2$) and $3N - 5$ for a linear molecule. Figure 4.8 illustrates some of the normal modes of N_2, H_2O, CO_2, O_3, and CH_4. If the motions are small amplitude, the quantum-mechanical result for the total vibrational energy is

$$E_{\rm v} = \sum_k h\nu_k(\upsilon_k + 1/2) \quad (\upsilon_k = 0, 1, 2, \ldots), \tag{4.47}$$

where the sum is over all modes denoted by the index k, $h\nu_k$ is the *vibrational constant* for that mode, ν_k is the mode frequency, and υ_k is an integer, the *vibrational quantum number*. The value of $h\nu_k$ will depend upon the molecule, as well as the particular electronic energy state, and is usually in the range 300–3,000 cm^{-1}. The constant $1/2$ is a quantum-mechanical feature associated with the "zero-point energy." The lowest vibrational energy levels are somewhat higher than thermal energy $\sim k_{\rm B}T \sim 200\ \text{cm}^{-1}$ for $T = 300$ K. For a classical simple harmonic oscillator, elementary analysis shows that $h\nu_k$ depends upon the square root of the "spring constant" $k_{\rm e}$ divided by the reduced mass. It is usually written in terms of a *vibrational constant* ($\omega_{\rm e}$ in cm^{-1}) as $h\nu_k = hc\omega_{\rm e}$. The intermolecular force for a diatomic molecule is given by the spatial derivative of the potential energy function $V(r)$, which for small-amplitude oscillations is given by $-k_{\rm e}(r - r_{\rm e})$. Here $r_{\rm e}$ is the equilibrium nuclear separation. Figure 4.9

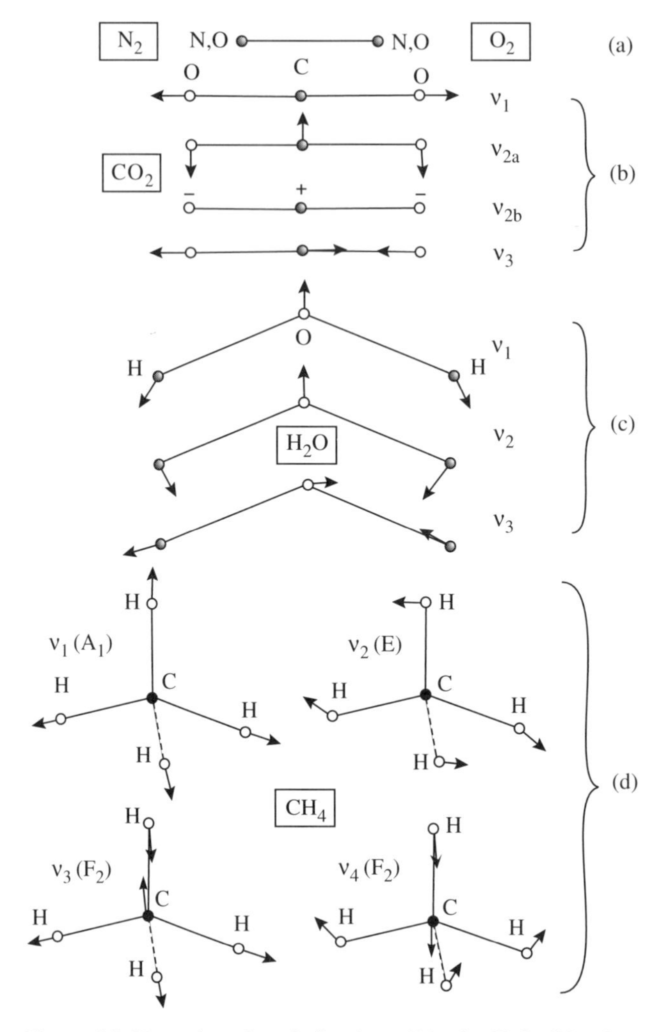

Figure 4.8 Normal modes of vibration of N_2, O_2, H_2O, CO_2, O_3, and CH_4. (a) N_2 and O_2 are diatomic, homonuclear molecules with only one mode of vibration. (b) CO_2 is a linear, triatomic molecule. Its ν_1 stretching mode is symmetric and therefore optically inactive. ν_{2a} and ν_{2b} are two separate modes but with the same energy and are said to be *degenerate*. The two modes differ only by a 90° rotation about the internuclear axis. ν_3 is the asymmetrical bending mode. (c) Both H_2O and O_3 (not shown) have three normal modes, all of which are optically active. (d) CH_4 has nine normal modes, but only ν_3 and ν_4 are active in the IR. ν_2 is doubly degenerate, and ν_3 is triply degenerate.

shows the function $V(r)$ for the H_2 molecule, along with the array of vibrational energy states.[20] Departures from strictly harmonic oscillations are described by higher-order terms.

In addition to being excited by molecular collisions, molecular vibrations may also be induced by absorption of radiation, provided the radiative energy is in

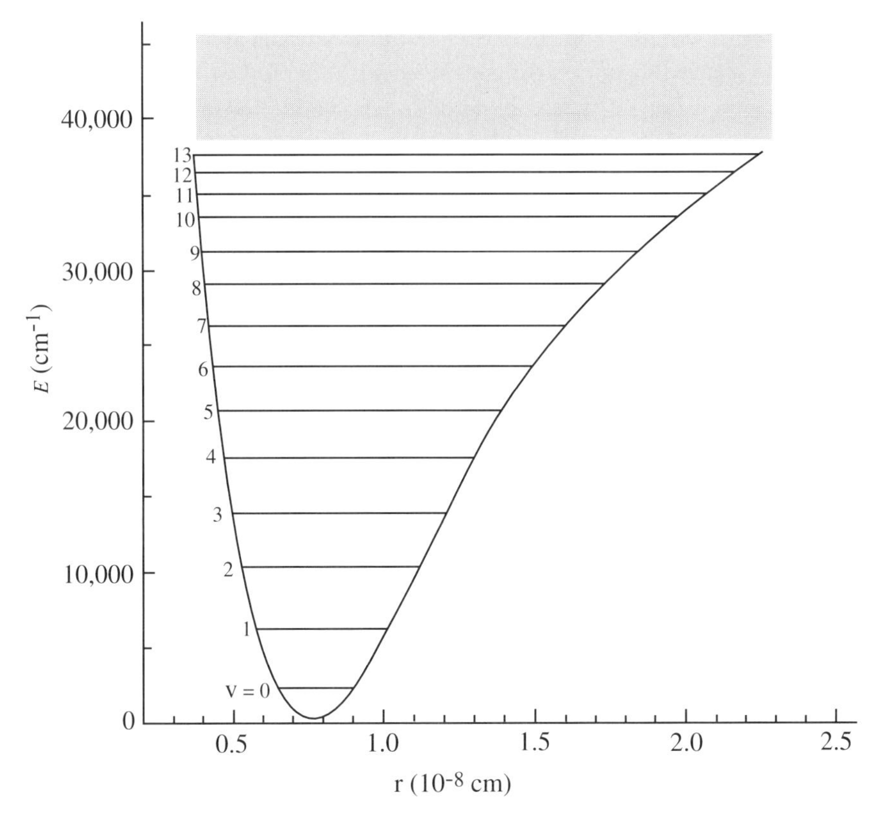

Figure 4.9 Potential curve of the H_2 ground state with vibrational levels and continuous term spectrum. The continuous term spectrum, above $v = 14$, is indicated by vertical hatching. The vibrational levels are drawn up to the potential curve, that is, their end points correspond to the classical turning points of the vibrational motion.

resonance with a normal mode. Classically we can think of this interaction as the temporary creation of an induced electric dipole moment by the incident electromagnetic field. This occurs if the new configuration results in an electron distribution whose first moment ("center of gravity") is displaced from its original position. In their ground, or lowest energy states, the dipole moment of symmetrical molecules, such as N_2, O_2, CO_2, and CH_4, is zero. However, there are asymmetrical stretching or bending modes of vibration (for example, the ν_2 state of CO_2) that result in an electric dipole. Radiative transitions between these states and the ground state are allowed because there is a change in the dipole moment. Note that the homonuclear molecules N_2 and O_2 are symmetrical in both their ground and (single) excited state, and therefore they have no vibrational spectra, that is, they are radiatively inactive.

In quantum theory, absorption takes place providing there is a finite dipole *matrix element*[21] between the initial and excited states. It happens that sometimes this matrix element is zero for certain combinations, and the transition is *forbidden*, at least for dipole transitions. Higher-order moments, such as *electric quadrupole* and *magnetic dipole* moments, may exist but their associated absorptions are much weaker than electric dipole transitions. *Selection rules* follow from considerations of whether a transition is "dipole allowed" or "dipole forbidden." The wavenumber of a vibrational transition is given by $hc\tilde{\nu} = E_v(v') - E_v(v'')$ with the selection rule $\Delta v = v' - v'' = \pm 1$, which is called the *fundamental*. Because of deviations from strict harmonic oscillator behavior of a real molecule, higher-order transitions (*overtone bands*) can occur where $\Delta v = \pm 2, \pm 3$, etc. If two (or more) modes change simultaneously during an absorption event, a *combination band* is said to occur.

Up to now we have ignored rotation, but rotational energy always accompanies vibrational energy. Rotation imposes a "fine structure" on the vibrational transitions, giving rise to a far richer absorption spectrum than Eq. 4.47 would imply.

4.5.1 Molecular Rotation: The Rigid Rotator

Molecular rotation is easy to understand in principle. For simplicity we assume that the molecule is a *rigid rotator*, that is, the internuclear separation is fixed, regardless of the rotation. A diatomic molecule will be characterized by one moment of inertia, I, expressed classically as $M_1 r_1^2 + M_2 r_2^2$, where M_j and r_j are the nuclear masses and distances along the principal axis from the center of gravity of the nuclei. The two radii are given by

$$r_1 = \frac{M_2}{M_1 + M_2} r \quad \text{and} \quad r_2 = \frac{M_1}{M_1 + M_2} r,$$

where $r = r_1 + r_2$ is the internuclear separation. The classical expression for the energy of rotation is $E_r = I\omega^2/2 = \mathcal{L}^2/2I$, where ω is the angular velocity of rotation about the principal axis, I is the corresponding moment of inertia, and $\mathcal{L}$ is the angular momentum. The usual quantum-mechanical "prescription" is to replace the classical angular momentum with the quantized quantity $(h/2\pi)$ times an integer, where h is Planck's constant. Since we are dealing with the square of the angular momentum, the quantum-mechanical equivalent is $\mathcal{L}^2 \rightarrow (h/2\pi)^2 J(J+1)$, where J is a positive integer, called the *rotational quantum number*. Thus, the rotational energy of a rigid rotator is given by

$$E_r(J) = \frac{1}{2I}\left(\frac{h}{2\pi}\right)^2 J(J+1) = hcB_v J(J+1) \quad (J = 0, 1, \ldots), \tag{4.48}$$

where $B_v \equiv h/(8\pi^2 cI)$ is the *rotational constant* corresponding to a particular electronic and vibrational state. Since B_v is inversely proportional to the moment of inertia, and therefore to the molecular mass, it follows that light molecules, such as H_2, will have more widely separated rotational energy levels (see Fig. 4.11 below) than heavier molecules.

How does rotation affect absorption and emission? Again, we invoke the principle that a changing electric dipole must be involved. In this case for radiative interaction it is necessary to have a permanent electric dipole moment. Since the dipole moment is a vector quantity, a change of the *direction* of this dipole moment would constitute a change in the dipole moment. This interaction leads to pure rotational transitions, whose energies occur in the far-infrared and microwave portion of the spectrum. The wavenumber of the emitted or absorbed photon is

$$\tilde{\nu} = B_v J'(J' + 1) - B_v J''(J'' + 1), \tag{4.49}$$

where the selection rule is $\Delta J = \pm 1$, that is, J may change or "jump" by only one unit. The pure rotational spectrum of a rigid rotator can be seen to be a sequence of equidistant lines. Linear molecules, such as N_2, O_2, or CO_2, are symmetrical in their ground states. They have no permanent dipole moment and thus no pure rotational spectrum. Finally, it should also be mentioned that pure rotational transitions prevail in the microwave spectrum. For example, H_2O exhibits intense microwave absorption at 22 and 183 GHz (1 gigahertz $= 1 \times 10^9$ Hz). Despite the fact that the ground state of O_2 possesses no electric dipole moment, it does have an unusually large magnetic dipole. Thus weak ("forbidden") magnetic dipole transitions occur in the microwave spectrum, which nevertheless are important for atmospheric absorption because of the very high abundance of O_2.

4.5.2 Molecular Vibration and Rotation: The Vibrating Rotator

Recognizing that vibration and rotation can occur simultaneously, we now consider the *vibrating rotator*. If there were no interaction between rotation and vibration, the energy would be simply the sum $E_\mathrm{v} + E_\mathrm{r}$. However, if the centrifugal force and the Coriolis force associated with the rotating frame are considered, the situation becomes more complicated. The energy levels can be written as *term values*

$$\frac{E(v, J)}{hc} = \omega_\mathrm{e}(v + 1/2) - \omega_\mathrm{e}x_\mathrm{e}(v + 1/2)^2 + B_v J(J + 1) - D_v J^2(J + 1)^2. \tag{4.50}$$

Here ω_e and $\omega_\mathrm{e}x_\mathrm{e}$ are vibrational constants, expressed in wavenumber units. The higher-order terms are the "interaction" terms for an *anharmonic oscillator*. Note the presence of two rotational constants, B_v and D_v, whose subscripts indicate their dependence on v. The term involving $\omega_\mathrm{e}x_\mathrm{e}$ is an anharmonic correction, which takes into account departures from simple harmonic oscillator motion.

The total molecular energy includes the electronic energy E_e. With this addition, the values of the rotational constants may also depend upon the particular electronic energy state. The wavenumber of a spectral line in a *vibration-rotation band* within a given electronic state is given by the difference of the *term values* of the two states defined by (v', J') and (v'', J''):

$$\tilde{\nu} = \tilde{\nu}_k + B'_v J'(J'+1) - B''_v J''(J''+1) \quad [\text{cm}^{-1}], \tag{4.51}$$

where $\tilde{\nu}_k$ is the basic wavenumber of the pure vibrational transition without taking into account any rotation (that is, when J' and J'' are set equal to zero). With $\Delta J = J' - J'' = +1$ and $\Delta J = J' - J'' = -1$, we obtain the wavenumbers of the *R-branch* and *P-branch*, respectively:

$$\tilde{\nu}_R = \tilde{\nu}_k + 2B'_v + (3B'_v - B''_v)J + (B'_v - B''_v)J^2 \quad (J = 0, 1, \ldots), \tag{4.52}$$

$$\tilde{\nu}_P = \tilde{\nu}_k - (B'_v + B''_v)J \quad + (B'_v - B''_v)J^2 \quad (J = 1, 2, \ldots). \tag{4.53}$$

(The notation for the ground-state term J'' has been changed to J.) Figure 4.10 shows the various transitions in a vibration-rotation band, illustrating the separation into two branches.[22]

The above description of a diatomic molecule is still approximate. The electrons (having small masses compared to the nuclei) have a small moment of inertia about the internuclear axis. Nevertheless their angular momenta are comparable to the nuclear value (which we now designate as $\mathbf{N}$), because they move much faster in their orbits. Only the component of this angular momentum along the axis, Λ, is constant (the other components average to zero). This occurs as a result of the electric field, which points along the axis of symmetry. The associated quantum number is Λ, a positive number. The total angular momentum $\mathbf{J}$ of the molecule is thus the vector sum of the nuclear angular momentum $\mathbf{N}$ (which points perpendicular to the axis) and the component of the electronic angular momenta $\mathbf{J}$ (which points along the axis). The magnitude of $\mathbf{J}$ is constant, and hence it is quantized according to $|\mathbf{J}| = \sqrt{(J(J+1)}h/2\pi$, where h is Planck's constant. J is greater than or equal to Λ and is given by $J = \Lambda, \Lambda + 1, \ldots$. For $\Lambda \neq 0$, there is a precession of $\mathbf{N}$ and $\boldsymbol{\Lambda}$ about the (constant) vector $\mathbf{J}$. Thus a more accurate picture of the diatomic molecule is a *symmetric top* nutating about the direction of the total angular momentum. The energy levels that result are thus the sum of the nuclear rotational energy and the nutational energy:

$$E_r/hc = B_v J(J+1) + (A_v - B_v)\Lambda^2,$$

$$\text{where } B_v = \frac{h}{8\pi^2 c I_B} \quad \text{and} \quad A_v = \frac{h}{8\pi^2 c I_A}.$$

The primary moment of inertia is I_B, and I_A is the much smaller moment about the internuclear axis. Generally, A_v is much larger than B_v. The quantum number Λ is usually a small (integral) value. Thus for a given electronic state, the levels of the symmetric top are the same as those of the simple rotator, except that there is a shift of magnitude $(A_v - B_v)\Lambda^2$. However, levels with $J < \Lambda$ are absent.

Ignoring electronic transitions, the selection rules are rather simple, since Λ does not change during the transition. Then for $\Lambda = 0$, $\Delta J = \pm 1$, and for $\Lambda \neq 0$, $\Delta J = 0, \pm 1$. In the first case, since the constant term $(A - B)\Lambda^2$ disappears when the two term values are subtracted, we obtain exactly the same branches as discussed for the simple

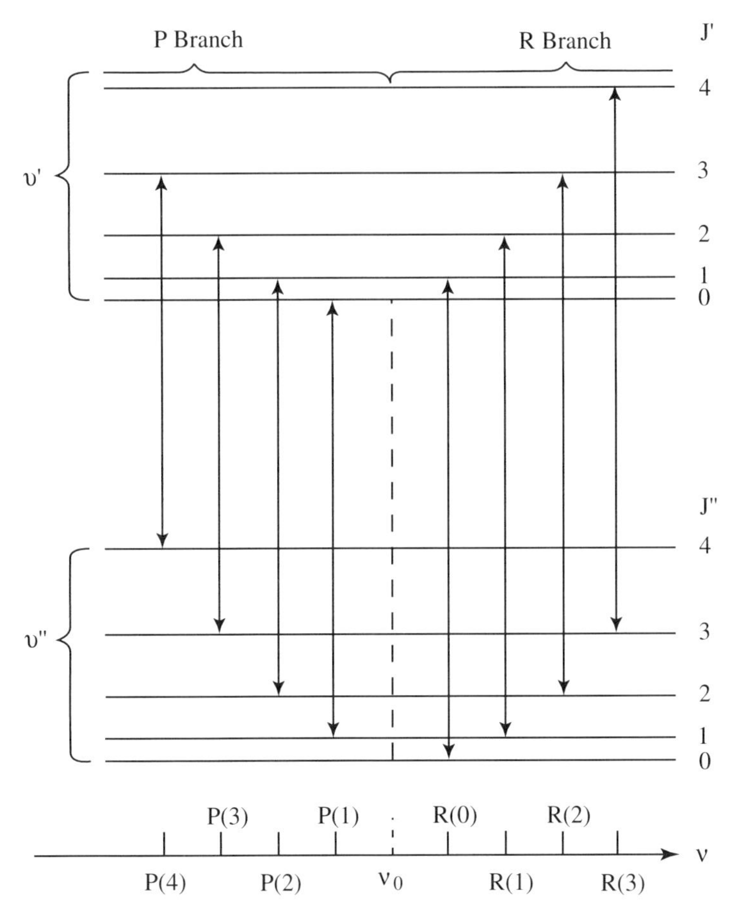

Figure 4.10 Energy levels of the vibrating rotator. v'' and J'' are the vibrational and rotational quantum numbers of the lower state. v', J' refer to the upper (excited) state. The vertical lines indicate allowed transitions ($\Delta J = \pm 1$). $R(0)$ denotes the R-branch ($\Delta J = +1$) ending in the $J = 0$ state, etc. In the lower part of the diagram, an idealized absorption is shown versus wave number. $P(1)$ denotes the P-branch ($\Delta J = -1$) ending in the $J = 1$ state, etc. The vertical dashed line indicates the band head, at $\nu = \nu_o$, which is missing in homonuclear diatomic molecules, because $v' = 0$ to $v'' = 0$ is forbidden.

rotator. In the second case there is a constant shift, but otherwise the term values are the same. However, more importantly, there arises a new branch, the *Q-branch*, with $\Delta J = 0$. The wavenumbers of the lines in this branch are

$$\tilde{\nu}_Q = \tilde{\nu}_k + (B_v'' - B_v')\Lambda^2 + (B_v' - B_v'')J + (B_v' - B_v'')J^2. \tag{4.54}$$

The only case of atmospheric interest for which a diatomic molecule has a Q-branch in its infrared spectrum is that of nitric oxide (NO), which has a nonzero Λ in its ground state. Its fundamental band at 5.3 μm is important for the energy budget of Earth's lower thermosphere.

Q-branches are more common in polyatomic molecular spectra, for example in the pure bending mode of the ν_2 mode of CO_2. The $\Delta v = 1$ transitions "pile up" at very nearly the same frequency, accounting for the very strong Q-branch in the 15-μm band (667–668 cm^{-1}; see Fig. 4.4).

More complex molecules are categorized in terms of the relationships of the various moments of inertias. For the most complicated molecule, all three moments of inertia are different and also unequal to zero. This is called the *asymmetric top* and is represented by the important molecules H_2O and O_3. If all three moments are equal, we have the *spherical top*, represented by CH_4. If two of the moments are equal, we have the *symmetric top*, already mentioned in the case of a diatomic molecule. It is represented by the molecule $CFCl_3$. Finally, we have the case where one of the moments is effectively zero, in which case we have a *linear* molecule, examples of which are CO_2, N_2O, CO, and NO. Equation 4.48 applies to the rotational energy for both the spherical top ($A_v = B_v$) and the linear molecule ($\Lambda = 0$). However, a more detailed analysis shows that linear molecules and spherical tops do not have the same rotational structure, as levels of equal J will "split" in different ways. This "splitting" occurs as the theory is made more precise to consider all the various couplings among electronic, vibrational, and rotational energies. Electron and nuclear *spin* are important for setting additional selection rules and in perturbing single energy levels into multiple levels. The *fine structure* results from the interaction of the magnetic dipole of the spinning electron with the electric field of the other electrons. *Hyperfine structure* results from a similar interaction of the nuclear spin.

4.5.3 Line Strengths

The rather formidable task of spectroscopy is to analyze the line frequencies of an absorption or emission spectrum in terms of the various quantum numbers, rotational and vibrational constants, etc. We will henceforth take a less formidable empirical approach and assume that we are given the complete set of *spectroscopic constants* (ω_e, $\omega_e x_e$, B, D, etc.) necessary to determine the frequency (or wavenumber) of all transitions within a specified frequency range. In addition to the spectroscopic constants, modern compilations of the *absorption line strengths* are also readily available. Knowledge of line strengths is necessary for determining the overall opacity of the atmosphere as a function of frequency. The strengths depend not only upon the nature of the individual transition, but also upon the equilibrium number of ground-state molecules. Thus it is necessary to return to the consideration of the Boltzmann distribution of energy states.

We will first consider a vibration-rotation band produced by a simple harmonic-oscillator rigid rotator. We will assume LTE conditions, so that the distribution of excited states is given by Boltzmann's formula. First consider molecular rotation only, in which the energy levels are denoted by the quantum number J and are given by Eq. 4.48. Since the quantum theory tells us that the statistical weight g_J (§4.3.4) of a rotational level J is $2J + 1$, we write for the ratio of state populations with different J-values

$$\frac{n(J)}{n(J')} = \frac{2J+1}{2J'+1}\exp\left\{-\frac{hcB}{k_B T}[J(J+1) - J'(J'+1)]\right\}. \tag{4.55}$$

A more convenient ratio is that of an excited-state population to the total number of states n within a given electronic and vibrational state,

$$\frac{n(J)}{n} = \frac{(2J+1)}{Q_r}\exp\left[-\frac{hcB_v}{k_B T}J(J+1)\right], \tag{4.56}$$

where Q_r is the *rotational partition function* given by

$$Q_r = \sum_{J'}(2J'+1)\exp\left[-\frac{hcB_v J'(J'+1)}{k_B T}\right]. \tag{4.57}$$

For sufficiently large T or small B_v, the spacing is very small compared with the total extent of the rotational energy. In this limiting case we may replace the sum with an integral, which is easily evaluated:

$$Q_r \approx \int_0^\infty dJ(2J+1)\exp[-hcB_v J(J+1)/k_B T] = \frac{k_B T}{hcB_v}.$$

The distribution of rotational energies with rotational quantum number J is shown in Fig. 4.11 for a number of molecules of atmospheric interest. This distribution is very important for the absorption coefficient, since the number of molecules in the ground (vibrational and electronic) state determines the rate of excitation. Note that in Fig. 4.11, when the average separation between states is relatively high, as in H_2, there are relatively few rotational states populated by collisions.

The LTE absorption cross section $\alpha^*_{\text{in}}(\nu)$ for an individual vibration-rotation line (denoted by i) can be written as the product of a numerical factor and a frequency-dependent line profile:

$$\alpha^*_{\text{in}}(\nu) = \mathcal{S}_i \Phi_i(\nu). \tag{4.58}$$

Here $\mathcal{S}_i$ is the *line strength* or *line intensity* of the ith line $(v'', J'') \rightarrow (v', J')$ given by

$$\mathcal{S}_i = \int d\nu \alpha^*_n(\nu) \quad [\text{m}^2 \cdot \text{s}^{-1}], \tag{4.59}$$

where the frequency integration is over the line breadth of a single line. In the above, double primes denote the lower state, and single primes denote the excited state. The absorption coefficient is given by $\alpha_i(\nu) = \alpha_{\text{in}}(\nu)n$, where n is the total density of radiatively active molecules (the sum over all ground and excited states). In tropospheric

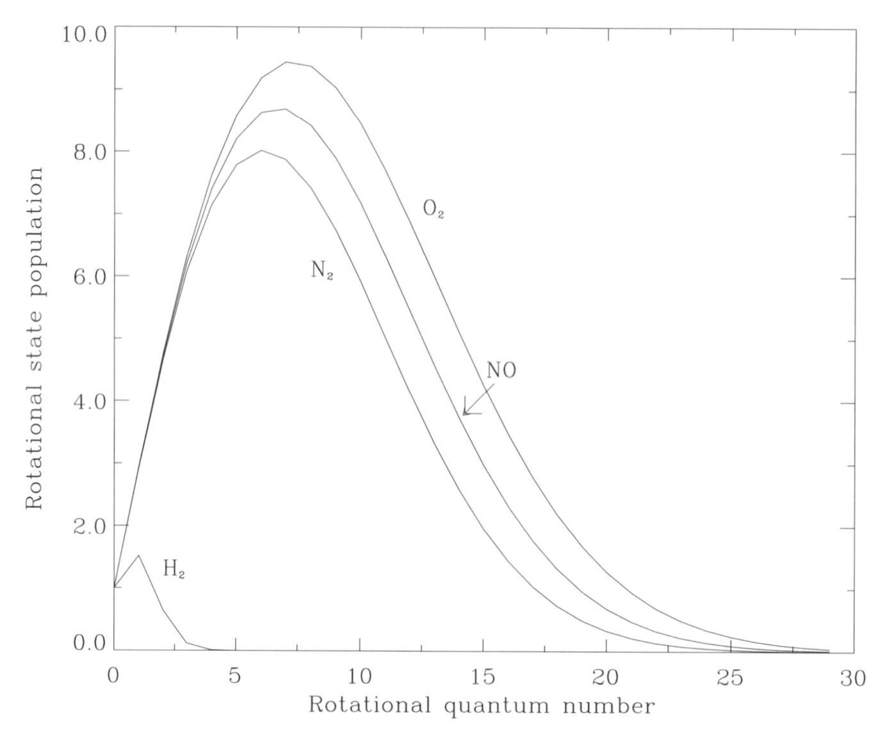

Figure 4.11 Distribution of rotational energy levels with rotational quantum number J for the molecules O_2, H_2, N_2, and NO, assuming $T = 250$ K.

radiation problems, it is permissible to assume LTE, so that Eq. 4.34 applies. In the present context, absorption and extinction are synononomous. In this equation, the initial (absorbing) state in the generalization to a multilevel molecule is $n_1^* \rightarrow n(v'', J'')$. Also, we let $B_{12} \rightarrow B_i$. Thus, equating Eqs. 4.34 and 4.58, we find the following expression for the LTE absorption coefficient:

$$k^*(\nu) \rightarrow \alpha^*(\nu) = \frac{h\nu_i}{4\pi} \Phi_i(\nu) n\left(v'', J''\right) B_i \left(1 - e^{-h\nu_i/k_B T}\right) \equiv \mathcal{S}_i n \Phi_i(\nu), \quad (4.60)$$

where ν_i denotes the central frequency of the line $h\nu_i = E(v', J') - E(v'', J'')$, the difference in energies of the two states connecting the transition i. The notation $\Phi_i(\nu)$ reminds us that the line profile may depend upon the particular transition, i, and differ from line to line and from band to band. This variation is usually small and slowly varying with frequency over lines within the same band. Solving for the line strength, we find

$$\mathcal{S}_i = \frac{h\nu_i}{4\pi} B_i \frac{n(v'', J'')}{n} \left(1 - e^{-h\nu_i/k_B T}\right). \quad (4.61)$$

Substitution from Eq. 4.55 for the population ratio of a rotational state yields

$$\mathcal{S}_i = \frac{h\nu_i (2J'' + 1)}{4\pi Q_i} B_i \exp\left[-hcB_v J''(J'' + 1)/k_B T\right]\left(1 - e^{-h\nu_i/k_B T}\right). \quad (4.62)$$

We should note here that the expressions above may differ from those found in the literature. First, our definition of the absorption coefficient in terms of the total density of molecules, as opposed to the number of molecules in the specific ground state, accounts for the factor $n(v'', J'')/n$ in the above equation. Second, the assumption of LTE allowed us to express stimulated emission as negative absorption, and the correction factor for this effect is $(1 - e^{-h\nu_i/k_B T})$. Not all authors include this factor, which is often near unity. Third, we have defined the Einstein coefficients in terms of the average intensity $\bar{I}_\nu$, instead of the energy density $\mathcal{U}_\nu$, which is often used in the astrophysical literature. This accounts for a factor of $4\pi/c$ between the two definitions.

In Eq. 4.62 we see the explicit dependence of the line strength on temperature through the Boltzmann distribution of initial-state populations. So far we have assumed that transitions connect only a ground state with an excited state. In fact, absorption can originate from a higher vibrational state. This is the case for so-called hot bands. Including the possibility of initial vibrational excitation, we have

$$\mathcal{S}_i = \mathcal{S}_{io} \frac{Q_v(T_o) Q_r(T_o)}{Q_v(T) Q_r(T)} \frac{e^{-E_i''/k_B T}}{e^{-E_i''/k_B T_o}} \frac{\left(1 - e^{-E_i/k_B T}\right)}{\left(1 - e^{-E_i/k_B T_o}\right)}. \tag{4.63}$$

The vibrational partition function, Q_v, is defined analogously to Q_r. E_i'' denotes the initial state energy. $\mathcal{S}_{io}$ is simply the line strength obtained from Eq. 4.62 evaluated at the reference temperature T_o. Fortunately the above result may be applied to any polyatomic molecule for which we know the various partition functions, line strengths, and central line frequencies. For standard tabulations, the temperature dependence of all the various terms is subsumed into the following semiempirical expression:

$$\mathcal{S}_i = \mathcal{S}_{io} \left(\frac{T_o}{T}\right)^m \exp\left[-\frac{E_i''}{k_B}\left(\frac{1}{T} - \frac{1}{T_o}\right)\right]. \tag{4.64}$$

Here m is a dimensionless quantity of order unity that serves as a fitting parameter.

The strength of a line can be determined in two basic ways: (1) from quantum theoretical calculations and (2) from laboratory measurements. The first method requires rather accurate knowledge of the wave functions, a very difficult problem for polyatomic molecules. In practice, laboratory results that rely upon the *Extinction Law* are used. The Air Force Geophysics Laboratory (now the Phillips Laboratory) provides an up-to-date listing of these parameters for atmospheric molecules. Specifically, it includes spectroscopic data for seven major atmospheric absorbers, O_2, H_2O, CO_2, O_3, N_2O, CO, and CH_4.

The 1992 HITRAN spectroscopic data base contains information for a total of 709,308 lines.[23] Included in the listing for each line are: ν_i, $\mathcal{S}_{io}$, width of the line at standard sea-level pressure and reference temperature, and energy E_i'' of the lower state, etc. A data base such as HITRAN is extremely useful to atmospheric radiative transfer practitioners, because it provides a well-accepted standard against which theory can be compared with data and with other theories.

4.6 Absorption Processes in the UV/Visible

As we proceed upward on the energy ladder from the infrared into the visible and UV, the spectroscopy becomes still more complex than previously discussed. At these higher energies, electronic excited states become accessible. Absorption of a photon causes an electron in an outer shell of the atom or molecule to be transferred to a higher electronic energy state. As in vibrational and rotational transitions, the transfer is accompanied by a change in the electric dipole moment of the atom or molecule. In the simple Bohr atom picture of an atom as a miniature solar system, the electron "jumps" from its initial orbit (around the massive nucleus) to one of larger radius. Thus, a third type of energy must be added to the vibrational and rotational types previously considered.

Electronic excited states are short lived in comparison to vibrational or rotational states. If the transition is electric dipole forbidden, the excited-state lifetimes are much longer than electric dipole states and are quenched by collisions well above the surface. For example, the $O_2(0, 1)$ A-band $(b\text{-}X)$ with band origin at 12,969 cm^{-1} (seen as the absorption feature at 771 nm in Fig. 5.8) is quenched at about 40-km altitude. Above this height, more of the absorbed radiative energy is promptly emitted as airglow emission, rather than ending up as thermal energy. The variable role of quenching produces absorption lines against the Rayleigh continuum background for tangent-ray heights below 40 km and emission lines above 40 km. High-spectral-resolution measurements from space[24] provide a sensitive probe of the temperature, density, and wind fields in the mesosphere and lower thermosphere. Neither O_2 nor N_2 absorb appreciably in the visible because of the absence of accessible electronic states with energies down to middle-UV wavelengths. In the UV, the *Herzberg continuum*, *Schumann–Runge* bands, and *Schumann–Runge* continuum (see Fig. 4.12) all result in a breakup of the oxygen molecule into free oxygen atoms. Absorption in the Herzberg continuum is a bound–free process that yields two ground-state oxygen atoms. The Schumann–Runge bands are due to a bound–bound transition to discrete upper levels. Electrons in this excited state are subject to a "level crossing" to a repulsive (unstable) electronic state. Thus, the upper state is very short lived, and because of the uncertainty principle, the absorption lines are broadened beyond that expected from ordinary pressure broadening. The unstable O_2 state almost instantaneously converts into two ground-state oxygen atoms. This process is known as *predissociation*. The Schumann–Runge continuum is a bound–free process resulting in dissociation into two O atoms: one in the ground state $O(^3P)$ and the other in an electronically excited state $O(^1D)$. Another important example of a bound–free process is the middle-UV *Hartley-band* absorption (see Fig. 4.12) of O_3 in which the molecule is fragmented into $O(^3P)$ and $O(^1D)$ products.[25]

In contrast to these bound–free continuum processes, more structured spectra result from transitions between two discrete electronic levels (bound–bound processes). Because molecules also have vibrational and rotational energy, the result will be a *band system*. The total term value can be written approximately as the sum of the

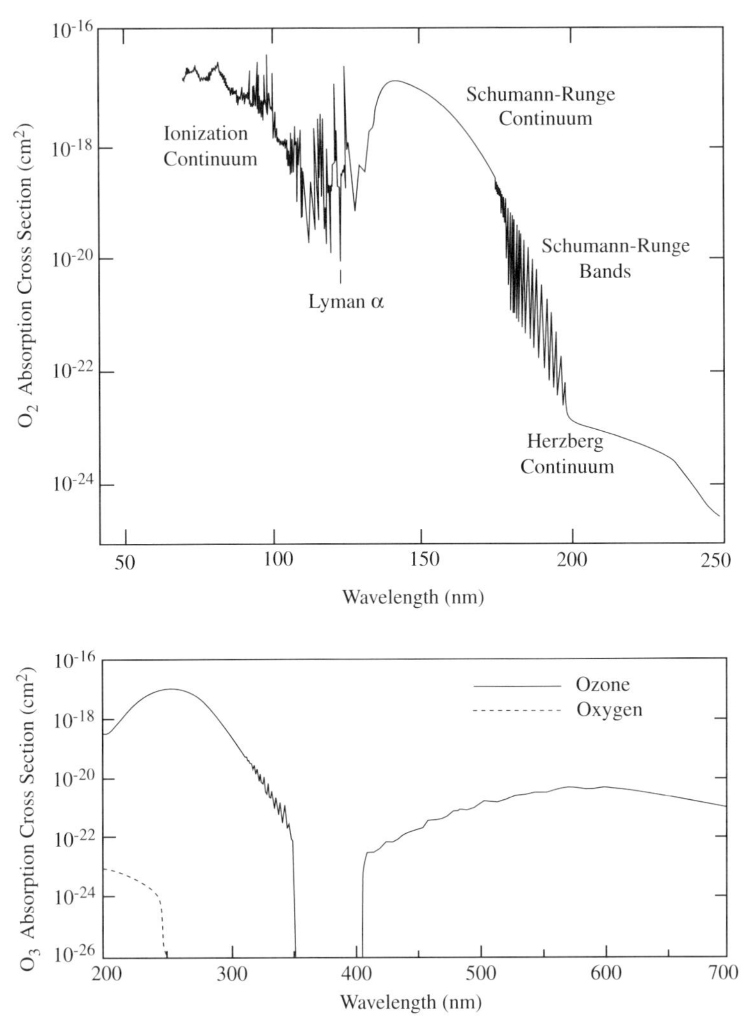

Figure 4.12 Top panel: O_2 absorption cross section illustrating the various band and continua (see text). The deep absorption line at 121.6 nm corresponds almost exactly with the strong solar hydrogen Lyman-α line. Bottom panel: Absorption cross section for O_3. The strong shortwave feature is the *Hartley continuum*, which blends into the band structure, called the *Hartley bands*. The weaker visible quasi-continuum is the *Chappuis bands*. Upper panel is from Brasseur, G. and S. Solomon (see Endnote 25).

electronic term value, T_e, the vibrational term value, G, and the rotational term value, F. Thus the frequency of a transition is written as the difference of two term values

$$\tilde{\nu} = T' - T'' = (T'_e - T''_e) + (G' - G'') + (F' - F''). \tag{4.65}$$

As usual we refer to the upper state with single-primed letters and the ground state with double-primed letters. For a given electronic transition, $\tilde{\nu}_e \equiv T'_e - T''_e$ is a constant. The remaining parts of Eq. 4.65 have forms similar to that for the vibration-rotation spectrum. The essential difference is that G' and G'' now belong to different vibrational term series with different values of ω_e and $\omega_e x_e$ (see Eq. 4.50). Similarly F' and F'' belong to two quite different rotational term series of different B_v-values (see Eq. 4.50).

Ignoring rotational "fine structure" for the moment, we can obtain a formula for the vibrational structure of an electronic band. Ignoring terms in T higher than v^2, we may write

$$\tilde{\nu} = \tilde{\nu}_e + \omega_e'(v' + 1/2) - \omega_e' x_e'(v' + 1/2)^2 - \omega_e''(v'' + 1/2) - \omega_e'' x_e''(v'' + 1/2)^2. \tag{4.66}$$

For electronic transitions there is no selection rule for the vibrational quantum number v. In principle, each upper vibrational state can be combined with each lower vibrational state.[26] Thus, a host of vibrational bands (v', v'') will exist within the entire electronic *band system*. The bands denoted by $(v', 0)$, $(v', 1)$, etc. for a fixed v' are called *v'-progressions*. The bands denoted by $(0, v'')$, $(1, v'')$, etc. are called *v''-progressions*. Another grouping of bands in which $\Delta v = v' - v''$ is a constant is called a *sequence*. Since ω_e and $\omega_e x_e$ are not very different between states, sequences are often grouped closely together in a spectrum.

At typical atmospheric temperatures, nearly all molecules will be in their ground $(v'' = 0)$ vibrational state. Thus absorption bands will exist normally only in a single v'-progression, with $v'' = 0$. The formula for the wavenumber of the absorption line progressions is

$$\tilde{\nu} = \tilde{\nu}_{00} + \omega_o v' - \omega_o' x_o' v'^2 + \cdots, \tag{4.67}$$

where $\tilde{\nu}_{00}$ is the term level of the (0, 0) band. With increasing v', the separations between successive vibrational levels approach the value zero, and then a *continuous* term spectrum joins onto the series of discrete vibrational levels. The energy where this occurs corresponds to the dissociation energy of the molecule. An example of this transition point is at 175 nm, where the structured Schumann–Runge band system converts to the smoothly varying Schumann–Runge continuum (see Fig. 4.12). This corresponds to the energy of dissociation plus the energy of the excited $O(^1D)$ oxygen atom.

Most molecules of interest to us (CO_2, H_2O, O_3, and CH_4) are polyatomic molecules. A discussion of their electronic spectra is beyond the scope of this book. Fortunately many of the concepts discussed above for the diatomic molecule may be carried over directly. Electronic transitions involving these molecules are not important for the energy balance of the lower atmosphere. However, they play critical roles in airglow, heating, and ionization processes in the upper atmosphere. The electronic absorption bands (the *Chappuis* and *Hartley bands*) of O_3 are semicontinuous over reasonably large frequency intervals and thus absorption obeys the Extinction Law. For example, calculating the irradiance-weighted cross sections over 10-nm bins may yield errors in the ozone photodissociation rate profile of only a few percent. Fortunately, unless one is interested in airglow spectra, it is generally not necessary to master the spectroscopy of polyatomic electronic band spectra. If necessary, one can fall back on the empirical method if a compilation of all the band positions and strengths is available from, for example, the HITRAN data base.

4.7 Summary

Absorption properties of aerosols depend upon the bulk optical properties, as well as the particle size and shape down to dimensions of order 0.1 μm, where quantum effects become important. Absorption by liquid water is governed by both the imaginary index of refraction of the pure state and by the effects of suspended particles. Molecular absorption is either discrete or continuous with frequency, depending upon the nature of the excited energy state. Of most importance to the IR are bound–bound (discrete) transitions. The strength and frequency width of these discrete features (absorption lines) are affected by the inherent properties (oscillator strengths), but they are also affected by the state of the ambient gas. The physical conditions of the gas (density, temperature, and chemical composition) affect the character of the radiative transitions through molecular collisions.

Various types of equilibria of matter were reviewed, including thermal, mechanical, chemical, and radiative. The most general, thermodynamic equilibrium, includes all of the above. It applies only to an isolated static system, a *hohlraum*. Despite the extreme departure of planetary media from these ideal situations, their study is quite useful in understanding less-restrictive equilibria, such as local thermodynamic equilibrium (LTE) and nonlocal thermodynamic (NLTE) equilibrium. The Planck Radiation Law and its Wien and Rayleigh–Jeans limits were reviewed, along with the Wien Displacement Law and the Stefan–Boltzmann Law. The Planck Law was derived from assuming that photons have discrete energies, whose states can be likened to standing waves in a closed box. The Planck Law distribution of light energy, as a function of frequency, was derived from finding the product of the mean energy per state and the density of states in an enclosure. The very important Kirchhoff's Law will be discussed in Chapter 5.

The idealized two-level atom permitted a simple visualization of the five basic processes of energy exchange between states: radiative absorption, radiative emission (spontaneous and induced), and collisional excitation and deexcitation. Detailed balance arguments allow the rates of processes to be related to their inverses. Reviews of the Maxwell–Boltzmann distribution of molecular velocities and of the Boltzmann distribution of discrete energy levels were given. A balance of radiative and collisional processes gives rise to the NLTE source function and the microscopic radiative transfer equation. The two extreme limits of gas density yield pure scattering in the low-density limit and the LTE limit for high density. The extension from a two-level atom to a multilevel molecule then allowed a realistic description of the line strength, in terms of temperature and density for a gas in LTE.

Infrared absorption and emission in molecular bands results from transfer among energy modes, or states, which arise from molecular vibration and rotation. These states are present in all molecules but are not always radiatively active. As a consequence of their linear symmetry, the dominant air species (N_2 and O_2, which both are homonuclear and diatomic) do not participate in IR radiative transfer. The way in which the various energies are described in terms of the basic vibrational and rotational quantum numbers was discussed. This was followed by a brief discussion of the spectroscopy of a diatomic molecule treated as a vibrating rotator. This simple model served as a prototype for more complex polyatomic molecules, which have more complicated vibrational and rotational modes. The temperature dependence of molecular line strengths, and more generally of the shape of molecular bands, was found to be a result of the Boltzmann distribution of energy states. Modern spectroscopic data bases were briefly discussed.

The more energetic shortwave absorption processes were discussed very briefly. Transitions in the visible and UV result from exchange between states of electronic excitation. These states have even greater complexity than IR absorption bands, because of the superposition of all three forms of energy: electronic, vibration, and rotation. For Earth atmospheric composition, these band systems are located primarily in the UV. Again, the spectroscopy of a diatomic molecule (such as O_2) serves to illustrate that of the more complex molecules, such as O_3. Thermal emission in these bands is negligible because collisional excitation rates at the low thermal energies of atmospheric molecules are very low.

Problems

4.1 The generalized Boltzmann distribution (Eq. 4.9) may be used to derive the equilibrium distribution of velocities. Setting $\beta = (k_B T)^{-1}$ for convenience, and allowing $\mathcal{E}_i$ to be the sum of kinetic and potential energies of the molecules, we have $\mathcal{E}_i = \frac{m}{2}(v_x^2 + v_y^2 + v_z^2) + V(x, y, z)$. Furthermore, the partition function Q may be converted from a sum to an integral, since the energy may be considered to be a continuous function of the spatial (x, y, z) and velocity-component (v_x, v_y, v_z) variables. Thus, we have $Q \propto \int \int \cdot \exp[-\beta\mathcal{E}(x, y, z, v_x, v_y, v_z)] dx dy dz dv_x dv_y dv_z$.

(a) Show by performing the integrations over the velocity components that the probability $p(\mathcal{E})$ that a molecule has position coordinates between x and $x + dx$, y and $y + dy$, and z and $z + dz$, and velocity components between v_x and $v_x + dv_x$, v_y and $v_y + dv_y$, and v_z and $v_z + dv_z$, is

$$p(\mathcal{E}) \propto f_{MB} e^{-\beta V(x,y,z)} dx dy dz,$$

where f_{MB} is given by Eq. 4.12.

(b) Assuming the atmosphere to be plane-parallel, and ignoring the variation of gravity and temperature with height, show that since $\beta V(x, y, z) = mgz/k_B T$, we recover the hydrostatic, or barometric law, Eq. 1.4.

(c) Explain why the atmosphere does not collapse, with all molecules falling to the ground under the influence of gravity.

(d) How is it possible for the molecules at each height to have the same average speed and kinetic energy?

4.2 Derive the equilibrium distribution of molecular *speeds* v, Eq. 4.17, beginning with the distribution of velocities, Eq. 4.12. (*Hint:* Transform the "volume element" $dv_x dv_y dv_z$ into polar coordinates, where the "radial" coordinate is dv, and integrate over the two angular coordinates.)

4.3 Show that if stimulated emission is neglected, leaving only two Einstein coefficients, an appropriate relation between the coefficients will be consistent with thermal equilibrium between the atom and a radiation field of a Wien spectrum, but not of a Planck spectrum.

4.4

(a) Calculate the rotational constant, B, for the electronic ground state of molecular hydrogen, H_2. The internuclear separation is $r_e = 1.059 \times 10^{-10}$ m. (Answer: $30\ \mathrm{cm}^{-1}$).

(b) Repeat for the ground state of O_2^{16} for which $r = 1.42 \times 10^{-10}$ m, and contrast this value with that for H_2.

4.5 The heating rate for the NLTE two-level atom problem is given by

$$\mathcal{H} = -h\nu_o \left(\frac{dn_1}{dt}\right)_{\mathrm{net}},$$

where $(dn_1/dt)_{\mathrm{net}}$ is the sum of the three radiative processes tending to change the population of the ground state.

(a) Show from the definition of $\mathcal{H} \equiv -\int d\omega d\nu (dI_\nu/ds)$ that

$$\mathcal{H} = 4\pi \mathcal{S}\epsilon_v (J - B_{\nu_o}) = 4\pi \mathcal{S}\epsilon_v \frac{(S - B_{\nu_o})}{1 - \epsilon_v},$$

where $\mathcal{S}$ is the line strength per molecule, ϵ_v is the coupling coefficient, S is the source function, B_{ν_o} is the Planck function evaluated at $\nu = \nu_o$, and

$$J = \int_0^\infty d\nu \Phi(\nu) \bar{I}_\nu.$$

(b) Let $\epsilon_v \to 0$. Assume that the radiation field is very dilute, so that $S \ll B_{\nu_o}$. Show that in this limit, $\mathcal{H} \approx -h\nu_o n_1 C_{12}$.

(c) Physically interpret the result (b).

Notes

1 Kyle, T. G., *Atmospheric Transmission, Emission and Scattering*, Pergamon Press, Oxford, 1991.

2 See Endnote 1.

3 See Endnote 1.

4 This may seem somewhat paradoxical. The high reflectance from a solid with a large absorption coefficient can be understood from the fact that the light is attenuated over a very small surface layer of a conductor, thereby "excluding" the incident light from the interior. The *Fresnel formula* (see Appendix E) for the normal reflectance of a metal in terms of the real and imaginary indices shows that when $m_i \to \infty$ then $\rho_s(0^\circ) \to 1$.

5 Jerlov, N. G., "Classification of sea water in terms of quanta irradiance," *Journal du Conseil-Conseil international pour l'exploration de la mer*, **37**, 281, 1977; see also Dera, J., *Marine Physics*, Elsevier, Amsterdam, 1992, Chapter 5.3, p. 275.

6 In the IR, we will often use wavenumber $(1/\lambda)$, instead of wavelength, where λ is the vacuum wavelength. Despite its being "non-SI," cm^{-1} remains the unit of choice by atmospheric spectroscopists. The SI unit of m^{-1} is never used!

7 Farmer, C. B. and R. H. Norton, *A High-Resolution Atlas of the Infrared Spectrum of the Sun and the Earth Atmosphere from Space, Vol. II. Stratosphere and Mesosphere, 650 to 3350 cm*$^{-1}$, NASA Reference Publication 1224, 684 pp., 1989.

8 See Endnote 7.

9 Goldman, A., F. J. Murcray, C. P. Rinsland, R. D. Blatherwick, F. H. Murcray, and D. G. Murcray, *Analysis of Atmospheric Trace Constituents from High Resolution Infrared Balloon-Borne and Ground-Based Solar Absorption Spectra*, Proc. of the Symposium on Remote Sensing of Atmospheric Chemistry, Society of Photo-Optical Instrumentation Engineers (SPIE), **1491**, 194–202, 1991.

10 The atmospheres of Venus, Earth, and Mars are in planetary radiative equilibrium, since internal energy sources are thought to be negligibly small for the terrestrial planets. Since planetary orbits are not circular, and radiative adjustment time scales are not short compared with an orbital period, the balance is more correctly stated as that existing between the solar energy received and the energy radiated over a complete orbit. For Jupiter and Saturn the atmospheres are also heated from energy sources within the planets themselves, presumably from an internal reservoir left over from their formation and/or from gravitational differentiation. Their atmospheres are definitely not in planetary radiative equilibrium.

11 This term in German means literally "empty space," but in the definition we have in mind it may be filled with matter of any type.

12 The physics of thermal radiation is described in M. Planck, *The Theory of Heat Radiation*, Dover, New York, 1991.

13 The real part of the index of refraction for gases for optical frequencies varies from unity only in the fourth decimal place. For example at STP, m_r in air varies from 1.000138 to 1.000142 over the visible spectrum. For CO_2, m_r varies from 1.000448 to 1.000454.

14 The question may be asked as to the role of the third body (M). Unless the collisional partners in an inelastic collision are fragmented, it is impossible for both total energy and momentum to be conserved without the participation of a third body, which acts as the carrier of the momentum/energy difference.

15 If inelastic cross sections were not much less than the elastic, there would be drastic consequences. Consider air at sea level, for which $\sim 10^9$ collisions occur per second. If a significant fraction of these collisions were inelastic – leading to an eventual radiative loss of heat – the air would cool down almost instantaneously!

16 Spontaneous emission may be thought of as being stimulated by fluctuations in the vacuum state of the electromagnetic field.

17 Since there is only one spectral line, the limits of integration can be safely extended over the entire spectrum.

18 The justification for why the line profiles for absorption and emission should be the same is given by Cooper, J., I. Hubeny, and J. Oxenius, "On the line profile coefficient for stimulated emission, *Astron. Astrophys.*, **127**, 224–6, 1983.

19 It might appear that we have cheated a bit by ignoring collisional processes. In fact, when we balance the radiative processes separately from the collisional processes, we are invoking the principle of *detailed balance*.

20 Herzberg, G., *Molecular Spectroscopy and Molecular Structure: I. Spectra of Diatomic Molecules*, van Nostrand, Princeton, 1950.

21 The electric dipole matrix element is the convolution of the ground- and excited-state quantum-mechanical wave functions with the electric dipole moment ex, where e is the electron charge and x is the displacement of the charge from the equilibrium position.

22 Adapted from Fig. 11.5 of Rybicki, R. G. and A. P. Lightman, *Radiative Processes in Astrophysics*, Wiley, New York, 1979.

23 An early version of the HITRAN data base is described by Rothman, L. S. et al., "AFRCL trace gas compilation:1982 edition," *Applied Optics*, **22**, 1616–27, 1983. An improved version was made available in 1986: Rothman, L. S. et al., "The HITRAN data base: 1986 edition," *Applied Optics*, **26**, 4058–97, 1987. The HITRAN data base is being updated continuously as more and better spectroscopic data become available. The latest published version is described by Rothman, L. S. et al., "The HITRAN molecular data base: Editions of 1991 and 1992," *J. Quant. Spectrosc. Radiative Transfer*, **48**, 469–508, 1992. There are currently three additional spectroscopic data bases that are widely used for line-by-line computations: the GEISA (Gestion et Etude des Informations Spectroscopiques Atmospheriques) data base: Husson, N. et al., "The GEISA spectroscopic line parameters data bank in 1984," *Ann. Geophys.*, **4**, 185–90, 1986; Husson, N., B. Bonnet, N. A., Scott, and A. Chedin, "Management and study of spectroscopic information: The GEISA program," *J. Quant. Spectrosc. Radiative Transfer*, **48**, 509–18, 1992; the ATMOS (Atmospheric Trace Molecule Spectroscopy) molecular line list compiled for the ATMOS experiment: Brown, L. R., C. B. Farmer, C. P. Rinsland, and R. A. Toth, "Molecular line parameters for the atmospheric trace molecule spectroscopy experiment," *Applied Optics*, **32**, 5154–82, 1987; and the JPL (Jet Propulsion Laboratory) compilation of microwave and submillimeter transmission: Poynter, R. L. and H. M. Pickett, "Submillimeter, millimeter, and microwave spectral line catalog," *Applied Optics*, **24**, 2235–2240, 1985.

24 Hays, P. et al., "The High-Resolution Doppler Imager on the Upper Atmosphere Research Satellite," *J. Geophys. Res.*, **98**, 10,713–24, 1993.

25 Brasseur, G. and S. Solomon, *Aeronomy of the Middle Atmosphere*, D. Reidel, Dordrecht, 1984.

26 In practice, the range of $v' - v''$ values is limited by the *Born–Oppenheimer Principle*.

Chapter 5

Principles of Radiative Transfer

5.1 Introduction

The radiation field in atmospheres and oceans is affected by the presence of a bounding surface, such as that of a solid land material, a liquid, such as an ocean, or an *effective surface*, such as a dense cloud deck. Similarly, the radiation field in the ocean is determined by the transmission of direct and diffuse light through the ocean–atmosphere interface. (A "surface" in our usage can even be the outer "edges" of a gaseous medium.) In this chapter we define emissive and reflective properties of surfaces applicable to their longwave and shortwave interactions, respectively. Infrared surface properties are generally simpler than UV/visible properties, since reflection of IR radiation is usually unimportant. Furthermore, except at millimeter wavelengths, surfaces emit thermal radiation approximately isotropically. Descriptions of surface interactions have historically been largely empirical, but in the past several decades, physically based models of surface reflection have gained in popularity with the need (for example) to relate land reflectance to crop yield and ocean color to water fertility. Somewhat analogous quantities for aerosols, the single-scattering albedo a and phase function p, describe the probability for scattering and the angular dependence of the scattered light, respectively. With these specifications, the radiative transfer equation including both scattering and absorption can be easily written down. We derive a closed-form solution for the local thermodynamic equilibrium intensity for the limiting case of zero scattering. The formal solution for the general case then immediately follows, although because it is expressed in terms of the unknown source function, it is not yet a usable result. The heating rate, photolysis rate, and dose rate are defined in terms of various integrals over the spectral intensity.

5.2 Boundary Properties of Planetary Media

A bounding surface[1] may have up to four distinct effects on the radiation environment: (1) reflection of a portion of the incident radiation back into the medium, (2) absorption of a portion of the incident radiation, (3) emission of thermal radiation, and (4) transmission of some of the incident radiation. All such processes modify the radiation field throughout the medium.

We refer to the *photometric* properties of surfaces when we deal with the unpolarized aspects of shortwave reflection and absorption. In the following sections we provide mathematical descriptions of these processes in terms of quantities that relate incoming and outgoing monochromatic intensities or fluxes. These photometric properties are specific to the type of surface, to the angular distribution of incoming radiation, and to the direction of outgoing radiation. Fortunately, the IR surface emittance depends only weakly upon its composition and texture.

Traditionally the study of diffuse reflection from solid surfaces has been an empirical science. However, in recent years the subject has been placed on a more physical basis. B. Hapke[2] has derived mathematical expressions for photometric quantities in terms of physically meaningful parameters. Some useful relationships among the various surface quantities are provided by *Kirchhoff's Law*, the *Principle of Reciprocity*, and the *Principle of Conservation of Energy*.

5.2.1 Thermal Emission from a Surface

As discussed previously, the small opening in a hohlraum (§4.3.1) is the ideal black "surface," emitting hemispherically isotropic blackbody radiation at all frequencies. However, emission from a real surface at the same temperature is usually quite different (and usually less efficient) than that from a blackbody. Generally, the emitted intensity will differ in the two polarization components, but as usual we ignore this complication.[3] Let $I^+_{\nu e}(\hat{\Omega}) \cos\theta d\omega$ be the emitted energy from a flat surface of temperature T_s within the solid angle $d\omega$ in the direction $\hat{\Omega}$. The corresponding energy emitted by a black surface at the same temperature is written $B_\nu(T_s) \cos\theta d\omega$. Here θ is the angle between the direction $\hat{\Omega}$ and the surface normal $\hat{n}$, so that $\cos\theta = |\hat{\Omega} \cdot \hat{n}|$. The spectral *directional emittance* is defined as the ratio of the energy emitted by a surface of temperature T_s to the energy emitted by a blackbody at the same frequency and temperature:

$$\epsilon(\nu, \hat{\Omega}, T_s) \equiv \frac{I^+_{\nu e}(\hat{\Omega}) \cos\theta \, d\omega}{B_\nu(T_s) \cos\theta \, d\omega} = \frac{I^+_{\nu e}(\hat{\Omega})}{B_\nu(T_s)}. \tag{5.1}$$

In general, ϵ depends upon the direction of emission, the surface temperature, and the frequency of the radiation, as well as other physical properties of the surface (index of refraction, chemical composition, texture, etc.). A surface for which ϵ is unity for all $\hat{\Omega}$ and ν is a *blackbody*, by definition. A hypothetical surface for which ϵ = constant < 1 for all frequencies is a *graybody*.

A second important surface quantity describes how the surface emits energy into a hemisphere (2π sr) relative to a blackbody at a particular frequency. The spectral *flux emittance* or *bulk emittance* relates the emitted flux to that of a blackbody at the same frequency and temperature:

$$\epsilon(\nu, 2\pi, T_s) \equiv \frac{\int_+ d\omega \cos\theta I_{\nu e}^+(\hat{\Omega})}{\int_+ d\omega \cos\theta B_\nu(T_s)} = \frac{\int_+ d\omega \cos\theta \epsilon(\nu, \hat{\Omega}, T_s) B_\nu(T_s)}{\pi B_\nu(T_s)}$$
$$= \frac{1}{\pi} \int_+ d\omega \cos\theta \epsilon(\nu, \hat{\Omega}, T_s). \qquad (5.2)$$

The flux emittance is usually measured directly, but in some cases it may be derived from first principles. (The directional emittance is almost never reported in the literature.) Experimentally it is easier to measure the reflectance ρ and to derive ϵ from the relationship $(1 - \rho)$. In the thermal IR, nearly all surfaces are efficient emitters, with ϵ generally exceeding 0.8. For pure substances, such as water ice, the imaginary index of refraction is needed to determine ϵ from theory.

5.2.2 Absorption by a Surface

Let a surface be illuminated by an angular beam of downward radiation with intensity $I_\nu^-(\hat{\Omega}')$ in a direction within a cone of solid angle $d\omega'$ around Ω'. Then the incident energy is $I_\nu^-(\hat{\Omega}') \cos\theta' d\omega'$. A certain amount $I_{\nu a}^-(\hat{\Omega}') \cos\theta' d\omega'$ of this energy is lost by absorption. Here θ' is the complement of the angle between the downward incident beam $\hat{\Omega}'$ and the upward surface normal $\hat{n}$, so that $\cos\theta' = |\hat{\Omega}' \cdot \hat{n}|$. (Our convention in using the negative sign in the notation $I_\nu^-(\hat{\Omega}')$ emphasizes the downward direction of the beam.) We define the *spectral directional absorptance* as the ratio of absorbed energy to incident energy of the beam,

$$\alpha(\nu, -\hat{\Omega}', T_s) \equiv \frac{I_{\nu a}^-(\hat{\Omega}') \cos\theta' d\omega'}{I_\nu^-(\hat{\Omega}') \cos\theta' d\omega'} = \frac{I_{\nu a}^-(\hat{\Omega}')}{I_\nu^-(\hat{\Omega}')}. \qquad (5.3)$$

Again, the minus sign in the notation $-\hat{\Omega}'$ emphasizes the downward direction of incidence. In analogy to the flux emittance, the spectral *flux absorptance* is the ratio of the absorbed flux to the incident flux,

$$\alpha(\nu, -2\pi, T_s) \equiv \frac{\int_- d\omega' \cos\theta' I_{\nu a}^-(\hat{\Omega}')}{\int_- d\omega' \cos\theta' I_\nu^-(\hat{\Omega}')}$$
$$= \frac{\int_- d\omega' \cos\theta' \alpha(\nu, -\hat{\Omega}', T_s) I_\nu^-(\hat{\Omega}')}{F_\nu^-}. \qquad (5.4)$$

If the illumination is from an isotropic blackbody, $I_\nu^-(\hat{\Omega}') = B_\nu(T_s)$, then

$$\alpha(\nu, -2\pi, T_s) = \frac{1}{\pi} \int_- d\omega' \cos\theta' \alpha(\nu, -\hat{\Omega}', T_s). \qquad (5.5)$$

5.2.3 Kirchhoff's Law for Surfaces

This famous law relates the emissive and absorptive abilities of a body in thermodynamic equilibrium. In the following we will apply heuristic arguments to show how this law follows from very simple physical considerations. We consider an opaque nonblack surface within a hohlraum, exposed to the isotropic radiance $I_\nu = B_\nu(T)$. Due to the isotropy of the radiation field, the upward radiation field emanating from the surface must also be uniform. However, in general not all this radiation will be *emitted* radiation. This is because we have allowed for values of emittance less than unity. (Since the surface material is assumed to be opaque, only reflection and absorption need to be considered.) The "deficit" of upward radiation due to the smaller amount of emitted radiation must be made up by a reflected component $I_{\nu\mathrm{r}}^+$. For each angular beam of light these two components must add to yield the Planck distribution, so that for all $\hat{\Omega}$

$$I_{\nu\mathrm{e}}^+(\hat{\Omega}) + I_{\nu\mathrm{r}}^+(\hat{\Omega}) = B_\nu(T_\mathrm{s}). \tag{5.6}$$

From conservation of energy, the sum of the reflected and absorbed energy must be equal to the incident energy, which in the hohlraum is also the Planck function. This yields

$$I_{\nu\mathrm{a}}^-(\hat{\Omega}) + I_{\nu\mathrm{r}}^+(\hat{\Omega}) = B_\nu(T_\mathrm{s}). \tag{5.7}$$

Using Eqs. 5.1, 5.3, 5.6, and 5.7 we find

Kirchhoff's Law for an Opaque Surface

$$\alpha(\nu, -\hat{\Omega}, T_\mathrm{s}) = \epsilon(\nu, \hat{\Omega}, T_\mathrm{s}). \tag{5.8}$$

Kirchhoff's Law describes the intimate connection between emission and absorption. The law is strictly valid only within an isothermal enclosure in thermodynamic equilibrium. However, in the form above it has much broader validity and for practical purposes may be considered to be an exact relationship for planetary surfaces.

Is there an equivalent form of this law relating the flux emittance and flux absorptance? Consider the expression for the flux absorptance in the special circumstance of hemispherically isotropic incidence. Comparing Eq. 5.5 with Eq. 5.2, and using Eq. 5.8, we find

$$\alpha(\nu, -2\pi, T_\mathrm{s}) = \epsilon(\nu, 2\pi, T_\mathrm{s}), \tag{5.9}$$

which states that the flux absorptance is indeed equal to the flux emittance. However, we stress that this form of Kirchhoff's Law is applicable *only if the incident intensity is uniform over a hemisphere*. If we were to seek relationships between the frequency-integrated emittance and absorptance, we would have to impose even more restrictive conditions on the radiation field. Therefore, *Kirchhoff's Law does not generally apply to angular-integrated and frequency-integrated quantities*.

5.2.4 Surface Reflection: The BRDF

The concepts of reflectance and transmittance are more complicated than those of emittance or absorptance, since they depend upon both the angles of incidence and reflection or transmission. Referring to Fig. 5.1, we consider a downward-moving angular beam of radiation with intensity $I_\nu^-(\hat{\Omega}')$ within a cone of solid angle $d\omega'$ around $\hat{\Omega}'$. Then the energy incident on a flat surface whose normal $\hat{n}$ is directed along the z axis (see Fig. 5.1) is $I_\nu^-(\hat{\Omega}')\cos\theta' d\omega'$. Denoting by $dI_{\nu r}^+(\hat{\Omega})$ the intensity of reflected light leaving the surface within a cone of solid angle $d\omega$ around the direction $\hat{\Omega}$, we define the *bidirectional reflectance distribution function*, or BRDF, as the ratio of the reflected intensity to the energy in the incident beam:

$$\rho(\nu, -\hat{\Omega}', \hat{\Omega}) \equiv \frac{dI_{\nu r}^+(\hat{\Omega})}{I_\nu^-(\hat{\Omega}')\cos\theta' d\omega'}. \tag{5.10}$$

We note that $dI_{\nu r}^+(\hat{\Omega})$ is a first-order differential quantity that balances the differential $d\omega'$ in the denominator, so that ρ is a finite quantity. Adding the contributions to the reflected intensity in the direction $\hat{\Omega}$ from beams incident on the surface in all downward directions, we obtain the total reflected intensity

$$I_{\nu r}^+(\hat{\Omega}) = \int_- dI_{\nu r}^+(\hat{\Omega}) = \int_- d\omega' \cos\theta' \rho(\nu, -\hat{\Omega}', \hat{\Omega}) I_\nu^-(\hat{\Omega}'). \tag{5.11}$$

Thus, the reflected intensity is the integral of the energy in each incident direction times the BRDF for that particular combination of incidence and observation angles

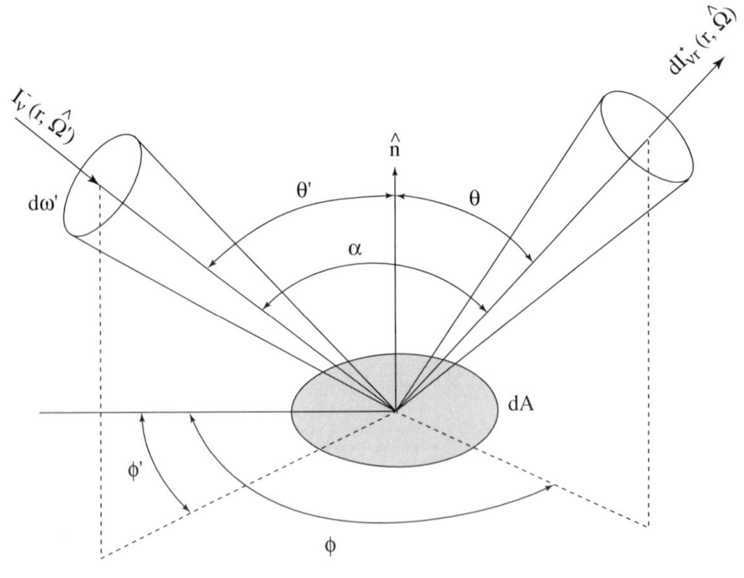

Figure 5.1 Geometry and symbols for the definition of the BRDF. The angle α is the backscattering angle.

under consideration. The fact that θ' can take on any value in the hemisphere in Fig. 5.1 illustrates the *diffuse* nature of the reflected light from most natural surfaces. The BRDF, or in short the *reflectance*, plays a central role in the remote sensing of planetary surfaces.

To help elucidate the physical significance of the BRDF, we consider the two extreme types of reflection: purely diffuse and purely specular. The former occurs at microscopically irregular surfaces, such as sand and snow. Such surfaces present a variety of angles to the incident light, and as a result the reflected light will have an approximately uniform angular distribution. Specular reflection occurs when the surface is perfectly smooth, like a mirror. A typical example is reflection of sunlight from a window pane.

If the reflected intensity from a diffuse surface is completely uniform with angle of observation, it is said to be a *Lambert surface*. Good examples of Lambert surfaces are ground glass and matte paper. *The BRDF for a Lambert surface is independent of both the direction of incidence and the direction of observation.* Then the reflectance simplifies to

$$\rho(\nu, -\hat{\Omega}', \hat{\Omega}) = \rho_{\rm L}(\nu), \tag{5.12}$$

where $\rho_{\rm L}$ is the *Lambert reflectance*, which may depend upon frequency. From Eq. 5.11, the reflected intensity is simply

$$I_{\nu r}^{+} = \rho_{\rm L}(\nu) \int_{-} d\omega' \cos\theta' I_{\nu}^{-}(\hat{\Omega}') = \rho_{\rm L}(\nu) F_{\nu}^{-}. \tag{5.13}$$

For this ideal surface, *the reflected intensity is proportional to the incident flux F_{ν}^{-} and is independent of the observation direction $\hat{\Omega}$.*[4]

Example 5.1 Collimated Incidence – Lambert Surface vs. Specular Reflection

Suppose that the incident light is direct (collimated) sunlight so that

$$I^{-}(\hat{\Omega}') = F^{\rm s}\delta(\hat{\Omega}' - \hat{\Omega}_0) = F^{\rm s}\delta(\cos\theta' - \cos\theta_0)\delta(\phi' - \phi_0), \tag{5.14}$$

where we have dropped the ν subscript.

What is the expression for the reflected intensity from a Lambert surface illuminated by direct (collimated) sunlight? Since the incident flux is given by $F^{-} = F^{\rm s}\cos\theta_0$, the result is simply $I_{\rm r}^{+}(\hat{\Omega}) = \rho_{\rm L}\mu_0 F^{\rm s}$, where $\mu_0 = \cos\theta_0$. Thus, *for a collimated light beam the intensity reflected from a Lambert surface is proportional to the cosine of the angle of incidence.*

The opposite extreme is specular reflection, for which the reflected intensity is directed along the angle of reflection (see Fig. 5.2). It is important for smooth, untextured surfaces, such as calm water, clean smooth ice, and the waxy surfaces of leaves. The specularly reflected intensity $I_{\rm r}^{+}$ is directly proportional to the incident intensity, but it is finite only in the direction of reflection θ, ϕ. The polar and azimuthal angles of reflection are given by $\theta = \theta'$ and $\phi = \phi' + \pi$. The proportionality constant is the *spectral reflection function* $\rho_{\rm s}(\nu, \theta)$, which depends upon reflection angle $\theta = \theta'$ and frequency. Thus, for an incident collimated beam in the direction θ_0, ϕ_0 given

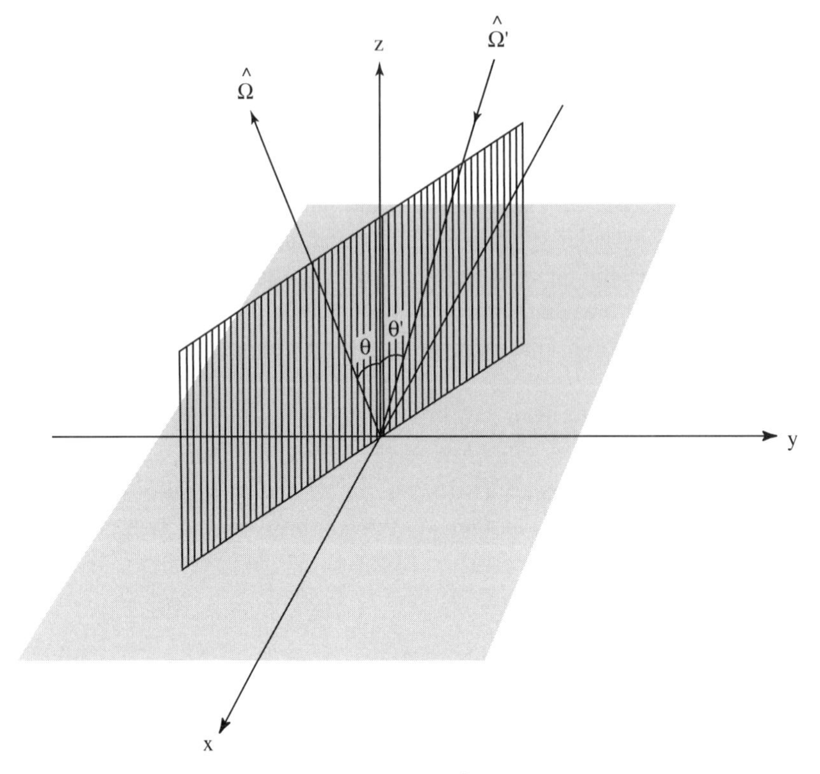

Figure 5.2 Intensity is incident in the direction $\hat{\Omega}'$ on the x–y plane. According to the law of reflection, the vertical plane through $\hat{\Omega}'$ and the surface normal $\hat{n}$ (the z direction) also contains the direction of reflection $\hat{\Omega}$. Furthermore, the angle of reflection θ is equal to the incident angle θ'.

by Eq. 5.14, the specularly reflected intensity is

$$I_r^+(\hat{\Omega}) = \rho_s(\theta) F^s \delta(\cos\theta_0 - \cos\theta)\delta(\phi - [\phi_0 + \pi]), \tag{5.15}$$

and the corresponding reflected flux becomes

$$\begin{aligned} F_r^+ &= \int_0^{2\pi} d\phi \int_0^{\pi/2} d\theta \sin\theta \cos\theta F^s \rho_s(\theta)\delta(\cos\theta_0 - \cos\theta)\delta(\phi - [\phi_0 + \pi]) \\ &= \rho_s(\theta_0) F^s \cos\theta_0. \end{aligned}$$

However, no natural surface is perfectly smooth. If the BRDF itself is of interest, the δ-function in Eq. 5.15 is to be replaced by ordinary functions that are sharply peaked in the reflection direction (see the parameterization in a following section).

In general, a surface will exhibit both a specular and a diffuse reflection component. Denoting the diffuse part of the BRDF as ρ_d and the specular part as ρ_s, we write ρ as a sum of specular and diffuse components,

$$\rho(\nu, -\hat{\Omega}', \hat{\Omega}) = \rho_s(\nu, -\hat{\Omega}', \hat{\Omega}) + \rho_d(\nu, -\hat{\Omega}', \hat{\Omega}), \tag{5.16}$$

so that the reflected intensity is given by

$$\begin{aligned} I_{\nu r}^{+}(\hat{\Omega}) &= \int_{-} d\omega' \cos\theta' \rho(\nu, -\hat{\Omega}', \hat{\Omega}) I_{\nu}^{-}(\hat{\Omega}') \\ &= \rho_s(\nu, \theta) I_{\nu}^{-}(\theta, \phi' + \pi) + \int_{-} d\omega' \cos\theta' \rho_d(\nu, -\hat{\Omega}', \hat{\Omega}) I_{\nu}^{-}(\hat{\Omega}'). \end{aligned} \tag{5.17}$$

Specular reflection from a smooth dielectric surface can be calculated from first principles, given the optical constants of air and the dielectric material. *Fresnel's equations* (Appendix E) yield the reflectance for an arbitrary direction of incidence. Figure 5.3 shows how the theoretical reflectance and transmittance of water varies with angle of incidence.

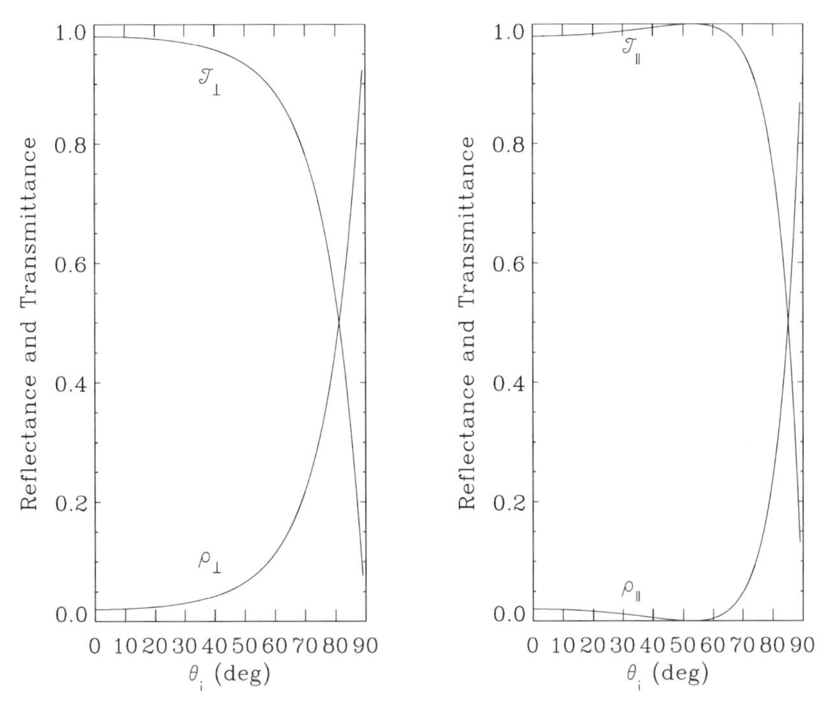

Figure 5.3 The bidirectional reflectance and transmittance at a smooth air–water interface based on Fresnel's equations (see Appendix E). The two curves show the reflectance and transmittance for the perpendicular and parallel components.

5.2.5 Albedo for Collimated Incidence

We are often interested in the reflected intensity and flux from direct (collimated) sunlight from the direction $\hat{\Omega}_0$. The incident intensity is given by Eq. 5.14, and from Eq. 5.11 we obtain the *diffusely reflected intensity*

$$I_{\nu r}^{+}(\hat{\Omega}) = F_\nu^s \int_{-} d\omega' \cos\theta' \rho_d(\nu, -\hat{\Omega}', \hat{\Omega}) \delta(\cos\theta' - \cos\theta_0)\delta(\phi' - \phi_0)$$
$$= F_\nu^s \cos\theta_0 \rho_d(\nu, -\hat{\Omega}_0, \hat{\Omega}). \tag{5.18}$$

The diffusely reflected flux is

$$F_{\nu r}^{+} = \int_{+} d\omega \cos\theta I_{\nu r}^{+}(\hat{\Omega}) = F_\nu^s \cos\theta_0 \int_{+} d\omega \cos\theta \rho_d(\nu, -\hat{\Omega}_0, \hat{\Omega}). \tag{5.19}$$

The ratio of the reflected flux to the incident solar flux is the *flux reflectance* or *plane albedo*[5]:

$$\rho(\nu, -\hat{\Omega}_0, 2\pi) \equiv \frac{F_{\nu r}^{+}}{F_\nu^s \cos\theta_0} = \int_{+} d\omega \cos\theta \rho(\nu, -\hat{\Omega}_0, \hat{\Omega}).$$

The 2π notation emphasizes that the integration is made over all reflected (outgoing) directions. In reality, $\rho(\nu, -\hat{\Omega}_0, 2\pi)$ is the sum of a diffuse and a specular component, that is,

$$\rho(\nu, -\hat{\Omega}_0, 2\pi) = \rho_s(\nu, -\hat{\Omega}_0, 2\pi) + \rho_d(\nu, -\hat{\Omega}_0, 2\pi)$$
$$= \int_{+} d\omega \cos\theta \rho_s(\nu, -\hat{\Omega}_0, \hat{\Omega}) + \int_{+} d\omega \cos\theta \rho_d(\nu, -\hat{\Omega}_0, \hat{\Omega}). \tag{5.20}$$

For most natural surfaces $\rho(\nu, -\hat{\Omega}_0, 2\pi)$ depends only upon the polar angle of the incident beam direction, θ_0.

If the BRDF is Lambertian, then Eq. 5.20 yields

$$\rho(\nu, -\hat{\Omega}_0, 2\pi) = \rho_L(\nu) \int_0^{2\pi} d\phi \int_0^{\pi/2} d\theta \sin\theta \cos\theta = \pi \rho_L(\nu).$$

This result is analogous to the relationship between the blackbody flux F_ν^{BB} (Eq. 4.1) and the blackbody intensity I_ν^{BB}, Eq. 4.4, that is, $F_\nu^{BB} = \pi I_\nu^{BB}$. Note that since $\rho_L \le \pi^{-1}$ the flux albedo is bounded from above by unity.

5.2.6 The Flux Reflectance, or Albedo: Diffuse Incidence

Sunlight reaching the surface of the Earth consists of two components, a *direct* (collimated) component and a *diffuse* (scattered) component. The diffuse radiation (skylight)

$I_\nu^-(\hat{\Omega}')$ is distributed over the entire downward hemisphere (§6.2). The reflected flux is

$$F_{\nu r}^+ = \int_+ d\omega \cos\theta I_{\nu r}^+(\hat{\Omega}) = \int_+ d\omega \cos\theta \int_- d\omega' \cos\theta' \rho(\nu, -\hat{\Omega}', \hat{\Omega}) I_\nu^-(\hat{\Omega}').$$

Interchanging the order of the integrations, we find

$$F_{\nu r}^+ = \int_- d\omega' \cos\theta' \left[\int_+ d\omega \cos\theta \rho(\nu, -\hat{\Omega}', \hat{\Omega}) \right] I_\nu^-(\hat{\Omega}').$$

The quantity in brackets [ ] is the flux reflectance for incidence from the direction $\hat{\Omega}'$,

$$\rho(\nu, -\hat{\Omega}', 2\pi) = \int_+ d\omega \cos\theta \rho(\nu, -\hat{\Omega}', \hat{\Omega}). \tag{5.21}$$

Therefore the reflected flux for a diffuse distribution of incident radiation is

$$F_{\nu r}^+ = \int_- d\omega' \cos\theta' \rho(\nu, -\hat{\Omega}', 2\pi) I_\nu^-(\hat{\Omega}'). \tag{5.22}$$

It is interesting to compare this result with our previous results for the emitted and absorbed fluxes,

$$F_{\nu e}^+ = \int_+ d\omega \cos\theta \epsilon(\nu, \hat{\Omega}, T_s) B_\nu(T_s), \tag{5.23}$$

$$F_{\nu a}^- = \int_- d\omega' \cos\theta' \alpha(\nu, -\hat{\Omega}', T_s) I_\nu^-(\hat{\Omega}'), \tag{5.24}$$

where we have used Eqs. 5.1 and 5.3. The three equations (5.22, 5.23, and 5.24) for the reflected, emitted, and absorbed fluxes show that the quantities $\rho(\nu, -\hat{\Omega}, 2\pi)$, $\epsilon(\nu, \hat{\Omega}, T_s)$, and $\alpha(\nu, -\hat{\Omega}, T_s)$ play analogous roles in "transforming" intensity distributions into fluxes.

We now derive an important relationship between the flux reflectance and the absorptance of an opaque surface. Consider an incident angular beam containing the energy $I_\nu^-(\hat{\Omega}') \cos\theta' d\omega'$, which is partially absorbed and partially reflected. Conservation of energy implies that the incident energy is equal to the sum of the absorbed and reflected energy. The absorbed energy (see Eq. 5.3) is $I_\nu^-(\hat{\Omega}') \cos\theta' d\omega' \alpha(\nu, -\hat{\Omega}', T_s)$. The energy reflected into the solid angle $d\omega$ around the direction $\hat{\Omega}$ is $dI_{\nu r}^+(\hat{\Omega}) \cos\theta d\omega$, where the reflected intensity is given by Eq. 5.10, that is, $dI_{\nu r}^+(\hat{\Omega}) = \rho(\nu, -\hat{\Omega}', \hat{\Omega}) I_\nu^-(\hat{\Omega}') \cos\theta' d\omega'$. The total energy reflected from the incident angular beam around $\hat{\Omega}'$ into all upward directions is given by the hemispherical sum of the contributions to the flux from all directions, that is,

$$\int_+ dI_{\nu r}^+(\hat{\Omega}) \cos\theta \, d\omega = I_\nu^-(\hat{\Omega}') \cos\theta' d\omega' \int_+ d\omega \cos\theta \rho(\nu, -\hat{\Omega}', \hat{\Omega}).$$

Equating the incident energy to the absorbed and total reflected energy, we have

$$I_\nu^-(\hat{\Omega}')\cos\theta' d\omega' = I_\nu^-(\hat{\Omega}')\cos\theta' d\omega' \alpha(\nu, -\hat{\Omega}', T_s) + I_\nu^-(\hat{\Omega}')\cos\theta' d\omega' \int_+ d\omega \cos\theta \rho(\nu, -\hat{\Omega}', \hat{\Omega}).$$

The integral in the above equation is the flux reflectance $\rho(\nu, -\hat{\Omega}', 2\pi)$. Canceling common factors, we find the

Relationship between the flux reflectance and directional absorptance for an opaque surface:

$$\alpha(\nu, -\hat{\Omega}', T_s) = 1 - \rho(\nu, -\hat{\Omega}', 2\pi). \tag{5.25}$$

Since this relationship follows from conservation of energy, it is generally valid. Its extension to a partially transmitting medium is considered in Problem 5.2.

5.2.7 Analytic Reflectance Expressions

It is convenient to describe surface reflection in terms of a smooth function involving a small number of adjustable parameters. Such a function may be used for interpolation if only a limited amount of reflectance data is available. Also, we may be interested in assessing how important the overall reflectance properties are in calculations of the radiation field. In such cases it would be useful to have a continuous variation of the parameters, for example between the two extremes of Lambert reflectance and specular reflectance. Below we discuss several analytic models that have appeared in the literature. In what follows we ignore any azimuthal dependence of the reflected light and express the input (θ_0) and output (θ) polar angles in terms of their cosines, $\mu_0 = \cos\theta_0$ and $\mu = \cos\theta$.

Example 5.2 The Minnaert Formula

A formula that has been extensively used in planetary astronomy is that of M. Minnaert:

$$\rho(\mu_0, \mu) = \rho_n \mu_0^{k-1} \mu^{k-1}. \tag{5.26}$$

Here k is a dimensionless parameter, which is adjusted to fit observations. When $k = 1$ we obtain the Lambert reflectance formula. For dark surfaces k is about 0.5. In the latter case, the *Minnaert formula* predicts that the greater the observation angle θ (also called the *off-nadir angle* or simply *nadir angle*), the brighter the surface. This prediction is in general accord with experiment. For brighter surfaces, k increases, and for very bright surfaces, $k \to 1$ and the reflectance approaches that of a Lambert surface, again in accord with experiment. Figure 5.4 shows the behavior of Eq. 5.26 for values of the parameters k and ρ_n chosen to fit the data for a number of prepared particulate surfaces, varying in the values of their normal reflectance, ρ_n.[6] The disadvantage of the Minnaert formula is that its parameters have no physical basis. An even more undesirable quality is that when fitting data, it is often necessary that parameters vary with the angle of incidence.

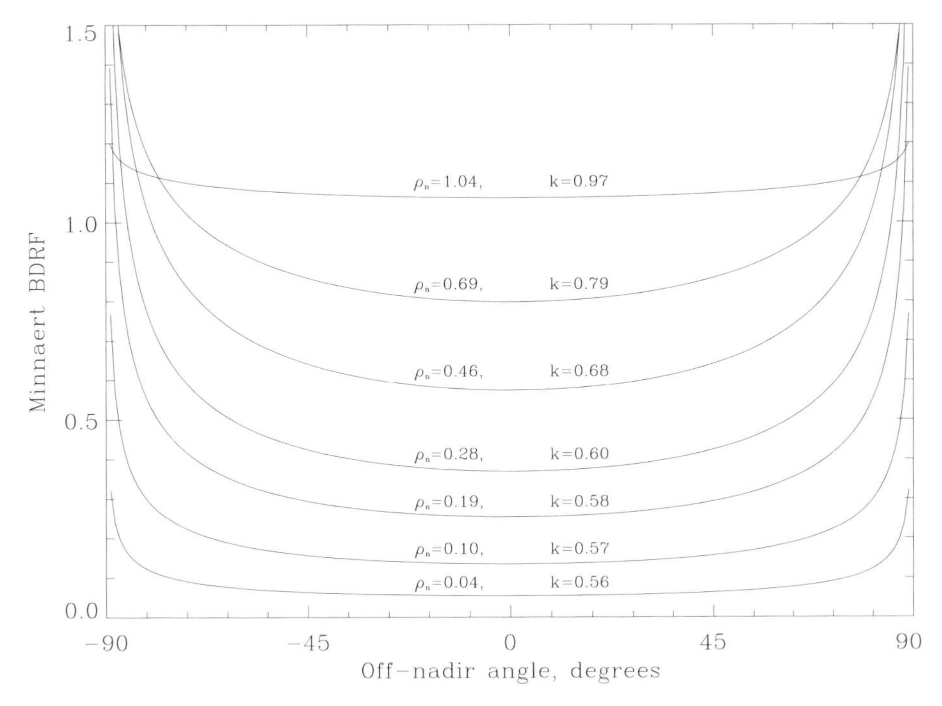

Figure 5.4 The BRDF for *Minnaert's formula* is plotted versus the off-nadir angle. Each curve corresponds to a different laboratory sample (see Endnote 6) ranging from very dark to very bright. For all curves the incident solar zenith angle is fixed at $\theta_0 = 60°$, so that $\mu_0 = 0.5$. If one is interested in a different solar zenith angle, each curve should be multiplied by $(2\mu_0)^{k-1}$.

As first pointed out by M. Minnaert in 1941, any satisfactory reflectance model must obey the *Principle of Reciprocity*,[8] according to which: *The reflectance is unaffected by an interchange of the directions of incidence and observation*, or in mathematical form

$$\rho(\theta', \phi'; \theta, \phi) = \rho(\theta, \phi; \theta', \phi').$$

Reciprocity is discussed in greater detail in Appendices M and P.

We next consider the *Lommel–Seeliger* formula for the reflectance

$$\rho(\mu_0, \mu) = \frac{2\rho_n}{(\mu_0 + \mu)}, \tag{5.27}$$

where $\rho_n \equiv \rho(1, 1)$ is called the *normal reflectance* in planetary astronomy. This form predicts that the reflectance should be minimum at normal viewing ($\mu = 1$) and maximum at a grazing view angle ($\mu \rightarrow 0$). This behavior is consistent with Minnaert's formula, as can be seen in Fig. 5.4. The Lommel–Seeliger formula also predicts that as the incident angle θ_0 increases from 0° to grazing (90°), ρ should exhibit a larger increase[9] with the off-nadir angle θ. Laboratory experiments confirm this behavior. Experience has shown that Eq. 5.27 applies rather well to most dark surfaces for which $\rho_n < 0.3$.

If we ignore the minor "peaking" effect (to be described shortly) the behavior of the Lommel–Seeliger formula is clearly seen in Fig. 5.5 for two land surface types.[10]

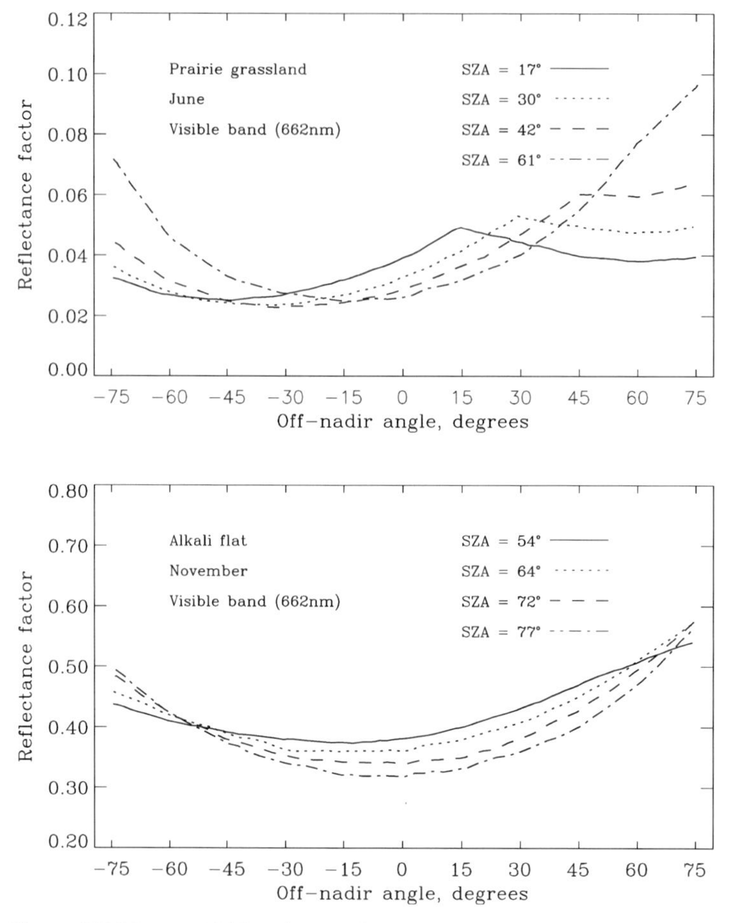

Figure 5.5 Measured bidirectional reflectance functions (BRDFs) for prairie grassland (upper plot) and for an alkali flat (lower plot). ρ is plotted versus the cosine of the off-nadir angle $\mu = \cos\theta$ in the plane of incidence. Each curve corresponds to a different solar zenith angle, θ_0.

5.2.8 The Opposition Effect

The *phase angle*, or *backscattering angle*, α, is defined as the complement of the angle (see Fig. 5.1) between the incident vector $\hat{\Omega}_0(\theta_0, \phi_0)$ and reflected vector $\hat{\Omega}(\theta, \phi)$: $\alpha \equiv \pi - \cos^{-1}(\hat{\Omega}_0 \cdot \hat{\Omega}) = \pi - \cos^{-1}[-\mu|\mu_0| + \sqrt{(1-\mu^2)(1-\mu'^2)}\cos\phi]$. Outer planets are generally viewed at very small values of α, whereas the inner planets are viewed over nearly the entire angular range.

A property of planetary surfaces that is not shared by atmospheres or oceans is the *opposition effect* (also called the *Heiligenschein* or the *hot-spot* phenomenon), an abrupt increase in brightness of the reflected light as $\alpha \to 0$. The lunar surface and many other natural surfaces possess this property. This effect is apparent even to the casual observer during the full moon. At lunar opposition when the Sun is behind the observer, the Moon is perceptably brighter than at other phases. All parts of the disk appear uniformly bright, despite the large spread in incidence and observation angles between 0° and 90°. This phenomenon was first observed by Seeliger in 1860 in reflection from the ring particles of Saturn. He correctly explained it as a result of particle shadowing. Since the scattering elements in solid surfaces are considerably larger than the wavelength of visible light, and are fairly densely packed, the elements cast shadows on each other. The interstices between the particles can be thought of as "tunnels" through which light can penetrate. At large α the interiors of most of the tunnels that the observer views are in shadow, because the light is blocked by the particles making up the walls of the tunnels. However, when α is small, the sides and bottoms of the tunnels are illuminated, resulting in the so-called opposition surge in brightness. Thus the phenomenon may be thought of as "shadow hiding." Note that the tunnels may be thought of as oriented in varying directions, depending upon how the surface is viewed. The effect is always largest when the Sun is "behind the shoulders" of the observer, that is, for small α. The surface can be inclined at any angle away from the observer, although in practice the effect is usually confined to solar incidence angles within 20°. The opposition effect is most pronounced for dark surface materials, since particle shadows are densest for strongly absorbing particles. Lighter materials containing translucent particles scatter light into shadowed regions, thus "softening" the shadows and diminishing the effect. Figure 5.5 illustrates the opposition effect from a prairie grassland.

An analytic formulation of the opposition phenomenon for the lunar soil was first given by B. Hapke in 1963.[11] This scheme has since been modified and extended to describe reflection from vegetative canopies. A simplified version[12] consists of a multiplicative correction to the Lommel–Seeliger formula, such that

$$\rho(\mu_0, \mu, \alpha) = \frac{A}{(\mu_0 + \mu)} \left\{ 1 + B \exp\left[-\frac{1}{h} \tan(\alpha/2) \right] \right\}, \tag{5.28}$$

where A is a normalization constant and $B \geq 0$ is the *opposition enhancement factor*. The *compaction parameter* h is a measure of the angular width of the opposition effect. It is related to the interparticle spacing in the surface layer. Low values of h imply a porous medium. Although h is usually in the range 0.2–0.6, it can be as high as unity. The correction results in a maximum at $\alpha = 0$, whose width is proportional to h. This peaking is clearly seen in Fig. 5.5 for prairie grassland.

5.2.9 Specular Reflection from the Sea Surface

A final example concerns specular reflection, most commonly observed as "sun glint." The delta-function behavior of the specular reflectance implied in Eq. 5.15 is never realized in practice. Even the smallest of surface irregularities in smooth, nonporous surfaces will spread out the peak. This property may be turned to advantage in remote sensing measurements of ocean roughness to infer surface wind speeds. In 1954 Cox and Munk[13] derived an analytic model for the sea-surface reflectance based on Fresnel's equations and the probability distribution for the orientation of scattering facets.

It has since been used to describe the specular features of snow and of the smooth lake beds of volcanic sites. Their expression is

$$\rho(\mu_0, \mu, \phi) = \frac{C}{\mu_0 \mu \mu_n^4} \rho_{\mathrm{F}}(\alpha/2) g(\tan\theta_n). \tag{5.29}$$

Here C is an empirical constant, and $\mu_n = \cos\theta_n$, where $\theta_n = \cos^{-1}[(\mu_0 + \mu)/2\cos(\alpha/2)]$. The angle θ_n is that particular angle between the scattering facet's surface normal and zenith for which specular reflection occurs. The term $\rho_{\mathrm{F}}(\alpha/2)$ is the *Fresnel reflectance* (see Appendix E), and $g(\tan\theta_n)$ is the probability of a specular contribution in the direction (μ, ϕ) given by[14]

$$g(\tan\theta_n) = \frac{1}{\pi\sigma^2} e^{-\tan\theta_n^2/\sigma^2}.$$

Here σ is the root mean square of the slope distribution of the surface, and $\tan\theta_n$ is the tangent of the wave slope whose components are $\partial Z/\partial x$ and $\partial Z/\partial y$.

Figure 5.6 shows the calculated reflected radiance above the surface of a calm ocean and above ocean surfaces that are disturbed by waves driven by winds of 5 knots (9.3 km $\cdot$ hr^{-1}) and 20 knots (37 km $\cdot$ hr^{-1}). Note the presence of the sharp specular peak for a calm sea and the broadening of this peak due to ocean roughness. At the higher wind speeds, there is no sign of the specular peak.[15]

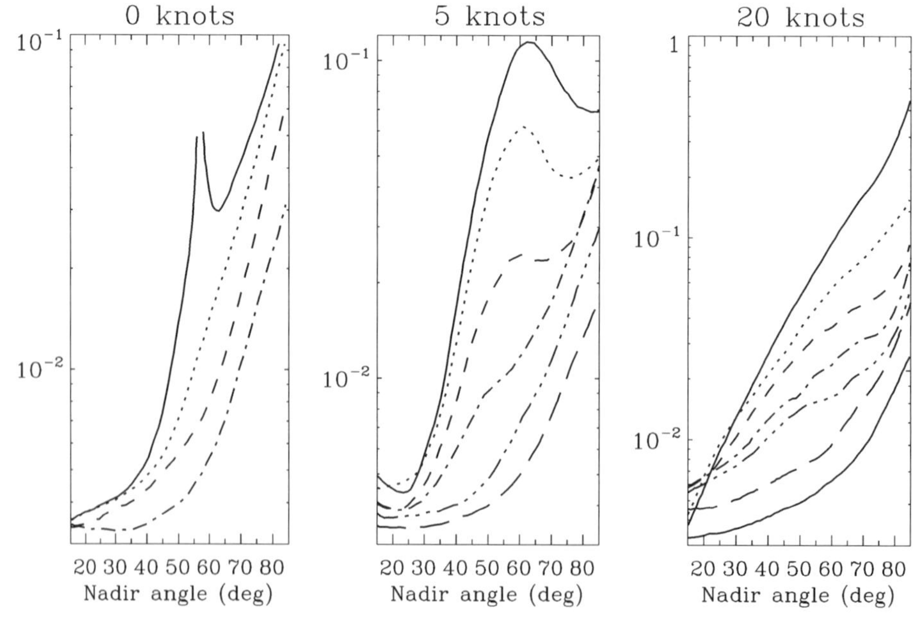

Figure 5.6 Upward intensity above the ocean surface for $\theta_0 = 57°$ at $\lambda = 0.46\ \mu$m, calculated using a Monte Carlo scheme and the Cox–Munk wave slope distribution. The three plots apply to three different assumed wind speeds. Each curve is for a different azimuthal plane, with $\phi = 0$ being the forward direction in the plane of incidence.

5.2.10 Transmission through a Slab Medium

If the surface is partially transparent (for example, a water or ice surface), the transmitted light can be further scattered or absorbed inside the medium. The description of these processes will be taken up in Chapters 6 and 8. If the medium has a finite vertical extent, then some of the scattered light contributes to the transmitted light. The *transmittance* is a measure of how much light reaches the lower surface, either in a direct, attenuated beam or as diffuse (scattered) radiation. For example, when we observe an optically thin cloud, we observe not only the directly transmitted sunlight but, by virtue of the diffuse light, we also view the bottomside of the cloud. Through the Extinction Law the direct or beam transmittance will depend upon the optical path and thus on the scattering and absorption properties of the medium.

We assume a homogeneous, horizontal slab so that the optical path τ_s depends only upon the polar angle θ and the vertical optical thickness (see Fig. 1.6). As in the case of surface reflection, we consider a downward-moving angular beam in the direction $-\hat{\Omega}'$, containing the energy $I_\nu^-(\hat{\Omega}')\cos\theta' d\omega'$, that is incident on a surface whose normal is directed along the positive z axis. The transmitted intensity leaving the bottom surface of the medium in direction $-\hat{\Omega}$ is $dI_{\nu t}^-(\hat{\Omega})$. The spectral *bidirectional transmittance function*, or simply *transmittance*, is defined as the ratio of the transmitted intensity $dI_{\nu t}^-(\hat{\Omega})$ to the incident energy, that is,

$$\mathcal{T}(\nu, -\hat{\Omega}', -\hat{\Omega}) \equiv \frac{dI_{\nu t}^-(\hat{\Omega})}{I_\nu^-(\hat{\Omega}')\cos\theta' d\omega'}.$$

Adding up the contributions to the transmitted intensity in the direction $-\hat{\Omega}$ from beams incident on the surface in all downward directions $-\hat{\Omega}'$, we obtain the total transmitted intensity

$$I_{\nu t}^-(\hat{\Omega}) = \int_- dI_{\nu t}^-(\hat{\Omega}) = \int_- d\omega' \cos\theta' \mathcal{T}(\nu, -\hat{\Omega}', -\hat{\Omega}) I_\nu^-(\hat{\Omega}'). \tag{5.30}$$

The transmitted intensity consists of two parts, the *direct* and *diffuse* components, analogous to the specular and diffuse components of the reflected radiation. The directly transmitted component of the intensity can be found immediately from the Extinction Law, $I_\nu^-(\hat{\Omega}')\mathcal{T}_b(\nu)$, where $\mathcal{T}_b(\nu) \equiv e^{-\tau_s(\nu)}$ is the *beam transmittance*. The diffuse part is given by Eq. 5.30 with $\mathcal{T}$ replaced by $\mathcal{T}_d$. The total transmitted intensity is written

$$I_{\nu t}^-(\hat{\Omega}) = I_\nu^-(\hat{\Omega})e^{-\tau_s(\nu)} + \int_- d\omega' \cos\theta' \mathcal{T}_d(\nu, -\hat{\Omega}', -\hat{\Omega}) I_\nu^-(\hat{\Omega}'),$$

where $\mathcal{T}_d(\nu, -\hat{\Omega}', \hat{\Omega})$ is the *diffuse transmittance*. If the incident beam is collimated so that $I_\nu^-(\hat{\Omega}')$ is given by Eq. 5.14, then

$$I_{\nu t}^-(\hat{\Omega}) = F_\nu^S \delta(\cos\theta - \cos\theta_0)\delta(\phi - \phi_0)e^{-\tau_s(\nu)} + F_\nu^S \cos\theta_0 \mathcal{T}_d(\nu, -\hat{\Omega}_0, -\hat{\Omega})$$

and the corresponding flux becomes

$$F_{\nu t}^{-} = \int_{-} d\omega \cos\theta I_{\nu t}^{-}(\hat{\Omega})$$

$$= F_{\nu}^{s} \cos\theta_0 \left[e^{-\tau_s(\nu)} + \int_{-} d\omega \cos\theta \mathcal{T}_{d}(\nu, -\hat{\Omega}_0, \hat{\Omega}) \right]. \quad (5.31)$$

We may use Eq. 5.31 to define the *flux transmittance*[16] as the ratio of the transmitted flux to the (collimated) incident flux:

$$\mathcal{T}(\nu, -\hat{\Omega}_0, -2\pi) \equiv \frac{F_{\nu t}^{-}}{F_{\nu}^{s} \cos\theta_0} = e^{-\tau_s(\nu)} + \int_{-} d\omega \cos\theta \mathcal{T}_{d}(\nu, -\hat{\Omega}_0, -\hat{\Omega}). \quad (5.32)$$

In IR radiative transfer one frequently encounters the flux transmittance of an isotropic incident radiation field $\mathcal{I}_{\nu}$. A frequency-integrated form of this quantity $\mathcal{T}_{F}$ is extensively used in the theory of transmission of radiative energy within molecular bands (§11.2.1). It is defined as

$$\mathcal{T}_{F}(\nu) \equiv \frac{\int_{2\pi} d\omega \cos\theta \mathcal{I}_{\nu} e^{-\tau(\nu)/\mu}}{\pi \mathcal{I}_{\nu}} = 2 \int_{0}^{1} d\mu \mu e^{-\tau(\nu)/\mu}. \quad (5.33)$$

5.2.11 Spherical, or Bond Albedo

An important property of a planet is its net ability to reflect, transmit, or absorb incident solar energy over its entire disk. To assess this ability we need to evaluate the emergent flux integrated over the planet. For simplicity, we assume that the planet's optical properties are uniform on any spherical surface. Consider the contribution from an annulus centered on the subsolar point, where the solar zenith angle is θ_0 (see Figure 5.7). If the planet's radius is R, then the radius of the annulus is $R \sin\theta_0$, and thus the area of the annulus presented to the Sun's rays is $2\pi R^2 \sin\theta_0 \cos\theta_0 d\theta_0$. The solar energy received by this annulus is therefore F_{ν}^{s} multiplied by the annular area. The energy reflected from this annulus is $2\pi R^2 F_{\nu}^{s} \rho(\nu, -\mu_0, 2\pi) \sin\theta_0 \cos\theta_0 d\theta_0$. Integrating over the entire planetary disk, we obtain the total spectral reflected energy. The total incoming solar energy at frequency ν is simply $\pi R^2 F_{\nu}^{s}$. The spectral *spherical albedo* (or *Bond albedo*) is defined to be the ratio of the disk-integrated reflected energy to the disk-integrated solar energy,[17]

$$\bar{\rho}(\nu) \equiv \frac{2\pi R^2 F_{\nu}^{s} \int_0^1 d\mu_0 \mu_0 \rho(\nu, -\mu_0, 2\pi)}{\pi R^2 F_{\nu}^{s}} = 2 \int_{0}^{1} d\mu_0 \mu_0 \rho(\nu, -\mu_0, 2\pi). \quad (5.34)$$

The *spherical transmittance* $\bar{\mathcal{T}}$ and *spherical absorptance* $\bar{\alpha}$ are also expressed as angular integrations over the corresponding flux transmittance and absorptance

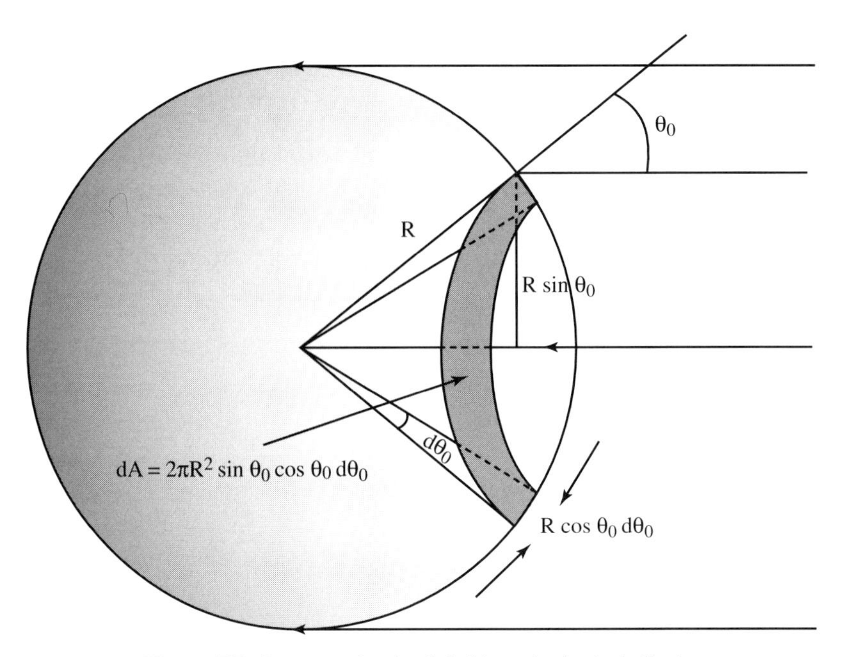

Figure 5.7 Geometry for the definition of spherical albedo.

quantities:

$$\bar{\mathcal{T}}(\nu) = 2\int_0^1 d\mu_0\mu_0\mathcal{T}(\nu, -\mu_0, -2\pi), \tag{5.35}$$

$$\bar{\alpha}(\nu) = 2\int_0^1 d\mu_0\mu_0\alpha(\nu, -\mu_0, 2\pi). \tag{5.36}$$

The frequency-integrated quantities are

$$\bar{\rho} = \int_0^\infty \bar{\rho}(\nu)d\nu, \qquad \bar{\mathcal{T}} = \int_0^\infty \bar{\mathcal{T}}(\nu)d\nu, \qquad \bar{\alpha} = \int_0^\infty \bar{\alpha}(\nu)d\nu.$$

The spherical albedo and spherical transmittance also appear in the problem of calculating the enhancement of the radiation field by a reflecting lower boundary (§6.11).

For planetary media the transmittance, reflectance, absorptance, and emittance are of fundamental importance. Given the photometric properties of the bounding surfaces, and the boundary conditions of illumination, these quantities must be evaluated by solving the radiative transfer equation. Obtaining their solution is one of the principal goals of this book. We now consider absorption and scattering in extended media.

5.3 Absorption and Scattering in Planetary Media

5.3.1 Kirchhoff's Law for Volume Absorption and Emission

To enable us to discuss radiative processes in an extended medium, rather than for a surface, we must define absorption and emission per unit volume rather than per unit area. We start by considering a hohlraum filled with matter throughout the volume, which both scatters and absorbs radiation at frequency ν. Kirchhoff's Law has a different form than that for a surface, since it relates the *thermal emission coefficient* $j_\nu^{\rm th}$ to the absorption coefficient and the Planck function.

Kirchhoff's Law Applied to Volume Emission:

$$j_\nu^{\rm th} = \alpha(\nu) B_\nu(T). \tag{5.37}$$

In accordance with Eq. 2.27 we define the thermal contribution to the source function by $j_\nu^{\rm th}/k(\nu)$:

$$S_\nu^{\rm th} = \frac{\alpha(\nu)}{k(\nu)} B_\nu(T). \tag{5.38}$$

In place of the spectral directional emittance for a surface, we define the spectral *volume emittance* $\epsilon_\nu(\nu, \hat{\Omega}, T)$ as the ratio of the thermal emission per unit volume of the matter under consideration to that of a perfectly "black" material of the same mass and temperature T, $S_\nu^{\rm BB} \equiv B_\nu(T)$. Mathematically,

$$\epsilon_\nu(\nu, \hat{\Omega}, T) = \frac{S_\nu^{\rm th}(\hat{\Omega}, T)}{B_\nu(T)}.$$

Most atmospheric and oceanic absorbers are isotropic emitters, so that the absorption coefficient is independent of angle. Dropping the angular dependence and using Eq. 5.38, we find

$$\epsilon_\nu(\nu, T) = \frac{\alpha(\nu, T)}{k(\nu)}. \tag{5.39}$$

The volume emittance ϵ_ν is proportional to the absorption coefficient α.

A planetary medium is of course far from an artificial closed system such as a hohlraum. It is therefore surprising that within spectral lines in the IR, planetary media radiate approximately as a blackbody, that is, the source function is equal to the Planck function. In addition, as previously discussed, all opaque surfaces (either solid or liquid) obey Kirchhoff's Law, Eq. 5.23. As discussed in the previous chapter, this situation prevails when the collisional rates of excitation/deexcitation of the quantum states are much larger than the corresponding radiative loss rates. Then the populations of these states are determined by the local kinetic temperature of the medium, rather than by the radiation field. In the troposphere thermal IR radiation is in local thermodynamic equilibrium. Note that LTE implies that Eqs. 5.37–5.39 are valid, but the intensity is not equal to the Planck function, as it would be in TE (see Eq. 4.4). At very low atmospheric densities LTE will break down even in the IR, and Eqs. 5.37–5.39 are no longer valid.

However, LTE does not apply to shortwave radiative processes, since the collisional excitation processes are completely ineffective. This is because the kinetic energy involved in thermal collisions ($\sim k_B T$) is much less than the excitation energies associated with visible and UV transitions (E_i). This is illustrated using the two-level atom relationship found in §4.4.2. If we set $h\nu_0 = E_i$ the excitation rate may be found from Eq. 4.38 to be proportional to $e^{-E_i/k_B T} \ll 1$. Thus the collisional quenching rates at shortwave energies greatly exceed the collisional excitation rates. Although sunlight heats the medium because of absorption, followed by collisional quenching, the opposite process (collisional energy converting to radiative energy) is absent. The imbalance of excitation versus quenching essentially uncouples the shortwave radiation from the thermal state of the gas.

As shown in Fig. 5.8 there is little overlap between the radiation spectra of the Sun and the Earth. Therefore, except for applications where the region of overlap (3–4 μm) is of special interest, we may treat the two spectra separately. Another important consideration is the absence of strong absorption by the major atmospheric gases throughout the visible spectrum. The major shortwave interaction is in the UV spectrum below 300 nm where sunlight never reaches the surface, being absorbed in the middle atmosphere by ozone.

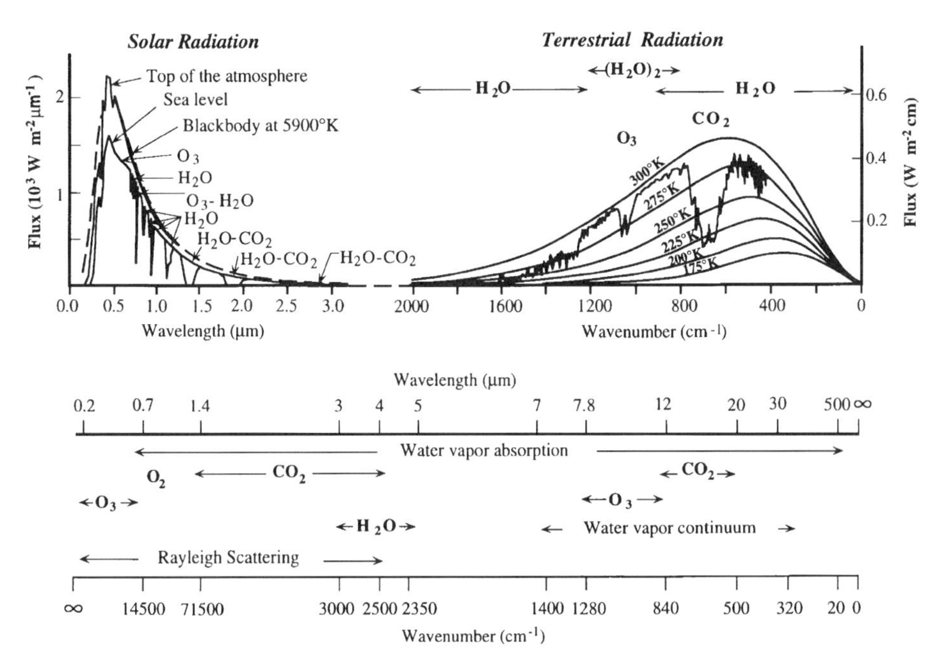

Figure 5.8 Spectral distribution of solar (shortwave) and terrestrial (longwave) radiation fields. Also shown are the approximate shapes and positions of the scattering and absorption features of the Earth's atmosphere.

5.3.2 Differential Equation of Radiative Transfer

We consider conservative scattering in which there is no change in frequency between the incident and the scattered radiation. Also, we assume that the incident radiation is collimated. We return to a formal description of the secondary emission that results from scattering. From Eq. 2.3 the radiative energy that is incident normally on the area dA in the direction $\hat{\Omega}'$ and within the solid angle $d\omega'$, centered, around $\hat{\Omega}'$, in time dt, and within frequency interval $d\nu$, is the fourth-order quantity

$$d^4E' = I_\nu(\hat{\Omega}')\, dA\, dt\, d\nu d\omega'.$$

The radiative energy scattered in *all* directions is $\sigma\, ds\, d^4E'$, where ds is the length of the scattering volume element in the direction normal to dA and σ is the scattering coefficient. We are interested in that fraction of the scattered energy that is directed into the solid angle $d\omega$ centered around the direction $\hat{\Omega}$. This fraction is proportional to $p(\hat{\Omega}', \hat{\Omega})\, d\omega/4\pi$, where $p(\hat{\Omega}', \hat{\Omega})$ is the scattering phase function defined in §3.4. If we multiply the scattered energy by this fraction and then integrate over all incoming directions, we find that the total scattered energy emerging from the volume element $dV = ds\, dA$ in the direction $\hat{\Omega}$ is

$$d^4E = \sigma(\nu) dV\, dt\, d\nu\, d\omega \int_{4\pi} d\omega' \frac{p(\hat{\Omega}', \hat{\Omega})}{4\pi} I_\nu(\hat{\Omega}').$$

We define the *emission coefficient for scattering* as

$$j_\nu^{\rm sc} \equiv \frac{d^4E}{dV dt\, d\nu\, d\omega} = \sigma(\nu) \int_{4\pi} \frac{d\omega'}{4\pi} p(\hat{\Omega}', \hat{\Omega}) I_\nu(\hat{\Omega}').$$

The source function for scattering is thus

$$S_\nu^{\rm sc}(\mathbf{r}, \hat{\Omega}) \equiv \frac{j_\nu^{\rm sc}}{k(\nu)} = \frac{\sigma(\nu)}{k(\nu)} \int_{4\pi} \frac{d\omega'}{4\pi} p(\hat{\Omega}', \hat{\Omega}) I_\nu(\hat{\Omega}'). \tag{5.40}$$

The quantity $\sigma(\nu)/k(\nu)$ appearing in Eq. 5.40 is given a special name, the *single-scattering albedo*, $a(\nu)$. Since $k(\nu) = \sigma(\nu) + \alpha(\nu)$, it is clear that $a(\nu) \leq 1$. We interpret a as the *probability that a photon will be scattered, given an extinction event*. Given that an interaction of radiation has occurred (either a scattering or an absorption, which we call an extinction event), the quantity $(1-a)$ is the *probability of absorption per extinction event*. The quantity $(1 - a)$ is also called the *co-albedo*. If thermal emission is involved, $(1 - a)$ is the *volume emittance* ϵ_ν (§4.4.2). The complete time-independent radiative transfer equation, which includes both scattering and absorption, is therefore

Radiative transfer equation including multiple scattering and absorption

$$\frac{dI_\nu}{d\tau_{\rm s}} = -I_\nu + [1 - a(\nu)]B_\nu(T) + \frac{a(\nu)}{4\pi} \int_{4\pi} d\omega'\, p(\hat{\Omega}', \hat{\Omega}) I_\nu(\hat{\Omega}'). \tag{5.41}$$

Equation 5.41 is a generalization of Eq. 2.28 to include both scattering (Eq. 5.40) and thermal emission (Eq. 5.38). An inspection of this equation reveals the major mathematical complexity of radiative transfer theory, namely, that it involves the solution of an *integro-differential equation*. We will show in the next section that there exists a formal solution to Eq. 5.41, which, in the absence of multiple scattering, represents a practical solution. This formal solution will help us understand the radiative transfer process and provide guidance for devising methods of numerical solution.

5.4 Solution of the Radiative Transfer Equation for Zero Scattering

In the limit of no scattering, the radiation is affected only by absorption and emission processes. The radiative transfer equation 5.41 simplifies to

$$\frac{dI_\nu}{d\tau_s} = -I_\nu + B_\nu(T), \tag{5.42}$$

where the source function $S_\nu = B_\nu(T)$ may be considered to be a known function of position through the temperature T. This problem is classified as a *local* one, since the source function does not depend upon the distant properties of the medium. This problem is much easier to solve than the more general *nonlocal* problem. Note that the optical path τ_s is measured along the beam direction, taken to be a straight line since we are ignoring refraction. A solution of Eq. 5.42, satisfying the appropriate boundary conditions, yields the radiation field I_ν at all positions τ_s along the beam direction. The solution will clearly vary with the frequency ν, the temperature T, and the optical properties of the medium, embodied in the absorption coefficient $\alpha(\nu)$, which in general may vary from point to point in the medium.

The problem of interest is that of an inhomogeneous medium in LTE, which at each point radiates thermal emission according to the Planck function at the local temperature. The medium may have arbitrarily shaped boundaries. It is illuminated by a beam of radiation in the direction $\hat{\Omega}$ at the boundary point P_1 (see Fig. 5.9). We want to find the solution for the intensity $I_\nu(P_2, \hat{\Omega})$ that emerges from the medium at point P_2 along the same direction. The solution for this particular direction of the incident beam may be considered to be an *elementary solution*. A completely general distribution of intensity in angle and frequency can be obtained by repeating the elementary solution $I_\nu(P_2, \hat{\Omega})$ for all incident beams and for all frequencies. The elementary solution will be found to be a sum of two terms: (a) the incident intensity $I_\nu(P_1, \hat{\Omega})$ attenuated by the intervening optical path along P_1P_2 and (b) a term consisting of the contributions from the internal (thermal) sources at all points P between P_1 and P_2 and attenuated by the intervening optical path along PP_2.

Equation 5.42 is readily integrated by using an *integrating factor*, which in this case is e^τ. For now we drop the subscripts s and ν to avoid a burdensome notation. After

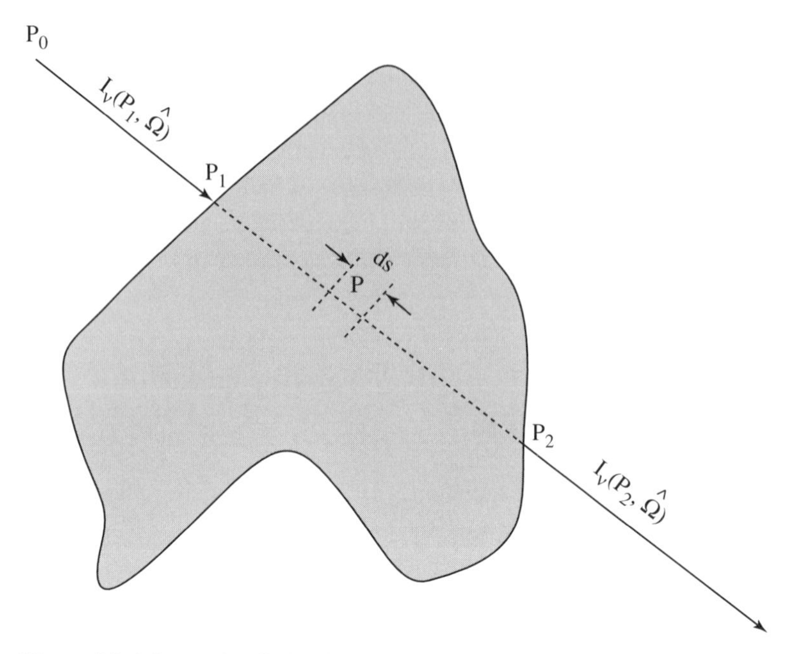

Figure 5.9 A beam of radiation is incident on an absorbing/emitting region at the boundary point P_1. It is attenuated along the path $P_1 P_2$ and emerges at the point P_2. The propagation direction of the beam is denoted by $\hat{\Omega}$. In addition, thermal emission adds to the beam at all points within the medium.

multiplying by e^τ, Eq. 5.42 may be written as a perfect differential

$$\frac{dI}{d\tau}e^\tau + Ie^\tau = \frac{d}{d\tau}(Ie^\tau) = Be^\tau. \tag{5.43}$$

Since we are ignoring refraction, we shall integrate along a straight path from the point P_1 to the point P_2, the latter point being at the boundary of the medium (see Fig. 5.9). The optical path from the point P_1 to an intermediate point P is given by

$$\tau(P_1, P) = \int_{P_1}^{P} \alpha\, ds = \int_{0}^{P} \alpha\, ds - \int_{0}^{P_1} \alpha\, ds \equiv \tau(P) - \tau(P_1).$$

Note that the optical path may be measured from an arbitrary reference point P_0 outside the medium, even though $\tau(P_0, P_1) = 0$. It increases monotonically with distance in the medium because of the addition of absorbing matter along the beam.

Integration of Eq. 5.43 along the straight line from P_1 to P_2 yields

$$\int_{\tau(P_1)}^{\tau(P_2)} dt \frac{d}{dt}(Ie^t) = I[\tau(P_2)]e^{\tau(P_2)} - I[\tau(P_1)]e^{\tau(P_1)} = \int_{\tau(P_1)}^{\tau(P_2)} dt B(t)e^t,$$

where the integration variable t stands for the optical path τ. Solving for the intensity at the point P_2, we find

$$I[\tau(P_2)] = I[\tau(P_1)]e^{-\tau(P_2)+\tau(P_1)} + \int_{\tau(P_1)}^{\tau(P_2)} dt B(t)e^{t-\tau(P_2)}$$

$$= I[\tau(P_1)]e^{-\tau(P_1,P_2)} + \int_{\tau(P_1)}^{\tau(P_2)} dt B(t)e^{-t(P,P_2)} \tag{5.44}$$

$$= I[\tau(P_1)]\mathcal{T}_b(P_1, P_2) + \int_{\mathcal{T}_b(P_1,P_2)}^{1} d\mathcal{T}_b(P, P_2)B(t). \tag{5.45}$$

In the last equation, we have expressed the result in terms of the beam transmittance, $\mathcal{T}_b$ (§5.2.10). This result has a direct physical interpretation. The intensity at the point P_2 emerging from the medium in the beam direction $P_1 P_2$ consists of two parts. The first term is the contribution from the intensity incident at the boundary point P_1, which has been attenuated by the beam transmittance. The surviving term is the contribution from thermal emission from those parts of the medium that lie along the beam, weighted by the appropriate beam transmittance $e^{-t(P,P_2)}$.

It is clear that the above solution is valid whether or not the points P_1 and P_2 lie at the boundaries of the medium (both P_1 and P_2 may be interior points). We chose these points to be at the boundaries so that $I[\tau(P_2)]$ would represent the intensity escaping from the medium at point P_2 due to the beam incident on the medium from the outside at point P_1 and the internal sources (due to thermal emission) between points P_1 and P_2. $I[\tau(P_2)]$ is called the *emergent intensity*.

Example 5.4 Isothermal Medium: Arbitrary Geometry

We assume that the temperature and absorption cross section α_n are constant throughout the medium. Then we can easily perform the integration in Eq. 5.45. We redefine the origin of the optical path scale so that it coincides with the point P_1 (i.e., $\tau(P_1) = 0$). Then

$$I[\tau(P_2)] = I[\tau(P_1)]e^{-\tau(P_2)} + B \int_0^{\tau(P_2)} dt\, e^{-[\tau(P_2)-t]}$$

$$= I[\tau(P_1) = 0]e^{-\tau(P_2)} + B[1 - e^{-\tau(P_2)}]. \tag{5.46}$$

Consider the behavior of the second term with optical path. If the medium is optically thin in a given direction (i.e., $\tau(P_2) \ll 1$), then the second term is $\approx B\tau(P_2)$, to first order in $\tau(P_2)$. In this limit the contribution from the medium depends linearly upon optical path, or in other words, upon the path number of emitting (or absorbing) molecules $\mathcal{N}(0, P_2)$:

$$\tau_s(P_2) = \int_0^{P_2} \alpha\, ds = \alpha_n \int_0^{P_2} n(s) ds \equiv \alpha_n \mathcal{N}(0, P_2),$$

where $\mathcal{N}(0, P_2)$ is (also) the path number of absorbing molecules. This dual interpetation of $\mathcal{N}(0, P_2)$ follows from our implicit use of Kirchhoff's Law in Eq. 5.42.

If we take both the absorption and scattering coefficient to be zero, then $\tau = 0$ and Eq. 5.46 shows that the intensity is constant along a given beam direction everywhere in the medium. A radiation field in such a transparent medium may be very anisotropic, since each beam may have its own intensity or constant of integration, while along each particular beam direction the intensity is constant. This is just *Theorem I* (see Chapter 2), which has already been proven in a more intuitive manner.

In the opposite case of an optically thick medium, for which $\tau(P_2) \gg 1$, the total intensity is $I_\nu(P_2) = B_\nu(T)$. In this case the medium radiates like a blackbody at all frequencies and in all directions; that is, it is in a state of thermodynamic equilibrium.

5.4.1 Solution with Zero Scattering in Slab Geometry

The most common geometry in the theory of radiative transfer is that of a *plane-parallel* medium, or a *slab*. This geometry is appropriate to both planetary atmospheres and oceans. The force of gravitation imposes a density stratification, so that the medium properties tend to vary primarily in the vertical direction. In many cases, we can ignore the horizontal variation in the medium. We will distinguish the vertical optical path τ (which we hereafter call the *optical depth*) from the slant optical path τ_s. It is convenient to measure the optical depth along the vertical direction downward from the "top" of the medium.[18] The relationship between the vertical and slant optical paths is

$$\tau(z) \equiv \int_z^\infty dz'\, k(z') = \tau_s|\cos\theta| = \tau_s|u|,$$

where $\theta = \cos^{-1} u$ is the polar angle of the beam direction. Since it is used so frequently, we assign a special symbol, u, to $\cos\theta$, so that $d\tau_s = -k\, dz/|u|$. The extinction optical depth can also be written in terms of the vertical column number, $\mathcal{N}$, and the extinction cross section,

$$\tau(z) = k_n \int_z^\infty dz'\, n(z') \equiv k_n \mathcal{N}(z),$$

or in terms of the mass extinction coefficient

$$\tau(z) = k_m \int_z^\infty dz' \rho(z') \equiv k_m \mathcal{M}(z),$$

where $\mathcal{M}(z)$ is the mass of the radiatively active material in a vertical column of unit cross-sectional area. Analogous expressions apply to the absorption and scattering optical depths.

If R is the radial distance from the center of the planet and H is the vertical scale length of the absorber, the slab approximation is valid if $H/R \ll 1$ and if θ is not too close to 90°. If these conditions are violated, it is necessary to take into account the curvature of the atmospheric layers.[19]

5.4.2 Half-Range Quantities in a Slab Geometry

The *half-range intensities* are defined by

$$\begin{aligned} I_\nu^+(\tau,\theta,\phi) &\equiv I_\nu(\tau,\theta \le \pi/2,\phi), \\ I_\nu^-(\tau,\theta,\phi) &\equiv I_\nu(\tau,\theta > \pi/2,\phi). \end{aligned} \tag{5.47}$$

These definitions may also be expressed in terms of $u = \cos\theta \ge 0$ and $u = \cos\theta < 0$. It will become apparent later that the variable $\mu = |u|$ makes the notation for slant optical depth simple and straightforward.

The radiative flux is also defined in terms of half-range quantities. From Eq. 2.5, we have

$$\begin{aligned} F_\nu^+(\tau) &= \int_+ d\omega \cos\theta\, I_\nu^+(\hat{\Omega}) = \int_0^{2\pi} d\phi \int_0^{\pi/2} d\theta \sin\theta \cos\theta\, I_\nu^+(\tau,\theta,\phi) \\ &= \int_0^{2\pi} d\phi \int_0^1 d\mu\, \mu I_\nu^+(\tau,\mu,\phi), \end{aligned} \tag{5.48}$$

$$\begin{aligned} F_\nu^-(\tau) &= -\int_- d\omega \cos\theta\, I_\nu^-(\hat{\Omega}) = -\int_0^{2\pi} d\phi \int_{\pi/2}^{\pi} d\theta \sin\theta \cos\theta\, I_\nu^-(\tau,\theta,\phi) \\ &= \int_0^{2\pi} d\phi \int_0^1 d\mu\, \mu I_\nu^-(\tau,\mu,\phi). \end{aligned} \tag{5.49}$$

The downward flux F^- is thus seen to be a positive quantity. From Eq. 2.6 the net flux is

$$\begin{aligned} F_\nu(\tau) &= \int_{4\pi} d\omega \cos\theta\, I_\nu(\hat{\Omega}) = \int_+ d\omega \cos\theta\, I_\nu^+(\hat{\Omega}) + \int_- d\omega \cos\theta\, I_\nu^-(\hat{\Omega}) \\ &= F_\nu^+(\tau) - F_\nu^-(\tau). \end{aligned} \tag{5.50}$$

Note that the net flux in slab geometry is positive if the net radiative energy flows in the upward (positive) direction, or toward increasing z and decreasing τ.

In the limit of no scattering the radiative transfer equations for the half-range intensities become

$$\mu \frac{dI_\nu^+(\tau,\mu,\phi)}{d\tau} = I_\nu^+(\tau,\mu,\phi) - B_\nu(\tau), \tag{5.51}$$

$$-\mu \frac{dI_\nu^-(\tau,\mu,\phi)}{d\tau} = I_\nu^-(\tau,\mu,\phi) - B_\nu(\tau). \tag{5.52}$$

Note that the independent variable is now the absorption optical depth, measured downwards from the "top" of the medium. This accounts for the difference in sign of the left-hand sides of Eqs. 5.51 and 5.52.

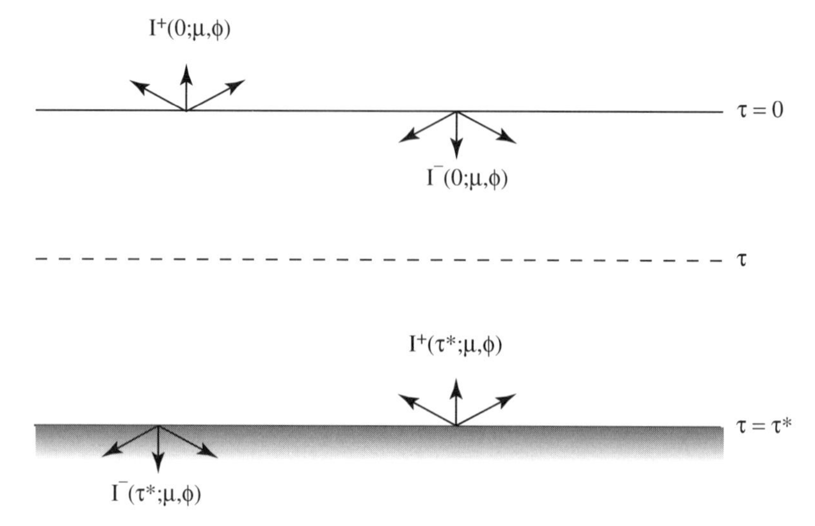

Figure 5.10 Half-range intensities in a slab geometry. The optical depth variable τ is measured downward from the "top" of the medium ($\tau = 0$) to the "bottom" ($\tau = \tau^*$). μ is equal to the absolute value of the cosine of the angle θ, the polar angle of the propagation vector $\hat{\Omega}$.

5.4.3 Formal Solution in a Slab Geometry

We first obtain a formal solution of Eq. 5.52. Choosing the integrating factor to be $e^{\tau/\mu}$, we obtain

$$\frac{d}{d\tau}\left(I_\nu^- e^{\tau/\mu}\right) = \left(\frac{dI_\nu^-}{d\tau} + \frac{1}{\mu}I_\nu^-\right)e^{\tau/\mu} = \frac{B_\nu(\tau)}{\mu}e^{\tau/\mu}. \tag{5.53}$$

In accordance with the physical picture in Fig. 5.10 of downgoing beams that start at the "top" and interact with the medium in the slab on their way downward, we integrate Eq. 5.53 along the vertical from the "top" ($\tau = 0$) to the "bottom" ($\tau = \tau^*$) of the medium to obtain

$$\int_0^{\tau^*} d\tau' \frac{d}{d\tau'}\left(I_\nu^- e^{\tau'/\mu}\right) = I_\nu^-(\tau^*, \mu, \phi)e^{\tau^*/\mu} - I_\nu^-(0, \mu, \phi)$$

$$= \int_0^{\tau^*} \frac{d\tau'}{\mu} e^{\tau'/\mu} B_\nu(\tau').$$

Solving for $I_\nu^-(\tau^*, \mu, \phi)$, we find

$$I_\nu^-(\tau^*, \mu, \phi) = I_\nu^-(0, \mu, \phi)e^{-\tau^*/\mu} + \int_0^{\tau^*} \frac{d\tau'}{\mu} B_\nu(\tau')e^{-(\tau^*-\tau')/\mu} \tag{5.54}$$

for the intensity emerging from the bottom of the slab. For an interior point, $\tau < \tau^*$, we integrate from 0 to τ. The solution is easily found by replacing τ^* by τ in Eq. 5.54:

$$I_\nu^-(\tau, \mu, \phi) = I_\nu^-(0, \mu, \phi)e^{-\tau/\mu} + \int_0^\tau \frac{d\tau'}{\mu} B_\nu(\tau')e^{-(\tau-\tau')/\mu}. \quad (5.55)$$

We now turn to the solution for the upper-half-range intensity. The integrating factor for Eq. 5.51 is $e^{-\tau/\mu}$, which yields

$$\frac{d}{d\tau}\left(I_\nu^+ e^{-\tau/\mu}\right) = \left(\frac{dI_\nu^+}{d\tau} - \frac{1}{\mu}I_\nu^+\right)e^{-\tau/\mu} = -\frac{B_\nu(\tau)}{\mu}e^{-\tau/\mu}.$$

In this case the physical picture in Fig. 5.10 involves upgoing beams that start at the bottom and interact with the medium on their way upward. Therefore, we integrate from the "bottom" to the "top" of the medium:

$$\int_{\tau^*}^0 d\tau' \frac{d}{d\tau'}\left(I_\nu^+ e^{-\tau'/\mu}\right) = I_\nu^+(0, \mu, \phi) - I_\nu^+(\tau^*, \mu, \phi)e^{-\tau^*/\mu}$$

$$= -\int_{\tau^*}^0 \frac{d\tau'}{\mu} e^{-\tau'/\mu} B_\nu(\tau') = \int_0^{\tau^*} \frac{d\tau'}{\mu} e^{-\tau'/\mu} B_\nu(\tau').$$

Solving for $I_\nu^+(0, \mu, \phi)$, we find

$$I_\nu^+(0, \mu, \phi) = I_\nu^+(\tau^*, \mu, \phi)e^{-\tau^*/\mu} + \int_0^{\tau^*} \frac{d\tau'}{\mu} e^{-\tau'/\mu} B_\nu(\tau').$$

To find the intensity at an interior point τ, we integrate from τ^* to τ, to obtain

$$I_\nu^+(\tau, \mu, \phi) = I_\nu^+(\tau^*, \mu, \phi)e^{-(\tau^*-\tau)/\mu} + \int_\tau^{\tau^*} \frac{d\tau'}{\mu} e^{-(\tau'-\tau)/\mu} B_\nu(\tau'). \quad (5.56)$$

On comparing Eq. 5.45 with Eq. 5.55 and Eq. 5.56, we find that the mathematical structures, and therefore the physical interpretations, are identical. Thus, the slab solutions could have been obtained directly from the general solution, Eq. 5.45. All that is necessary is to replace τ_s with τ/μ, and change variables from τ_s to τ. Note also that our dummy variable in the slab geometry is τ', the (variable) optical depth, to distinguish it clearly from the slant optical path t. To obtain a good physical understanding of the radiative transfer process the integration should be thought of as directed along the beam direction. Note, however, that from a mathematical point of view, the integration direction is irrelevant: Either direction gives the same answer.

It is interesting to consider the situation where $\mu \to 0$, that is, where the line of sight traverses an infinite distance parallel to the slab. It is intuitively clear that since

$B_\nu(\tau)$ is constant in Eq. 5.55 and 5.56

$$I_\nu^{\pm}(\tau, \mu = 0, \phi) = B_\nu(\tau). \tag{5.57}$$

5.5 Gray Slab Medium in Local Thermodynamic Equilibrium

It is instructive to study the role of infrared radiation in a simple climate model, in which the rather drastic assumption is made that the optical depth is independent of frequency. This is the *gray approximation*, which has played a very important historical role in the subject and is still of great pedagogic value. If scattering is ignored, the source function is simply the Planck function B_ν, and the solutions to the monochromatic transfer equation are given by Eqs. 5.55, 5.56, and 5.57. The frequency-integrated intensity is given by $I(\tau, \mu, \phi) = \int d\nu I_\nu(\tau, \mu, \phi)$, where the limits of the integration include the thermal IR wavelengths longer than $\sim 3\ \mu$m (see Figs. 4.6 and 5.8). From Eqs. 4.6 and 4.7, $\int d\nu B_\nu(\tau, \mu, \phi) = \sigma_B T^4/\pi$. Because the atmosphere receives no appreciable IR radiation from space, we set $I^-(0, \mu, \phi) = 0$. If we assume that the lower boundary intensity is determined by an isotropically emitting black surface of temperature T_s, then we set $I^+(\tau^*, \mu, \phi) = \sigma_B T_s^4/\pi$. The frequency-integrated intensities are then given by

$$I^+(\tau, \mu, \phi) = \frac{\sigma_B T_s^4}{\pi} e^{-(\tau^* - \tau)/\mu} + \int_\tau^{\tau^*} \frac{d\tau'}{\mu} \frac{\sigma_B T^4(\tau')}{\pi} e^{-(\tau' - \tau)/\mu},$$

$$I^-(\tau, \mu, \phi) = \int_0^\tau \frac{d\tau'}{\mu} \frac{\sigma_B T^4(\tau')}{\pi} e^{-(\tau - \tau')/\mu}. \tag{5.58}$$

The corresponding fluxes are obtained by integrating the (cosine-weighted) intensity over solid angle in each hemisphere (see Eqs. 5.48 and 5.49):

$$F^+(\tau) = 2\sigma_B T_s^4 E_3(\tau^* - \tau) + 2 \int_\tau^{\tau^*} d\tau' \sigma_B T^4(\tau') E_2(\tau' - \tau), \tag{5.59}$$

$$F^-(\tau) = 2 \int_0^\tau d\tau' \sigma_B T^4(\tau') E_2(\tau - \tau'). \tag{5.60}$$

Here we define the *exponential integral* of order n for $n > 0$ and $x \geq 0$ as

$$E_n(x) \equiv \int_0^1 d\mu \mu^{n-2} e^{-x/\mu} = \int_1^\infty \frac{dt}{t^n} e^{-tx}. \tag{5.61}$$

These functions cannot be analytically integrated, except in certain limits. However, fast computational software programs are available for evaluating the E_n functions.

In this regard, they can be considered no more difficult to handle than the trigonometric functions. For more information we refer the interested reader to the literature.[20] In this book we will make use of only two of their interesting mathematical properties, which we list below:

$$E_n(0) = \begin{cases} +\infty & \text{for } n = 1, \\ \frac{1}{n-1} & \text{for } n = 2, 3, 4, \ldots, \end{cases}$$
$$\frac{d}{dx}E_n(x) = \begin{cases} -e^{-x}/x & \text{for } n = 1, \\ -E_{n-1}(x) & \text{for } n = 2, 3, 4, \ldots. \end{cases} \tag{5.62}$$

5.6 Formal Solution Including Scattering and Emission

If we cannot neglect scattering, the source function is written (see Eq. 5.41)

Source function due to thermal emission and multiple scattering

$$S(\tau, \hat{\Omega}) = [1 - a(\tau)]B(\tau) + \frac{a(\tau)}{4\pi} \int_{4\pi} d\omega' p(\tau, \hat{\Omega}', \hat{\Omega}) I(\tau, \hat{\Omega}'). \tag{5.63}$$

We note that the independent variable is the extinction optical depth, which is the sum of the absorption and scattering optical depths. The source function is generally a function of the direction $\hat{\Omega}$ of the "emitted" beam and also a function of the local intensity distribution. The general radiative transfer equation is

$$\mu \frac{dI(\tau, \hat{\Omega})}{d\tau} = -I(\tau, \hat{\Omega}) + S(\tau, \hat{\Omega}). \tag{5.64}$$

The method of the integrating factor may be used in exactly the same manner as in the previous section. We can therefore write the formal solution of Eq. 5.64 by replacing B with S in Eq. 5.45:

Intensity in terms of an integration over the source function

$$I[\tau(P_2), \hat{\Omega}] = I[\tau(P_1), \hat{\Omega}]e^{-\tau(P_1, P_2)} + \int_{\tau(P_1)}^{\tau(P_2)} dt\, S(t, \hat{\Omega}) e^{-t(P, P_2)}. \tag{5.65}$$

We stress that this solution is only a *formal* solution, since (in contrast to Eq. 5.45) the source function is unknown, and in fact it depends upon the radiation field, as seen in Eq. 5.63. The importance of this solution is that it emphasizes that, apart from boundary terms, *a knowledge of the source function is tantamount to knowledge of the complete solution of the radiative transfer problem.* This follows from the fact that finding I is reduced to evaluating the integral in Eq. 5.65 by numerical quadrature.[21] Furthermore, there is an important class of radiative transfer problems in which the source function is approximately independent of angle. Then we need solve for the unknown $S(\tau)$ having only one independent variable (τ), rather than two or even three variables. Finally, there are some problems in which we are not interested in the intensity, and the source function is itself the unknown of interest.

In slab geometry the solutions to Eqs. 5.63 and 5.65 are given by Eqs. 5.55, 5.56, and 5.57:

$$I^-(\tau,\mu,\phi) = I^-(0,\mu,\phi)e^{-\tau/\mu} + \int_0^{\tau} \frac{d\tau'}{\mu} S(\tau',\mu,\phi)e^{-(\tau-\tau')/\mu}, \tag{5.66}$$

$$I^+(\tau,\mu,\phi) = I^+(\tau^*,\mu,\phi)e^{-(\tau^*-\tau)/\mu} + \int_{\tau}^{\tau^*} \frac{d\tau'}{\mu} S(\tau',\mu,\phi)e^{-(\tau'-\tau)/\mu}, \tag{5.67}$$

$$I^{\pm}(\tau,\mu=0,\phi) = S(\tau,\mu=0,\phi). \tag{5.68}$$

The source function in slab geometry is easily derived from Eq. 5.63, where we express the radiation field in terms of the half-range intensities:

$$S(\tau,\mu,\phi) = (1-a)B(\tau) + \frac{a}{4\pi}\int_0^{2\pi} d\phi' \int_0^1 d\mu' p(\mu',\phi';\mu,\phi)I^+(\tau,\mu',\phi') + \frac{a}{4\pi}\int_0^{2\pi} d\phi' \int_0^1 d\mu' p(-\mu',\phi';\mu,\phi)I^-(\tau,\mu',\phi'). \tag{5.69}$$

This very important set of equations constitutes the basis for much of the analysis in future chapters. An important point is that the convergence of the integrals in the above equations requires the following condition at large optical depths:

$$\lim_{\tau\to\infty} S(\tau)e^{-\tau} \to 0. \tag{5.70}$$

Equations 5.63 and 5.65 are two coupled equations in the unknowns S and I. If we were to substitute one equation into the other we would arrive at two new equations for either S or I separately. The rather complicated results are usually simplified by various assumptions and serve as a starting point for many radiative transfer techniques.[22] An example of such an *integral equation* is discussed in §7.2.1.

Example 5.5 Physical Basis for the Lommel–Seeliger Law

Here we illustrate the use of the formal solution of the radiative transfer equation (Eqs. 5.66, 5.67, and 5.69) in deriving the reflectance of an idealized flat surface illuminated by a collimated beam. The solid is modeled as a homogeneous, semi-infinite ($\tau^* \to \infty$) slab medium. Even though the medium is optically infinite, because of the dense packing of the scatterers, the penetration of photons is actually quite shallow in real space. The surface is of course located at an optical depth $\tau = 0$. The particles of the medium are assumed to be independent scatterers of single-scattering albedo a. The particles scatter according to a phase function p. The absorption is assumed to dominate the scattering (a is small), so that only the singly scattered photons leave the surface.

Under the above assumptions, the reflected intensity is a simplified form of Eq. 5.67,

$$I^+(0,\mu,\phi) = \int_0^\infty \frac{d\tau'}{\mu} S(\tau',\mu,\phi) e^{-\tau'/\mu}, \tag{5.71}$$

where the source function S is obtained from Eq. 5.69. To evaluate S and thus I^+, we note that we need to determine the two unknown half-range intensities, I^+ and I^-, throughout the medium. This circularity is always present in scattering theory, and to solve the problem generally requires methods not yet discussed in the book. However, in this simple example, we have simplified the problem in order to specify the radiation field explicitly. The single-scattering assumption permits us to replace the intensities on the right-hand side of Eq. 5.69 with the incident radiation field. In this case I^- is the solar intensity, evaluated within the medium from the extinction law, $I^-(\tau,\mu',\phi') = F^s\delta(\mu'-\mu_0)\delta(\phi'-\phi_0)e^{-\tau/\mu'}$. The term involving I^+ within the medium can be neglected, since it represents higher-order scattering. Inserting this result into Eq. 5.69 and ignoring thermal emission, we find

$$S(\tau,\mu,\phi) = \frac{a}{4\pi} p(-\mu_0,\phi_0;\mu,\phi) F^s e^{-\tau/\mu_0}. \tag{5.72}$$

Inserting this result into Eq. 5.71 we find

$$\begin{aligned} I^+(0,\mu,\phi) &= \frac{a}{4\pi} p(-\mu_0,\phi_0;\mu,\phi) F^s \int_0^\infty \frac{d\tau'}{\mu} e^{-\tau'/\mu_0} e^{-\tau'/\mu} \\ &= \frac{a}{4\pi(1+\mu/\mu_0)} p(-\mu_0,\phi_0;\mu,\phi) F^s. \end{aligned} \tag{5.73}$$

The reflectance (BRDF) is given by

$$\rho(-\mu_0,\phi_0;\mu,\phi) = \frac{I^+(0,\mu,\phi)}{\mu_0 F^s} = \frac{a}{4\pi} \frac{p(-\mu_0,\phi_0;\mu,\phi)}{(\mu+\mu_0)}. \tag{5.74}$$

On setting $\rho_n = a/8\pi$ and assuming isotropic scattering ($p = 1$) we recover the Lommel–Seeliger Law (Eq. 5.27). Thus this originally empirical law is seen to have an underlying physical basis, applying to a dark surface whose particles scatter isotropically. It is shown in Problem 9.7 that the above result may be generalized to include multiple scattering. This results in a more general expression for ρ, which applies equally well to dark and bright surfaces.

5.7 Radiative Heating Rate

We now examine the rate at which radiation exchanges energy with matter. From Eq. 2.16, the differential change of energy over the distance ds along a beam is

$$\delta(d^4E) = dI_\nu \, dA \, dt \, d\nu \, d\omega.$$

Dividing this expression by $ds dA = dV$, and also by $d\nu dt$, we obtain the time rate of change in radiative energy per unit volume per unit frequency, due to the change in intensity for beams within the solid angle $d\omega$. Since there is (generally) incoming radiation from all directions, the total change in energy per unit frequency per unit time per unit volume is

$$\int_{4\pi} d\omega \frac{dI_\nu}{ds} = \int_{4\pi} d\omega (\hat{\Omega} \cdot \nabla I_\nu),$$

where we have used the alternate form for the directional derivative (see Eq. 2.28). The *spectral radiative heating rate* $\mathcal{H}_\nu$ is (minus) the rate of change of the radiative energy per unit volume,

$$\mathcal{H}_\nu = -\int_{4\pi} d\omega (\hat{\Omega} \cdot \nabla I_\nu).$$

The net radiative heating rate $\mathcal{H}$ is

$$\mathcal{H} = -\int_0^\infty d\nu \int_{4\pi} d\omega (\hat{\Omega} \cdot \nabla I_\nu). \tag{5.75}$$

In a slab geometry the radiative heating rate is written[23]

$$\mathcal{H} = -\int_0^\infty d\nu \frac{\partial F_\nu}{\partial z} = -2\pi \int_0^\infty d\nu \int_{-1}^{+1} du\, u \frac{\partial I_\nu}{\partial z}, \tag{5.76}$$

where $F_\nu = F_\nu^+ - F_\nu^-$ is the radiative flux in the z direction (Eq. 5.50).

5.7.1 Generalized Gershun's Law

Substituting for the term $dI_\nu/d\tau_s = -u dI_\nu/d\tau$ in Eq. 5.76 from the radiative transfer equation (Eq. 5.41) and using $k(\nu) = \sigma(\nu) + \alpha(\nu)$, we obtain

$$\mathcal{H}_\nu = 4\pi\alpha(\nu)\bar{I}_\nu + 4\pi\sigma(\nu)\bar{I}_\nu - 4\pi\alpha(\nu)B_\nu(T) \\ - \sigma(\nu) \int_{4\pi} d\omega' \int_{4\pi} \frac{d\omega}{4\pi} p(\hat{\Omega}', \hat{\Omega}) I_\nu(\hat{\Omega}'),$$

where $\bar{I}_\nu$ is the angular average of the intensity (Eq. 2.7). We also used the isotropic property of $B_\nu(T)$. Employing the normalization condition for the phase function (Eq. 3.24), we see that the inner integral is unity. Thus the two scattering terms cancel. Integration over frequencies therefore yields the result that

the radiative heating rate is the rate at which radiative energy is absorbed, less the rate at which radiative energy is emitted.

$$\mathcal{H} = -\nabla \cdot \mathbf{F} = 4\pi \int_0^\infty d\nu \alpha(\nu) \bar{I}_\nu - 4\pi \int_0^\infty d\nu \alpha(\nu) B_\nu(T). \tag{5.77}$$

When internal emission is absent, Eq. 5.77 is known as *Gershun's Law* in ocean optics.

When $\mathcal{H} = 0$, the volume absorption rate exactly balances the volume emission rate. This may happen locally at points where the net heating rate happens to change sign, but if the entire medium experiences this balance, we have the condition of *planetary radiative equilibrium*. Clearly, if there is no absorption anywhere in the medium, then $\mathcal{H} = 0$ everywhere. In a slab medium, radiative equilibrium implies that $\partial F/\partial z = 0$ and thus $F =$ constant.

5.7.2 Warming Rate, or the Temperature Tendency

When $\mathcal{H} \neq 0$ an imbalance of heating occurs. Then the second law of thermodynamics dictates a change in the temperature of the medium, which will act to eliminate the heating excess or deficit. To examine the response of the medium to the heating imbalance we will ignore all other diabatic heating mechanisms except radiative processes.

We consider the instantaneous rate of change of temperature of a fluid parcel containing absorbing particles. In general, an imbalance of heating and cooling results in a gain or loss of thermal energy. We ignore any changes of internal (chemical) energy and consider only changes in the thermal energy $\rho c T$, where c is the specific heat per unit mass. The question is whether to use c_v (applicable to a process in which the volume is held constant) or c_p (applicable to a process in which pressure is held constant). Clearly, if a fluid parcel[24] is heated there will be no constraining walls to maintain a fixed volume. The parcel will expand against its environment and do work on its surroundings to maintain a constant pressure. The rate of change in the thermal energy per unit volume of a gas that is free to expand against its surroundings is therefore given by $\rho c_p(\partial T/\partial t)_p$. Thus,

> *in the absence of other heating or cooling processes, the rate of change of temperature of a parcel at constant pressure is*
>
> $$\left(\frac{\partial T}{\partial t}\right)_p = -\frac{\mathcal{H}}{\rho c_p}. \tag{5.78}$$

The above discussion motivates the use of c_p instead of c_v, since the former applies to a parcel that is free to expand as it is heated rather than being constrained to a fixed volume. Since $c_p > c_v$, the heating rate $\mathcal{H} > 0$ is less than it would be for a fixed volume, because some of the absorbed energy goes into work of expansion. However, if $\mathcal{H} < 0$ the rate of temperature reduction is also less than at constant volume. This is because the environmental pressure compresses and heats the parcel, thus canceling some of the radiative cooling. It is conventional in atmospheric science to describe this quantity in units of kelvins per day. In the field of atmospheric dynamics, it is called the *temperature tendency*. We prefer to call it the *warming rate*[25] $\mathcal{W}$, given by the formula (since there are 86,400 seconds per day)

$$\mathcal{W} = 86{,}400\frac{\mathcal{H}}{\rho c_p} \quad \text{[K per day]}. \tag{5.79}$$

5.7.3 Actinic Radiation, Photolysis Rate, and Dose Rate

A quantity of key importance to atmospheric photochemistry is the *photolysis rate* J, defined as the local rate of a photoabsorption process leading to a specific photodissociation event. The radiative energy absorbed at a frequency ν may not be 100% efficient in producing the desired event. Defining this efficiency or *quantum yield* by $\eta^i(\nu)(0 \leq \eta^i \leq 1)$, we express the photolysis rate for the photodissociation of a particular species (denoted by superscript i) as follows:

The photolysis rate

$$J_i \equiv \int_{\nu_c}^{\infty} d\nu 4\pi(\bar{I}_\nu/h\nu)\alpha^i(\nu)\eta^i(\nu) \quad [\mathrm{m}^{-3}\cdot\mathrm{s}^{-1}]. \tag{5.80}$$

Here ν_c is the minimum frequency corresponding to the threshold energy for the photoabsorption. According to Eq. 2.9, $4\pi\bar{I}_\nu/h\nu = \mathcal{U}_\nu c/h\nu$ is the density of photons at a given frequency. Photochemists sometimes use the term *actinic flux* for the quantity $4\pi\bar{I}_\nu$, but this is to be avoided because the term *flux* should be reserved for the hemispherical flow of radiative energy or momentum. We may, however, refer to $4\pi\bar{I}_\nu$ as the *actinic radiation field* without introducing confusion.

Of considerable biological interest is the rate at which a flat horizontal surface, or a spherical surface, receives radiative energy capable of initiating certain biological processes. For example, UV-B radiation between 280 and 320 nm is harmful to human skin and damaging to the DNA molecule, the carrier of genetic information in all living cells. The received radiation is weighted by a specific spectral function $A(\nu) < 1$ called the *action spectrum*. $A(\nu)$ gives the efficiency as a function of frequency of a particular process, for example, the UV "kill rate." The rate at which a flat surface is "exposed" is called the *dose rate*,

$$D \equiv \int_0^{\infty} d\nu A(\nu) F_\nu \quad [\mathrm{W}\cdot\mathrm{m}^{-2}], \tag{5.81}$$

where F_ν is the net flux. Alternatively, we may be interested in the rate at which a spherical particle receives energy from the radiation field. This describes the dose rate received by small ocean creatures or the heating rate of a cloud water droplet. Note that in the case of heating of a spherical droplet, the appropriate quantity is $4\pi\bar{I}_\nu$, not F_ν. The dose rate in this case is

$$D = 4\pi \int_0^{\infty} d\nu A(\nu)\bar{I}_\nu \quad [\mathrm{W}\cdot\mathrm{m}^{-2}]. \tag{5.82}$$

The *radiation dose* is defined to be the total time-integrated amount of energy received (usually over one day), $\int dt D(t)$.

5.8 Summary

In this chapter we have laid the groundwork for the radiative transfer theory to be elaborated on in succeeding chapters. The interaction properties of matter and radiation (emittance, absorptance, surface reflectance, scattering and absorption coefficients, emission coefficients, etc.) were considered as specified functions of frequency, angle, temperature, etc. The fundamental principle underlying the theory is the phenomenological *Extinction Law* (Eqs. 2.17 or 2.23). Radiative emission is governed by *Kirchhoff's Law*, according to which the directional emittance equals the directional absorptance in the case of surfaces (Eq. 5.8); and in the case of extended media the emission coefficient is proportional to the absorption coefficient (Eq. 5.37). The *Principle of Reciprocity* equates reflectances evaluated for the direct ray path and for the time-reversed ray path. Several analytic representations of the reflectance (commonly referred to as the *bidirectional reflectance distribution function*) are *Minnaert's formula* and the *Lommel–Seeliger formula*. Physically based models were used to describe the *Opposition Effect* and also specular reflection from a wind-roughened sea surface. The time-independent radiative transfer equation (Eq. 5.41) describes how the intensity is affected by extinction, emission, and scattering processes. The solution of this equation requires the specification of external radiation sources and of the absorption, emission, and scattering properties of the medium and its bounding surface. The latter properties include the efficiencies of reflection, emission, absorption, and transmission as functions of frequency, temperature, and direction. These properties illustrate the equivalent status of the source function S_ν and the intensity I_ν as information carriers of the radiation field. From the energetics point of view, the radiative flux divergence (Eq. 5.76) is the key to determining the heating rate (Eq. 5.77). For atmospheric photochemistry, the quantity of interest is the photolysis rate, which depends upon the product of the average intensity and the effective absorption coefficient (Eq. 5.80). UV exposure of the biosphere was quantified in the dose rate D or the time-integrated radiation dose. In the next chapter we will discuss how these matter–radiation descriptions enter into the mathematical formulation of radiative transfer problems.

Problems

5.1 Two gray surfaces 1 and 2, at temperatures T_1 and T_2, have flux emittances $\epsilon_1(2\pi)$ and $\epsilon_2(2\pi)$ and flux reflectances $\rho_1(2\pi)$ and $\rho_2(2\pi)$. The surfaces are flat, parallel, and separated by a transparent medium. Surface 2 is above surface 1.

(a) Show that the upward and downward frequency-integrated fluxes are given by

$$F^+ = \frac{\epsilon_1 \sigma_B T_1^4 + \epsilon_2 \rho_1 \sigma_B T_2^4}{1 - \rho_1 \rho_2}, \qquad F^- = \frac{\epsilon_2 \sigma_B T_2^4 + \epsilon_1 \rho_2 \sigma_B T_1^4}{1 - \rho_1 \rho_2}.$$

(b) Show that the net rate of energy transfer between the surfaces per unit area is

$$F = F^+ - F^- = \frac{\sigma_B\left(T_2^4 - T_1^4\right)}{\epsilon_2^{-1} + \epsilon_1^{-1} - 1}.$$

5.2 Prove that for a partially transparent surface, Eq. 5.25 is generalized to

$$\alpha(\nu, -\hat{\Omega}') = 1 - \rho(\nu, -\hat{\Omega}', 2\pi) - \mathcal{T}(\nu, -\hat{\Omega}', -2\pi).$$

5.3 Suppose that a surface reflects partially by Lambert reflection and partially by specular reflection. Show from conservation of energy that

$$\pi\rho_L + \rho_s(-\hat{\Omega}_o) \le 1.$$

5.4 The Earth's atmosphere, plus its underlying surface, can be considered to have an effective absorptance and reflectance. Suppose the atmosphere is purely absorbing, so that the total shortwave atmospheric transmittance is determined by the beam transmittance, $\mathcal{T}_b$. Let the diffusely reflecting surface be characterized by a BRDF with no azimuthal dependence of the reflection, such that $\rho_d(-\hat{\Omega}', \hat{\Omega}) = \rho_d(\theta', \theta)$.

(a) Show that the net flux absorptance of the combined atmosphere–surface for solar radiation incident at the polar angle θ_0 is given by

$$\alpha_{net}(-2\pi, -\mu_0) = 1 - 2\pi\mathcal{T}_b(-\mu_0)\int_0^1 d\mu\mu\rho_d(-\mu_0, \mu)\mathcal{T}_b(\mu),$$

where $\mu = \cos\theta$ and $\mu_0 = \cos\theta_0$.

(b) Suppose that there is also a specular component to the surface reflection, $\rho_s(\theta)$. Show that the net flux absorptance is

$$\alpha_{net}(-2\pi, -\mu_0) = 1 - \mathcal{T}_b(\mu_o)\rho_s(-\mu_o)\mu_o\mathcal{T}_b(-\mu_o) - 2\pi\mathcal{T}_b(-\mu_0)\int_0^1 d\mu\mu\rho_d(-\mu_0, \mu)\mathcal{T}_b(\mu).$$

5.5 A plane-parallel planetary atmosphere is heated by a collimated solar beam whose flux normal to the beam at the top of the atmosphere is $-F_\nu^s$ and is incident at the angle $\cos^{-1}\mu_0$. An underlying Lambertian surface has a flux reflectance ρ_L. Show from the radiative transfer equation that the heating rate due to solar absorption by a single species is given by

$$\mathcal{H}_\nu(z) = -\frac{\partial F_\nu}{\partial z} = \alpha(z)F_\nu^s\left[e^{-\tau/\mu_0} + \rho_L\mu_0 e^{-\tau^*/\mu_0}2E_2(\tau^* - \tau)\right].$$

Here F_ν is the net flux (downwelling solar flux combined with the upward reflected solar flux) and $\alpha(z)$ is the absorption coefficient of the absorbing species. E_n is the exponential integral of order n (see Eqs. 5.62). Ignore scattering processes. (*Hint*: Pay particular attention to the signs of the direct and reflected fluxes.)

5.6

(a) Show that for a semi-infinite, purely absorbing isothermal plane-parallel atmosphere in LTE, the net flux at frequency ν and at optical depth τ is given by

$$F_\nu(\tau) \equiv \pi B_\nu(T)\mathcal{T}(\nu, -2\pi, -2\pi) = 2\pi B_\nu(T)E_3(\tau).$$

(b) Evaluate this expression at $\tau = 0$ and as $\tau \to \infty$. Interpret the results.

(c) Show by differentiation that the spectral heating rate is given by

$$\mathcal{H}_\nu = -2\pi\alpha(\nu)B_\nu E_2(\tau).$$

Interpret this result, using the concept of "cooling to space."

(d) Show that the above result is compatible with Eq. 5.77 ($\mathcal{H}_\nu = 4\pi\alpha(\nu)(\bar{I}_\nu - B_\nu)$) by evaluating $\bar{I}_\nu$ explicitly. In other words use the solutions for $I^+(\tau)$ and $I^-(\tau)$ and integrate these results over solid angle.

5.7 Show that the column-integrated heating rate due to direct solar absorption in the atmosphere is given by

$$\int_0^\infty dz'\mathcal{H}(z') = \int_0^\infty d\nu F_\nu^{\mathrm{s}}\mu_0\alpha(\nu, \mu_0, 2\pi),$$

where μ_0 is the cosine of the incident solar zenith angle, α is the flux absorptance, and F^{s} is the solar flux.

Notes

1 An excellent reference for boundary properties of materials is Siegel, R., and J. R. Howell, *Thermal Radiation Heat Transfer*, 3rd ed., Hemisphere Publ., Washington, DC, 1992.

2 Hapke, B., *Theory of Reflectance and Emittance Spectroscopy*, Cambridge University Press, Cambridge, 1993.

3 The formalism is readily generalized to apply when polarization is important. The boundary quantities defined in this section may be defined separately for the perpendicular and parallel components.

4 Note that the maximum value of ρ_L is $1/\pi$. For this reason, some authors define the BRDF as $r \equiv \pi\rho$. However, it is then necessary to define the albedo as $1/\pi \int d\omega \cos\theta r$.

5 $\rho(\nu, -\hat{\Omega}, 2\pi)$ is also called the *directional-hemispherical reflectance*. A similar quantity, the *hemispherical-directional reflectance* $\rho(\nu, -2\pi, \hat{\Omega})$ is the ratio of reflected intensity to incident flux for uniform illumination. The term *albedo* originated in a book published in 1760 by Johann Lambert. Writing in Latin, Lambert used the word *albedo,* meaning "whiteness" for the fraction of light reflected diffusely by a body.

6 The data for ρ_n and k are taken from Veverka, J., J. Goguen, S. Yange, and J. Elliot, "Scattering of light from particulate surfaces I. A laboratory assessment of multiple-scattering effects," *Icarus*, **34**, 406–14, 1978.

7 See Endnote 6.

8 Papers still appear in which this important principle is ignored.

9 In the language of planetary photometry, there is a weak "limb brightening" from a surface obeying the *Lommel–Seeliger* formula. "Limb darkening/brightening" refers to the variation of brightness of an extended object, such as the moon or a planet, with observation angle, that is, from the center of the disk to the edge ("limb"). Usually it refers to observations near zero phase angle.

10 The smooth curves are fits of the *Lommel–Seeliger formula*, modified to include an opposition effect. Taken from Ahmad, S. P., and D. W. Deering, "A simple analytical function for bidirectional reflectance," *J. Geophys. Res.*, **97**, 18,867–86, 1992.

11 Hapke, B., "A theoretical photometric function for the lunar surface," *J. Geophys. Res.*, **68**, 4571–86, 1963.

12 Ahmad, S. P., and D. W. Deering, "A simple analytical function for bidirectional reflectance," *J. Geophys. Res.*, **97**, 18,867–86, 1992.

13 Cox, C., and W. Munk, "Measurement of the roughness of the sea surface from photographs of the sun glint," *J. Opt. Soc. Am.*, **44**, 838–50, 1954.

14 Nakajima, T., and M. Tanaka, "Effect of wind-generated waves on the transfer of solar radiation in the atmosphere–ocean system," *J. Quant. Spectr. Radiative Transfer*, **29**, 521–37, 1983.

15 Figure 5.6 is taken from Plass, G. N., G. W. Kattawar, and J. A. Guinn, Jr., "Radiative transfer in the Earth's atmosphere and ocean: Influence of ocean waves," *Applied Optics*, **14**, 1924–36, 1975.

16 This quantity is also called the *directional-hemispherical transmittance*. From reciprocity it may be shown to be equal to the total transmittance $\mathcal{T}(\nu; -2\pi, -\hat{\Omega})$, which describes the transmission of light in the direction $-\hat{\Omega}$ for a hemispherically isotropic incoming radiation field. This is called the *hemispherical-directional transmittance* (see Appendix M).

17 Spherical albedoes for the planet Venus for a variety of wavelengths are found in Irvine, W. M., "Monochromatic phase curves and albedoes for Venus," *J. Atmos. Sci.*, 610–16, 1968.

18 This convention is universal in the astrophysical literature but has not been accepted by all atmospheric radiative transfer workers. Caution is in order in reading the older literature. It is a natural convention in oceanic applications.

19 An additional restriction must be imposed even if the above condition is met: The angle that incident sunlight makes with the local vertical, θ_0, must not be greater than about 82° (for the Earth). Otherwise the horizontal distance through which the sunlight penetrates before being absorbed or reflected is large enough for curvature effects to become important.

20 A compact discussion is found in the appendix of the text by M. N. Ozisik, *Radiative Transfer and Interactions with Conduction and Convection*, Wiley-Interscience, New York, 1973. A more exhaustive discussion is found in M. Abramowitz and I. A. Stegun, *Handbook of Mathematical Functions*, Dover Publications, New York, 1965.

21 Equation 5.65 is also useful as an interpolation formula in angle, once the approximate solution is known at a set of discrete angles (see Chapter 8).

22 An excellent modern review of analytic solution methodology is given in the book by E. G. Yanovitskij, *Light Scattering in Inhomogeneous Atmospheres*, Springer-Verlag, Berlin, 1997.

23 Equation 5.76 assumes all absorbed energy results in heat. This is not the case if collisions are not capable of quenching the upper excited states before they have a chance to radiate. This NLTE situation applies to rarefied upper levels of planetary atmospheres. In NLTE, Eq. 5.76 must contain an efficiency factor $e(\nu)$ $(0 \leq e(\nu) \leq 1)$.

24 A fluid parcel could apply to either an atmosphere or ocean and defines a fixed number of molecules or particles. The particles move together, more or less like a separate entity. The volume of the parcel is compressed when it sinks and expands when it rises. It is clear that a parcel will not maintain its identity forever but will eventually be distorted and eventually fragmented by dynamical and diffusive processes. However, it is a useful concept for air motions over time scales that are short compared to a diffusion time scale (hours to days, depending upon the height).

25 Since the warming rate is negative when only the thermal infrared part of the radiative budget is considered, this quantity is sometimes called the *cooling rate*.

Chapter 6

Formulation of Radiative Transfer Problems

6.1 Introduction

In this chapter, we further refine the mathematical description of the radiative transfer process. We will find that it is as important to be able to set up a problem correctly as it is to solve it. Experience has shown that an investment of attention at the "front end" is well rewarded when it comes time to submit the problem to analytic or numerical solutions. To quote A. Einstein and L. Infeld,[1]

> The formulation of a problem is often more essential than its solution, which may be merely a matter of mathematical or experimental skill.

For example, some applications are more amenable to an integral-equation approach. In other cases, a transformation can convert a problem that might involve hundreds of terms in the expansion of the phase function to one involving just a few terms. Also, in scattering problems it is usually advisable to separate the direct solar component from the diffuse component. Finally, we will introduce several prototype problems, which are invaluable as tools for learning various solution techniques. Since accurate solutions to these problems are readily available, they provide a practical means of testing numerical techniques, which can then be applied to more realistic problems.

6.2 Separation into Diffuse and Direct (Solar) Components

There are two distinctly different components of the shortwave radiation field.[2] The first one is the direct or *solar* component I_s, which is that part of the solar radiation field that has survived extinction:

$$I_s^-(\tau, \mu, \phi) = F^s e^{-\tau/\mu_0} \delta(\hat{\Omega} - \hat{\Omega}_0) = F^s e^{-\tau/\mu_0} \delta(\mu - \mu_0)\delta(\phi - \phi_0), \quad (6.1)$$

where we have suppressed the ν subscripts. This is sometimes called the "uncollided" component. The second part of the radiation is the *diffuse* component I_d, which consists of light that has been scattered at least once. This part is also called the *multiple-scattering* component, which may be thought of as the medium's "self-illumination." A particular volume element of the medium can be said to be illuminated by two sources: by the Sun and by the rest of the medium (including the planetary surface or the bottom of the ocean). Since the direct component is described by the Extinction Law, it is useful to isolate this part from the total radiation field, so that the difficult part remains:

$$I^-(\tau, \mu, \phi) = I_d^-(\tau, \mu, \phi) + I_s^-(\tau, \mu, \phi). \quad (6.2)$$

For the present discussion we let the lower surface be black so that no solar radiation is reflected back into the medium. We also ignore thermal emission from the surface, but include thermal radiation from the medium itself. Then $I_s^+(\tau^*, \mu, \phi) = 0$, where τ^* denotes the total optical depth of the medium, and $I^+(\tau, \mu, \phi) = I_d^+(\tau, \mu, \phi)$.

We now recall the form of the radiative transfer equation for the two half-range intensities. We can write Eq. 5.41 in terms of the half-range intensities for slab geometry as

$$-\mu \frac{dI^-(\tau, \hat{\Omega})}{d\tau} = I^-(\tau, \hat{\Omega}) - (1-a)B - \frac{a}{4\pi} \int_+ d\omega' p(+\hat{\Omega}', -\hat{\Omega}) I^+(\tau, \hat{\Omega}')$$
$$- \frac{a}{4\pi} \int_- d\omega' p(-\hat{\Omega}', -\hat{\Omega}) I^-(\tau, \hat{\Omega}'), \quad (6.3)$$

$$\mu \frac{dI^+(\tau, \hat{\Omega})}{d\tau} = I^+(\tau, \hat{\Omega}) - (1-a)B - \frac{a}{4\pi} \int_+ d\omega' p(+\hat{\Omega}', +\hat{\Omega}) I^+(\tau, \hat{\Omega}')$$
$$- \frac{a}{4\pi} \int_- d\omega' p(-\hat{\Omega}', +\hat{\Omega}) I^-(\tau, \hat{\Omega}'). \quad (6.4)$$

With regard to the notation $I^-(\tau, \mu, \phi) \equiv I^-(\tau, \hat{\Omega}) \equiv I(\tau, -\hat{\Omega})$ and $p(-\hat{\Omega}', +\hat{\Omega})$ (for example) indicates that a photon is moving downward before the scattering $(-\hat{\Omega}')$ and upward $(+\hat{\Omega})$ after the scattering. We now substitute for the total intensity field the sum of the direct and diffuse components (Eqs. 6.1 and 6.2) into Eq. 6.3 to obtain

$$-\mu \frac{dI_d^-(\tau, \hat{\Omega})}{d\tau} - \mu \frac{dI_s^-(\tau, \hat{\Omega})}{d\tau}$$
$$= I_d^-(\tau, \hat{\Omega}) + I_s^-(\tau, \hat{\Omega}) - (1-a)B - \frac{a}{4\pi} \int_- d\omega' p(-\hat{\Omega}', -\hat{\Omega}) I_s^-(\tau, \hat{\Omega}')$$
$$- \frac{a}{4\pi} \int_+ d\omega' p(+\hat{\Omega}', -\hat{\Omega}) I_d^+(\tau, \hat{\Omega}') - \frac{a}{4\pi} \int_- d\omega' p(-\hat{\Omega}', -\hat{\Omega}) I_d^-(\tau, \hat{\Omega}'). \quad (6.5)$$

The two nonintegral terms involving the direct component cancel, because $-\mu dI_s^-/d\tau = I_s^-$. If we substitute for I_s^- from Eq. 6.1 in the first integral term, we obtain the result

$$-\mu\frac{dI_d^-(\tau,\hat{\Omega})}{d\tau} = I_d^-(\tau,\hat{\Omega}) - (1-a)B - S^*(\tau,-\hat{\Omega}) - \frac{a}{4\pi}\int_+ d\omega' p(\hat{\Omega}',-\hat{\Omega})I_d^+(\tau,\hat{\Omega}') - \frac{a}{4\pi}\int_- d\omega' p(-\hat{\Omega}',-\hat{\Omega})I_d^-(\tau,\hat{\Omega}'), \tag{6.6}$$

where

$$S^*(\tau,-\hat{\Omega}) = \frac{a}{4\pi}\int_- d\omega' p(-\hat{\Omega}',-\hat{\Omega})F^s e^{-\tau/\mu_0}\delta(\hat{\Omega}'-\hat{\Omega}_0) = \frac{a}{4\pi}p(-\hat{\Omega}_0,-\hat{\Omega})F^s e^{-\tau/\mu_0}. \tag{6.7}$$

We repeat this procedure for the upward component to obtain

$$\mu\frac{dI_d^+(\tau,\hat{\Omega})}{d\tau} = I_d^+(\tau,\hat{\Omega}) - (1-a)B - S^*(\tau,+\hat{\Omega}) - \frac{a}{4\pi}\int_+ d\omega' p(+\hat{\Omega}',+\hat{\Omega})I_d^+(\tau,\hat{\Omega}') - \frac{a}{4\pi}\int_- d\omega' p(-\hat{\Omega}',+\hat{\Omega})I_d^-(\tau,\hat{\Omega}'), \tag{6.8}$$

where

$$S^*(\tau,+\hat{\Omega}) = \frac{a}{4\pi}\int_- d\omega' p(-\hat{\Omega}',+\hat{\Omega})F^s e^{-\tau/\mu_0}\delta(\hat{\Omega}'-\hat{\Omega}_0) = \frac{a}{4\pi}p(-\hat{\Omega}_0,+\hat{\Omega})F^s e^{-\tau/\mu_0}. \tag{6.9}$$

The equations of transfer for the total field and for the diffuse field differ by the presence of an extra "source" term $S^*(\tau,\pm\hat{\Omega})$ (compare Eqs. 6.6 and 6.8 with Eq. 5.41). This *single-scattering source function* "drives" the diffuse radiation field. Without $S^*(\tau,\pm\hat{\Omega})$ there would be no diffuse radiation in the absence of thermal emission ($B=0$). Note also that the azimuthal dependence of the radiation field can be traced to that of $S^*(\tau,\pm\hat{\Omega})$ through the phase function $p(-\hat{\Omega}_0,\pm\hat{\Omega})$. If the external source had a more general angular dependence, then $S^*(\tau,\pm\hat{\Omega})$ would be expressed in terms of the angular integration over this external radiation field, weighted by the phase function.

We note for completeness and later reference that in full-range slab geometry the radiative transfer equation for the diffuse intensity may be written more compactly as

$$u\frac{dI(\tau,u,\phi)}{d\tau} = I(\tau,u,\phi) - \frac{a}{4\pi}\int_0^{2\pi} d\phi' \int_{-1}^{1} du' p(u',\phi';u,\phi) I(\tau,u',\phi') - (1-a)B - S^*(\tau,u,\phi), \tag{6.10}$$

where $S^*(\tau,u,\phi)$ denotes the solar beam driving term derived above.

6.2.1 Lower Boundary Conditions

For a lower surface at temperature T_s that emits thermal radiation with emittance ϵ and reflects the incident radiation field $I^-(\tau^*,\hat{\Omega}')$ with reflectance ρ, it is straightforward to show that the following source terms must be added to Eqs. 6.7 and 6.9:

$$S_b^*(\tau,\pm\hat{\Omega}) = \frac{a}{4\pi}\int_+ d\omega' p(+\hat{\Omega}',\pm\hat{\Omega}) e^{-(\tau^*-\tau)/\mu} \times \left[\underbrace{\int_- d\omega'' \rho(-\hat{\Omega}'',\hat{\Omega}')\cos\theta' I^-(\tau^*,\hat{\Omega}'')}_{\text{reflected component}} + \underbrace{\epsilon(\hat{\Omega})B(T_s)}_{\text{emitted}} \right] \tag{6.11}$$

where

$$I^-(\tau^*,\hat{\Omega}'') = \underbrace{F^s e^{-\tau^*/\mu_0}\delta(\hat{\Omega}_0 - \hat{\Omega}'')}_{\text{solar component}} + \underbrace{I_d^-(\tau^*,\hat{\Omega}'')}_{\text{diffuse component}}. \tag{6.12}$$

Finally if we add the unscattered component of the intensity reflected from the lower boundary (analogous to the solar component for the downward intensity) the total upward intensity is given by

$$I^+(\tau,\mu,\phi) = \int_0^{2\pi} d\phi' \int_0^1 d\mu\mu' \rho(-\mu',\phi';\mu,\phi) I^-(\tau^*,\mu',\phi') e^{-(\tau^*-\tau)/\mu)} + \int_\tau^{\tau^*} \frac{d\tau'}{\mu} S_{\rm tot}(\tau',\mu,\phi) e^{-(\tau'-\tau)/\mu} + \epsilon(\mu,\phi)B(T_s)e^{-(\tau'-\tau)/\mu}, \tag{6.13}$$

where $S_{\rm tot} \equiv (1-a)B + S^* + S + S_b^*$. We have rewritten the angular integration in polar coordinates. It is instructive to sketch a number of photon trajectories, each having different reflection and scattering histories, to determine which of the six terms in Eq. 6.13 contains that trajectory. Unfortunately, $I_d^-(\tau^*, \hat{\Omega}'')$, the downward diffuse component, is unknown. However, there are ways to incorporate this boundary condition into various solution techniques. An example of this approach is found in §6.6, where the effects of specular reflection from a calm sea surface are included in a term derived from the reflected component above.

A common way of dealing with the unknown term begins with the solution of the *standard problem* (in which the underlying surface is assumed to be nonreflecting, i.e., black) and proceeds by adding to this an analytic correction to obtain the solution for the full problem with a reflecting surface, known as the *planetary problem*. This procedure, which is discussed in §6.11, would be the preferred method if we were interested solely in the total energy balance of the atmosphere–surface system, that is, ρ, α, and $\mathcal{T}$. Unfortunately this approach does not take into account an emitting surface and it is limited to Lambert surfaces. Nor can it be used if we are interested in the internal radiation field. We will generally deal with the standard problem in this book. For Lambert surfaces the reflection can be included as an analytic correction (§6.11) with the caveats mentioned above.

Methods that transform the exact radiative transfer equation to a system of coupled partial differential equations can incorporate the above boundary condition straightforwardly into the solution. An example is given in Chapter 8, where the *discrete-ordinate method* is worked out in detail.

6.2.2 Multiple Scattering

The interpretation of Eqs. 6.6–6.9 for the diffuse intensity is straightforward. The extra term S^* is an "imbedded source" of radiation, which has been scattered once within the medium. The integral terms constitute the source of multiply scattered radiation.[3] Since I_d is the radiation arising from scattering by the medium itself (as contrasted with boundary or external sources), it adds to the source function through additional scattering. Thus, the total source function consists of the following sum:

$$S(\tau, \hat{\Omega}) = \underbrace{[1 - a]B(T)}_{\textit{thermal emission}} + \underbrace{S^*(\tau, \hat{\Omega})}_{\textit{first-order scattering}} + \underbrace{\frac{a}{4\pi} \int_{4\pi} d\omega' p(\hat{\Omega}', \hat{\Omega}) I_d(\hat{\Omega}')}_{\textit{multiple scattering}}. \tag{6.14}$$

It should be emphasized that $S(\tau, \hat{\Omega})$ refers to the sources of all *internal* radiation and includes boundary sources if S^* includes the boundary term S_b^* (Eq. 6.11).

Example 6.1 Isotropic Scattering in Slab Geometry

It is sometimes permissible to assume that the scattering is isotropic, so that $p = 1$. The source term is therefore also isotropic: $S^{*\pm}(\tau, \hat{\Omega}) = S^*(\tau)$, and the radiative transfer equations for the half-range diffuse intensity fields, Eqs. 6.6 and 6.8, are greatly simplified because the integrals are independent of the azimuthal angle, ϕ. Dropping the subscript d, and assuming a black lower boundary, we find

$$\mu \frac{dI^+(\tau,\mu)}{d\tau} = I^+(\tau,\mu) - (1-a)B - S^*(\tau) - \frac{a}{2}\int_0^1 d\mu' I^+(\tau,\mu') - \frac{a}{2}\int_0^1 d\mu' I^-(\tau,\mu'), \tag{6.15}$$

$$-\mu \frac{dI^-(\tau,\mu)}{d\tau} = I^-(\tau,\mu) - (1-a)B - S^*(\tau) - \frac{a}{2}\int_0^1 d\mu' I^+(\tau,\mu') - \frac{a}{2}\int_0^1 d\mu' I^-(\tau,\mu'), \tag{6.16}$$

where $S^*(\tau) = \frac{a}{4\pi} F^s e^{-\tau/\mu_0}$.

Because S^* is isotropic in this case, the intensities are independent of the angle ϕ, which is an enormous simplification over the anisotropic scattering case. A great deal of the early work in the field was performed on this type of problem, because of its analytic simplicity rather than its resemblance to real problems. Nevertheless, there are a few practical problems for which the isotropic approximation is valid, and these will be mentioned later in the book.

The source function is, from Eq. 6.14,

$$S(\tau) = (1-a)B + S^* + \frac{a}{2}\int_0^1 d\mu \left[I_d^+(\tau,\mu) + I_d^-(\tau,\mu) \right]. \tag{6.17}$$

Given the source function $S(\tau)$, the diffuse intensities follow from Eqs. 5.66 and 5.67:

$$I_d^-(\tau,\mu) = \int_0^\tau \frac{d\tau'}{\mu} S(\tau') e^{-(\tau-\tau')/\mu}, \tag{6.18}$$

$$I_d^+(\tau,\mu) = \int_\tau^{\tau^*} \frac{d\tau'}{\mu} S(\tau') e^{-(\tau'-\tau)/\mu}. \tag{6.19}$$

6.2.3 Azimuth Independence of Flux and Mean Intensity

We now prove the important result that, in slab geometry, the fluxes and mean intensity depend only on the azimuthally averaged intensity. By averaging Eqs. 6.6 and 6.8 over azimuth, we obtain the following pair of equations for the azimuthally averaged half-range diffuse intensities:

$$\mu \frac{dI^+(\tau,\mu)}{d\tau} = I^+(\tau,\mu) - (1-a)B - \frac{a}{2}\int_0^1 d\mu' p(+\mu',+\mu) I^+(\tau,\mu')$$
$$- \frac{a}{2}\int_0^1 d\mu' p(-\mu',+\mu) I^-(\tau,\mu') - S^*(\tau,\mu), \quad (6.20)$$

$$-\mu \frac{dI^-(\tau,\mu)}{d\tau} = I^-(\tau,\mu) - (1-a)B - \frac{a}{2}\int_0^1 d\mu' p(+\mu',-\mu) I^+(\tau,\mu')$$
$$- \frac{a}{2}\int_0^1 d\mu' p(-\mu',-\mu) I^-(\tau,\mu') - S^*(\tau,-\mu), \quad (6.21)$$

where

$$I^{\pm}(\tau,\mu) \equiv \frac{1}{2\pi}\int_0^{2\pi} d\phi' I^{\pm}(\tau,\mu,\phi'), \quad (6.22)$$

$$p(\pm\mu',\pm\mu) \equiv \frac{1}{2\pi}\int_0^{2\pi} d\phi' p(\pm\mu',\phi';\pm\mu,\phi'), \quad (6.23)$$

and

$$S^*(\tau,\pm\mu) = \frac{1}{2\pi}\int_0^{2\pi} d\phi' S^*(\tau,\pm\mu,\phi') = \frac{a}{4\pi} p(-\mu_0,\pm\mu) F^{s} e^{-\tau/\mu_0}. \quad (6.24)$$

Note that we have dropped the subscript d since we are mainly concerned from this point onward with the diffuse intensity. The presence of S^* is always the clue that we are referring to the diffuse intensity. By definition

$$F^{\pm}(\tau) \equiv \int_0^{2\pi} d\phi' \int_0^1 d\mu' \mu' I^{\pm}(\tau,\mu',\phi') = 2\pi \int_0^1 d\mu' \mu' \frac{1}{2\pi}\int_0^{2\pi} d\phi' I^{\pm}(\tau,\mu',\phi')$$
$$= 2\pi \int_0^1 d\mu' \mu' I^{\pm}(\tau,\mu'). \quad (6.25)$$

To simplify the notation, we have used the absence of ϕ arguments to indicate independence of azimuth angle. Hence, we see that *in a slab geometry the flux depends only on the azimuthally averaged intensity.*

The above result shows that if we are interested only in radiative flux (as opposed to angular-dependent intensities) then all we have to consider is the azimuthally independent component of the intensity. Similarly, we find that the mean intensity depends only on the azimuthally averaged intensity. Switching back to the u coordinate, we have

$$\bar{I}(\tau) = \frac{1}{4\pi}\int_0^{2\pi} d\phi' \int_{-1}^{1} du' I(\tau, u', \phi') = \frac{1}{2}\int_{-1}^{1} du' \frac{1}{2\pi}\int_0^{2\pi} d\phi' I(\tau, u', \phi')$$
$$= \frac{1}{2}\int_{-1}^{1} du' I(\tau, u'). \tag{6.26}$$

Finally, we may integrate Eqs. 6.3 and 6.4 over 4π steradians so that we consider the total (diffuse plus direct) radiation field. Adding the two terms yields

$$\frac{dF}{d\tau} = 4\pi(1-a)(\bar{I} - B) = (1-a)F^{s}e^{-\tau/\mu_0} + 4\pi(1-a)(\bar{I}_{\mathrm{d}} - B), \tag{6.27}$$

which shows that a constant flux is obtained if either there is no absorption in the medium ($a = 1$) or the slab is in *monochromatic radiative equilibrium* ($\bar{I} = B$). We note that Eq. 6.27 is proportional to the *spectral heating rate* (see Eqs. 5.76 and 5.77).

6.3 Azimuthal Dependence of the Radiation Field

We have seen that if only fluxes or heating rates are desired, we need to solve a radiative transfer problem involving only two variables, τ and u. However, if we desire the intensity or the source function, we are faced with having to solve for a function of three variables, τ, u, and ϕ. We will show below that by using an *Addition Theorem* we can reduce the latter problem to solving for a small number of functions of only two variables. We will describe a transformation that reduces the problem to one of solving a finite set of uncoupled radiative transfer equations, each of which depends on only two variables, τ and u.

We start by expanding the phase function in a finite series of $2N$ *Legendre polynomials*[4] as follows:

$$p(\tau, \cos\Theta) \approx \sum_{l=0}^{2N-1} (2l+1)\chi_l(\tau)P_l(\cos\Theta), \tag{6.28}$$

where P_l is the lth *Legendre polynomial*. The lth expansion coefficient is given by

$$\chi_l(\tau) = \frac{1}{2}\int_{-1}^{1} d(\cos\Theta)P_l(\cos\Theta)p(\tau, \cos\Theta). \tag{6.29}$$

It is common to denote the first moment of the phase function by the symbol $g \equiv \chi_1$. (Another notation is $\langle\cos\Theta\rangle$.) The first moment represents the degree of asymmetry of the angular scattering and is therefore called the *asymmetry factor*. Special values for the asymmetry factor are given below:

$g = 0$ (isotropic scattering, or symmetric about $\cos\Theta = 0$),

$g = -1$ (complete backscattering),

$g = 1$ (complete forward scattering).

The Legendre polynomials $P_l(\cos\Theta)$ comprise a natural basis set of orthogonal polynomials over the angular domain ($0^\circ \leq \Theta \leq 180^\circ$). The first five Legendre polynomials are

$$P_0(u) = 1, \qquad P_1(u) = u, \qquad P_2(u) = \frac{1}{2}(3u^2 - 1),$$

$$P_3(u) = \frac{1}{2}(5u^3 - 3u), \qquad P_4(u) = \frac{1}{8}(35u^4 - 30u^2 + 3).$$

An important property of the Legendre polynomials is their *orthogonality*:

$$\frac{1}{2}\int_{-1}^{+1} du\, P_l(u) P_k(u) = \frac{1}{2l+1}\delta_{lk}.$$

Here, δ_{lk} is the *Kronecker delta* ($\delta_{lk} = 1$ for $l = k$ and $\delta_{lk} = 0$ for $l \neq k$). The number of terms, $2N$, in the expansion required for an accurate representation of $p(\tau, \cos\Theta)$ depends on how asymmetric the phase function is. Obviously, for isotropic scattering ($p = 1$) only one term is needed. In this case $\chi_0 = 1$ and $\chi_l = 0$ for $l = 1, 2, 3 \ldots 2N$. In general, the more asymmetric the phase function the more terms are required for an accurate representation. Figure 6.1 illustrates this for an increasingly anisotropic *Henyey–Greenstein* phase function (§6.7).

The above representation is of little use in the transfer equation, which is described in terms of the angular coordinates measured with respect to the vertical axis, that is, θ and ϕ. The relationship of these two angles to the scattering angle, Θ, is given by Eq. 3.22. However, the substitution of this relationship into the Legendre polynomials yields a rather complicated and useless form. The key in simplifying the expansion of the phase function is the *Addition Theorem for Spherical Harmonics*,[5] which we may write as follows:

$$P_l(\cos\Theta) = P_l(u')P_l(u) + 2\sum_{m=1}^{l} \Lambda_l^m(u')\Lambda_l^m(u)\cos m(\phi' - \phi). \tag{6.30}$$

To simplify the formulas we have introduced the *normalized associated Legendre polynomial* defined by[6]

$$\Lambda_l^m(u) \equiv \sqrt{\frac{(l-m)!}{(l+m)!}}\, P_l^m(u), \tag{6.31}$$

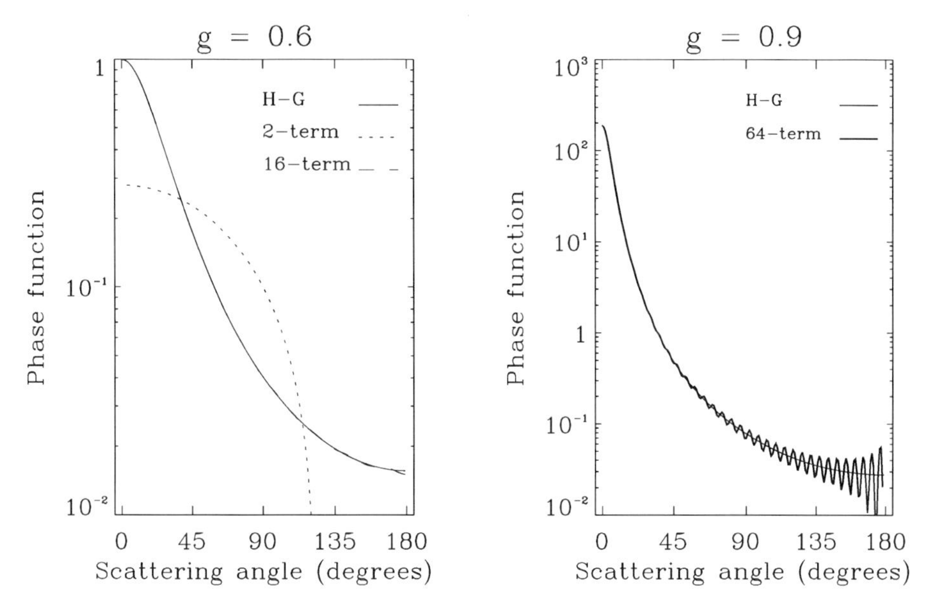

Figure 6.1 Illustration of Legendre polynomial fit to a synthetic phase function for varying degrees of asymmetry as indicated by the parameter g. The larger the value of g the more anisotropic the phase function. Less anisotropic phase functions require fewer terms in the expansion of the phase function in Legendre polynomials to obtain a reasonable fit.

where $P_l^m(u)$ is the *associated Legendre polynomial*. The following orthogonality properties apply:

$$\frac{1}{2}\int_{-1}^{1} du\, P_l^m(u) P_k^m(u) = \frac{(l+m)!}{(2l+1)(l-m)!}\delta_{lk}$$

or

$$\frac{1}{2}\int_{-1}^{1} du\, \Lambda_l^m(u) \Lambda_k^m(u) = \frac{\delta_{lk}}{2l+1}.$$

If $m = 0$, $\Lambda_l^0(u) = P_l^0(u) \equiv P_l(u)$, the Legendre polynomial. Note that we have defined $\Lambda_l^m(u)$ in such a way that it satisfies the same orthogonality condition as $P_l(u)$. Thus, the function $\sqrt{(2l+1)/2}\Lambda^m(u)$ is orthonormal with respect to the polar angle θ. The first few associated Legendre polynomials are

$$P_1^1(u) = \sqrt{1-u^2}, \qquad P_1^2(u) = 3u\sqrt{1-u^2}, \qquad P_2^2(u) = 3\sqrt{1-u^2},$$

$$P_3^2(u) = 15u(1-u^2), \qquad P_3^1(u) = \frac{3}{2}\sqrt{1-u^2}(5u^2-1).$$

The addition theorem allows us to express the phase function in terms of products of functions of θ (or $u = \cos\theta$) and ϕ separately as follows:

$$p(\cos\Theta) = p(u', \phi'; u, \phi) = \sum_{l=0}^{2N-1} (2l+1)\chi_l \times \left\{ P_l(u')P_l(u) + 2\sum_{m=1}^{l} \Lambda_l^m(u')\Lambda_l^m(u)\cos m(\phi' - \phi) \right\}.$$

Inverting the order of the summation, we have

$$p(u', \phi'; u, \phi) = \sum_{m=0}^{2N-1} p^m(u', u)\cos m(\phi' - \phi), \tag{6.32}$$

where

$$p^m(u', u) = (2 - \delta_{0m}) \sum_{l=m}^{2N-1} (2l+1)\chi_l \Lambda_l^m(u')\Lambda_l^m(u). \tag{6.33}$$

Since the expansion of the phase function in Eq. 6.32 is essentially a *Fourier cosine series*, we should expand the intensity in a similar way:

$$I(\tau, u, \phi) = \sum_{m=0}^{2N-1} I^m(\tau, u)\cos m(\phi_0 - \phi). \tag{6.34}$$

As explained in Appendix L, for spherical geometry we need to consider sine terms too, because the derivative operator contains such terms. In slab geometry sine terms are absent. If we substitute Eqs. 6.32 and 6.34 into the full-range version of the radiative transfer equation (Eq. 5.41), we obtain, after some tedious but straightforward manipulation, an equation for each of the Fourier components (see Appendix S for details):

$$\begin{aligned} u\frac{dI^m(\tau, u)}{d\tau} &= I^m(\tau, u) \\ &\quad - \frac{a(\tau)}{2}\int_{-1}^{1} du' p^m(\tau, u', u) I^m(\tau, u') - X_0^m(\tau, u)e^{-\tau/\mu_0} \\ &\quad - (1 - a)B(\tau)\delta_{0m} \quad (m = 0, 1, 2, \ldots, 2N-1), \end{aligned} \tag{6.35}$$

where $p^m(\tau, u', u)$ is given by Eq. 6.33, and

$$X_0^m(\tau, u) = \frac{a}{4\pi} F^s (2 - \delta_{0m}) p^m(\tau, -\mu_0, u), \tag{6.36}$$

where we have used the relation $\Lambda_l^m(-u) = (-1)^{l+m}\Lambda_l^m(u)$.

We have effectively "eliminated" the azimuthal dependence from the radiative transfer equation in the sense that the various Fourier components in Eqs. 6.35 are entirely uncoupled. Thus, in slab geometry independent solutions for each m give the azimuthal components, and the sum in Eq. 6.34 yields the complete azimuthal dependence of the intensity. It is important to note that the azimuthal dependence is forced upon us by the boundary conditions. Hence, when there is no beam source and no azimuth-dependent reflection at either of the slab boundaries, the sum in Eq. 6.34 reduces to the $m = 0$ term. Then the angles μ_0 and ϕ_0 are irrelevant and there is no azimuthal dependence of the diffuse intensity. We also note that if the particles scatter isotropically, there is no azimuthal dependence, as we already showed in Eqs. 6.15–6.16. This follows from Eq. 6.36, since $X_0^m = 0$ for $m > 0$ if the phase function is set to unity. Finally, we note that Eq. 6.35 is of the same mathematical form for all azimuth components m. This means that a method available for solving the $m = 0$ azimuth-independent equation can be readily applied to solve the equation for all $m > 0$.

6.4 Spherical Shell Geometry

So far in this chapter we have been dealing with slab geometry, which is appropriate for a planetary atmosphere if we ignore horizontal inhomogeneities associated with clouds and aerosols, and if the Sun is high enough above the horizon that the non-flat shape of the planet is unimportant. Although the remainder of this book deals almost exclusively with slab geometry for both practical and pedagogical reasons, we shall briefly discuss here the complications arising when plane-parallel geometry is no longer appropriate. We shall distinguish between a stratified, *spherical shell medium*, in which the optical properties are assumed to depend only on the distance from the center of the planet, and a *nonstratified medium*, in which the optical properties are truly inhomogeneous (both horizontally and vertically). The former case applies to a clear planetary medium assumed to be inhomogeneous only in the vertical, whereas the latter applies to a situation in which horizontal inhomogeneities are also introduced, for example by the presence of particles in the atmosphere or ocean.

For solar zenith angles greater than about 82° and twilight situations, we have to take the curvature of the Earth into account and solve the radiative transfer equation appropriate for a spherical shell atmosphere.[7]

In spherical shell geometry, the derivative of the intensity consists of three terms in addition to the one term occurring for slab geometry. These extra terms express the change in the intensity associated with changes in polar angle, azimuthal angle, and solar zenith angle. For a spherical shell medium illuminated by a collimated beam of radiation, the appropriate radiative transfer equation for the diffuse intensity may be expressed as (see §2.8)

$$\hat{\Omega} \cdot \nabla I(r, u, \phi, \mu_0) = -k(r)[I(r, u, \phi, \mu_0) - S(r, u, \phi, \mu_0)]. \quad (6.37)$$

Here r is the distance from the center of the planet and k is the extinction coefficient, while u and ϕ are the cosine of the polar angle and the azimuthal angle, respectively. The symbol $\hat{\Omega} \cdot \nabla$ denotes the derivative operator or the "streaming term" appropriate for this geometry. To arrive at this term we must use spherical geometry. If we map the intensity from a set of global spherical coordinates to a local set with reference to the local zenith direction, then the streaming term[8] may be written (see Appendix O)

$$\hat{\Omega} \cdot \nabla \equiv u \frac{\partial}{\partial r} + \frac{1-u^2}{r} \frac{\partial}{\partial u} + \frac{1}{r} f(u, \mu_0) \left[\cos(\phi - \phi_0) \frac{\partial}{\partial \mu_0} + \frac{\mu_0}{1-\mu_0^2} \sin(\phi - \phi_0) \frac{\partial}{\partial(\phi - \phi_0)} \right], \tag{6.38}$$

where the factor f is given by

$$f(u, \mu_0) \equiv \sqrt{1-u^2}\sqrt{1-\mu_0^2}. \tag{6.39}$$

For slab geometry, only the first term contributes. Thus, for spherically symmetric geometry, the second term must be added, while the third and fourth terms are required for a spherical shell medium illuminated by collimated beam radiation. The source function in Eq. 6.37 is

$$S(r, u, \phi, \mu_0) \equiv \frac{a(r)}{4\pi} \int_0^{2\pi} d\phi' \int_{-1}^{1} du' p(r, u', \phi'; u, \phi) I(r, u', \phi', \mu_0) + \frac{a(r)}{4\pi} p(r, -\mu_0, \phi_0; u, \phi) F^{s} e^{-\tau Ch(r, \mu_0)}. \tag{6.40}$$

The first term in Eq. 6.40 is due to multiple scattering and the second term is due to first-order scattering. We have used the diffuse/direct splitting so that Eq. 6.37 describes the diffuse radiation field only. We note that for isotropic scattering, the primary scattering "driving term" becomes isotropic, which implies that the intensity becomes azimuth independent. The argument in the exponential, $Ch(r, \mu_0)$, is the *air-mass factor* or the *Chapman function*: the quantity by which the vertical optical depth must be multiplied to obtain the slant optical path. For a slab geometry, $Ch(r, \mu_0) = 1/\mu_0 = \sec\theta_0$. Hence $\exp[-\tau Ch(r, \mu_0)]$ yields the attenuation of the incident solar radiation of flux F^{s} (normal to the beam) along the solar beam path.

In a stratified planetary atmosphere, spherical effects become important around sunrise and sunset. It has been shown that in a stratified atmosphere, mean intensities may be calculated with sufficient accuracy for zenith angles less than 90° by ignoring all angle derivatives in the streaming term, but using spherical geometry to compute the direct beam attenuation. Then, we may simply write the streaming term as $\hat{\Omega} \cdot \nabla \cong uk\partial/\partial\tau$.

Although this pseudo-spherical approach works adequately for the computation of intensities in the zenith- and nadir-viewing directions and mean intensities (for zenith angles less than 90°), it may be inadequate for computation of intensities in directions off-zenith (or off-nadir) unless it can be shown that the angle derivative terms are indeed small.

6.5 Nonstratified Media

The Earth's atmosphere and ocean are inhomogeneous, both in the vertical and the horizontal. The horizontal variation is caused by nonuniform distributions of aerosols and broken cloud fields in the atmosphere and of particles in the ocean. Thus, while it is a good approximation for a clear atmosphere or pure ocean water to ignore horizontal inhomogeneities, this is, in general, not the case for an aerosol-loaded and partly overcast atmosphere or a "turbid" ocean. We have seen above that, in certain circumstances, it is possible to reduce the three-dimensional radiative transfer equation valid for spherical shell geometry to a one-dimensional equation. This is not a valid approach if we want to treat the horizontal inhomogeneity of the atmosphere and ocean. Because inhomogeneities occur over a range of scales, a realistic treatment should be three dimensional. Cuboidal cloud forms have been adopted in three-dimensional models to arrive at the somewhat obvious result that a finite cloud has lower albedo than the plane-parallel counterpart due to "leakage" of radiation out of the sides of the cloud. More general, multidimensional formulations for the treatment of broken cloud fields have also been developed that allow for arbitrary variability over many scales. Fractal descriptions of clouds are popular in the current research literature. Fractals make pictures of clouds that look like clouds, but there are, at the present time, no convincing physical arguments leading to a fractal description of clouds. To quote W. Wiscombe[9]:

> ... measurements are the acid test of any model; it is not enough that a model simply "looks" better. Perhaps a plane-parallel model taking proper account of vertical inhomogeneity will agree better with measurements than typical cubic cloud models with their spatially-invariant liquid water and drop distributions. Perhaps weighting plane-parallel albedos by the proper measure of cloudiness fraction will correctly predict the albedo of a patchy cloud field. But more importantly, our job is not to make our models as complicated as nature herself; it is to simplify and idealize, in order to gain understanding. Plane-parallel cloud modeling is an entirely acceptable way to do this. And, on a practical level, (a) we will never know, or want to know, the shape and size of every single cloud on Earth, and (b) plane-parallel clouds can be modeled with a level of spectral and angular detail unreachable in finite cloud models. Our job is to learn how to make simple *adjustments* to plane-parallel predictions to mimic patchiness, not to reject this very valuable modeling approach out of hand.

A better understanding of radiative transfer in inhomogeneous media is clearly needed. This is an active research area in which many questions remain to be formulated and answered. For the development of practical models to compute quantities that depend on the radiation field, such as warming/cooling and photolysis rates, reliable and fast computational schemes are required. At the present time, only radiative transfer models for stratified media have reached a stage of development to be suitable for incorporation into large-scale models.

6.6 Radiative Transfer in the Atmosphere–Ocean System

In Chapter 2 and in Problem 2.2 it was shown that the invariant intensity I/m_r^2 is invariant along a beam path in the absence of scattering and absorption. We have assumed that we are dealing with media for which the index of refraction is constant throughout the medium. The coupled atmosphere–ocean system provides an important exception to this situation because we have to consider the change in the index of refraction across the interface between the atmosphere (with $m_r \approx 1$) and the ocean (with $m_r \approx 1.33$). In this section, we shall discuss the modifications of the framework provided previously that are required to properly describe the radiative transfer process throughout a system consisting of two adjacent strata.[10]

Radiative transfer in aquatic media is a mature discipline in itself with its own nomenclature, terminology, and methodology. To help readers more familiar with oceanic rather than atmospheric applications (and vice versa) we provide a list of nomenclature in Appendix D. We should also point out that radiative transfer in aquatic media is similar in many respects to radiative transfer in gaseous media. In pure aquatic media, density fluctuations lead to Rayleigh-like scattering phenomena. *Turbidity* (which formally is a ratio of the scattering from particles to the scattering from the pure medium) in an aquatic medium is caused by dissolved organic and inorganic matter acting to scatter and absorb radiation in much the same way aerosol and cloud "particles" do in the atmosphere. In the following, we shall focus on the transfer of solar radiation in the atmosphere–ocean system as illustrated in Fig. 6.2.

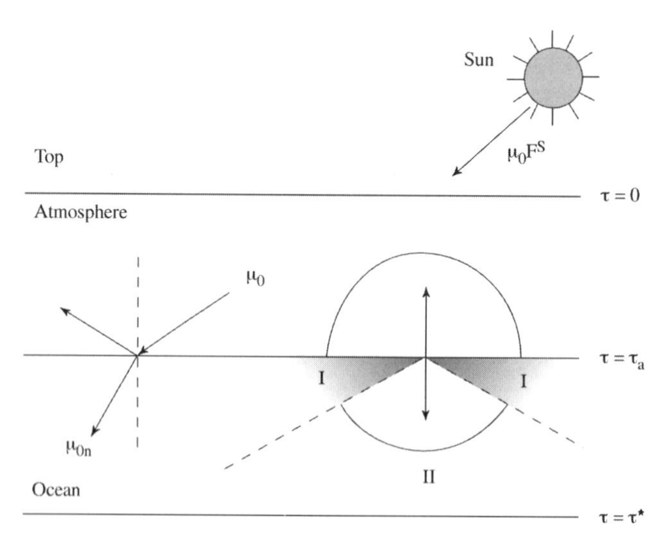

Figure 6.2 Schematic illustration of two adjacent media with a flat interface such as the atmosphere overlying a calm ocean. The atmosphere has a different index of refraction ($m_r \approx 1$) than the ocean ($m_r = 1.33$). Therefore, radiation in the atmosphere distributed over 2π sr will be confined to a cone of less than 2π sr in the ocean (region II). Radiation in the ocean within region I will be totally reflected when striking the interface from below.

To simplify the situation, we shall assume that the ocean surface is calm (i.e., perfectly flat). In principle, the radiative coupling of the two media is very simple because it is described by the well-known laws of reflection and refraction that apply at the interface as expressed mathematically by *Snell's Law* and *Fresnel's Equations*. The practical complications that arise are due to multiple scattering and total internal reflection. As illustrated in Fig. 6.2, the downward radiation distributed over 2π sr in the atmosphere will be restricted to an angular cone less than 2π sr (hereafter referred to as region II; see Fig. 6.2) after being refracted across the interface into the ocean. Beams outside the refractive region in the ocean are in the total reflection region (referred to as region I hereafter; see Fig. 6.2). The demarcation between the refractive and total reflective region in the ocean is given by the critical angle (see Eq. 6.45), schematically illustrated by the dashed line separating regions I and II in Fig. 6.2. Upward-traveling beams in region I in the ocean will be reflected back into the ocean upon reaching the interface. Thus, beams in region I cannot reach the atmosphere directly (and vice versa); they must be scattered into region II in order to be returned to the atmosphere.

6.6.1 Two Stratified Media with Different Indices of Refraction

Since the radiation field in the ocean is driven by solar radiation passing through the atmosphere, we may use the same radiative transfer equation in the ocean as in the atmosphere as long as we properly incorporate the changes occurring at the atmosphere–ocean interface. Thus, the appropriate radiative transfer equation in either medium is Eq. 6.10,

$$u\frac{dI(\tau,u,\phi)}{d\tau} = I(\tau,u,\phi) - \frac{a(\tau)}{4\pi}\int_0^{2\pi} d\phi' \int_{-1}^{1} du' p(\tau,u',\phi';u,\phi)I(\tau,u',\phi') - S^*(\tau,u,\phi),$$

where $S^*(\tau,u,\phi)$ represents the solar driving term. This term is different in the atmosphere and the ocean. In the atmosphere, we have

$$S^*_{\text{air}}(\tau,u,\phi) = \frac{a(\tau)F^{\text{s}}}{4\pi}p(\tau,-\mu_0,\phi_0;u,\phi)e^{-\tau/\mu_0} + \frac{a(\tau)F^{\text{s}}}{4\pi}\rho_{\text{s}}(-\mu_0;m_{\text{rel}})p(\tau,\mu_0,\phi_0;u,\phi)e^{-(2\tau_{\text{a}}-\tau)/\mu_0},$$

where as usual μ_0, ϕ_0, and F^{s} refer to the incident solar beam at the top of the atmosphere; $m_{\text{rel}} \equiv m_{\text{ocn}}/m_{\text{atm}}$ is the real index of refraction in the ocean relative to the atmospher; and τ_{a} is the optical depth of the atmosphere. To simplify the notation we have written $\rho_{\text{s}}(-\mu_0,m_{\text{rel}}) \equiv \rho_{\text{s}}(-\mu_0,\phi_0;\mu_0,\phi_0+\pi;m_{\text{rel}})$ for the specular reflection by the atmosphere–ocean interface caused by the change in index of refraction between air and water. The first term is due to the usual solar beam source, whereas the second term is due to specular reflection by the atmosphere–ocean interface.

The source term in the ocean consists of the attenuated solar beam refracted through the interface:

$$S^*_{\text{ocn}}(\tau, u, \phi) = \frac{a(\tau)F^{\text{s}}}{4\pi}\frac{\mu_0}{\mu_{0m}}\mathcal{T}_{\text{b}}(-\mu_0; m_{\text{rel}})$$
$$\times\, p(\tau, -\mu_{0m}, \phi_0; u, \phi)e^{-\tau_{\text{a}}/\mu_0}e^{-(\tau-\tau_{\text{a}})/\mu_{0m}},$$

where $\mathcal{T}_{\text{b}}(-\mu_0; m_{\text{rel}}) \equiv \mathcal{T}_{\text{b}}(-\mu_0, \phi_0; -\mu_{0m}, \phi_0; m_{\text{rel}})$ is the beam transmittance through the interface, and μ_{0m} is the cosine of the solar zenith angle in the ocean, related to μ_0 by *Snell's Law*

$$\mu_{0m} \equiv \mu_{0m}(\mu_0, m_{\text{rel}}) = \sqrt{1 - (1-\mu_0^2)/m_{\text{rel}}^2}.$$

Isolating the azimuth dependence is now accomplished as usual. The source terms become

$$S^*_{\text{air}}(\tau, u) = X_0^m(\tau, u)e^{-\tau/\mu_0} + X_{01}^m(\tau, u)e^{\tau/\mu_0},$$

$$S^*_{\text{ocn}}(\tau, u) = X_{02}(\tau, u)e^{-\tau/\mu_{0m}},$$

where $X_0^m(\tau, u)$ is given by Eq. 6.36, and

$$X_{01}^m(\tau, u) = \frac{a(\tau)F^{\text{s}}}{4\pi}\rho_{\text{s}}(-\mu_0; m_{\text{rel}})e^{-2\tau_{\text{a}}/\mu_0}(2-\delta_{0m}) \cdot \sum_{l=0}^{2N-1}(2l+1)\chi_l(\tau)\Lambda_l^m(u)\Lambda_l^m(\mu_0), \tag{6.41}$$

$$X_{02}^m(\tau, u) = \frac{a(\tau)F^{\text{s}}}{4\pi}\frac{\mu_0}{\mu_{0m}}\mathcal{T}_{\text{b}}(-\mu_0, -\mu_{0m}; m_{\text{rel}})e^{-\tau_{\text{a}}(1/\mu_0 - 1/\mu_{0m})}(2-\delta_{0m}) \cdot \sum_{l=0}^{2N-1}(-1)^{l+m}(2l+1)\chi_l(\tau)\Lambda_l^m(u)\Lambda_l^m(\mu_{0m}). \tag{6.42}$$

In addition to applying boundary conditions at the top of the atmosphere and the bottom of the ocean, we must properly account for the reflection from and transmission through the interface. Here the following conditions apply:

$$I_{\text{a}}^+(\tau_{\text{a}}, \mu^{\text{a}}) = \rho_{\text{s}}(-\mu_0; m_{\text{rel}})I_{\text{a}}^-(\tau_{\text{a}}, \mu^{\text{a}}) + \mathcal{T}_{\text{b}}(\mu^{\text{o}}; m_{\text{rel}})\left[I_{\text{o}}^+(\tau_{\text{a}}, \mu^{\text{o}})/m_{\text{rel}}^2\right], \tag{6.43}$$

$$\frac{I_{\text{o}}^-(\tau_{\text{a}}, \mu^{\text{o}})}{m_{\text{rel}}^2} = \rho_{\text{s}}(\mu^{\text{o}}; m_{\text{rel}})\frac{I_{\text{o}}^+(\tau_{\text{a}}, \mu^{\text{o}})}{m_{\text{rel}}^2} + \mathcal{T}_{\text{b}}(-\mu^{\text{a}}; m_{\text{rel}})I_{\text{a}}^-(\tau_{\text{a}}, \mu^{\text{a}}) \quad (\mu^{\text{o}} > \mu_{\text{c}}),$$
$$I_{\text{o}}^-(\tau_{\text{a}}, \mu^{\text{o}}) = I_{\text{o}}^+(\tau_{\text{a}}, \mu^{\text{o}}) \quad (\mu^{\text{o}} < \mu_{\text{c}}). \tag{6.44}$$

Here $I_{\text{a}}(\tau_{\text{a}}, \mu^{\text{a}})$ refers to the intensity in the atmosphere evaluated at the interface, while $I_{\text{o}}(\tau_{\text{a}}, \mu^{\text{o}})$ refers to the intensity in the ocean evaluated at the interface. The first of these equations states that the upward intensity at the interface in the atmosphere consists of the specularly reflected downward atmospheric radiation plus the transmitted upward oceanic radiation. Similarly, the second equation states that the downward intensity at the interface in the ocean consists of the reflected component of oceanic origin plus a transmitted component originating in the atmosphere. Finally, the last equation ensures that radiation in the total reflection region is properly taken into account.

The demarcation between the refractive and the total reflective region in the ocean is given by the critical angle, whose cosine is

$$\mu_c = \sqrt{1 - 1/m_{rel}^2}. \tag{6.45}$$

μ^o and μ^a are connected through the relation

$$\mu^o = \mu^o(\mu^a) = \sqrt{1 - [1 - (\mu^a)^2]/m_{rel}^2}.$$

Note that we have defined $\rho_s(\mu; m_{rel})$ and $\mathcal{T}_b(\mu; m_{rel})$ as the specular reflectance and transmittance of the invariant intensity, I/m_r^2, where m_r is the local index of the real part of the refractive index. The reflectance and transmittance are derived from Fresnel's equations (see Appendix E).

The solution of the above equations requires techniques to be described briefly in Chapter 8. Alternative treatments are given in texts on marine optics.[11]

6.7 Examples of Phase Functions

The action of scattering particles (including molecules) on the intensity and the state of polarization of an incident radiation field can be represented as a linear operator, called the *scattering* or *phase matrix*. As explained in Appendix I the elements of the phase matrix depend upon the optical properties of the particles. The 4×4 phase matrix connects the Stokes vector of the incident radiation to the scattered radiation. In this book we are mainly concerned with the intensity of light (i.e., the first component of the Stokes vector I) since it conveys the information on the energy carried by the light field. In the *scalar approximation* we require only one element of the phase matrix, which is usually referred to as the *phase function*.[12] In many applications, such as heating/cooling of the medium, photodissociation of molecules, and biological dose rates, it is often permissible to ignore polarization effects. This is because the error incurred by doing so is very small compared to errors caused by uncertainties in the input parameters to the computation, which determine the optical properties of the medium. We limit our attention to the phase function here, although we should caution that in some remote sensing applications, describing the state of polarization may be absolutely necessary.

6.7.1 Rayleigh Phase Function

As explained in Chapter 3, if the light frequency is not close to a resonant frequency, the scattering of light by molecules is similar to that of an induced dipolar oscillator. The classical model that fits observations quite well considers the (unpolarized) incident wave to induce a motion of the bound electrons, which is in phase with the wave. Its interaction with an unpolarized wave results in the molecule extracting energy from the wave and then reradiating it in all directions. If we assume that the molecule is isotropic (which is not quite true for linear molecules such as N_2 and O_2) and the incident radiation is unpolarized, then it was shown in §3.4.1 that the normalized phase function is given by

$$p_{RAY}(\cos\Theta) = \frac{3}{4}(1+\cos^2\Theta). \tag{6.46}$$

Expanding p_{RAY} in terms of the incident and scattered polar and azimuthal angles, we find using Eq. 3.22

$$p_{RAY}(u',\phi';u,\phi) = \frac{3}{4}[1 + u'^2u^2 + (1-u'^2)(1-u^2)\cos^2(\phi'-\phi) + 2u'u(1-u'^2)^{1/2}(1-u^2)^{1/2}\cos(\phi'-\phi)]. \tag{6.47}$$

The azimuthally averaged phase function is found to be

$$p_{RAY}(u',u) = \frac{1}{2\pi}\int_0^{2\pi} d\phi'\, p_{RAY}(u',\phi';u,\phi) = \frac{3}{4}\left[1 + u'^2u^2 + \frac{1}{2}(1-u'^2)(1-u^2)\right]. \tag{6.48}$$

Expressing the above in terms of Legendre polynomials, one may show that

$$p_{RAY}(u',u) = 1 + \frac{1}{2}P_2(u)P_2(u').$$

The asymmetry factor for the Rayleigh phase function is therefore

$$g = \chi_1 = \frac{1}{2}\int_{-1}^{+1} du'\, p_{RAY}(u',u)P_1(u') = \frac{1}{2}\int_{-1}^{+1} du'u'\, p_{RAY}(u',u) = 0$$

because of the orthogonality of the Legendre polynomials. This result can be proven for any *even* function of $\cos\Theta$, that is, for any terms in the phase function symmetric around $\Theta = 90°$, and should be obvious from symmetry arguments.

6.7.2 The Mie–Debye Phase Function

Scattering in planetary media is caused by molecules and particulate matter. If the size of the scatterer is small compared to the wavelength as is the case for scattering of solar radiation by molecules, then the scattering phase function is only mildly anisotropic. Such a phase function poses no special problem for solving the radiative transfer equation. However, scattering of solar radiation by larger particles is characterized by

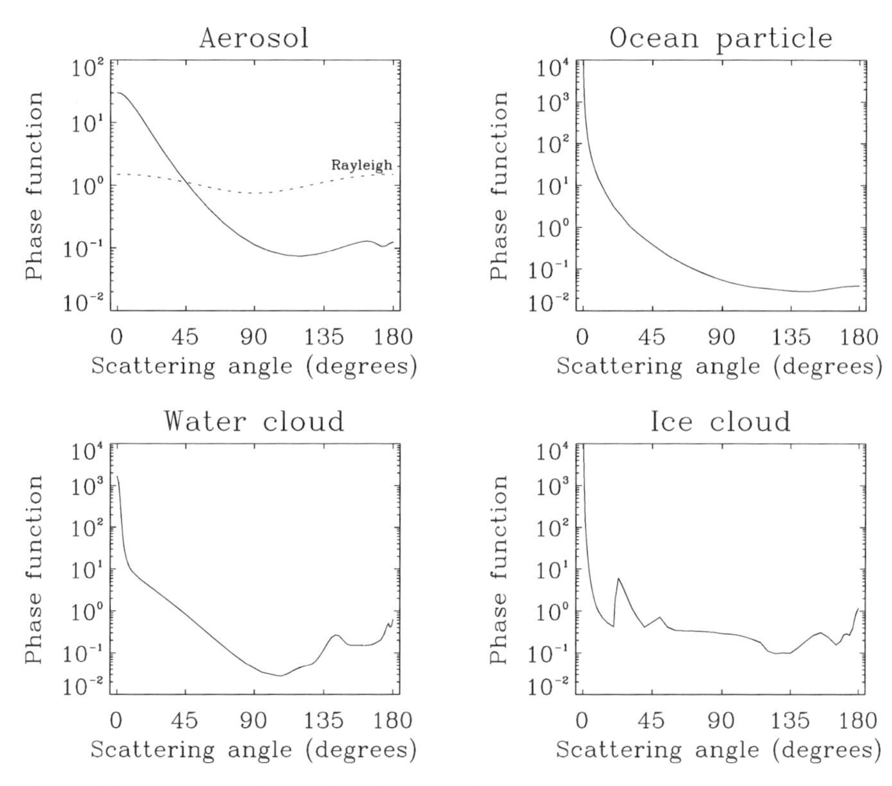

Figure 6.3 Illustration of phase functions occurring in planetary media. Shown are phase functions for: molecular (Rayleigh) scattering and aerosol particles (upper left); hydrosols (upper right); cloud droplets (lower left); and ice crystals (lower right).

strong forward scattering with a *diffraction peak* in the forward direction. Examples of phase functions for four different media are shown in Fig. 6.3.

The *Mie–Debye theory* (discussed in §3.5) has been refined and developed by hundreds of investigators. Although the mathematical foundation is complete, its numerical implementation has proven to be very challenging. Progress in developing fast and accurate computer algorithms continues to the present day. Although it is beyond the scope of this book to explore this byway, every serious student of radiative transfer should be familiar with the subject, for which there are many excellent references.[13]

Example 6.2 The Henyey–Greenstein Phase Function

A one-parameter phase function proposed by L. Henyey and J. Greenstein in 1941 is

$$p_{\mathrm{HG}}(\cos\Theta) = \frac{1-g^2}{(1+g^2-2g\cos\Theta)^{3/2}}. \tag{6.49}$$

This function has no physical basis and should be considered as a one-parameter analytic fit to an actual phase function. It should not be used except when the fit is reasonably good. However, as far as the radiative transfer is concerned, the requirement of "reasonableness" is not very strict, because the multiple-scattering process tends to smooth out irregularities present in the more accurate function. A remarkable feature of the *H–G function* (proven in Problem 6.1) is the fact that the Legendre polynomial coefficients are simply

$$\chi_l = (g)^l.$$

This partly explains its popularity, because only the first moment of the phase function (i.e., the asymmetry factor g) must be specified. Thus, the Legendre polynomial expansion of the H–G phase function is given simply by

$$p_{\mathrm{HG}}(\cos\Theta) = 1 + 3g\cos\Theta + 5g^2 P_2(cos\Theta) + 7g^3 P_3(cos\Theta) + \cdots.$$

Also, the Henyey–Greenstein phase function has the desirable feature that it yields complete forward scattering for $g = 1$, isotropic scattering for $g = 0$, and complete backward scattering for $g = -1$. The linear combination

$$p(\cos\Theta) = b p_{\mathrm{HG}}(g, \cos\Theta) + (1 - b) p_{\mathrm{HG}}(g', \cos\Theta)$$

is sometimes used to simulate a phase function with both a forward- and a backward-scattering component ($g > 0$ and $g' < 0$). Here $0 < b < 1$ and g and g' are usually different.

6.8 Scaling Transformations Useful for Anisotropic Scattering

The solution of the radiative transfer equation for strongly forward-peaked scattering is notoriously difficult. An accurate expansion of the phase function may require several hundred terms for a typical cloud phase function.

As we shall see, most methods of solving Eq. 6.35 start by approximating the integral term by a finite sum that is usually of the same order ($2N$) as the number of terms necessary to get a good Legendre polynomial representation of the phase function. This may lead to a large system of equations that requires such inordinate amounts of computer storage space and time as to render the solution impractical even on modern computers.

To circumvent this numerical difficulty associated with strongly forward-peaked scattering, *scaling transformations* have been invented. The motivation for scaling in this case is to transform a transfer equation with a strongly peaked phase function into a more tractable problem with a phase function that is much less anisotropic.

The pronounced forward scattering by cloud droplets becomes even more extreme if we plot the phase function as a function of the cosine of the scattering angle (instead of the scattering angle). In fact, the forward scattering peak takes on the resemblance of a Dirac δ-function when plotted versus cosine of the scattering angle. This suggests that it would be useful to treat photons scattered within the sharp forward peak as unscattered, and truncate this peak from the phase function. In ocean optics, this has been referred to as the *quasi-single-scattering approximation*.[14]

We start by assuming that the forward-scattering peak can be represented by a Dirac δ-function, while the remainder of the phase function is expanded in Legendre polynomials as usual. Thus, we set

$$\begin{aligned}
\hat{p}_{\delta-N}(\cos\Theta) &= \hat{p}_{\delta-N}(u',\phi';u,\phi) \\
&\equiv 2f\delta(1-\cos\Theta) + (1-f)\sum_{l=0}^{2N-1}(2l+1)\hat{\chi}_l P_l(\cos\Theta) \\
&= 4\pi f\delta(u'-u)\delta(\phi'-\phi) + (1-f)\sum_{l=0}^{2N-1}(2l+1)\hat{\chi}_l \\
&\quad\times\left\{\sum_{m=0}^{l}\Lambda_l^m(u')\Lambda_l^m(u)\cos m(\phi'-\phi)\right\},
\end{aligned} \tag{6.50}$$

where f $(0 \leq f \leq 1)$ is a dimensionless parameter to be determined by a fit to an actual phase function. We shall refer to this transformation as the δ-N method as indicated by the subscript. Note that if $f = 0$ we retain the usual Legendre polynomial expansion and $\hat{\chi}_l \equiv \chi_l$. Illustrations of actual phase functions occurring in nature and approximations using the δ-N method are provided in Fig. 6.4. This figure suggests that it would be problematic to obtain accurate intensities with low-order scaled phase functions, although such phase functions give accurate flux and mean intensity.

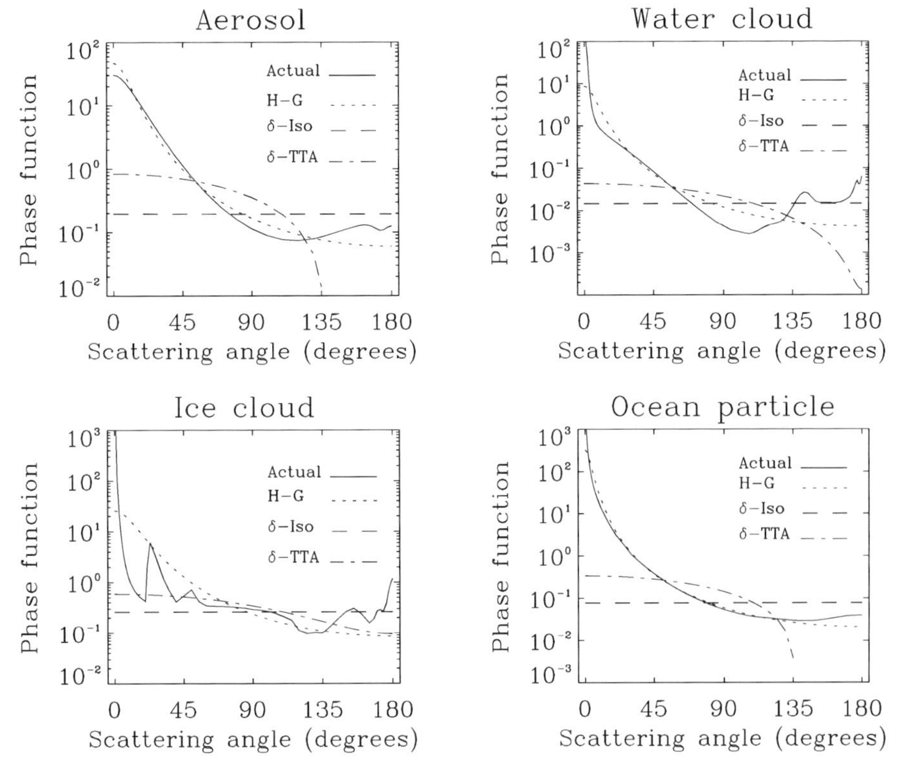

Figure 6.4 Illustration of actual and δ-N scaled phase functions for aerosol particles, water cloud droplets, ice particles, and hydrosols.

For simplicity we shall consider the azimuthally averaged radiative transfer equation below (see Problem 6.3 for a generalization to azimuthal dependence). We first find a general expression for the azimuthally averaged scaled phase function:

$$\hat{p}(u', u) = \frac{1}{2\pi} \int_0^{2\pi} d\phi \hat{p}(\cos\Theta)$$

$$= 2f\delta(u' - u) + (1 - f) \sum_{l=0}^{2N-1} (2l + 1)\hat{\chi}_l P_l(u') P_l(u). \tag{6.51}$$

The second part of the scaling problem concerns how we approximate the remaining (nonpeaked) part of the phase function. In the following subsections, we shall start with the simplest case, which emphasizes the principles involved, and then generalize this method to achieve better accuracy. A more formal mathematical approach (see Appendix J) to the scaling procedure is provided elsewhere.[15]

6.8.1 The δ-Isotropic Approximation

The crudest way of handling the problem with strong forward scattering is to approximate the remainder of the phase function by a constant (isotropic scattering). The azimuthally averaged phase function becomes

$$\hat{p}_{\delta\text{-iso}}(u', u) \equiv \frac{1}{2\pi} \int_0^{2\pi} d\phi \hat{p}(\cos\Theta) = 2f\delta(u' - u) + (1 - f). \tag{6.52}$$

Substitution of this phase function into the azimuthally averaged radiative transfer equation (Eq. 6.35) yields

$$u\frac{dI(\tau, u)}{d\tau} = I(\tau, u) - \frac{a}{2} \int_{-1}^{1} du' p(u', u) I(\tau, u') \tag{6.53}$$

$$= I(\tau, u) - afI(\tau, u) - \frac{a(1 - f)}{2} \int_{-1}^{1} du' I(\tau, u') \tag{6.54}$$

or

$$u\frac{dI(\hat{\tau}, u)}{d\hat{\tau}} = I(\hat{\tau}, u) - \frac{\hat{a}}{2} \int_{-1}^{1} du' I(\hat{\tau}, u'), \tag{6.55}$$

where

$$d\hat{\tau} \equiv (1 - af)d\tau, \qquad \hat{a} \equiv \frac{(1 - f)a}{1 - af}. \tag{6.56}$$

For simplicity we ignored the source term $Q(\tau, u) \equiv S^*(\tau, u) + (1 - a)B(\tau)$. Because we have divided by $1 - af$ this term would simply become $\hat{Q}(\hat{\tau}, u) = Q(\tau, u)/(1 - af)$ (see also Problem 4.3).

Finally, to complete the scaling, we ask: How do we specify f, the strength of the forward-scattering peak? Clearly there is no unique choice, but it should depend in some simple way on the asymmetry factor, g. Since $\chi_1 = g$ is the first moment of the unscaled phase function, this suggests that we equate the first moment of p_{acc} (the accurate phase function) to the first moment of the scaled phase function $\hat{p}$. This is

$$\hat{\chi}_1 = \frac{1}{2} \int_{-1}^{1} du' u' \hat{p}_{\delta\text{-iso}}(u', u) = f. \tag{6.57}$$

The above result follows from substitution of Eq. 6.52 and carrying out the integration. Our matching requirement sets $f = \chi_1 = g$.

The δ-isotropic approximation is sometimes referred to as the *transport approximation*. Use of it in the radiative transfer equation leads to the transformation of Eq. 6.53 to an equation with isotropic scattering, but with a scaled optical depth $d\hat{\tau} = (1 - ag)d\tau$ and a scaled single-scattering albedo $\hat{a} = (1 - g)a/(1 - ag)$. Thus, the radiative transfer equation with strongly anisotropic scattering is reduced to an equation with isotropic scattering, which is easier to handle numerically. These particular scaling transformations of the optical depth and the single-scattering albedo are sometimes referred to as *similarity relations*.[16]

6.8.2 The δ-Two-Term Approximation

A better approximation results from representing the remainder of the phase function by two terms[17] as follows (setting $N = 1$ in Eq. 6.51):

$$\hat{p}_{\delta\text{-TTA}}(u', u) = 2f\delta(u' - u) + (1 - f)\sum_{l=0}^{l=1}(2l + 1)\hat{\chi}_l P_l(u')P_l(u). \tag{6.58}$$

Substitution of this phase function into the azimuthally averaged radiative transfer equation yields

$$u\frac{dI(\hat{\tau}, u)}{d\hat{\tau}} = I(\hat{\tau}, u) - \frac{\hat{a}}{2}\sum_{l=0}^{l=1}(2l + 1)\hat{\chi}_l P_l(u) \int_{-1}^{1} du' P_l(u')I(\hat{\tau}, u'), \tag{6.59}$$

where $d\hat{\tau}$ and $\hat{a}$ are defined in Eq. 6.56. Again, by matching moments of the approximate and accurate phase functions we find

$$\hat{\chi}_1 \equiv \hat{g} = \frac{\chi_1 - f}{1 - f} = \frac{g - f}{1 - f}, \quad f = \chi_2.$$

6.8.3 Remarks on Low-Order Scaling Approximations

We may now ask what (if anything) has been gained by the scaling? First, we note that the two-term approximation (TTA) makes the replacement $\chi_l = 0$ for $l \geq 2$. Thus, all the higher-order moments, which may contribute substantially to the phase function if it happens to be strongly anisotropic, have been brutally set to zero. From the discussion of the general case (arbitrary N) below, it will be shown that the δ-TTA approximation is equivalent to making the replacement $\chi_l = \chi_2$ for $l \geq 2$. Thus, for strongly anisotropic phase functions, $\chi_3 = \chi_2$, which demonstrates the advantage of the TTA scaling transformation (i.e., the third moment and higher moments are set equal to the second moment, which is generally expected to be better than setting them to zero).

The two-term approximation is commonly used in connection with the *two-stream* and *Eddington approximations* to be discussed in Chapter 7. As we shall see, in these approximations the transfer equation is replaced by two coupled, first-order differential equations, which are easily solved analytically. In the two-stream case, the two coupled equations are obtained by replacing the integral (multiple-scattering term) by just two terms or "streams" (this essentially amounts to adopting a two-point quadrature). In the Eddington approximation, one expands the intensity in Legendre polynomials, keeping only the first two terms (i.e., we set $I(\tau, u) = I_0(\tau) + uI_1(\tau)$). Inserting this approximation into Eqs. 6.20–6.21 and using the first two moments, we arrive at a set of two coupled equations. This will be discussed in some detail in Chapter 7. Here it suffices to note that, as a rule of thumb, it is customary to keep the number of terms in the expansion of the phase function equal to the number of quadrature terms (or expansion terms for the intensity) in the approximation. Hence, a two-term expansion of the phase function leads naturally to the two-stream (or Eddington) approximation, although in principle one could retain more terms in the expansion of the phase function even in the two-stream approximation to the radiative transfer equation. In fact, as explained in Chapter 7, it is possible to use the exact phase function even in the two-stream approximation. The rule of thumb referred to above merely reflects the philosophy that *there may not be much to gain* from using a very accurate representation of the phase function if the level of approximation for solving the radiative transfer equation is much cruder.

Finally, if we use the H–G phase function in the δ-TTA approximation, then $f = g^2$, and therefore $\hat{g} = g/(1+g)$. Then $0 \leq \hat{g} \leq 0.5$ when $0 \leq g \leq 1$. This implies that the δ-TTA approximation applies to a range of $\hat{g}$ ($\hat{g} < 0.5$) for which the two-stream approximation has been shown to be reasonably accurate. We note, however, that we must require $\hat{g} \equiv \hat{\chi}_1 < 1/3$ to guarantee that the truncated phase function [i.e., $\hat{p}_{\text{trunc}}(\cos\Theta) = (1-f)(1+3\hat{g}\cos\Theta)$] is positive for all scattering angles. This should alert us to the possibility that we may obtain unphysical results (e.g., negative reflectance) unless $\hat{g} < 1/3$ or $g < 1/2$.

6.8.4 The δ-N Approximation: Arbitrary N

We generalize the method outlined above to include an arbitrary number of terms to represent the remainder of the phase function in Eq. 6.50. The resulting transformation, referred to as the *δ-N method*,[18] is designed to scale phase functions that are sharply peaked in the forward direction. It has proven to be particularly useful in solving practical radiative transfer problems involving sharply forward-peaked phase functions associated with scattering from cloud droplets and particles.

By substituting Eq. 6.51 into Eq. 6.53, we find

$$u\frac{dI(\hat{\tau},u)}{d\hat{\tau}} = I(\hat{\tau},u) - \frac{\hat{a}}{2}\sum_{l=0}^{2N-1}(2l+1)\hat{\chi}_l P_l(u)\int_{-1}^{1}du' P_l(u')I(\hat{\tau},u'), \quad (6.60)$$

where $d\hat{\tau}$ and $\hat{a}$ are defined in Eq. 6.56.

As before, we set the expansion coefficients $\hat{\chi}_l$ equal to the moments, χ_l, of the accurate (unscaled) phase function by equating moments of the accurate and approximate phase functions:

$$\chi_l = \frac{1}{2}\int_{-1}^{1} d(\cos\Theta)p_{\rm acc}(\cos\Theta)P_l(\cos\Theta),$$

$$\hat{\chi}_l = \frac{1}{2}\int_{-1}^{1} d(\cos\Theta)\hat{p}_{\delta\text{-}N}(\cos\Theta)P_l(\cos\Theta),$$

where $p_{\rm acc}$ denotes the accurate value for p. This leads to

$$\chi_l = f + (1-f)\hat{\chi}_l \quad \text{or} \quad \hat{\chi}_l = \frac{\chi_l - f}{1-f}.$$

It is easy to see that if we set $f = 0$, then $d\hat{\tau} = d\tau$, $\hat{a} = a$, and $\hat{\chi}_l = \chi_l$, implying that the *scaled* equation reduces to the *unscaled* one as it should. We determine f by setting $f = \chi_{2N}$ (truncation), which is clearly a generalization of the procedure used for $N = 1$. We note that setting $\hat{\chi}_l = 0$ for $l \geq 2N$ is equivalent to replacing χ_l with χ_{2N} for $l \geq 2N$. Thus, whereas the ordinary Legendre polynomial expansion of order $2N$ sets $\chi_l = 0$ for $l \geq 2N$, the δ-N method makes the replacement $\chi_l = \chi_{2N}$ for $l \geq 2N$. Finally we note that the error in the phase function representation incurred by using the δ-N method is

$$p_{\rm acc}(\cos\Theta) - \hat{p}_{\delta\text{-}N}(\cos\Theta) = \sum_{l=2N+1}^{\infty}(2l+1)(\chi_l - \chi_{2N})P_l(\cos\Theta).$$

Example 6.3 The δ-Henyey–Greenstein Approximation (δ-HG)

In this case we have $\hat{p}(\cos\Theta) = 2f\delta(1-\cos\Theta) + (1-f)p_{\rm HG}(\cos\Theta)$, where $p_{\rm HG}(\cos\Theta) = \sum_{l=0}^{\infty}(2l+1)g^l P_l(\cos\Theta)$. By matching the first two moments of this phase function ($\hat{\chi}_1$ and $\hat{\chi}_2$)

to the actual phase function we want to approximate, we find

$$\hat{\chi}_1 = f + (1 - f)g = \chi_1,$$
$$\hat{\chi}_2 = f + (1 - f)g^2 = \chi_2.$$

Solving for g and f we find

$$g = \frac{\chi_1 - \chi_2}{1 - \chi_1}, \qquad f = \frac{\chi_2 - \chi_1^2}{1 - 2\chi_1 + \chi_2}.$$

6.8.5 Mathematical and Physical Meaning of the Scaling

It is worth noting that an essential feature of the scaling is to turn the unscaled problem into one in which the optical depth is reduced ($d\hat{\tau} < d\tau$), while the absorption is artificially increased ($\hat{a} < a$). In addition the scattering phase function appears considerably less anisotropic. It should be stressed, however, that Eq. 6.60 is of identical mathematical form to Eq. 6.35. This implies that whatever tools we have available for solving the unscaled equation can be applied to the scaled equation. From a pragmatic point of view, what has been gained by the scaling is to render the problem more tractable numerically because the new equation has a phase function that is much less anisotropic due to the truncation of the forward-scattering peak. Hence, we expect that considerably fewer terms are needed to obtain an adequate Legendre polynomial expansion of the phase function. This, in turn, implies that the scaled equation is easier to solve by numerical means, as will be illustrated in Chapters 7 and 8. Thus, the δ-N method is not a method of solution but rather a way of making the transfer equation easier to solve by whatever analytical and/or numerical techniques we have at our disposal.

From the physical point of view the δ-N approximation relies on the following premise: *Those beams that are scattered through the small angles contained within the forward peak are not scattered at all.* These beams are in fact "added back" to the original radiation field. This explains why the scaled optical depth $\hat{\tau}$ is smaller than the original τ. The effective asymmetry factor is also less than the original (unscaled) value, since the angular distribution of those beams scattered outside the forward peak is (by definition) less extreme in its angular dependence. Consider the transmitted flux in the scaled problem

$$F(\hat{\tau}^*) = F_{\rm d}(\hat{\tau}^*) + \mu_0 F^{\rm s} e^{-\hat{\tau}^*/\mu_0},$$

where the d subscript denotes the diffuse flux. Since $\hat{\tau}^* < \tau^*$, this means that the scaled directly transmitted solar flux is greater than it is in the unscaled problem. Because of the phase-function truncation, the "direct" flux actually contains some scattered beams of radiation traveling in very nearly the same direction as the incident beam. As an example, the Sun's rays shining through a hazy or dusty atmosphere are spread out into a very bright blurry disk, somewhat greater than the Sun's disk itself. This is called the Sun's *aureole*, which has been used as a means of inferring the mean particle size in tropospheric haze. A substantial fraction of the solar aureole would be included in the δ-N direct flux. The scattered flux in this approximation would apply to those beams scattered largely outside the aureole. Finally, we note that since the total downward flux must be the same whether we use scaling or not, that is, $F_{\text{tot}}^-(\hat{\tau}) = F_{\text{tot}}^-(\tau)$, or

$$F_{\text{d}}^-(\hat{\tau}) + \mu_0 F^{\text{s}} e^{-\hat{\tau}/\mu_0} = F_{\text{d}}^-(\tau) + \mu_0 F^{\text{s}} e^{-\tau/\mu_0},$$

we can always recover the unscaled downward diffuse flux by solving for $F_{\text{d}}^-(\tau)$:

$$F_{\text{d}}^-(\tau) = F_{\text{d}}^-(\hat{\tau}) + \mu_0 F^{\text{s}} \big(e^{-\hat{\tau}/\mu_0} - e^{-\tau/\mu_0}\big), \tag{6.61}$$

where all the quantities on the right are known. No such "correction" for the upward flux is necessary.

6.9 Prototype Problems in Radiative Transfer Theory

In this section, we describe a number of standard radiative transfer problems, which have received much attention in the research literature. By concentrating on these special cases, we may compare the numerical results from approximate solutions to exact solutions. Because of the many idealizations used in these problems, it is possible to study the transfer process in detail, without the distraction of the many complexities that always accompany real-life transfer problems. In addition, the methods of attack we find successful for these prototype problems can be applied with more confidence to the more realistic problems.

Each prototype problem is for a slab geometry and an optically uniform (homogeneous) medium. The radiation is considered to be monochromatic and unpolarized. The complete specification of a prototype problem requires five input variables: (1) τ^*, the vertical optical depth of the slab; (2) $S^*(\tau, \hat{\Omega})$, the internal or external sources; (3) $p(\hat{\Omega}', \hat{\Omega})$, the phase function; (4) a, the single-scattering albedo; and (5) $\rho(-\hat{\Omega}', \hat{\Omega})$, the bidirectional reflectance of the underlying surface ($\rho_{\text{L}} =$ constant, for a Lambert surface). For $\tau^* \to \infty$, a condition on the source function is that $\lim_{\tau\to\infty} S(\tau)e^{\tau} \to 0$. The analytic or numerical solution of the radiative transfer equation provides the following output variables: the reflectance, transmittance, absorptance, and emittance; the source function; the internal intensity field; and the heating rate and net flux throughout the medium. A cartoon illustration of the standard problems we describe below is provided in Fig. 6.5.

6.9.1 Prototype Problem 1: Uniform Illumination

Here the incident radiation field is taken to be constant ($=\mathcal{I}$) in the downward hemisphere. Because of the azimuthal symmetry of the incident radiation, the radiation field depends only upon τ and μ. Furthermore, the source function depends only upon τ.

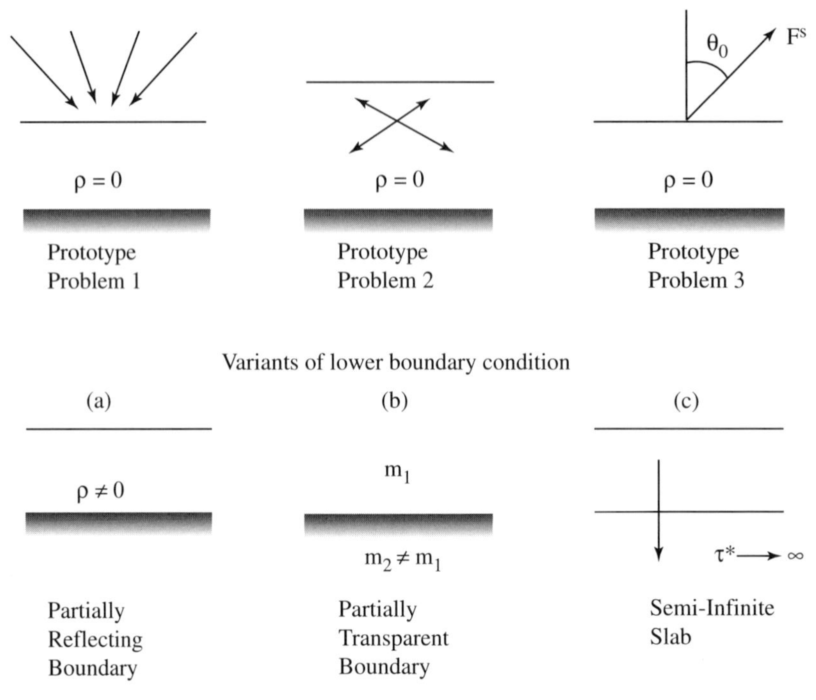

Figure 6.5 Illustration of Prototype Problems in radiative transfer.

It so happens that the frequency-integrated problem for conservative and isotropic scattering reduces mathematically to that of a simple *greenhouse problem*. This describes the enhancement of the surface temperature over that expected from planetary radiative equilibrium. This application is discussed in Chapter 12.

In addition, *Prototype Problem 1* approximately reproduces the illumination conditions provided by an optically thick cloud overlying an atmosphere. The source for the diffuse emission is

$$S^*(\tau) = \frac{a}{4\pi}\int_0^{2\pi} d\phi \int_0^1 d\mu\, \mathcal{I} e^{-\tau/\mu} = \frac{a}{2}\mathcal{I} E_2(\tau),$$

where $E_2(\tau)$ is the *second exponential integral* defined by Eq. 5.61. However, this problem is solved more efficiently by setting $S^* = 0$ in Eqs. 6.6 and 6.8 and applying a uniform boundary condition to the *total* intensity $I^-(\tau = 0, \mu) = \mathcal{I}$ at the upper boundary. It is straightforward to add the effects of surface reflection later, as described at the end of this section.

6.9.2 Prototype Problem 2: Constant Imbedded Source

For thermal radiation problems, the term $(1 - a)B$ is the "driver" of the scattered radiation. This is an "imbedded source" equal to the local rate at which inelastic collisions are producing thermal emission. Note that in general, this term will be a strong function of frequency, and of course it depends upon the temperature, through Eq. 4.4. In our *Prototype Problem 2* we assume that the product $(1 - a)B$ is constant with depth.

Example 6.4 The Conservative Limit for *Prototype Problem 2*: The Milne Problem

The classical *Milne Problem* can be thought of as the limit of the imbedded-source problem, in which there is no absorption so that $a \to 1$. Then the source term is zero, and one is left with the simple radiative transfer equation

$$u\frac{dI(\tau,u)}{d\tau} = I(\tau,u) - \frac{1}{2}\int_{-1}^{+1} du' I(\tau,u').$$

Note that we consider the full-range intensity ($-1 \le u \le +1$) because no particular simplifications follow from using half-range intensities, and because this is the traditional form of the Milne problem. The medium is usually taken to be semi-infinite, and a source of radiation is assumed to be placed at an infinite distance away from the boundary ($\tau = 0$). The radiation therefore trickles upward and escapes without any losses into the half-space above the boundary. As we shall see in Chapter 7 the upward radiative flux is constant, since there is no absorption in the medium.

6.9.3 Prototype Problem 3: Diffuse Reflection Problem

In this problem we consider collimated incidence at $\tau = 0$, and a lower boundary that may be partly reflecting as explained below. The case of collimated incidence as opposed to uniform incidence can be considered to be *the* classical planetary problem. For shortwave applications, the term $(1 - a)B$ can be ignored and the only source term is

$$S^*(\tau, \pm\mu, \phi) = \frac{aF^{\text{s}}}{4\pi} p(-\mu_0, \phi_0; \pm\mu, \phi)e^{-\tau/\mu_0}. \tag{6.62}$$

Note that in contrast to *Prototype Problems 1* and *2*, the radiation field depends upon both μ and the azimuthal coordinate ϕ. The lower boundary condition appropriate for this problem is described below.

An illustration of the total source functions, S given by Eq. 6.14, in *Prototype Problems 1* through *3* is provided in Fig. 6.6. For simplicity we assume isotropic scattering so that S depends only on optical depth in all the *Prototype Problem* solutions displayed in Fig. 6.6.

6.9.4 Boundary Conditions: Reflecting and Emitting Surface

We first consider a Lambertian surface (BRDF $= \rho_{\text{L}}$), which also emits thermal IR with an emittance ϵ_{s} and temperature T_{s}. The upward intensity at the surface is given

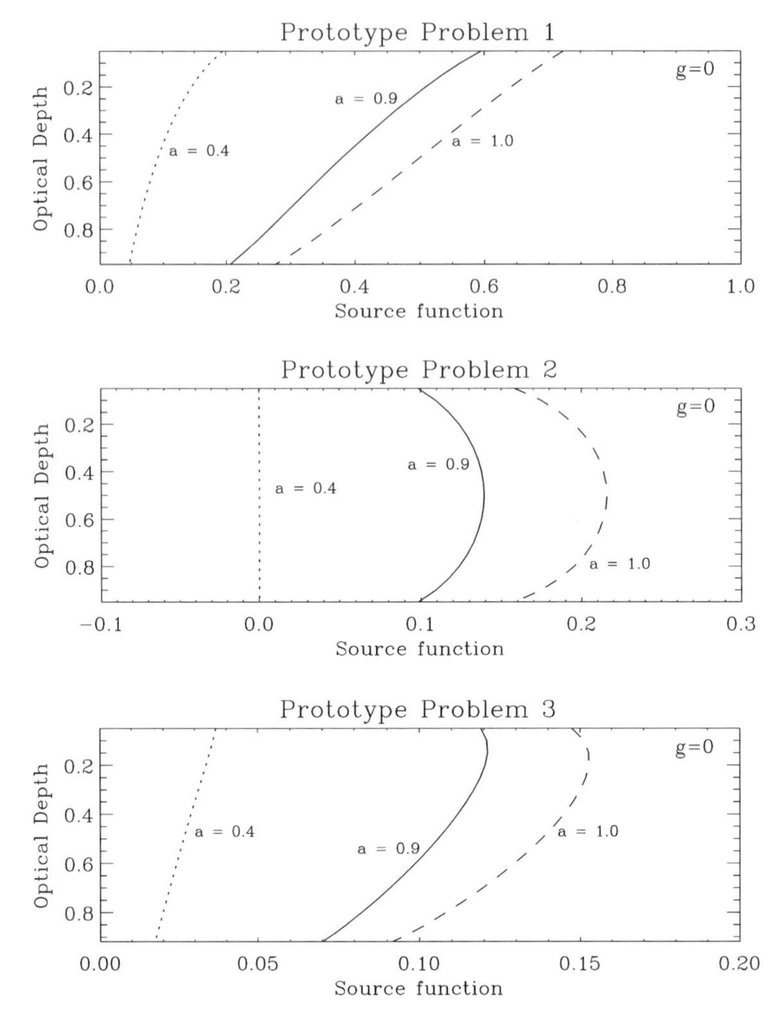

Figure 6.6 Source functions for isotropic scattering for *Prototype Problems 1* through *3* considered in this book. The upper panel pertains to uniform illumination ($\mathcal{I} = 1$), the middle panel to a constant internal source ($B = 1$), and the lower one to an incident collimated beam ($F^{\rm s} = 1$). A vertical optical depth $\tau^* = 1$ is assumed. The lower boundary is assumed to be black. These source functions were computed using the DISORT code (see Chapter 8) with eight streams.

by (see Eq. 5.11)

$$I^+(\tau^*, \mu, \phi) = \rho_{\rm L} F_{\rm d}^-(\tau^*) + \rho_{\rm L} \int_0^{2\pi} d\phi' \int_0^1 d\mu' \mu' I^-(0, \mu', \phi') e^{-\tau^*/\mu'} + \epsilon_{\rm s} B(T_{\rm s}). \quad (6.63)$$

Here $B(T_{\rm s})$ denotes the Planck function at the appropriate frequency and $F_{\rm d}^-(\tau^*)$ is the downward diffuse intensity at the lower boundary. The upper and lower boundary conditions for the three prototype problems can now be written down immediately:

Prototype Problem 1

$$I^-(0,\mu) = \mathcal{I}, \qquad I^+(\tau^*,\mu) = \rho_L\left[F_d^-(\tau^*) + 2\pi\mathcal{I}E_3(\tau^*)\right] + \epsilon_s B(T_s), \tag{6.64}$$

Prototype Problem 2

$$I^-(0,\mu) = 0, \qquad I^+(\tau^*,\mu) = \rho_L F_d^-(\tau^*); \tag{6.65}$$

Prototype Problem 3

$$\begin{aligned} I^-(0,\mu,\phi) &= F^s\delta(\mu-\mu_0)\delta(\phi-\phi_0), \\ I^+(\tau^*,\mu,\phi) &= \rho_L\left[F_d^-(\tau^*) + \mu_0 F^s e^{-\tau^*/\mu_0}\right]. \end{aligned} \tag{6.66}$$

The above equations are expressed in terms of the unknown quantity $F_d^-(\tau^*)$. Although this would not appear to be very helpful in the ultimate solution, it is not a difficulty for the methods of solution described in this book (Chapters 7 and 8). These techniques rely upon the intensity (and flux) being expressed as a finite sum (say of N terms) of known functions of angle and optical depth. These functions are weighted by unknown coefficients. Thus the problem becomes one of solving for the coefficients, which are determined by a set of algebraic equations of $N-2$ equations in the N unknowns. The two boundary conditions provide two necessary constraints, which then provide the necessary N equations.

For *Prototype Problem 3* we may encounter situations where the surface is described by a more general reflectance law. In this case the upward intensity at the surface is given by

$$\begin{aligned} I^+(\tau^*,\mu,\phi) = &\int_0^{2\pi} d\phi' \int_0^1 d\mu'\mu'\rho_d(-\mu',\phi';\mu,\phi)I^-(\tau^*,+\mu',\phi') \\ &+ \mu_0 F^s e^{-\tau^*/\mu_0}\rho_d(-\mu_0,\phi_0;+\mu,\phi) + \epsilon(\mu)B(T_s), \end{aligned} \tag{6.67}$$

where we have assumed no ϕ dependence of the thermal emission. Here ρ_d is the diffuse BRDF, assumed to be a function only of the *difference* between the azimuthal angles of the incident and the reflected radiation. This will enable us to separate the Fourier components by expanding the BRDF in a Fourier cosine series of $2N$ terms. Thus, we write

$$\begin{aligned} \rho_d(-\mu',\phi';\mu,\phi) &= \rho_d(-\mu',\mu;\phi-\phi') \\ &= \sum_{m=0}^{2N-1} \rho_d^m(-\mu',\mu)\cos m(\phi-\phi'), \end{aligned} \tag{6.68}$$

where the expansion coefficients are computed from

$$\rho_d^m(-\mu',\mu) \equiv \frac{1}{\pi}\int_{-\pi}^{\pi} d(\phi-\phi')\rho_d(-\mu',\mu;\phi-\phi')\cos m(\phi-\phi'). \tag{6.69}$$

Substituting Eq. 6.68 into Eq. 6.67 and using Eq. 6.34, we find that each Fourier component must satisfy the bottom boundary condition (see Appendix N for a derivation)

$$I^{m+}(\tau^*, \mu) = \delta_{m0}\epsilon(\mu)B(T_s) + (1 + \delta_{m0}) \int_0^1 d\mu' \mu' \rho_d^m(-\mu', \mu) I^{m-}(\tau^*, \mu')$$
$$+ \frac{\mu_0}{\pi} F^s \rho_d^m(-\mu_0, \mu) e^{-\tau^*/\mu_0} \quad m = 0, 1, \ldots, 2N - 1. \quad (6.70)$$

Finally, for an atmosphere overlying a body of calm water, we must use the interface conditions provided in §6.6 (Eqs. 6.43–6.44) to account for the reflection and transmission taking place at the interface between the two strata with differing indices of refraction. Then the coupled atmosphere–ocean system should be considered as described in §6.6.

6.10 Reciprocity, Duality, and Inhomogeneous Media

We have noted previously (see Appendices J, P, and Eq. 9) that the reciprocity relationships satisfied by the BRDF and flux reflectance are

$$\rho(-\hat{\Omega}', \hat{\Omega}) = \rho(-\hat{\Omega}, \hat{\Omega}'), \qquad \rho(-\hat{\Omega}', 2\pi) = \rho(-2\pi, \hat{\Omega}'), \quad (6.71)$$

where we have suppressed the ν argument. The *Reciprocity Principle* states that in any linear system, the pathways leading from a cause (or action) at one point to an effect (or response) at another point can be equally well traversed in the opposite direction.[19] Equation 6.71 states that the BRDF is unchanged upon a reversal of the direction of the light rays. Equation 6.71 follows from the reciprocity of the BRDF and implies that the flux reflectance of *Prototype Problems 2* and *3* are related (see §6.9). This is an example of the more general *Principle of Duality*, in which the solution of one problem is related to the solution of a second problem.

A similar relationship exists for the transmittance. It is important to point out that our previous discussions are applicable only to homogeneous media. For plane-parallel media, this means that the optical properties, such as a and p, are uniform with optical depth. It is not difficult to show that for an inhomogeneous slab the reflectance and transmittance for a slab illuminated from the top are in general different from those of a slab illuminated from the bottom. To emphasize this fact, we denote properties for illumination from the bottom with the symbol $\tilde{}$. The more general reciprocity relationships for transmittance and flux transmittance are

$$\mathcal{T}(-\hat{\Omega}', -\hat{\Omega}) = \tilde{\mathcal{T}}(+\hat{\Omega}, +\hat{\Omega}'), \qquad \mathcal{T}(-2\pi, -\hat{\Omega}) = \tilde{\mathcal{T}}(+\hat{\Omega}, +2\pi). \quad (6.72)$$

Therefore, the principle of duality connects the transmittances of *Prototype Problems 1* and *3*, in a similar way as for reflectances. The remaining relationships can now be listed:

$$\tilde{\rho}(+\hat{\Omega}', -\hat{\Omega}) = \tilde{\rho}(+\hat{\Omega}, -\hat{\Omega}'), \qquad \tilde{\rho}(+\hat{\Omega}', -2\pi) = \tilde{\rho}(+2\pi, -\hat{\Omega}'). \tag{6.73}$$

The practical implications of the above results can be illustrated by one example. Suppose we are interested in only boundary quantities, such as flux reflectance and transmittance, and we want to solve a problem involving collimated radiation for many values of the incoming solar direction. Then it is much more efficient to consider the problem of uniform illumination and solve for the reflected intensity from

$$\begin{aligned} I^+(0, \hat{\Omega}) &= \int_+ d\omega' \cos\theta' \rho(-\hat{\Omega}', +\hat{\Omega})\mathcal{I} = \mathcal{I}\rho(-2\pi, +\hat{\Omega}) \\ &= \mathcal{I}\rho(-\hat{\Omega}, +2\pi). \end{aligned} \tag{6.74}$$

The last result follows from Eq. 6.71 and is the desired flux reflectance for collimated incidence. We can find $\rho(-\hat{\Omega}, +2\pi)$, the flux reflectance for collimated light incident in direction $-\hat{\Omega}$, for every value of $-\hat{\Omega}$ of interest, by applying uniform illumination with $\mathcal{I} = 1$ and solving for the intensity $I^+(0, \hat{\Omega})$. Moreover, by integrating Eq. 6.74 we find

$$F^+ = 2\pi \int_0^1 d\mu\mu I^+(0, \mu) = 2\pi\mathcal{I} \int_0^1 d\mu\mu\rho(-\mu, +2\pi) = \pi\mathcal{I}\bar{\rho}, \tag{6.75}$$

which shows that we can obtain the spherical albedo, $\bar{\rho}$ (§5.2.11), by computing the flux reflectance, $F^+/\pi\mathcal{I}$, resulting from uniform illumination.

6.11 Effects of Surface Reflection on the Radiation Field

We now consider the effects of a reflecting lower boundary on the reflectance and transmittance of a homogeneous plane-parallel slab overlying a partially reflecting surface. The principle can be established by simplifying the boundary reflectance to that of an idealized Lambert surface. We will show that we can express the solutions for the emergent intensities algebraically in terms of the solutions derived above for a completely black or nonreflecting lower boundary, which we refer to as the *standard problem*. In other words, we express the solutions for the partly reflecting lower boundary, referred to as the *planetary problem*, in terms of those for the standard problem.[20]

From Fig. 6.7, the total reflected flux from the combined slab plus lower boundary is the sum of the following components: (1) the reflection from the slab itself; (2) the flux that reaches the surface, is reflected, and then transmitted; (3) that part of (2) which is reflected back to the surface and is then reflected a second time and transmitted; and (4) all higher-order terms, reflected three, four, etc. times from the

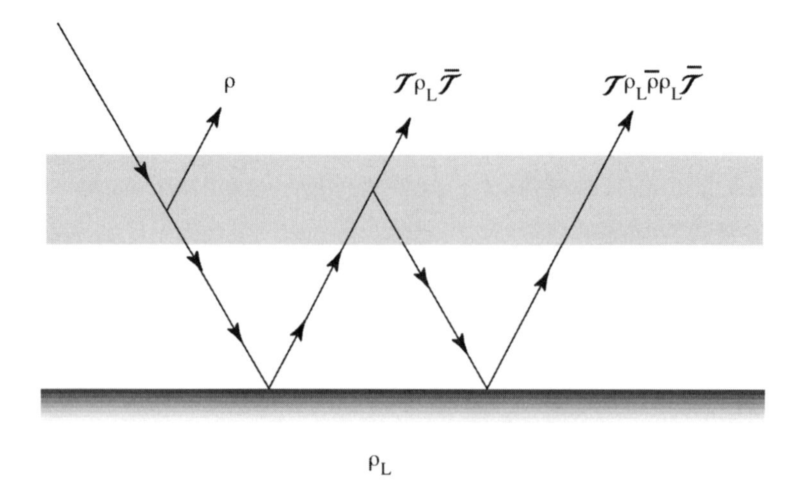

Figure 6.7 Addition of a reflecting surface leads to a geometric (binomial) series.

surface before being transmitted. The first term is just the ordinary flux reflectance $\rho(-\hat{\Omega}, +2\pi)$. The second term is proportional to the flux transmittance (including both the direct and diffuse components) $\mathcal{T}(-\hat{\Omega}, -2\pi)$. After reflection, it is proportional to $\mathcal{T}(-\hat{\Omega}, -2\pi)\rho_L$. Upon transmission through the slab, this term is multiplied by the spherical transmittance $\bar{\mathcal{T}}$. Thus, the second term is $\mathcal{T}(-\hat{\Omega}, -2\pi)\rho_L\bar{\mathcal{T}}$. The third term takes the part of term (2) up to where it was reflected from the surface, but then instead of multiplying by the flux transmittance, we multiply by the spherical reflectance $\bar{\rho}$, thus bringing it back for a second surface reflection. It is then multiplied by ρ_L, and finally gets transmitted, bringing in the term $\bar{\mathcal{T}}$. Thus the third term is $\mathcal{T}(-\hat{\Omega}, -2\pi)\rho_L\bar{\rho}\rho_L\bar{\mathcal{T}}$. Proceeding in a similar way with the higher-order components, we find that the sum can be written as (see Fig. 6.7)

$$\begin{aligned}
&\rho(-\hat{\Omega}, +2\pi) + \mathcal{T}(-\hat{\Omega}, -2\pi)\rho_L\bar{\mathcal{T}}[1 + \bar{\rho}\rho_L + (\bar{\rho}\rho_L)^2 + (\bar{\rho}\rho_L)^3 + \cdots] \\
&\quad = \rho(-\hat{\Omega}, +2\pi) + \frac{\mathcal{T}(-\hat{\Omega}, -2\pi)\rho_L\bar{\mathcal{T}}}{1 - \bar{\rho}\rho_L},
\end{aligned}$$

where we have used the fact that the infinite sum is the binomial expansion of $(1 - \bar{\rho}\rho_L)^{-1}$. Thus,

$$\rho_{\text{tot}}(-\hat{\Omega}, +2\pi, \rho_L) = \rho(-\hat{\Omega}, +2\pi) + \frac{\mathcal{T}(-\hat{\Omega}, -2\pi)\rho_L\bar{\mathcal{T}}}{1 - \bar{\rho}\rho_L}, \tag{6.76}$$

where the quantities on the right-hand side (ρ, $\bar{\rho}$, $\mathcal{T}$, and $\bar{\mathcal{T}}$) are evaluated for a black surface, $\rho_L = 0$.

A final step is to include the possibility of inhomogeneity. When the transmission or reflection is from below, it is necessary to replace $\bar{\rho}$ with $\bar{\bar{\rho}}$ and $\bar{\mathcal{T}}$ with $\bar{\bar{\mathcal{T}}}$. Thus the total flux reflectance is written

$$\rho_{\rm tot}(-\hat{\Omega}, +2\pi, \rho_{\rm L}) = \rho(-\hat{\Omega}, +2\pi) + \frac{\mathcal{T}(-\hat{\Omega}, -2\pi)\rho_{\rm L}\tilde{\tilde{\mathcal{T}}}}{1 - \tilde{\tilde{\rho}}\rho_{\rm L}}. \tag{6.77}$$

The total flux transmittance can be determined in a similar fashion:

$$\mathcal{T}_{\rm tot}(-\hat{\Omega}, -2\pi, \rho_{\rm L}) = \mathcal{T}(-\hat{\Omega}, -2\pi) + \frac{\mathcal{T}(-\hat{\Omega}, -2\pi)\rho_{\rm L}\tilde{\tilde{\rho}}}{1 - \tilde{\tilde{\rho}}\rho_{\rm L}}$$
$$= \frac{\mathcal{T}(-\hat{\Omega}, -2\pi)}{1 - \tilde{\tilde{\rho}}\rho_{\rm L}}. \tag{6.78}$$

It is straightforward, but a bit tedious, to show that these results also follow from the conservation of energy. Also, the method can be extended to derive a similar relationship for the directional transmittance and reflectance (see Appendix P for details). The results for these quantities are

$$\rho_{\rm tot}(-\hat{\Omega}_0, \hat{\Omega}, \rho_{\rm L}) = \rho(-\hat{\Omega}_0, \hat{\Omega}) + \frac{\rho_{\rm L}\mathcal{T}(-\hat{\Omega}_0, -2\pi)\tilde{\mathcal{T}}(\hat{\Omega}, +2\pi)}{\pi[1 - \tilde{\tilde{\rho}}\rho_{\rm L}]}, \tag{6.79}$$

$$\mathcal{T}_{\rm tot}(-\hat{\Omega}_0, -\hat{\Omega}, \rho_{\rm L}) = \mathcal{T}(-\hat{\Omega}_0, -\hat{\Omega}) + \frac{\rho_{\rm L}\mathcal{T}(-\hat{\Omega}_0, -2\pi)\tilde{\rho}(\hat{\Omega}, +2\pi)}{\pi[1 - \tilde{\tilde{\rho}}\rho_{\rm L}]}. \tag{6.80}$$

We have shown that the bidirectional reflectance and transmittance of a slab overlying a reflecting surface with a Lambertian reflectance are given by the sums and products of quantities evaluated for a black surface. This means that we only have to solve one radiative transfer problem involving a nonreflecting lower boundary. If the slab is inhomogeneous, we should, however, apply uniform illumination from both the top and the bottom to allow for rapid computation of $\tilde{\mathcal{T}}(-\hat{\Omega}, -2\pi)$, $\tilde{\rho}(-\hat{\Omega}, +2\pi)$, and $\tilde{\tilde{\rho}}$ as discussed above. Thus, as long as we are interested in only the transmitted and reflected intensities an analytic correction allows us to find the solutions pertaining to reflecting (Lambert) surfaces.

6.12 Integral Equation Formulation of Radiative Transfer

We have seen that the general radiative transfer equation (Eq. 5.41) is an *integro-differential equation*, which relates the spatial derivative of the intensity to an angular integral over the local intensity. This approach emphasizes the intensity, I_ν, as the fundamental quantity of interest. However, as discussed in §5.3, we could as well consider the source function, S_ν, to be the desired quantity, since a knowledge of either I_ν or S_ν (together with boundary conditions) constitutes a complete solution. We will consider below an alternate formulation of the radiative transfer problem,[21] where the source function is the principal dependent variable.

We proceed by making the same approximations as in Example 6.1 – isotropic scattering, lower absorbing boundary, and a homogeneous medium. We may obtain an equation for the source function by eliminating the unknowns I_d^+ and I_d^- from Eq. 6.17. Substituting the Eqs. (6.19 and 6.18) for I_d^+ and I_d^- we find

$$S(\tau) = \frac{a}{4\pi}\int_0^{2\pi} d\phi' \left[\int_0^1 d\mu' \overbrace{\int_0^{\tau} \frac{d\tau'}{\mu'} S(\tau') e^{-(\tau-\tau')/\mu'}}^{I_d^-(\tau',\mu)} \right.$$
$$\left. + \int_0^1 d\mu' \underbrace{\int_{\tau}^{\tau^*} \frac{d\tau'}{\mu'} S(\tau') e^{-(\tau'-\tau)/\mu'}}_{I_d^+(\tau',\mu)} \right] + (1-a)B(\tau) + S^*(\tau). \quad (6.81)$$

The two terms in brackets may be combined into a single term by noting that because the two quantities $(\tau' - \tau)$ and $(\tau - \tau')$ are positive, they are therefore left unchanged if we substitute their absolute value. The integrands are now identical in form, and we may combine the two integrals into a single integral over the entire medium. Interchanging orders of integration, we find

$$S(\tau) = (1-a)B(\tau) + S^*(\tau) + \frac{a}{2}\int_0^{\tau^*} d\tau' \left[\int_0^1 \frac{d\mu'}{\mu'} e^{-|\tau-\tau'|/\mu'} \right] S(\tau'). \quad (6.82)$$

The quantity in brackets is the *first exponential integral* (see Eq. 5.61),

$$E_1(|\tau - \tau'|) = \int_0^1 \frac{d\mu}{\mu} e^{-|\tau-\tau'|/\mu} = \int_1^{\infty} \frac{dx}{x} e^{-|\tau-\tau'|x}. \quad (6.83)$$

Defining the sum of the two "internal" sources as $Q = (1-a)B + S^*$, we arrive at the final form of our integral equation:

$$S(\tau) = Q(\tau) + \frac{a}{2}\int_0^{\tau^*} d\tau' E_1(|\tau - \tau'|) S(\tau'). \quad (6.84)$$

This has come to be known as the *Milne–Schwarzschild integral equation.* $E_1(|\tau - \tau'|)$ is called the *kernal* of the integral equation. The quantity $(a/2)E_1(|\tau - \tau'|)d\tau'$ is the probability that a photon that has been emitted within the plane layer between τ' and $\tau' + d\tau'$ travels a net vertical distance to the plane at τ and is scattered at τ. The physical significance of the integral term is therefore clear. It is the contribution to the source function from multiply scattered photons occurring within the medium. (In other words it is the contribution of the "self-illumination" of the medium.)

In the Milne–Schwarzschild formulation, we are faced with solving an integral equation for the source function, $S(\tau)$, instead of solving an integro-differential equation for the intensity, $I(\tau, \mu)$. It is clear that in this specific application (isotropic scattering) the integral approach is simpler, since it involves finding a solution for a function of only one variable, rather than a function of two variables.

There are other favorable features of the integral formulation that can be mentioned. The multiple-scattering term can be ignored if $a\tau^* \ll 1$. Then the integral equation solution provides the explicit solution $S(\tau) \approx Q(\tau)$. This result is easily generalized to anisotropic scattering and thermal emission:

$$S(\tau, \mu, \phi) \approx [(1-a)B(\tau) + aF^{s}\, p(-\mu_0, \phi_0; \mu, \phi)/4\pi]e^{-\tau/\mu_0} \quad (a\tau^* \ll 1). \tag{6.85}$$

Once the solution for $S(\tau)$ is available, the intensity follows by evaluating Eqs. 6.19 and 6.18 by numerical quadrature. In the more general case of anisotropic scattering, the integral equation formulation loses its primary advantage, in that the source function now depends upon not only position but also the direction, $\hat{\Omega}$, of emission.

6.13 Probabilistic Aspects of Radiative Transfer

In this section we consider an alternate formulation of the radiative transfer process. This approach will focus on properties of the scattering medium, which are independent of the distribution of sources of radiation, either external or internal. The *point-direction gain* and the *escape probability* are the basic quantities of interest. These quantities incorporate all the basic scattering properties of the medium, through the single-scattering albedo and the phase function, plus the knowledge of the total optical depth. Through the point-direction gain and its angular moments it is possible to solve problems differing in their sources of radiation. In the days before computers, this approach provided a considerable advantage, since in principle, these fundamental quantities could be calculated once and for all and presented in tables for general use. This approach is no longer necessary, since computers make it a relatively simple matter to alter the boundary conditions or internal sources of radiation in the program code. However, these probabilistic concepts are still of great pedagogical interest. We will therefore discuss only the essential elements of the method.

We begin with a "thought experiment," which will illustrate the concept. We assume a slab geometry and an isotropic scattering law. The medium may be inhomogeneous, so that the single-scattering albedo depends upon the optical depth. Consider an interior point at the optical depth τ' within the medium. Within the thin layer between τ' and $\tau' + d\tau'$ is contained an isotropic source of radiation, given by $Q(\tau')$, whose detailed specification will not concern us. The source Q will generally consist of the sum of a thermal source, $(1-a)B$, plus an imbedded source of first-order scattered photons, S^*. (We will continue to suppress the frequency subscript.) The emergent photons will execute a variety of scattering trajectories, depending upon the specific emission direction and upon the random nature of the angular scattering process (see Fig. 6.8).

We are interested in those photons that eventually reach the surface ($\tau = 0$) and that leave the medium in the direction given by μ. (There is no dependence on the

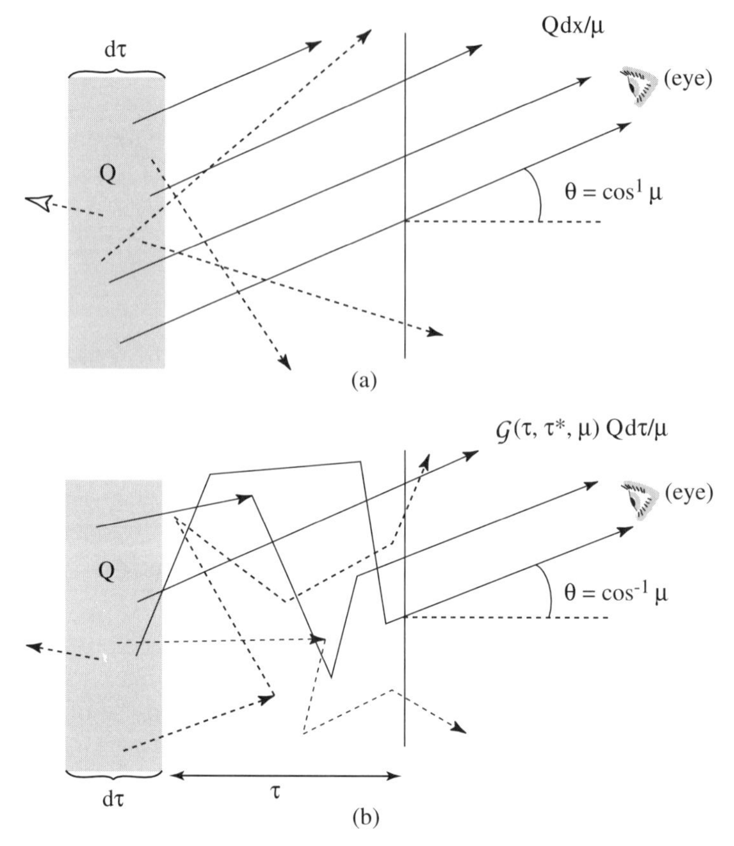

Figure 6.8 Illustration of the point-direction gain concept. (a) Without an interacting medium, the intensity emitted by the region within the thin layer of thickness $d\tau$, in direction $\cos^{-1}\mu$, is given by $Q\frac{d\tau}{\mu}$. (b) When the thin layer is imbedded in an absorbing/scattering medium, the intensity emitted in the same direction is $\mathcal{G}Q\frac{d\tau}{\mu}$, where $\mathcal{G}$ is the point-direction gain.

azimuthal angle ϕ, because of the assumption of an isotropic source and an isotropic phase function.) The differential contribution to the emergent intensity due to this thin layer may be written

$$dI^+(0,\mu) \equiv \mathcal{G}(\tau',\mu;a,\tau^*)Q(\tau')\frac{d\tau'}{\mu},$$

where $\mathcal{G}$ is the (dimensionless) *point-direction gain*. To understand the meaning of this quantity, consider the contribution to the emergent intensity from the same source in the absence of a medium. This is just

$$dI^+(0,\mu) = Q(\tau')\frac{d\tau'}{\mu} \quad (\mathcal{G}=1).$$

The presence of the term $d\tau'/\mu$ is explained by the fact that the column of emitting material is in the direction μ. The change in the radiation field, $\mathcal{G}$, is a result of the intervening medium. We may now write down an expression for the emergent intensity from the entire medium, since this is just a linear sum over all layers (an integral in the limit of very small layer thicknesses):

$$I^{+}(0,\mu) = \int_0^{\tau^*} \frac{d\tau'}{\mu} \mathcal{G}(\tau', \mu; a, \tau^*) Q(\tau'). \tag{6.86}$$

The intensity emerging from the bottom of the medium (assuming no reflection from the boundary) is given by

$$I^{-}(\tau^*,\mu) = \int_0^{\tau^*} \frac{d\tau'}{\mu} \tilde{\mathcal{G}}(\tau^* - \tau', \mu; a, \tau^*) Q(\tau'). \tag{6.87}$$

For an inhomogeneous medium, the point-direction gain is different for downward emitted radiation than for upward radiation, and hence we use the different symbol, $\tilde{\mathcal{G}}$. For a homogeneous medium, $\tilde{\mathcal{G}} = \mathcal{G}$.

The above experiment is called *direct*, in that we seek the external radiation field derived from a certain internal source. We now consider the *inverse* experiment, that is, we are given an external radiation field, $I^{-}(0,\mu)$, and we seek to determine the internal source function. We assume that no thermal sources of radiation are present. We postulate that the source function arising from this incident radiation field is given by

$$S(\tau) = \frac{a}{4\pi} \int_0^{2\pi} d\phi \int_0^1 d\mu \mathcal{G}(\tau, \mu; a, \tau^*) I^{-}(0, \mu, \phi).$$

In the particular case of an incident collimated beam

$$I^{-}(0,\mu,\phi) = F^{\mathrm{S}} \delta(\mu - \mu_0) \delta(\phi - \phi_0),$$

the source function is

$$S_3(\tau) = \frac{a}{4\pi} F^{\mathrm{S}} \mathcal{G}(\tau, \mu_0; a, \tau^*), \tag{6.88}$$

where the subscript 3 refers to *Prototype Problem 3* (i.e., collimated incidence). Since S_3 satisfies the Milne–Schwarzschild equation (Eq. 6.84)

$$S_3(\tau) = \frac{a}{4\pi} F^{\mathrm{S}} e^{-\tau/\mu_0} + \frac{a}{2} \int_0^{\tau^*} d\tau' E_1(|\tau - \tau'|) S_3(\tau')$$

then $\mathcal{G}$ satisfies the following equation:

$$\mathcal{G}(\tau, \mu_0; a, \tau^*) = e^{-\tau/\mu_0} + \frac{a}{2} \int_0^{\tau^*} d\tau' E_1(|\tau - \tau'|) \mathcal{G}(\tau', \mu_0; a, \tau^*). \tag{6.89}$$

We now return to the direct experiment and assume that the internal source is just $Q = S^* = [a/4\pi]F^{\rm s} e^{-\tau/\mu_0}$, that is, it is provided by an external collimated beam with flux $F^{\rm s}$. From Eq. 6.86, the emergent intensity from this source is

$$I^+(0, \mu) = F^{\rm s} \int_0^{\tau^*} \frac{d\tau'}{\mu} \frac{a}{4\pi} \mathcal{G}(\tau', \mu; a, \tau^*) e^{-\tau'/\mu_0}.$$

We set the angle $\mu = \mu_0$, so that

$$I^+(0, \mu_0) = F^{\rm s} \int_0^{\tau^*} \frac{d\tau'}{\mu_0} \frac{a}{4\pi} \mathcal{G}(\tau', \mu_0; a, \tau^*) e^{-\tau'/\mu_0},$$

and use the relationship between intensity and the source function (where we set $S(\tau') = S_3(\tau')$)

$$I^+(0, \mu_0) = \int_0^{\tau^*} \frac{d\tau'}{\mu_0} e^{-\tau'/\mu_0} S_3(\tau').$$

The above two equations for $I^+(0, \mu_0)$ are valid for all μ_0 and τ^*, if and only if

$$S_3(\tau) = \frac{a}{4\pi} F^{\rm s} \mathcal{G}(\tau, \mu_0; a, \tau^*), \tag{6.90}$$

which is the same result as in the inverse experiment, Eq. 6.88. It is another illustration of the Principle of Duality, discussed in §6.9.

Other relationships connecting $\mathcal{G}$ and its angular moments may be established with the classical functions of radiative transfer theory. It suffices to mention only one set of relationships to the X-, Y-, and H-functions, described by V. A. Ambartsumyan and S. Chandrasekhar. These functions have played a key role in the development of the classical theory of radiative transfer.[22] It may be shown that these functions are equal to the point-direction gain evaluated at the upper and lower bounds of the medium (assumed homogeneous),

$$X(\mu; a, \tau^*) \equiv \mathcal{G}(0, \mu; a, \tau^*), \qquad Y(\mu; a, \tau^*) \equiv \mathcal{G}(\tau^*, \mu; a, \tau^*), \tag{6.91}$$

$$H(\mu; a) \equiv \mathcal{G}(0, \mu; a, \tau^* \to \infty). \tag{6.92}$$

It should be apparent that knowledge of $\mathcal{G}$ allows one to solve entire classes of radiative transfer problems. This is the case for the source function for an arbitrary incident intensity on either or both faces, and also for the emergent intensities for any arbitrary disposition of internal sources.

6.13.1 The Escape Probability

As discussed earlier, the quantity $(a/4\pi)\mathcal{G}$ has a dual interpretation. In the inverse problem, it is the source function for the problem of an isotropically scattering slab of optical depth τ^* illuminated by a collimated beam of radiation of flux unity (see Eq. 6.90). For the direct problem, we can interpret it in terms of a *directional escape probability*. The formal relationship is

$$\mathcal{P}(\tau, \mu; a, \tau^*) \equiv \frac{a}{4\pi}\mathcal{G}(\tau, \mu; a, \tau^*). \tag{6.93}$$

The quantity $\mathcal{P}d\omega$ is interpreted as the joint probability of two successive events: First, $a/4\pi$ is the probability per unit solid angle that a photon, having suffered an extinction event at τ, will emerge as a scattered photon; and second, $\mathcal{G}d\omega$ is the probability that a photon will emerge from the upper surface of the medium in the direction μ within the solid angle $d\omega$. If we then integrate the product of these two probabilities over all emergent angles in the upper hemisphere, we obtain the *hemispherical escape probability*,

$$\mathcal{P}(\tau; a, \tau^*) \equiv \int_0^{2\pi} d\phi \int_0^1 d\mu \mathcal{P}(\tau, \mu; a, \tau^*) = \frac{a}{2}\int_0^1 d\mu \mathcal{G}(\tau, \mu; a, \tau^*). \tag{6.94}$$

Here $\mathcal{P}(\tau; a, \tau^*)$ is the probability of a photon emerging from the medium in *any* (upward) direction, and it is seen to be proportional to the zeroth-order moment of the point-direction gain. The escape probability for a homogeneous medium is proportional to the source function for *Prototype Problem 1* (uniform incidence), which is illustrated in Fig. 6.6. Suppose the medium is optically thin, so that $\mathcal{G} \approx e^{-\tau/\mu}$, from Eq. 6.89. Then $\mathcal{P} \to P$ where

$$P(\tau; a, \tau^*) \equiv \frac{a}{2}\int_0^1 d\mu e^{-\tau/\mu} = \frac{a}{2}E_2(\tau) = S_1(\tau), \tag{6.95}$$

where $S_1(\tau)$ is the source function for *Prototype Problem 1*. The above quantity, the *single-flight escape probability*, is given a special symbol because of its importance. It describes the probability of direct escape of all photons emerging at that point either from thermal emission or from a scattering. It is clear that those photons which have been scattered one or more times are counted in the general escape probability $\mathcal{P}$. Note that the single-flight escape probability out the *bottom* of the medium is $P(\tau^* - \tau)$.

6.14 Summary

As stated at the beginning the purpose of this chapter is to take a close look at the formulation of radiative transfer problems. The main idea is that we can save a lot of resources by properly analyzing the problem to see if there are possible simplifications making it easier to find a solution.

As a first example we demonstrated that a separation into diffuse and solar components is very useful for problems with collimated beam incidence. A second example is the simplification that is possible in slab geometry if we need to compute the azimuthal dependence of the radiation field. In this case it turns out to be sufficient to solve a series of uncoupled azimuth-independent equations. Each of these equations is similar to the azimuthally averaged equation, which we can solve, by methods discussed in subsequent chapters. We referred to this as "the isolation of the azimuthal dependence"; it allows us to compute very efficiently the complete azimuthal dependence of the radiation field.

Many problems in radiative transfer involve scattering phase functions that are strongly peaked in the forward direction. The one-parameter Henyey–Greenstein phase function was introduced as a prototype for large-particle scattering. Scaling transformations in which the forward-scattering peak is truncated have proven to be extremely useful for making the transfer equation more amenable to solution. This provided a third example of the usefulness of formulating the problem properly prior to attempting a solution.

It is also useful to try to relate a problem for which solution is sought to one that is easier to solve. This issue was addressed in §6.10, where we discussed how certain duality relations that can be traced to the Reciprocity Principle allow us to relate problems pertaining to collimated incidence to an equivalent problem pertaining to uniform incidence. The practical consequence is that we can save considerable amounts of computing resources because the latter problem requires substantially less computation than the former one. The use of uniform incidence allows us to compute very efficiently bulk quantities such as flux reflectance and transmittance and the corresponding "spherical" quantities. These in turn can be used to provide analytic corrections for surface effects if the surface can be approximated by a Lambert reflector.

Three prototype radiative transfer problems were introduced in this chapter. These problems will be solved by approximate methods in Chapter 7 and accurately in Chapter 8. Their solutions provide physical insight as well as serving to check the performance of solution methods. We discussed briefly the integral equation formulation approach to radiative transfer and the use of probabilistic concepts in radiative transfer theory. The point-direction gain $\mathcal{G}$ is similar to a *Green's function* – if we are given the basic distribution of photon sources, we can find the emergent intensities from an integration of the sources weighted by $\mathcal{G}$. It is interpreted as the change in the emergent intensity due to the intervening medium over that obtained from the sources in the absence of a medium. It is also interpreted as the probability of a photon emerging from a medium in either a specified direction (in which case we refer to the *directional* escape probability) or over a hemisphere (the *hemispherical* escape probability). A knowledge of $\mathcal{G}$ evaluated at the two boundaries of a slab medium provides us with the Chandrasekhar X-, Y- and H-functions. These will be shown in the next chapter to be useful in finding emergent intensities from a slab medium.

Problems

6.1 The Henyey–Greenstein phase function is given by Eq. 6.49.
(a) Verify that it is correctly normalized so that

$$\frac{1}{4\pi}\int_{4\pi} d\omega p_{\mathrm{HG}}(\cos\Theta) = 1.$$

(b) Derive the following expression for the azimuthally averaged phase function:

$$p_{\mathrm{HG}}(u, u') = \frac{1}{2\pi}\int_0^{2\pi} d\phi p_{\mathrm{HG}}(u, \phi; u', \phi') = \sum_{l=0}^{\infty} g^l (2l+1) P_l(u) P_l(u'),$$

where $P_l(u)$ are the Legendre polynomials. Show explicitly that $\chi_l = g^l$ by expanding Eq. 6.49 in a binomial series. (*Hint*: Expand p_{HG} as a power series expansion in the quantity $g^2 - 2g\cos\Theta$ and collect the terms multiplying the various powers of g.)

6.2
(a) Show that the delta-Legendre polynomial representation for the phase function

$$\hat{p}(\cos\Theta) = 2f\delta(1-\cos\Theta) + (1-f)\sum_{l=0}^{2N-1}(2l+1)\,\hat{\chi}_l P_l(\cos\Theta)$$

leads to the usual equation (as outlined in §6.3, Eq. 6.35) for the mth Fourier component of the intensity, $I^m(\hat{\tau}, \mu)$, except that the optical depth, single-scattering albedo, and source term are scaled according to

$$d\hat{\tau} = (1-af)d\tau, \qquad \hat{a} = \frac{(1-f)a}{1-af}, \qquad \hat{Q}^m(\hat{\tau}, \mu) = \frac{Q^m(\tau, \mu)}{1-af}.$$

(b) Verify that $\hat{p}(\cos\Theta)$ is correctly normalized.

6.3 Prove Eq. 6.47.

6.4 A disadvantage to the Henyey–Greenstein phase function P_{HG} is its failure in the limit of $g \to 0$ (small-particle limit) to approach the Rayleigh phase function p_{RAY}. (Instead it approaches $p \to 1$.) This problem describes a more satisfactory phase function parameterization for small particles.
(a) Show that the following parameterization approaches p_{RAY} as $g \to 0$:

$$p'_{\mathrm{HG}}(\Theta) = \frac{3(1-g^2)(1+\cos^2\Theta)}{2(2+g^2)(1+g^2-2g\cos\Theta)^{3/2}}.$$

(b) Show that p'_{HG} is properly normalized by using the property $\chi_l = \frac{1}{2}\int_0^{\pi} d\Theta \sin\Theta p_{\mathrm{HG}}(\Theta) P_l(\Theta) = g^l$. (*Hint*: $\cos^2\Theta = \frac{1}{3}(2P_2(\Theta)+1)$.)

(c) Show that the asymmetry factor is

$$\langle\cos\Theta\rangle \equiv \frac{1}{2}\int_0^{\pi} d\Theta \sin\Theta \cos\Theta p'_{\rm HG}(\Theta) = \frac{3g(g^2+4)}{5(g^2+2)}.$$

6.5

(a) Prove that the escape probability (§6.13, Eq. 6.94) for a semi-infinite medium is related to the source function for *Prototype Problem 2* (B = constant) through

$$\mathcal{P}(\tau, \tau^* \to \infty) = 1 - \frac{S_2(\tau, \tau^* \to \infty)}{B}.$$

(b) Prove that $\mathcal{P}(\tau, \tau^* \to \infty) = S_1(\tau, \tau^* \to \infty)/\mathcal{I}$, where S_1 is the solution to *Prototype Problem 1*.

6.6 Show that the spherical albedo, $\bar{\rho}$, and the spherical emittance, $\bar{\mathcal{T}}$, for a collimated beam problem (*Prototype Problem 3*) are equal to the flux reflectance, $\rho(-2\pi, +2\pi)$, and flux transmittance, $\mathcal{T}(-2\pi, +2\pi)$, respectively, for *Prototype Problem 1*.

6.7

(a) Show that the flux emittance in *Prototype Problem 2* is equal to the flux absorptance in *Prototype Problem 1*, in the case of a semi-infinite slab.
(b) How should the boundary conditions be altered in order for the above relationship to apply to a finite slab?
(c) Explain in physical terms the basis of this relationship.

6.8 A homogeneous, conservative-scattering medium of spherical reflectance $\bar{\rho}_{\rm a}$ overlies a Lambert reflecting surface of albedo $\rho_{\rm L}$.

(a) Use Eqs. 6.77 and 6.78 to show that the sum of the total spherical reflectance $\bar{\rho}_{\rm tot}$ and the total flux transmittance $\bar{\mathcal{T}}_{\rm tot}$ of the medium is given by

$$\bar{\rho}_{\rm tot} + \bar{\mathcal{T}}_{\rm tot} = \frac{1 + \rho_{\rm L} - 2\rho_{\rm L}\bar{\rho}_{\rm a}}{1 - \rho_{\rm L}\bar{\rho}_{\rm a}}.$$

(b) Show that $\bar{\rho}_{\rm tot} + \bar{\mathcal{T}}_{\rm tot} = 2$ in the limit $\rho_{\rm L} = 1$.
(c) What is the physical interpretation of a surface that has both perfect reflectance and perfect transmittance? (Compare with a transparent surface imbedded in a hohlraum.)

6.9 The *spherical transmittance* $\bar{\bar{\mathcal{T}}}$ has been previously defined (Eq. 5.35) as

$$\bar{\bar{\mathcal{T}}} = 2\int_0^1 d\mu\mu\left[\tilde{\mathcal{T}}_{\rm d}(+\mu, +2\pi) + e^{-\tau^*/\mu}\right] = \bar{\bar{\mathcal{T}}}_{\rm d} + 2\int_0^1 d\mu\mu e^{-\tau^*/\mu}, \qquad (6.96)$$

where $\tilde{\mathcal{T}}_{\rm d}$ is the diffuse flux transmittance and $\bar{\bar{\mathcal{T}}}_{\rm d}$ is the corresponding *spherical* diffuse transmittance. Show that if we integrate the above expressions for the reflectance and transmittance over the hemisphere to obtain the flux reflectance and flux transmittance, we obtain Eqs. 6.77 and 6.78.

Notes

1 Einstein, A. and L. Infeld, *The Evolution of Physics*, Simon and Shuster, New York, 1961.

2 The separation of the radiation field into diffuse and direct ("solar") components is a "standard" procedure used to deal with boundary conditions in mathematical physics, see, e.g., Morse, P. M. and H. Feschbach, *Methods of Theoretical Physics*, Dover, New York, 1950. In radiative transfer theory, it is described for example in Chandrasekhar, S., *Radiative Transfer*, Clarendon, Oxford, 1950 and Liou, K.-N., *An Introduction to Atmospheric Radiation*, Academic Press, Orlando, FL, 1980.

3 Collections of key papers on radiative transfer in multiple-scattering and absorbing media is contained in *Selected Papers on Scattering in the Atmosphere*, SPIE Milestone Series, Volume MS07, 1989, edited by C. F. Bohren, and *Selected Papers on Multiple Scattering in Plane Parallel Atmospheres and Oceans: Methods*, SPIE Milestone Series, Volume MS42, 1991, edited by G. W. Kattawar.

4 The expansion of the phase function in Legendre polynomials and the isolation of the azimuth dependence for the slab problem are "standard" procedures described in Chandrasekhar's book. Note that if $N \to \infty$, Eq. 6.28 becomes an equality.

5 See, for example, Arfken, G., *Mathematical Methods of Physics*, 3rd ed., Academic Press, Boston, 1985.

6 A practical reason for using the Λ_l^m is that they remain bounded whereas the P_l^m can become quite large and cause computer overflow (see NASA report cited as Endnote 5 of Chapter 8).

7 The treatment of spherical geometry is described in Sobolev, V. V., *Light Scattering in Planetary Atmospheres* (Transl. by W. M. Irvine), Pergamon, New York, 1975.

8 A derivation of the "streaming" term is provided in Kylling, A., *Radiation Transport in Cloudy and Aerosol Loaded Atmospheres*, Ph.D. Thesis, University of Alaska, Fairbanks, 1992, and a discussion of the azimuthally averaged equation is provided in Dahlback, A. and K. Stamnes, "A new spherical model for computing the radiation field available for photolysis and heating at twilight," *Planet. Space Sci.*, **39**, 671–83, 1991.

9 Wiscombe, W. J., "Atmospheric radiation: 1975–1983," *Rev. Geophys.*, **21**, 997–1021, 1983.

10 Several authors have considered the transfer of radiation within the coupled atmosphere–ocean system. A comparison of different approaches including numerical results is provided in Mobley, C. D., B. Gentili, H. R. Gordon, Z. Jin, G. W. Kattawar, A. Morel, P. Reinersman, K. Stamnes, and R. H. Stavn, "Comparison of numerical models for computing underwater light fields," *Applied Optics*, **32**, 7484–504, 1993. Our presentation follows the treatment of Z. Jin and K. Stamnes, "Radiative transfer in nonuniformly refracting media such as the atmosphere/ocean system," *Applied Optics.*, **33**, 431–42, 1993.

11 Radiative transfer in aquatic media is described in some detail in the classical text by N. G. Jerlov, *Marine Optics*, Elsevier, Amsterdam, 1976, and more briefly in the text by J. Dera, *Marine Physics*, Elsevier, Amsterdam, 1992. A recent text is that of Mobley, C. D., *Light and Water: Radiative Transfer in Natural Waters*, Academic Press, San Diego, 1995.

12 Scattering of radiation by independent particles and the calculation of phase functions are described in considerable detail in the classic treatise by H. C. van de Hulst, *Scattering by Small Particles*, Wiley, New York, 1957, in the more specialized treatment by

D. Deirmendjian, *Electromagnetic Scattering on Spherical Polydispersions*, American Elsevier, New York, 1969, in the more elementary book by E. J. McCartney, *Optics of the Atmosphere: Scattering by Molecules and Particles*, Wiley, New York, 1976, as well as in the comprehensive treatment by C. F. Bohren and D. R. Huffman, *Absorption and Scattering of Light by Small Particles*, John Wiley & Sons, New York, 1983.

13 A practical reference that discusses numerical algorithms is Wiscombe, W. J., "Improved Mie scattering algorithms," *Applied Optics*, **19**, 1505–9, 1980.

14 Gordon, H. R., "Modeling and simulating radiative transfer in the ocean," in *Ocean Optics*, ed. R. W. Spinrad, K. L. Carder, and M. J. Perry, Oxford University Press, New York, 1994.

15 This approach is due to A. F. McKellar and M. A. Box, "Scaling group of the radiative transfer equation," *J. Atmos. Sci.*, **38**, 1063–8, 1981.

16 Similarity transformations were first introduced by H. C. van de Hulst and are discussed in his book, *Multiple Light Scattering: Tables, Formulas, and Applications*, Academic Press, Orlando, FL, 1980.

17 The δ-TTA scaling transformation is described in J. H. Joseph, W. J. Wiscombe, and J. A. Weinman, "The delta-Eddington approximation for radiative flux transfer," *J. Atmos. Sci.*, **33**, 2452–9, 1976.

18 The δ-N approximation is described in W. J. Wiscombe, "The delta-M method: Rapid yet accurate radiative flux calculations for strongly asymmetric phase functions," *J. Atmos. Sci.*, **34**, 1408–22, 1977.

19 It is closely connected to the *Principle of Detailed Balance* (see §4.4), which states that in thermodynamic equilibrium any detailed process which we choose to consider has a reverse process, and the rates of these processes are in exact balance. It has been proven for any system governed by a Hamiltonian that possesses time-reversal invariance, both in classical systems and in quantum mechanics. For more details, see van de Hulst, H. C., *Scattering by Small Particles*, Wiley, New York, 1957.

20 The inclusion of surface reflection as an "analytic correction" is described in Chandrasekhar's book (pages 271–3) for a homogeneous slab. The extension to an inhomogeneous slab outlined in Appendix P is taken from K. Stamnes, "Reflection and transmission by a vertically inhomogeneous planetary atmosphere," *Planet. Space Sci.*, **30**, 727–32, 1982.

21 The classical literature is replete with solutions of radiative transfer integral equations, usually in an idealized scattering/absorbing medium. A prime example is E. Milne, "Thermodynamics of the stars" in *Handbuch Astrophys.*, **3** (1), 65, reprinted by Dover Press in the volume edited by D. H. Menzel, *Selected Papers on the Transfer of Radiation*, 1966. More modern treatments are given by V. Kourganoff, *Basic Methods in Transfer Problems*, Clarendon Press, Oxford, and Heaslet, M. A. and R. F. Warming, "Radiative source predictions for finite and semi-infinite non-conservative atmospheres," *Astrophys. Space Science*, **1**, 460–98, 1968.

22 Chandrasekhar, S., *Radiative Transfer*, Dover Publications, New York, 1960. For a readable exposition of classical methods of the Russian school, see V. V. Sobolev, *A Treatise on Radiative Transfer*, English translation by S. I. Gaposchkin, D. Van Nostrand, Princeton, 1963.

Chapter 7

Approximate Solutions of Prototype Problems

7.1 Introduction

We now describe several approximate methods of solution of the radiative transfer equation. Approximate methods play an important role in the subject, because they usually provide more insight than the more accurate methods. Indeed their simple mathematical forms help clarify physical aspects that are not easily discerned from the numerical output of a computer code. Another redeeming feature of approximate methods is that they are often sufficiently accurate that no further effort is necessary. Unlike some of the more sophisticated numerical techniques, these methods also yield approximations for the internal radiation field, including the source function. Of central importance is the two-stream approximation. This class of solutions has been given various names in the past (Schuster–Schwarzschild, Eddington, two-stream, diffusion approximation, two-flow analysis, etc.). In all variations of the method, the intractable integro-differential equation of radiative transfer is replaced with representations of the angular dependence of the radiation field in terms of just two functions of optical depth. These two functions obey two linear, coupled ordinary differential equations. When the medium is homogeneous, the coefficients of these equations are constants, and analytic closed-form solutions are possible. The mathematical forms of these solutions are exponentials in optical depth, depending on the total optical depth of the medium, the single-scattering albedo, one or two moments of the phase function, and the boundary intensities. Some disadvantages are that two-stream solutions maintain acceptable accuracy over a rather restricted range of the parameters; there is no useful a priori method to estimate the accuracy; and one generally needs an accurate solution to obtain a useful estimate of the accuracy.

In recent years two-stream methods have been applied to multilayer problems and to the computation of photolysis and heating rates in the presence of scattering. Although computationally intensive radiative transfer schemes are capable of accuracy better than 1%, it is unlikely that the inputs to the models are nearly this accurate. In the present era of fast computers, there is still a need for fast computational radiation algorithms (such as the two-stream method), particularly in three-dimensional photochemical/dynamical models. For example, the NCAR CCM3 General Circulation Model uses an algorithm for calculating shortwave radiative heating that is based on the two-stream method.

In this chapter we first consider so-called optically thin problems, or more precisely, first-order scattering solutions. The results are simple analytic expressions, which nevertheless are sufficiently general to make this method an important part of one's arsenal of weapons. We first discuss the single-scattering approximation for solar beam incidence in which the multiple-scattering contribution to the source function is ignored. In this case the source function for the diffuse radiation is given by a simple analytic expression and the integration may be carried out directly. We then discuss the multiple-scattering series and the condition of validity of the single-scattering approximation. The effect of a reflecting surface is considered next. We will then concentrate on the two-stream approach in solving the radiative transfer equation. We find approximate solutions of the three prototype problems introduced in Chapter 5 in the simplifying case of isotropic scattering. We introduce the exponential-kernal method, based on the integral-equation approach (§4.11). *Prototype Problem 1* is solved as an example of this technique to illustrate its equivalence to the traditional differential-equation method.

We then look at anisotropic scattering and compare the approximating differential equations for the classical Eddington approach with those of the two-stream approach. The two methods will be found to be equivalent, provided the mean ray inclinations for both the internal and boundary radiation are chosen to be the same. The backscattering coefficients, which play a vital role in the two-stream method, are defined as angular integrations over the phase function. Solutions are obtained for the case when these coefficients differ between the two hemispheres, a situation that approximates radiative transfer in the ocean. The accuracy of several of these approximate solutions is assessed by comparing with the results of more accurate techniques. These latter methods will be considered in Chapter 8.

7.2 Separation of the Radiation Field into Orders of Scattering

We showed in Chapter 5 that if the source function is known then we may integrate the radiative transfer equation directly. This is the case for thermal emission in the absence of scattering. A second important case where the source function is known is when multiple scattering is negligible. The solution to the radiative transfer equation in the absence of multiple scattering is usually referred to as the *single-scattering approximation.* We will first derive these solutions for the diffuse intensity and then discuss situations where this approximation is valid. In slab geometry the radiative transfer equations for the half-range intensities are

$$\mu \frac{dI^+(\tau, \mu, \phi)}{d\tau} = I^+(\tau, \mu, \phi) - S^+(\tau, \mu, \phi), \tag{7.1}$$

$$-\mu \frac{dI^-(\tau, \mu, \phi)}{d\tau} = I^-(\tau, \mu, \phi) - S^-(\tau, \mu, \phi). \tag{7.2}$$

We have dropped the ν subscripts for convenience. As usual, the independent variable is the optical depth, measured downward from the "top" of the slab. We showed in Chapter 6 that formal solutions to the above equations are a sum of direct (I_s) and diffuse (I_d) contributions:

$$I^-(\tau, \mu, \phi) = \overbrace{I^-(0, \mu, \phi) e^{-\tau/\mu}}^{I_s^-(\tau,\mu,\phi)} + \overbrace{\int_0^\tau \frac{d\tau'}{\mu} e^{-(\tau-\tau')/\mu} S^-(\tau', \mu, \phi)}^{I_d^-(\tau,\mu,\phi)}, \tag{7.3}$$

$$I^+(\tau, \mu, \phi) = \overbrace{I^+(\tau^*, \mu, \phi) e^{-(\tau^*-\tau)/\mu}}^{I_s^+(\tau,\mu,\phi)} + \overbrace{\int_\tau^{\tau^*} \frac{d\tau'}{\mu} e^{-(\tau'-\tau)/\mu} S^+(\tau', \mu, \phi)}^{I_d^+(\tau,\mu,\phi)}, \tag{7.4}$$

$$I(\tau, \mu = 0, \phi) = S(\tau). \tag{7.5}$$

In the single-scattering approximation, we assume that the multiple-scattering contribution (the integral terms) to the source function is negligible. From Eqs. 6.7 and 6.9, the source function simplifies to

$$\begin{aligned} S^{\pm}(\tau', \mu, \phi) &\approx (1-a)B + S^*(\tau', \pm\mu, \phi) \\ &= (1-a)B + \frac{aF^s}{4\pi} p(-\mu_0, \phi_0; \pm\mu, \phi) e^{-\tau'/\mu_0}. \end{aligned} \tag{7.6}$$

Here B is the Planck function, $(1-a)$ is the volume emittance, $S^{\pm}$ are the *half-range source functions* analogous to the half-range *intensities* $I^{\pm}(\tau, \mu)$, and S^* is the contribution of singly scattered solar photons to the source function. Since this latter term varies exponentially with optical depth the integration is easily carried out. Substituting the simplified form (Eq. 7.6) into Eqs. 7.3 and 7.4, and carrying out the integrations, we obtain the following analytic results for the first-order scattered intensity:

$$\begin{aligned} I_d^-(\tau, \mu, \phi) &= (1-a)B\left[1 - e^{-\tau/\mu}\right] \\ &\quad + \frac{a\mu_0 F^s p(-\mu_0, \phi_0; -\mu, \phi)}{4\pi(\mu_0 - \mu)}\left[e^{-\tau/\mu_0} - e^{-\tau/\mu}\right], \end{aligned} \tag{7.7}$$

$$I_d^+(\tau, \mu, \phi) = (1 - a)B\left[1 - e^{-(\tau^* - \tau)/\mu}\right] + \frac{a\mu_0 F^s p(-\mu_0, \phi_0; +\mu, \phi)}{4\pi(\mu_0 + \mu)}\left[e^{-\tau/\mu} - e^{-[(\tau^* - \tau)/\mu + \tau^*/\mu_0]}\right]. \tag{7.8}$$

To obtain the total intensity, we must add to the diffuse intensity the boundary terms in Eqs. 7.3 and 7.4. Since $I^-(0, \mu, \phi) = F^s\delta(\mu - \mu_0)\delta(\phi - \phi_0)$ and $I^+(\tau^*, \mu, \phi) = 0$ (if the lower boundary is assumed to be perfectly absorbing), the total intensity is given by

$$I^-(\tau, \mu, \phi) = I^-(0, \mu, \phi)e^{-\tau/\mu} + I_d^-(\tau, \mu, \phi) = F^s e^{-\tau/\mu}\delta(\mu - \mu_0)\delta(\phi - \phi_0) + I_d^-(\tau, \mu, \phi), \tag{7.9}$$

$$I^+(\tau, \mu, \phi) = \overbrace{I^+(\tau^*, \mu, \phi)e^{-(\tau^* - \tau)/\mu}}^{0} + I_d^+(\tau, \mu, \phi) = I_d^+(\tau, \mu, \phi), \tag{7.10}$$

where the diffuse terms are given by Eqs. 7.7 and 7.8.

Favorable aspects of the single-scattering approximation are:

1. The solution is valid for any phase function.
2. It is easily generalized to include polarization.
3. It applies to any geometry, as long as we replace the slab optical path τ/μ with the expression appropriate to the incident ray path. For example, in spherical geometry, τ/μ_0 is replaced with $\tau Ch(\mu_0)$, where $Ch(\mu_0)$ is the Chapman function (see §6.4).
4. It is useful when an approximate solution is available for the multiple scattering, for example, from the two-stream approximation. In this case the diffuse intensity is given by the sum of single-scattering and (approximate) multiple-scattering contributions.
5. It serves as a starting point for expanding the radiation field in a sum of contributions from first-order, second-order scattering, etc. The latter expansion technique, known as *Lambda iteration*, allows us to evaluate more precisely the validity of the first-order scattering approximation.

7.2.1 Lambda Iteration: The Multiple-Scattering Series

We have previously shown that the source function is the sum of single-scattering and multiple-scattering contributions. The latter is just the sum of contributions from photons scattered twice, three times, etc. To explicitly reveal the structure of this *multiple-scattering series*, we consider the integral equation formulation, Eq. 6.81. To avoid unnecessary complexity, we will assume isotropic scattering in a homogeneous medium, for which Eq. 6.84 is valid (a = constant):

$$S(\tau) = (1 - a)B(\tau) + S^*(\tau) + \frac{a}{2}\int_0^{\tau^*} d\tau' E_1(|\tau - \tau'|)S(\tau'). \tag{7.11}$$

This integral equation forms the basis for an iterative solution, in which we first approximate the integrand $S(\tau')$ with the first-order scattering source function, $S^{(1)}(\tau') = (1 - a)B(\tau') + S^*(\tau')$. Then an improved result $S^{(2)}$ is obtained by inserting $S^{(1)}(\tau')$ into the integrand of Eq. 7.11,

$$S^{(2)}(\tau) \approx S^{(1)}(\tau) + \frac{a}{2}\int_0^{\tau^*} d\tau' E_1(|\tau' - \tau|)S^{(1)}(\tau'). \tag{7.12}$$

This procedure is then repeated *ad infinitum* to improve the approximation. Defining the dummy integration variables as $\tau_1, \tau_2, \ldots, \tau_n$, we obtain the infinite series

$$S(\tau) = \sum_{n=0}^{\infty} \Lambda^{(n)}(\tau, \tau_n)S^{(1)}(\tau_n) = S^{(1)}(\tau) + \frac{a}{2}\int_0^{\tau^*} d\tau_1 E_1(|\tau - \tau_1|)S^{(1)}(\tau_1)$$
$$+ \left(\frac{a}{2}\right)^2 \int_0^{\tau^*} d\tau_1 \int_0^{\tau^*} d\tau_2 E_1(|\tau - \tau_1|)E_1(|\tau_1 - \tau_2|)S^{(1)}(\tau_2) + \cdots, \tag{7.13}$$

where the *Lambda operator* is defined as

$$\Lambda^{(0)}(\tau, \tau_0) \equiv \delta(\tau - \tau_0),$$

$$\Lambda^{(1)}(\tau, \tau_1) \equiv \frac{a}{2}\int_0^{\tau^*} d\tau_1 E_1(|\tau - \tau_1|),$$

$$\Lambda^{(2)}(\tau, \tau_2) \equiv \left(\frac{a}{2}\right)^2 \int_0^{\tau^*} d\tau_1 \int_0^{\tau^*} d\tau_2 E_1(|\tau - \tau_1|)E_1(|\tau_1 - \tau_2|),$$

$$\Lambda^{(n)}(\tau, \tau_n) \equiv \left(\frac{a}{2}\right)^n \int_0^{\tau^*} d\tau_1 E_1(|\tau - \tau_1|) \int_0^{\tau^*} d\tau_2 E_1(|\tau_1 - \tau_2|)$$
$$\cdots \int_0^{\tau^*} d\tau_n E_1(|\tau_{n-1} - \tau_n|). \tag{7.14}$$

Equation 7.13 is called the *Neumann series expansion* of the source function, and it is easily interpreted as the sum of the first-order, second-order, etc. scattering contributions. It is not obvious that the above series converges, that is, whether it is the desired solution to Eq. 7.11. A proof that it is indeed absolutely convergent is obtained by evaluating a second series expansion in closed form, whose terms are upper bounds to those in the above series expansion.

The proof of convergence also provides us with an approximate solution that offers more insight into the general nature of the multiple-scattering series. We assume the conditions of *Prototype Problem 2*, in which the only source of radiation is thermal emission, and for which the "imbedded source" is constant with τ. For simplicity we temporarily set this source equal to unity, $S^{(1)} = (1-a)B = 1$. Consider the second-order contribution to the source function

$$S^{(2)}(\tau) \equiv \Lambda^{(1)}(\tau, \tau_1)S^{(1)}(\tau_1) = \frac{a}{2}\int_0^{\tau^*} d\tau_1 E_1(|\tau - \tau_1|)$$

$$= \frac{a}{2}\left[\int_0^{\tau} d\tau_1 E_1(\tau - \tau_1) + \int_{\tau}^{\tau^*} d\tau_1 E_1(\tau_1 - \tau)\right]. \tag{7.15}$$

Using the property $dE_2(t)/dt = -E_1(t)$, and noting that $E_2(0) = 1$, we find

$$S^{(2)}(\tau) = \int_0^{\tau} d\tau_1 \frac{dE_2(\tau - \tau_1)}{d\tau_1} + \int_{\tau}^{\tau^*} d\tau_1 (-)\frac{dE_2(\tau_1 - \tau)}{d\tau_1}$$

$$= a\,[1 - 1/2E_2(\tau) - 1/2E_2(\tau^* - \tau)]$$

$$\equiv a\,[1 - P(\tau) - P(\tau^* - \tau)]. \tag{7.16}$$

The functions $P(\tau) \equiv \frac{1}{2}E_2(\tau)$ and $P(\tau^* - \tau) \equiv \frac{1}{2}E_2(\tau^* - \tau)$ are the hemispherical *single-flight escape probabilities* for a photon released at the optical depth τ (see Eq. 6.95). They describe the probability of escape without further scattering through the top ($\tau = 0$) or bottom ($\tau^* - \tau$) of the slab, respectively. As usual, we are assuming a black lower boundary, so that a photon is lost when it reaches either boundary.

Since $1 - P(\tau) - P(\tau^* - \tau)$ is the probability of photon capture upon emission at τ, the interpretation of Eq. 7.16 is clear. The source function of second-order scattered photons is the product of two factors: (a = probability of a photon being scattered following an extinction event) × (the probability that a photon is "captured"). It defines the contribution from photons that are emitted from a unit volume,[1] suffer *one extinction*, and survive the extinction as a scattering event.

Since we are interested in an upper bound to the source function, this occurs where the escape probability is a minimum, that is, at the mid-point in the slab, $\tau = \tau^*/2$. Let us replace the equality Eq. 7.16 with the inequality

$$S^{(2)}(\tau) < a\,[1 - P(\tau^*/2)].$$

Continuing the procedure, we find that the third-order term is

$$S^{(3)}(\tau) = \left(\frac{a}{2}\right)^2 \int_0^{\tau^*} d\tau_1 E_1(|\tau - \tau_1|) \int_0^{\tau^*} d\tau_2 E_1(|\tau_1 - \tau_2|). \tag{7.17}$$

An upper limit follows from evaluating both the integrals at $\tau^*/2$:

$$S^{(3)}(\tau) < \left(\frac{a}{2}\right)^2 \int_0^{\tau^*} d\tau_1 E_1(|\tau_1 - \tau^*/2|) \int_0^{\tau^*} d\tau_2 E_1(|\tau_2 - \tau^*/2|)$$
$$= a^2[1 - P(\tau^*/2)]^2.$$

Repeating this process for every order of scattering, we find the general upper bound

$$S^{(n)}(\tau) < a^{n-1}[1 - P(\tau^*/2)]^{n-1}.$$

The total source function thus obeys the following inequality:

$$S(\tau) = \sum_{n=1}^{\infty} S^{(n)} < \sum_{n=0}^{\infty} x^n, \quad \text{where } x \equiv a[1 - P(\tau^*/2)].$$

The above series is easily evaluated by recognizing that the geometric series expansion of $(1-x)^{-1}$ is $1 + x + x^2 + \cdots$. Thus

$$S(\tau) < \frac{(1-a)B}{1 - a[1 - P(\tau^*/2)]}. \tag{7.18}$$

We have replaced the numerator (1) with the original thermal source $(1-a)B$. Since $0 < P(\tau) < 1$, and $0 < a \leq 1$, this result is finite. Note that in the limit of an optically thick, conservative-scattering medium where the escape probability goes to zero, $S \to (1-a)B/(1-a) = B$, as it should. Thus, we have shown that *the Neumann series converges for any first-order scattering source that is finite everywhere in the medium*. Equation 7.18 is also interesting in its own right, since it provides in some circumstances a useful estimate for the actual source function. If we use the more general first-order source function, Eq. 7.15, we can replace the upper limit with the approximation

$$S(\tau, \mu, \phi) \approx \frac{(1-a)B + (aF^{\mathrm{s}}/4\pi)p(-\mu_0, \phi_0; \mu, \phi)e^{-\tau/\mu_0}}{1 - a[1 - P(\tau^*/2)]}. \tag{7.19}$$

In astrophysics, the above equation is called the *Sobolev approximation*, and in infrared atmospheric physics it goes by the name of the *cooling-to-space approximation* (§10.2.6).

Now let $\tau^* \ll 1$. We note that $1 - P(\tau^*/2) = 1 - E_2(\tau^*/2) \approx 1 - e^{-\tau^*/2\bar{\mu}} \approx \tau^*$ (assuming $\bar{\mu} = 1/2$). The multiple-scattering series becomes

$$S(\tau, \mu, \phi) \approx \left[(1-a)B + (aF^{\mathrm{s}}/4\pi)p(-\mu_0, \phi_0; \mu, \phi)e^{-\tau/\mu_0}\right]$$
$$\times [1 + a\tau^* + (a\tau^*)^2 + \cdots].$$

This shows that it is permissible to use first-order scattering provided that $a\tau^* \ll 1$. This is reasonable because if absorption is large (a is small), then even if τ^* is large, the contribution of multiply scattered photons may still be negligible. Thus, *an optically thin medium ($\tau^* \ll 1$) is not necessarily required for multiple scattering to be negligible.*

The multiple-scattering series method is practical only when the series converges rapidly. Otherwise, the necessity of performing multiple integrals makes the computation prohibitively expensive. The second-order scattering term may be used to estimate the error in the first-order scattering approximation, although because of the monotonic nature of the series, this only provides a lower bound to the error, which is not very helpful. A practical example where single scattering is the dominant contribution is that of a thin Rayleigh-scattering atmosphere overlying a dark surface. The angular distribution of near-infrared skylight as seen by a ground observer is described with reasonable accuracy by the single-scattered component (Problem 5.1). The next most important contribution is usually that of sunlight reflected from the surface and then scattered once in the atmosphere. We next consider this effect.

7.2.2 Single-Scattered Contribution from Ground Reflection: The Planetary Problem

The radiation reflected upward from a surface is often comparable to the direct solar radiation, and the first-order scattering from this source may appreciably augment the source function. From the expression 6.11, the boundary contribution to the source function is given by

$$S_{\mathrm{b}}^{*}(\tau, \pm\mu, \phi) = \frac{a}{4\pi} \int_{0}^{2\pi} d\phi' \int_{0}^{1} d\mu' p(+\mu', \phi'; \pm\mu, \phi) I_{\mathrm{r}}^{+}(\tau, \mu', \phi'), \tag{7.20}$$

where I_{r}^{+} is the upward beam of reflected solar radiation received at the optical depth τ. We will assume that the reflected light is primarily the component arriving directly from the surface. From the definition of the BRDF (Eq. 5.10),

$$I_{\mathrm{r}}^{+}(\tau, \mu', \phi') = \mu_0 F^{\mathrm{s}} e^{-(\tau^*-\tau)/\mu'} \rho(-\mu_0, \phi_0; \mu', \phi')\, e^{-\tau^*/\mu_0}.$$

Inserting the above result into Eq. 7.20, we find

$$S_{\mathrm{b}}^{*}(\tau, \pm\mu, \phi) = \frac{a\mu_0 F^{\mathrm{s}}}{4\pi} e^{-\tau^*/\mu_0} \int_{0}^{2\pi} d\phi' \cdot \int_{0}^{1} d\mu' p(\mu', \phi'; \pm\mu, \phi) \rho(\mu_0, \phi_0; \mu', \phi')\, e^{-(\tau^*-\tau)/\mu'}. \tag{7.21}$$

The single-scattering source function for the planetary problem is therefore $S^* + S_{\mathrm{b}}^*$, where S^* is given by Eq. 7.6, and S_{b}^* by Eq. 7.21. Integration over a line of sight yields the intensity, using Eqs. 7.3 and 7.4. The effects of ground reflection should always be considered in any first-order scattering formulation. As we will show, its influence is comparable to the direct first-order scattering (Eqs. 7.15 and 7.16), unless the surface is quite dark.

We will estimate the first-order surface contribution to the Rayleigh-scattered skylight distribution. For this purpose, it is sufficient to replace the BRDF with a Lambertian reflectance function ($\rho = \rho_{\rm L}$ = constant). In addition, the integration of the reflected radiation over 2π sr will "wash out" the signature of the weak angular dependence of the Rayleigh phase function. Thus, it is permissible to replace the function $p_{\rm RAY}$ with an isotropic phase function, $p = 1$. The error is expected to be less than that of ignoring the multiple scattering. The simplified result is

$$S_{\rm b}^* \approx \frac{a\mu_0\rho_{\rm L}F^{\rm s}}{2} e^{-\tau^*/\mu_0} E_2(\tau^* - \tau). \tag{7.22}$$

The ratio of $S_{\rm b}$ to S^* is therefore $2\pi\mu_0\rho_{\rm L}E_2(\tau^* - \tau)p^{-1}$. For small τ^*, this ratio is $\sim 2\pi\mu_0\rho_{\rm L}p^{-1}$, which can actually exceed unity. For an average solar incidence, $\mu_0 = 0.5$, this term may be ignored only if $\rho_{\rm L} \ll (\pi)^{-1}$. Thus even for the dark ocean surface ($\pi\rho_{\rm L} \approx 0.1$), the skylight is influenced by surface reflection. Note that this result is independent of the wavelength.

7.3 The Two-Stream Approximation: Isotropic Scattering

7.3.1 Approximate Differential Equations

Before considering the more realistic anisotropic scattering problem, we will spend some time solving problems in which the scattering is isotropic ($p = 1$). This assumption greatly reduces the algebraic complexity without sacrificing any essential aspects. We will solve the three prototype problems for homogeneous media and for a black lower boundary. We begin with the governing integro-differential equations and later show that the integral-equation approach yields identical results.

The radiative transfer equations for the half-range intensity fields are given by (see Eqs. 6.3 and 6.4)

$$\mu\frac{dI^+(\tau,\mu)}{d\tau} = I^+(\tau,\mu) - \frac{a}{2}\int_0^1 d\mu' I^+(\tau,\mu') - \frac{a}{2}\int_0^1 d\mu' I^-(\tau,\mu') - (1-a)B,$$

$$-\mu\frac{dI^-(\tau,\mu)}{d\tau} = I^-(\tau,\mu) - \frac{a}{2}\int_0^1 d\mu' I^+(\tau,\mu') - \frac{a}{2}\int_0^1 d\mu' I^-(\tau,\mu') - (1-a)B.$$

Because the scattering is isotropic, the radiation field has no azimuthal dependence as explained previously in Chapter 6. In the two-stream approximation we replace the angularly dependent quantities $I^{\pm}$ by their averages over each hemisphere, $I^{+}(\tau)$ and $I^{-}(\tau)$. This leads to the following pair of coupled differential equations[2], which are called

The two-stream equations:

$$\bar{\mu}^{+}\frac{dI^{+}(\tau)}{d\tau} = I^{+}(\tau) - \frac{a}{2}I^{+}(\tau) - \frac{a}{2}I^{-}(\tau) - (1-a)B, \tag{7.23}$$

$$-\bar{\mu}^{-}\frac{dI^{-}(\tau)}{d\tau} = I^{-}(\tau) - \frac{a}{2}I^{+}(\tau) - \frac{a}{2}I^{-}(\tau) - (1-a)B. \tag{7.24}$$

Here $\bar{\mu}^{\pm}$ is the cosine of the average polar angle $\bar{\theta}$ made by a beam, which generally differs in the two hemispheres. The precise values are described in the next section. These linear, coupled, ordinary differential equations allow for analytic solutions by standard methods if the medium is homogeneous so that $a(\tau) = a =$ constant. Before embarking on this course of action, we note that the two-stream approximation will be most accurate when the radiation field is nearly isotropic. This situation should occur deep inside the medium, far away from any boundary or from sources or sinks of radiation. However, we will often find, somewhat surprisingly, that it is accurate even at the boundaries themselves. Thus, far from being simply an "asymptotic" theory, the method can teach us about radiative transfer over a large range of variables, from optically thin to optically thick conditions, and for both scattering- and emission-dominated problems. We shall return to the accuracy of the two-stream approximation in §7.6.

The approximate two-stream expressions for the source function, the flux, and the heating rate are

$$S(\tau) = \frac{a}{2}\int_0^1 d\mu[I^{+}(\tau,\mu) + I^{-}(\tau,\mu)] + (1-a)B$$

$$\approx \frac{a}{2}[I^{+}(\tau) + I^{-}(\tau)] + (1-a)B, \tag{7.25}$$

$$F(\tau) = 2\pi\int_0^1 d\mu\mu[I^{+}(\tau,\mu) - I^{-}(\tau,\mu)]$$

$$\approx 2\pi[\bar{\mu}^{+}I^{+}(\tau) - \bar{\mu}^{-}I^{-}(\tau)], \tag{7.26}$$

and

$$\mathcal{H}(\tau) = -\frac{\partial F}{\partial z} \approx 2\pi\alpha[I^{+}(\tau) + I^{-}(\tau)] - 4\pi\alpha B. \tag{7.27}$$

In Eq. 7.27, α is the absorption coefficient. We have used Eq. 5.76 for the heating rate.

7.3.2 The Mean Inclination: Possible Choices for $\bar{\mu}$

The two-stream approach does not provide us with a prescription for determining the value of $\bar{\mu}^{\pm}$, but some options are discussed below. We could obviously define $\bar{\mu}^{\pm}$ formally as the intensity-weighted angular means

$$\bar{\mu}^{\pm} = \langle\mu\rangle^{\pm} \equiv \frac{2\pi \int_0^1 d\mu \mu I^{\pm}(\tau,\mu)}{2\pi \int_0^1 d\mu I^{\pm}(\tau,\mu)} = \frac{F^{\pm}}{2\pi I^{\pm}},$$

but since we do not know the intensity distribution a priori this definition is of little use. It demonstrates, however, that $\bar{\mu}$ will in general vary with optical depth and takes on a different value in the two hemispheres. Hence picking the same constant value for this quantity in both hemispheres ($\bar{\mu} = \bar{\mu}^{+} = \bar{\mu}^{-} = \text{constant}$) is clearly an approximation. If the intensity field were strictly hemispherically isotropic, this formula would yield $\bar{\mu} = 1/2$ for all depths and for both hemispheres. This is identical to the result obtained in evaluating the hemispherical integral numerically using a one-point Gaussian quadrature (explained in more detail in Chapter 8).

If the intensity distribution were approximately linear in μ, say $I(\mu) \approx C\mu$, where C is a constant, then $\bar{\mu} = 2/3$. In the solution for *Prototype Problem 2*, we find that in order to get the optimum accuracy in calculating the emittance of a highly absorbing slab, $\bar{\mu}$ must lie between 2/3 and 1/2.

Alternatively, we could use the root-mean-square value

$$\bar{\mu} \equiv \mu_{\text{rms}} = \sqrt{\langle\mu^2\rangle} = \sqrt{\frac{\int_0^1 d\mu \mu^2 I(\tau,\mu)}{\int_0^1 d\mu I(\tau,\mu)}}.$$

If the radiation field were isotropic this definition would yield $\bar{\mu} = 1/\sqrt{3}$, which happens to be identical to the value obtained from a two-point Gaussian quadrature for the complete range of $u = \cos\theta$ ($-1 \leq u \leq 1$). A linear variation of the radiation field would yield $\bar{\mu} = 1/\sqrt{2} = 0.71$.

Thus, these possible choices yield $\bar{\mu}$ values ranging from 0.5 to 0.71. There is really no certain way to decide categorically and a priori which choice is optimal or if another definition would be even better. It appears that we are faced with having to pick the optimal $\bar{\mu}$ value on a trial-and-error basis for each type of problem. In most of what follows, we assume a single value for $\bar{\mu}$ but leave its value undetermined to remind us that it represents some sort of average over a hemisphere. We shall return to this issue in §7.5.5.

7.3.3 Prototype Problem 1: Differential-Equation Approach

In this problem, we ignore the thermal emission term. We will manipulate the two-stream equations to uncouple the quantities I^{+} and I^{-}. By first adding Eqs. 7.23 and

7.24 and then subtracting Eq. 7.24 from Eq. 7.23 we obtain

$$\bar{\mu}\frac{d(I^+ - I^-)}{d\tau} = (1-a)(I^+ + I^-), \tag{7.28}$$

$$\bar{\mu}\frac{d(I^+ + I^-)}{d\tau} = (I^+ - I^-). \tag{7.29}$$

Differentiating Eq. 7.29 with respect to τ, and substituting for $d(I^+ - I^-)/d\tau$ from Eq. 7.28, we find

$$\frac{d^2(I^+ + I^-)}{d\tau^2} = \frac{(1-a)}{\bar{\mu}^2}(I^+ + I^-).$$

This provides us with an equation involving only the sum of the intensities. Similarly, differentiating the first of the above equations, and substituting for $d(I^+ + I^-)/d\tau$ from the second equation, we find

$$\frac{d^2(I^+ - I^-)}{d\tau^2} = \frac{(1-a)}{\bar{\mu}^2}(I^+ - I^-),$$

which involves only the difference of the intensities. We see that we have the same differential equation to solve for both quantities. Calling the unknown Y, we obtain a simple second-order *diffusion equation*

$$\frac{d^2Y}{d\tau^2} = \Gamma^2 Y, \quad \text{where } \Gamma \equiv \sqrt{1-a}/\bar{\mu}, \tag{7.30}$$

for which the general solution is a sum of positive and negative exponentials

$$Y = A'e^{\Gamma\tau} + B'e^{-\Gamma\tau}.$$

Here A' and B' are arbitrary constants to be determined. Since the sum and difference of the two intensities are both expressed as sums of exponentials, each intensity component must also be expressed in the same way:

$$I^+(\tau) = Ae^{\Gamma\tau} + Be^{-\Gamma\tau}, \qquad I^-(\tau) = Ce^{\Gamma\tau} + De^{-\Gamma\tau}, \tag{7.31}$$

where A, B, C, and D are additional arbitrary constants.

We now introduce boundary conditions at the top and the bottom of the medium. We begin with *Prototype Problem 1* for which

$$I^-(\tau = 0) = \mathcal{I} = \text{constant}, \qquad I^+(\tau^*) = 0. \tag{7.32}$$

We choose this as our first example, as the two-stream solution to this problem has the simplest analytic form of the three considered. Furthermore it has several interesting aspects:

1. It bears a resemblance to the important problem of solar beam incidence (*Prototype Problem 3*), particularly at large optical depths.
2. Its solution is that of the simplest greenhouse problem (Chapter 12).
3. The source function is proportional to the escape probability $\mathcal{P}$ (see §6.13 and Problems 6.4 and 7.6).
4. Through the duality property, the directional reflectance and transmittance for the direction μ are simply related to the flux reflectance and flux transmittance for *Prototype Problem 3*, where μ is replaced by the cosine of the angle of incidence of sunlight μ_0.
5. The flux reflectance and transmittance for this problem are the spherical reflectance and transmittance for *Prototype Problem 3* (see Problem 6.5).

Equations 7.31 display four constants of integration, but the two boundary conditions, Eqs. 7.32, and the fact that the differential equation is of degree two, suggest that there are only two independent constants. To obtain the two necessary relationships between A, B, C, and D, we substitute Eqs. 7.31 into Eqs. 7.23–7.24. We find that

$$\frac{C}{A} = \frac{B}{D} = \frac{a}{2 - a + 2\bar{\mu}\Gamma} = \frac{1 - \bar{\mu}\Gamma}{1 + \bar{\mu}\Gamma} = \frac{1 - \sqrt{1 - a}}{1 + \sqrt{1 - a}} \equiv \rho_\infty. \tag{7.33}$$

An explanation of the physical meaning of the above ratio defined as ρ_∞, where $0 \le \rho_\infty \le 1$, is provided in Example 7.2. We now substitute into the general solutions, Eqs. 7.31, to obtain

$$I^+(\tau) = Ae^{\Gamma\tau} + \rho_\infty De^{-\Gamma\tau}, \tag{7.34}$$

$$I^-(\tau) = \rho_\infty Ae^{\Gamma\tau} + De^{-\Gamma\tau}. \tag{7.35}$$

We now apply the boundary conditions (Eqs. 7.32), which yield

$$I^-(\tau = 0) = \rho_\infty A + D = \mathcal{I}, \qquad I^+(\tau = \tau^*) = Ae^{\Gamma\tau^*} + \rho_\infty De^{-\Gamma\tau^*} = 0.$$

Solving for A and D we find

$$A = \frac{-\rho_\infty \mathcal{I} e^{-\Gamma\tau^*}}{e^{\Gamma\tau^*} - \rho_\infty^2 e^{-\Gamma\tau^*}}, \qquad D = \frac{\mathcal{I} e^{\Gamma\tau^*}}{e^{\Gamma\tau^*} - \rho_\infty^2 e^{-\Gamma\tau^*}}.$$

The solutions are

$$I^+(\tau) = \frac{\mathcal{I}\rho_\infty}{\mathcal{D}}\left[e^{\Gamma(\tau^*-\tau)} - e^{-\Gamma(\tau^*-\tau)}\right], \tag{7.36}$$

$$I^-(\tau) = \frac{\mathcal{I}}{\mathcal{D}}\left[e^{\Gamma(\tau^*-\tau)} - \rho_\infty^2 e^{-\Gamma(\tau^*-\tau)}\right], \tag{7.37}$$

where the denominator is

$$\mathcal{D} \equiv e^{\Gamma\tau^*} - \rho_\infty^2 e^{-\Gamma\tau^*}. \tag{7.38}$$

The solutions for the source function, flux, and heating rate which follow from Eqs. 7.25–7.27 are

$$S(\tau) = \frac{a\mathcal{I}}{2\mathcal{D}}(1+\rho_\infty)\left[e^{\Gamma(\tau^*-\tau)} - \rho_\infty e^{-\Gamma(\tau^*-\tau)}\right], \tag{7.39}$$

$$F(\tau) = -2\bar{\mu}\frac{\pi\mathcal{I}}{\mathcal{D}}(1-\rho_\infty)\left[e^{\Gamma(\tau^*-\tau)} + \rho_\infty e^{-\Gamma(\tau^*-\tau)}\right], \tag{7.40}$$

$$\mathcal{H}(\tau) = \frac{2\pi\alpha\mathcal{I}}{\mathcal{D}}(1+\rho_\infty)\left[e^{\Gamma(\tau^*-\tau)} - \rho_\infty e^{-\Gamma(\tau^*-\tau)}\right]. \tag{7.41}$$

Note that Eq. 7.26 yields $F^-(0) = 2\pi\bar{\mu}I^-(0) = 2\pi\bar{\mu}\mathcal{I}$ for the incoming flux at the top of the slab. We might be tempted to set $\bar{\mu} = 0.5$ so that this expression would yield the exact value, $\pi\mathcal{I}$. However, to remain consistent with the two-stream approximation, it is important to use the approximate expression, Eq. 7.26. The flux reflectance $\rho(-2\pi, 2\pi)$, flux transmittance $\mathcal{T}(-2\pi, -2\pi)$, and flux absorptance $\alpha(-2\pi)$ become

$$\rho(-2\pi, 2\pi) = \frac{2\pi\bar{\mu}I^+(0)}{2\pi\bar{\mu}\mathcal{I}} = \frac{\rho_\infty}{\mathcal{D}}[e^{\Gamma\tau^*} - e^{-\Gamma\tau^*}], \tag{7.42}$$

$$\mathcal{T}(-2\pi, -2\pi) = 2\pi\bar{\mu}\frac{I^-(\tau^*)}{2\pi\bar{\mu}\mathcal{I}} = \frac{1-\rho_\infty^2}{\mathcal{D}}, \tag{7.43}$$

and

$$\alpha(-2\pi) = 1 - \rho(-2\pi, 2\pi) - \mathcal{T}(-2\pi, -2\pi)$$
$$= \frac{(1-\rho_\infty)}{\mathcal{D}}[e^{\Gamma\tau^*} + \rho_\infty e^{-\Gamma\tau^*} - 1 - \rho_\infty]. \tag{7.44}$$

It is important to note that the flux transmittance includes the "beam" transmittance, which in this problem is

$$\mathcal{T}_b(-2\pi, -2\pi) = \frac{\int_0^1 d\mu\mu\mathcal{I}e^{-\tau^*/\mu}}{\int_0^1 d\mu\mu\mathcal{I}} = 2E_3(\tau^*).$$

Thus, the diffuse flux transmittance is

$$\mathcal{T}_d(-2\pi, -2\pi) = \mathcal{T}(-2\pi, -2\pi) - \mathcal{T}_b(-2\pi, -2\pi)$$
$$= \frac{1-\rho_\infty^2}{\mathcal{D}} - 2E_3(\tau^*). \tag{7.45}$$

In the upper left panel of Fig. 7.1 we show two-stream values for the upward and downward fluxes and the mean intensity as a function of optical depth for a specific set of optical parameters, τ^*, a, $\bar{\mu}$, and $\mathcal{I}$. The errors incurred by using the two-stream approximation are shown in the upper right panel. Note that the upward flux is zero at the bottom boundary ($\tau = \tau^* = 1.0$), and the downward flux is $2\pi\bar{\mu}\mathcal{I}$ at the top boundary ($\tau = 0$). The error in these fluxes grows almost monotonically away from

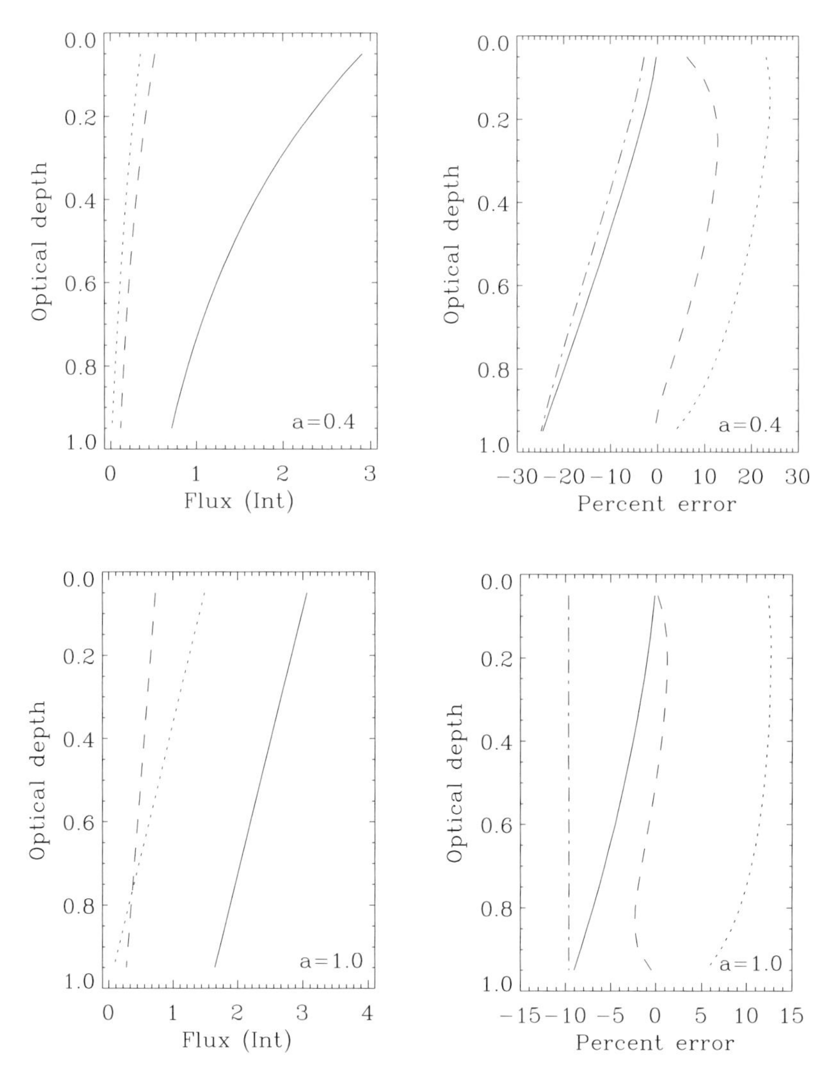

Figure 7.1 Two-stream solutions for uniform illumination (*Prototype Problem 1*). Upper left panel: The parameters are $\mathcal{I} = 1$, $\tau^* = 1$, $a = 0.4$, $\bar{\mu} = 1/2$, and $p = 1$. The downward flux F^- (solid line), the upward flux F^+ (dotted line), and the mean intensity $\bar{I}$ (dashed line). Upper right panel: Errors (ϵ) in the two-stream approximation, $\epsilon_- = F^- - F^-_{\rm ac}$ (solid line), $\epsilon_+ = F^+ - F^+_{\rm ac}$ (dotted line), $\epsilon_{\rm net} = F - F_{\rm ac}$ (dot-dashed line), and $\epsilon_{\rm div} = (\Delta F - \Delta F_{\rm ac})/\Delta\tau$ (long-dashed line). The subscript "ac" denotes the accurate solution, computed from an eight-stream discrete-ordinates solution (Chapter 8). $\Delta F/\Delta\tau$ is the flux divergence, and $\epsilon_{\rm div}$ is the error in the flux divergence. Lower panels: Same as upper panels except that $a = 1.0$.

the boundaries and becomes as large as 25%, in this particular example. In lower panels of Fig. 7.1 we show similar results for conservative scattering ($a = 1.0$). The heating rate (flux divergence) is proportional to the total (direct plus diffuse) mean intensity. The error here is less than that of the hemispherical fluxes, suggesting that energy deposition may be estimated accurately in the two-stream approximation.

Example 7.1 Thermalization Length and Random Walk

Equations 7.36 and 7.37 show that the $(1/e)$-depth of the penetration of photons is

$$\Gamma^{-1} = \frac{\bar{\mu}}{\sqrt{1-a}}.$$

The quantity Γ^{-1} is called the *thermalization length*. It is interpreted to be the mean optical depth of a photon penetration following repeated scatterings before it suffers an absorption (thermalization). The dependence on $\bar{\mu}$ is straightforward, since the steeper the mean inclination of the rays, the more shallow the penetration. The inverse dependence on $\sqrt{1-a}$ is qualitatively reasonable; when $a \to 1$ (conservative scattering), the penetration depth becomes infinite, as one would expect for no photon loss.

Why should the *square root* of the volume emittance $(1-a)$ be the relevant dependence? To answer this we will use the *random walk* picture of "diffusion" of a multiply scattered photon over distance. Imagine that photons are released repeatedly into an unbounded scattering medium and are randomly scattered on the average $\langle N \rangle$ times before being destroyed (absorbed). Since the probability of being destroyed *per collision* is $(1-a)$, then it is clear that $\langle N \rangle(1-a) = 1$. According to the random-walk theory, the mean total distance through which an average photon "wanders" after $\langle N \rangle$ collisions is $\sqrt{\langle N \rangle} mfp$, where mfp is the photon mean free path. Now mfp is just one optical depth times the cosine of the mean ray inclination. Thus, the mean total distance covered before being absorbed is just $\sqrt{\langle N \rangle} \cdot \bar{\mu} = \bar{\mu}/\sqrt{1-a}$.

Example 7.2 Semi-Infinite Slab

In this example we consider the limit of $\tau^* \to \infty$. This is an approximation to a very thick planetary atmosphere (such as that of Venus or Jupiter) or a deep ocean. Invoking the condition $S(\tau)e^{-\tau} \to 0$, we must exclude the positive exponential terms. The solutions simplify as follows:

$$I^-(\tau) = \mathcal{I} e^{-\Gamma\tau}, \qquad I^+(\tau) = \mathcal{I} \rho_\infty e^{-\Gamma\tau}, \tag{7.46}$$

$$S(\tau) = \frac{a}{2}\mathcal{I}(1+\rho_\infty)e^{-\Gamma\tau}, \tag{7.47}$$

$$F(\tau) = -2\pi\bar{\mu}\mathcal{I}(1-\rho_\infty)e^{-\Gamma\tau}, \tag{7.48}$$

$$\mathcal{H}(\tau) = \frac{\alpha\mathcal{I}(1+\rho_\infty)}{2\bar{\mu}}e^{-\Gamma\tau}. \tag{7.49}$$

Note that the sign of $F(\tau)$ is negative, indicating that the net flow of energy is downward. We also note that $F(\tau) \to 0$ as τ exceeds many thermalization lengths. The flux reflectance of the semi-infinite slab is

$$\rho(-2\pi, 2\pi) = \frac{I^+(\tau=0)}{\mathcal{I}} = \frac{\mathcal{I}\rho_\infty}{\mathcal{I}} = \rho_\infty = \frac{1-\sqrt{1-a}}{1+\sqrt{1-a}}. \tag{7.50}$$

The meaning of the notation ρ_∞ should now be clear. The above expression is an exact result for the reflected intensity (see Problem 7.9). Equation 7.50 also provides us with the flux absorptance,

$$\alpha(-2\pi) = 1 - \rho(-2\pi, 2\pi) = \frac{2\sqrt{1-a}}{1+\sqrt{1-a}}. \tag{7.51}$$

Example 7.3 The Conservative-Scattering Limit

There are two ways to find this solution. The first is to take the limit $a \to 1$ of the expressions valid for nonconservative scattering using *L'Hôspital's Rule* to handle the 0/0 limits. The second way is to return to the (simplified) set of coupled differential equations and solve them afresh. Then only a first-order differential equation must be solved. Both methods yield the following results (see Problem 7.2):

$$I^+(\tau) = \frac{\mathcal{I}(\tau^* - \tau)}{2\bar{\mu} + \tau^*}, \qquad I^-(\tau) = \frac{\mathcal{I}[2\bar{\mu} + (\tau^* - \tau)]}{2\bar{\mu} + \tau^*}, \tag{7.52}$$

$$S(\tau) = \frac{\mathcal{I}[\bar{\mu} + (\tau^* - \tau)]}{2\bar{\mu} + \tau^*}, \qquad F(\tau) = -\frac{4\pi\bar{\mu}^2\mathcal{I}}{2\bar{\mu} + \tau^*}, \qquad \mathcal{H}(\tau) = 0. \tag{7.53}$$

Note that the flux is constant throughout the atmosphere and that the heating rate is zero, as should be the case for a conservatively scattering slab. The above results, derived by Schuster in 1905, were one of the first published solutions of the radiative transfer equation.[3] Assuming $\bar{\mu} = 1/2$, Schuster found the reflectance and transmittance of the medium to be

$$\rho(-2\pi, 2\pi) = \frac{I^+(0)}{\mathcal{I}} = \frac{\tau^*}{1 + \tau^*}, \qquad \mathcal{T}(-2\pi, -2\pi) = \frac{I^-(\tau^*)}{\mathcal{I}} = \frac{1}{1 + \tau^*}.$$

Schuster also pointed out the following remarkable property of a conservatively scattering slab: It transmits light in an *arithmetic ratio* to the thickness, whereas an absorbing slab transmits light in a *geometric ratio* to the thickness. This effect is caused by the augmentation of the exponentially attenuated beam by the multiply scattered light.

Example 7.4 Highly Absorbing Medium

In this limit, it is easily shown by expanding the previous expressions in a power series in a, and maintaining only the lowest-order terms, that the reflectance for a semi-infinite slab, in the limit $a \to 0$, is $\rho_\infty \approx a/4$. The source function and the flux are given by

$$S(\tau) \approx \frac{a\mathcal{I}}{2} e^{-\tau/\bar{\mu}}, \qquad F(\tau) \approx -2\pi\bar{\mu}\mathcal{I}e^{-\tau/\bar{\mu}}.$$

These results show that only first-order scattering contributes to the source function, and only the directly transmitted radiation contributes to the net flux. The reflectance, transmittance, and absorptance are

$$\rho(-2\pi, 2\pi) \approx \frac{a}{4}\left(1 - e^{-2\tau^*/\bar{\mu}}\right), \qquad \mathcal{T}(-2\pi, 2\pi) \approx e^{-\tau^*/\bar{\mu}},$$
$$\alpha(-2\pi) \approx 1 - e^{-\tau^*/\bar{\mu}}.$$

Note in particular the presence of the term $e^{-2\tau^*/\bar{\mu}}$ in the reflectance. A reflected photon will travel twice as far through the slab, on the average, as a transmitted photon.

A further interesting limit is for $\tau^* \ll 1$ and $a \ll 1$, the optically thin, highly absorbing case. Then,

$$\rho(-2\pi, 2\pi) \approx \frac{a\tau^*}{2\bar{\mu}} = \frac{\tau_s^*}{2\bar{\mu}}, \qquad \mathcal{T}(-2\pi, 2\pi) \approx 1 - \frac{\tau_a^*}{\bar{\mu}}, \qquad \alpha(-2\pi) \approx \frac{\tau_a^*}{\bar{\mu}}.$$

In this limit, the reflectance depends linearly upon the mean slant scattering optical path $\tau_s^*/\bar{\mu}$. The factor of 1/2 accounts for the fact that for isotropic scattering, half the photons are scattered in the forward direction. It is also reasonable that the slab transmittance and slab absorptance are linearly dependent upon the mean slant absorption optical path, $\tau_a^*/\bar{\mu}$.

Example 7.5 Angular Distribution of the Radiation Field

To find the intensity $I^{\pm}(\tau,\mu)$ in the two-stream approximation, it is necessary to integrate the (approximate) source function, using Eqs. 7.3 and 7.4. This method yields a closed-form solution for the angular dependence of the intensity and may provide sufficient accuracy for some problems. We proceed by considering the expressions for the upward and downward intensity:

$$I^+(\tau,\mu) = \int_{\tau}^{\tau^*} \frac{d\tau'}{\mu} S(\tau') e^{-(\tau'-\tau)/\mu},$$
$$I^-(\tau,\mu) = \int_{0}^{\tau} \frac{d\tau'}{\mu} S(\tau') e^{-(\tau-\tau')/\mu} + \mathcal{I} e^{-\tau/\mu}. \tag{7.54}$$

Inserting the approximate (two-stream) approximations for the source function, Eq. 7.39, and performing the integration, we find, after some algebraic manipulation using the relationships between ρ_∞, Γ, and $\bar{\mu}$ (Eq. 7.33),

$$I^+(\tau,\mu) = \frac{\mathcal{I}\rho_\infty}{\mathcal{D}}\{C^+(\mu)e^{\Gamma(\tau^*-\tau)} - C^-(\mu)e^{-\Gamma(\tau^*-\tau)} + [C^-(\mu) - C^+(\mu)]e^{-(\tau^*-\tau)/\mu}\}, \tag{7.55}$$

$$I^-(\tau,\mu) = \frac{\mathcal{I}}{\mathcal{D}}\{C^-(\mu)e^{\Gamma(\tau^*-\tau)} - C^+(\mu)\rho_\infty^2 e^{-\Gamma(\tau^*-\tau)} + [1 - C^-(\mu)]e^{\Gamma\tau^*-\tau/\mu} - \rho_\infty^2[1 - C^+(\mu)]e^{-\Gamma\tau^*-\tau/\mu}\}, \tag{7.56}$$

where $C^{\pm}(\mu) \equiv (1 \pm \Gamma\bar{\mu})/(1 \pm \Gamma\mu)$. This form is convenient because it shows explicitly that when $\mu = \bar{\mu}$, then $C^{\pm}(\bar{\mu}) = 1$ and the results become identical to Eqs. 7.36–7.37.

It is interesting that for a viewing angle parallel to the slab ($\mu = 0$), the above results both approach the source function, $I^{\pm}(\tau, \mu \to 0) = S(\tau)$, Eq. 7.39, which is a property shared by the exact result. It is easily verified that the above results satisfy the boundary conditions for all values of μ, that is, $I^-(0,\mu) = \mathcal{I}$ and $I^+(\tau^*,\mu) = 0$.

The expressions for the hemispherical-directional reflectance $I^+(0,\mu)/\mathcal{I}$ and the transmittance $I^-(\tau^*,\mu)/\mathcal{I}$ are

$$\rho(-2\pi,\mu) = \frac{\rho_\infty}{\mathcal{D}}\{C^+(\mu)e^{\Gamma\tau^*} - C^-(\mu)e^{-\Gamma\tau^*} + [C^-(\mu) - C^+(\mu)]e^{-\tau^*/\mu}\}, \tag{7.57}$$

$$\mathcal{T}(-2\pi,\mu) = \frac{1}{\mathcal{D}}\{C^-(\mu) - \rho_\infty^2 C^+(\mu) + [1 - C^-(\mu)]e^{\Gamma\tau^*-\tau^*/\mu} - \rho_\infty^2[1 - C^+(\mu)]e^{-\Gamma\tau^*-\tau^*/\mu}\}. \tag{7.58}$$

In the special case $\mu = \bar{\mu}$, since $C^{\pm}(\bar{\mu}) = 1$, the results for the flux reflectance and transmittance agree with the two-stream results, Eqs. 7.42 and 7.43. From the duality principle, the above equations are equal to the directional-hemispherical reflectance $\rho(-\mu, 2\pi)$ and transmittance $\mathcal{T}(-\mu, 2\pi)$. The latter quantities may also be derived from the solution of the problem with collimated incidence (*Prototype Problem 3*; see §7.3.5), but with considerably more difficulty.

Example 7.6 The Exponential-Kernal Approximation

An alternate method of solving radiative transfer problems is to begin with the Milne–Schwarzschild integral equation for the source function. The source function yields the diffuse intensity through an integration, which may then be added to the direct (solar) intensity. We illustrate this approach by again solving *Prototype Problem 1* and comparing to the previous results. Equation 6.84 is written for isotropic scattering for a general internal (thermal) source $S^{\rm int}$ and a boundary contribution S^*:

$$S(\tau) = S^{\rm int}(\tau) + S^*(\tau) + \frac{a}{2}\int_0^{\tau^*} d\tau' E_1(|\tau' - \tau|)S(\tau'). \tag{7.59}$$

The boundary contribution may be written in terms of a general distribution of intensity $I^-(0, \mu, \phi)$ falling on the top of the medium. From Eq. 5.63

$$S^*(\tau) = \frac{a}{4\pi}\int_{4\pi} d\omega' I^-(0, \mu', \phi')e^{-\tau/\mu}, \tag{7.60}$$

and for hemispherically isotropic radiation of intensity $\mathcal{I}$

$$S^*(\tau) = \frac{a}{2}\mathcal{I}\int_0^1 d\mu' e^{-\tau/\mu'} = \frac{a}{2}\mathcal{I}E_2(\tau), \tag{7.61}$$

where E_2 is the exponential integral of order 2.

The *exponential-kernal approximation* consists of the following replacement:

$$E_1(|\tau' - \tau|) = \int_0^1 \frac{d\mu'}{\mu'}e^{-|\tau'-\tau|/\mu'} \approx \frac{1}{\bar{\mu}}e^{-|\tau'-\tau|/\bar{\mu}}, \tag{7.62}$$

where $\bar{\mu}$ has its usual meaning. This may be thought of as a one-point quadrature evaluation of the integral or as a replacement of the angular integral with the integrand evaluated at the mean angle of the inclination of the rays. Note that E_2 in the expression for S^* becomes $\sim e^{-\tau/\bar{\mu}}$. The nth exponential integral is approximated by

$$E_n(\tau) \approx e^{-\tau/\bar{\mu}}\bar{\mu}^{n-2}. \tag{7.63}$$

Substituting the above results, we find that Eq. 7.59 becomes

$$S(\tau) = \frac{a}{2}\mathcal{I}e^{-\tau/\bar{\mu}} + \frac{a}{2}\int_0^{\tau^*} \frac{d\tau'}{\bar{\mu}}e^{-|\tau'-\tau|/\bar{\mu}}S(\tau'). \tag{7.64}$$

This equation can be shown to have a solution consisting of positive and negative exponentials. Substituting the trial solution $S(\tau) = Ae^{\Gamma\tau} + Ce^{-\Gamma\tau}$, where Γ is given by Eq. 7.30, we carry out the integrations to yield

$$Ae^{\Gamma\tau} + Ce^{-\Gamma\tau} = \frac{a}{2}\mathcal{I}e^{-\tau/\bar{\mu}} + \frac{a}{2}\int_0^\tau \frac{d\tau'}{\bar{\mu}}e^{-(\tau-\tau')/\bar{\mu}}[Ae^{\Gamma\tau'} + Ce^{-\Gamma\tau'}]$$

$$+ \int_\tau^{\tau^*} \frac{d\tau'}{\bar{\mu}}e^{-(\tau'-\tau)/\bar{\mu}}[Ae^{\Gamma\tau'} + Ce^{-\Gamma\tau'}]$$

$$= Ae^{\Gamma\tau} + Ce^{-\Gamma\tau} + e^{-\tau/\bar{\mu}}\left[\frac{a}{2}\left(1 - \frac{A}{(1-\Gamma\bar{\mu})} - \frac{C}{(1+\Gamma\bar{\mu})}\right)\right]$$

$$- \frac{a}{2}\left[\frac{Ae^{-\Gamma\tau^*}}{(1+\Gamma\bar{\mu})} + \frac{Ce^{\Gamma\tau^*}}{(1-\Gamma\bar{\mu})}\right].$$

After cancellation of equal terms on the left and right, we obtain an equation in which the left-hand side is zero. For this equation to be correct *for all values of* τ, it is necessary that the coefficients of the two linearly independent terms (one of which is proportional to $e^{-\tau/\bar{\mu}}$ and the other being just a constant) be separately equal to zero, that is,

$$1 - \frac{A}{(1-\Gamma\bar{\mu})} - \frac{C}{(1+\Gamma\bar{\mu})} = 0, \qquad \frac{Ae^{-\Gamma\tau^*}}{(1+\Gamma\bar{\mu})} + \frac{Ce^{\Gamma\tau^*}}{(1-\Gamma\bar{\mu})} = 0.$$

These are easily solved to yield

$$A = \frac{-\mathcal{I}(1-\Gamma\bar{\mu})e^{\Gamma\tau^*}}{\mathcal{D}}, \qquad C = \frac{-\rho_\infty\mathcal{I}(1-\Gamma\bar{\mu})e^{-\Gamma\tau^*}}{\mathcal{D}},$$

where we have used the definition of ρ_∞, Eq. 7.33, and $\mathcal{D}$, Eq. 7.38. Noting that $1 - \Gamma\bar{\mu} = (a/2)(1+\rho_\infty)$, we find the same solution for $S(\tau)$ obtained earlier, Eq. 7.39. Given the source function, the diffuse intensity is given by

$$I_d^+(\tau) \approx I^+(\tau,\bar{\mu}) = \int_\tau^{\tau^*} \frac{d\tau'}{\bar{\mu}}S(\tau')e^{-(\tau'-\tau)/\bar{\mu}}, \tag{7.65}$$

$$I_d^-(\tau) \approx I^-(\tau,\bar{\mu}) = \int_0^\tau \frac{d\tau'}{\bar{\mu}}S(\tau')e^{-(\tau-\tau')/\bar{\mu}}. \tag{7.66}$$

The total radiation field is the sum of the diffuse and direct ("solar") terms, $I_d + I_s$, where the direct term is

$$I_s(\tau) = \frac{a}{2}\mathcal{I}E_2(\tau) \approx \frac{a}{2}\mathcal{I}e^{-\tau/\bar{\mu}}.$$

Carrying out the integrations, one finds that the total intensities $I^\pm(\tau)$ agree with the earlier results, Eqs. 7.34–7.35. If $\mu \neq \bar{\mu}$, the intensity $I(\tau,\mu)$ agrees with the result of Example 7.5, Eq. 7.54–7.55.

We have shown that, at least for *Prototype Problem 1*, the exponential-kernal methods yields the same solution as the traditional two-stream differential-equation approach. It should be obvious that the two methods are equivalent, since they both rely upon the same approximation replacing the angular variation of the radiation with a constant value.

7.3.4 Prototype Problem 2: Imbedded Source

We now return to the differential equation approach and solve *Prototype Problem 2*, where the only source of radiation is thermal emission within the slab. We further assume that the slab is isothermal, isotropically scattering, and homogeneous. The two-stream equations are

$$\bar{\mu}\frac{dI^+(\tau)}{d\tau} = I^+(\tau) - (1-a)B - \frac{a}{2}I^+(\tau) - \frac{a}{2}I^-(\tau), \tag{7.67}$$

$$-\bar{\mu}\frac{dI^-(\tau)}{d\tau} = I^-(\tau) - (1-a)B - \frac{a}{2}I^+(\tau) - \frac{a}{2}I^-(\tau), \tag{7.68}$$

with the boundary conditions $I^-(0) = I^+(\tau^*) = 0$. These equations differ from the previous set by having an extra *inhomogeneous term* on the right-hand side. To handle this complication, it is standard practice to first seek a solution to the *homogeneous equation* for which the imbedded source $(1-a)B$ is set equal to zero. Next, we find a *particular solution* that satisfies the full equation. The general solution is then the sum of the homogeneous and particular solutions. For the former, we seek solutions of the form

$$I^+(\tau) = Ae^{\Gamma\tau} + \rho_\infty De^{-\Gamma\tau}, \qquad I^-(\tau) = \rho_\infty Ae^{\Gamma\tau} + De^{-\Gamma\tau},$$

where A and D are to be determined from the boundary conditions, and where Γ and ρ_∞ were defined in the previous section.

The particular solution is obtained by guessing that $I^+ = B$ and $I^- = B$ are solutions. This is easily verified by substituting these solutions into the governing equations. Imposing the boundary conditions, we arrive at the following two simultaneous equations:

$$Ae^{\Gamma\tau^*} + \rho_\infty De^{-\Gamma\tau^*} + B = 0, \qquad \rho_\infty A + D + B = 0.$$

Solving for A and D we find

$$A = \frac{-B(1-\rho_\infty e^{-\Gamma\tau^*})}{\mathcal{D}}, \qquad D = \frac{-B(e^{\Gamma\tau^*} - \rho_\infty)}{\mathcal{D}},$$

where $\mathcal{D}$ was defined previously (Eq. 7.38). Using these results, we substitute into the general solution to find

$$I^+(\tau) = \frac{B}{\mathcal{D}}\left\{\rho_\infty^2 e^{-\Gamma\tau} - e^{\Gamma\tau} + \rho_\infty\left[e^{-\Gamma(\tau^*-\tau)} - e^{\Gamma(\tau^*-\tau)}\right]\right\} + B, \tag{7.69}$$

$$I^-(\tau) = \frac{B}{\mathcal{D}}\left\{\rho_\infty^2 e^{-\Gamma(\tau^*-\tau)} - e^{\Gamma(\tau^*-\tau)} + \rho_\infty\left[e^{-\Gamma\tau} - e^{\Gamma\tau}\right]\right\} + B. \tag{7.70}$$

The expression for the flux is, from Eq. 7.26,

$$\begin{aligned} F(\tau) = {} & 2\pi\bar{\mu}\frac{B}{\mathcal{D}}\left\{\rho_\infty^2\left[e^{-\Gamma\tau} - e^{-\Gamma(\tau^*-\tau)}\right] + \left[e^{\Gamma(\tau^*-\tau)} - e^{\Gamma\tau}\right]\right\} \\ & + 2\pi\bar{\mu}\frac{B}{\mathcal{D}}\rho_\infty\left[e^{-\Gamma(\tau^*-\tau)} - e^{-\Gamma\tau} - e^{\Gamma(\tau^*-\tau)} + e^{\Gamma\tau}\right], \end{aligned} \tag{7.71}$$

and the source function is, from Eq. 7.25,

$$S(\tau) = \frac{a}{2}(I^+ + I^-) + (1-a)B, \tag{7.72}$$

$$\frac{S(\tau)}{B} = 1 - \frac{a(1+\rho_\infty)}{2\mathcal{D}}\left[e^{\Gamma\tau} - \rho_\infty e^{-\Gamma(\tau^*-\tau)} + e^{\Gamma(\tau^*-\tau)} - \rho_\infty e^{-\Gamma\tau}\right]. \tag{7.73}$$

The slab (or bulk) emittance is, from Eq. 5.2,

$$\epsilon(2\pi) = \frac{I^+(0)}{B} = \frac{I^-(\tau^*)}{B} = \frac{1}{\mathcal{D}}\left[\rho_\infty^2 - 1 - \rho_\infty(e^{\Gamma\tau^*} - e^{-\Gamma\tau^*})\right] + 1. \tag{7.74}$$

Examples of the depth variation of the flux and average intensity are provided in Fig. 7.2 for a highly absorbing medium ($a = 0.4$, upper panels), and a conservative case ($a = 1.0$, lower panels), respectively. The error in upward and downward fluxes decreases from about 10% for the absorptive case ($a = 0.4$), to 1–2% for the conservative case, whereas the error in the mean intensity is larger: about 10% for $a = 0.4$ and 15% for $a = 1.0$.

Example 7.7 Symmetry Properties of the Solutions

Figure 7.2 shows that $S(\tau)$ and $F(\tau)$ are symmetric and antisymmetric about $\tau^*/2$, respectively. This can be proven by transforming τ into $\tau^* - \tau$ and $\tau^* - \tau$ into τ. We obtain

$$S(\tau) = S(\tau^* - \tau), \qquad F(\tau) = -F(\tau^* - \tau).$$

The above results can be restructured by taking advantage of this symmetry. We will change variables from τ to t, where $\tau = T + t$ and $T \equiv \tau^*/2$. The new variable, t, can be negative, varying over the range $-T \le t \le +T$. After some manipulation, we obtain the following compact expressions (see Problem 7.4):

$$I^+(t) = B\left[1 - \frac{e^{\Gamma t} + \rho_\infty e^{-\Gamma t}}{e^{\Gamma T} + \rho_\infty e^{-\Gamma T}}\right], \qquad I^-(t) = B\left[1 - \frac{e^{-\Gamma t} + \rho_\infty e^{\Gamma t}}{e^{\Gamma T} + \rho_\infty e^{-\Gamma T}}\right], \tag{7.75}$$

$$S(t)/B = 1 - a\frac{(1+\rho_\infty)(e^{\Gamma t} + e^{-\Gamma t})}{2(e^{\Gamma T} + \rho_\infty e^{-\Gamma T})}, \qquad F(t) = 2\pi\bar{\mu}B(1-\rho_\infty)\frac{(e^{-\Gamma t} - e^{\Gamma t})}{(e^{\Gamma T} + \rho_\infty e^{-\Gamma T})}. \tag{7.76}$$

A plot of S/B versus τ/τ^* is shown in the left panel of Fig. 7.3 for several values of τ^* and for $a = 0.95$. This diagram shows that for $\tau^* > 15$, the source function "saturates," that is, it approaches the Planck function in the center of the medium. This is a consequence of the average intensity approaching the Planck function when the optical depth greatly exceeds the thermalization depth. The scattering contribution to the source function, $a\bar{I} \approx aB$, then makes up for the deficit in the thermal contribution, $(1-a)B$. The variation with optical depth of $S(\tau)/B$ is shown in the right panel of Fig. 7.3, where the optical depth of the medium is held constant at $\tau^* = 2$ and the single-scattering albedo, a, takes on the values $a = 0.2$, 0.6, 0.8, and 0.95. The tendency for $S \to B$ as $a \to 0$ is clear.

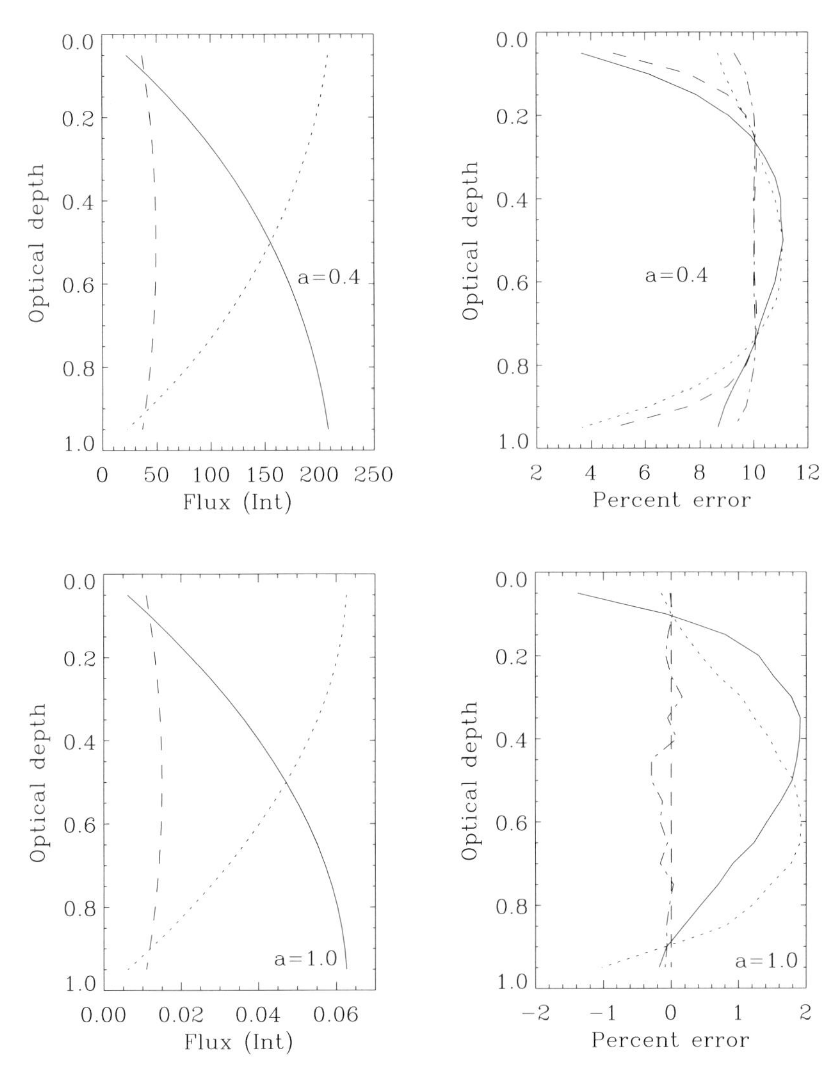

Figure 7.2 Two-stream solutions for an imbedded source (*Prototype Problem 2*). Upper panels: The parameters are $B = 100$, $\tau^* = 1$, $a = 0.4$, $\bar{\mu} = 1/2$, and $p = 1$. Lower panels: Same as upper panels except that $a = 1.0$. The line symbols are explained in Fig. 7.1.

The thermal radiation emitted by the slab relative to that of a blackbody is described by the bulk emittance

$$\epsilon(2\pi) = \frac{F(T)}{2\pi\bar{\mu}B} = (1 - \rho_\infty)\frac{(e^{\Gamma\mathrm{T}} - e^{-\Gamma\mathrm{T}})}{e^{\Gamma\mathrm{T}} + \rho_\infty e^{-\Gamma\mathrm{T}}}.$$

Note that for consistency with the two-stream approximation we have used the expression for the blackbody flux $2\pi\bar{\mu}B$, rather than the exact value πB.

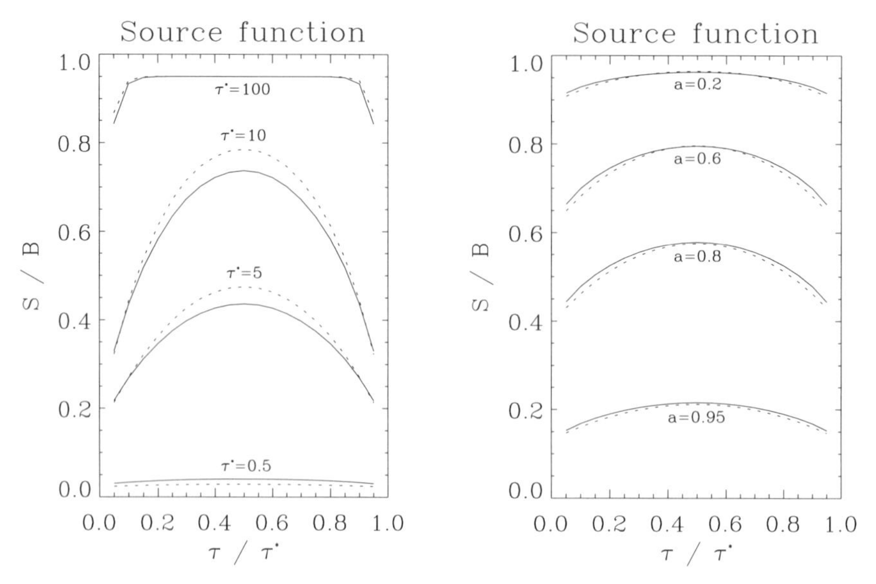

Figure 7.3 Source function S divided by the imbedded source B versus optical depth (scaled by the total optical thickness of the slab) for *Prototype Problem 2*. Dashed lines are two-stream results; solid lines are accurate results. Left panel: Single-scattering albedo of 0.95 and several values of total optical thickness. Right panel: Total optical thickness $\tau^* = 2$ and several values of single-scattering albedo.

Example 7.8 Semi-Infinite Slab

Using the previous results (Eqs. 7.73), and taking the limit $\tau^* \to \infty$, we find

$$S/B = 1 - \left(1 - \sqrt{1-a}\right) e^{-\Gamma\tau}.$$

Similarly the net flux (Eq. 7.71) becomes

$$F(\tau) = 2\pi\bar{\mu}B(1 - \rho_\infty)\, e^{-\Gamma\tau}.$$

The flux is positive, as it should be for radiation flowing upward toward the single boundary at $\tau = 0$. Notice also that the flux approaches zero at a large optical depth, where $S \to B$. More specifically, this occurs when $\Gamma\tau \gg 1$ or when the optical depth greatly exceeds the thermalization depth. The heating rate for a semi-infinite medium is given by

$$\mathcal{H}(\tau) = -\frac{4\pi\alpha B}{1+\sqrt{1-a}} e^{-\Gamma\tau}.$$

Note that there is a net *cooling* everywhere, with the maximum occurring at $\tau = 0$. The source function at $\tau = 0$ deserves special attention. This is

$$S(\tau = 0) = \sqrt{1-a}\, B,$$

which turns out to be an exact result. For small values of $(1 - a)$, the surface value of the source function falls well below the Planck function. This is a result of "leakage" of photons from the boundary region at a rate that cannot be sustained by the radiation sources.

Example 7.9 Limit of Conservative Scattering for a Semi-Infinite Slab

In the limit $a \to 1$, the imbedded source vanishes and the solution might appear to be undetermined. However, it can be derived by considering the ratio $S(\tau)/S(0)$, which, for the semi-infinite slab, is simply

$$\frac{S(\tau)}{S(0)} = \frac{1}{\sqrt{1-a}}\left[1 - \left(1 - \sqrt{1-a}\right) e^{-\Gamma\tau}\right].$$

We expand the exponential and keep only the first-order term

$$e^{-\Gamma\tau} \approx 1 - \Gamma\tau = 1 - \tau\sqrt{1-a}/\bar{\mu}.$$

Substituting into the above expression we find, in the limit $a \to 1$,

$$S(\tau) = S(0)(1 + \tau/\bar{\mu}).$$

This is the two-stream solution to the *Milne problem*.[4] It describes the problem of a conservative-scattering semi-infinite medium, in which radiation flows upward from a radiation source located deeply within the medium. It approximates the flow of radiation through the Sun's optically thick photosphere, an application first made by K. Schwarzschild in 1906.[5]

It is possible to relate the source function to the upward flux F^+ by noting that in the two-stream approximation, $F^+(\tau = 0) = 2\pi\bar{\mu}I^+(\tau = 0) = 4\pi\bar{\mu}S(\tau = 0)$. If we set $\bar{\mu} = 1/\sqrt{3}$, which we will justify later, we find

$$S(\tau = 0) = \frac{F}{4\pi\bar{\mu}} = \frac{\sqrt{3}}{4\pi}F,$$

which is also an exact result (the *Hopf–Bronstein relationship*).

The two-stream solution for the source function in the Milne problem in which the upwelling flux is F = constant is

$$S(\tau) = \frac{F}{4\pi\bar{\mu}}(1 + \tau/\bar{\mu}) = \frac{3F}{4\pi}\left[\tau + \frac{1}{\sqrt{3}}\right] \quad \left(\bar{\mu} = 1/\sqrt{3}\right).$$

We now compare this solution with the *exact* solution

$$S(\tau) = \frac{3F}{4\pi}[\tau + q(\tau)],$$

where $q(\tau)$ is the *Hopf function*. This exact solution was first derived by E. Hopf in 1934, who showed that $q(\tau = 0) = 0.577350$, that $q(\tau)$ increases slowly with optical depth, and that at great depth, $q(\infty) = 0.710446$. The two-stream approximation sets $q \approx q_a$ = constant. If $\bar{\mu} = 1/\sqrt{3}$, then $q_a = \bar{\mu} = 0.577350$. The best overall value for q_a is $2/3 = 0.666667$.

Example 7.10 Limit of Pure Absorption

This is a useful limit since we have exact solutions at our disposal. For $a \to 0$, the solution for the isothermal slab becomes (see Example 7.8)

$$I^+(t) = B\left[1 - e^{-(\mathrm{T}-t)/\bar{\mu}}\right], \qquad I^-(t) = B\left[1 - e^{-(\mathrm{T}+t)/\bar{\mu}}\right], \qquad S(t) = B,$$
$$F(t) = 2\pi\bar{\mu}B\left[e^{-(\mathrm{T}+t)/\bar{\mu}} - e^{-(\mathrm{T}-t)/\bar{\mu}}\right], \qquad \epsilon(2\pi) = \left(1 - e^{-2\mathrm{T}/\bar{\mu}}\right) = \left(1 - e^{-\tau^*/\bar{\mu}}\right).$$

The two-stream result for the emittance differs from the exact result, which is

$$\epsilon(2\pi) = 1 - 2E_3(\tau^*).$$

Equating these two results, we require

$$e^{-\tau^*/\bar{\mu}} = 2E_3(\tau^*) = 2\int_0^1 d\mu\mu e^{-\tau^*/\mu}.$$

We can find a "best value" for $\bar{\mu}$ by requiring that

$$\int_0^\infty dt\left[E_3(t) - \frac{1}{2}e^{-t/\bar{\mu}}\right] = 0.$$

The solution to this equation can be easily verified to be $\bar{\mu} = 2/3$. However, this choice is not necessarily the best one for all circumstances, as previously discussed.

When the medium is semi-infinite, ($\tau^* \to \infty$), it is easier to deal with the flux in the $(0 - \tau^*)$ coordinate, Eq. 7.73. The emergent flux and emittance are easily seen to be

$$F(0) = 2\bar{\mu}\pi B \quad \text{and} \quad \epsilon(2\pi) = 1,$$

which requires that $\bar{\mu} = 1/2$ for the result to agree with the exact solution for $F(0) = \pi B$.

7.3.5 Prototype Problem 3: Beam Incidence

We now consider the most important scattering problem in planetary atmospheres – that of a collimated solar beam of flux F^{s}, incident from above on a planetary atmosphere. This is *Prototype Problem 3* as described in Chapter 5. We will find it convenient to deal with the *diffuse* intensity in this problem. We simplify to an isotropically scattering, homogeneous atmosphere and, as usual, assume a black lower boundary. (Both these restrictions will be removed later.) Setting the angle of incidence to be $\cos^{-1}\mu_0$, we find that the appropriate two-stream equations are

$$\bar{\mu}\frac{dI_{\mathrm{d}}^+}{d\tau} = I_{\mathrm{d}}^+ - \frac{a}{2}\left(I_{\mathrm{d}}^+ + I_{\mathrm{d}}^-\right) - \frac{a}{4\pi}F^{\mathrm{s}}e^{-\tau/\mu_0} \tag{7.77}$$

and

$$-\bar{\mu}\frac{dI_{\mathrm{d}}^-}{d\tau} = I_{\mathrm{d}}^- - \frac{a}{2}\left(I_{\mathrm{d}}^+ + I_{\mathrm{d}}^-\right) - \frac{a}{4\pi}F^{\mathrm{s}}e^{-\tau/\mu_0}, \tag{7.78}$$

where $I_{\rm d}^{+}$ and $I_{\rm d}^{-}$ are the diffuse intensities. As before, we take the sum and difference of the above equations

$$\bar{\mu}\frac{d\left(I_{\rm d}^{+} - I_{\rm d}^{-}\right)}{d\tau} = (1-a)\left(I_{\rm d}^{+} + I_{\rm d}^{-}\right) - \frac{a}{2\pi}F^{\rm S}e^{-\tau/\mu_0}, \tag{7.79}$$

$$\bar{\mu}\frac{d\left(I_{\rm d}^{+} + I_{\rm d}^{-}\right)}{d\tau} = \left(I_{\rm d}^{+} - I_{\rm d}^{-}\right). \tag{7.80}$$

Differentiating Eq. 7.80 and substituting into Eq. 7.79, we find

$$\bar{\mu}^2\frac{d^2\left(I_{\rm d}^{+} + I_{\rm d}^{-}\right)}{d\tau^2} = (1-a)\left(I_{\rm d}^{+} + I_{\rm d}^{-}\right) - \frac{a}{2\pi}F^{\rm S}e^{-\tau/\mu_0}.$$

Similarly, if we differentiate Eq. 7.79 and substitute into Eq. 7.80 we get

$$\bar{\mu}^2\frac{d^2\left(I_{\rm d}^{+} - I_{\rm d}^{-}\right)}{d\tau^2} = (1-a)\left(I_{\rm d}^{+} - I_{\rm d}^{-}\right) + \frac{a\bar{\mu}}{2\pi\mu_0}F^{\rm S}e^{-\tau/\mu_0}.$$

We may use the same solution method used earlier for *Prototype Problem 2*. We first consider the homogeneous solution. As was shown previously, this can be written as follows:

$$I_{\rm d}^{+} = Ae^{\Gamma\tau} + \rho_\infty De^{-\Gamma\tau}, \qquad I_{\rm d}^{-} = \rho_\infty Ae^{\Gamma\tau} + De^{-\Gamma\tau},$$

where Γ and ρ_∞ have their usual meanings. We guess that the particular solution is proportional to $e^{-\tau/\mu_0}$. We set

$$I_{\rm d}^{+} = Ae^{\Gamma\tau} + \rho_\infty De^{-\Gamma\tau} + Z^{+}e^{-\tau/\mu_0},$$

and

$$I_{\rm d}^{-} = \rho_\infty Ae^{\Gamma\iota} + De^{-\Gamma\tau} + Z^{-}e^{-\tau/\mu_0},$$

where Z^{+} and Z^{-} are constants to be determined. Substituting into Eqs. 7.77–7.78, we find

$$Z^{+} + Z^{-} = -\frac{aF^{\rm S}\mu_0^2}{2\pi\bar{\mu}^2\left(1-\Gamma^2\mu_0^2\right)}, \qquad Z^{+} - Z^{-} = \frac{aF^{\rm S}\mu_0\bar{\mu}}{2\pi\bar{\mu}^2\left(1-\Gamma^2\mu_0^2\right)}. \tag{7.81}$$

The above two equations may be solved for Z^{+} and Z^{-} separately:

$$Z^{+} = \frac{aF^{\rm S}\mu_0(\bar{\mu}-\mu_0)}{4\pi\bar{\mu}^2\left(1-\Gamma^2\mu_0^2\right)}, \qquad Z^{-} = -\frac{aF^{\rm S}\mu_0(\mu_0+\bar{\mu})}{4\pi\bar{\mu}^2\left(1-\Gamma^2\mu_0^2\right)}. \tag{7.82}$$

We are now ready to apply boundary conditions for the diffuse intensity: $I_{\rm d}^{-}(\tau=0)=0$ and $I_{\rm d}^{+}(\tau^*)=0$. From these two conditions, we obtain two simultaneous equations for A and D. After some manipulation we find

$$A = -\frac{aF^s\mu_0}{4\pi\bar{\mu}^2(1-\Gamma^2\mu_0^2)\mathcal{D}}\left[\rho_\infty(\bar{\mu}+\mu_0)e^{-\Gamma\tau^*} + (\bar{\mu}-\mu_0)e^{-\tau^*/\mu_0}\right],$$

$$D = \frac{aF^s\mu_0}{4\pi\bar{\mu}^2(1-\Gamma^2\mu_0^2)\mathcal{D}}\left[(\bar{\mu}+\mu_0)e^{\Gamma\tau^*} + \rho_\infty(\bar{\mu}-\mu_0)e^{-\tau^*/\mu_0}\right],$$

where $\mathcal{D}$ is defined in Eq. 7.38.

We may now solve for the source function, flux, etc. For example, the source function is

$$S(\tau) = \frac{a}{2}\left(I_\mathrm{d}^+ + I_\mathrm{d}^-\right) + \frac{aF^s}{4\pi}e^{-\tau/\mu_0}. \tag{7.83}$$

Rather than display the rather complicated solution for a finite medium, we will consider the simpler situation of a semi-infinite medium. With the condition on the boundedness of the solution $S(\tau)e^\tau \to 0$, the positive exponentials must be discarded, so that $A = 0$. The constant D reduces to

$$D = \frac{aF^s\mu_0(\bar{\mu}+\mu_0)}{4\pi\bar{\mu}^2(1-\Gamma^2\mu_0^2)}.$$

The diffuse intensities are

$$\begin{aligned} I_\mathrm{d}^+(\tau) &= \rho_\infty De^{-\Gamma\tau} + Z^+e^{-\tau/\mu_0} \\ &= \frac{aF^s\mu_0}{4\pi\bar{\mu}^2(1-\Gamma^2\mu_0^2)}\left[\rho_\infty(\bar{\mu}+\mu_0)e^{-\Gamma\tau} + (\bar{\mu}-\mu_0)\,e^{-\tau/\bar{\mu}}\right], \end{aligned} \tag{7.84}$$

$$\begin{aligned} I_\mathrm{d}^-(\tau) &= De^{-\Gamma\tau} + Z^-e^{-\Gamma/\mu_0} \\ &= \frac{aF^s\mu_0}{4\pi\bar{\mu}^2(1-\Gamma^2\mu_0^2)}\left[(\bar{\mu}+\mu_0)e^{-\Gamma\tau} - (\bar{\mu}+\mu_0)e^{-\tau/\mu_0}\right], \end{aligned} \tag{7.85}$$

and the source function becomes (Eq. 7.83)

$$\begin{aligned} S(\tau) = \frac{aF^s}{4\pi}&\left\{\frac{a\mu_0}{\bar{\mu}^2(1-\Gamma^2\mu_0^2)}\right. \\ &\times\left.\left[\frac{1}{2}(\bar{\mu}+\mu_0)(1+\rho_\infty)e^{-\Gamma\tau} - \mu_0e^{-\tau/\mu_0}\right] + e^{-\tau/\mu_0}\right\}. \end{aligned} \tag{7.86}$$

We may ask: What happens if the denominator $(1-\Gamma^2\mu_0^2)$ is zero in the equations for $I_\mathrm{d}^\pm$? This can occur if the Sun is at a specific location in the sky. This *removable singularity* can be "cured"[6] by the application of L'Hôspital's rule, which leads to a new algebraic form that varies as $\tau\,\exp(-\tau/\mu_0)$.

The total net flux and heating rate are

$$F(\tau) = 2\pi\bar{\mu}\left(I_d^+ - I_d^-\right) - \mu_0 F^s e^{-\tau/\mu_0}$$
$$= \frac{aF^s\mu_0(\bar{\mu}+\mu_0)}{2\bar{\mu}\left(1-\Gamma^2\mu_0^2\right)}\left[\rho_\infty(\bar{\mu}+\mu_0)e^{-\Gamma\tau} - 2\mu_0 e^{-\tau/\mu_0}\right] - \mu_0 F^s e^{-\tau/\mu_0} \tag{7.87}$$

and

$$\mathcal{H}(\tau) = 2\pi\alpha\left(I_d^+ + I_d^-\right) + \alpha F^s e^{-\tau/\mu_0}$$
$$= 2\pi\alpha\frac{aF^s\mu_0(\bar{\mu}+\mu_0)}{4\pi\bar{\mu}^2\left(1-\Gamma^2\mu_0^2\right)}\left[(1+\rho_\infty)(\bar{\mu}+\mu_0)e^{-\Gamma\tau} - 2\mu_0 e^{-\tau/\mu_0}\right]$$
$$+\alpha F^s e^{-\tau/\mu_0}. \tag{7.88}$$

Note that we have added the terms $-\mu_0 F^s e^{-\tau/\mu_0}$ in the flux equation and $\alpha F^s e^{-\tau/\mu_0}$ in the heating equation to include the contributions from the solar component.

Profiles of computed fluxes and average intensity are shown in Fig. 7.4 for an absorbing slab ($a = 0.4$, upper panels) as well as for a nonabsorbing slab ($a = 1.0$, lower panels). The error incurred by using the two-stream approximation is also shown.

Example 7.11 Point-Direction Gain for a Semi-Infinite Medium

From Eq. 6.90 the point-direction gain for a semi-infinite medium is

$$\mathcal{G}(\tau, \mu_0; \tau^* \to \infty) = \frac{S(\tau; \tau^* \to \infty; -\mu_0)}{(aF^s/4\pi)}, \tag{7.89}$$

where S is given by Eq. 7.86. The *H-function* is given by Eq. 6.92:

$$H(\mu) = \mathcal{G}(0, \mu; \tau^* \to \infty) = 1 + \frac{a\mu}{\bar{\mu}^2\left(1-\Gamma^2\mu_0^2\right)}\left[\frac{1}{2}(\bar{\mu}+\mu)(1+\rho_\infty) - \mu\right].$$

The above expression may be reduced to a much simpler algebraic form (Problem 7.3)

$$H(\mu) = \frac{\bar{\mu}+\mu}{\bar{\mu}+\mu\sqrt{1-a}}. \tag{7.90}$$

The directional-hemispherical albedo is

$$\rho(-\mu_0, 2\pi) = \frac{2\pi\bar{\mu}I_d^+(0)}{\mu_0 F^s} = \frac{a\mu_0}{2\bar{\mu}\left(1-\Gamma^2\mu_0^2\right)}[\rho_\infty(\bar{\mu}+\mu_0) + (\bar{\mu}-\mu_0)], \tag{7.91}$$

which may also be simplified to

$$\rho(-\mu_0, 2\pi; \tau^* \to \infty) = \frac{\bar{\mu}-\bar{\mu}\sqrt{1-a}}{\bar{\mu}+\mu_0\sqrt{1-a}} = 1 - \sqrt{1-a}H(\mu_0). \tag{7.92}$$

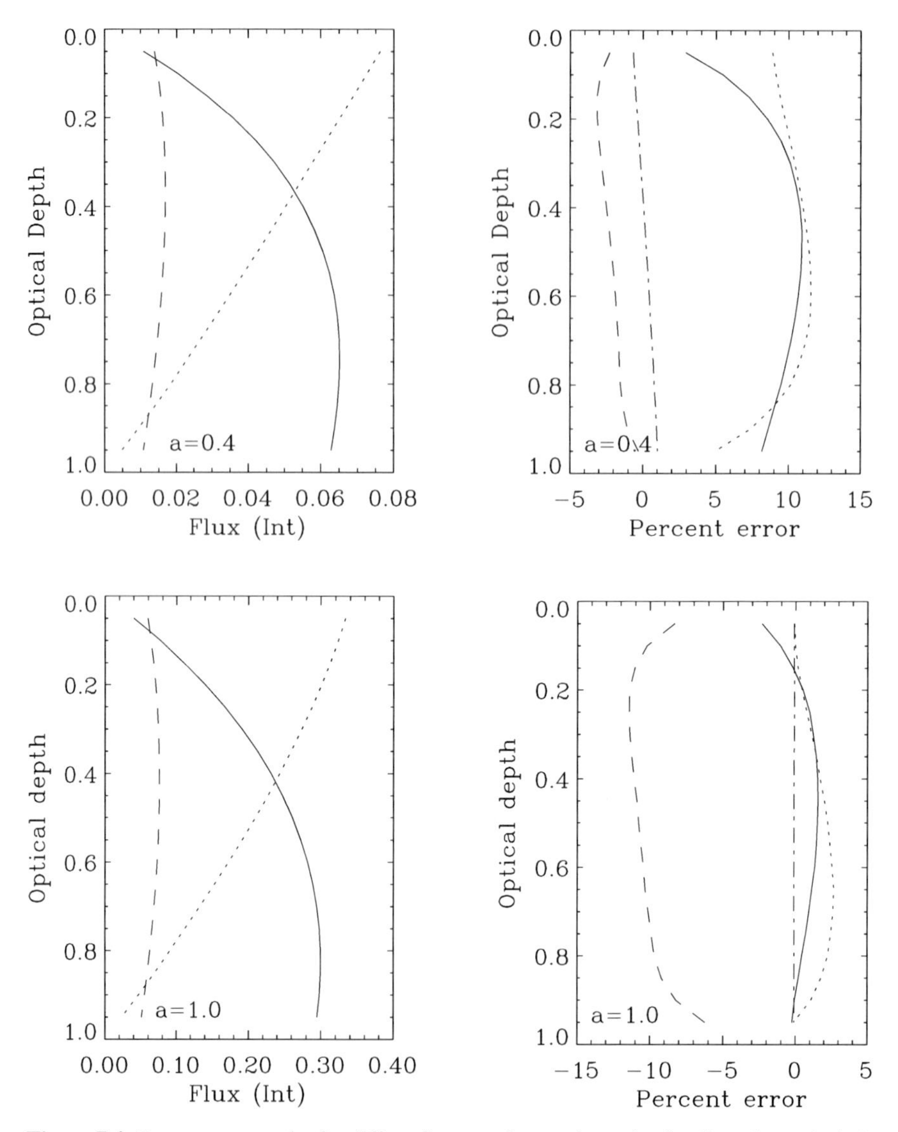

Figure 7.4 Two-stream results for diffuse fluxes and mean intensity for direct beam (solar) illumination (*Prototype Problem 3*). Upper panels: The parameters are $F^s = 1$, $\mu_0 = 1.0$, $\tau^* = 1$, $a = 0.4$, $\bar{\mu} = 1/2$, and $p = 1$. Lower panels: Same as upper panels except that $a = 0.4$. The line symbols are explained in Fig. 7.1.

An interesting relationship exists between the above quantity and the two-stream solution for the hemispherical-directional albedo in *Prototype Problem 1* (Eq. 7.57), which in the limit of $\tau^* \to \infty$ becomes

$$\rho(-2\pi; \mu; \tau^* \to \infty) = \rho_\infty C^+(\mu) = \frac{1 - \sqrt{1-a}}{1 + \sqrt{1-a}} \frac{1 + \Gamma\bar{\mu}}{1 + \Gamma\mu} = \frac{\bar{\mu} - \bar{\mu}\sqrt{1-a}}{\bar{\mu} + \mu\sqrt{1-a}}, \tag{7.93}$$

which is the same result as Eq. 7.92. *Thus, the two-stream approximation for the directional albedo obeys the Duality Principle*, $\rho(-\mu, 2\pi; \tau^* \to \infty) = \rho(-2\pi, \mu; \tau^* \to \infty)$.

7.4 Conservative Scattering in a Finite Slab

It is instructive to solve the radiative transfer problem anew in this case, since a new feature occurs, namely that the homogeneous solution for $I^+ + I^-$ is now a linear function of τ (say, $B\tau + C$). We will not show the details of the solution (this is Problem 7.5). The results are given below:

$$I_d^+(\tau) = \frac{F^s m}{4\pi}\left[\frac{(m+1)(\tau^* - \tau) + (m-1)(\tau + 2\bar{\mu})\,e^{-\tau^*/\mu_0}}{(\tau^* + 2\bar{\mu})} - (m-1)e^{-\tau/\mu_0}\right], \tag{7.94}$$

$$I_d^-(\tau) = \frac{F^s m}{4\pi}\left[\frac{(m+1)(\tau^* - \tau + 2\bar{\mu}) + (m-1)\tau e^{-\tau^*/\mu_0}}{(\tau^* + 2\bar{\mu})} - (m+1)\,e^{-\tau/\mu_0}\right], \tag{7.95}$$

$$S(\tau) = \frac{F^s m}{4\pi(\tau^* + 2\bar{\mu})}\left[(m+1)(\tau^* - \tau + \bar{\mu}) + (m-1)(\tau + \bar{\mu})e^{-\tau^*/\mu_0} - m(\tau^* + 2\bar{\mu})e^{-\tau/\mu_0}\right] + \frac{F^s}{4\pi}e^{-\tau/\mu_0}, \tag{7.96}$$

$$F(\tau) = -\frac{F^s \mu_0 \bar{\mu}}{(\tau^* + 2\bar{\mu})}\left[(1+m) + (1-m)e^{-\tau^*/\mu_0}\right], \tag{7.97}$$

$$\rho(-\mu_0, 2\pi) = \frac{F^+(0)}{\mu_0 F^s} = \frac{\tau^* + (\bar{\mu} - \mu_0)(1 - e^{-\tau^*/\mu_0})}{\tau^* + 2\bar{\mu}}, \tag{7.98}$$

$$\mathcal{T}(-\mu_0, -2\pi) = \frac{F^-(\tau^*)}{\mu_0 F^s} = \frac{\bar{\mu} + \mu_0 + (\bar{\mu} - \mu_0)e^{-\tau^*/\mu_0}}{\tau^* + 2\bar{\mu}}, \tag{7.99}$$

where $m \equiv \mu_0/\bar{\mu}$. Since Eq. 7.95 is the downward diffuse intensity, the total transmittance given by Eq. 7.99 is

$$\mathcal{T}(-\mu_0, -2\pi) = \mathcal{T}_d(-\mu_0, -2\pi) + e^{-\tau^*/\mu_0} = \left(2\pi\bar{\mu}I_d^-(\tau^*)/\mu_0 F^s\right) + e^{-\tau^*/\mu_0}.$$

It is easily verified that $\rho(-\mu_0, 2\pi) + \mathcal{T}(-\mu_0, -2\pi) = 1$.

Consider the limit of large optical depth. For $\tau^* \to \infty$, $F(\tau^*) \to 0$, $\rho(-\mu_0, 2\pi) \to 1$, and $\mathcal{T}(-\mu_0, 2\pi) \to 0$. In this limit, the source function is

$$S(\tau) = \frac{F^s}{4\pi}\left[(1 - m^2)e^{-\tau/\mu_0} + m(m+1)\right]. \tag{7.100}$$

From this expression, we see that our two-stream approximation yields $S(\tau \to \infty)/S(0) = m = \sqrt{3}\mu_0$ when $\bar{\mu} = 1/\sqrt{3}$, a result that is *exact*.[7]

Example 7.12 Angular Distribution of the Intensity

The angular distribution of the intensity for *Prototype Problem 3* can be obtained in the same fashion as in *Prototype Problem 1*. The procedure is straightforward, but the algebra is rather daunting. However, a short cut is possible, provided we are interested only in the emergent intensities. We may use the closed-form results for the emergent intensities given by S. Chandrasekhar[8]:

$$I^+(0, \mu; \tau^*) = \frac{aF^s\mu_0}{4\pi(\mu + \mu_0)}[X(\mu)X(\mu_0) - Y(\mu)Y(\mu_0)], \tag{7.101}$$

$$I^-(\tau^*, \mu; \tau^*) = \frac{aF^s\mu_0}{4\pi(\mu - \mu_0)}[Y(\mu)X(\mu_0) - X(\mu)Y(\mu_0)], \tag{7.102}$$

where the X- and Y-functions are obtained directly from the point-direction gain (Eqs. 6.91), $X(\mu) = \mathcal{G}(0, \mu; \tau^*)$ and $Y(\mu) = \mathcal{G}(\tau^*, \mu; \tau^*)$. In Problem 7.3, the X- and Y-functions are obtained in the two-stream approximation.

For a semi-infinite medium, Chandrasekhar's result is written

$$\begin{aligned} I^+(0, \mu; \tau^* \to \infty) &= \frac{aF^s\mu_0}{4\pi(\mu + \mu_0)}H(\mu)H(\mu_0) \\ &\approx \frac{aF^s\mu_0}{4\pi(\mu + \mu_0)}\left[\frac{(\bar{\mu} + \mu)}{(\bar{\mu} + \mu\sqrt{1-a})}\right]\left[\frac{(\bar{\mu} + \mu_0)}{(\bar{\mu} + \mu_0\sqrt{1-a})}\right]. \end{aligned} \tag{7.103}$$

7.5 Anisotropic Scattering

7.5.1 Two-Stream Versus Eddington Approximations

Two-stream types of approximations are used primarily to compute fluxes and mean intensities in plane geometry.[9] As we showed in Chapter 5, flux and mean intensity depend only on the azimuthally averaged radiation field. We are therefore interested in simple solutions to the azimuthally averaged radiative transfer equation valid for anisotropic scattering:

$$u\frac{dI_d(\tau, u)}{d\tau} = I_d(\tau, u) - \frac{a}{2}\int_{-1}^{1} du' p(u', u)I_d(\tau, u') - S^*(\tau, u). \tag{7.104}$$

For now, we ignore thermal emission, but the method is by no means restricted to scattering alone. To obtain approximate solutions, we proceed by integrating Eq. 7.104 over each hemisphere to find two coupled, first-order differential equations for hemispherically averaged upward and downward intensity "streams." As we have seen, this leads to the usual two-stream approximation. We can obtain a similar result by replacing the integral in Eq. 7.104 by a two-term quadrature. This quadrature can be full range (based on the complete interval $-1 \leq u \leq +1$) or half range in which the intervals $-1 \leq u \leq 0$ and $0 \leq u \leq +1$ are considered separately. The latter is similar to the hemispheric averaging mentioned above.

We may alternatively proceed by approximating the angular dependence of the intensity by a polynomial in u. By choosing a linear polynomial, $I(\tau, u) = I_0(\tau) + uI_1(\tau)$, and taking angular moments of Eq. 7.104, we arrive at two coupled equations for the zeroth and first moments of the intensity, I_0 and I_1. This approach is usually referred to as the *Eddington approximation*.

In the following, we examine both the Eddington and the two-stream approximation. We shall be particularly interested in exposing the similarities and differences between these two approaches.

Assuming collimated incidence, $S^*(\tau, u) = (aF^s/4\pi)p(-\mu_0, u)e^{-\tau/\mu_0}$, we approximate the angular dependence of the intensity as a constant plus a term linear in u, $I(\tau, u) \approx [I_0(\tau) + uI_1(\tau)]$, which upon substitution into Eq. 7.104 yields

$$u\frac{d(I_0 + uI_1)}{d\tau} = (I_0 + uI_1) - \frac{a}{2}\int_{-1}^{1} du' p(u', u)(I_0 + u'I_1) - \frac{aF^s}{4\pi}p(-\mu_0, u)e^{-\tau/\mu_0}. \tag{7.105}$$

We expand the phase function in Legendre polynomials as usual and find that the azimuthally averaged phase function is

$$p(u', u) = \sum_{l=0}^{\infty}(2l + 1)\chi_L P_l(u)P_l(u'),$$

where the moments of the phase function are given by

$$\chi_l = \frac{1}{2}\int_{-1}^{+1} du' p(u', u)P_l(u').$$

In the two-stream approximation, we normally retain only two terms: the zeroth moment, which is unity because of the normalization of the phase function ($\chi_0 = 1$), and the first moment, which we refer to as the *asymmetry factor*, χ_1 (more commonly denoted by g). Then

$$\frac{a}{2}\int_{-1}^{1} du' p(u', u)(I_0 + u'I_1) = a(I_0 + 3gu\langle u\rangle_2 I_1),$$

where the $\langle\ \rangle$ symbol denotes an angular average over the sphere

$$\langle u\rangle_2 \equiv \frac{1}{2}\int_{-1}^{1} du u^2.$$

Since $p(-\mu_0, u) = 1 - 3gu\mu_0$, Eq. 7.105 becomes

$$u\frac{d(I_0 + uI_1)}{d\tau} = I_0 + uI_1 - a(I_0 + 3gu\langle u\rangle_2 I_1) - \frac{aF^{\mathrm{s}}}{4\pi}(1 - 3gu\mu_0)e^{-\tau/\mu_0}. \tag{7.106}$$

We first integrate Eq. 7.106 over u (from -1 to 1). This yields the first equation below. We then multiply Eq. 7.106 by u, and integrate again, to obtain the second equation below. Thus, we are left with the following pair of coupled equations for the moments of intensity, I_0 and I_1:

$$\frac{dI_1}{d\tau} = \frac{1}{\langle u\rangle_2}(1 - a)I_0 - \frac{aF^{\mathrm{s}}}{4\pi\langle u\rangle_2}e^{-\tau/\mu_0}, \tag{7.107}$$

$$\frac{dI_0}{d\tau} = (1 - 3ga\langle u\rangle_2)I_1 + \frac{3aF^{\mathrm{s}}}{4\pi}g\mu_0 e^{-\tau/\mu_0}. \tag{7.108}$$

Rather than solve these coupled equations immediately, we consider a slightly different approach.

We start by writing Eq. 7.104 in terms of the half-range intensities

$$\begin{aligned}\mu\frac{dI_{\mathrm{d}}^{+}(\tau,\mu)}{d\tau} &= I_{\mathrm{d}}^{+}(\tau,\mu) - \frac{a}{2}\int_0^1 d\mu' p(-\mu',\mu)I_{\mathrm{d}}^{-}(\tau,\mu') \\ &\quad - \frac{a}{2}\int_0^1 d\mu' p(\mu',\mu)I_{\mathrm{d}}^{+}(\tau,\mu') - \frac{aF^{\mathrm{s}}}{4\pi}p(-\mu_0,\mu)e^{-\tau/\mu_0} \\ &\equiv I_{\mathrm{d}}^{+}(\tau,\mu) - S^{+}(\tau,\mu)\end{aligned} \tag{7.109}$$

and

$$\begin{aligned}-\mu\frac{dI_{\mathrm{d}}^{-}(\tau,\mu)}{d\tau} &= I_{\mathrm{d}}^{-}(\tau,\mu) - \frac{a}{2}\int_0^1 d\mu' p(-\mu',-\mu)I_{\mathrm{d}}^{-}(\tau,\mu') \\ &\quad - \frac{a}{2}\int_0^1 d\mu' p(\mu',-\mu)I_{\mathrm{d}}^{+}(\tau,\mu') - \frac{aF^{\mathrm{s}}}{4\pi}p(-\mu_0,-\mu)e^{-\tau/\mu_0} \\ &\equiv I_{\mathrm{d}}^{-}(\tau,\mu) - S^{-}(\tau,\mu).\end{aligned} \tag{7.110}$$

The above equations are exact. We proceed by integrating both equations over the hemisphere by applying the operator $\int_0^1 d\mu$. If the $I^{\pm}(\tau,\mu)$ are replaced by their averages over each hemisphere, $I^{\pm}(\tau)$, and the explicit appearance of μ is replaced by some average value $\bar{\mu}$, this leads to the following pair of coupled equations for $I^{\pm}$ (dropping the d subscript):

$$\bar{\mu}\frac{dI^{+}}{d\tau} = I^{+} - a(1 - b)I^{+} - abI^{-} - S^{*+}, \tag{7.111}$$

$$-\bar{\mu}\frac{dI^{-}}{d\tau} = I^{-} - a(1 - b)I^{-} - abI^{+} - S^{*-}, \tag{7.112}$$

where

$$S^{*+} \equiv \frac{aF^{\mathrm{s}}}{2\pi} b(\mu_0) e^{-\tau/\mu_0} \equiv X^{+} e^{-\tau/\mu_0},$$
$$S^{*-} \equiv \frac{aF^{\mathrm{s}}}{2\pi} [1 - b(\mu_0)] e^{-\tau/\mu_0} \equiv X^{-} e^{-\tau/\mu_0}. \tag{7.113}$$

Here

$$X^{+} \equiv \frac{a}{2\pi} F^{\mathrm{s}} b(\mu_0), \qquad X^{-} \equiv \frac{a}{2\pi} F^{\mathrm{s}} [1 - b(\mu_0)]. \tag{7.114}$$

The *backscattering coefficients* are defined as

$$b(\mu) \equiv \frac{1}{2} \int_0^1 d\mu' p(-\mu', \mu) = \frac{1}{2} \int_0^1 d\mu' p(\mu', -\mu), \tag{7.115}$$

$$b \equiv \int_0^1 d\mu b(\mu) = \frac{1}{2} \int_0^1 d\mu \int_0^1 d\mu' p(-\mu', \mu)$$
$$= \frac{1}{2} \int_0^1 d\mu \int_0^1 d\mu' p(\mu', -\mu), \tag{7.116}$$

$$1 - b = \frac{1}{2} \int_0^1 d\mu \int_0^1 d\mu' \, p(\mu', \mu) = \frac{1}{2} \int_0^1 d\mu \int_0^1 d\mu' \, p(-\mu', -\mu). \tag{7.117}$$

We have used the *Reciprocity Relations* satisfied by the phase function, $p(-\mu', \mu) = p(\mu', -\mu)$; $p(-\mu', -\mu) = p(\mu', \mu)$, as well as the normalization property.

Example 7.13 Alternative Derivation Using Quadrature

We could have derived these two-stream equations by using quadrature. We proceed as explained above by applying the operator $\int_0^1 d\mu$ to the half-range equations for the azimuthally averaged intensities, but instead of replacing $I^{\pm}(\tau, \mu)$ with their average values in each hemisphere, we now evaluate these integrals by approximating them with the value of the integrand at one particular direction (or "stream") $\mu = \mu_1$. This will yield a pair of two-stream equations identical to Eqs. 7.111 and 7.112 if we choose $\mu_1 = \bar{\mu}$. The difference is that we interpret $I^{\pm}(\tau, \mu_1)$ as the value of the intensity in the particular direction $\mu = \mu_1$ rather than the hemispheric average. We shall return to the use of quadrature in Chapter 8 where we will generalize it to include an arbitrary number of "streams."

In terms of two-stream quadrature we have

$$b = \frac{1}{2} p(-\bar{\mu}, \bar{\mu}) = \frac{1}{2}(1 - 3g\bar{\mu}^2) = \frac{1}{2} p(\bar{\mu}, -\bar{\mu}), \tag{7.118}$$

$$1 - b = \frac{1}{2} p(\bar{\mu}, \bar{\mu}) = \frac{1}{2}(1 - 3g\bar{\mu}^2) = \frac{1}{2} p(-\bar{\mu}, -\bar{\mu}), \tag{7.119}$$

$$b(\mu') = \frac{1}{2}p(-\mu', \bar{\mu}) = \frac{1}{2}(1 - 3g\bar{\mu}\mu'), \tag{7.120}$$

$$1 - b(\mu') = \frac{1}{2}p(-\mu', -\bar{\mu}) = \frac{1}{2}(1 + 3g\bar{\mu}\mu'). \tag{7.121}$$

As before, we have kept only two terms in the expansion of the phase function in Legendre polynomials.

Equations 7.111 and 7.112 are the two-stream equations for anisotropic scattering. In the limit of isotropic scattering, ($p = 1$ or $b = 1/2$), they reduce to the equations considered in the previous section, as they should. We note that if we choose $\bar{\mu} = 1/\sqrt{3}$, then the backscattering coefficient and the asymmetry factor are related through $b = \frac{1}{2}(1 - g)$.

We have derived two sets of differential equations (Eqs. 7.107–7.108 and Eqs. 7.111–7.112), both of which are derived from similar assumptions. What is the relationship, if any, between them? To answer this question, we will attempt to bring Eqs. 7.111 and 7.112 into a form similar to Eqs. 7.107 and 7.108. We do so by using the change of variable

$$I^{\pm}(\tau) = I_0 \pm \bar{\mu} I_1,$$

consistent with the Eddington approximation. By first adding Eqs. 7.111 and 7.112, and then subtracting 7.111 from 7.112, we find after some manipulation that Eqs. 7.111 and 7.112 are equivalent to

$$\frac{dI_1}{d\tau} = \frac{1-a}{\bar{\mu}^2} I_0 - \frac{a}{4\pi\bar{\mu}^2} F^{\mathrm{s}} e^{-\tau/\mu_0},$$

$$\frac{dI_0}{d\tau} = (1 - a + 2ab) I_1 + \frac{a}{4\pi\bar{\mu}} F^{\mathrm{s}} [1 - 2b(\mu_0)] e^{-\tau/\mu_0}.$$

Since $1 - a + 2ab = 1 - a + a(1 - 3g\bar{\mu}^2) = 1 - 3ag\bar{\mu}^2$ and $1 - 2b(\mu_0) = 1 - (1 - 3g\bar{\mu}\mu_0) = 3g\bar{\mu}\mu_0$, these last two equations become

$$\frac{dI_1}{d\tau} = \frac{1-a}{\bar{\mu}^2} I_0 - \frac{a}{4\pi\bar{\mu}^2} F^{\mathrm{s}} e^{-\tau/\mu_0}, \tag{7.122}$$

$$\frac{dI_0}{d\tau} = (1 - 3ga\bar{\mu}^2) I_1 + \frac{3a}{4\pi} g\mu_0 F^{\mathrm{s}} e^{-\tau/\mu_0}. \tag{7.123}$$

By comparing Eqs. 7.107–7.108 and 7.122–7.123, we conclude that the *equations describing the Eddington and two-stream approximations are identical provided* $\langle u \rangle_2 = \bar{\mu}^2$. Thus, the choices $\langle u \rangle_2 = 1/3$ and $\bar{\mu} = 1/\sqrt{3}$ make the governing equations for the two methods the same. Therefore any remaining difference between the two must stem from different boundary conditions. This is readily seen as follows: A homogeneous boundary condition for the downward diffuse intensity consistent with the two-stream approximation leads to the boundary condition

$$I^-(0) = I_0 - \bar{\mu} I_1 = 0.$$

If, however, we require the downward diffuse *flux* to be zero at the upper boundary (common practice in the Eddington approximation), then we find

$$I_0 - \frac{2}{3} I_1 = 0.$$

As we shall see later (Chapter 8), the value $\bar{\mu} = 1/\sqrt{3}$ for the average cosine follows from applying full-range Gaussian quadrature whereas a half-range Gaussian quadrature would lead to $\bar{\mu} = 1/2$. We now consider in more detail the useful concept of the backscattering coefficient.

7.5.2 The Backscattering Coefficients

The backscattering coefficients $b(\mu_0)$ and b were defined previously in Eqs. 7.115 and 7.116. These define the fraction of the energy that is scattered into the backward hemisphere. Of course, $1 - b$ or $1 - b(\mu_0)$ is the fraction that is forward scattered. We showed in the previous subsection that if we use the approximate expression (Eq. 7.120) and choose $\bar{\mu} = 1/\sqrt{3}$, then the backscattering coefficient is related to the asymmetry factor through $g = (1 - 2b)$ where $-1 \leq g \leq 1$.

Next, let us take a closer look at the relationship between the backscattering coefficients[10] and p, the azimuthally averaged phase function as expressed by Eqs. 7.115 and 7.116. As explained in Chapter 5, we normally do not use the phase function itself, but rather its expansion in Legendre polynomials

$$p(u', u) = \sum_{l=0}^{2N-1} (2l + 1)\chi_l P_l(u') P_l(u),$$

where we have retained $2N$ terms but have left the value of N unspecified for the moment. Substituting this phase function expansion into Eq. 7.115 and 7.116, we find

$$\begin{aligned} b(\mu) &= \frac{1}{2} \int_0^1 d\mu' p(-\mu', \mu) \\ &= \frac{1}{2} \sum_{l=0}^{2N-1} (-1)^l (2l + 1)\chi_l P_l(\mu) \int_0^1 d\mu' P_l(\mu') \equiv \sum_{l=0}^{2N-1} b_l(\mu), \end{aligned} \tag{7.124}$$

where we have used the relation $P_l(-\mu) = (-1)^l P_l(\mu)$ satisfied by the Legendre polynomials and defined $b_l(\mu) \equiv \frac{1}{2}(-1)^l (2l + 1)\chi_l P_l(\mu) \int_0^1 d\mu' P_l(\mu')$. Using Eqs. 7.115 and 7.124, we obtain

$$b = \int_0^1 d\mu b(\mu) = \frac{1}{2} \sum_{l=0}^{2N-1} (-1)^l (2l + 1)\chi_l \left[\int_0^1 d\mu P_l(\mu) \right]^2. \tag{7.125}$$

For $N = 1$ these formulas yield $b(\mu) = \frac{1}{2}(1 - \frac{3}{2}g\mu)$ and $b = \frac{1}{2}(1 - \frac{3}{4}g)$, which are identical with the results obtained from Eqs. 7.118 and 7.120 using quadrature with the choice $\bar{\mu} = 1/2$. This is to be expected because this corresponds to the use of half-range Gaussian quadrature (based on the Legendre polynomials), and it is consistent with the half-range integration of the same polynomial in the above formula. We note that whereas the use of $\bar{\mu} = 1/\sqrt{3}$ in Eqs. 7.118 and 7.120 yields the sensible results $b = 0$ for complete forward scattering ($g = 1$) and $b = 1$ for complete backscattering ($g = -1$), we obtain the unphysical results $b = 1/8$ and $b = 7/8$, respectively, for $\bar{\mu} = 1/2$. This provides a strong incentive for adopting $\bar{\mu} = 1/\sqrt{3}$ in Eqs. 7.118 and 7.120 if we want to use them to compute the backscattering coefficients from the (presumably known) asymmetry factor. We can, however, determine these coefficients more accurately from Eqs. 7.115 and 7.116 or the above "summation" formulas by numerically integrating the double integrals. The summation formulas are preferable because they do not require knowledge of the complete azimuthally averaged phase function; the moments are sufficient. Moreover, from a computational point of view the summation formulas are expected to be more efficient because the integrals over the Legendre polynomials can be precomputed once and for all. It can be shown, however, that the use of $N \to \infty$ in the above formula for b yields the following exact result involving only a single integral:

$$b = \frac{1}{2\pi} \int_0^{\pi} d\Theta\, \Theta \sin\Theta p(\cos\Theta) = \frac{1}{2\pi} \int_{-1}^{1} du \cos^{-1} u p(u).$$

Thus, it is possible to compute the backscattering coefficients directly from the phase function to any desired accuracy by numerical integration. This would allow us, at least in principle, to use an "exact" backscattering coefficient in the two-stream approximation instead of the commonly used approximate formulas (Eqs. 7.118 and 7.120) relating the backscattering coefficient to the asymmetry factor. If the phase function is strongly peaked in the forward direction, we should apply the δ-N-scaling (discussed in §6.8) prior to solving the two-stream equations. The exact scaled backscattering coefficients ($\hat{b}(\mu)$ and $\hat{b}$) could then be obtained by replacing the original phase function in the single integral for b by the scaled one and replacing the moments in the summation formula for $b(\mu)$ by their scaled counterparts. Alternatively, we could use the summation formulas to compute both coefficients, and this may be preferable, not only for consistency, but also because the scaling transformation discussed in §6.8 presumes that the moments of the phase function are known. In addition, we need these moments to compute $\hat{b}(\mu)$ exactly from the summation formula given above in any case. Finally, if we use the Henyey–Greenstein phase function, then all we need to know is the first moment of the phase function – the asymmetry factor – because all the higher moments are just powers of the first, as explained in Chapter 6.

An illustration of the exact angular backscattering coefficient $b(\mu)$ is provided in Fig. 7.5 for a Henyey–Greenstein phase function with several values of the asymmetry factor between 0 and 0.95. Figure 7.6 provides an indication of the number of terms

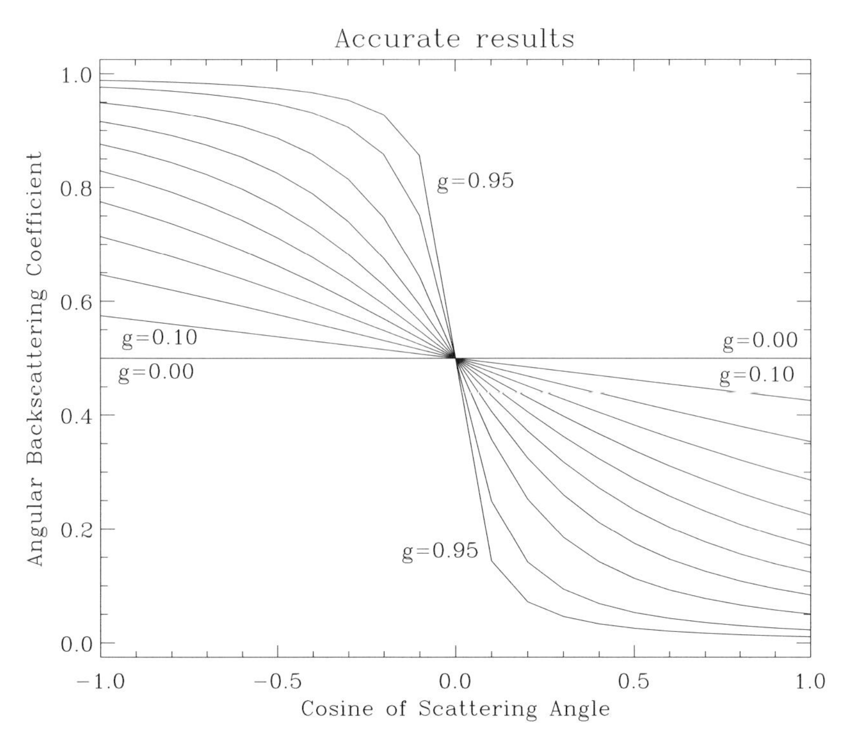

Figure 7.5 Angular backscattering coefficient: "Exact" values of the angular backscattering coefficient for the Henyey–Greenstein phase function with asymmetry factors between 0 and 0.95.

needed to obtain accurate representations of the angular backscattering coefficient computed from the above summation formula for a Henyey–Greenstein phase function with asymmetry factor 0.95. For an asymmetry factor $g = 0.4$ (not shown), four terms ($N = 2$ in Eq. 7.124) are sufficient to obtain very accurate results. For $g = 0.95$ (left panel), more than thirty terms (not shown) are required for the summation formula to converge. If we truncate the summation at $N = 1$ (two-term approximation), the backscatter coefficient becomes negative for $\mu > 0.7$. This range corresponds to solar elevations greater than about 45°. Equation 7.120 implies that S^{*+} in Eq. 7.113 becomes negative for solar elevation angles such that $\mu_0 > 1/3g\bar{\mu}$. For an optically thin medium this may lead to negative reflectance, which can be avoided by a suitable scaling as shown below.

The advantage of using δ-N scaling of the phase function is illustrated in the bottom panel of Fig. 7.6, which shows the δ-N scaled backscattering coefficient for a Henyey–Greenstein phase function with $g = 0.95$. Here, we have used the scaled moments $\hat{\chi}_l = (\chi_l - f)/(1 - f)$ in the above summation formulas with $f = \chi_{N+1}$ as discussed in Chapter 6. We see that $N = 8$ is sufficient to obtain a reasonably accurate representation of the scaled backscatter coefficient for all μ. It is also clear that a two-term expansion of the phase function ($N = 1$) may give reasonably accurate results. In particular we note that no negative values appear in Fig. 7.6. For a specified solar elevation

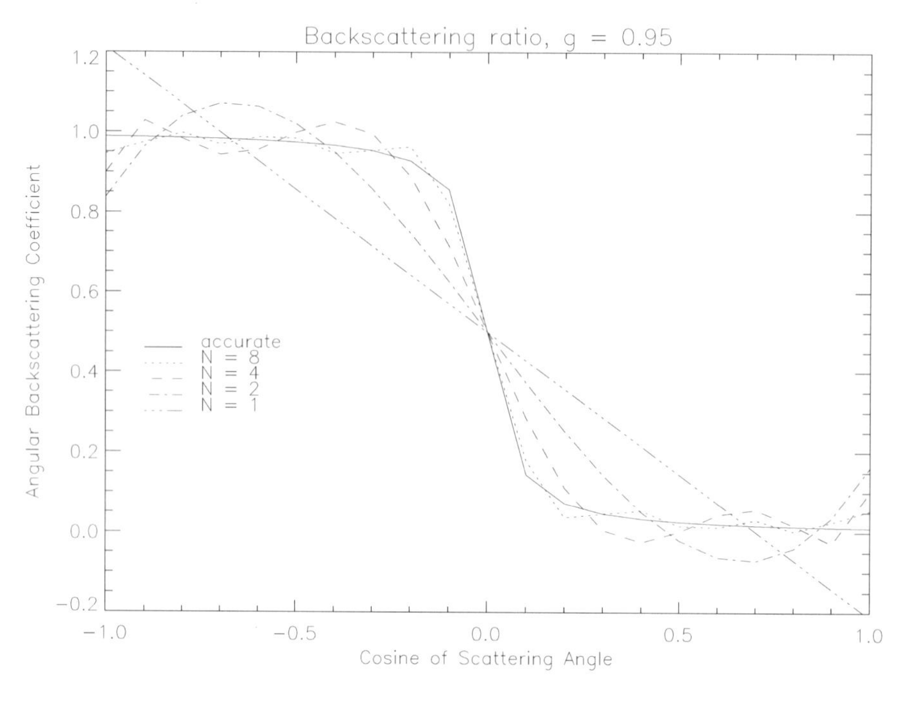

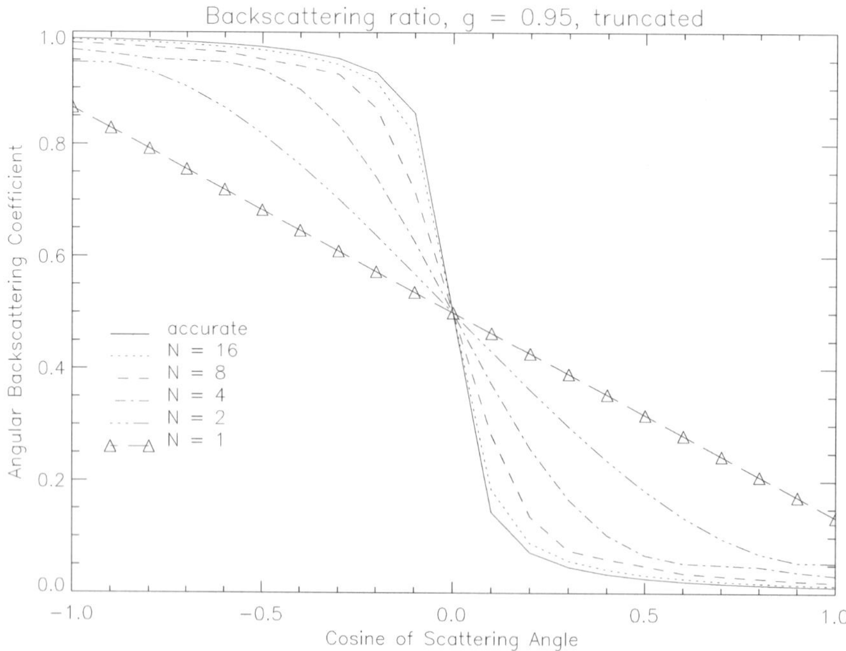

Figure 7.6 Approximate representations of the angular backscattering coefficient obtained from the summation formula. The value of N indicates half the number of terms used (i.e., $N = 1$ yields two terms, $N = 2$ four terms, and so on). Both panels are for $g = 0.95$ (before scaling). In the top panel no scaling is applied, while the bottom panel pertains to the δ-N scaled phase function.

(fixed μ_0) the (unscaled) asymmetry factor must satisfy the condition $g < 1/3\bar{\mu}\mu_0$ to avoid negative values of the backscattering coefficient. For $\bar{\mu} = 1/\sqrt{3}$, and $\mu_0 = 1.0$ (overhead Sun), we have $1/3\bar{\mu}\mu_0 = 0.5773$. However, we showed in §6.8 that if the δ-N scaling is invoked, then $|g'| < 0.5$ for the Henyey–Greenstein phase function. Hence, the δ-N scaling will overcome the problem with negative reflectance referred to above. The total (integrated) backscattering coefficient (Eq. 7.125) for the Henyey–Greenstein phase function varies smoothly but nonlinearly with g between 1 ($g = 0$) and 0 ($g = 1$).

It is customary to use only two terms in the expansion of the backscatter coefficients (Eqs. 7.124 and 7.125), although (as mentioned previously) it is possible to compute "exact" values for these coefficients. To justify the use of only two terms we examine the first few terms in the expansion coefficients of Eq. 7.124. The first few Legendre polynomials are $P_0(u) = 1$, $P_1(u) = u$, $P_2(u) = \frac{1}{2}(3u^2 - 1)$, $P_3(u) = \frac{1}{2}(5u^3 - 3u)$, $P_4(u) = \frac{1}{8}(35u^4 - 30u^2 + 3)$, and $P_5(u) = \frac{1}{8}(63u^5 - 70u^3 + 15u)$. The values for the expansion coefficients, $b_l(\mu)$, in Eq. 7.124 are displayed in Fig. 7.7 for a Henyey–Greenstein phase function with asymmetry factor g. If we choose $\bar{\mu} = 1/\sqrt{3}$, then $P_2(\bar{\mu}) = 0$. Thus, there is no contribution from the second term in Eq. 7.124. Furthermore, the contribution from the third term is negligible. The fourth term involving $P_4(\bar{\mu})$ contributes for $|g| > 0.5$. However, if we use the δ-N transformation, then $0 \le g \le 0.5$ for the Henyey–Greenstein phase function. Finally, the fifth term involving $P_5(\bar{\mu})$ contributes negligibly for all values of g. Hence, in the two-stream approximation it is sufficient to include only the first two terms in the expansion of the backscattering coefficient, so that $b(\mu) = \frac{1}{2}(1 - 3g\mu\bar{\mu})$.

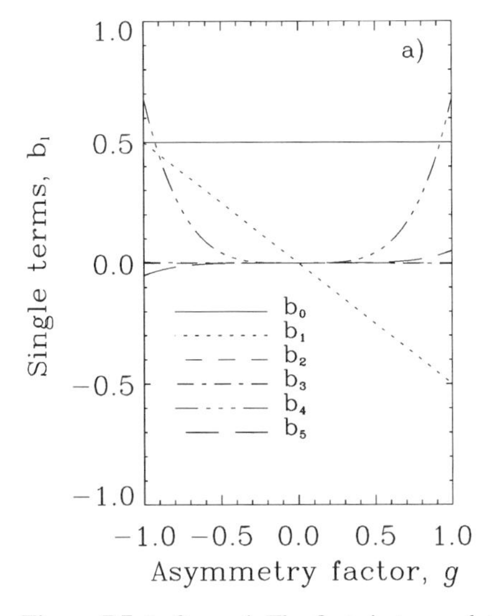

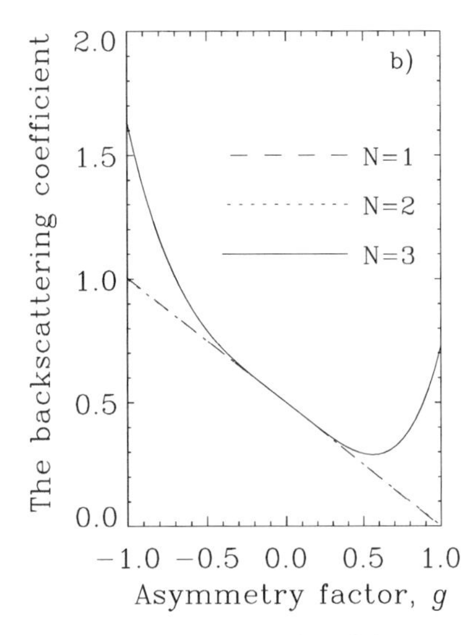

Figure 7.7 Left panel: The first six terms, $b_l(\mu)$ ($l = 0, \ldots, 5$), in the expansion of the backscattering coefficient $b(\mu)$. Right panel: The backscattering coefficient obtained for different choices of N. Note that the curves for $N = 1$ and $N = 2$ are indistinguishable.

7.5.3 Two-Stream Solutions for Anisotropic Scattering

Focusing first on the homogeneous solution, we add and subtract Eqs. 7.111 and 7.112 to obtain

$$\frac{d\left(I_{\mathrm{d}}^{+}+I_{\mathrm{d}}^{-}\right)}{d\tau}=-(\alpha-\beta)\left(I_{\mathrm{d}}^{+}-I_{\mathrm{d}}^{-}\right),$$

$$\frac{d\left(I_{\mathrm{d}}^{+}-I_{\mathrm{d}}^{-}\right)}{d\tau}=-(\alpha+\beta)\left(I_{\mathrm{d}}^{+}+I_{\mathrm{d}}^{-}\right),$$

where we have defined $\alpha \equiv -[1-a(1-b)]/\bar{\mu}$ and $\beta \equiv ab/\bar{\mu}$. By differentiating one equation and substituting into the second, we obtain the following uncoupled equations to solve:

$$\frac{d^2\left(I_{\mathrm{d}}^{+}+I_{\mathrm{d}}^{-}\right)}{d\tau^2}=\Gamma^2\left(I_{\mathrm{d}}^{+}+I_{\mathrm{d}}^{-}\right), \qquad \frac{d^2\left(I_{\mathrm{d}}^{+}-I_{\mathrm{d}}^{-}\right)}{d\tau^2}=\Gamma^2\left(I_{\mathrm{d}}^{+}-I_{\mathrm{d}}^{-}\right),$$

where

$$\Gamma=\sqrt{(\alpha-\beta)(\alpha+\beta)}=(1/\bar{\mu})\sqrt{(1-a)(1-a+2ab)}.$$

As in the case of isotropic scattering, the homogeneous solutions are

$$I_{\mathrm{d}}^{+}(\tau)=Ae^{\Gamma\tau}+Be^{-\Gamma\tau}=Ae^{\Gamma\tau}+\rho_{\infty}De^{-\Gamma\tau}, \tag{7.126}$$

$$I_{\mathrm{d}}^{-}(\tau)=Ce^{\Gamma\tau}+De^{-\Gamma\tau}=\rho_{\infty}Ae^{\Gamma\tau}+De^{-\Gamma\tau}. \tag{7.127}$$

The coefficients A, B, C, and D are not all independent, as pointed out in §7.3.3. The relation between them is found by substituting Eqs. 7.126 and 7.127 into Eqs. 7.111 and 7.112, yielding

$$\frac{C}{A}=\frac{B}{D}=\frac{\sqrt{1-a+2ab}-\sqrt{1-a}}{\sqrt{1-a+2ab}+\sqrt{1-a}}\equiv\rho_{\infty}.$$

Equations 7.111 and 7.112 suggest seeking a particular solution of the form

$$I_{\mathrm{d}}^{\pm}=Z^{\pm}e^{-\tau/\mu_0}. \tag{7.128}$$

Substitution of Eq. 7.128 into Eqs. 7.111 and 7.112 yields

$$Z^{\pm}=\frac{abX^{\mp}+[1-a+ab\mp\bar{\mu}/\mu_0]X^{\pm}}{(1-a)(1-a+2ab)-(\bar{\mu}/\mu_0)^2},$$

where $X^{\pm}$ are given by Eqs. 7.114. Before proceeding with the solution, we note the resemblance to the equations for isotropic scattering. If we set $b=1/2\,(g=0)$ and observe that in this case $X^{+}=X^{-}$, it can be verified that Γ and $Z^{\pm}$ are identical to those terms for the corresponding isotropic scattering (Eqs. 7.30 and 7.82). It is also clear that for $b=1/2$ we recover the earlier result for ρ_{∞} (see Eq. 7.33).

Returning to the solution of Eqs. 7.111 and 7.112, we determine the constants A and D in Eqs. 7.126–7.127 from the homogeneous radiation boundary conditions appropriate for the diffuse intensities. The reader may verify that the following solutions are the desired forms:

$$A = \frac{\left(-Z^+ e^{-\tau^*/\mu_0} + Z^- \rho_\infty e^{-\Gamma\tau^*}\right)}{\mathcal{D}}, \qquad D = \frac{\left(Z^+ \rho_\infty e^{-\tau^*/\mu_0} - Z^- e^{\Gamma\tau^*}\right)}{\mathcal{D}},$$

where $\mathcal{D}$ is defined by Eq. 7.38.

As can be easily shown, the above solutions satisfy the differential Eqs. 7.111 and 7.112 and also obey homogeneous boundary conditions. It is easy to show that in the limit of isotropic scattering the expressions for A and D above reduce to those following Eqs. 7.82 as they should. The solutions for the diffuse intensities are

$$I_{\rm d}^+ = \frac{1}{\mathcal{D}}\Big[\left(-Z^+ e^{-\tau/\mu_0} + Z^- \rho_\infty e^{-\Gamma\tau^*}\right)e^{\Gamma\tau} + \rho_\infty\left(Z^+ \rho_\infty e^{-\tau^*/\mu_0} - Z^- e^{\Gamma\tau^*}\right)e^{-\Gamma\tau}\Big] + Z^+ e^{-\tau/\mu_o},$$

$$I_{\rm d}^- = \frac{1}{\mathcal{D}}\Big[\left(-Z^+ e^{-\tau/\mu_0} + Z^- \rho_\infty e^{-\Gamma\tau^*}\right)\rho_\infty e^{\Gamma\tau} + \left(Z^+ \rho_\infty e^{-\tau^*/\mu_0} - Z^- e^{\Gamma\tau^*}\right)e^{-\Gamma\tau}\Big] + Z^- e^{-\tau/\mu_0}.$$

We can now solve for the half-range source functions, the flux, and the heating rate:

$$S^+(\tau) = a(1-b)I_{\rm d}^+(\tau) + abI_{\rm d}^-(\tau) + \frac{aF^{\rm s}e^{-\tau/\mu_0}}{2\pi}b(\mu_0),$$

$$S^-(\tau) = a(1-b)I_{\rm d}^-(\tau) + abI_{\rm d}^+(\tau) + \frac{aF^{\rm s}e^{-\tau/\mu_0}}{2\pi}[1-b(\mu_0)],$$

$$F(\tau) = 2\pi\bar{\mu}\left[I_{\rm d}^+(\tau) - I_{\rm d}^-(\tau)\right] - \mu_0 F^{\rm s}e^{-\tau/\mu_0},$$

$$\mathcal{H}(\tau) = 2\pi\alpha\left[I_{\rm d}^+(\tau) + I_{\rm d}^-(\tau)\right] + \alpha F^{\rm s}e^{-\tau/\mu_0}.$$

In Fig. 7.8, we show computed fluxes and average intensities for single-scattering albedo 0.4 (upper panel) and 1.0 (lower panel). A Henyey–Greenstein phase function with asymmetry factor 0.9 was used in these calculations, and the δ-N scaling was used to truncate the forward-scattering peak of the phase function.

Example 7.14 Addition of a Reflecting Boundary: The Planetary Problem

For a two-stream approximation, it is sufficiently accurate to assume a Lambert reflecting surface. Ignoring surface thermal emission, we obtain from Eq. 5.13 the boundary condition that replaces $I_{\rm d}^+(\tau^*) = 0$.

$$I_{\rm d}^+(\tau^*) = \rho_{\rm L}\left[\mu_0 F^{\rm s}e^{-\tau^*/\mu_0} + 2\pi\bar{\mu}I_{\rm d}^-(\tau^*)\right],$$

where $\rho_{\rm L}$ is the Lambert BRDF. Together with $I_{\rm d}^-(0) = 0$, these relationships allow us to once again solve for the coefficients A and D. The results are rather complicated. Problem 5.1 provides a simpler example.

Alternatively, if we were interested only in the reflectance, absorptance, and transmittance, then we could use the solutions for the black surface (which we derived above) and add the boundary "correction terms," described in §6.11 (Eqs. 6.79 and 6.80). It was shown in Problem 4.5 that the spherical albedo and spherical transmittance for *Prototype Problem 3* are equal to the flux albedo

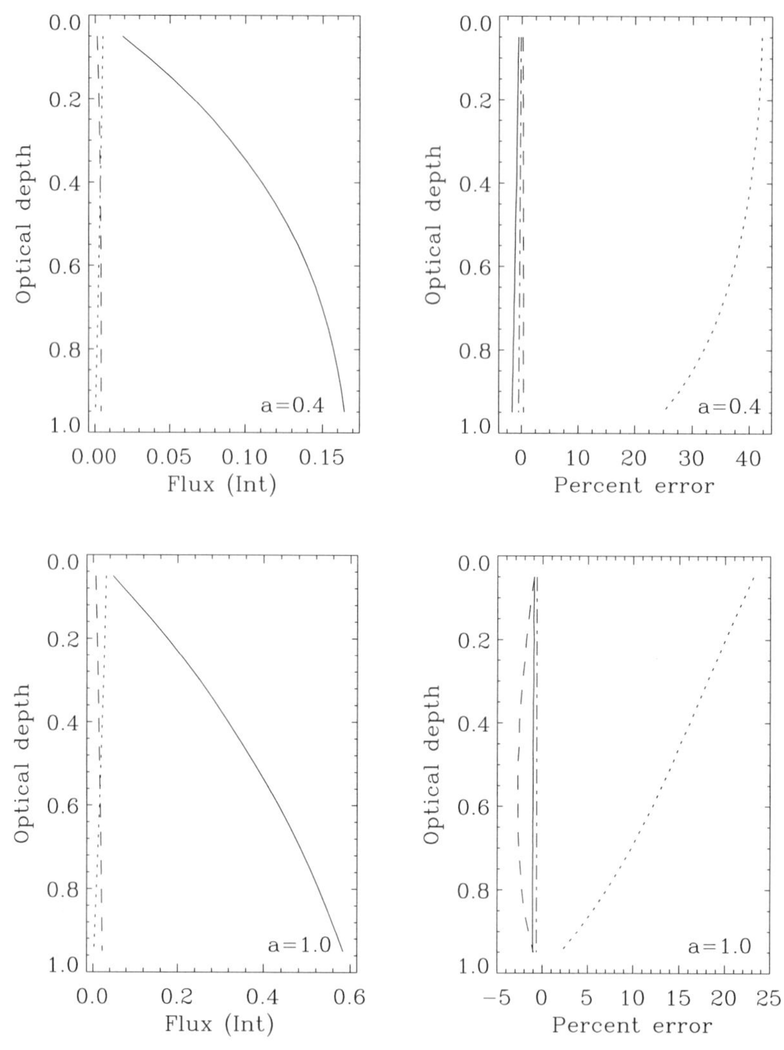

Figure 7.8 Two-stream results for direct beam (solar) illumination. Upper panel: The parameters are $F^s = 1$, $\mu_0 = 1.0$, $\tau^* = 1$, $a = 0.4$, $\bar{\mu} = 1/2$, and Henyey–Greenstein phase function with $g = 0.9$. Lower panel: Same as upper panel except that $a = 1.0$. The line symbols are explained in Fig. 7.1.

and flux transmittance (respectively) for *Prototype Problem 1*. Therefore, no additional effort is needed to complete the solution for the planetary problem.

7.5.4 Scaling Approximations for Anisotropic Scattering

In §6.7 we noted that accurate representation of sharply peaked phase functions typically requires several hundred terms in a Legendre polynomical expansion. If we make the approximation that photons scattered within this peak are not scattered at all, we find that the radiative transfer equation becomes more tractable, while we lose only a small amount of accuracy. This artifice is known as a *scaling approximation*, and it takes on various forms depending upon the choice of the truncation. We found in the δ-isotropic approximation that the scaled radiative transfer equation corresponds to an isotropic scattering problem, but with a different optical depth $\hat{\tau} = (1 - af)\tau$ and a different single-scattering albedo $\hat{a} = (1 - f)a/(1 - af)$. Here f is the fraction of the phase function within the forward peak. The value of f is somewhat arbitrary, but a good choice is $f = g$, where g is the asymmetry factor χ_1 (Eq. 6.28). If the remainder of the phase function is constant, the radiative transfer equation to be solved is

$$\mu \frac{dI^{\pm}(\hat{\tau}, \mu)}{d\hat{\tau}} = I^{\pm}(\hat{\tau}, \mu) - \frac{\hat{a}}{2} \int_0^1 d\mu' [I^{+}(\hat{\tau}, \mu') + I^{-}(\hat{\tau}, \mu')].$$

Since we have solved the above equation in the two-stream approximation for three prototype problems, it is a trivial matter to rewrite the solutions in terms of the scaled parameters, $\hat{a}$ and $\hat{\tau}$. We will write the asymmetry factor in terms of the backscattering coefficient, $b = (1/2)(1 - g)$. We use as an example the conservative-scattering limit, $\hat{a} = 1$ and $\hat{\tau} = 2b\tau$. For *Prototype Problem 3* the scaled solutions for the reflectance and transmittance are taken from Eqs. 7.98 and 7.99:

$$\rho(-\mu_0, 2\pi) = \frac{2b\tau^* + (\bar{\mu} - \mu_0)\left(1 - e^{-2b\tau^*/\mu_0}\right)}{2b\tau^* + 2\bar{\mu}}, \tag{7.129}$$

$$\mathcal{T}(\mu_0, 2\pi) = \frac{\bar{\mu} + \mu_0 + (\bar{\mu} - \mu_0)e^{-2b\tau^*/\mu_0}}{2b\tau^* + 2\bar{\mu}}. \tag{7.130}$$

Because isotropic scattering problems have much simpler closed-form solutions than those for anisotropic scattering, scaling makes it possible to obtain approximate solutions with considerably less algebra. For example, the above solutions may be used (see Problem 7.11) to simulate the effects of an optically thick cloud consisting of conservatively and anisotropically scattering water droplets.[11]

7.5.5 Generalized Two-Stream Equations

As mentioned in §7.3.2 the mean inclination (average cosine) will in general take on a different value in the two hemispheres unless the radiation field is isotropic. Similarly, the efficiency of scattering from one hemisphere to the other (defined as $r^{\pm}$ below) will be hemispherically dependent, because it depends on the angular distribution of the radiation field, which in general will be different in the two hemispheres.

We now generalize Eqs. 7.111 and 7.112 to apply for mean inclinations and scattering efficiencies that are different in the two hemispheres. Integrating Eq. 7.109 over the upper hemisphere, we find

$$\bar{\mu}^+ \frac{dI^+}{d\tau} = I^+ - \frac{a}{2}\int_0^1 d\mu \int_0^1 d\mu' p(-\mu', \mu) I^-(\tau, \mu')$$
$$- \frac{a}{2}\int_0^1 d\mu \int_0^1 d\mu' p(\mu', \mu) I^+(\tau, \mu')$$
$$- \frac{aF^s}{4\pi}\int_0^1 d\mu p(-\mu_0, \mu) e^{-\tau/\mu_0}. \tag{7.131}$$

Defining

$$r^+(\tau) \equiv \frac{1}{(1-b)I^+}\frac{1}{2}\int_0^1 d\mu \int_0^1 d\mu' p(\mu', \mu) I^+(\tau, \mu'), \tag{7.132}$$

$$r^-(\tau) \equiv \frac{1}{bI^-}\frac{1}{2}\int_0^1 d\mu \int_0^1 d\mu' p(-\mu', \mu) I^-(\tau, \mu'), \tag{7.133}$$

we may rewrite Eq. 7.131 as

$$\bar{\mu}^+ \frac{dI^+}{d\tau} = I^+(\tau, \mu) - a(1-b)r^+ I^+ - abr^- I^- - \frac{aF^s}{2\pi} b(\mu_0) e^{-\tau/\mu_0}. \tag{7.134}$$

The backscattering coefficient b, defined previously (Eq. 7.116), is a single-scattering property that is independent of the radiation field. In a similar way, we integrate Eq. 7.110 over the lower hemisphere to obtain

$$-\bar{\mu}^- \frac{dI^-}{d\tau} = I^-(\tau, \mu) - a(1-b)r^- I^-$$
$$- abr^+ I^+ - \frac{aF^s}{2\pi}[1 - b(\mu_0)] e^{-\tau/\mu_0}. \tag{7.135}$$

Equations 7.134 and 7.135 relate the downward (I^-) and upward (I^+) radiances in the corresponding hemispheres. In ocean optics terminology $2\pi I^-$ and $2\pi I^+$ are referred to as downward and upward *scalar irradiance*, respectively. If we define the coefficients α and β,

$$\alpha^{\pm} \equiv [(1 - a(1-b)r^{\pm}]/\bar{\mu}^{\pm}, \qquad \beta^{\pm} \equiv abr^{\mp}/\bar{\mu}^{\pm}, \tag{7.136}$$

we can rewrite the above Eqs. 7.134 and 7.135 as

$$\bar{\mu}^+ \frac{dI^+}{d\tau} = \alpha^+ I^+ - \beta^+ I^- - \frac{aF^s b(\mu_0)}{2\pi} e^{-\tau/\mu_0}, \tag{7.137}$$

$$-\bar{\mu}^- \frac{dI^-}{d\tau} = \alpha^- I^- - \beta^- I^+ - \frac{aF^s[1 - b(\mu_0)]}{2\pi} e^{-\tau/\mu_0}. \tag{7.138}$$

We note that Eqs. 7.134 and 7.135 are similar to Eqs. 7.111 and 7.112. In fact, they become identical to Eqs. 7.111 and 7.112 if $r^+ = r^- = 1$ and $\bar{\mu}^+ = \bar{\mu}^-$.

The product $(1 - b)r^+$ defined by Eq. 7.132 gives the probability that upward-traveling photons continue traveling upward, whereas br^- gives the probability that photons traveling in the downward direction are backscattered into the upward direction. The coefficients $r^\pm$ take into account the angular distribution of the radiation field. Thus, br gives a true backscatter efficiency as the product of the (single-scattering) backscattering coefficient b and the coefficient r, which properly accounts for the anisotropy of the radiation field. Similarly, the product $(1 - b)r$ gives a true forward-scatter probability. Of course, to compute the angular distribution of the radiation field accurately we need to apply more sophisticated methods (see Chapter 8). Using such methods we could compute accurate (τ-dependent) values of $r^\pm$ and $\bar{\mu}^\pm$, and then apply them in Eqs. 7.137 and 7.138 to provide accurate flux values. In fact, Eqs. 7.137 and 7.138 can be considered to be exact, because no approximations have been invoked in their derivation.

Combining the homogeneous versions of Eqs. 7.137 and 7.138, we arrive at the following equations:

$$\frac{d^2F^\pm}{d\tau^2} = (\alpha^+ - \alpha^-)\frac{dF^\pm}{d\tau} + (\alpha^+\alpha^- - \beta^+\beta^-)F^\pm. \tag{7.139}$$

Solutions to two-stream equations similar to Eq. 7.139 have been applied to radiative transfer in the ocean.[12]

We have now provided detailed two-stream solutions for the three prototype problems of Chapter 5. Of course, we have not completely shown all the various limiting forms ($a \to 1$, $\tau^* \to \infty$, etc.). Because the algebra is considerably simplified, the student is encouraged to consider these limiting cases, which can provide further insight. We now investigate how well the method performs in practice.

7.6 Accuracy of the Two-Stream Method

Two-stream (Eddington) approximations[13] using the δ-TTA approximation for the Henyey–Greenstein phase function to deal with sharply forward-peaked scattering may be compared with accurate (doubling) computations of reflectance, transmittance, and absorptance for a homogeneous slab. The comparisons show that the accuracy is remarkably good, being better than 2.5% in most cases for reflected and transmitted fluxes. They are accurate to better than 2% when the solar zenith angle satisfies $\mu_0 \geq 0.4$. The errors increase as μ_0 decreases, with a maximum error in the reflectance increasing to 15% as $\mu_0 \to 0$.

We should note, however, that the solutions presented in this chapter pertaining to a homogeneous slab contain exponentials with arguments $\pm k\tau$. From a practical computational point of view this represents a flaw, because "overflow" problems are encountered when $k\tau$ becomes too large. Also, to solve realistic problems involving

Table 7.1. *Two-stream results compared to accurate computations for the diffuse upward and downward fluxes at the top and bottom of the layer.*[a]

Case	μ_0	τ^*	a	g	$F^+(0)$ Exact	$F^+(0)$ Twostr	Error (%)	$F^-(\tau^*)$ Exact	$F^-(\tau^*)$ Twostr	Error (%)
1	1.000	1.00	1.0000	0.7940	0.173	0.174	0.65	1.813	1.812	−0.07
2	1.000	1.00	0.9000	0.7940	0.124	0.133	7.03	1.516	1.522	0.38
3	0.500	1.00	0.9000	0.7940	0.226	0.221	−2.14	0.803	0.864	7.59
4	1.000	64.00	1.0000	0.8480	2.662	2.683	0.81	0.480	0.454	−5.50
5	1.000	64.00	0.9000	0.8480	0.376	0.376	−0.05	0.000	0.000	0.00

Note: [a]The surface albedo was set to $\rho_L(2\pi) = 0.0$.

radiative transfer in the atmosphere and ocean, we must be able to deal with the vertical inhomogeneity of the medium. Thus, the solutions should be extended to apply to multilayered media in which the optical properties are allowed to change from layer to layer. Both these shortcomings will be removed in the next chapter, where we generalize the two-stream method to an arbitrary number of streams and an arbitrary number of layers.

In Tables 7.1–7.3 we show results for beam illumination of a homogeneous slab (*Prototype Problem 3*). Table 7.1 shows upward flux at the top and downward flux at the bottom of a slab of optical thickness $\tau^* = 1$ and 64. The error varies from being negligible to as large as 7%. Table 7.2 shows the net flux and the flux divergence at several levels in an optically thick slab ($\tau^* = 64$). For a conservative slab ($a = 1.0$) the errors are small (4–5%) for the net flux and negligible for the flux divergence. For a moderately absorbing slab ($a = 0.9$) the error is of similar magnitude for the net flux and flux divergence; it is relatively small closer to the top of the medium ($\tau < 12$) but becomes as large as 80% deep within the medium. Results for the mean intensity at the top and bottom of a Rayleigh-scattering slab are shown in Table 7.3. The slab overlies a Lambert reflector with albedo values $\rho_L = 0$ (nonreflecting), $\rho_L = 0.25$, and $\rho_L = 0.80$. The error varies depending on the angle of illumination (μ_0), the optical thickness of the slab (τ^*), and surface reflectance (ρ_L), but it is typically small (several percent) except for a few cases where it is 10% or larger.

Tables 7.4 and 7.5 show results for an imbedded (thermal) source (*Prototype Problem 2*). Table 7.4 displays exiting fluxes (top and bottom) and the flux divergence. The temperature was assumed to vary linearly across the slab from 270 K at the top to 280 K at the bottom. The lower boundary ("surface") temperature was taken to be 0 K for the absorbing case $a = 0$ and 300 K for the nonabsorbing case ($a = 1$). The total integrated Planck function ($\sigma_B T^4/\pi$) was used to drive the radiation field. The cases with ($a = 0$) are "extreme" because the two-stream approximation is known to have problems in this limit. These cases may therefore be considered as less favorable situations. We note that the error is never larger than about 11% for exiting fluxes,

Table 7.2. *Two-stream results compared to accurate computations for the net flux,* $F = F^{+} - F^{-}$, *and the flux divergence,* $F_{\text{layer},i+1} - F_{\text{layer},i}$.[a]

Case 1: $\mu_0 = 1.000$, $a = 1.0000$, $\tau^* = 64.00$, $g = 0.8480$

τ	Net flux Exact	Net flux Twostr	Error (%)	Divergence Exact	Divergence Twostr	Error (%)
.000	0.48000	0.45804	−4.6	0.00000	0.00026	0.0
3.200	0.48000	0.45778	−4.6	0.00000	0.00035	0.0
6.400	0.48000	0.45743	−4.7	0.00000	0.00070	0.0
12.800	0.48000	0.45674	−4.8	0.00000	0.00180	0.0
32.000	0.48000	0.45493	−5.2	0.00000	0.00092	0.0
48.000	0.48000	0.45401	−5.4	0.00000	0.00039	0.0
64.000	0.48000	0.45362	−5.5	0.00000	0.00000	0.0

Case 2: $\mu_0 = 1.000$, $a = 0.9000$, $\tau^* = 64.00$, $g = 0.8480$

τ	Net flux Exact	Net flux Twostr	Error (%)	Divergence Exact	Divergence Twostr	Error (%)
.000	2.77600	2.76577	−0.4	1.16100	1.16050	0.0
3.200	1.60500	1.60526	0.0	0.77500	0.78321	1.1
6.400	0.83000	0.82205	−1.0	0.63400	0.63931	0.8
12.800	0.19600	0.18274	−6.8	0.19400	0.18143	6.5
32.000	0.00220	0.00131	−40.2	0.00210	0.00130	−38.3
48.000	0.00010	0.00002	−81.0	0.00010	0.00002	−81.3
64.000	0.00000	0.00000	0.0	0.00000	0.00000	0.0

Note: [a]The surface albedo was set to $\rho_L(2\pi) = 0.0$.

whereas the error is negligible for the flux divergence. In the conservative case the two-stream approximation yields slightly nonzero values for the flux divergence. In Table 7.5 two-stream results are compared with accurate multistream results for slabs of optical thicknesses $\tau^* = 0.1$, 1.0, 10.0, and 100.0, single-scattering albedo $a =$ 0.1 and 0.95, and asymmetry factors $g = 0.05$ and 0.75. The internal source was taken to be (isotropic) thermal radiation. The "surface" temperature was taken as 0 K, the top temperature as 200 K, and the bottom temperature 300 K. Again, the temperature was assumed to vary linearly across the slab, and the Planck function was integrated between wavenumbers 300 cm^{-1} and 800 cm^{-1}, which includes the main portion of the thermal radiation. The maximum error in the two-stream results is about 12%.

A two-stream algorithm designed to work for a multilayer medium has been used to investigate the adequacy of two-stream approximations for computations of surface ultraviolet and visible solar fluxes, atmospheric photolysis rates, and warming/cooling rates due to solar and thermal infrared radiation under clear-sky, overcast, and hazy

Table 7.3. *Two-stream results compared to accurate computation for the mean intensity for conservative Rayleigh scattering ($a = 1$, $g = 0$).*

		$\rho_L(2\pi) = 0.00$		Error	$\rho_L(2\pi) = 0.25$		Error	$\rho_L(2\pi) = 0.80$		Error
μ_0	τ^*	Exact	Twostr	(%)	Exact	Twostr	(%)	Exact	Twostr	(%)
					$4\pi \bar{I}(0)/F^s$					
0.10	0.02	1.045	1.016	−2.8	1.089	1.061	−2.6	1.187	1.161	−2.2
0.10	0.25	1.170	1.085	−7.3	1.189	1.107	−6.9	1.239	1.165	−6.0
0.10	1.00	1.212	10119	−7.7	1.220	1.128	−7.5	1.247	1.166	−6.5
0.40	0.02	1.047	1.017	−2.9	1.235	1.210	−2.1	1.653	1.640	−0.8
0.40	0.25	1.284	1.164	−9.3	1.402	1.296	−7.5	1.707	1.645	−3.7
0.40	1.00	1.534	1.374	−10.4	1.584	1.431	−9.7	1.778	1.650	−7.2
0.92	0.02	1.040	1.017	−2.2	1.477	1.467	−0.7	2.453	2.471	0.7
0.92	0.25	1.279	1.191	−6.9	1.597	1.542	−3.4	2.404	2.466	2.6
0.92	1.00	1.691	1.572	−7.1	1.851	1.754	−5.2	2.398	2.457	2.5
					$4\pi \bar{I}(\tau^*)/F^s$					
0.10	0.02	0.864	0.834	−3.4	0.912	0.881	−3.4	1.018	0.985	−3.3
0.10	0.25	0.192	0.156	−18.6	0.224	0.188	−16.1	0.307	0.271	−3.3
0.10	1.00	0.057	0.054	−4.4	0.082	0.081	−1.1	0.186	0.183	9.2
0.40	0.02	0.998	0.968	−3.0	1.203	1.168	−2.9	1.661	1.613	−2.9
0.40	0.25	0.787	0.693	−11.9	0.988	0.883	−10.6	1.502	1.382	−8.0
0.40	1.00	0.385	0.344	−10.6	0.540	0.500	−7.4	1.071	1.099	2.6
0.92	0.02	1.018	0.996	−2.2	1.495	1.461	−2.3	2.561	2.500	−2.4
0.92	0.25	1.028	0.950	−7.6	1.560	1.453	−6.9	2.982	2.776	−5.2
0.92	1.00	0.881	0.822	−6.7	1.384	1.320	−4.6	3.019	3.241	4.2

conditions. The defining formulas are given in Chapter 5 and the procedures used for integration over the solar and thermal infrared spectrum are described in Chapter 9. Here we summarize briefly the performance of the two-stream approximation for such computations.

For clear-sky situations the error in photolysis rates incurred by using the two-stream approximation is insignificant compared to other sources of errors in these computations associated with uncertainties in cross sections and quantum yields. For overcast and hazy atmospheres the error in photolysis rates incurred by use of the two-stream approach may become as large as 20–40% for the worst cases. For surface ultraviolet (UV-B: 280–320 nm and UV-A: 320–400 nm) and photosynthetically active radiation (PAR: 400–700 nm) the error is relatively small for clear, hazy, and overcast conditions if the surface albedo is less than 0.8. For surface albedos larger than 0.8 (snow-covered areas) the error may become larger than 10%. The two-stream approximation overestimates PAR under overcast conditions by up to 50% for low solar elevations. Finally, it is encouraging to note that the two-stream approximation yields very accurate results for the computation of warming/cooling rates except for layers containing clouds and aerosols, where the errors are about 10%.

Table 7.4. *Upward and downward fluxes and the flux divergence for a single layer in the limits $a = 0$ and $a = 1$ for isotropic scattering.*[a]

τ^*	a	$F^+(0)$ Exact	$F^+(0)$ Twostr	Error (%)	$F^-(\tau^*)$ Exact	$F^-(\tau^*)$ Twostr	Error (%)
1.0	0.0	248.2	274.2	10.5	259.1	286.9	10.8
1000.0	0.0	301.4	301.4	0.0	348.5	348.5	0.0
10.0	1.0	53.623	41.791	−22.1	417.674	417.553	2.9

τ^*	a	$dF(0)/d\tau$ Exact	$dF(0)/d\tau$ Twostr	Error (%)	$dF(\tau^*)/d\tau$ Exact	$dF(\tau^*)/d\tau$ Twostr	Error (%)
1.0	0.0	−669.4	−657.0	−1.8	−823.4	−820.2	−0.4
1000.0	0.0	−602.6	−602.6	0.0	−697.1	−697.1	0.0
10.0	1.0	0.000	−0.013	0.0	0.000	0.004	0.0

Note: [a]The temperature at the top of layer is 270 K and it is 280 K at the bottom. It is assumed to vary linearly across the layer. The surface temperature is 0 K for the $a = 0$ cases and 300 K for the $a = 1$ case. The Planck function was integrated over the interval 0.0–10,000.0 cm^{-1}. Exact results are from 16-stream calculations by the DISORT algorithm (described in Chapter 8).

Table 7.5. *Two-stream results compared to accurate computations for thermal radiation for different optical thicknesses, single-scattering albedos, and asymmetry factors.*

τ^*	a	g	$F^+(0)$ Exact	$F^+(0)$ Twostr	Error (%)	$F(0) - F(\tau^*)$ Exact	$F(0) - F(\tau^*)$ Twostr	Error (%)
0.10	0.10	0.05	19.271	20.919	8.55	−39.839	−42.721	7.23
0.10	0.95	0.75	1.264	1.282	1.40	−2.575	−2.581	0.23
1.00	0.10	0.05	80.164	87.250	8.84	−195.348	−214.882	10.00
1.00	0.95	0.75	11.317	11.615	2.64	−24.135	−24.577	1.83
10.0	0.10	0.05	63.725	61.501	−3.49	−265.788	−265.497	−0.11
10.0	0.95	0.75	53.001	48.385	−8.71	−136.632	−130.601	−4.41
100.00	0.10	0.05	56.541	55.965	−1.02	−270.983	−269.922	−0.39
100.00	0.95	0.75	39.423	34.325	−12.93	−172.113	−153.005	−11.10

7.7 Final Comments on the Two-Stream Method

A bewildering variety of seemingly different two-stream methods have been proposed over the years to solve the radiative transfer equation for anisotropic scattering.[14] To avoid confusion, it is useful to distinguish between the approximation utilized to represent the phase function in the problem and the subsequent approximate solution of the radiative transfer equation for a given representation of the phase function. As an example, we refer to the Eddington and two-stream methods considered above to solve for the azimuthally averaged intensity. For a two-term representation of the phase function, we found that the governing equations for the Eddington and two-stream approximations were identical provided that the value of $\bar{\mu}$ was taken to be $1/\sqrt{3}$. Thus, the only remaining difference must then be attributed to the application of boundary conditions. Below we shall discuss what options are available.

To help relate the following discussion to previous published papers on the two-stream approximation, we shall rewrite the two-stream equations in the following form, which is frequently found in the literature[15]:

$$\frac{dI_{\rm d}^{+}}{d\tau} = \gamma_1 I_{\rm d}^{+} - \gamma_2 I_{\rm d}^{-} - \gamma_3 \frac{aF^{\rm s}}{2\pi} e^{-\tau/\mu_0},$$

$$-\frac{dI_{\rm d}^{-}}{d\tau} = -\gamma_2 I_{\rm d}^{+} + \gamma_1 I_{\rm d}^{-} - \gamma_4 \frac{aF^{\rm s}}{2\pi} e^{-\tau/\mu_0}.$$

In terms of our notation, $\gamma_1 \equiv [1 - a(1-b)]/\bar{\mu}$, $\gamma_2 \equiv ab/\bar{\mu}$, $\gamma_3 \equiv b(\mu_0)/\bar{\mu}$, and $\gamma_4 \equiv [1 - a(1 - b(\mu_0))]/\bar{\mu}$. Since $F_{\rm d}^{\pm} = 2\pi\bar{\mu}I_{\rm d}^{\pm}$ in the two-stream approximation, we could have derived virtually identical equations for $F_{\rm d}^{\pm}$. In fact, the generalized two-stream equations (Eqs. 7.139) also reduce to those given above for $r^{\pm} = 1$ and $\bar{\mu}^{\pm} = 1$).

Let's first consider the simple case of isotropic scattering. In this case, $b(\mu) = b = 1/2$ and therefore $\gamma_1 = (1 - a/2)/\bar{\mu}$, $\gamma_2 = a/2\bar{\mu}$, $\gamma_3 = \gamma_4 = 1/2\bar{\mu}$. Thus, only a change in $\bar{\mu}$ can affect the γ's. A change of $\bar{\mu}$ will obviously affect the solution, but it will not affect the method of solution. Next, let us consider anisotropic scattering. The common use of a two-term expansion of the phase function in the Eddington and two-stream approximations was justified in §7.5.2. The advantage of adopting δ-N scaling was also emphasized, because it avoids problems with negative reflectances that are inevitable for strongly forward-peaked phase functions. Also, the use of scaled values for the backscattering coefficients leaves only $\bar{\mu}$ to be chosen.

It has been customary to classify two-stream methods according to the values adopted for the γ's. Obviously, these constants will depend on (1) the choice of $\bar{\mu}$ and (2) for anisotropic scattering, the representation of the phase function (including scaling if necessary). This practice has led to a variety of different two-stream methods and much energy has been expended on searching for values of these coefficients that give optimal accuracy. Reading the voluminous literature on this subject can be somewhat frustrating, because there has been a tendency to confuse solution methodology with preparation of the equations prior to solution. This preparation includes adequate treatment of the angular scattering such as the truncation of the forward-scattering peak and the selection of the number of terms to be used in the expansion of the phase function. The confusion is partly due to the fact that too little attention has been paid to the "front end" of the problem concerning the choice of γ values that are physically reasonable, compared to the effort spent on comparing solutions for a variety of values of the coefficients. To alleviate this state of confusion and provide some guidance to readers who are entering this field, we offer the following suggestions: (i) Use scaling to truncate the forward scattering peak (this is mandatory if the asymmetry factor is greater than 0.5); (ii) for a beam source, $\bar{\mu} = 1/\sqrt{3}$ is expected to yield the overall best results for quantities integrated over both hemispheres, such as the mean intensity and the flux divergence, as well as for quantities integrated over a single hemisphere such as upward and downward fluxes; (iii) for imbedded, isotropic sources such as thermal radiation, $\bar{\mu} = 0.5$ yields the best overall accuracy.

Finally we may ask: Is the two-stream approximation consistent with energy conservation? To provide a partial answer to this question we note that if we ignore the solar "pseudo-source" so that we are considering the total radiation field, and then add the pair of two-stream equations, we obtain

$$\frac{dF}{d\tau} = 4\pi(1 - a)\bar{I}.$$

Conservation of energy follows from the fact that this equation is identical to the one obtained by integrating the full radiative transfer equation over 4π sr. Of course, this is just a consequence of the two-stream phase function (embodied in the backscattering ratio) being correctly normalized.

7.8 Summary

In this chapter, we have discussed simple approximate solutions to the radiative transfer equation. The first method discussed was that of single scattering, for which we showed that the condition for validity is the smallness of the quantity $a\tau^*$. This method provides convenient analytic forms, which are valid even for inhomogeneous media, anisotropic scattering, and nonslab geometry. The two-stream method, in its differential form, replaces the complicated integro-differential equation with two coupled differential equations for two functions of optical depth. When the medium is homogeneous, the resulting equations have constant coefficients and analytic solutions are obtained. These simple analytic solutions have great value from a pedagogical point of view, but they are also of considerable practical value in large-scale photochemical and dynamical models, which require repeated solutions of the radiative transfer equation for the purpose of computing photolysis and/or warming/cooling rates across the temporal and spatial domains. We also showed that for *Prototype Problem 1*, a method equivalent to the differential two-stream method is the exponential-kernal approximation. It is clear that both approaches are valid provided the scattering is isotropic.

We applied the two-stream approximation to the solution of the prototype problems defined in Chapter 5. By focusing first on the simple case of isotropic scattering, we were able to expose the essential aspects of the solutions, discuss a variety of limiting cases, and introduce several important physical concepts such as the thermalization length. We then generalized the two-stream solutions to include anisotropic scattering and compared the two-stream approach to the Eddington approximation. We also provided generalized two-stream equations that are valid for coefficients (mean inclinations and scattering efficiencies) that are different in the two hemispheres. We noted that, in principle, these latter equations can be considered to be exact, because no approximations were invoked in their derivation. In practice, however, an accurate method is needed to derive the position-dependent coefficients for each hemisphere.

Finally, we note that the literature is replete with papers on various two-stream and Eddington approximations that have been proposed over the years. Reading these papers can be somewhat frustrating. To avoid confusion, we note that for isotropic scattering there is essentially just one "tuning" parameter, $\bar{\mu}$. For anisotropic scattering, it is useful to distinguish the approximate representation of the phase function from the subsequent method of solution. Ever since the invention of the scaling of the phase function discussed in §6.8, there has been a tendency to confuse the scaling issue with the solution technique. As discussed in §6.8, it is always advisable to use scaling if the phase function is strongly forward peaked regardless of solution technique. We reiterate: *The representation of the phase function* (including scaling) *does not change the mathematical form of the radiative transfer equation.* Scaling, in particular, merely changes the optical properties of the medium to make the scattering appear less anisotropic. This in turn makes the scaled equation easier to solve by whatever analytic or numerical technique we choose to apply. Thus, contrary to common misconception, a two-stream or Eddington approximation and subsequent solution of the scaled radiative transfer equation, which is often referred to as a "δ-two-stream" or "δ-Eddington" method, does not really qualify as a new method of solution. It merely uses the same old methods to solve the scaled equation.

So then, what are the differences between the various methods? There are really only three ways in which two-stream and Eddington type of methods can differ: (i) the choice of $\bar{\mu}$, (ii) the application of boundary conditions, and (iii) the representation of the phase function for anisotropic scattering. The first item is related to how we integrate over polar angle or apply quadrature to obtain the approximate equations. In §7.5.1, we showed that the governing equations for the Eddington and two-stream approximations are identical provided we choose $\bar{\mu} = 1/\sqrt{3}$. So if we use the same representation of the phase function (including the scaling), the only remaining difference must be attributed to the application of boundary conditions.

Problems

7.1 Assume that the intensity distribution of the clear daytime sky is described by photons that have been scattered only once.

(a) Ignoring surface reflection, show that the downward intensity of skylight at the lower boundary is

$$I_{\mathrm{d}}^{-}(\tau^{*}, \mu, \phi) = \frac{a p_{\mathrm{RAY}}(\cos\Theta)}{4\pi} F^{\mathrm{s}} \frac{\mu_0}{\mu_0 - \mu}\left(e^{-\tau^{*}/\mu_0} - e^{-\tau^{*}/\mu}\right).$$

(b) What does this distribution reduce to when $\mu \to \mu_0$? (*Hint*: Assume τ^*/μ_0 and τ^*/μ are both $\ll 1$.)

(c) Let $\theta_0 = 45°$. Plot the relative intensity of skylight at sea level in the principal plane of the Sun ($\phi = 0°$ and $\phi = 180°$ for four different clear-sky optical depths, $\tau^* = 1.0$, 0.25, 0.15, and 0.05. (These correspond to the wavelengths 316, 440, 498 and 654 nm.)

(d) Assume there is a lower reflecting Lambertian surface with BRDF ρ_L. Find the new expression for the skylight intensity by integrating over the source function $S^* + S_b^*$ (S_b^* is given by Eq. 7.21). Show that, in the limit of small τ^*, the sky intensity at the surface is given by

$$I_d^-(\tau^*, \mu, \phi) = \frac{F^S \tau^*}{4\pi\mu}[p_{RAY}(-\mu_0, \phi_0; -\mu, \phi) + 2\pi \mu_0 \rho_L].$$

(Use the exponential-kernal approximation $E_2(\tau^* - \tau) \approx e^{-(\tau^* - \tau)/\bar{\mu}}$.)

(e) Plot $I_d^-(\tau^*, \mu, \phi)$ for the same conditions as part (c), but including the effects of a reflecting surface with $\pi\rho_L = 0.3$ and 0.7.

7.2 Derive Eqs. 7.52 and 7.53 by two different methods: (1) by taking the limit $a \to 1$ and using L'Hôspital's Rule, and (2) by solving the relevant set of first-order differential equations.

7.3

(a) Find the two-stream expression for the point-direction gain $\mathcal{G}$ for nonconservative, isotropic scattering. Use Eqs. 7.82 for $S_3(\tau)$, and use the relationship of $S_3(\tau)$ with $\mathcal{G}$ in Eq. 6.94.

(b) From part (a) derive the two-stream expressions for the X- and Y-functions, using Eqs. 6.94, 6.95, and 6.96. Assume $a = 1$ and use Eq. 7.92 for $S_3(\tau)$. Show that the resulting expressions are

$$X(\mu) = 1 + \frac{\mu\left[(\bar{\mu} - \mu)\left(1 - e^{-\tau^*/\bar{\mu}}\right) + \tau^*\right]}{\bar{\mu}(\tau^* + 2\bar{\mu})},$$

$$Y(\mu) = e^{-\tau^*/\bar{\mu}} + \frac{\mu\left[(\bar{\mu} + \mu) - (\tau^* + \mu + \bar{\mu})e^{-\tau^*/\bar{\mu}}\right]}{\bar{\mu}(\tau^* + 2\bar{\mu})}.$$

(c) Show that $X(\mu) \to H(\mu)$ and $Y(\mu) \to 0$, where $H(\mu)$ is given by Eq. 7.90 in the limit $\tau^* \to \infty$.

7.4 Derive Eqs. 7.75 through 7.76.

7.5 Derive Eqs. 7.95 through 7.100.

7.6 The escape probability for an isotropically scattering semi-infinite slab satisfies the following integral equation:

$$\mathcal{P}(\tau) = \frac{a}{2}E_2(\tau) + \frac{a}{2}\int_0^\infty d\tau' \mathcal{P}(\tau') E_1(|\tau - \tau'|).$$

We will consider an approximate form of this equation by using the kernal approximation, $E_1(|\tau - \tau'|) \approx e^{-|\tau-\tau'|/\bar{\mu}}/\bar{\mu}$.

(a) Solve the above equation by λ-iteration out to the third term; that is, find the first three terms in the Neumann-series expansion

$$\mathcal{P}(\tau) = P_1(\tau) + P_2(\tau) + P_3(\tau) + \cdots.$$

(b) Show that the same results are obtained by expanding the two-stream result for *Prototype Problem 1* (see Problem 6.4b and Eqs. 7.46–7.47) in a Taylor-series expansion in powers of a, the single-scattering albedo, that is,

$$P(\tau) = f(\tau, a = 0) + a\frac{\partial f(\tau, a = 0)}{\partial a} + \frac{a^2}{2!}\frac{\partial^2 f(\tau, a = 0)}{\partial^2 a} + \cdots,$$

where $f(\tau, a) = a(1 + \rho_\infty)e^{-\Gamma\tau}/2$.

(c) Plot the first three terms for $\mathcal{P}(\tau)$ versus optical depth for a number of values of a, and plot and compare their sum with the "exact result," $\mathcal{P} = f(\tau, a)$. Show that, for those photons escaping from near the surface, $\mathcal{P}$ consists largely of the first few orders of scattering, at least for some range of a.

(d) Infer by mathematical induction the expression for the nth term, $P_n(\tau)$. (*Hint*: Use the property that $\mathcal{P}(\tau, a = 1) = 1$ for all τ.) Plot $P_n(\tau)$ for some large value of n and interpret the behavior with τ.

7.7 The two-stream solution to the flux reflectance for *Prototype Problem 1* for an isotropically scattering slab overlying a black surface is given by Eq. 7.42. We want to include a reflecting lower boundary, which acts as a Lambertian reflector with flux reflectance ρ_L.

(a) Show, using the boundary condition $I^+(\tau^*) = \rho_L I^-(\tau^*)$, that

$$\rho_{tot}(-2\pi, 2\pi) = \frac{\rho_\infty(1 - \rho_\infty\rho_L)\,e^{\Gamma\tau^*} + (\rho_L - \rho_\infty)\,e^{-\Gamma\tau^*}}{(1 - \rho_\infty\rho_L)\,e^{\Gamma\tau^*} + \rho_\infty(\rho_L - \rho_\infty)\,e^{-\Gamma\tau^*}}.$$

(b) What is the expression for $\mathcal{T}_{tot}(-2\pi, -2\pi)$?

(c) Show that if ρ_L is set equal to ρ_∞, then $\rho_{tot} = \rho_\infty$. How is this result related to Ambartsumyan's *Principle of Invariance* (see book by K.-N. Liou, pp. 203–6, op cit., Endnote 3 of Chapter 2)?

(d) We derived an *exact* result for the total reflectance and transmittance of an atmosphere/reflecting surface in terms of the solutions for an underlying black surface (Eqs. 6.77 and 6.78). What are the two-stream versions of these equations? Show that if one assumes that the spherical albedo and transmittance are the same as the flux albedo and flux transmittance for this problem, then the results are identical to those derived in parts (a) and (b).

7.8

(a) Repeat Problem 7.7, except use beam boundary conditions, or in other words, solve the planetary problem for *Prototype Problem 3*, assuming a Lambert reflecting surface. Assume isotropic scattering.

(b) Find ρ_{tot} when $\rho_{\text{L}} = 1$.
(c) Find ρ_{tot} when $a = 1$.
(d) Find the limit for ρ_{tot} when $\tau^* \to \infty$.

7.9 Consider monochromatic light incident on a homogeneous, semi-infinite medium that scatters isotropically. For a collimated beam of incident light of intensity $I^-(\tau, \mu) = F^{\text{s}}\delta(\mu - \mu_0)\delta(\phi - \phi_0)$, the angular distribution of the reflected intensity can be expressed in terms of the H-function as (see Eq. 7.103)

$$I^+(0, +\mu, \mu_0) = \frac{a}{4\pi} F^{\text{s}} \frac{\mu_0}{\mu + \mu_0} H(\mu) H(\mu_0), \tag{7.140}$$

where a is the single-scattering albedo and μ_0 is the cosine of the beam incidence angle.

(a) Show that the total reflected intensity ($I^+(0) = 2\pi \int_0^1 d\mu I^+(\tau = 0, \mu)$) divided by F^{s} is $I^+(0)/F^{\text{s}} = [H(\mu_0) - 1]$ and that the directional-hemispherical albedo becomes

$$\rho(-\mu_0, 2\pi) = \frac{F^+(0)}{\mu_0 F^{\text{s}}} = [1 - \sqrt{1 - a} H(\mu_0)].$$

Assume now that instead of a collimated beam the incident light has an angular distribution $\mathcal{I} f(\mu_0)$, where $\mathcal{I}$ = constant and $2 \int_0^1 d\mu \mu f(\mu) = 1$, so that the incident flux is $\pi \mathcal{I}$.

(b) Show that the angular distribution of the reflected light can now be expressed as

$$\frac{I^+(0, \mu)}{\mathcal{I}} = \frac{H(\mu)}{H_{\text{f}}(\mu)} - H(\mu) \left[1 - \int_0^1 d\mu f(\mu) \right]^{\frac{1}{2}}, \tag{7.141}$$

where $H_{\text{f}}(\mu)$ satisfies the equation

$$\frac{a}{2} \int_0^1 \frac{d\mu' f(\mu') H_f(\mu')}{\mu + \mu'} = \frac{1}{H_{\text{f}}(\mu)} - \left[1 - a \int_0^1 d\mu f(\mu) \right]^{\frac{1}{2}}.$$

(c) What is the shape of the angular distribution if the scattering is conservative ($a = 1$) and the incident illumination is uniform ($f(\mu_0) = 1$)? Justify your answer.

(d) Derive the following expression for the total reflected intensity (scalar irradiance) divided by $\mathcal{I}$:

$$\frac{I^+(0)}{\mathcal{I}} = \int_0^1 d\mu \frac{H(\mu)}{H_f(\mu)} - \frac{2}{a} \left[1 - \sqrt{1 - a} \right] \left[1 - a \int_0^1 d\mu f(\mu) \right]^{\frac{1}{2}}. \tag{7.142}$$

(e) Show that for uniform incidence the following exact expressions are valid:

$$\frac{I^+(0,\mu=0)}{\mathcal{I}} = \left[1-\sqrt{1-a}\right], \qquad \frac{I^+(0)}{\mathcal{I}} = \frac{1-\sqrt{1-a}}{1+\sqrt{1-a}}.$$

(f) Derive an expression for the flux reflectance or albedo in terms of H-functions, and show that for uniform illumination this expression reduces to one that is consistent with the result obtained in (a) for the directional-hemispherical albedo.

7.10 Show that the coefficients $r^{\pm}(\tau)$ defined by Eqs. 7.132–7.133 can be expressed in terms of Legendre polynomials, $P_l(\mu)$, as

$$r^+(\tau) = \frac{1}{bI^+} r_1^+(\mu) \int_0^1 d\mu' P_l(\mu') I^+(\tau,\mu'),$$

$$r^-(\tau) = \frac{1}{(1-b)I^-} r_1^-(\mu) \int_0^1 d\mu' P_l(\mu') I^-(\tau,\mu'),$$

where

$$r_l^{\pm}(\mu) = \frac{1}{2}\sum_{l=0}^{2N-1} (\pm 1)^l (2l+1)\chi_l P_l(\mu)$$

and χ_l are the expansion coefficients of the phase function in Legendre polynomials (see Eq. 6.29).

7.11 A cloud begins to form in the sky overhead. As the cloud thickens, the visual brightness of the cloud's bottom side will brighten, reach a maximum, and then begin to decrease as the cloud becomes optically thick. Ignore effects of ground reflection.

(a) Explain this behavior in physical terms.

(b) Assume the cloud is a plane-parallel slab that scatters visible radiation conservatively ($a = 1$). The cloud particles have an asymmetry factor g and a probability of backscatter $b = (1-g)/2$. Show that the approximate (two-stream) equations of radiative transfer are

$$\bar{\mu}\frac{d(I^+ - I^-)}{d\tau} = 0 \quad \text{and} \quad \bar{\mu}\frac{d(I^+ + I^-)}{d\tau} = (1-g)(I^+ - I^-).$$

(c) Solve the above equations with the boundary conditions of *Prototype Problem 1*. Separate the solution into solar and diffuse components, and show that the transmitted diffuse radiation $I_d^-(\tau^*)$ has the properties described above. (This means solving for the total transmitted intensity, $I^-(\tau^*)$, and subtracting the solar transmitted radiation, $\mathcal{I}e^{-\tau^*/\mu_0}$.)

(d) Plot the solution for $I_d^-(\tau^*)$ versus τ^*, assuming that $g = 0.75$ and $\bar{\mu} = 0.6$. Is your result compatible with the behavior of the cloud described above? How important is the assumption of a dark surface to your conclusion?

(e) Repeat the above analysis for beam boundary conditions (*Prototype Problem 3*) by solving for $I_d^-(\tau^*)$ directly. Include the effect of a reflecting boundary by using

Eq. 6.78. Use for the spherical albedo the flux albedo for *Prototype Problem 1*. (The equality of these two quantities was proven in Problem 4.5.) Plot the result and compare with part (d).

7.12 The radiative transfer equation for a semi-infinite anisotropically scattering medium such as the deep ocean with an internal source of radiation is given by

$$u\frac{dI}{d\tau} = I - \frac{a}{2}\int_{-1}^{1} du' p(u', u)I(\tau, u') - \frac{G(u)}{4\pi}e^{-\gamma\tau},$$

where $G(u)$ is an arbitrary function of u and the parameter $\gamma > 0$.

(a) Show that the two-stream solution to the above equation can be written for $\gamma \neq \Gamma$:

$$I^{\pm}(\tau) = A^{\pm}e^{-\Gamma\tau} + B^{\pm}e^{-\gamma\tau},$$

where

$$A^- = \mathcal{I} - B^-, \qquad \mathcal{I} \equiv I^-(\tau = 0) \text{ is the boundary condition,}$$

$$A^+ = \rho_\infty A^-, \qquad \rho_\infty \equiv \frac{\sqrt{1 - a + 2ab} - \sqrt{1 - a}}{\sqrt{1 - a + 2ab} + \sqrt{1 - a}},$$

$$B^{\pm} = \frac{(\Gamma\bar{\mu})^2}{1 - a}\frac{G^{\pm}}{[(\Gamma\bar{\mu})^2 - (\gamma\bar{\mu})^2]}\left\{1 \mp \frac{\gamma\bar{\mu}(1 - a)}{(\Gamma\bar{\mu})^2} - \frac{1}{2(\Gamma\bar{\mu})^2}[(\Gamma\bar{\mu})^2 - (1 - a)^2]\left(1 - \frac{G^{\mp}}{G^{\pm}}\right)\right\},$$

$$G^{\pm} \equiv \frac{1}{4\pi}\int_0^1 d\mu G(\pm\mu), \qquad \Gamma\bar{\mu} = \sqrt{(1 - a)(1 - a + 2ab)}.$$

(b) Show that the two-stream approximation to the source function

$$S(\tau, u) = \frac{a}{2}\int_{-1}^{1} du' p(u', u)I(\tau, u') + \frac{G(u)}{4\pi}e^{-\gamma\tau}$$

can be expressed as

$$S^{\pm}_{\text{tsa}}(\tau) = \mathcal{I}C^{\pm}e^{-\Gamma\tau} + B^- D^{\pm}\left(e^{-\gamma\tau} - \frac{C^{\pm}}{D^{\pm}}e^{-\Gamma\tau}\right) + G^{\pm}e^{-\gamma\tau},$$

where

$$C^+ = a[\rho_\infty + b(1 - \rho_\infty)], \qquad C^- = a[1 - b(1 - \rho_\infty)],$$

$$D^+ = a\left[b + (1 - b)\frac{B^+}{B^-}\right], \qquad D^- = a\left[1 - b\left(1 - \frac{B^+}{B^-}\right)\right].$$

(c) Show that for isotropic scattering ($b = 1/2$) and isotropic internal source ($G^+ = G^- \equiv \bar{G} = G/4\pi$) the above equation reduces to

$$S^+_{\text{tsa}}(\tau) = S^-_{\text{tsa}}(\tau) = \mathcal{I}\left[1 - \sqrt{1-a}\right] e^{-\Gamma\tau} + \bar{G} H_{\text{tsa}}\left(\frac{1}{\gamma}\right) H_{\text{tsa}}\left(-\frac{1}{\gamma}\right)\left[e^{-\gamma\tau} - \frac{1-\sqrt{1-a}}{1-\gamma\bar{\mu}} e^{-\Gamma\tau}\right],$$

where $H_{\text{tsa}}(x)$ is given by Eq. 7.90 and is the two-stream approximation to the H-function.

(d) Use the above equation to derive an approximate solution for the emergent intensity for $\mathcal{I} = 0$ (i.e., no incident radiation) given by

$$I^+_{\text{tsa}}(\tau = 0, \mu) = \bar{G} \frac{H_{\text{tsa}}\left(\frac{1}{\gamma}\right)}{1 + \gamma\mu} H_{\text{tsa}}(\mu).$$

Discuss how the emergent intensity varies with the spatial variation of the internal source, that is, with varying γ.

7.13 A plane-parallel planetary atmosphere is transparent to solar radiation and is thus subjected to heating from only the surface (assumed to be black). Assume the atmosphere is gray in the infrared, optically thick, and purely absorbing. The frequency-integrated source function is thus

$$B(\tau) = \int_0^\infty d\nu B_\nu(\tau) = \frac{\sigma_{\text{B}}}{\pi} T^4(\tau),$$

where B_ν is the Planck function, ν is (IR) frequency, σ_{B} is the Stefan–Boltzmann constant, T is temperature, and τ is optical depth.

(a) Show that an approximate relationship between the frequency-integrated IR flux F and the source function is obtained from the Eddington approximation

$$F(\tau) \approx \frac{4\pi}{3} \frac{\partial B(\tau)}{\partial \tau}.$$

(b) Assume that the atmosphere is in radiative equilibrium. Show that the source function at any optical depth τ is

$$B(\tau) = B(\tau = 0) + \frac{3F}{4\pi}\tau.$$

(c) Assume that $\bar{\mu} = 1/2$. Show from the boundary conditions that this implies $F = 2\pi B(\tau = 0)$.

(d) Assume that the absorbing gas is well mixed in the atmosphere and obeys the hydrostatic equation, $(\partial p_{\text{a}}/\partial z) = -\rho_{\text{a}} g$, where p_{a} is the partial pressure of the absorbing gas and ρ_a is its mass density. Define the optical depth as

$$\tau = \int_z^\infty dz' \alpha_m \rho_a(z'),$$

where α_m is the (gray) absorption coefficient per unit mass. Show that

$$T(\tau) = T(\tau = 0)\left(1 + \frac{3}{2}\alpha_m p_a/g\right)^{1/4}.$$

(e) Define the effective temperature of the planet, T_e as $F = \sigma_B T_e^4$. Find the relationship between T_e and the "skin" temperature $T(\tau = 0)$.

(f) Show that the radiative equilibrium temperature lapse rate is given by

$$\frac{\partial T}{\partial z} = \frac{-g}{R} \frac{\frac{3}{8}\tau}{\left(1 + \frac{3}{2}\tau\right)}.$$

From this result show that an atmosphere becomes convectively unstable when the optical depth is large, and only if $c_p > 4R$. Here c_p and R are the specific heat and gas constant per unit mass.

7.14 A gigantic volcanic eruption causes a worldwide layer of sulfuric acid particles to form in the Earth's stratosphere. Widespread reports of ultrabright skies and unprecedented cases of sunburn are reported by the world's hospitals. Scientists find that their measurements of the visible-light sky brightness have increased substantially over their pre-eruption values. After a late-autumn snowfall and the return of cloudless skies, their spectrometers record even more downward sky radiation than is available in the extraterrestrial flux!

Explain this behavior using the two-stream approximation for *Prototype Problem 3*. Assume the following: $a = 1$ for the aerosols (an excellent approximation for sulfuric acid droplets); the aerosol layer is horizontally homogeneous; and Lambertian reflectance for the snow cover, ρ_L. Ignore the influence of the atmosphere, either Rayleigh scattering or atmospheric absorption. This problem involves the determination of the total flux transmittance of the diffuse radiation, $\mathcal{T}_{tot}(-\mu_0, -2\pi)$, to include the anisotropically scattering aerosol layer and the reflecting surface.

(a) First, assume $\rho_L = 0$. Show that the two-stream scaled expressions for the directional-hemispherical transmittance and reflectance for *Prototype Problem 3* obey the relationship

$$\mathcal{T}(-\mu_0, -2\pi) + \rho(-\mu_0, 2\pi) = 1.$$

(b) Show that the spherical albedo $\bar{\rho}_3$ for *Prototype Problem 3* is equal to the flux reflectance for *Prototype Problem 1*, which we denote as $\rho(-2\pi, 2\pi)$.

(c) Using the relationship (Eq. 6.78) relating the solution for the directional-hemispherical transmittance $\mathcal{T}_0$ of a medium with a completely absorbing lower boundary to that for a reflecting lower boundary (with a Lambert reflectance ρ_L), find the total (diffuse plus direct) directional-hemispherical transmittance $\mathcal{T}_{tot}$ appropriate to this problem.

(d) Show that, for an appropriate combination of parameters, μ_0, $\bar{\mu}$, and ρ_L, the transmittance can *exceed* unity. Explain how "photon trapping" explains this remarkable result and thus accounts for the observed sky brightness following the volcanic eruption.

(e) Find the corresponding expression for the total diffuse reflectance of the atmosphere/surface system and show that

$$\rho_{tot}(-\mu_0, 2\pi) + (1 - \rho_L)\mathcal{T}_{tot}(-\mu_0, -2\pi) = 1.$$

Notes

1 A unit volume in "tau space" is not the same as in geometrical space. Imagine a small cylindrical volume whose length is $d\tau$ and whose cross-sectional area dA has units m^2. Then the volume will be $dAd\tau$, whose units are m^2, in contrast to the geometrical volume $dzdA$, which has units of m^3.

2 The two-stream approximation dates back to A. Schuster, "Radiation through a foggy atmosphere," *Astrophysical Journal*, **21**, 1–22, 1905, and K. Schwarzschild, "On the equilibrium of the Sun's atmosphere," *Nachrichten von der Königlichen Gesellschaft der Wissenschaften zu Göttingen, Math.-Phys. Klasse*, **195**, 41–53, 1905. The Eddington approximation originated with A. S. Eddington, "On the radiative equilibrium of the stars," *Monthly Notices of the Royal Astronomical Society*, **77**, 16–35, 1916. These three classical papers have been reprinted in *Selected Papers on the Transfer of Radiation*, ed. D. H. Menzel, Dover, New York, 1966.

3 C. Mobley has pointed out that E. Lommel was the first to derive the radiative transfer equation. See E. Lommel's "Die Photometrie der diffusen Zurückwefung," *Ann. Phys. U. Chem. (N.F.)*, **36**, 473–502, 1889.

4 It is also called the *Eddington approximation* solution of the Milne problem when $\bar{\mu} = 2/3$ and $S(0)$ is equal to $F/2\pi$.

5 Schwarszchild assumed that radiative equilibrium applied in the Sun's outer atmosphere. Taking a gray absorption for the photospheric material, he related the source function to $\sigma_B T^4$. From this, he compared the limb darkening to that predicted by adiabatic equilibrium. Observations of limb darkening agreed much better with the radiative equilibrium assumption.

6 In computational work it is usually sufficient to use numerical "dithering" by which μ_0 is changed slightly away from the "singular value." This artifice produces satisfactory results and avoids the inconvenience of having to deal with a special case involving a different solution.

7 See S. Chandrasekhar, *Radiative Transfer*, Dover, New York, 1960, Eq. 131, p. 87.

8 Chandrasekhar, S., *Radiative Transfer*, p. 209, his Eqs. 3, 4, and 5.

9 The relationship between the two-stream and the Eddington approximation was discussed by D. R. Lyzenga, "Note on the modified two-stream approximation of Sagan and Pollack," *Icarus*, **19**, 240–3, 1973.

10 The backscatter coefficients are discussed in some detail in Wiscombe, W. J. and G. W. Grams, "The Backscattered Fraction in Two-Stream Approximations," *J. Atmos. Sci.*, **33**, 2440–451, 1976.

11 An entertaining description of how the two-stream approximation may be used to explain numerous radiative transfer phenomena is found in Bohren, C. F., "Multiple scattering of light and some of its observable consequences," *Am. J. Physics*, **55**(6), 524-33, 1987.

12 See, for example Aas, E., "Two-stream irradiance model for deep waters," *Applied Optics*, **26**, 2095-101.

13 The accuracy of the Eddington and two-stream methods for anisotropic scattering was explored by W. J. Wiscombe and J. H. Joseph, "The range of validity of the Eddington approximation," *Icarus*, **32**, 362–77, 1977, who found that it was accurate for values of g less than 0.5. This explains why the δ-Eddington and δ-two-stream methods are so valuable: The scaled asymmetry factor is always less than 0.5. Tables 7.1–7.5 are taken from Kylling el al., 1995 (the full reference is provided at the end of the next note).

14 Attempts to combine two-stream solutions for several adjacent slabs with different optical properties date back more than twenty-nine years – Shettle, E. P. and J. A. Weinman, "The transfer of solar irradiance through inhomogeneous turbid atmospheres evaluated by Eddington's approximation," *J. Atmos. Sci.*, **27**, 1048–55, 1970. These solutions were intrinsically ill-conditioned, because the matrix (that had to be inverted to determine the constants of integration in the problem) contained a combination of very small and very large elements resulting from the negative and positive arguments of the exponential solutions. In the δ-Eddington method, subdivision of layers was employed to circumvent the ill-conditioning, but at the expense of increasing the computational burden substantially for thick layers (Wiscombe, W. J., NCAR Tech. Note NCAR/TN-121+STR). The ill-conditioning problem was eliminated by a scaling transformation that removed the positive arguments of the exponential solutions (Stamnes, K. and P. Conklin, "A new multilayer discrete ordinate approach to radiative transfer in vertically inhomogeneous atmospheres," *J. Quant. Spectrosc. Radiative Transfer*, **31**, 273–82, 1984.) This scaling transformation was eventually implemented into a general-purpose multistream (including two-stream) radiative transfer algorithm by Stamnes, K., S.-C. Tsay, W. J. Wiscombe and K. Jayaweera, "Numerically stable algorithm for discrete-ordinate-method radiative transfer in multiple scattering and emitting layered media," *Applied Optics*, **27**, 2502–9, 1988. The resulting code has been made generally available to interested users and will be briefly described in the next chapter. A specific two-stream code that made use of this scaling transformation to remove the ill-conditioning has been developed (Toon, O. B., C. P. McKay, T. P. Ackerman, and K. Santhanam, "Rapid calculation of radiative heating rates and photodissociation rates in inhomogeneous multiple scattering atmospheres," *J. Geophys. Res.*, **94**, 16287–301, 1989). Finally, a two-stream algorithm derived from the general-purpose multistream algorithm mentioned above has been extended for application to spherical geometry and to layers in which the internal source may vary rapidly (Kylling, A., K. Stamnes, and S.-C. Tsay, "A reliable and efficient two-stream algorithm for radiative transfer: Documentation of accuracy in realistic layered media," *J. Atmos. Chem.*, **21**, 115–50, 1995.) This two-stream code is also generally available to interested users.

15 Modern discussions of the two-stream method are due to Meador, W. E. and W. R. Weaver, "Two-stream approximations to radiative transfer in planetary atmospheres: A unified

description," *J. Atmos. Sci.*, **37**, 630–43, 1980; Zdunkowski, W. G., R. M. Welch, and G. Korb, "An investigation of the structure of typical two-stream-methods for the calculation of solar fluxes and heating rates in clouds," *Contrib. Atmos. Phys.*, **53**, 147–66, 1980; King, M. D. and Harshvardhan, "Comparative accuracy of selected multiple scattering approximations," *J. Atmos. Sci.*, **43**, 784–801, 1986; and Harshvardhan and M. D. King, "Comparative accuracy of diffuse radiative properties computed using selected multiple scattering approximations," *J. Atmos. Sci.*, **50**, 247–59, 1993. Its accuracy for a single, homogeneous slab was explored in these (and other) papers for a variety of combinations of the single-scattering albedo, asymmetry factor, and slab optical thickness, and for several solar elevations.

Chapter 8

Accurate Numerical Solutions of Prototype Problems

8.1 Introduction

We now consider a class of more sophisticated approximation techniques that are capable of approaching the exact solution as closely as desired. This class includes the discrete-ordinate method,[1] the spherical-harmonic method, and the doubling-adding method. It will be seen that in lowest order the first two methods become the two-stream and Eddington approximations, respectively. The discrete-ordinate and doubling-adding methods are also closely related, as will be discussed.

8.2 Discrete-Ordinate Method – Isotropic Scattering

8.2.1 Quadrature Formulas

The solution of the isotropic-scattering problem involves the following integral over angle:

$$\int_{-1}^{1} du\, I(\tau, u) = \int_{0}^{1} d\mu\, I^{+}(\tau, \mu) + \int_{0}^{1} d\mu\, I^{-}(\tau, \mu).$$

In the two-stream method we replaced the integration over u with the simple formula

$$\int_{-1}^{1} du\, I \approx I^{+}(\tau) + I^{-}(\tau).$$

This is obviously a crude approximation. We could improve the accuracy by including more points in a *numerical quadrature formula*

$$\int_{-1}^{1} du\, I(\tau, u) \approx \sum_{j=1}^{m} w'_j I(\tau, u_j).$$

Here w'_j is a *quadrature weight* and u_j is the *discrete ordinate*. The simplest example is the *trapezoidal rule*

$$\int_{-1}^{1} du\, I \approx \Delta u \left(\tfrac{1}{2} I_1 + I_2 + I_3 + \cdots + I_{m-1} + \tfrac{1}{2} I_m\right),$$

and the more accurate Simpson's rule is

$$\int_{-1}^{1} du\, I \approx \frac{\Delta u}{3}(I_1 + 4I_2 + 2I_3 + 4I_4 + \cdots + I_m),$$

where Δu is the (equal) spacing between the adjacent points, u_j, and the I_j denotes $I(\tau, u_j)$.

If we have m points at which we evaluate $I(\tau, u)$, we can replace I with its *approximating polynomial* $\phi(u)$, which is a polynomial of degree $(m-1)$. Consider the following form for $\phi(u)$, for $m = 3$:

$$\phi(u) = I(u_1)\frac{(u-u_2)(u-u_3)}{(u_1-u_2)(u_1-u_3)} + I(u_2)\frac{(u-u_1)(u-u_3)}{(u_2-u_1)(u_2-u_3)} + I(u_3)\frac{(u-u_1)(u-u_2)}{(u_3-u_1)(u_3-u_2)}.$$

It is easily verified that $\phi(u)$ is a second-degree polynomial, which, when evaluated at the points u_1, u_2, and u_3, yields $I(u_1)$, $I(u_2)$, and $I(u_3)$, respectively. The above is an example of *Lagrange's interpolation formula*. We can write this in abbreviated form, if we use the notation $\prod$ to indicate products of terms. For example, we may define

$$F(u) \equiv \prod_{j=1}^{m}(u - u_j) = (u-u_1)(u-u_2)\cdots(u-u_m).$$

Then, since the polynomial $(u-u_1)(u-u_2)\cdots(u-u_{j-1})(u-u_{j+1})\cdots(u-u_m)$ becomes $F(u)/(u-u_j) = \prod_{k\neq j}^{m}(u-u_k)$, we can write the polynomial $\phi(u)$ in a shorthand form

$$\phi(u) = \sum_{j=1}^{m} I(u_j)\frac{F(u)}{(u-u_j)F'(u_j)},$$

where $F'(u_j)$ is defined as $dF/du\rfloor_{u=u_j}$. We see that the derivative will give a long string of polynomials of degree $(m-1)$; however, when it is evaluated at $u = u_j$, all terms become zero except the term $(u-u_1)(u-u_2)\cdots(u-u_{j-1})(u-u_{j+1})\cdots(u-u_m)$.

Hence, the quadrature formula arising from the assumption that the intensity is a polynomial of degree $(m-1)$ is

$$\int_{-1}^{1} du\, I(u) = \sum_{j=1}^{m} w'_j I(u_j), \qquad w'_j = \frac{1}{F'(u_j)} \int_{-1}^{1} \frac{du F(u)}{(u-u_j)}.$$

The quadrature points u_j are, so far, arbitrary.

It can be shown that the error incurred by using the Lagrange interpolation formula is proportional to the mth derivative of the functions $[I(u)]$ being approximated.[2] Thus, it is clear that if $I(\tau, u)$ happens to be a polynomial of degree $(m-1)$ or smaller, then the m-point quadrature formula is exact.

Example 8.1 Simple Demonstration of Quadrature

Let's assume that the intensity is a polynomial of degree 3,

$$I(u) = a_0 + a_1 u + a_2 u^2 + a_3 u^3, \tag{8.1}$$

where a_i $(i = 0, \ldots, 3)$ are constants. Evaluating the function at the points $u_1 = -1$, $u_2 = 0$, and $u_3 = 1$ yields three evenly spaced points in the interval $[-1, 1]$. We find $I(u_1) = a_0 - a_1 + a_2 - a_3$, $I(u_2) = a_0$, and $I(u_3) = a_0 + a_1 + a_2 + a_3$. Thus, the approximating polynomial becomes

$$\begin{aligned}\phi(u) &= \frac{1}{2}(a_0 - a_1 + a_2 - a_2)(u^2 - u) \\ &\quad + a_0(u^2 - 1) + \frac{1}{2}(a_0 + a_1 + a_2 + a_3)(u^2 + u), \\ F(u) &= (u - u_1)(u - u_2)(u - u_3) \\ &= u^3 - (u_1 + u_2 + u_3)u^2 + (u_1 u_2 + u_1 u_3 + u_2 u_3)u - u_1 u_2 u_3, \\ F'(u) &= (u - u_2)(u - u_3) + (u - u_1)(u - u_3) + (u - u_1)(u - u_2)\end{aligned}$$

and the quadrature weights become

$$w'_1 = \frac{1}{F'(u_1)} \int_{-1}^{1} \frac{du F(u)}{u - u_1} = \frac{1}{(u_1 - u_2)(u_1 - u_3)} \int_{-1}^{1} (u - u_2)(u - u_3) du = \frac{1}{3}$$

and similarly

$$w'_2 = \frac{1}{F'(u_2)} \int_{-1}^{1} \frac{du F(u)}{u - u_2} = \frac{4}{3} \quad \text{and} \quad w'_3 = \frac{1}{F'(u_3)} \int_{-1}^{1} \frac{du F(u)}{u - u_3} = \frac{1}{3}.$$

We have just derived *Simpson's rule*.

So clearly

$$\int_{-1}^{1} du\, I(u) = \sum_{i=1}^{3} w'_i I(u_i) = \frac{1}{3}[I(u_1) + 4I(u_2) + I(u_3)] = 2a_o + \frac{2}{3}a_2,$$

which is the same as the exact result

$$\int_{-1}^{1} du\, I(u) = \int_{-1}^{1} du(a_0 + a_1 u + a_2 u^2 + a_3 u^3)$$

$$= \left[a_0 u + a_1 \frac{u^2}{2} + a_2 \frac{u^3}{3} + a_3 \frac{u^4}{4}\right]_{-1}^{1} = 2a_0 + \frac{2}{3} a_2.$$

Thus, we have demonstrated that Lagrange's three-point formula integrates exactly a polynomial of degree 3 or less.

As already mentioned, the error in the Lagrange interpolation polynomial of degree $(m-1)$ is proportional to the mth derivative of the function being approximated. The resulting quadrature schemes (usually referred to as the *Newton–Cotes formulas*) rely on using even spacing between the points at which the function is evaluated. We may ask: Is it possible to obtain higher accuracy? This would appear to be so if we were to choose the quadrature points in an optimal manner.

Gauss showed that one can, in fact, do better, even in the absence of information about the function $I(u)$. He showed that if $F(u)$ is a certain polynomial, and the u_j are the roots of that polynomial, then we get the accuracy of a polynomial of degree $(2m-1)$. *This polynomial is the Legendre polynomial* $P_m(u)$. As we have seen earlier, they have the special property of being orthogonal to every power of u less than m, that is,

$$\int_{-1}^{1} du\, P_m(u) u^l = 0 \quad (l = 0, 1, 2, \ldots, m-1).$$

Note that if u_j is a root of an even Legendre polynomial, then $-u_j$ is also a root. Also, all m roots are real.

8.2.2 The Double-Gauss Method

We will proceed using a variant of the standard discrete-ordinate method, which will in general turn out to be the most accurate solution for a given order of approximation. It is customary to choose the even-order Legendre polynomials as the approximating polynomials. This choice is made because the roots of the even orders appear in pairs: If we use a negative index to label points in the downward hemisphere and a positive index for points in the upper hemisphere, then $u_{-i} = -u_{+i}$. The quadrature weights are the same in each hemisphere, that is, $w'_i = w'_{-i}$. The "full-range" approach has certain problems because it assumes that $I(\tau, u)$ is a smoothly varying function of $u(-1 \le u \le +1)$ with no sharp corners for all values of τ. We noted earlier that, in the absence of any information about the integrand, the Legendre polynomial yields optimum accuracy. However, let's introduce some information by noting that, at least for small τ, the intensity changes rather rapidly as u passes through zero, that is, as the line of sight passes through the horizontal. In fact at $\tau = 0$, this change is quite abrupt. Since there is no incoming diffuse radiation, $I(\tau,u)$ (for u slightly negative) is zero; and for slightly positive u values it will generally have a finite value. It will clearly be difficult to fit such a discontinuous distribution with a small number of terms involving polynomials that span continuously the full range between $u = -1$ and $u = 1$. Because the region near the surface is the most troublesome in terms of getting accurate solutions, this is obviously the region to which we should pay the most attention.

To remedy this situation, the *Double-Gauss method* was devised.[3] In this method, the hemispheres are treated separately. Instead of approximating $\int_{-1}^{1} du I(u)$ by the sum $\sum_{i=-N}^{+N} w_i' I(u_i)$, where w_i' and u_i are the weights and roots of the even-ordered Legendre polynomial P_{2N}, we break the angular integration into two hemispheres, and approximate each integral separately,

$$\int_{-1}^{1} du\, I = \int_{0}^{1} d\mu\, I^+ + \int_{0}^{1} d\mu\, I^- \approx \sum_{j=1}^{M} w_j I^+(\mu_j) + \sum_{j=1}^{M} w_j I^-(\mu_j),$$

where the w_j and μ_j are the weights and roots of the approximating polynomial for the half range. Note that we have used the same set of weights and roots for both hemispheres. Again, this is not necessary, but it is obviously convenient. Now within each hemisphere, if we are to obtain the highest accuracy, we must again use Gaussian quadrature. However, our new interval is $(0 \le \mu \le 1)$ instead of $(-1 \le u \le 1)$. This is easily arranged by defining the variable $u = 2\mu - 1$ so that the orthogonal polynomial is $P_M(2\mu - 1)$. The new quadrature weight is given by

$$w_j = \frac{1}{P_M'(2\mu_j - 1)} \int_{0}^{1} d\mu \frac{P_M(2\mu - 1)}{(\mu - \mu_j)}, \tag{8.2}$$

and the μ_j are the roots of the half-range polynomials. It is easy to find the weights for $M = 1$ from the above formula (see Example 6.2 below).

Algorithms to compute the roots and weights are usually based on the full range. It is therefore useful to relate the half-range quadrature points and weights to those for the full range. Fortunately, it turns out that the new half-range weights and roots can be found easily in terms of the weights w_j' and points u_j for the full range. Since the linear transformation $t = (2x - x_1 - x_2)/(x_2 - x_1)$ will map any interval $[x_1, x_2]$ into $[-1, 1]$ provided $x_2 > x_1$, Gaussian quadrature can be used to approximate

$$\int_{x_1}^{x_2} dx\, I(x) = \int_{-1}^{1} dt\, I\left[\frac{(x_2 - x_1)t + x_2 + x_1}{2}\right] \frac{(x_2 - x_1)}{2}.$$

Choosing $x_1 = 0$, $x_2 = 1$, $x = \mu$, and $t = u$, we find

$$\int_{0}^{1} d\mu\, I(\mu) = \frac{1}{2} \int_{-1}^{1} du\, I\left(\frac{u+1}{2}\right),$$

and by applying Gaussian quadrature to each integral, we find on setting $M = 2N$ for the half range

$$\int_0^1 d\mu, I(\mu) = \sum_{j=1}^{2N} w_j I(\mu_j) = \frac{1}{2}\int_{-1}^{1} du\, I\left(\frac{u+1}{2}\right)$$
$$= \frac{1}{2}\sum_{\substack{j=-N\\ j\neq 0}}^{N} w'_j I\left(\frac{u_j+1}{2}\right). \tag{8.3}$$

Thus, in even orders the half-range points and weights are related to the full-range ones by

$$\mu_j = \frac{u_j + 1}{2}, \qquad w_j = \frac{1}{2}w'_j, \tag{8.4}$$

showing that the new double-Gauss weights in even orders are half the Gaussian weights in half the order (see Example 8.2). According to Eqs. 8.4 each pair of roots $\pm|u_j|$ for any order N (full range) generates two positive roots $\mu_j = (-|u_j| + 1)/2$ and $\mu_{2N+1-j} = (|u_j| + 1)/2$ of order $2N$ (half range). We have chosen to label the roots so that they appear in ascending order, that is, $\mu_1 < \mu_2 < \mu_3 < \cdots < \mu_{2N}$ (see also Table 8.1).

Finally, we note that Eq. 8.3 implies that $\sum w_j = 1$ (obtained by setting $I(\mu) = 1$), which is confirmed by inspection of Table 8.1 provided in the following example.

Example 8.2 Low-Order Quadrature

Let's examine the $M = 1$ approximation to see if we retrieve the two-stream approximation. Consider μ_1, which is $(1 + u_1)/2$. Now u_1 is the root of $P_1(u) = u$. This gives $u_1 = 0$, and hence $\mu_1 = \frac{1}{2}$. The weight w_1 is easily determined from its definition in Eq. 8.2:

$$w_1 = \frac{1}{P'_1(2\mu_1 - 1)}\int_0^1 \frac{d\mu(2\mu - 1)}{\left(\mu - \frac{1}{2}\right)} = 2/P'_1(\mu_1).$$

Since $P_1 = 2\mu - 1$, $P'_1 = 2$, and hence $w_1 = 1$. Therefore we retrieve, in the lowest-order double-Gauss formula, the same equations as the two-stream *Schuster–Schwarzschild* equations, in which $\bar{\mu} = 1/2$.

Following the same equations for the lowest *even-order* Gauss formula, we obtain the same expressions except that $\bar{\mu} = 1/\sqrt{3}$, rather than $1/2$. This follows since the lowest-order even Gauss formula refers to the $P_2(u) = \frac{1}{2}(3u^2 - 1)$ Legendre polynomial for which $P_2(u) = 0$ for $u_1 = \pm 1/\sqrt{3}$. In summary, the lowest-order double-Gauss formula leads to the half-range two-stream Schuster–Schwarzschild equations; and the lowest-order (even) Gauss formula leads to the full-range two-stream or Eddington approximation.

We may now use the formulas given above to find the half-range roots and weights for $N = 1$. Since the corresponding full-range roots and weights are $u_{\pm 1} = \pm 1/\sqrt{3}$ and $w'_{\pm 1} = 1$, respectively, we find $\mu_1 = \frac{1}{2}(1 - 1/\sqrt{3})$, $\mu_2 = \frac{1}{2}(1 + 1/\sqrt{3})$, $w_1 = \frac{1}{2}$, and $w_2 = \frac{1}{2}$ for the half-range roots and weights for $0 < \mu < 1$. For $-1 < \mu < 0$ the weights are the same and $\mu_{-i} = -\mu_i$. Gaussian and corresponding double-Gaussian points and weights are listed in Table 8.1.

Table 8.1. *Gaussian and double-Gaussian points and weights.*

N	j	$2N+1-j$	u_j	w'_j	μ_j	w_j	μ_{2N+1-j}	w_{2N+1-j}
1	1	2	0.57735	1.00000	0.21132	0.50000	0.78868	0.50000
2	1	4	0.33998	0.65215	0.06943	0.17393	0.93057	0.17393
	2	3	0.86114	0.34785	0.33001	0.32607	0.66999	0.32607
3	1	6	0.23862	0.46791	0.03377	0.08566	0.96623	0.08566
	2	5	0.66121	0.36076	0.16940	0.18038	0.83060	0.18038
	3	4	0.93247	0.17132	0.38069	0.23396	0.61931	0.23396
4	1	8	0.18343	0.36268	0.01986	0.05061	0.98014	0.05061
	2	7	0.52553	0.31371	0.10167	0.11119	0.89833	0.11119
	3	6	0.79667	0.22238	0.23723	0.15685	0.76277	0.15685
	4	5	0.96029	0.10123	0.40828	0.18134	0.59172	0.18134
5	1	10	0.14887	0.29552	0.01305	0.03334	0.98695	0.03334
	2	9	0.43340	0.26927	0.06747	0.07473	0.93253	0.07473
	3	8	0.67941	0.21909	0.16030	0.10954	0.83970	0.10954
	4	7	0.86506	0.14945	0.28330	0.13463	0.71670	0.13463
	5	6	0.97391	0.06667	0.42556	0.14776	0.57444	0.14776
6	1	12	0.12523	0.24915	0.00922	0.02359	0.99078	0.02359
	2	11	0.36783	0.23349	0.04794	0.05347	0.95206	0.05347
	3	10	0.58732	0.20317	0.11505	0.08004	0.88495	0.08004
	4	9	0.76990	0.16008	0.20634	0.10158	0.79366	0.10158
	5	8	0.90412	0.10694	0.31608	0.11675	0.68392	0.11675
	6	7	0.98156	0.04718	0.43738	0.12457	0.56262	0.12457

8.3 Anisotropic Scattering

8.3.1 General Considerations

We shall now generalize the discrete-ordinate method to situations including anisotropic scattering in finite inhomogeneous (layered) media. In doing so we shall introduce a matrix formulation[4] that not only allows for a compact notation but also greatly facilitates the numerical implementation of the method. This formulation is valid for isotropic scattering considered in the previous section as well as for any phase function that depends only on the angle between the incoming direction $\hat{\Omega}'$ and the scattered direction $\hat{\Omega}$ through $\hat{\Omega}' \cdot \hat{\Omega} = \cos\Theta$, where Θ is the scattering angle.

For simplicity we start by considering a homogeneous slab. As we discussed in Chapter 5, it is possible to "eliminate" the azimuth dependence in the sense that the intensity may be written as a Fourier cosine series (Eq. 6.34) in which each Fourier component of the series satisfies a radiative transfer equation mathematically identical to the azimuthally averaged equation. Thus, we may focus on Eq. 6.35 for $m = 0$ or Eq. 6.59 if we want to utilize the δ-N scaling to handle sharply forward-peaked phase functions. Mathematically these two equations are identical, and so it does not matter which one we choose. In fact, scaling only influences the optical properties of the medium and will not affect the mathematical solution. Therefore, in the following it suffices to consider the following pair of equations for the azimuthally averaged (or $m = 0$ component) half-range diffuse intensities (cf. Eqs. 6.20–6.21):

$$\mu \frac{dI^+(\tau,\mu)}{d\tau} = I^+(\tau,\mu) - \frac{a}{2}\int_0^1 d\mu' p(\mu',\mu) I^+(\tau,\mu') \\ - \frac{a}{2}\int_0^1 d\mu' p(-\mu',\mu) I^-(\tau,\mu') - X_0^+ e^{-\tau/\mu_0}, \tag{8.5}$$

$$-\mu \frac{dI^-(\tau,\mu)}{d\tau} = I^-(\tau,\mu) - \frac{a}{2}\int_0^1 d\mu' p(\mu',-\mu) I^+(\tau,\mu') \\ - \frac{a}{2}\int_0^1 d\mu' p(-\mu',-\mu) I^-(\tau,\mu') - X_0^- e^{-\tau/\mu_0}, \tag{8.6}$$

where (see Eqs. 6.35–6.36)

$$p(\mu',\mu) = \sum_{l=0}^{2N-1} (2l+1)\chi_l P_l(\mu) P_l(\mu') \tag{8.7}$$

and

$$X_0^\pm \equiv X_0(\pm\mu) = \frac{a}{4\pi} F^s p(-\mu_0, \pm\mu). \tag{8.8}$$

We consider the collimated beam case first. This is the most difficult of the three prototype problems defined in Chapter 6 because we need to deal with the full azimuthal dependence to arrive at the intensity distribution. The internal source problem, *Prototype Problem 2*, is used as an example below (see Example 8.4).

As before, the discrete-ordinate approximation to Eqs. 8.5–8.8 is obtained by replacing the integrals by quadrature sums and thus transforming the pair of coupled integro-differential equations 8.5 and 8.6 into a system of coupled differential equations as follows:

$$\mu_i \frac{dI^+(\tau,\mu_i)}{d\tau} = I^+(\tau,\mu_i) - \frac{a}{2}\sum_{j=1}^{N} w_j p(\mu_j,\mu_i) I^+(\tau,\mu_j) \\ - \frac{a}{2}\sum_{j=1}^{N} w_j p(-\mu_j,\mu_i) I^-(\tau,\mu_j) - X_{0i}^+ e^{-\tau/\mu_0}, \tag{8.9}$$

$$-\mu_i \frac{dI^-(\tau,\mu_i)}{d\tau} = I^-(\tau,\mu_i) - \frac{a}{2}\sum_{j=1}^{N} w_j p(\mu_j,-\mu_i) I^+(\tau,\mu_j) \\ - \frac{a}{2}\sum_{j=1}^{N} w_j p(-\mu_j,-\mu_i) I^-(\tau,\mu_j) - X_{0i}^- e^{-\tau/\mu_0}. \tag{8.10}$$

8.3.2 Quadrature Rule

As noted previously it is convenient to use the same quadrature in each hemisphere so that $\mu_{-i} = -\mu_i$ and $w_{-i} = w_i$. There are many quadrature rules that satisfy this requirement, but the use of Gaussian quadrature is essential because it ensures that the phase function is correctly normalized, that is,

$$\sum_{\substack{j=-N \\ j\neq 0}}^{N} w_j p(\tau, \mu_i, \mu_j) = \sum_{\substack{i=-N \\ i\neq 0}}^{N} w_i p(\tau, \mu_i, \mu_j) = 1, \tag{8.11}$$

implying that energy is conserved in the computation. The reason for this is simply that the Gaussian rule is based on the zeros of the Legendre polynomials, which we have also used for our expansion of the phase function. The normalization property (Eq. 8.11) guarantees that there will be no spurious absorption in problems with conservative scattering ($a = 1$). The big advantage of using the expansion of the phase function in Legendre polynomials (in addition to the isolation of the azimuth dependence discussed previously) is that this *normalization holds in all orders of approximation* (i.e., for arbitrary values of N).

The quadrature points and weights of the "double-Gauss" scheme adopted here satisfy $\mu_{-j} = -\mu_j$ and $w_{-j} = w_j$. As pointed out previously (see §8.2), "double-Gauss" simply refers to a quadrature rule in which the Gaussian formula is applied separately to the half ranges $-1 < u < 0$ and $0 < u < 1$. The main advantage of this scheme is that the quadrature points (in even orders) are distributed symmetrically around $|u| = 0.5$ and clustered both toward $|u| = 1$ and $|u| = 0$, whereas, in the Gaussian scheme for the complete range, $-1 < u < 1$, they are clustered toward $u = \pm 1$. The clustering toward $|u| = 0$ will give superior results near the boundaries where the intensity varies rapidly around $|u| = 0$. A half-range scheme is also preferable since the intensity is discontinuous at the boundaries. Another advantage is that half-range quantities such as upward and downward fluxes and average intensities are obtained immediately without any further approximations. Computation of half-range quantities using a full-range quadrature scheme is obviously not self-consistent.

8.4 Matrix Formulation of the Discrete-Ordinate Method

8.4.1 Two- and Four-Stream Approximations

Before we consider the general multistream solution, we shall first describe the two- and four-stream cases ($N = 1$ and 2) for pedagogical reasons. It will then become obvious how the multistream case works.

The two-stream approximation is obtained by setting $N = 1$ in Eqs. 8.9 and 8.10, which yields two coupled differential equations as usual,

$$\mu_1 \frac{dI^+(\tau)}{d\tau} = I^+(\tau) - \frac{a}{2} p(-\mu_1, \mu_1) I^-(\tau) - \frac{a}{2} p(\mu_1, \mu_1) I^+(\tau) - Q'^+(\tau), \tag{8.12}$$

$$-\mu_1 \frac{dI^-(\tau)}{d\tau} = I^-(\tau) - \frac{a}{2}p(-\mu_1, -\mu_1)I^-(\tau) - \frac{a}{2}p(\mu_1, -\mu_1)I^+(\tau) - Q'^-(\tau), \quad (8.13)$$

where

$$I^\pm(\tau) \equiv I^\pm(\tau, \mu_1),$$
$$Q'^\pm(\tau) \equiv \frac{a}{4\pi} F^s p(-\mu_0, \pm\mu_1) e^{-\tau/\mu_0},$$
$$\frac{a}{2}p(\mu_1, -\mu_1) \equiv \frac{a}{2}(1 - 3g\mu_1^2) \equiv ab = \frac{a}{2}p(-\mu_1, \mu_1),$$
$$\frac{a}{2}p(\mu_1, \mu_1) \equiv \frac{a}{2}(1 + 3g\mu_1^2) \equiv a(1 - b) = \frac{a}{2}p(-\mu_1, -\mu_1).$$

We note first that Eqs. 8.12–8.13 are identical to Eqs. 7.111–7.112 as they should be. Also, we recall that $b \equiv \frac{1}{2}(1 - 3g\mu_1^2)$ is called the backscatter ratio and that g is the first moment of the phase function as defined in Eq. 6.29 and is commonly referred to as the asymmetry factor. If we take $\mu_1 = 3^{-\frac{1}{2}}$, then for $g = -1$ we have complete backscattering ($b = 1$), for $g = 1$ complete forward scattering ($b = 0$), and for $g = 0$ isotropic scattering ($b = 1/2$). As shown above the value $\mu_1 = 3^{-\frac{1}{2}}$ corresponds to Gaussian quadrature for the full range $[-1, 1]$, while Gaussian quadrature for the half range [0, 1] (referred to as double-Gauss; see §8.2 above) yields $\mu_1 = 1/2$.

We may rewrite Eqs. 8.12 and 8.13 in matrix form as

$$\frac{d}{d\tau}\begin{bmatrix} I^+ \\ I^- \end{bmatrix} = \begin{bmatrix} -\alpha & -\beta \\ \beta & \alpha \end{bmatrix}\begin{bmatrix} I^+ \\ I^- \end{bmatrix} - \begin{bmatrix} Q^+ \\ Q^- \end{bmatrix}, \quad (8.14)$$

where

$$Q^\pm \equiv \pm\mu_1^{-1} Q'^\pm,$$
$$\alpha \equiv \mu_1^{-1}\left[\frac{a}{2}p(\mu_1, \mu_1) - 1\right] = \mu_1^{-1}\left[\frac{a}{2}p(-\mu_1, -\mu_1) - 1\right]$$
$$= \mu_1^{-1}[a(1 - b) - 1],$$
$$\beta \equiv \mu_1^{-1}\frac{a}{2}p(\mu_1, -\mu_1) = \mu_1^{-1}\frac{a}{2}p(-\mu_1, \mu_1) = \mu_1^{-1}ab.$$

Example 8.3 Four-Stream Approximation ($N = 2$)

In this case we obtain four coupled differential equations from Eqs. 8.9 and 8.10 as follows (again by assuming that we have chosen a quadrature satisfying $\mu_{-i} = -\mu_i$, $w_{-i} = w_i$):

$$\mu_1 \frac{dI^+(\tau, \mu_1)}{d\tau} = I^+(\tau, \mu_1) - Q'^+(\tau, \mu_1)$$
$$- w_2 \frac{a}{2}p(-\mu_2, \mu_1)I^-(\tau, \mu_2) - w_1 \frac{a}{2}p(-\mu_1, \mu_1)I^-(\tau, \mu_1)$$
$$- w_1 \frac{a}{2}p(\mu_1, \mu_1)I^+(\tau, \mu_1) - w_2 \frac{a}{2}p(\mu_2, \mu_1)I^+(\tau, \mu_2),$$

$$\mu_2 \frac{dI^+(\tau, \mu_2)}{d\tau} = I^+(\tau, \mu_2) - Q'^+(\tau, \mu_2)$$
$$- w_2 \frac{a}{2} p(-\mu_2, \mu_2) I^-(\tau, \mu_2) - w_1 \frac{a}{2} p(-\mu_1, \mu_2) I^-(\tau, \mu_1)$$
$$- w_1 \frac{a}{2} p(\mu_1, \mu_2) I^+(\tau, \mu_1) - w_2 \frac{a}{2} p(\mu_2, \mu_2) I^+(\tau, \mu_2),$$

$$-\mu_1 \frac{dI^-(\tau, \mu_1)}{d\tau} = I^-(\tau, \mu_1) - Q'^-(\tau, \mu_1)$$
$$- w_2 \frac{a}{2} p(-\mu_2, -\mu_1) I^-(\tau, \mu_2) - w_1 \frac{a}{2} p(-\mu_1, -\mu_1) I^-(\tau, \mu_1)$$
$$- w_1 \frac{a}{2} p(\mu_1, -\mu_1) I^+(\tau, \mu_1) - w_2 \frac{a}{2} p(\mu_2, -\mu_1) I^+(\tau, \mu_2),$$

$$-\mu_2 \frac{dI^-(\tau, \mu_2)}{d\tau} = I^-(\tau, \mu_2) - Q'^-(\tau, \mu_2)$$
$$- w_2 \frac{a}{2} p(-\mu_2, -\mu_2) I^-(\tau, \mu_2) - w_1 \frac{a}{2} p(-\mu_1, -\mu_2) I^-(\tau, \mu_1)$$
$$- w_1 \frac{a}{2} p(\mu_1, -\mu_2) I^+(\tau, \mu_1) - w_2 \frac{a}{2} p(\mu_2, -\mu_2) I^+(\tau, \mu_2).$$

We may rewrite these equations in matrix form as

$$\frac{d}{d\tau} \begin{bmatrix} I^+(\tau, \mu_1) \\ I^+(\tau, \mu_2) \\ I^-(\tau, \mu_1) \\ I^-(\tau, \mu_2) \end{bmatrix} = \begin{bmatrix} -\alpha_{11} & -\alpha_{12} & -\beta_{11} & -\beta_{12} \\ -\alpha_{21} & -\alpha_{22} & -\beta_{21} & -\beta_{22} \\ \beta_{11} & \beta_{12} & \alpha_{11} & \alpha_{12} \\ \beta_{21} & \beta_{22} & \alpha_{21} & \alpha_{22} \end{bmatrix} \begin{bmatrix} I^+(\tau, \mu_1) \\ I^+(\tau, \mu_2) \\ I^-(\tau, \mu_1) \\ I^-(\tau, \mu_2) \end{bmatrix} - \begin{bmatrix} Q^+(\tau, \mu_1) \\ Q^+(\tau, \mu_2) \\ Q^-(\tau, \mu_1) \\ Q^-(\tau, \mu_2) \end{bmatrix}, \tag{8.15}$$

where

$$Q^\pm(\tau, \mu_i) = \pm \mu_i^{-1} Q'^\pm(\tau, \mu_i), \quad i = 1, 2,$$
$$\alpha_{11} = \mu_1^{-1} \left[w_1 \frac{a}{2} p(\mu_1, \mu_1) - 1 \right] = \mu_1^{-1} \left[w_1 \frac{a}{2} p(-\mu_1, -\mu_1) - 1 \right],$$
$$\alpha_{12} = \mu_1^{-1} w_2 \frac{a}{2} p(\mu_1, \mu_2) = \mu_1^{-1} w_2 \frac{a}{2} p(-\mu_1, -\mu_2),$$
$$\alpha_{21} = \mu_2^{-1} w_1 \frac{a}{2} p(\mu_2, \mu_1) = \mu_2^{-1} w_1 \frac{a}{2} p(-\mu_2, -\mu_1),$$
$$\alpha_{22} = \mu_2^{-1} \left[w_2 \frac{a}{2} p(\mu_2, \mu_2) - 1 \right] = \mu_2^{-1} \left[w_2 \frac{a}{2} p(-\mu_2, -\mu_2) - 1 \right],$$
$$\beta_{11} = \mu_1^{-1} w_1 \frac{a}{2} p(\mu_1, -\mu_1) = \mu_1^{-1} w_1 \frac{a}{2} p(-\mu_1, \mu_1),$$
$$\beta_{12} = \mu_1^{-1} w_2 \frac{a}{2} p(\mu_1, -\mu_2) = \mu_1^{-1} w_2 \frac{a}{2} p(-\mu_1, \mu_2),$$
$$\beta_{21} = \mu_2^{-1} w_1 \frac{a}{2} p(\mu_2, -\mu_1) = \mu_2^{-1} w_1 \frac{a}{2} p(-\mu_2, \mu_1),$$
$$\beta_{22} = \mu_2^{-1} w_2 \frac{a}{2} p(\mu_2, -\mu_2) = \mu_2^{-1} w_2 \frac{a}{2} p(-\mu_2, \mu_2).$$

By introducing the vectors

$$\mathbf{I}^{\pm} = \{I^{\pm}(\tau, \mu_i)\}, \qquad \mathbf{Q}^{\pm} = \{Q^{\pm}(\tau, \mu_i)\}, \quad i = 1, 2,$$

we may write Eq. 8.15 in a more compact form as

$$\frac{d}{d\tau}\begin{bmatrix}\mathbf{I}^+\\ \mathbf{I}^-\end{bmatrix} = \begin{bmatrix}-\tilde{\alpha} & -\tilde{\beta}\\ \tilde{\beta} & \tilde{\alpha}\end{bmatrix}\begin{bmatrix}\mathbf{I}^+\\ \mathbf{I}^-\end{bmatrix} - \begin{bmatrix}\mathbf{Q}^+\\ \mathbf{Q}^-\end{bmatrix}, \tag{8.16}$$

where all the elements of the matrices $\tilde{\alpha}$ and $\tilde{\beta}$ are defined above. It should be noted that this equation is very similar to the one obtained in the two-stream approximation except that the scalars α and β have become 2×2 matrices. It should also be obvious from Eqs. 8.14 and 8.16 that $\tilde{\alpha}$ and $\tilde{\beta}$ may be interpreted as local transmission and reflection operators, respectively. This will become more evident when we discuss the doubling method.

8.4.2 Multistream Approximation (N Arbitrary)

From the preceding description of the two- and four-stream cases it should be obvious how to generalize this scheme. We may now write Eqs. 8.9 and 8.10 in matrix form as

$$\frac{d}{d\tau}\begin{bmatrix}\mathbf{I}^+\\ \mathbf{I}^-\end{bmatrix} = \begin{bmatrix}-\tilde{\alpha} & -\tilde{\beta}\\ \tilde{\beta} & \tilde{\alpha}\end{bmatrix}\begin{bmatrix}\mathbf{I}^+\\ \mathbf{I}^-\end{bmatrix} - \begin{bmatrix}\mathbf{Q}^+\\ \mathbf{Q}^-\end{bmatrix}, \tag{8.17}$$

where

$$\begin{aligned}
\mathbf{I}^{\pm} &= \{I^{\pm}(\tau, \mu_i)\}, \quad i = 1, \ldots, N,\\
\mathbf{Q}^{\pm} &= \pm\mathbf{M}^{-1}\mathbf{Q}'^{\pm} = \{Q^{\pm}(\tau, \mu_i)\}, \quad i = 1, \ldots, N,\\
\mathbf{M} &= \{\mu_i \delta_{ij}\}, \quad i, j = 1, \ldots, N,\\
\tilde{\alpha} &= \mathbf{M}^{-1}\{\mathbf{D}^+\mathbf{W} - \mathbf{1}\},\\
\tilde{\beta} &= \mathbf{M}^{-1}\mathbf{D}^-\mathbf{W},\\
\mathbf{W} &= \{w_i \delta_{ij}\}, \quad i, j = 1, \ldots, N,\\
\mathbf{1} &= \{\delta_{ij}\}, \quad i, j = 1, \ldots, N,\\
\mathbf{D}^+ &= \frac{a}{2}\{p(\mu_i, \mu_j)\} = \frac{a}{2}\{p(-\mu_i, -\mu_j)\}, \quad i, j = 1, \ldots, N,\\
\mathbf{D}^- &= \frac{a}{2}\{p(-\mu_i, \mu_j)\} = \frac{a}{2}\{p(\mu_i, -\mu_j)\}, \quad i, j = 1, \ldots, N.
\end{aligned}$$

We note that the structure of the $(2N \times 2N)$ matrix

$$\begin{bmatrix}-\tilde{\alpha} & -\tilde{\beta}\\ \tilde{\beta} & \tilde{\alpha}\end{bmatrix}$$

in Eq. 8.17 can be traced to the fact that the phase function depends only on the scattering angle (i.e., the angle between $\hat{\Omega}(\mu, \phi)$ and $\hat{\Omega}'(\mu', \phi')$; cf. Chapter 3). This special structure is also a consequence of having chosen a quadrature rule satisfying $\mu_{-i} = -\mu_i$, $w_{-i} = w_i$. As we shall see below, because of this structure, Eq. 8.17 permits eigensolutions with eigenvalues occurring in positive/negative pairs. In particular we will find that it can be used to reduce the order of the resulting algebraic eigenvalue problem by a factor of 2, which leads to a decrease of the computational burden by a factor of 8. This follows because the timing for eigensolution solvers is proportional to the cube of the matrix dimension.

8.5 Matrix Eigensolutions

8.5.1 Two-Stream Solutions (N = 1)

Seeking solutions to the homogeneous version of Eq. 8.14 ($Q^{\pm} = 0$) of the form $I^{\pm} = g^{\pm}e^{-\lambda\tau}$, $g^{\pm} = g(\pm\mu_1)$, we find that this leads to the following algebraic eigenvalue problem:

$$\begin{bmatrix} \alpha & \beta \\ -\beta & -\alpha \end{bmatrix} \begin{bmatrix} g^+ \\ g^- \end{bmatrix} = \lambda \begin{bmatrix} g^+ \\ g^- \end{bmatrix}. \tag{8.18}$$

Writing this matrix equation as

$$\alpha g^+ + \beta g^- = \lambda g^+,$$
$$-\beta g^+ - \alpha g^- = \lambda g^-$$

and adding and subtracting these two equations, we find

$$(\alpha - \beta)(g^+ - g^-) = \lambda(g^+ + g^-), \tag{8.19}$$

$$(\alpha + \beta)(g^+ + g^-) = \lambda(g^+ - g^-). \tag{8.20}$$

Substitution of the last equation into Eq. 8.19 yields

$$(\alpha - \beta)(\alpha + \beta)(g^+ + g^-) = \lambda^2(g^+ + g^-),$$

which has the solutions $\lambda_1 = k$, $\lambda_{-1} = -k$ with

$$k = \sqrt{\alpha^2 - \beta^2} = \frac{1}{\mu_1}\sqrt{(1-a)(1-a+2ab)} > 0 \quad (a < 1) \tag{8.21}$$

and

$$g^+ + g^- = \text{arbitrary constant}, \tag{8.22}$$

which we may set equal to unity.

For $\lambda_1 = k$ Eq. 8.20 yields

$$g^+ - g^- = (\alpha + \beta)/k \tag{8.23}$$

(assuming $k \neq 0$ or $a \neq 1$). Combining Eqs. 8.22 and 8.23 we find

$$\frac{g_1^+}{g_1^-} = \frac{k + (\alpha + \beta)}{k - (\alpha + \beta)} = \frac{\sqrt{1-a+2ab} - \sqrt{1-a}}{\sqrt{1-a+2ab} + \sqrt{1-a}} \equiv \rho_\infty \tag{8.24}$$

and thus

$$\begin{bmatrix} g_1^+ \\ g_1^- \end{bmatrix} = \begin{bmatrix} \rho_\infty \\ 1 \end{bmatrix}, \tag{8.25}$$

which is the eigenvector belonging to eigenvalue $\lambda_1 = k$. Repeating this for $\lambda_{-1} = -k$, we find $g_{-1}^-/g_{-1}^+ = \rho_\infty$, and

$$\begin{bmatrix} g_{-1}^+ \\ g_{-1}^- \end{bmatrix} = \begin{bmatrix} 1 \\ \rho_\infty \end{bmatrix}. \tag{8.26}$$

The complete homogeneous solution becomes a linear combination of the exponential solutions for eigenvalues $\lambda_1 = k$ and $\lambda_{-1} = -k$, that is,

$$\begin{aligned} I^+(\tau) = I(\tau, +\mu_1) &= C_{-1}g_{-1}(+\mu_1)e^{+k\tau} + C_1 g_1(+\mu_1)e^{-k\tau} \\ &= C_{-1}g_{-1}(+\mu_1)e^{+k\tau} + \rho_\infty C_1 g_1(-\mu_1)e^{-k\tau}, \end{aligned} \tag{8.27}$$

$$\begin{aligned} I^-(\tau) = I(\tau, -\mu_1) &= C_{-1}g_{-1}(-\mu_1)e^{+k\tau} + C_1 g_1(-\mu_1)e^{-k\tau} \\ &= \rho_\infty C_{-1}g_{-1}(+\mu_1)e^{+k\tau} + C_1 g_1(-\mu_1)e^{-k\tau}, \end{aligned} \tag{8.28}$$

where C_1 and C_{-1} are constants of integration. We note that these solutions are identical to Eqs. 7.126 and 7.127 given previously for the two-stream approximation, as they should be. In anticipation of the extension to more than two streams we rewrite the solution in the following somewhat artificial form:

$$I^\pm(\tau, \mu_i) = \sum_{j=1}^{1} C_{-j}g_{-j}(\pm\mu_i)e^{k_j\tau} + \sum_{j=1}^{1} C_j g_j(\pm\mu_i)e^{-k_j\tau}, \quad i = 1, 1 \tag{8.29}$$

with $k_1 = k$, given by Eq. 8.21.

8.5.2 Multistream Solutions (N Arbitrary)

Equation 8.17 is a system of $2N$ coupled, ordinary differential equations with constant coefficients. These coupled equations are linear and our goal is to uncouple them by using well-known methods of linear algebra. From the discussion of the two- and four-stream cases it is now obvious how we should proceed. Seeking solutions to the homogeneous version ($\mathbf{Q} = \mathbf{0}$) of Eq. 8.17 of the form

$$\mathbf{I}^\pm = \mathbf{g}^\pm e^{-k\tau}, \tag{8.30}$$

we find

$$\begin{bmatrix} \tilde{\alpha} & \tilde{\beta} \\ -\tilde{\beta} & -\tilde{\alpha} \end{bmatrix} \begin{bmatrix} \mathbf{g}^+ \\ \mathbf{g}^- \end{bmatrix} = k \begin{bmatrix} \mathbf{g}^+ \\ \mathbf{g}^- \end{bmatrix}, \tag{8.31}$$

which is a standard algebraic eigenvalue problem of order $2N \times 2N$ determining the eigenvalues k and the eigenvectors $\mathbf{g}^\pm$.

As noted previously, because of the special structure of the matrix in Eq. 8.31, the eigenvalues occur in positive/negative pairs and the order of the algebraic eigenvalue problem (Eq. 8.31) may be reduced as follows. Rewriting the homogeneous version of Eq. 8.17 as

$$\frac{d\mathbf{I}^+}{d\tau} = -\tilde{\alpha}\mathbf{I}^+ - \tilde{\beta}\mathbf{I}^-,$$
$$\frac{d\mathbf{I}^-}{d\tau} = \tilde{\alpha}\mathbf{I}^- + \tilde{\beta}\mathbf{I}^+$$

and then adding and subtracting these two equations, we find

$$\frac{d(\mathbf{I}^+ + \mathbf{I}^-)}{d\tau} = -(\tilde{\alpha} - \tilde{\beta})(\mathbf{I}^+ - \mathbf{I}^-) \tag{8.32}$$

and

$$\frac{d(\mathbf{I}^+ - \mathbf{I}^-)}{d\tau} = -(\tilde{\alpha} + \tilde{\beta})(\mathbf{I}^+ + \mathbf{I}^-). \tag{8.33}$$

Combining Eqs. 8.32 and 8.33, we obtain

$$\frac{d^2(\mathbf{I}^+ + \mathbf{I}^-)}{d\tau^2} = (\tilde{\alpha} - \tilde{\beta})(\tilde{\alpha} + \tilde{\beta})(\mathbf{I}^+ + \mathbf{I}^-),$$

or in view of Eq. 8.30

$$(\tilde{\alpha} - \tilde{\beta})(\tilde{\alpha} + \tilde{\beta})(\mathbf{g}^+ + \mathbf{g}^-) = k^2(\mathbf{g}^+ + \mathbf{g}^-), \tag{8.34}$$

which completes the reduction of the order. To proceed we solve Eq. 8.34 to obtain eigenvalues and eigenvectors $(\mathbf{g}^+ + \mathbf{g}^-)$. We then use Eq. 8.33 to determine $(\mathbf{g}^+ - \mathbf{g}^-)$ and proceed as in the four-stream case to construct a complete set of eigenvectors.

8.5.3 Inhomogeneous Solution

It is easily verified that a particular solution for collimated beam incidence (*Prototype Problem 3*) is

$$I(\tau, u_i) = Z_0(u_i)e^{-\tau/\mu_0}, \tag{8.35}$$

where the $Z_0(u_i)$ are determined by the following system of linear algebraic equations:

$$\sum_{\substack{j=-N \\ j\neq 0}}^{N} \left[(1 + u_j/\mu_0)\delta_{ij} - w_j \frac{a}{2} p(u_i, u_j)\right] Z_0(u_j) = X_0(u_i). \tag{8.36}$$

Equation 8.36 is obtained by simply substituting the "trial" solution (Eq. 8.35) into Eqs. 8.9–8.10.

In the two-stream case Eq. 8.36 reduces to a system of two algebraic equations with two unknowns, which is easily solved analytically, and the solutions were provided in Chapter 7. The four-stream case involves four algebraic equations and may also be solved analytically, but this may not be worth the effort, since standard linear equation solvers have built-in features like pivoting, implying that such a software package is, in general, likely to produce numerical results superior to those obtained from the analytic solutions.

Example 8.4 Thermal Source

For thermal sources the emitted radiation is isotropic so that the source term is azimuthally independent,

$$Q'(\tau) = (1 - a)B(\tau).$$

To account for the temperature variation in the slab we may approximate the Planck function for each layer by a polynomial in optical depth τ:

$$B[T(\tau)] = \sum_{l=0}^{K} b_l \tau^l.$$

Then if we insist that the solution should also be a polynomial in τ, that is,

$$I(\tau, u_i) = \sum_{l=0}^{K} Y_l(u_i)\tau^l,$$

it can be shown (see Problem 6.1) that the coefficients $Y_l(u_i)$ are determined by solving the following system of linear algebraic equations:

$$Y_K(u_i) = (1 - a)b_K, \tag{8.37}$$

$$\sum_{j=-N}^{N} \left(\delta_{ij} - w_j \frac{a}{2} p(u_i, u_j)\right) Y_l(u_j) = (1 - a)b_l - (l + 1)u_i Y_{l+1}(u_i),$$
$$l = K - 1, K - 2, \ldots, 0. \tag{8.38}$$

In practice it is popular to use a linear approximation ($K = 1$), which only requires knowledge of the temperature at layer interfaces to compute the Planck function there. Noting that the Planck function depends linearly on temperature in the long-wavelength (Rayleigh–Jeans) limit but exponentially in the short-wavelength (Wien's) limit, we expect an exponential times a linear dependence of the Planck function on τ to work well under most circumstances (see Problem 6.1).

8.5.4 General Solution

The general solution to Eqs. 8.9 and 8.10 consists of a linear combination, with coefficients C_j, of all the homogeneous solutions, plus the particular solution,

$$I^{\pm}(\tau, \mu_i) = \sum_{j=1}^{N} C_{-j} g_{-j}(\pm\mu_i) e^{k_j \tau} + \sum_{j=1}^{N} C_j g_j(\pm\mu_i) e^{-k_j \tau}$$
$$+ Z_0(\pm\mu_i) e^{-\tau/\mu_0}, \quad i = 1, \ldots, N. \tag{8.39}$$

The k_j and $g_j(\pm\mu_i)$ are the eigenvalues and eigenvectors obtained as described above. The $\pm\mu_i$ are the quadrature angles, and the $C_{\pm j}$ the constants of integration.

8.6 Source Function and Angular Distributions

For a slab of thickness τ^*, we may solve Eqs. 8.9 and 8.10 formally to obtain ($\mu > 0$)

$$I^+(\tau, \mu) = I^+(\tau^*, \mu)e^{-(\tau^*-\tau)/\mu} + \int_\tau^{\tau^*} \frac{dt}{\mu} S^+(t, \mu)e^{-(t-\tau)/\mu}, \tag{8.40}$$

$$I^-(\tau, \mu) = I^-(0, \mu)e^{-\tau/\mu} + \int_0^\tau \frac{dt}{\mu} S^-(t, \mu)e^{-(\tau-t)/\mu}. \tag{8.41}$$

As previously discussed in Chapter 5, these two equations show that if we know the source function $S^\pm(t, \mu)$, we can find the intensity at arbitrary angles by integrating the source function. Below we shall use the discrete-ordinate solutions to derive explicit expressions for the source function that can be integrated analytically. Although this procedure is sometimes referred to as the "iteration of the source-function technique," it essentially consists of an interpolation.

In view of Eqs. 8.9–8.10 the discrete-ordinate approximation to the source function may be written as

$$\begin{aligned} S^\pm(\tau, \mu) = {} & \frac{a}{2} \sum_{i=1}^N w_i p(-\mu_i, \pm\mu) I^-(\tau, \mu_i) \\ & + \frac{a}{2} \sum_{i=1}^N w_i p(+\mu_i, \pm\mu) I^+(\tau, \mu_i) + X_0^\pm(\mu)e^{-\tau/\mu_0}. \end{aligned} \tag{8.42}$$

Substituting the general solution of Eq. 8.39 into Eq. 8.42, we find

$$\begin{aligned} S^\pm(\tau, \mu) = {} & \sum_{j=1}^N C_{-j} \tilde{g}_{-j}(\pm\mu)e^{k_j\tau} + \sum_{j=1}^N C_j \tilde{g}_j(\pm\mu)e^{-k_j\tau} \\ & + \tilde{Z}_0^\pm(\mu)e^{-\tau/\mu_0}, \end{aligned} \tag{8.43}$$

where

$$\tilde{g}_j(\pm\mu) = \frac{a}{2} \sum_{i=1}^N \{w_i p(-\mu_i, \pm\mu) g_j(-\mu_i) + w_i p(+\mu_i, \pm\mu) g_j(+\mu_i)\}, \tag{8.44}$$

$$\begin{aligned} \tilde{Z}_0^\pm(\mu) = {} & \frac{a}{2} \sum_{i=1}^N \{w_i p(-\mu_i, \pm\mu) Z_0(-\mu_i) \\ & + w_i p(+\mu_i, \pm\mu) Z_0(+\mu_i)\} + X_0(\pm\mu). \end{aligned} \tag{8.45}$$

Equations 8.44 and 8.45 are simply convenient analytic interpolation formulas for the $\tilde{g}_j(\pm\mu)$ and the $\tilde{Z}_0(\pm\mu)$. They clearly reveal the interpolatory nature of Eq. 8.43 for the source function. The fact that they are derived from the basic radiative transfer equation to which we are seeking solutions indicates that these expressions may be superior to any other standard interpolation scheme, as demonstrated in Example 8.5 below.

Using Eqs. 8.43 in Eqs. 8.40–8.41, we find that for a layer of thickness τ^*, the intensities become

$$I^+(\tau,\mu) = I^+(\tau^*,\mu)e^{-(\tau^*-\tau)/\mu} + \sum_{j=-N}^{N} C_j \frac{\tilde{g}_j(+\mu)}{1+k_j\mu}\left\{e^{-k_j\tau} - e^{-[k_j\tau^*+(\tau^*-\tau)/\mu]}\right\}, \tag{8.46}$$

$$I^-(\tau,\mu) = I^-(0,\mu)e^{-\tau/\mu} + \sum_{j=-N}^{N} C_j \frac{\tilde{g}_j(-\mu)}{1-k_j\mu}\left\{e^{-k_j\tau} - e^{-\tau/\mu}\right\}, \tag{8.47}$$

where we have for convenience included the particular solution as the $j = 0$ term in the sum so that $C_0\tilde{g}_0(\pm\mu) \equiv \tilde{Z}_0(\pm\mu)$ and $k_0 \equiv 1/\mu_0$. The basic soundness and merit of the intensity expressions given above will be demonstrated in the following example. First, we note that Eqs. 8.46–8.47, when evaluated at the quadrature points, yield results identical to Eqs. 8.39 (see Problem 8.2). Second, Eqs. 8.46–8.47 satisfy the boundary conditions for all μ values (even though we have imposed such conditions only at the quadrature points!). Third, the more complicated expressions (i.e., Eqs. 8.46–8.47 as compared to Eq. 8.39) have the merit of "correcting" the simpler expression (Eq. 8.39) for μ values not coinciding with the quadrature points.

Example 8.5 The Merit of the Interpolation Scheme

Equations 8.46 and 8.47 provide a convenient means of computing the intensities for arbitrary angles, at any desired optical depth. However, the merit of these expressions depends crucially on the ability to compute efficiently the eigenvectors $\tilde{g}_j(\pm\mu)$ and the particular solution vector $\tilde{Z}_0(\pm\mu)$. Since the $g_j(\mu)$ are known at the quadrature points ($\mu = \mu_i, i = \pm1, \ldots, \pm N$), this information can be used as a basis for interpolation using any standard interpolation scheme. To illustrate the problems one might encounter in interpolation using standard technique, we show in Fig. 8.1 the eigenvector corresponding to the smallest eigenvalue for a phase function typical of atmospheric aerosols with single-scattering albedo $a = 0.9$. This illustrates the typical behavior of some of the eigenvectors. A 16-stream computation ($N = 8$) was used in this example. The values at the quadrature points to be interpolated are indicated by the filled circles. We notice that there is a pronounced dip close to $\mu = 0$. It is obviously difficult to fit a polynomial to a function with such a pronounced dip. A cubic spline interpolation also performs poorly on both sides of the dip as illustrated, whereas the analytic expression (Eq. 8.44) yields quite adequate results.

To illustrate the consequence of using different interpolation schemes we computed diffuse intensities for a slab of optical thickness $\tau^* = 1$. The slab was illuminated by a collimated beam with $F^s = 1$ at an angle of incidence so that $\mu_0 = 0.5$. The results are based on Eqs. 8.46 and 8.47. Figure 8.2 shows the azimuthally averaged diffuse intensity at optical depth $\tau = 0.05$. The solid line results from using the analytic expressions (Eqs. 8.44 and 8.45) to compute the eigenvectors and the particular solution vector. The results obtained by using cubic spline interpolation of the eigenvectors are shown by the dashed line. We notice that the cubic spline interpolation leads to erroneous results for $-0.6 < u < -0.1$.

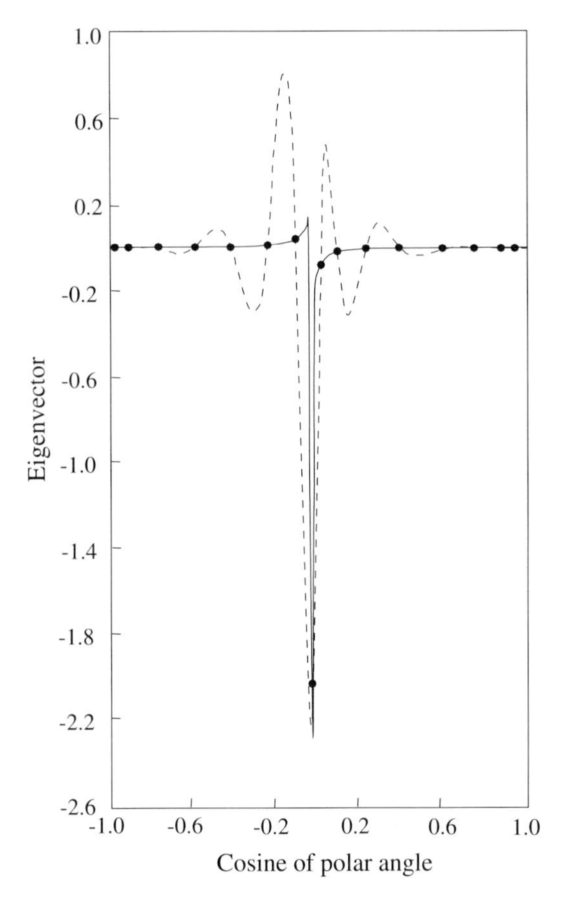

Figure 8.1 Interpolated eigenvector. The solid line pertains to the use of the analytic formula, the dashed line is a cubic spline interpolation, and the filled circles refer to values at the quadrature points. Note that the spline fails to produce accurate results, whereas the analytic formula gives adequate results.

This example illustrates that an interpolation scheme that interpolates the eigenvectors is perhaps best suited as a general purpose interpolation scheme since it can provide intensities at any desired angle and depth. As we have seen, the analytic expressions (Eqs. 8.44 and 8.45) yield adequate results.

8.7 Boundary Conditions – Removal of Ill-Conditioning

8.7.1 Boundary Conditions

We considered the inclusion of boundary conditions in §6.11 when we discussed *Prototype Problem 3*. We noted that if the diffuse bidirectional reflectance, $\rho_d(\mu, \phi; -\mu', \phi')$, is a function only of the difference between the azimuthal angles before and after

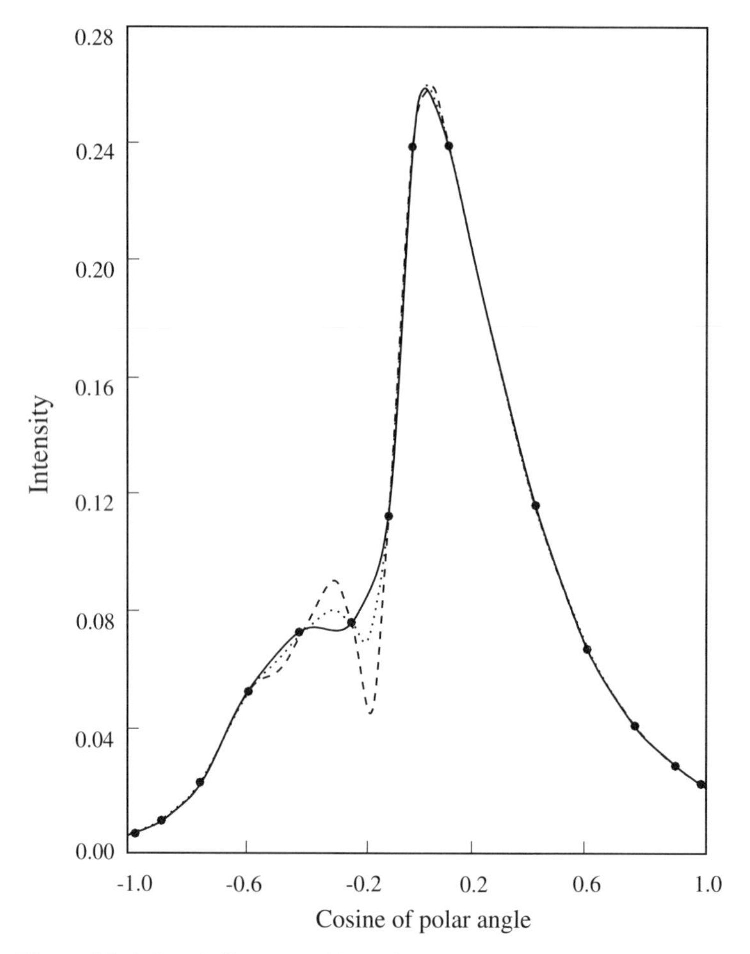

Figure 8.2 Azimuthally averaged intensity at $\tau = 0.05$ for oblique incidence ($\mu_0 = 0.5$) of collimated light of unit flux on a slab of optical thickness $\tau^* = 1$. The slab consists of particles that scatter radiation according to an anisotropic phase function (aerosol particles; see Fig. 6.3) with single-scattering albedo $a = 0.9$. The solid line is obtained from the analytic expressions, the dashed line from cubic spline interpolation of the eigenvectors, and the dotted line from cubic spline interpolation of the intensities at the quadrature points. The filled circles denote the values of the intensities at the quadrature points.

reflection, then we may expand it in a cosine series as follows:

$$\rho_d(-\mu', \phi'; \mu, \phi) = \rho_d(-\mu', \mu; \phi - \phi') = \sum_{m=0}^{2N-1} \rho_d^m(-\mu', \mu) \cos m(\phi - \phi'),$$

where the expansion coefficients are computed from

$$\rho_d^m(-\mu', \mu) = \frac{1}{\pi} \int_{-\pi}^{\pi} d(\phi - \phi') \rho_d(-\mu', \mu; \phi - \phi') \cos m(\phi - \phi').$$

Here the superscript m refers to the azimuthal component. The advantage of this expansion is that we again are able to isolate the azimuthal components. In fact, it was shown (see §6.9) that each Fourier component must satisfy the bottom boundary condition

$$
\begin{aligned}
I^m(\tau^*, +\mu) &= \delta_{m0}\epsilon(\mu)B(T_s) + (1+\delta_{m0})\int_0^1 d\mu'\,\mu'\rho_d^m(-\mu', \mu)I^m(\tau^*, -\mu') \\
&\quad + \frac{\mu_0 F^s}{\pi}e^{-\tau^*/\mu_0}\rho_d^m(-\mu_0, \mu) \\
&\equiv I_s^m(\mu),
\end{aligned}
\tag{8.48}
$$

where T_s and $\epsilon(\mu)$ are the temperature and the emittance of the lower boundary surface, respectively. Thus, Eqs. 8.39 must satisfy boundary conditions

$$
I^m(0, -\mu_i) = \mathcal{I}^m(-\mu_i), \quad i = 1, \ldots, N, \tag{8.49}
$$

$$
I^m(\tau^*, +\mu_i) = I_s^m(\mu_i), \quad i = 1, \ldots, N, \tag{8.50}
$$

where

$$
\begin{aligned}
I_s^m(\mu_i) &= \delta_{m0}\epsilon(\mu_i)B(T_s) + (1+\delta_{m0})\sum_{j=1}^{N} w_j\mu_j\rho_d^m(\mu_j, -\mu_i)I^m(\tau^*, -\mu_j) \\
&\quad + \frac{\mu_0 F^s}{\pi}e^{-\tau^*/\mu_0}\rho_d^m(\mu_0, -\mu_i).
\end{aligned}
\tag{8.51}
$$

$\mathcal{I}^m(-\mu_i)$ is the radiation incident at the top boundary. Note that for *Prototype Problem 1* we would have $\mathcal{I}^m(-\mu_i) =$ constant (the same for all μ_i) for $m = 0$, and $\mathcal{I}^m(-\mu_i) = 0$ for $m \neq 0$ (uniform illumination). For *Prototype Problems 2* and *3* we have, of course, $\mathcal{I}^m(-\mu_i) = 0$ since there is by definition no diffuse radiation incident in *Prototype Problem 3* and *Prototype Problem 2* is assumed to be driven entirely by internal radiation sources. Since Eqs. 8.49 and 8.50 introduce a fundamental distinction between downward directions (denoted by $-$) and upward directions (denoted by $+$), one should select a quadrature rule that integrates separately over the downward and upward directions. As noted previously, the double-Gauss rule that we have adopted satisfies this requirement.

For the discussion of boundary conditions, it is convenient to write the discrete ordinate solution in the following form ($k_j > 0$ and $k_{-j} = -k_j$):

$$
I^{\pm}(\tau, \mu_i) = \sum_{j=1}^{N}\left[C_j g_j(\pm\mu_i)e^{-k_j\tau} + C_{-j}g_{-j}(\pm\mu_i)e^{+k_j\tau}\right] + U^{\pm}(\tau, \mu_i),
\tag{8.52}
$$

where the sum contains the homogeneous solution involving the unknown coefficients (the C_j) and $U^{\pm}(\tau, \mu_i)$ is the particular solution given by Eq. 8.35. Insertion of Eq. 8.52

into Eqs. 8.49–8.51 yields (omitting the m superscript)

$$\sum_{j=1}^{N}\{C_j g_j(-\mu_i) + C_{-j} g_{-j}(-\mu_i)\} = \mathcal{I}(-\mu_i) - U^-(0, \mu_i),$$
$$i = 1, \ldots, N, \tag{8.53}$$

$$\sum_{j=1}^{N}\left\{C_j r_j(\mu_i) g_j(+\mu_i) e^{-k_j\tau^*} + C_{-j} r_{-j}(\mu_i) g_{-j}(+\mu_i) e^{k_j\tau^*}\right\} = \Gamma(\tau^*, \mu_i),$$
$$i = 1, \ldots, N, \tag{8.54}$$

where

$$r_j(\mu_i) = 1 - (1 + \delta_{m0}) \sum_{n=1}^{N} \rho_d(\mu_i, -\mu_n) w_n \mu_n g_j(-\mu_n)/g_j(+\mu_i) \tag{8.55}$$

and

$$\begin{aligned}\Gamma(\tau^*, \mu_i) = {} & \delta_{m0}\epsilon(\mu_i) B(T_s) - U^+(\tau^*, \mu_i) \\ & + (1 + \delta_{m0}) \sum_{j=1}^{N} \rho_d(\mu_i, -\mu_j) w_j \mu_j U^-(\tau^*, \mu_j) \\ & + \frac{\mu_0 F^s}{\pi} e^{-\tau^*/\mu_0} \rho_d(-\mu_0, \mu_i).\end{aligned} \tag{8.56}$$

Equations 8.53 and 8.54 constitute a $2N \times 2N$ system of linear algebraic equations from which the $2N$ unknown coefficients, the C_j ($j = \pm 1, \ldots, \pm N$), are determined. The numerical solution of this set of equations is seriously hampered by the fact that Eqs. 8.53 and 8.54 are intrinsically ill-conditioned. Fortunately, this ill-conditioning may be entirely eliminated by a simple scaling transformation discussed in the next subsection. Suffice it to state here: By "ill-conditioning" we mean that when Eqs. 8.53 and 8.54 are written in matrix form the resulting matrix cannot be successfully inverted by existing computers that work with "finite-digit" arithmetic. As we shall see in the next subsection, if τ^* is sufficiently large, some of the elements of the matrix become huge while others become tiny, and it is this situation that leads to ill-conditioning.

8.7.2 Removal of Numerical Ill-Conditioning

Attempts to solve Eqs. 8.53 and 8.54 as they stand reveal that they are notoriously ill-conditioned. In fact, this problem explains why the discrete-ordinate method has not been used very frequently by researchers in the past. We shall now show that this ill-conditioning can be completely eliminated, thereby rendering the method very useful for solving practical problems. The root of the ill-conditioning problem lies in the occurrence of exponentials with positive arguments in Eqs. 8.53 and 8.54 (recall that $k_j > 0$ by convention), which must be removed. This is achieved by the scaling transformation

$$C_{+j} = C'_{+j} e^{k_j \tau_t} \quad \text{and} \quad C_{-j} = C'_{-j} e^{-k_j \tau_b}, \tag{8.57}$$

where we have written τ_t and τ_b for the optical depths at the top and the bottom of the layer, respectively. This was done deliberately in anticipation of generalizing this scaling scheme to apply to a multilayered medium. In the present one-layer case we have, of course, $\tau_t = 0$ and $\tau_b = \tau^*$.

Inserting Eqs. 8.57 into Eqs. 8.53 and 8.54 and solving for the C'_j instead of the C_j, we find that all the exponential terms in the coefficient matrix have negative arguments ($k_j > 0$, $\tau_b > \tau_t$). Consequently, numerical ill-conditioning is avoided, implying that the system of algebraic equations determining the C'_j will be unconditionally stable for arbitrary layer thickness.

As stated above, the merit of the scaling transformation is to remove all positive arguments of the exponentials occurring in the matrix elements of the coefficient matrix. To demonstrate how this scheme works we shall use the two-stream case as an example.

Example 8.6 Removal of Ill-Conditioning – Two-Stream Case ($N = 1$)

In this simple case, Eqs. 8.53 and 8.54 reduce to

$$C_1 g_1(-\mu_1) e^{-k\tau_t} + C_{-1} g_{-1}(-\mu_1) e^{k\tau_t} = C_1 g_1^- e^{-k\tau_t} + C_{-1} g_{-1}^- e^{k\tau_t} = (\text{RHS})_t$$

and

$$\begin{aligned} r_1 C_1 g_1(+\mu_1) e^{-k\tau_b} + r_{-1} C_{-1} g_{-1}(+\mu_1) e^{k\tau_b} &= r_1 C_1 g_1^+ e^{-k\tau_b} + r_{-1} C_{-1} g_{-1}^+ e^{k\tau_b} \\ &= (\text{RHS})_b, \end{aligned}$$

where we have used Eqs. 8.27 and 8.28. The left-hand side may be written in matrix form as

$$\begin{bmatrix} g_1^- e^{-k\tau_t} & g_{-1}^- e^{k\tau_t} \\ r_1 g_1^+ e^{-k\tau_b} & r_{-1} g_{-1}^+ e^{k\tau_b} \end{bmatrix} \begin{bmatrix} C_1 \\ C_{-1} \end{bmatrix}.$$

This matrix is ill-conditioned because one element becomes very large while another one becomes very small as $k\tau_b$ (the product of the eigenvalue and the optical depth) becomes large. In practice this limits solutions to problems for which $k\tau_b < 3$ or 4. As we go beyond the two-stream case the problem becomes more severe because some of the eigenvalues become large. We recall that for isotropic scattering the eigenvalues are flanked by the values $1/\mu_1$, $1/\mu_2$, etc., showing that the larger N is, the larger the biggest eigenvalue. This is the case also for anisotropic scattering. Hence, it is clear that for the method to be of any practical value this problem must be overcome.

Using the scaling transformation we find that the above matrix becomes

$$\begin{bmatrix} g_1^- & g_{-1}^- e^{-k(\tau_b - \tau_t)} \\ r_1 g_1^+ e^{-k(\tau_b - \tau_t)} & r_{-1} g_{-1}^+ \end{bmatrix} \begin{bmatrix} C'_1 \\ C'_{-1} \end{bmatrix}.$$

In the limit of large values of $k(\tau_b - \tau_t)$ this matrix becomes

$$\begin{bmatrix} g_1^- & 0 \\ 0 & r_{-1} g_{-1}^+ \end{bmatrix},$$

which shows that the ill-conditioning problem has been entirely eliminated.

8.8 Inhomogeneous Multilayered Media

8.8.1 General Solution – Boundary and Layer Interface Conditions

So far we have considered only a homogeneous slab in which the optical properties specified by the single-scattering albedo and the phase function were assumed to be constant throughout the slab. We shall now allow for these properties to be a function of optical depth. To approximate the behavior of a vertically inhomogeneous slab we will divide it into a number of layers. We will assume that the optical properties are constant within each layer, but we will allow them to be different from layer to layer. Thus, the slab is assumed to consist of L adjacent layers in which the single-scattering albedo and the phase function are taken to be constant within each layer (but allowed to vary from layer to layer), as illustrated in Fig. 8.3. For an emitting slab we assume that we know the temperature at the layer boundaries. The idea is that by using enough layers we can approximate the actual variation in optical properties and temperature as closely as desired.

The advantage of this approach is that we can use the solutions derived previously because each of the layers by assumption is homogeneous. This implies that we may

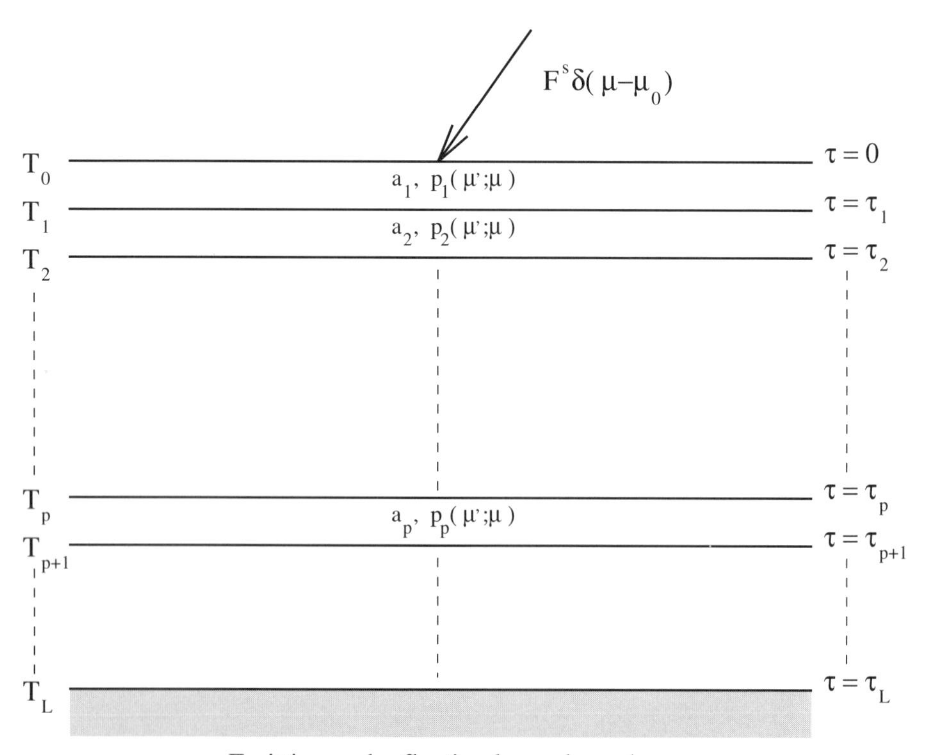

Figure 8.3 Schematic illustration of a multilayered, inhomogeneous medium overlying an emitting and partially reflecting surface.

write the solution for the pth layer as ($k_{jp} > 0$ and $k_{-jp} = -k_{jp}$)

$$I_p^{\pm}(\tau, \mu_i) = \sum_{j=1}^{N} \left[C_{jp} g_{jp}(\pm\mu_i) e^{-k_{jp}\tau} + C_{-jp} g_{-jp}(\pm\mu_i) e^{+k_{jp}\tau}\right] + U_p^{\pm}(\tau, \mu_i),$$
$$p = 1, 2, \ldots, L, \tag{8.58}$$

where the sum contains the homogeneous solution involving the unknown coefficients (the C_{jp}) and $U_p^{\pm}(\tau, \mu_i)$ is the particular solution given by Eq. 8.35. We note that except for the layer index p, Eq. 8.58 is identical to Eq. 8.52, as it should be. The solution contains $2N$ constants per layer, yielding a total of $2N \times L$ unknown constants. In addition to boundary conditions we must now require the intensity to be continuous across layer interfaces. As we shall see this will lead to a set of algebraic equations from which the $2N \times L$ unknown constants can be determined. Thus, Eq. 8.39 must now satisfy boundary and continuity conditions as follows:

$$I_1^m(0, -\mu_i) = \mathcal{I}^m(-\mu_i), \quad i = 1, \ldots, N; \tag{8.59}$$

$$I_p^m(\tau_p, \mu_i) = I_{p+1}^m(\tau_p, \mu_i), \quad i = \pm 1, \ldots, \pm N, \; p = 1, \ldots, L-1; \tag{8.60}$$

$$I_L^m(\tau_L, +\mu_i) = I_s^m(\mu_i), \quad i = 1, \ldots, N; \tag{8.61}$$

where $I_s^m(\mu_i)$ is given by Eq. 8.51 with τ^* replaced by τ_L.

Equation 8.60 is included to ensure that the intensity is continuous across layer interfaces. Insertion of Eq. 8.58 into Eqs. 8.59–8.61 yields (omitting the m superscript)

$$\sum_{j=1}^{N} \{C_{j1} g_{j1}(-\mu_i) + C_{-j1} g_{-j1}(-\mu_i)\} = \mathcal{I}(-\mu_i) - U_1(0, -\mu_i),$$
$$i = 1, \ldots, N; \tag{8.62}$$

$$\sum_{j=1}^{N} \Big\{C_{jp} g_{jp}(\mu_i) e^{-k_{jp}\tau_p} + C_{-jp} g_{-jp}(\mu_i) e^{k_{jp}\tau_p} - \big[C_{j,p+1} g_{j,p+1}(\mu_i) e^{-k_{j,p+1}\tau_p}$$
$$+ C_{-j,p+1} g_{-j,p+1}(\mu_i) e^{k_{j,p+1}\tau_p}\big]\Big\} = U_{p+1}(\tau_p, \mu_i) - U_p(\tau_p, \mu_i),$$
$$i = \pm 1, \ldots, \pm N, \quad p = 1, \ldots, L-1; \tag{8.63}$$

$$\sum_{j=1}^{N} \Big\{C_{jL} r_j(\mu_i) g_{jL}(\mu_i) e^{-k_{jL}\tau_L} + C_{-jL} r_{-j}(\mu_i) g_{jL}(\mu_i) e^{k_{jL}\tau_L}\Big\} = \Gamma(\tau_L, \mu_i),$$
$$i = 1, \ldots, N; \tag{8.64}$$

where r_j is given by Eq. 8.55 with g_j replaced by g_{jL}, and Γ is given by Eq. 8.56 with $U^{\pm}$ replaced by $U_L^{\pm}$ and τ^* by τ_L.

Equations 8.62–8.64 constitute a $(2N \times L) \times (2N \times L)$ system of linear algebraic equations from which the $2N \times L$ unknown coefficients, the C_{jp} ($j = \pm 1, \ldots, \pm N$; $p = 1, \ldots, L$) are determined. We note that Eqs. 8.62 and 8.64 constitute the boundary conditions and are therefore identical to Eqs. 8.53 and 8.54 (again except for the layer indices). As in the one-layer case we must deal with the fact that Eqs. 8.62–8.64 are intrinsically ill-conditioned. Again, this ill-conditioning may be entirely eliminated by the scaling transformation introduced previously (Eqs. 8.57). To illustrate how this scheme works for a multilayered slab it suffices to consider two layers in the two-stream approximation (see Problem 8.5).

8.8.2 Source Functions and Angular Distributions

In a multilayered medium we may evaluate the integral in Eqs. 8.40 and 8.41 by integrating layer by layer as follows ($\tau_{p-1} \le \tau \le \tau_p$ and $\mu > 0$):

$$\int_{\tau}^{\tau_L} \frac{dt}{\mu} S^+(t,\mu) e^{-(t-\tau)/\mu} = \int_{\tau}^{\tau_p} \frac{dt}{\mu} S_p^+(t,\mu) e^{-(t-\tau)/\mu} + \sum_{n=p+1}^{L} \int_{\tau_{n-1}}^{\tau_n} \frac{dt}{\mu} S_n^+(t,\mu) e^{-(t-\tau)/\mu}, \tag{8.65}$$

$$\int_{0}^{\tau} \frac{dt}{\mu} S^-(t,\mu) e^{-(\tau-t)/\mu} = \sum_{n=1}^{p-1} \int_{\tau_{n-1}}^{\tau_n} \frac{dt}{\mu} S_n^-(t,\mu) e^{-(\tau-t)/\mu} + \int_{\tau_{p-1}}^{\tau} \frac{dt}{\mu} S_p^-(t,\mu) e^{-(\tau-t)/\mu}. \tag{8.66}$$

Using Eq. 8.43 for $S_n^{\pm}(t,\mu)$ in each layer (properly indexed) in Eqs. 8.65 and 8.66, we find

$$I_p^+(\tau,\mu) = I^+(\tau_L,\mu) e^{-(\tau_L-\tau)/\mu} + \sum_{n=p}^{L} \sum_{j=-N}^{N} C_{jn} \frac{\tilde{g}_{jn}(+\mu)}{1+k_{jn}\mu} \times \left[e^{-[k_{jn}\tau_{n-1}+(\tau_{n-1}-\tau)/\mu]} - e^{-[k_{jn}\tau_n+(\tau_n-\tau)/\mu]} \right], \tag{8.67}$$

with τ_{n-1} replaced by τ for $n = p$, and

$$I_p^-(\tau,\mu) = I^-(0,\mu) e^{-\tau/\mu} + \sum_{n=1}^{p} \sum_{j=-N}^{N} C_{jn} \frac{\tilde{g}_{jn}(-\mu)}{1-k_{jn}\mu} \times \left[e^{-[k_{jn}\tau_n+(\tau-\tau_n)/\mu]} - e^{-[k_{jn}\tau_{n-1}+(\tau-\tau_{n-1})/\mu]} \right], \tag{8.68}$$

with τ_n replaced by τ for $n = p$. It is easily verified that for a single layer ($\tau_{n-1} = \tau$, $\tau_n = \tau_L = \tau^*$ in Eq. 8.67; $\tau_n = \tau$, $\tau_{n-1} = 0$ in Eq. 8.68), Eqs. 8.67 and 8.68 reduce to Eqs. 8.46 and 8.47, as they should.

Equations 8.58, 8.67 and 8.68 contain exponentials with positive arguments, which will eventually lead to numerical overflow for large enough values of these arguments. Fortunately, we can remove all these positive arguments by introducing the scaling transformation into our solutions. Since only the homogeneous solution is affected, it suffices to substitute Eqs. 8.57 into the homogeneous version of Eq. 8.58, ignoring the particular solution $U_p^{\pm}(\tau, \mu_i)$, that is,

$$I_p^{\pm}(\tau, \mu_i) = \sum_{j=1}^{N} \{C'_{jp} g'_{jp}(\pm\mu_i) e^{-k_{jp}(\tau - \tau_{p-1})} + C'_{-jp} g_{-jp}(\pm\mu_i) e^{-k_{jp}(\tau_p - \tau)}\}. \tag{8.69}$$

Since $k_{jp} > 0$ and $\tau_{p-1} \leq \tau \leq \tau_p$, all exponentials in Eq. 8.69 have negative arguments, as they should to avoid overflow in the numerical computations.

Introducing the scaling into Eqs. 8.67 and 8.68 is straightforward but leads to somewhat cumbersome expressions.[5]

8.8.3 Numerical Implementation of the Discrete-Ordinate Method

The solution of the radiative transfer equation described in previous sections has been implemented numerically into a code written in FORTRAN. This code applies to vertically inhomogeneous, nonisothermal, plane-parallel media and it includes all the physical processes discussed previously, namely thermal emission, scattering, absorption, bidirectional reflection, and thermal emission at the lower boundary. The medium may be forced at the top boundary by direct (collimated) or diffuse illumination and by internal and boundary sources as well.[6]

As discussed in §6.8 for strongly forward-peaked scattering, it is difficult to obtain accurate solutions to the radiative transfer equation. The δ-N method, which replaces the forward-scattering peak of the phase function by a δ-function (see §6.8), is useful and improves the accuracy significantly, especially for fluxes and mean intensity. The intensity computation is also generally improved by using δ-N but further improvements are desirable and essential if one desires to use low-order discrete-ordinate approximations (say $N < 10$) to reduce the computational burden. Special algorithms have been invented to correct the intensity computation for strongly forward-peaked scattering. The development of such algorithms starts with the notion that the single-scattering solution can be computed exactly and used to improve the accuracy. Such an algorithm, described in §8.9, is implemented in the DISORT code and used in conjunction with the δ-N method to provide acceptable accuracy for as little as 10 streams. Without these algorithms similar accuracy will typically require a quintupling of the number of streams, which implies that they provide computational savings of the order of $5^3 = 125$ since the most time-consuming computation in DISORT is the solution of the algebraic eigenvalue problem in which the computation time varies as the cube of half the number of streams (i.e., N^3).

In §6.11 we derived simple expressions for the flux reflectance and transmittance for media without internal sources and showed that an analytic correction allows us to find the solutions pertaining to reflecting (Lambert) surfaces. These expressions, which are implemented in DISORT, offer substantial computational advantages when only integrated quantities such as flux reflectance and transmittance are required.

Example 8.7 Intensities Computed with and without δ-N Scaling

To illustrate the merit of invoking the δ-N scaling of the phase function, we show in Fig. 8.4 plots of intensity versus cosine of the polar angle for three different optical depths and for azimuthal angles $\Delta\phi = 0, 90, 180°$. The solid curve is for 48 streams ($2N = 48$) and the dashed and dotted–dashed curves are for 16 streams with and without δ-N scaling, respectively. For the azimuthal plane of the Sun ($\Delta\phi = 0$) the largest deviation between the 16-stream and the accurate 48-stream results occurs in the forward direction (i.e., for $\mu \approx \mu_0$). The 16-stream approximation underestimates the intensity near the forward direction, but the unscaled intensity deviates less

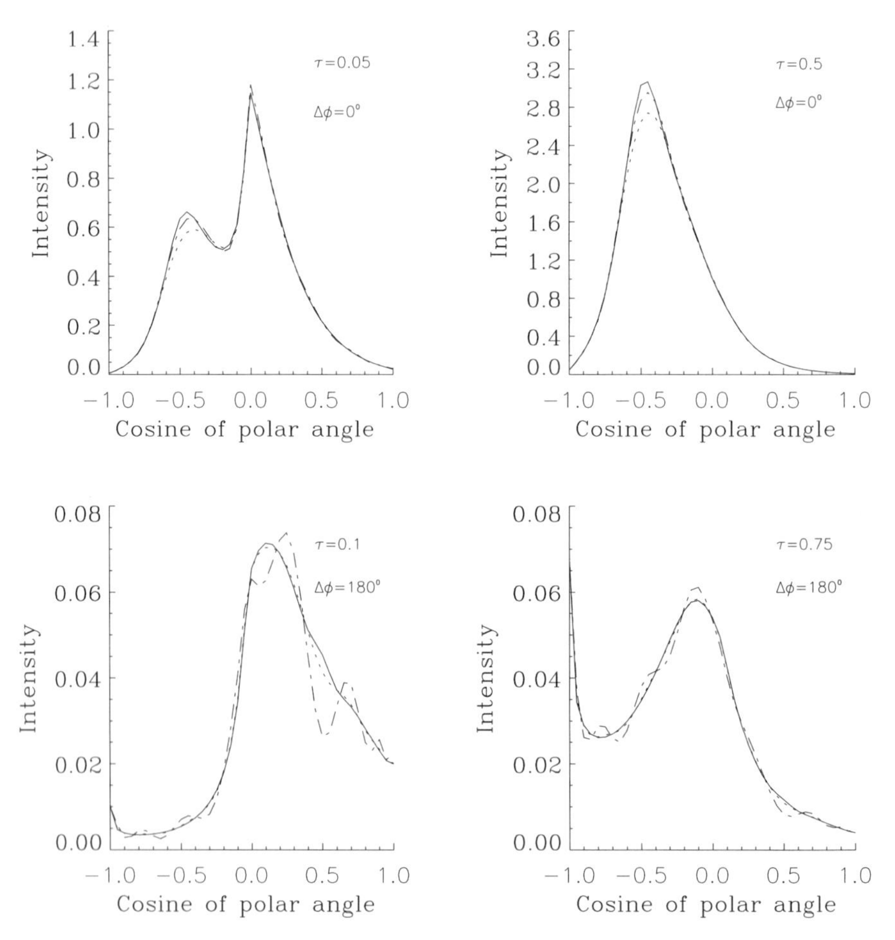

Figure 8.4 Comparison of accurate (48-stream) and approximate (16-stream) diffuse intensities computed with and without δ-N scaling at several optical depths within an aerosol layer of total optical thickness $\tau_N = 1$ for $\Delta\phi = 0$, 90, and 180°; $a = 0.9$, and $\mu_0 = -0.5$. Note that the ordinate scale is not the same in the various diagrams.

than the result obtained by invoking δ-N scaling. This is expected because of the truncation of the forward-scattering peak of the phase function incurred in the scaling. We note, however, that for angles not close to the forward direction better results are obtained with scaling than without. This is particularly noticeable for $\Delta\phi = 180°$, where the unscaled 16-stream results exhibit an oscillatory behavior with appreciable deviations, whereas the scaled computations yield quite satisfactory results. Below we shall discuss procedures invented to improve the speed and accuracy of the intensity computation.

8.9 Correction of the Truncated Intensity Field

The selection of the truncation, or, in other words the choice of the number of streams used to compute the intensity field, has been a vexing problem. As we have pointed out previously the difficulty stems from the strong forward peaking of the phase function encountered whenever the scattering "particle" is large compared to the wavelength of the incident radiation. In Chapter 6 we showed that a large number of terms is required to obtain an accurate representation of such phase functions when expanded in Legendre polynomials. As a rule of thumb it is advisable to use a similar number of streams in the discretization of the integro-differential equation to turn it into a set of ordinary differential equations for which analytic solutions are feasible. Thus we need to adopt a large value of N, say several hundred, in any discretization scheme to fully resolve the scattering pattern. This consumes enormous amounts of computing resources and makes routine usage impossible. Therefore, truncation procedures that provide accurate intensities when N is kept reasonably small become highly desirable.

8.9.1 The Nakajima–Tanaka Correction Procedure

The δ-N transformation of the phase function has proven to be a most reliable means for truncation in flux computations. As we mentioned in §6.8, application of the δ-N will artificially enhance the direct flux component at the expense of the diffuse component, but the sum will be computed accurately. Since the sum must be independent of the δ-N scaling, it is trivial to "unscale" the diffuse flux since the true direct component is easily obtained (cf. Eq. 6.61). This unscaling of the diffuse flux is done in DISORT. As shown in Fig. 8.4 the accuracy of the intensity computation is generally improved by the use of δ-N except in the forward direction. This example pertains to a phase function for aerosols that is far from extreme (see upper left panel of Fig. 6.3). For more strongly forward-peaked phase functions (e.g., those associated with hydrosols or clouds illustrated in Fig. 6.3) the error incurred by straightforward application of the δ-N method becomes unacceptable for practical purposes (i.e., small N), especially in the region around the Sun's disk referred to as the solar aureole. Fortunately, it can be shown[7] that by combining the δ-N method with exact computation of low orders

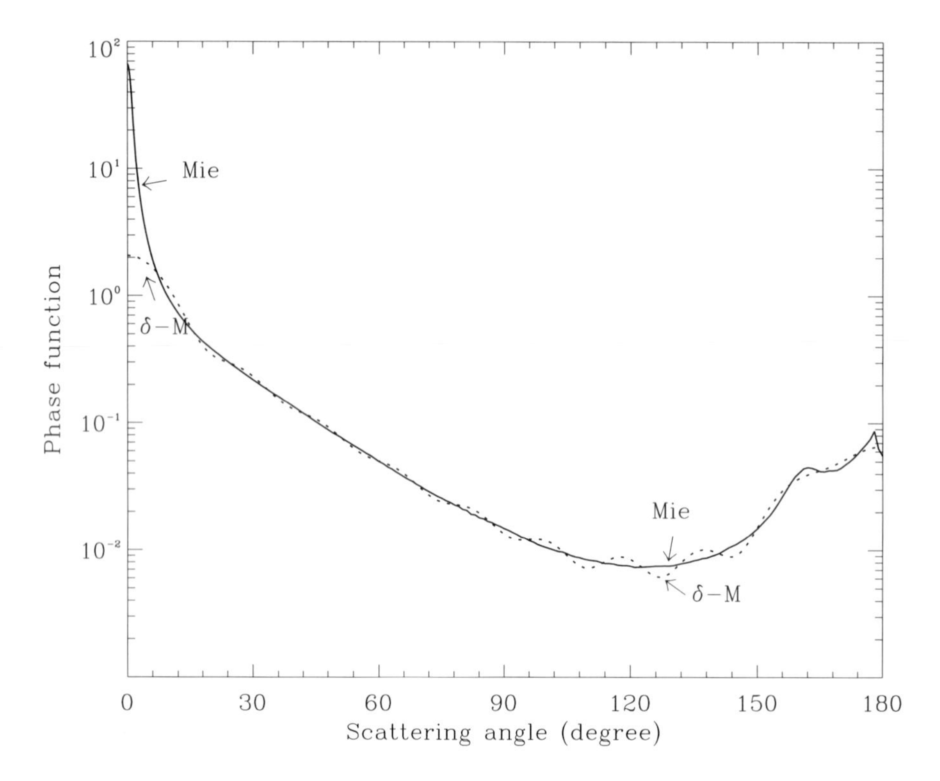

Figure 8.5 Scattering phase function computed by Mie theory and the δ-N representation for $N = 10$.

of scattering (notably first- and second-order scattering), the error in the intensity computation can be reduced to less than 1%.

Application of the δ-N transformation results in a truncated phase function that oscillates systematically around the original phase function as illustrated in Fig. 8.5. The magnitude of the oscillation depends on the size of the truncation, f, which in δ-N is set equal to χ_{2N}, where $2N$ is the number of terms used in the expansion of the phase function in Legendre polynomials. The χ_l are the expansion coefficients (see §6.8). The intensity obtained by the application of the δ-N transformation also oscillates around the exact value. This suggests that the singly scattered intensity, which resembles the scattering phase function, contributes most to the fluctuation. Furthermore, it is expected and well established that multiple scattering tends to produce an isotropic intensity distribution. Thus, we should first construct the exact solution for the singly scattered intensity and subsequently apply correction procedures to account for errors introduced by the truncation as described below.

In §7.2.1 we showed that it is easy to derive exact solutions for the singly scattered intensity field. The expressions derived there for a single homogeneous slab may easily be extended to apply to a multilayered medium. In fact, by ignoring the multiple-scattering term in Eqs. 6.3 and 6.4, we may easily arrive at simple analytic expressions for the half-range intensity fields. Thus, for the upward intensity we find (assuming there is no radiation emanating from the lower boundary) the following expression for a multilayered slab consisting of $L - 1$ layers as shown in Fig. 8.3:

$$I^+(\tau, \mu, \phi) = \frac{F^s}{4\pi(1 + \mu/\mu_0)} \sum_{n=p}^{L} a_n p_n(-\mu_0, \phi_0; +\mu, \phi) \cdot \left[e^{-(\tau_{n-1} - \tau/\mu + \tau_{n-1}/\mu_0)} - e^{-(\tau_n - \tau/\mu + \tau_n/\mu_0)}\right], \tag{8.70}$$

with τ_{n-1} replaced by τ for $n = p$. We note that since $\tau_L > \cdots > \tau_n > \tau_{n-1} \geq \tau$, all the exponentials in Eq. 8.70 have negative arguments. Similarly, for the downward intensity we find on setting $I(\tau = 0) = 0$

$$I^-(\tau, \mu, \phi) = \frac{F^s}{4\pi(1 - \mu/\mu_0)} \sum_{n=1}^{p} a_n p_n(-\mu_0, \phi_0; -\mu, \phi) \cdot \left[e^{-(\tau - \tau_n/\mu + \tau_n/\mu_0)} - e^{-(\tau - \tau_{n-1}/\mu + \tau_{n-1}/\mu_0)}\right], \tag{8.71}$$

with τ_n replaced by τ for $n = p$. Again, since $\tau > \cdots > \tau_n > \tau_{n-1} > 0$, all the exponentials in Eq. 8.71 have negative arguments, as they should to avoid computer overflow problems.

To improve the accuracy of the intensity computation resulting from the straightforward application of the δ-N method we may add a correction term obtained from single-scattering computations. For example, if the single-scattering computation is done for the original unscaled optical depth and single-scattering albedo, the correction is simply taken to be the difference between the singly scattered intensity computed from the exact phase function and that from the δ-N truncated phase function (which is what remains after the forward-scattering peak is removed). The improvement obtained by application of this correction is shown in Fig. 8.6. Clearly, this procedure acts to suppress the fluctuation of the intensity field outside of the solar aureole region.

Since this correction does not completely remove the error, a refinement is necessary. Thus, we may consider the possible improvement obtained by including the effect of the τ-scaling in the single-scattering computation: We subtract the single-scattering contribution resulting from the δ-N method and add the single-scattering contribution obtained by using the exact phase function and the unscaled single-scattering albedo, but with the δ-N scaled optical depth. This leads to a substantial improvement except in the forward direction where significant error remains. This remaining error in the solar aureole region is caused by secondary and higher order scattering within the forward-scattering peak. A further improvement is obtained by accounting for secondary scattering effects to correct the transmitted intensity in the solar aureole region. The algebra involved in this correction is somewhat cumbersome and will not be reproduced here. We note, however, that except for the solar aureole region a simple single-scattering correction of the δ-N results is sufficient to obtain the accuracy required for most practical purposes.

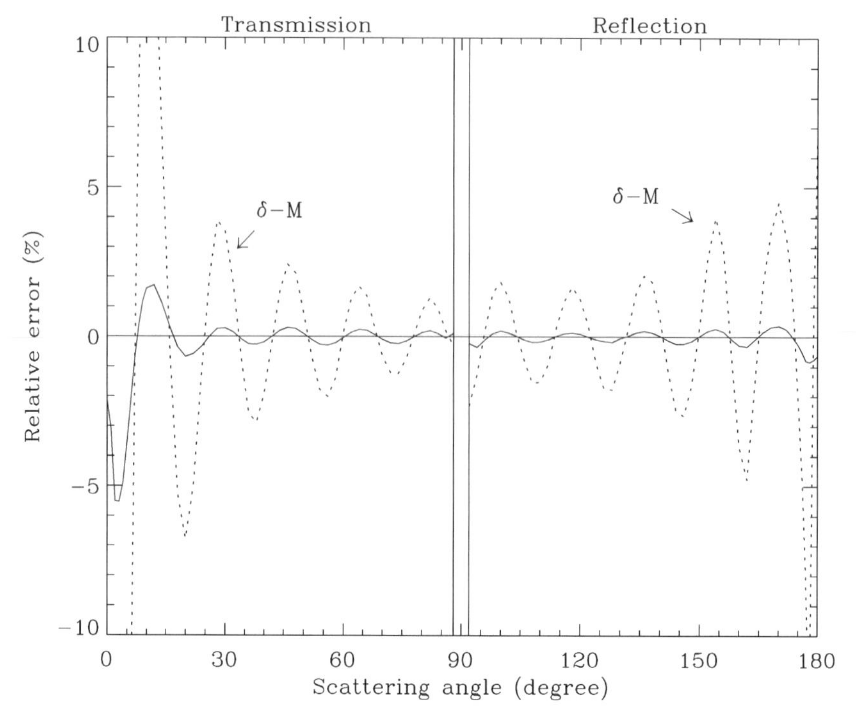

Figure 8.6 Relative error of the reflected and transmitted intensities computed by strict application of δ-N and by applying a correction to the δ-N method (solid line), which is simply the difference between the singly scattered intensity computed from the exact phase function and that from the δ-N-scaled phase function. This example pertains to vertical (collimated) illumination of a homogeneous slab of total optical thickness 0.8 consisting of particles with scattering properties defined in the previous figure.

8.9.2 Computed Intensity Distributions for the Standard Problem

To demonstrate the use of the solutions presented above we shall use *Prototype Problem 3* as an example. Thus, we consider a direct (collimated) beam of light incident on a slab of particles that may scatter and absorb radiation. We assume a beam of radiation incident in direction $\theta_0 = \cos^{-1}\mu_0 = 60°$ so that $\mu_0 F^s = \mu_0(F^s = 1) = 0.5$ on a slab of optical thickness $\tau^* = \tau_L = 1.0$. Our first example is for partly absorbing aerosol particles with single-scattering albedo $a = 0.9$ and phase function labeled "Aerosol" (illustrated in Fig. 6.3). The intensity as a function of polar and azimuthal angles for several optical depths ranging from $\tau = 0$ (top) to $\tau^* = 1$ (bottom) is shown in Fig. 8.7. The most striking feature observed in these plots is the rapid decrease in intensity with increasing azimuthal angle. We note also that the reflected intensity exhibits limb brightening, while for optical depths ≤ 0.2 two distinct maxima occur, one close to $\theta = 90°$ and another one close to the direction of incidence $\theta \approx \theta_0 = 60°$. For optical depths >0.2 the two peaks combine to yield a single maximum in the forward

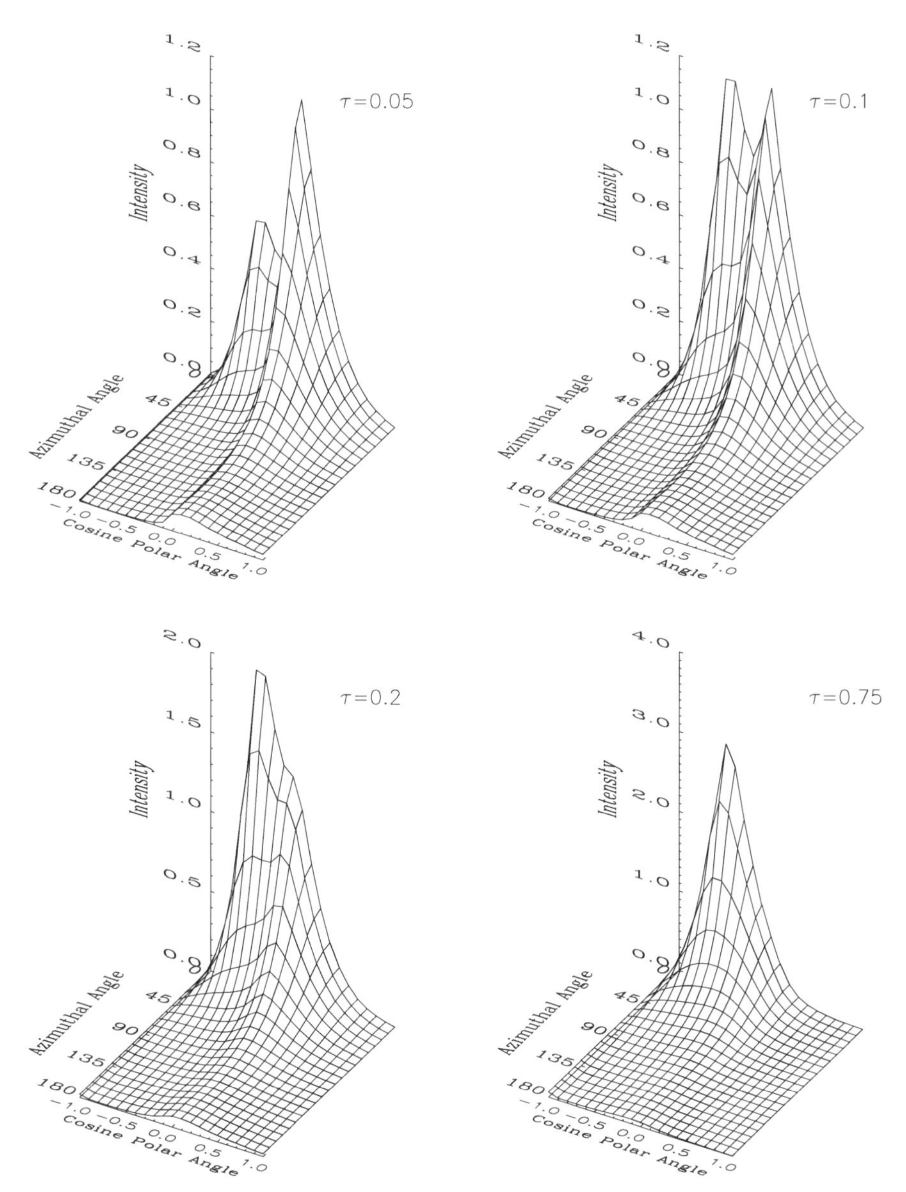

Figure 8.7 Three-dimensional display of diffuse intensity versus polar and azimuthal angles for several optical depths within a layer consisting of aerosol particles of optical thickness $\tau^* = 1$, single-scattering albedo $a = 0.9$, and cosine of solar zenith angle $\mu_0 = 0.5$.

direction (i.e., $\mu \approx \mu_0$). The magnitude of the peak intensity generally increases and the peak becomes narrower with optical depth. This behavior is a direct consequence of the forward-peaked aerosol phase function.

In our second example we use the phase function for hydrosols, which is even more strongly peaked than the aerosol phase function, as is evident from Fig. 6.3. As a result

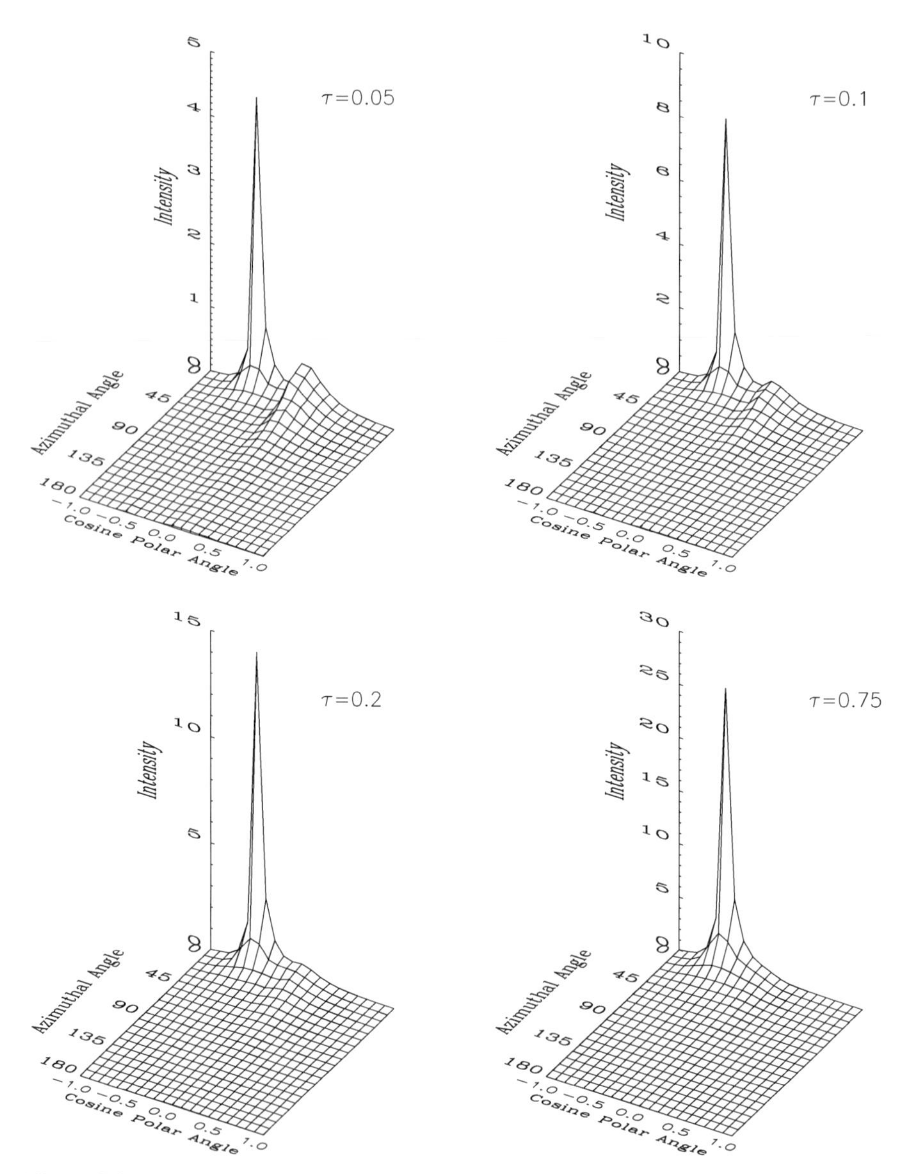

Figure 8.8 Three-dimensional display of diffuse intensity versus polar and azimuthal angles for several optical depths within a layer consisting of oceanic particles (hydrosols) of optical thickness $\tau^* = 1$, single-scattering albedo $a = 0.9$, and cosine of solar zenith angle $\mu_0 = 0.5$.

the intensity pattern becomes more extreme than in the previous case (see Fig. 8.8). As the beam penetrates into the medium the intensity is small everywhere except for the region around the Sun, where the scattered photon intensity becomes increasingly dominant with penetration depth due to forward scattering.

We repeated the calculation above and kept everything the same except that we replaced the phase function for hydrosols with the water cloud phase function shown in Fig. 6.3. The intensity pattern (not shown) is very similar to that for hydrosols, which is understandable because the common feature of the two phase function is the strong forward peaking, which tends to dominate the intensity pattern.

8.10 The Coupled Atmosphere–Ocean Problem

In the previous section we discussed how to deal with changes in optical properties in a vertically inhomogeneous slab. We did not, however, consider any changes in the index of refraction such as that occurring at the interface between the atmosphere and the ocean. In §6.6 we described how to deal with two strata with different indices of refraction. Below we shall describe how to solve the pertinent equations in the discrete-ordinate approximation.[8] We consider an interface that is perfectly flat and exhibits only specular reflection.

8.10.1 Discretized Equations for the Atmosphere–Ocean System

The atmosphere and the ocean may each be divided into a sufficient number of layers to resolve their optical properties as described in the previous section. We start as usual by replacing the integral term in Eqs. 6.20–6.21 by a quadrature sum. Thus, for each layer in the atmosphere, we obtain

$$\mu_i^{\rm a}\frac{dI^+(\tau,\mu_i^{\rm a})}{d\tau} = I^+(\tau,\mu_i^{\rm a}) - \frac{a}{2}\sum_{j=1}^{N_1} w_j^{\rm a} p(\mu_j^{\rm a},\mu_i^{\rm a}) I^+(\tau,\mu_j^{\rm a}) - \frac{a}{2}\sum_{j=1}^{N_1} w_j^{\rm a} p(-\mu_j^{\rm a},\mu_i^{\rm a}) I^-(\tau,\mu_j^{\rm a}) - S_{\rm air}^*(\tau,\mu_i^{\rm a}), \tag{8.72}$$

$$-\mu_i^{\rm a}\frac{dI^-(\tau,\mu_i^{\rm a})}{d\tau} = I^-(\tau,\mu_i^{\rm a}) - \frac{a}{2}\sum_{j=1}^{N_1} w_j^{\rm a} p(\mu_j^{\rm a},-\mu_i^{\rm a}) I^+(\tau,\mu_j^{\rm a}) - \frac{a}{2}\sum_{j=1}^{N_1} w_j^{\rm a} p(-\mu_j^{\rm a},-\mu_i^{\rm a}) I^-(\tau,\mu_j^{\rm a}) - S_{\rm air}^*(\tau,-\mu_i^{\rm a}). \tag{8.73}$$

Similarly, for layers in the ocean we find

$$\mu_i^{\rm o}\frac{dI^+(\tau,\mu_i^{\rm o})}{d\tau} = I^+(\tau,\mu_i^{\rm o}) - \frac{a}{2}\sum_{j=1}^{N_2} w_j^{\rm o} p(\mu_j^{\rm o},\mu_i^{\rm o}) I^+(\tau,\mu_j^{\rm o}) - \frac{a}{2}\sum_{j=1}^{N_2} w_j^{\rm o} p(-\mu_j^{\rm o},\mu_i^{\rm o}) I^-(\tau,\mu_j^{\rm o}) - S_{\rm ocn}^*(\tau,\mu_i^{\rm o}), \tag{8.74}$$

$$-\mu_i^{\mathrm{o}} \frac{dI^-(\tau,\mu_i^{\mathrm{o}})}{d\tau} = I^-(\tau,\mu_i^{\mathrm{o}}) - \frac{a}{2}\sum_{j=1}^{N_2} w_j^{\mathrm{o}} p(\mu_j^{\mathrm{o}}, -\mu_i^{\mathrm{o}}) I^+(\tau,\mu_j^{\mathrm{o}})$$

$$- \frac{a}{2}\sum_{j=1}^{N_2} w_j^{\mathrm{o}} p(-\mu_j^{\mathrm{o}}, -\mu_i^{\mathrm{o}}) I^-(\tau,\mu_j^{\mathrm{o}}) - S_{\mathrm{air}}^*(\tau, -\mu_i^{\mathrm{o}}). \tag{8.75}$$

Here μ_i^{a}, w_i^{a} and μ_i^{o}, w_i^{o} are quadrature points and weights for atmosphere and ocean, respectively, and $\mu_{-i} = -\mu_i$, $w_{-i} = w_i$ (Fig. 8.9). Note that we have used different numbers of streams for the atmosphere ($2N_1$) and the ocean ($2N_2$). In the refractive region of the ocean (region II), which communicates directly with the atmosphere, we use the same number of streams ($2N_1$) as in the atmosphere. This will properly account for the shrinking caused by refraction of the angular domain in the ocean. In region I of the ocean, total reflection occurs at the ocean–atmosphere interface for photons moving in the upward direction. In this region we invoke additional streams ($2N_2 - 2N_1$) to accomodate the scattering interaction between regions I and II in the ocean.

8.10.2 Quadrature and General Solution

We use the double-Gauss rule to determine quadrature points and weights μ_i^{a} and w_i^{a} ($i = 1, \ldots, N_1$) in the atmosphere, as well as the quadrature points and weights μ_i^{o} and w_i^{o} ($i = N_1 + 1, \ldots, N_2$) in the total reflection region of the ocean. The quadrature points in the refractive region of the ocean are obtained by simply "refracting" the downward streams in the atmosphere, $[\mu_1^{\mathrm{a}}, \ldots, \mu_{N_1}^{\mathrm{a}}]$, into the ocean as shown schematically in Fig. 8.9. Thus, in this region, μ_i^{o} is related to μ_i^{a} by Snell's Law

$$\mu_i^{\mathrm{o}} = S(\mu_i^{\mathrm{a}}) = \sqrt{1 - \left[1 - (\mu_i^{\mathrm{a}})^2\right]/m_{\mathrm{rel}}^2},$$

and from this relation the weights for this region are derived as

$$w_i^{\mathrm{o}} = w_i^{\mathrm{a}} \left[\frac{dS(\mu^{\mathrm{a}})}{d\mu^{\mathrm{a}}}\right]_{\mu^{\mathrm{a}}=\mu_i^{\mathrm{a}}} = \frac{\mu_i^{\mathrm{a}}}{m_{\mathrm{rel}}^2 S(\mu_i^{\mathrm{a}})} w_i^{\mathrm{a}}, \quad i = 1, \ldots, N_1.$$

The advantage of this choice of quadrature is that the points are clustered toward $\mu = 0$ both in the atmosphere and the ocean and, in addition, toward the critical angle direction in the ocean. This clustering will give superior results near these directions where the intensities vary rapidly. Also, the phase function is still correctly normalized.

The solution of the homogeneous version of Eqs. 8.72–8.75 was presented in §8.5. Following the same procedure, we may write the homogeneous solution as follows. In the atmosphere it is (see Eq. 8.39)

$$I^{\pm}(\tau,\mu_i^{\mathrm{a}}) = \sum_{j=1}^{N_1}\left[C_{-j} g_{-j}(\pm\mu_i^{\mathrm{a}}) e^{k_j^{\mathrm{a}}\tau} + C_j g_j(\pm\mu_i^{\mathrm{a}}) e^{-k_j^{\mathrm{a}}\tau}\right], \quad i = 1, \ldots, N_1,$$

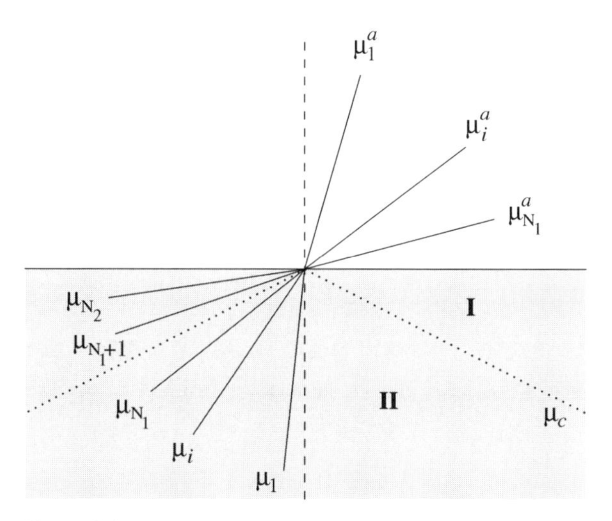

Figure 8.9 Schematic illustration of the quadrature adopted for the coupled atmosphere–ocean system. The dotted line labeled μ_c marks the separation between region II, in which photons can communicate directly with the atmosphere, and the total reflection region (I). The quadrature angles in region II are obtained from those in the atmosphere by using Snell's law. Additional quadrature points are then added to represent region I as indicated.

and similarly, in the ocean

$$I^{\pm}(\tau, \mu_i^{\mathrm{o}}) = \sum_{j=1}^{N_2} \left[C_{-j} g_{-j}(\pm\mu_i^{\mathrm{o}}) e^{k_j^{\mathrm{o}}\tau} + C_j g_j(\pm\mu_i^{\mathrm{o}}) e^{-k_j^{\mathrm{o}}\tau} \right], \quad i = 1, \ldots, N_2,$$

where the k_j and g_j are eigenvalues and eigenvectors determined by solving an algebraic eigenvalue problem, and the C_j are unknown constants of integration to be determined by the application of boundary and continuity conditions as discussed below.

The particular solution in the atmosphere can be expressed as

$$U^{\pm}(\tau, \mu_i^{\mathrm{a}}) = Z_0(\pm\mu_i^{\mathrm{a}}) e^{-\tau/\mu_0} + Z_{01}(\pm\mu_i^{\mathrm{a}}) e^{\tau/\mu_0},$$

where $i = 1, \ldots, N_1$, and the coefficients $Z_0(\pm\mu_i^{\mathrm{a}})$ and $Z_{01}(\pm\mu_i^{\mathrm{a}})$ are determined by the following system of linear algebraic equations (see Eq. 8.36):

$$\sum_{\substack{j=-N_1 \\ j\neq 0}}^{N_1} \left[\left(1 + \frac{u_j^{\mathrm{a}}}{\mu_0}\right) \delta_{ij} - w_j^{\mathrm{a}} \frac{a}{2} p(u_i^{\mathrm{a}}, u_j^{\mathrm{a}}) \right] Z_0(u_j^{\mathrm{a}}) = X_0(u_i^{\mathrm{a}}),$$

$$\sum_{\substack{j=-N_1 \\ j\neq 0}}^{N_1} \left[\left(1 - \frac{u_j^{\mathrm{a}}}{\mu_0}\right) \delta_{ij} - w_j^{\mathrm{a}} \frac{a}{2} p(u_i^{\mathrm{a}}, u_j^{\mathrm{a}}) \right] Z_{01}(u_j^{\mathrm{a}}) = X_{01}(u_i^{\mathrm{a}}).$$

Similarly, in the ocean the particular solution can be expressed as

$$U^{\pm}(\tau, \mu_i^{\rm o}) = Z_{02}(\pm\mu_i^{\rm o}) e^{-\tau/\mu_{0m}},$$

where $i = 1, \ldots, N_2$ and $Z_{02}(\pm\mu_i^{\rm o})$ is determined by the following system of linear algebraic equations:

$$\sum_{\substack{j=-N_2 \\ j\neq 0}}^{N_2} \left[\left(1 + \frac{u_j^{\rm o}}{\mu_{0m}}\right)\delta_{ij} - w_j^{\rm o}\frac{a}{2} p(u_i^{\rm o}, u_j^{\rm o})\right] Z_{02}(u_j^{\rm o}) = X_{02}(u_i^{\rm o}).$$

As usual the general solution is just the sum of the homogeneous and particular solutions.

8.10.3 Boundary, Continuity, and Atmosphere–Ocean Interface Conditions

The vertically inhomogeneous medium is represented by multiple adjacent homogeneous layers in the atmosphere and the ocean, respectively. The solutions derived previously pertain in each layer. We assume that the system consists of L_1 layers of atmosphere and L_2 layers of ocean. Then we may write the solution for the pth layer as (cf. Eq. 8.58)

$$I_p^{\pm}(\tau, \mu_i^{\rm a}) = \sum_{j=1}^{N_1} \left[C_{-jp} g_{-jp}(\pm\mu_i^{\rm a}) e^{k_{jp}^{\rm a}\tau} + C_{jp} g_{jp}(\pm\mu_i^{\rm a}) e^{-k_{jp}^{\rm a}\tau}\right] + U_p^{\pm}(\tau, \mu_i^{\rm a}), \quad i = 1, \ldots, N_1 \text{ and } p \leq L_1, \tag{8.76}$$

$$I_p^{\pm}(\tau, \mu_i^{\rm o}) = \sum_{j=1}^{N_2} \left[C_{-jp} g_{-jp}(\pm\mu_i^{\rm o}) e^{k_{jp}^{\rm o}\tau} + C_{jp} g_{jp}(\pm\mu_i^{\rm o}) e^{-k_{jp}^{\rm o}\tau}\right] + U_p^{\pm}(\tau, \mu_i^{\rm o}), \quad i = 1, \ldots, N_2 \text{ and } L_1 < p \leq L_2. \tag{8.77}$$

In total there are $(2N_1 \times L_1) + (2N_2 \times L_2)$ unknown coefficients C_{jp} in Eqs. 8.76 and 8.77. They will be determined by (i) the boundary conditions to be applied at the top of the atmosphere and the bottom of the ocean, (ii) the continuity conditions at each interface between layers in the atmosphere and ocean, and finally (iii) the reflection and refraction occurring at the atmosphere–ocean interface where we require Fresnel's equations to be satisfied. These conditions are implemented as follows.

At the top, we require

$$I_1(0, -\mu_i^{\rm a}) = \mathcal{I}(-\mu_i^{\rm a}), \quad i = 1, \ldots, N_1, \tag{8.78}$$

at the interfaces between atmospheric layers,

$$I_p(\tau_p, \mu_i^{\rm a}) = I_{p+1}(\tau_p, \mu_i^{\rm a}), \quad i = 1, \ldots, N_1 \text{ and } p = 1, \ldots, L_1 - 1, \tag{8.79}$$

at the interface between the atmosphere and the ocean (cf. Eqs. 6.43–6.44),

$$I_{L_1}(\tau_a, \mu_i^a) = I_{L_1}(\tau_a, -\mu_i^a)\rho_s(-\mu_i^a, m_{rel}) + \left[\frac{I_{L_1+1}(\tau_a, \mu_i^o)}{m_{rel}^2}\right]\mathcal{T}_s(+\mu_i^o, m_{rel}), \quad i = 1, \ldots, N_1, \tag{8.80}$$

$$\frac{I_{L_1+1}(\tau_a, -\mu_i^o)}{m_{rel}^2} = \left[\frac{I_{L_1+1}(\tau_a, \mu_i^o)}{m_{rel}^2}\right]\rho_s(+\mu_i^o, m_{rel}) + I_{L_1}(\tau_a, -\mu_i^a)\mathcal{T}_s(-\mu_i^a, m_{rel}), \quad i = 1, \ldots, N_1, \tag{8.81}$$

$$I_{L_1+1}(\tau_a, -\mu_i^o) = I_{L_1+1}(\tau_a, \mu_i^o), \quad i = N_1 + 1, \ldots, N_2, \tag{8.82}$$

at the interfaces between ocean layers,

$$I_p(\tau_p, \mu_i^o) = I_{p+1}(\tau_p, \mu_i^o), \quad i = 1, \ldots, N_2 \quad \text{and} \quad p = L_1 + 1, \ldots, L_1 + L_2 - 1, \tag{8.83}$$

and finally at the bottom boundary,

$$I_{L_1+L_2}(\tau^*, \mu_i^o) = I_g(\mu_i^o), \quad i = 1, \ldots, N_2. \tag{8.84}$$

We defined $\rho_s(\pm\mu_i, m_{rel})$ and $\mathcal{T}_s(\pm\mu_i, m_{rel})$ as the specular reflectance and transmittance of the invariant intensity, I/m^2 (where m is the index of refraction at the location where I is measured). The minus sign applies for the downward intensity, and the positive for the upward intensity. Formulas for ρ_s and $\mathcal{T}_s$ can be derived from the basic Fresnel equations. The results are (see Appendix E)

$$\rho_s(-\mu_i^a, m_{rel}) = \frac{1}{2}\left[\left(\frac{\mu_i^a - m_{rel}\mu_i^o}{\mu_i^a + m_{rel}\mu_i^o}\right)^2 + \left(\frac{\mu_i^o - m_{rel}\mu_i^a}{\mu_i^o + m_{rel}\mu_i^a}\right)^2\right],$$

$$\rho_s(+\mu_i^o, m_{rel}) = \rho_s(-\mu_i^a, m_{rel}),$$

$$\mathcal{T}_s(-\mu_i^a, m_{rel}) = 2m_{rel}\mu_i^a\mu_i^o\left[\left(\frac{1}{\mu_i^a + m_{rel}\mu_i^o}\right)^2 + \left(\frac{1}{\mu_i^o + m_{rel}\mu_i^a}\right)^2\right],$$

$$\mathcal{T}_s(-\mu_i^o, m_{rel}) = \mathcal{T}_s(-\mu_i^a, m_{rel}).$$

Equations 8.80 and 8.81 ensure that, by satisfying Fresnel's equations, the radiation fields in the atmosphere and the ocean are properly coupled through the interface, whereas Eq. 8.82 represents the total reflection in region I of the ocean. The total optical depth of the entire medium (atmosphere and ocean) is denoted by τ^* in Eq. 8.84. $\mathcal{I}(-\mu_i^a)$ is the intensity incident at the top of the atmosphere, and $I_g(\mu_i^o)$ is determined by the bidirectional reflectance of the underlying surface at the bottom of the ocean. Substitution of Eqs. 8.76 and 8.77 into Eqs. 8.78–8.84 leads to a system of $(2N_1 \times L_1) + (2N_2 \times L_2)$ linear algebraic equations for the same number of unknown coefficients, the $C_{\pm jp}$. Matrix inversion of this system of equations yields the desired coefficients and thereby completes the solution for the coupled atmosphere–ocean system. Once we have obtained the solutions for all Fourier components using Eqs. 8.76 and 8.77, we may compute the intensity at the quadrature directions from Eq. 6.34. Fluxes and mean intensity can now be easily computed from the zeroth-order Fourier component of the intensity given above by using the quadrature to convert the appropriate integrals into summations, as usual.

Removal of ill-conditioning is just as important in the coupled atmosphere–ocean problem (two strata with different indices of refraction) as in the one-stratum case. How to accomplish this removal is addressed in Problem 8.7 for the case of two streams in the atmosphere and four streams in the ocean.

Example 8.8 Computational Illustration

Here we shall provide a sample application of this model to a simplified atmosphere–ocean system consisting of a clear model atmosphere (only molecular scattering) and pure seawater. Therefore, the Rayleigh-scattering phase function is assumed to apply both in the atmosphere and the ocean. Due to the nonconstant mixing ratio of absorbing gases (notably ozone in the Chappuis band for the wavelength considered here) with height, a multilayered approach is required to account properly for the change in optical properties with altitude in the atmosphere, whereas the optical properties for pure seawater are constant with depth in the ocean if the effects of ocean impurities (particles) are ignored. Accordingly, a midlatitude model atmosphere divided into 24 layers is used in conjunction with a one-layer ocean, which is sufficient in view of the assumed homogeneity of pure seawater. Since the absorbed energy at any level is proportional to the mean intensity there (cf. the definition of heating rate in Chapter 5), we show in Fig. 8.10 the mean intensity (total scalar irradiance$/4\pi$ in ocean terminology) as a function of height in the atmosphere and depth in the ocean. The same results obtained by ignoring refraction are also shown as well as the relative error, which may be as large as 20% just below the ocean surface. Although the error is large in the deep ocean, the radiation field is already significantly attenuated there.

The azimuthally averaged intensity distribution just above and just below the ocean surface is shown in Fig. 8.11. Again, results obtained by ignoring refraction are also shown. We note that the change in refraction between the atmosphere and the ocean significantly alters the radiative intensity distribution. Just below the ocean surface, the downward intensity discontinuity position shifts from the horizontal direction for the case of no change in refraction to the critical angle direction when refraction is included. The refraction also significantly changes the upward radiation field just above the ocean surface. Knowledge of the intensity distribution here is important for correct interpretations of intensity measurements in remote sensing applications.

8.11 The Doubling-Adding and the Matrix Operator Methods

We shall now discuss a method that has been widely used to solve radiative transfer problems in planetary atmospheres. In this method *doubling* refers to how one finds the reflection and transmission matrices of two layers with *identical* optical properties from those of the individual layers, while *adding* refers to the combination of two or more layers with *different* optical properties.

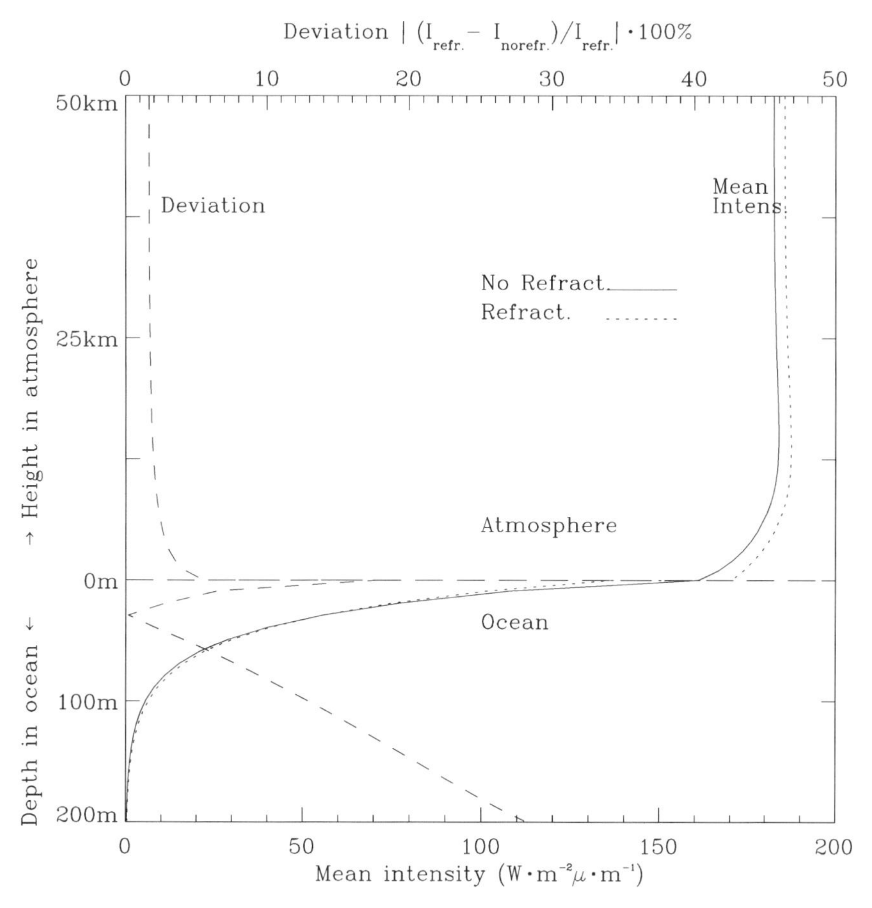

Figure 8.10 Distribution of the total mean intensity (total scalar irradiance/4π in ocean terminology) with height in the atmosphere and depth in the ocean. The result of neglecting the refraction occurring at the atmosphere–ocean interface is also shown. The computation was done for a solar zenith angle of 30° and wavelength of 500 nm.

The doubling concept is rather old and seems to have originated with Stokes. It was rediscovered and put to practical use in atmospheric science by van de Hulst and others.[9] The doubling method as commonly practiced today uses the known reflection and transmission properties of a single homogeneous layer to derive the resulting properties of two identical layers. To start the doubling procedure the initial layer is frequently taken to be thin enough that its reflection and transmission properties can be computed from single scattering. Repeated "doublings" are then applied to reach the desired optical thickness. The division of an inhomogeneous slab into a series of adjacent sublayers, each of which is homogeneous, but in principle different from all the others, is usually taken to be identical to that discussed previously for the discrete-ordinate method. The solution proceeds by first applying doubling to find the reflection and transmission matrices for each of the homogeneous layers, whereupon adding is subsequently used to find the solution for all the different layers combined.

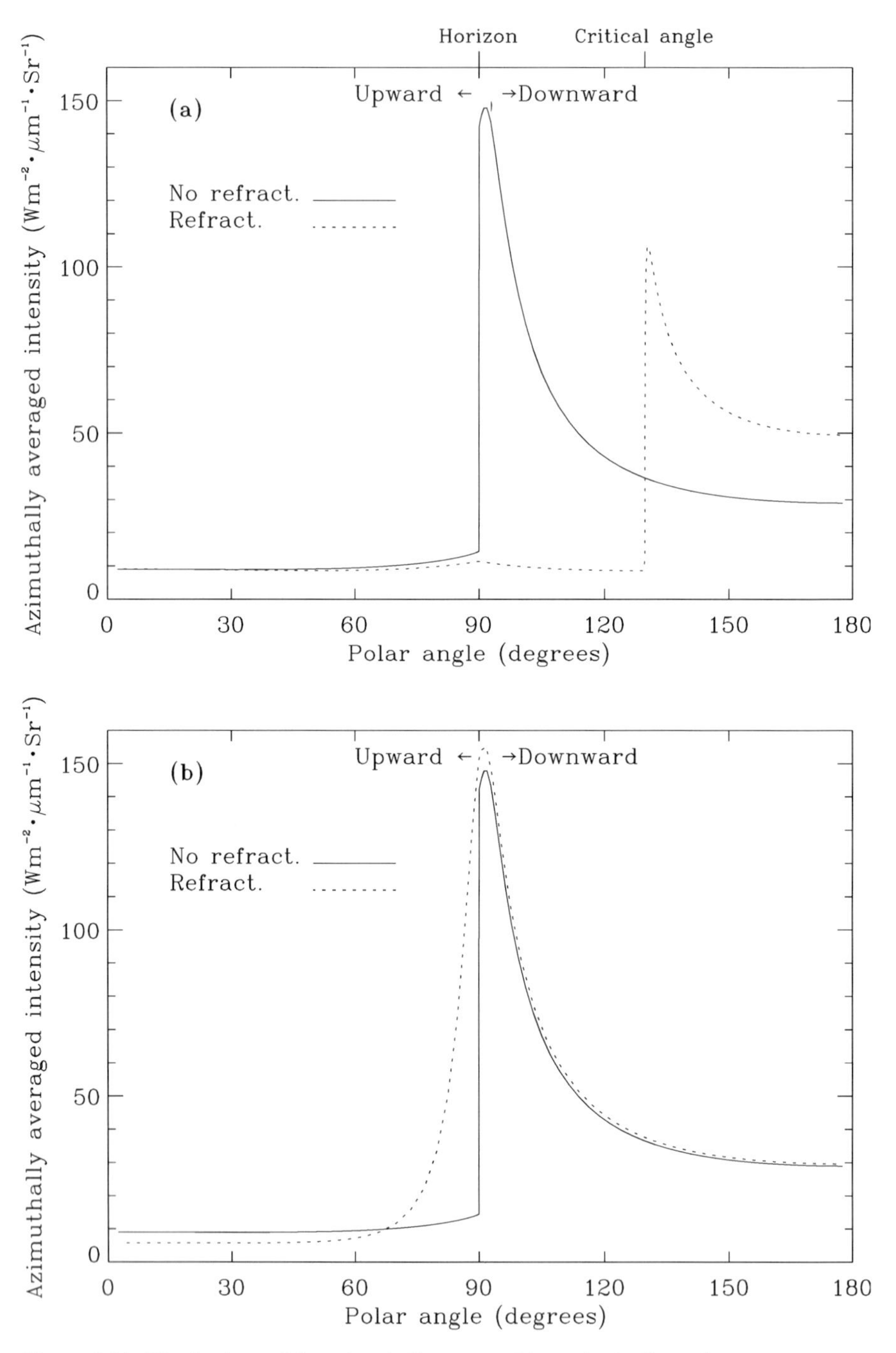

Figure 8.11 Distributions of the azimuthally averaged intensity (radiance in ocean terminology) including refraction in the ocean ($m_r = 1.33$), and ignoring it ($m_r = 1.0$). The computation was done for a solar zenith angle of 30° and wavelength of 500 nm. (a) Just below the ocean surface; (b) just above the ocean surface.

8.11.1 Matrix-Exponential Solution – Formal Derivation of Doubling Rules

In the doubling-adding method one also usually adopts the same discretization in angle as in the discrete-ordinate method. The basic problem is therefore to find solutions to the homogeneous version of Eq. 8.17, which we may rewrite as

$$\frac{d}{d\tau}\begin{bmatrix}\mathbf{I}^-\\ \mathbf{I}^+\end{bmatrix} = -\tilde{\Gamma}\begin{bmatrix}\mathbf{I}^-\\ \mathbf{I}^+\end{bmatrix}, \tag{8.85}$$

where the $2N \times 2N$ matrix $\tilde{\Gamma}$ consists of the $N \times N$ matrices $\tilde{\alpha}$ and $\tilde{\beta}$ defined after Eq. 8.17. Although Eq. 8.30 represents an explicit solution to Eq. 8.85, we may solve this equation formally for a homogeneous slab to obtain

$$\begin{bmatrix}\mathbf{I}^-(\tau^*)\\ \mathbf{I}^+(\tau^*)\end{bmatrix} = e^{-\tilde{\Gamma}\tau^*}\begin{bmatrix}\mathbf{I}^-(0)\\ \mathbf{I}^+(0)\end{bmatrix}, \tag{8.86}$$

where τ^* is the optical thickness of the slab. We emphasize that this matrix-exponential solution is only a formal solution since we are dealing with a two-point boundary value problem in which the *incident* intensities $\mathbf{I}^-(0)$ and $\mathbf{I}^+(\tau^*)$ constitute the boundary conditions and the *emergent* intensities $\mathbf{I}^+(0)$ and $\mathbf{I}^-(\tau^*)$ are to be determined. We notice the formal similarity with an *initial-value problem* in which the left side could be computed if the "initial" values (i.e., the right side) were known (which they are not in our problem).

Basically, the doubling concept starts with the notion that the emergent intensities $\mathbf{I}^+(0)$ and $\mathbf{I}^-(\tau^*)$ in Eq. 8.86 are determined by a reflection matrix, ρ, and a transmission matrix, $\mathcal{T}$, through the relations

$$\mathbf{I}^-(\tau^*) = \mathcal{T}\mathbf{I}^-(0) + \rho\mathbf{I}^+(\tau^*),$$
$$\mathbf{I}^+(0) = \mathcal{T}\mathbf{I}^+(\tau^*) + \rho\mathbf{I}^-(0)$$

for a homogeneous slab of thickness τ^*. These two relations, which are frequently referred to as *the Interaction Principle*, may be rewritten in matrix form as

$$\begin{bmatrix}\mathbf{I}^-(\tau^*)\\ \mathbf{I}^+(\tau^*)\end{bmatrix} = \begin{bmatrix}\mathcal{T} - \rho\mathcal{T}^{-1}\rho & \rho\mathcal{T}^{-1}\\ -\mathcal{T}^{-1}\rho & \mathcal{T}^{-1}\end{bmatrix}\begin{bmatrix}\mathbf{I}^-(0)\\ \mathbf{I}^+(0)\end{bmatrix}, \tag{8.87}$$

where the superscript -1 denotes matrix inversion. By comparing Eqs. 8.86 and 8.87, we find

$$e^{-\tilde{\Gamma}\tau^*} = \begin{bmatrix}\mathcal{T} - \rho\mathcal{T}^{-1}\rho & \rho\mathcal{T}^{-1}\\ -\mathcal{T}^{-1}\rho & \mathcal{T}^{-1}\end{bmatrix}. \tag{8.88}$$

If we now let $\mathcal{T}_1 = \mathcal{T}(\tau^*)$, $\rho_1 = \rho(\tau^*)$ and $\mathcal{T}_2 = \mathcal{T}(2\tau^*)$, $\rho_2 = \rho(2\tau^*)$, then the identity $e^{-\tilde{\Gamma}2\tau^*} = (e^{-\tilde{\Gamma}\tau^*})^2$ implies (using Eq. 8.88)

$$\begin{bmatrix}\mathcal{T}_2 - \rho_2\mathcal{T}_2^{-1}\rho_2 & \rho_2\mathcal{T}_2^{-1}\\ -\mathcal{T}_2^{-1}\rho_2 & \mathcal{T}_2^{-1}\end{bmatrix} = \begin{bmatrix}\mathcal{T}_1 - \rho_1\mathcal{T}_1^{-1}\rho_1 & \rho_1\mathcal{T}_1^{-1}\\ -\mathcal{T}_1^{-1}\rho_1 & \mathcal{T}_1^{-1}\end{bmatrix}^2.$$

Solving for $\mathcal{T}_2$ and ρ_2, we find

$$\mathcal{T}_2 = \mathcal{T}_1\left(\mathbf{1} - \rho_1^2\right)^{-1}\mathcal{T}_1, \tag{8.89}$$

$$\rho_2 = \rho_1 + \mathcal{T}_1\rho_1\left(\mathbf{1} - \rho_1^2\right)^{-1}\mathcal{T}_1, \tag{8.90}$$

where $\mathbf{1}$ is the identity matrix. Equations 8.89 and 8.90 constitute the basic *doubling rules* from which the reflection and transmission matrices for a layer of thickness $2\tau^*$ are obtained from those of half the thickness, τ^*.

8.11.2 Connection between Doubling and Discrete-Ordinate Methods

As already mentioned, in practical numerical implementations of the doubling method it is common to start with an infinitesimally thin layer so that multiple scattering can be ignored. The starting values for the ρ and $\mathcal{T}$ matrices are then simply determined from single scattering. Since the computational time is directly proportional to the number of doublings required to obtain results for a given thickness, it would be useful to start the procedure with thicker layers. This could be achieved by evaluating the left side of Eq. 8.88, requiring the eigenvalues and eigenvectors of $\tilde{\Gamma}$ to be determined by standard procedures as discussed previously. Inspection of the right side of Eq. 8.88 shows that inversion of the lower right quadrant of the resulting matrix yields $\mathcal{T}$ and postmultiplication of the upper right quadrant with $\mathcal{T}$ yields ρ.[10]

The discussion above shows that the discrete-ordinate method and the doubling method, which are conceptually very different, are, in fact, intimately related. In particular we have seen that the formal matrix-exponential (discrete-ordinate) solution can be used to derive the doubling rules. Moreover, the eigenvalues and eigenvectors that were used to construct the basic solutions in the discrete-ordinate method can also be used to compute the reflection and transmission matrices occurring in the doubling method. We noted previously that we may interpret the matrices $\tilde{\alpha}$ and $\tilde{\beta}$ occurring in Eq. 8.17 as *local* or *differential* transmission and reflection operators. The ρ and $\mathcal{T}$ matrices are the analogous *global* operators for a finite layer. The relationship between the local and global operators is given by Eq. 8.88.

8.11.3 Intuitive Derivation of the Doubling Rules – Adding of Dissimilar Layers

The doubling concept is illustrated in Fig. 8.12, in which the two sublayers are taken to be identical. From this figure the doubling rules can be derived in a more intuitive way. Writing ρ_1 and $\mathcal{T}_1$ for the individual layers and ρ_2 and $\mathcal{T}_2$ for the combined layers,

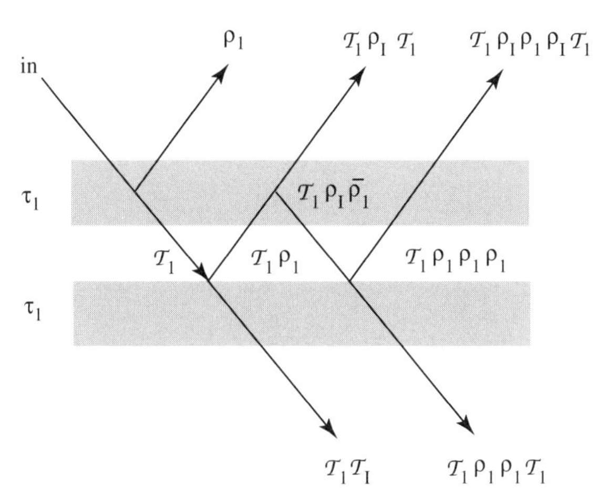

Figure 8.12 Illustration of the doubling concept.

we find

$$\begin{aligned}\rho_2 &= \rho_1 + \mathcal{T}_1\rho_1\mathcal{T}_1 + \mathcal{T}_1\rho_1\rho_1\rho_1\mathcal{T}_1 + \mathcal{T}_1\rho_1\rho_1\rho_1\rho_1\rho_1\mathcal{T}_1 + \cdots \\ &= \rho_1 + \mathcal{T}_1\rho_1(\mathbf{1} + \rho_1\rho_1 + \rho_1\rho_1\rho_1\rho_1 + \cdots)\mathcal{T}_1 \\ &= \rho_1 + \mathcal{T}_1\rho_1\left(\mathbf{1} - \rho_1^2\right)^{-1}\mathcal{T}_1,\end{aligned}$$

$$\begin{aligned}\mathcal{T}_2 &= \mathcal{T}_1\mathcal{T}_1 + \mathcal{T}_1\rho_1\rho_1\mathcal{T}_1 + \mathcal{T}_1\rho_1\rho_1\rho_1\rho_1\mathcal{T}_1 + \cdots \\ &= \mathcal{T}_1(\mathbf{1} + \rho_1\rho_1 + \rho_1\rho_1\rho_1\rho_1 + \cdots)\mathcal{T}_1 \\ &= \mathcal{T}_1\left(\mathbf{1} - \rho_1^2\right)^{-1}\mathcal{T}_1,\end{aligned}$$

where we have used matrix inversion to sum the infinite series, that is,

$$\mathbf{1} + \rho\rho + \rho\rho\rho\rho + \cdots = (\mathbf{1} - \rho\rho)^{-1}.$$

We note that these expressions are identical to Eqs. 8.89 and 8.90, as they should be.

For two dissimilar vertically inhomogeneous layers we must generalize the expressions to account for the fact that the transmission and reflection matrices will in general be different for illumination from the bottom of the layer than that from the top. This is illustrated in Fig. 8.13, where the reflection and transmission matrices pertinent for illumination from below are denoted by the symbol ˜, the individual layers are denoted by subscripts 1 and 2, and the combined layers are without subscript. Referring to Fig. 8.13 we have

$$\begin{aligned}\rho &= \rho_1 + \mathcal{T}_1\rho_2\tilde{\mathcal{T}}_1 + \mathcal{T}_1\rho_2\tilde{\rho}_1\rho_2\tilde{\mathcal{T}}_1 + \cdots \\ &= \rho_1 + \mathcal{T}_1\rho_2(\mathbf{1} - \tilde{\rho}_1\rho_2)^{-1}\tilde{\mathcal{T}}_1,\end{aligned}$$

$$\begin{aligned}\mathcal{T} &= \mathcal{T}_1\mathcal{T}_2 + \mathcal{T}_1\rho_2\tilde{\rho}_1\mathcal{T}_2 + \mathcal{T}_1\rho_2\tilde{\rho}_1\rho_2\tilde{\rho}_1\mathcal{T}_2 + \cdots \\ &= \mathcal{T}_1(\mathbf{1} - \rho_2\tilde{\rho}_1)^{-1}\mathcal{T}_2,\end{aligned}$$

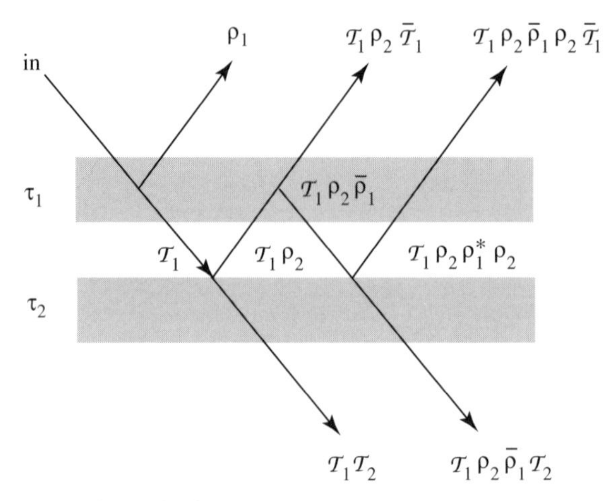

Figure 8.13 Illustration of the adding concept.

which are the reflection and transmission matrices for the combined layers for illumination from the top boundary as illustrated in Fig. 8.13.

Illumination from the bottom boundary leads to the following expressions:

$$\begin{aligned}\tilde{\rho} &= \tilde{\rho}_2 + \tilde{\mathcal{T}}_2\tilde{\rho}_1\mathcal{T}_2 + \mathcal{T}_2\tilde{\rho}_1\rho_2\tilde{\rho}_1\mathcal{T}_2 + \cdots \\ &= \tilde{\rho}_2 + \mathcal{T}_2\tilde{\rho}_1(\mathbf{1} - \rho_2\tilde{\rho}_1)^{-1}\mathcal{T}_2,\end{aligned}$$

$$\begin{aligned}\tilde{\mathcal{T}} &= \tilde{\mathcal{T}}_2\tilde{\mathcal{T}}_1 + \tilde{\mathcal{T}}_2\tilde{\rho}_1\rho_2\tilde{\mathcal{T}}_1 + \tilde{\mathcal{T}}_2\tilde{\rho}_1\rho_2\tilde{\rho}_1\rho_2\tilde{\mathcal{T}}_1 + \cdots \\ &= \tilde{\mathcal{T}}_2(\mathbf{1} - \tilde{\rho}_1\rho_2)^{-1}\tilde{\mathcal{T}}_1.\end{aligned}$$

This completes the derivation of doubling-adding formulas for external illumination. To deal with internal sources, additional formulas are required. Readers who are interested in further pursuing this topic should consult the references given at the end of this chapter.

8.12 Other Accurate Methods

A variety of techniques have been developed to solve the radiative transfer equation. We shall not attempt here to describe these various methods in any detail. Several excellent reviews and books exist and are referred to in the notes at the end of this chapter. Our aim here is rather to briefly mention some of the more powerful and general methods that have proven useful for solving practical radiative transfer problems.

8.12.1 The Spherical-Harmonics Method

We have already discussed the doubling-adding method and how it is closely related to the discrete-ordinate method despite being seemingly quite different in concept.

Another method that is closely related to the discrete-ordinate method is the spherical-harmonics method,[11] which starts by expanding the intensity in Legendre polynomials. As already mentioned, this expansion was first suggested by Eddington in 1916 and led to the widely used Eddington approximation by retaining only two terms in the expansion. In Chapter 7 we showed that the Eddington and two-stream methods are closely related. Since the generalization of the two-stream and Eddington methods leads to the discrete-ordinate method and the spherical-harmonics method, respectively, it is perhaps not surprising to learn that these two latter methods are also closely related. One reason for this similarity is that the spherical-harmonics method relies on an expansion of the intensity in Legendre polynomials, while the discrete-ordinate method relies on using quadrature, which in turn is based on approximating the intensity with an interpolating polynomial that makes essential use of the Legendre polynomials to achieve optimum accuracy. The difference between the two methods lies mainly in the implementation of boundary conditions. As we have seen, this is quite straightforward in the discrete-ordinate approach, but it appears somewhat more cumbersome in the spherical-harmonics method in which moments of the intensity are specified at each boundary instead of specifying the intensity in discrete directions as is done in the discrete-ordinate method. In spite of this difficulty the spherical-harmonics method has been developed into a reliable and efficient solution technique.

8.12.2 Invariant Imbedding

The radiative transfer equation may be classified as a linear transport equation that must be solved subject to boundary conditions at the top and the bottom of the medium. Mathematically, we refer to this problem as a linear two-point boundary value problem. There is another class of methods in which the two-point boundary value problem is converted into a set of initial-value problems. This is usually referred to as *invariant imbedding*.[12] From a mathematical point of view this essentially amounts to the transformation of a "difficult" yet linear two-point boundary value problem into a set of "simpler" but partly nonlinear initial-value problems. Invariant imbedding has been used extensively to solve astrophysical radiative transfer problems as well as in radiation dosimetry calculations and radiative transfer in the ocean.[13]

8.12.3 Iteration Methods

There are several methods of an iterative nature, such as "successive orders of scattering," "lambda iteration" (or Neumann series), and "Gauss–Seidel iteration" that have been important both for the development and understanding of multiple-scattering theories.[14] The advantage is that these approaches are physically based and this allows for easy intuitive interpretation of the results. The disadvantage is that they apply only under restrictive conditions such as optically thin media and nonconservative scattering. We have already discussed these types of methods in Chapter 7 and shall not discuss them further here.

8.12.4 The Feautrier Method

Another class of methods was first introduced by P. Feautrier[15] in 1964. Feautrier's approach, which has gained prominence and popularity in astrophysics, is based on using symmetric and antisymmetric averages of the radiation field as the dependent variable. The resulting equations are discretized in both angle and optical depth and solved numerically using finite-difference techniques. The method was originally used mainly for problems with isotropic scattering, but it has been generalized to apply also to problems with anisotropic scattering.

8.12.5 Integral Equation Approach

As shown in Chapter 6 we may convert the integro-differential radiative transfer equation into a Fredholm-type integral equation[16] commonly referred to as the Schwarzschild–Milne integral equation. This approach is particularly appealing for line transfer problems occurring in astrophysics in which isotropic scattering and complete frequency redistribution prevail, since the resulting integral equation becomes angle and frequency independent. Although it is readily generalized to anisotropic scattering, this approach has not received much attention in the non-astrophysical literature.

8.12.6 Monte Carlo Methods

For geometries other than plane-parallel and/or media with irregular boundaries, Monte Carlo methods[17] become attractive. In essence the Monte Carlo approach consists of simulating the trajectories of individual photons using probabilistic methods and concepts such as those discussed in §6.13. In order to get good statistics a large number of trajectories must be simulated. Such simulations can, in principle, yield very precise results. The accuracy is primarily limited by computer resources. Monte Carlo methods have been developed to a high degree of sophistication and used to solve a variety of radiative transfer problems in plane-parallel media as well as media with complicated geometries. This approach has also been widely used to solve radiative transfer problems in the ocean including the coupled ocean–atmosphere problem in the presence of a nonplanar (wavy) interface.

8.13 Summary

The main objective of this chapter was to discuss accurate solution methods. To this end we have provided a detailed description of one accurate method for solving the radiative transfer equation and a more cursory description of other powerful methods. For the more detailed description we picked the discrete-ordinate method, with which we are intimately familiar and which has been developed into a sophisticated, robust, and versatile general-purpose numerical tool suitable for a variety of applications. Other methods were briefly described, their relationship to the discrete-ordinate was pointed out, and a list of pertinent references is provided at the end of the chapter.

The discrete-ordinate method may be considered to be the obvious or natural extension of the two-stream method described in detail in Chapter 7. To accomplish this generalization it was necessary to discuss the topic of numerical quadrature because the method relies on approximating integrals by finite sums, which in turn converts an integro-differential equation into a set of coupled ordinary differential equations. Thus, in §8.2 we started by introducing the concept of numerical quadrature using the simple case of isotropic scattering for demonstration purposes. The advantage of using Gaussian quadrature, which makes essential use of the Legendre polynomials, was stressed. Likewise, we discussed in some detail the advantage of using half-range quadrature, in which each hemisphere is treated separately, instead of full-range quadrature.

In §8.3 we applied the discrete-ordinate approximation to the more general problem of anisotropic scattering and we showed in §8.4 how this problem may be written in matrix form. The matrix formulation is essential because it allows us to apply powerful numerical methods from linear algebra. In §8.5 the solutions were derived in some detail starting with the simple two- and four-stream cases, which we then generalized to apply to an arbitrary number of streams. In §8.6 we derived an expression for the source function, which was used to generate analytic formulas for the upward and downward intensities, allowing us to compute the intensity at arbitrary angles (not only the quadrature points).

The inclusion of boundary conditions was discussed in some detail (§8.7) and the removal of ill-conditioning problems that arise when solving for the constants of integration was accomplished, both for a single layer (§8.7) and for a multilayered medium (§8.8). This last point is essential; *the ill-conditioning must be removed for the method to work.* The numerical aspects of implementing the discrete-ordinate method on the computer were briefly discussed at the end of §8.8. A more comprehensive description of the computer code called DISORT (DIScrete Ordinate Radiative Transfer) is provided in a NASA report cited in the notes below. A discussion of improved intensity computations as well as concrete examples of such computations were provided in §8.9.

In §8.10 we discussed the solution of the coupled atmosphere–ocean problem. Particular attention was paid to the selection of quadrature to allow effective solution of the coupled problem as well as the implementation of the conditions properly describing reflection and refraction at the atmosphere–ocean interface.

The doubling-adding method has been intensively used to solve radiative transfer problems. We therefore discussed this method in some detail (§8.11) and showed in particular how it is related to the discrete-ordinate method. Finally, in §8.12 we provided a very brief discussion of other accurate methods that are currently used to solve radiative transfer problems in atmospheres and oceans.

Problems

8.1

(a) In Example 8.4 the thermal source was approximated by a polynomial. Verify that the solution is given by Eqs. 8.37 and 8.38.

Assume that the Planck function across any layer can be approximated by an exponential times a linear function of optical depth τ, that is,

$$B(T(\tau)) = e^{-\sigma\tau}(b_0 + b_1\tau).$$

(b) Derive an expression for σ, b_0, and b_1 by assuming that the temperature is known at the top and the bottom of the layer. Assume that the temperature at the center of the layer is given by $T(\tau_{\text{center}}) = [T(\tau_{\text{top}}) + T(\tau_{\text{bottom}})]/2$.

(c) Assume that the particular solution varies as $I(\tau, u_i) = e^{-\alpha\tau}[Y_0(u_i) + Y_1(u_i)\tau]$ and derive suitable expressions for α, Y_0, and Y_1.

8.2

(a) Verify Eqs. 8.43–8.47 of §8.6.

(b) Show that the solutions for the angular distributions given by Eqs. 8.46 and 8.47 reduce to the results given in Chandrasekhar's book (pp. 82–83) for isotropic scattering.

(c) Prove that the interpolated intensities given by Eqs. 8.46 and 8.47 coincide with the basic solutions (Eq. 8.39) at the quadrature points.

(d) Derive Eqs. 8.67–8.69.

8.3 Prove that the interpolated intensities given by Eqs. 8.46 and 8.47 satisfy boundary conditions for all angles.

8.4

(a) Derive Eq. 8.48 and verify that the boundary conditions can be written as in Eqs. 8.53–8.56 for a homogeneous slab (§8.7).

(b) For multilayered slab verify that the boundary and continuity conditions expressed by Eqs. 8.62–8.64 are correct.

8.5 Proceed as in the one-layer case considered in Example 8.6 to prove that the ill-conditioning problem can be removed in the two-stream, two-layer case ($N = 1$, $L = 2$).

8.6 Compute (use the DISORT code) and plot fluxes (upward diffuse, downward diffuse, total downward, and net flux) for direct beam incidence of radiation on a homogeneous layer of optical thickness $\tau^* = 1$. Use (solar) zenith angles $\theta = 10^\circ$ and $\theta = 60^\circ$.

(a) Assume a nonreflecting lower boundary and consider the following cases:
  (i) Rayleigh phase function with conservative scattering (use $a = 0.9999$).
  (ii) Henyey–Greenstein phase function with asymmetry factor, $g = 0.75$, and $a = 0.9999$ and $a = 0.8$.

(b) Repeat the above computations for a Lambertian lower boundary with albedo 0.8.
(c) Discuss the physical meaning of the results and potential applications.

8.7 For the coupled atmosphere-ocean problem show that the ill-conditioning can be removed by considering two streams in the atmosphere and four streams in the ocean ($N_1 = 1$, $N_2 = 2$). Use one layer each in the atmosphere and ocean ($L_1 = L_2 = 1$).

8.8 Consider the doubling-adding rules in §8.11.
(a) Verify that the formal derivation of the doubling rules from the interaction principle is correct.
(b) Repeat for the adding rules.

8.9 Use the DISORT code to compute the sky intensity in the principal plane, for the same conditions as in Problem 7.1. Assume a Lambert surface, with $\pi\rho_L = 0$, $\pi\rho_L = 0.3$, and $\pi\rho_L = 0.7$, and compare with the approximate results of Problem 7.1.

8.10 Consider a cumulus cloud containing conservatively scattering water drops whose angular scattering is approximated by a Henyey–Greenstein phase function with $g = 0.85$.
(a) Using DISORT, compute the sky intensity in the principal plane for the same solar illumination as Problem 7.1. Assume $\pi\rho_L = 0$, $\pi\rho_L = 0.3$, and $\pi\rho_L = 0.7$.
(b) Compare and contrast with the results from Problem 8.9.

Notes

1 The discrete ordinate method was first introduced by G. C. Wick, in "Über ebene Diffusionsprobleme," *Zeit. f. Phys.*, **121**, 702, 1943, and further developed by Chandrasekhar in the 1940s as described in his classic text *Radiative Transfer* first published in 1950. The material in §8.2 follows the earlier development except that we have emphasized the advantage of using double-Gauss quadrature and elucidated the connection between full range Gaussian quadrature, applicable for the interval [−1, 1], and half range (i.e., double-Gauss) quadrature in which the Gaussian formula is applied separately for the intervals [−1, 0] and [0, 1].

2 See, e.g., Burden, R. L., and J. D. Faires, *Numerical Analysis*, 3rd. ed., Prindle, Weber and Schmidt, Boston, 1985, p. 153.

3 Sykes, J. B., "Approximate integration of the equation of transfer," *Monthly Notices of the Royal Astronomical Society*, **111**, 377–86, 1951, is generally given the credit

for this suggestion. Actually, this method was first proposed by J. Yvon, according to Kourganoff, V., Basic Methods in Transfer Problems, Dover, 1963, p. 101.

4 The matrix formulation and solution applicable to anisotropically scattering, multilayered media (§8.4–8.8) are based on developments described in a series of papers by Stamnes and coworkers. A summary of these developments is provided in Stamnes, K., "The theory of multiple scattering of radiation in plane parallel atmospheres," *Rev. Geophys.*, **24**, 299–310, 1986. The numerical and computational implementation of a comprehensive, multiple-scattering radiative transfer code based on the discrete-ordinate method (DISORT) is described and documented in Stamnes, K., S.-C. Tsay, W. J. Wiscombe, and K. Jayaweera, "Numerically stable algorithm for discrete-ordinate-method radiative transfer in multiple scattering and emitting layered media," *Applied Optics*, **27**, 2,502–9, 1988; and Stamnes, K., S.-C. Tsay, and W. J. Wiscombe, "A general-purpose, numerically stable computer code for discrete-ordinate-method radiative transfer in scattering and emitting layered media," to appear as NASA report, 1999.

5 For details see Stamnes, K. and P. Conklin, "A new multilayer discrete ordinate approach to radiative transfer in vertically inhomogeneous atmospheres," *J. Quant. Spectrosc. Radiative Transfer*, **31**, 273–82, 1984.

6 A description of the code called DISORT is provided in the NASA report cited in Endnote 4.

7 The improved intensity computation (§8.9) is based on Nakajima, T. and M. Tanaka, "Algorithms for radiative intensity calculations in moderately thick atmospheres using a truncation approximation," *J. Quant. Spectrosc. Radiative Transfer*, **40**, 51–69, 1988. Figures 8.5 and 8.6 are based on this work. The computational illustrations in §8.9 are generated with the DISORT code in which the improved intensity computations have been incorporated.

8 This material is based on Z. Jin and K. Stamnes,"Radiative transfer in nonuniformly refracting media such as the atmosphere/ocean system," *Applied Optics*, **33**, 431–42, 1993.

9 The doubling concept seems to have originated in 1862 (Stokes, G., "On the intensity of the light reflected from or transmitted through a pile of plates," *Proc. R. Soc. London*, **11**, 545–56, 1862). It was introduced into atmospheric physics one century later (Twomey, S., H. Jacobowitz, and J. Howell, "Matrix methods for multiple scattering problems," *J. Atmos. Sci.*, **23**, 289–96, 1966; van de Hulst, H. C. and K. Grossman, "Multiple light scattering in planetary atmospheres," in *The Atmospheres of Venus and Mars*, edited by J. C. Brandt and M. V. McElroy, Gordon and Breach, New York, 1968; Hunt, G. E. and I. P. Grant, "Discrete space theory of radiative transfer and its application to problems in planetary atmospheres," *J. Atmos. Sci.*, **26**, 963–72, 1969). The theoretical aspects as well as the numerical techniques have since been developed by a number of investigators; for references see, e.g., Wiscombe, W. J., "Atmospheric radiation: 1975–1983," *Rev. Geophys.*, **21**, 997–1021, 1983. These methods or slight variants thereof are also referred to as discrete space theory (Preisendorfer, R. W., *Radiative Transfer on Discrete Spaces*, Pergamon, New York, 1965; Stephens, G. L., "Radiative transfer on a linear lattice: Application to anisotropic ice crystals clouds," *J. Atmos. Sci.*, **37**, 2095–104, 1980) or matrix operator theory (Plass, G. N., G. W. Kattawar, and F. E. Catchings, "Matrix operator theory of radiative transfer, I, Rayleigh scattering," *Applied Optics*, **12**, 314–29, 1973; Tanaka, T. and M. Nakajima, "Matrix formulations for the transfer of

solar radiation in a plane-parallel scattering atmosphere," *J. Quant. Spectrosc. Radiative Transfer*, **28**, 13–21, 1986; Lenoble J. (ed.), *Radiative Transfer in Scattering and Absorbing Atmospheres: Standard Computational Procedures*, A. Deepak, Hampton, VA, 1985).

10 See Waterman, P. C., "Matrix-exponential description of radiative transfer," *J. Opt. Soc. Am.*, **71**, 410–22, 1981, and Stamnes, K., "The theory of multiple scattering in plane parallel atmospheres," *Rev. Geophys.*, **24**, 299–310, 1986.

11 The spherical-harmonic method has been developed into an advanced technique for solving the radiative transfer equation. The most recent developments are by Karp, A. H., J. J. Greenstadt, and J. A. Filmore, "Radiative transfer through an arbitrarily thick, scattering atmosphere," *J. Quant. Spectrosc. Radiative Transfer*, **24**, 391–406, 1980.

12 The invariant imbedding method is described by several authors: Bellman, R., R. Kalaba, and M. Prestrud, *Invariant Imbedding and Radiative Transfer in Slabs of Finite Thickness*, Elsevier, New York, 1963; Bellman, R., H. Kawigada, R. Kalaba, and S. Ueno, "Invariant imbedding equations for the dissipation functions of an inhomogeneous finite slab with anisotropic scattering," *J. Math. Phys.*, **8**, 2137–42, 1967; Hummer, D. G. and G. Rybicki, "Computational methods for NLTE line-transfer problems", in *Methods in Computational Physics*, edited by A. Alder, S. Fernbach, and M. Rotheberg, Academic, Orlando, FL, 1967; Lenoble J. (ed.), op. cit., 1985.

13 For applications in the ocean, see Mobley, C. D., *Light and Water: Radiative Transfer in Natural Waters*, Academic Press, San Diego, 1994.

14 For a review of iteration methods, see Irvine, W. M., "Multiple scattering in planetary atmospheres," *Icarus*, **25**, 175–204, 1975; and Lenoble, op. cit., 1985.

15 Feautrier's method was first introduced thirty five years ago (Feautrier, P., "Sur la resolution numerique de l'equation de transfert," *C. R. Acad. Sci. Paris*, **258**, 3189–91, 1964), and it has since been used extensively (e.g., Mihalas, D., *Stellar Atmospheres*, W. H. Freeman, San Francisco, CA, 1978; Mihalas, D., "The computation of radiation transport using Feautrier variables, 1, Static media," *J. Comput. Phys.*, **57**, 1–25, 1985; see also articles in Kalkofen, W. (ed.), *Methods in Transfer Problems*, Cambridge University Press, New York, 1984).

16 The integral equation approach is described by Chandrasekhar (1960, op. cit.) and by Cheyney, H. and A. Arking, "A new formulation for anisotropic radiative transfer problems, I, Solution with a variational technique," *Astrophys. J.*, **207**, 808–19, 1976, and several investigators have applied this method to solve a variety of problems (e.g., Anderson, D. E., "The troposphere–stratosphere radiation field at twilight: A spherical model," *Planet. Space Sci.*, **31**, 1517–23, 1983; Hummer, D. F. and Rybicki, G. B. op. cit., 1967; Strickland, D. J., "The transport of the resonance radiation in a non-isothermal medium: The effect of a varying Doppler width," *J. Geophys. Res.*, **84**, 5890–96, 1979; Strickland, D. J. and T. M. Donahue, "Excitation and radiative transport of OI 1304 Å radiation, 1, The dayglow," *Planet. Space Sci.*, **18**, 661–89, 1970).

17 The Monte Carlo method is well documented (Collins, D. G., W. G. Blattner, M. B. Wells, and H. G. Horak, "Backward Monte-Carlo calculations of the polarization characteristics of the radiation field emerging from spherical shell atmospheres," *Applied Optics*, **11**, 2684–705, 1972; Lenoble, op. cit., 1985) and has been extensively utilized in a variety of applications (see, e.g., Plass, G. N. and G. W. Kattawar, "Monte-Carlo calculations of light scattering from clouds," *Applied Optics*, **7**, 669–704, 1968; Danielson

R. E., D. R. Moore, and H. C. van de Hulst, "The transfer of visible radiation through clouds," *J. Atmos. Sci.*, **26**, 1078–87, 1969; Collins et al., op. cit., 1972; Appleby, J. F. and W. M. Irvine, "Path length distributions of photons diffusely reflected from a semi-infinite atmosphere," *Astrophys. J.*, **183**, 337–46, 1973; Kattawar, G. W., G. N. Plass, J. A. Quinn, "Monte-Carlo calculation of polarization of radiation in the earth's atmosphere–ocean system," *J. Phys. Oceanogr.*, **3**, 353–72, 1973; Meier, R. R. and J.-S. Lee, "Angle-dependent frequency redistribution: Internal source case," *Astrophys. J.*, **250**, 376–83, 1981; Lee, J.-S. and R. R. Meier, "Angle-dependent frequency distribution in a plane-parallel medium: External source case," *Astrophys. J.*, **240**, 185–95, 1980; Lenoble, op. cit., 1985; Davies, R., W. L. Ridgeway, and K.-E. Kim, "Spectral absorption of solar radiation in cloudy atmospheres: A 20 cm^{-1} model," *J. Atmos. Sci.*, **41**, 2126–37, 1984).

Chapter 9

Shortwave Radiative Transfer

9.1 Introduction

There are currently two prominent problems in atmospheric and environmental science that have received much attention: the possibility of widespread ozone depletion and the potential for global warming. The primary concerns of public debate and scientific research have focused on (i) to what extent ozone depletion and global warming are, in fact, occurring and (ii) if so, to what extent these phenomena are due to natural rather than anthropogenic causes. There is growing evidence relating ozone depletion directly to the release of man-made trace gases, notably chlorofluorocarbons used in the refrigeration industry and as propellants in spray cans. Since ozone provides an effective shield against damaging ultraviolet radiation from the Sun, there is indeed good reason to be concerned, because a thinning of the ozone layer could have serious biological ramifications. The most harmful ultraviolet radiation reaching the Earth's surface, commonly referred as UV-B, lies in the wavelength range between 280 and 320 nm (see Table 1.1). UV-B radiation, which has enough energy to damage the DNA molecule, is strongly absorbed by ozone. Radiation with wavelengths between 320 and 400 nm, referred to as UV-A, is relatively little affected by ozone. UV-A radiation can mitigate some of the damage inflicted by UV-B (this is known as "photo-repair"), but it causes sunburn and is therefore believed to be a partial cause of skin cancer. In addition to the harmful effects on humans, too much ultraviolet radiation has deleterious effects on terrestrial animals and plants, as well as aquatic life forms.

Ozone is a trace gas, whose bulk content resides in the stratosphere. Its abundance is determined by a balance between production and loss processes. Chemical reactions as well as photolysis are responsible for the destruction of atmospheric ozone. Its formation in the stratosphere relies on the availability of atomic oxygen, which is produced by photodissociation of molecular oxygen. Ozone is then formed when an oxygen atom (O) and an oxygen molecule (O_2) combine to yield O_3. It is produced mainly high in the atmosphere at low latitudes where light is abundant, and it is subsequently transported to higher latitudes by the equator-to-pole circulation.[1] Thus, the distribution of ozone in the atmosphere, vertically and globally, results from a subtle interplay between radiation, chemistry, and dynamics.

Ozone interacts with ultraviolet/visible radiation as well as with thermal infrared (terrestrial) radiation. A thinning of the ozone layer renders the stratosphere more transparent in the 9.6-μm region, thereby allowing more transmission and less backwarming of surface emission. Thus ozone depletion cools the surface and tends to partially mitigate warming from increased greenhouse gases. In addition to ozone, several other atmospheric trace gases, notably water vapor, carbon dioxide, chlorofluorocarbons, and methane, are infrared active. These so-called greenhouse gases strongly absorb and emit infrared radiation and thereby trap radiative energy that would otherwise escape to space. The global warming issue is concerned with the effects enhanced abundances of these gases (due mainly to human activities) may have on the radiative energy balance of the Earth and hence on climate.

Life began with light. Solar radiation illuminating the primordial atmosphere gave rise to chemical reactions that constituted the basis for biological evolution and eventually led to the establishment of photosynthesis, which is a prerequisite for life as we know it. Three groups of organisms use light for photosynthesis: photosynthetic bacteria, blue-green algae, and green plants. The green color of plants and biologically productive waters is due to chlorophyll, which absorbs blue and red light. The end result of photosynthesis is that light, water, and carbon dioxide combine to yield carbohydrates and oxygen, which are the building blocks of life. Photosynthesis is driven by light with wavelengths between 400 and 700 nm. Therefore, this region of the spectrum is referred to as the photosynthetically active radiation (PAR).

In this chapter we shall consider the radiative output from the Sun, and the ultraviolet and visible radiation in particular, which drive photochemistry and photobiology. Solar near-IR and terrestrial IR radiation will be discussed in Chapter 11, and the role of radiation in climate will be discussed in Chapter 12. In §9.2 we discuss the solar radiative output and the basic notion of penetration of solar radiation into the atmosphere and ocean. The optical properties of the atmosphere, ocean, snow, and ice are briefly discussed in §9.3. The modeling of shortwave radiation in these regimes is considered in §9.4.

9.2 Solar Radiation

The spectral input of solar radiation falling on a planetary atmosphere is governed roughly by the Sun's effective radiating temperature, which is about 5,780 K. More precisely it is governed by the abundance and temperature of the absorbing and emitting gaseous species in the Sun's outer atmosphere. Until the space age our knowledge of the solar spectrum was largely confined to visible wavelengths longward of the ozone cutoff at $\lambda = 300$ nm. The extraterrestrial irradiance $F^s(\nu)$ when integrated over all wavelengths, and referenced to the mean distance of the Earth from the Sun (1 AU, or 1.50×10^8 km), is known as the *solar constant*.[2] Since we know that the Sun's output is variable, the modern parlance is *total solar irradiance*.

Pre-space-age determinations of $F^s(\nu)$ relied upon the cosecant law for the dependence of the atmospheric optical path on zenith angle. To correct for atmospheric extinction the ground-based measurements were extrapolated to their extraterrestrial values through the *Bouguer–Langley method*. This consists of taking direct solar measurements at several solar elevations as the Sun is rising or setting, preferably from a high-altitude location. Since the direct solar radiation attenuates according to the Extinction Law, a plot of the solar irradiance versus the secant of the solar zenith angle on a semilogarithmic scale yields a straight line, if the atmosphere is horizontally homogeneous and the solar elevation is not too small. This straight line is then extrapolated to the (fictitious) point where the secant of the solar zenith angle becomes zero (at "zero air mass"). Limitations are due primarily to the presence of atmospheric aerosols, which generally are not homogeneously distributed in the horizontal. Nevertheless, ground-based measurements over the first half of the twentieth century yielded a surprisingly accurate result (about 1,350 $\text{W} \cdot \text{m}^{-2}$) in comparison with the currently accepted value derived from satellite measurements of $1{,}368 \pm 5$ $\text{W} \cdot \text{m}^{-2}$.

In Fig. 1.1 we showed the extraterrestrial solar spectrum in the region 200–700 nm. Some of the more important aspects of the UV/Visible spectrum should be mentioned:

1. Most of the emission arises within the *photosphere*, where the Sun's visible optical depth reaches unity. The finer structure is due to *Fraunhofer absorption* lines caused by gases in the cooler (higher) portions of the photosphere.
2. For $125 < \lambda < 380$ nm, the effective radiating temperature falls to values as low as 4,500 K, due to increased numbers of overlapping absorbing lines (so-called line blanketing). At still shorter wavelengths, some of the emission originates in the hotter *chromosphere*, which overlies the photosphere, and the effective temperature increases.
3. The UV irradiance is noticeably dependent upon the solar cycle, being more intense at high solar activity than at low solar activity. Roughly speaking, this variability is due to the chromospheric component and begins at about the aluminum absorption edge at 210 nm, where the solar cycle modulation is $\sim$5% (peak to trough). At the wavelength of the hydrogen resonance line (Lyman-alpha at 121.6 nm) it is $\sim$70%, and in the extreme UV and X-ray region the modulation can be factors of ten or more.

9.3 Optical Properties of the Earth–Atmosphere System

9.3.1 Gaseous Absorption and Penetration Depth

For wavelengths longer than 200 nm ozone is the most important species affecting the penetration of UV radiation through the atmosphere (see Fig. 4.12). Unlike the well-mixed gases, whose densities fall off approximately exponentially with height according to the hydrostatic balance (Eq. 1.4), the ozone density typically peaks at about 20 km altitude. Most of the ozone in the atmosphere resides in the stratosphere. For the U.S. Standard Atmosphere (tabulated in Appendix C) the column amount is 350 *Dobson units* (DU).[3] Figure 9.1 shows the annual variation of the ozone amount over Antarctica for 1987. From January until September the column ozone amount varies between 250 and 300 DU. The severe ozone depletion referred to as the *ozone hole* started in September this particular year. In October ozone column amounts less than 140 DU were measured. In recent years column amounts less than 100 DU have been observed over Antarctica, which is about 65% less than average for the rest of the year. As shown in Fig. 9.2 during the austral spring ozone hole over Antarctica the ozone concentration is most severely depleted in the altitude range between 12 and 25 km, where the depletion is nearly 100%. It is now well established that the ozone hole over Antarctica is due to chlorine photochemistry. Aircraft observations have demonstrated a clear anticorrelation between ClO and O_3 over Antarctica (see Fig. 9.2).[4] High concentrations of ClO are observed only in air that earlier in winter had experienced cold temperatures (less than −80°C; see Fig. 9.1), allowing the formation of polar stratospheric clouds consisting of nitric acid and water condensed on sulfuric acid particles. On these particles, inactive chlorine compounds are converted to ClO molecules, which lead to catalytic destruction of ozone in the presence of sunlight. The observed anticorrelation of ClO and ozone is considered to be the "smoking gun" for ozone loss by chlorine compounds.

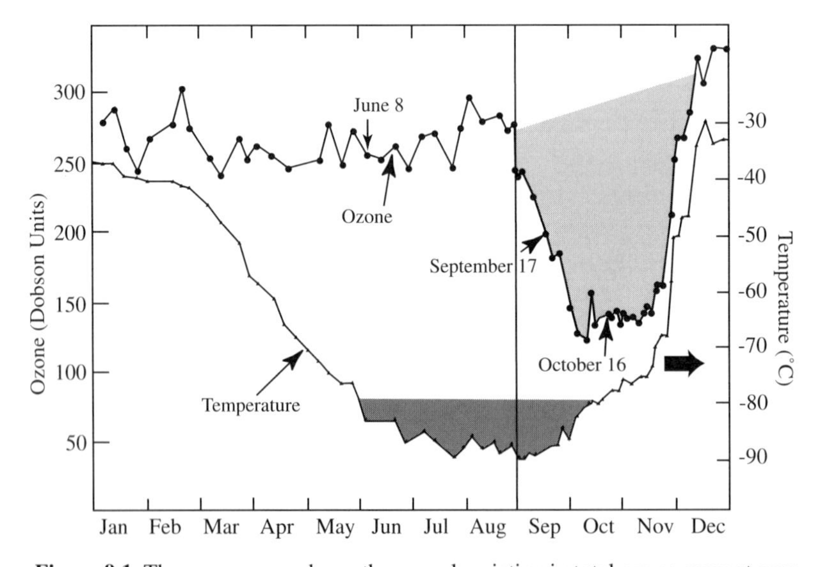

Figure 9.1 The upper curve shows the annual variation in total ozone amount over Antarctica in 1987. The shaded area indicates the *ozone hole*. The lower curve shows the annual mean temperature between 14 and 19 km. The dark-shaded area indicates that the temperature was less than −80°C from June into October. The vertical distribution on three days marked in this figure are shown in Fig. 9.2.

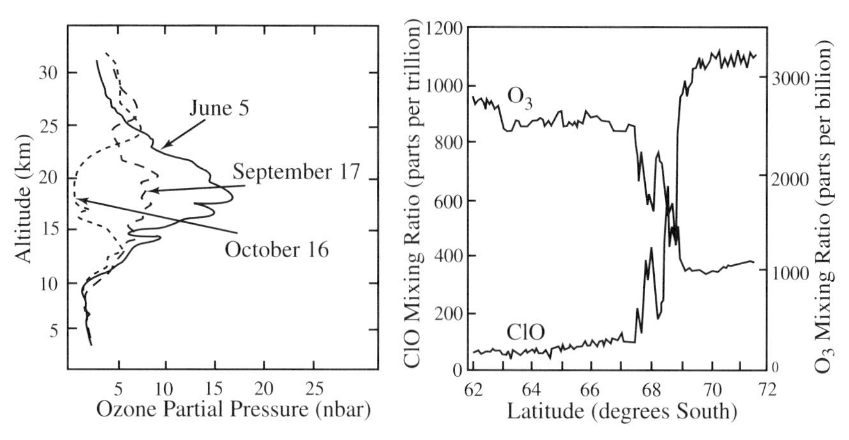

Figure 9.2 Left panel: Ozone vertical distributions over Antarctica on three days in 1987 marked in the previous figure. The solid curve shows the vertical distribution on June 5 when the total ozone amount was about 260 DU. The dashed line is for September 17 when the ozone depletion was already substantial, and the dotted line shows the extreme depletion in the 12 to 25 km region on October 16. Right panel: Aircraft observations of ClO and ozone over Antarctica on September 16, 1987. As the aircraft moves south into the Antarctic vortex, the abundance of ClO increases dramatically, while the ozone amount decreases.

It is of interest to note that the main instrument to measure ozone from the ground was designed by Dobson in the late 1920s.[5] The fact that it is still in use attests to its robustness and ingenious design. The key feature behind its success is that it was designed to measure ratios of spectral intensities (rather than absolute intensities; see Problem 9.5), thus avoiding problems with absolute calibrations, which still pose significant challenges in spectrophotometry. A network of these instruments has been deployed around the world for several decades, long before the launch of instruments for ozone measurements on satellites starting in the 1970s. In fact, the discovery of the Antarctic ozone hole was based on measurements with the Dobson spectrophotometer.[6]

The continuum in the O_3 absorption longward of 310 nm is due to the *Huggins bands*. The weak but spectrally broad *Chappuis bands* of O_3 at $450 < \lambda < 750$ nm produce significant solar absorption for low solar elevation angle because of the large amount of solar irradiance present at these wavelengths. An instructive way of looking at the effects of gaseous absorption on incoming radiation is through the concept of *penetration depth*.[7] This is defined as the height at which the solar radiation reaches optical depth unity, for a clear atmosphere exposed to an overhead Sun. The UV penetration depth below 320 nm is shown in Fig. 9.3. The smallest penetration depths occur in the thermosphere at the most opaque wavelengths. They arise from the very high absorption coefficients of air in the X-ray ($\lambda < 10$ nm) and extreme-UV ($10 < \lambda < 100$ nm) regions. In the far-UV between 100 and 200 nm (see Table 1.1), solar radiation is absorbed in the lower thermosphere and mesosphere (50–120 km). Middle-UV, or UV-C ($200 < \lambda < 280$ nm), atmospheric absorption is dominated by the intense *Hartley bands* due to O_3. In the absence of O_3, the Earth's land and ocean surface would

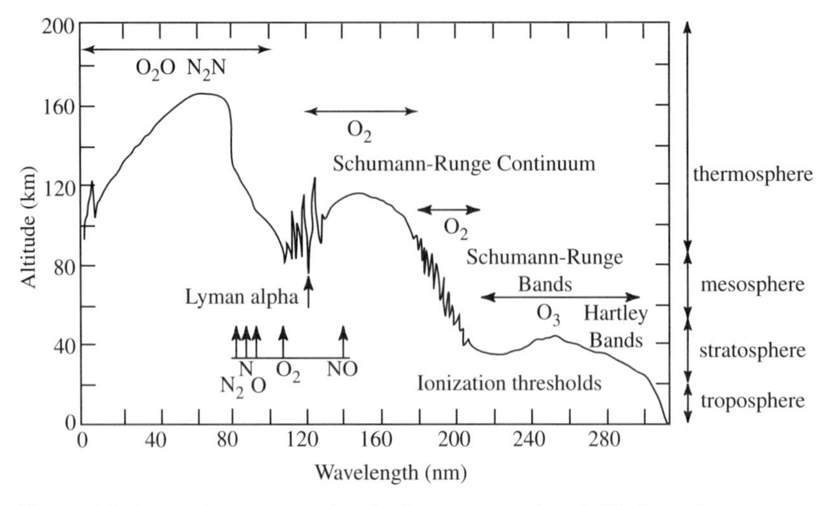

Figure 9.3 Atmospheric penetration depth versus wavelength. Horizontal arrows indicate the molecule (and band) responsible for absorption in that spectral region. Vertical arrows indicate the ionization thresholds of the various species.

be irradiated directly by UV-C to the detriment, if not the total eradication, of surface life. The absorption of this important energy band causes the inversion in the Earth's stratospheric (15–50 km) temperature profile. Solar radiation near 250 nm, at the peak of the Hartley band absorption, is absorbed near 50-km height where the warming rate maximizes (see Problem 9.7). Living organisms are several orders of magnitude more sensitive to damage by UV-B (280–320 nm) radiation than by radiation in the more benign UV-A (320–400 nm) region. UV-B penetration is very sensitive to the total column abundance of O_3. The photosynthetically active radiation (PAR) spanning the spectral range between 400 and 700 nm has the greatest clear-air transparency (except for radio waves), because it contains only the weak absorption features of O_2 and O_3. The eyes of humans (and animals) are most sensitive to light in this spectral range, a fact of extreme importance for evolutionary adaptation.

9.3.2 Optical Properties of Atmospheric Aerosols

Small particles injected into the atmosphere may stay in the air (be suspended) for some time depending on their size. Such suspended particles of a variety of origins are generically referred to as "aerosols" or "haze" because of their impact on visibility. The particles in the lowest part of the atmosphere are generally due to wind-blown dust and industrial pollution. Over the ocean, sea-salt spray from whitecaps is believed to be a major source of aerosols. In addition there is evidence that dimethylsulfide (DMS) gas of biogenic origin is another source of optically significant aerosol particles over the ocean. Tropospheric aerosol abundance may be conveniently related to surface visibility. Based on available information on typical particle mass loading, composition, and size distribution, aerosol models have been constructed for a variety of atmospheric conditions. Using these models we may compute the optical properties from electromagnetic theory (the *Mie–Debye theory*), if we assume that the aerosol particles have spherical shapes.

In the stratosphere the sources of aerosol particles are primarily volcanic eruptions and combustion from fossil fuels. The stratospheric aerosol layer is typically situated between 15 and 25 km. The impact of such aerosols on the radiation budget has been the subject of several climate-related investigations. To derive the optical properties we must know the mean particle size and refractive index; the latter depends on composition. Then if we assume that the particles have spherical shapes we may compute their optical properties from theory. Data from in situ optical sampling devices carried in sounding balloons following the El Chichón (1982) and Pinatubo (1991) volcanic eruptions indicate that the perturbed stratospheric aerosol layer consisted primarily of 0.1–0.3 μm radius spherical liquid droplets with a $\sim$75% concentration (by weight) of sulfuric acid. The quiet-time layer (the *Junge layer*) consists of smaller ($<$0.1 μm) sulfuric acid particles. At peak aerosol loading, the transmission of direct solar radiation is appreciably altered. The optical properties of the stratospheric aerosol layer change with time and location due to advective transport, gas-to-particle conversion, and gravitational sedimentation. The single-scattering albedo and the asymmetry factor are shown in Fig. 9.4 for stratospheric aerosols of fresh and aged volcanic origin as well as background conditions.

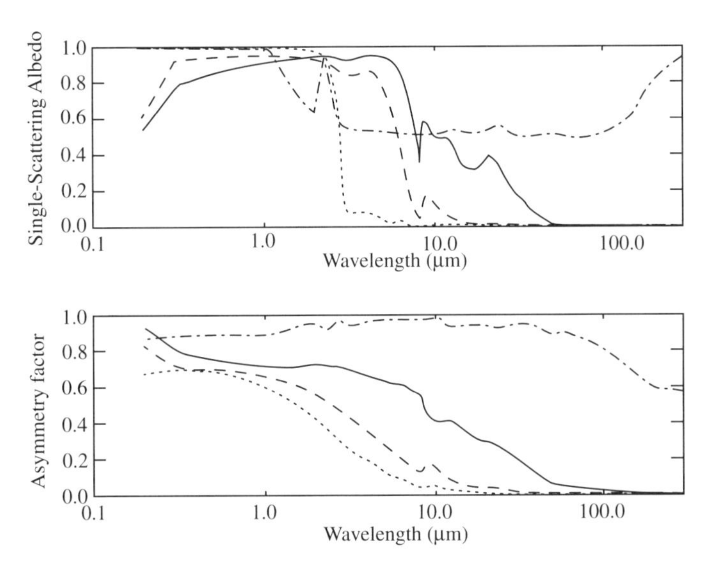

Figure 9.4 Wavelength dependence of the single-scattering albedo and the asymmetry factor for aerosols in the stratosphere. The solid line pertains to fresh volcanic conditions, the dotted line to background, and the dashed line to aged volcanic aerosol. For comparison the dashed-dotted line shows the optical properties of a cirrus cloud assumed to consist of spherical ice particles.

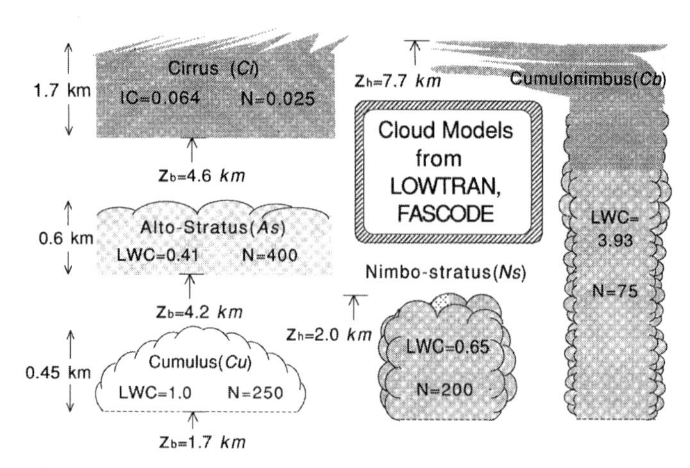

Figure 9.5 Schematic illustration of cloud models made available by the Phillips Laboratory. Typical number concentrations (N), liquid water (LWC), or ice (IWC) contents as well as altitude ranges are indicated for several cloud types.

9.3.3 Optical Properties of Warm (Liquid Water) Clouds

Observations from ground and from space show that a variety of cloud forms cover about 60% of the Earth's surface, on average. Low clouds strongly reflect solar radiation, while higher clouds tend to reduce the longwave radiation emitted to space. They achieve this by absorbing terrestrial radiation and reemitting it at the colder cloud-top temperature. Clouds therefore have a strong influence on the climate of the planet by reflecting solar radiation (which cools the Earth) and trapping terrestrial radiation (which tends to warm the Earth). Results from the *Earth Radiation Budget Experiment*[8] indicate that clouds presently cool the Earth. A variety of cloud types occur in the atmosphere. A summary of cloud types for which models exist is provided in Fig. 9.5.

Assuming that cloud droplets are spherical, we may calculate their optical properties from theory. The important output quantities are the scattering and absorption cross sections and the asymmetry factor of the individual cloud droplets. Knowing these, we may determine the bulk optical properties by an integration over the size distribution of the cloud droplets. (Unfortunately, the size distribution is usually unknown.) These computations are very time consuming and therefore not affordable in large-scale models. Fortunately, it is possible to obtain accurate parameterizations of the Mie computations in terms of the effective radius and the liquid water content of the cloud droplets. It turns out that as far as the radiative properties are concerned these are the only parameters that matter. Detailed information about the droplet size distribution, such as skewness, width, and shape, is not required for this purpose.

The *effective radius* is defined as

$$\langle r \rangle = \int_0^\infty dr\, n(r) r^3 \Big/ \int_0^\infty dr\, n(r) r^2, \tag{9.1}$$

where $n(r)$ is the cloud droplet size distribution $[\text{m}^{-3} \cdot \mu\text{m}^{-1}]$. The numerator is proportional to the liquid water content (LWC), while the denominator is related to the extinction coefficient

$$k_c = \int_0^\infty dr\, Q_{\text{ext}}(r) n(r) r^2, \tag{9.2}$$

where the *extinction efficiency*, Q_{ext}, is defined as the ratio of the extinction cross section to the geometrical cross section (πr^2) of the spherical droplet. If the size of the droplet is large compared to the wavelength λ (i.e., $2\pi r/\lambda \gg 1$), the extinction efficiency asymptotically approaches 2. Therefore, in the visible range, where this inequality applies, we find from Eqs. 9.1 and 9.2 that the extinction coefficient of the cloud may be approximated by[9]

$$k_c \approx \frac{3}{2}\frac{LWC}{\langle r \rangle} \quad \text{(cgs units)}. \tag{9.3}$$

This equation suggests that a parameterization of cloud optical properties in terms of just the effective radius and the liquid water content might be expected to work well. It turns out, in fact, that we can replace the lengthy Mie–Debye computations with a small table of coefficients to derive the optical properties from the following simple algebraic expressions:[10]

$$k_c/LWC = a_1 \langle r \rangle^{b_1} + c_1, \tag{9.4}$$

$$1 - a_c = a_2 \langle r \rangle^{b_2} + c_2, \tag{9.5}$$

$$g_c = a_3 \langle r \rangle^{b_2} + c_3. \tag{9.6}$$

Here k_c is the cloud extinction coefficient, a_c is the single-scattering albedo, and g_c is the asymmetry factor. The coefficients (the a_is, the b_is, and the c_is) are determined by fitting these expressions to exact computations. Comparison with exact computations shows that this parameterization yields very accurate results. Optical properties of water clouds are shown in the left panel of Fig. 9.6.

9.3.4 Optical Properties of Ice Clouds

Cirrus clouds are high-altitude ice clouds occurring over land and ocean. They are present worldwide, covering 20–30% of the globe. These clouds consist of nonspherical ice crystals of various shapes, including hexagonal plates and columns. In the polar regions ice clouds occur closer to the surface and, because of their small sizes, are frequently referred to as "diamond dust."

A major difficulty in determining reliable and efficient radiative properties of ice particles in clouds is associated with the nonspherical, irregular shape of ice crystals. Attempts to use either area-equivalent or volume-equivalent ice spheres, so that Mie–Debye theory could be applied to compute ice crystal scattering and absorption

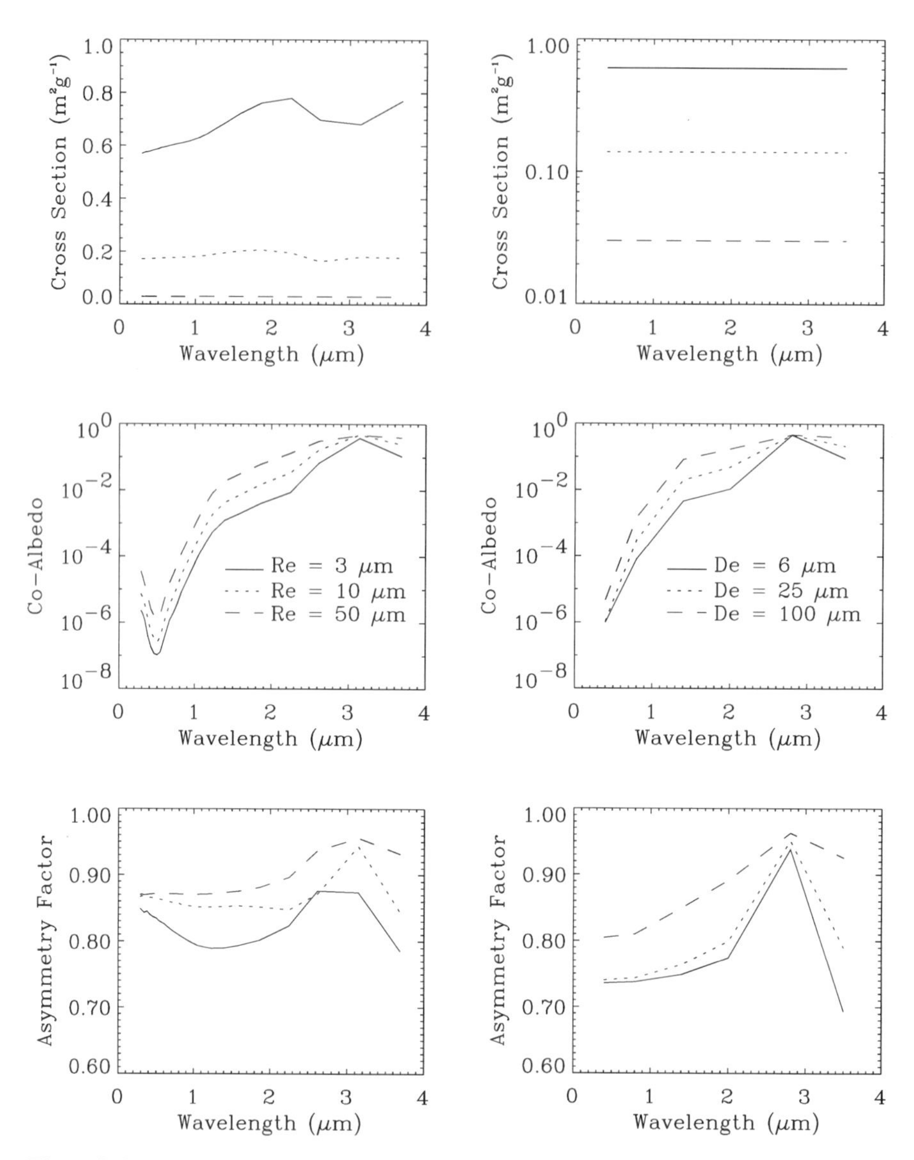

Figure 9.6 The three panels to the left show from top to bottom the volume extinction coefficient (divided by liquid water content), the single-scattering co-albedo, and the asymmetry factor versus wavelength for equivalent radii 3, 10, and 50 μm. The panels to the right show the same properties for ice clouds for equivalent diameters 10, 50, and 100 μm.

properties, have been shown to be inadequate. It has been shown that[11] (i) use of an equivalent sphere model to infer cirrus cloud properties from remote sensing data leads to significant underestimation of cloud optical depth and underestimation of cloud height, (ii) interpretation of polarization measurements for ice clouds cannot be based on spherical models, (iii) correct interpretation of broadband albedo of cirrus clouds must rely on correct nonspherical scattering properties, and (iv) the climatic forcing computed by spherical models will be seriously misleading. We should also expect that estimates of photolysis rates and UV exposure based on spherical aerosol/ice crystal particle models will be unrealistic.

Nonspherical ice crystal size distributions are usually expressed in terms of the maximum dimension (or length) L. Assuming that light scattering is proportional to the cross-sectional area of nonspherical particles, we may define an *effective size* (analogous to the effective radius for spherical droplets, Eq. 9.1) as follows:[12]

$$D_e \equiv \int_{L_{min}}^{L_{max}} dL\, D^2 \cdot L \cdot n(L) \Bigg/ \int_{L_{min}}^{L_{max}} dL\, D \cdot L \cdot n(L). \tag{9.7}$$

Here $D(L)$ denotes the ice crystal width, $n(L)$ is the ice particle distribution, and L_{max} and L_{min} are the maximum and minimum lengths of ice particles. Thus, the effective width (or size) is defined solely in terms of the ice particle size distribution. The numerator is related to the ice water content (IWC) given by

$$IWC = \frac{3\sqrt{3}}{8}\rho_i \int_{L_{min}}^{L_{max}} dL\, D^2 \cdot L \cdot n(L), \tag{9.8}$$

where the volume of an hexagonal ice crystal, $3\sqrt{3}D^2L/8$, is used and ρ_i is the bulk density of ice. The extinction coefficient is defined by

$$k_c \equiv \int_{L_{min}}^{L_{max}} dL\, k(D, L) n(L). \tag{9.9}$$

Here $k(D, L)$ is the extinction cross section for a single crystal. In the geometric optics limit, the extinction cross section is twice the effective cross-sectional area and may be expressed as

$$k(D, L) = \frac{3}{2}D\left[\frac{\sqrt{3}}{4}D + L\right]. \tag{9.10}$$

This expression for $k(D, L)$ is based on the assumption that the ice crystals are hexagonal and randomly oriented in space. In this case it can be shown that k_c/IWC is linearly related to $1/D_e$, that is,

$$k_c \approx IWC(a + b/D_e), \tag{9.11}$$

where a and b are constants. By further assuming that the absorption is small, one may express the absorption cross section as the product of the imaginary part of the refractive index, m_i, and the particle volume:

$$\alpha_n(\lambda) = \frac{3\sqrt{3}\pi m_i(\lambda)}{2\lambda} D^2 L. \tag{9.12}$$

The extinction and absorption cross sections defined above, combined with the notion that D and L are related, allow us to express the single-scattering co-albedo approximately as

$$1 - a_c \approx c + d \cdot D_e, \tag{9.13}$$

where c and d are best-fit constants.

For cases in which the geometric optics assumption (underlying Eq. 9.11) and the assumption of small absorption (underlying Eq. 9.13) are not valid, higher-order expansions may be used to represent the single-scattering properties of ice crystals more accurately:

$$k_c = IWC \sum_{n=0}^{N} a_n / D_e^n, \qquad 1 - a_c = \sum_{n=0}^{N} b_n / D_e^n, \tag{9.14}$$

where a_n and b_n are coefficients to be determined by fitting accurate values to the above expressions, and N is the total number of terms required to achieve a desired accuracy. Note that Eqs. 9.14 reduce to Eqs. 9.11 and 9.13 for $N = 1$. Finally, the moments of the phase function may be determined in a similar fashion. For example, the first moment (i.e., the asymmetry factor, g_c) may be expressed in terms of coefficients, c_n, as follows:

$$g_c = \sum_{n=0}^{N} c_n / D_e^n. \tag{9.15}$$

To determine the unknown coefficients one may use ray-tracing techniques developed for hexagonally shaped particles if $D_e > 30$. For smaller D_e, solutions developed for spheroidally shaped particles may be adopted in the absence of solutions valid for hexagonally shaped crystals. Since the smaller sizes are usually associated with IR wavelengths where ice is highly absorbing, the detailed shape may not be critical in scattering and absorption computations. Optical properties of ice clouds are shown in the right panel of Fig. 9.6.

In spite of a considerable effort by several groups worldwide to compute the optical properties of nonspherical particles the situation is less than satisfactory in that exact and approximate solutions seem to be generally lacking for nonspherical particles with size parameters between 20 and 30, in particular for shapes deviating substantially from that of the sphere.[13] For size parameters larger than 30, ray tracing based on geometrical optics may be used, while a variety of approaches have been tried for size parameters comparable to the wavelength of light. Because these approaches are generally based on theories valid for relatively small departures from the spherical shape, they suffer from numerical ill-conditioning[14] when they are applied to particles with (i) large real and/or imaginary refractive index, (ii) large size compared with the wavelength of incident light, and (iii) extreme shapes that deviate substantially from that of a sphere.

9.3.5 Optical Properties of the Ocean

The underwater spectral radiation field depends on the spectral and angular distribution of the solar radiation arriving at the water surface and on the optical properties of the water body. The latter are determined by Rayleigh scattering and absorption by water molecules (pure water) as well as scattering and absorption by suspended particles of organic and inorganic origin. As explained previously it is important to take into account the change in refractive index occurring at the atmosphere–ocean interface, in order to describe accurately the transfer of radiation in the coupled atmosphere–ocean system. In addition, modelers generally separate the atmosphere and ocean into a sufficient number of equivalent layers to resolve adequately the optical properties of each medium. Vertical profiles of atmospheric ozone, density, cloud cover, aerosol loading, as well as suspended particles and dissolved organic materials in the ocean are used to generate optical properties (optical depth, single-scattering albedo, and phase function) for each layer of both the atmosphere and the ocean. Based on these inputs, models are used to compute the radiation field throughout the coupled system.

Spectral absorption and scattering coefficients may be computed from a three-component model[15] with contributions from phytoplankton, nonchlorophyllous particles, and dissolved organic matter. The absorption coefficient, $a_{\mathrm{case1}}(\lambda)$, in open oceanic waters, which is commonly referred to as *Case 1* waters, is

$$\alpha_{\mathrm{case1}}(\lambda) = \left[\alpha_{\mathrm{w}}(\lambda) + 0.06\alpha^*_{\mathrm{C}}(\lambda)C^{0.65}\right][1.0 + 0.2 * Y(\lambda)] \quad [\mathrm{m}^{-1}]. \tag{9.16}$$

Here $\alpha_{\mathrm{w}}(\lambda)$ is the absorption coefficient for pure seawater, $\alpha^*_{\mathrm{C}}(\lambda)$ is the specific chlorophyll-*a* absorption with units $[\mathrm{m}^{-1}\ (\mathrm{mg}\cdot\mathrm{m}^{-3})^{-0.65}]$, and C is the pigment concentration with units $[\mathrm{mg}\cdot\mathrm{m}^{-3}]$. Absorption by locally formed yellow substance correlated with the chlorophyll-*a* concentration is included in the second factor in Eq. 9.16 with a spectral variation $Y(\lambda)$ given as

$$Y(\lambda) = e^{\Gamma(\lambda-\lambda_0)}, \tag{9.17}$$

where $\lambda_0 = 440$ [nm] and $\Gamma = -0.014$ $[\mathrm{nm}^{-1}]$. For turbid coastal waters, commonly referred to as *Case 2* waters, we have

$$\alpha(\lambda) = \alpha_{\mathrm{case1}}(\lambda) + b_{\mathrm{S}}(\lambda_{\mathrm{S}})\alpha_{\mathrm{S}}(\lambda) + \alpha_{\mathrm{Y}}(\lambda_0)Y(\lambda), \tag{9.18}$$

where $\alpha_{\mathrm{S}}(\lambda)$ is the specific absorption by suspended matter, and $b_{\mathrm{S}}(\lambda_{\mathrm{S}})$ (with $\lambda_{\mathrm{S}} =$ 550 nm) and $\alpha_{\mathrm{Y}}(\lambda_0)$ are measures of the concentrations of suspended matter and yellow substance, respectively. $\alpha_{\mathrm{w}}(\lambda)$, $\sigma_{\mathrm{w}}(\lambda)$ (see below), $\alpha^*_{\mathrm{C}}(\lambda)$, and $\alpha_{\mathrm{S}}(\lambda)$ are tabulated.[16]

The total scattering coefficient is

$$\sigma(\lambda) = \sigma_{\mathrm{w}}(\lambda) + \sigma_{\mathrm{C}}(\lambda) + \sigma_{\mathrm{S}}(\lambda) \quad [\mathrm{m}^{-1}], \tag{9.19}$$

where $\sigma_{\mathrm{w}}(\lambda)$ is the scattering coefficient of pure seawater. The scattering by phytoplankton $\sigma_C(\lambda)$ is computed from the pigment concentration C by

$$\sigma_C(\lambda) = \Lambda\ C^{0.62}\frac{\lambda_S}{\lambda}, \tag{9.20}$$

where Λ is 0.3 m^{-1} $(\text{mg}\cdot\text{m}^{-3})^{-0.62}$ and $\lambda_S = 550$ nm. This empirical expression is deduced from measurements in the near-UV and visible part of the spectrum (above 350 nm). Data for wavelengths shorter than 350 nm are generally lacking, but absorption and scattering by chlorophyll and dissolved organic matter seem to be the principal contributors to extinction in the UV spectral range. However, the concentration of dissolved organic matter seems to be influenced much more by natural (or anthropogenic) land sources than by marine biological activity. Thus, in regions where land sources are deemed to be of lesser importance – such as the open ocean or the waters surrounding the Antarctic continent – it may be sufficient to include absorption and scattering by chlorophyllous pigment. The scattering coefficient for chlorophyll may then be computed from Eq. 9.20. Variations in the specific spectral absorption coefficient are shown in Fig. 9.7.[17]

For more general situations, including coastal waters, we may use the total scattering coefficient $\sigma(\lambda_S)$ to characterize the scattering by nonchlorophyllous particles. The contribution $\sigma_S(\lambda)$ from nonchlorophyllous particles to the total scattering coefficient $\sigma(\lambda_S)$ is found by subtracting the contributions from algae and pure water,

$$\sigma_S(\lambda_S) = \sigma(\lambda_S) - \sigma_C(\lambda_S) - \sigma_w(\lambda_S). \tag{9.21}$$

The spectral variation of $\sigma_S(\lambda)$ is given by

$$\sigma_S(\lambda) = \sigma_S(\lambda_S)\left(\frac{\lambda_S}{\lambda}\right)^{-n}. \tag{9.22}$$

Here n is a number between 0 and 2, which depends on the specific type of sediment. Dissolved organic material (yellow substance) is assumed not to contribute to the total scattering coefficient.

The total asymmetry factor is computed from the individual contributions as follows:

$$g(\lambda) = \frac{g_w\sigma_w(\lambda) + g_C\sigma_C(\lambda) + g_S\sigma_S(\lambda)}{\sigma(\lambda)}. \tag{9.23}$$

If we adopt the Henyey–Greenstein phase function, this description is sufficient. For Rayleigh scattering by water we have $g_w = 0$. Chlorophyllous as well as nonchlorophyllous particles are large compared to the wavelength of light and therefore have large g values (for example, $g_C \sim 0.99$, and $g_S \sim 0.97$). However, the numerical values of g_C and g_S will vary with the water type.

9.3.6 Optical Properties of Snow and Ice

Making the somewhat dubious assumption that snow consists of spherical particles, we may obtain their optical properties from Mie–Debye computations, which require

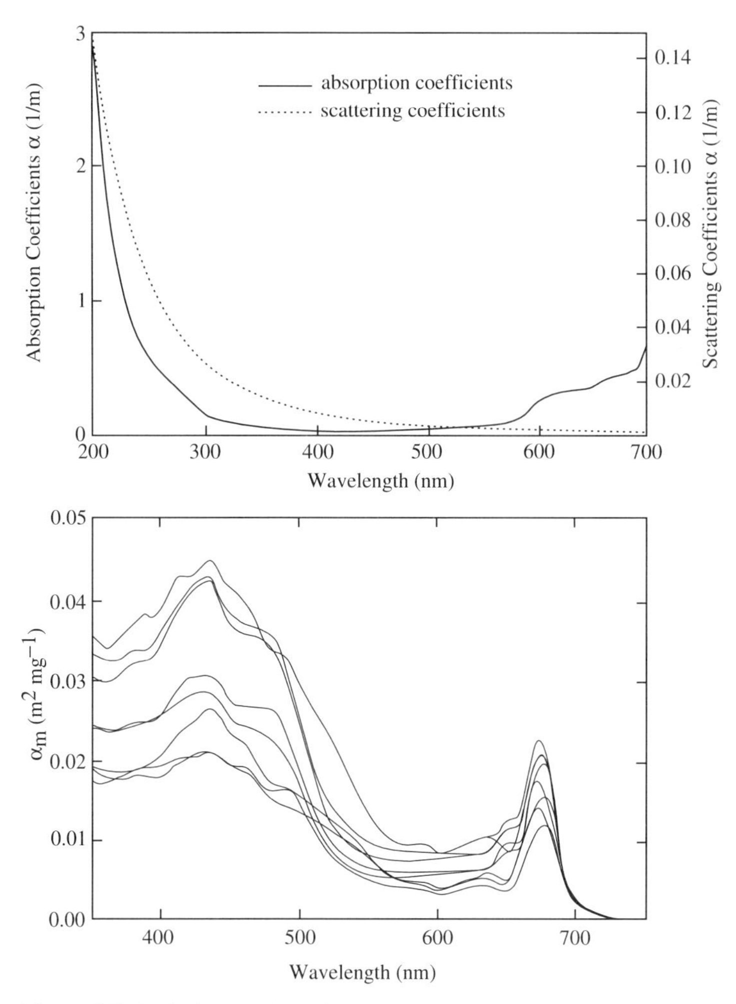

Figure 9.7 Optical properties of seawater. Upper panel: Scattering and absorption coefficients for pure seawater. Lower panel: Chlorophyll-specific spectral absorption coefficient for eight species of phytoplankton.

the refractive index of ice and the mean radius as input. Surprisingly this approach leads to results for predicted snow albedo that agree well with available observations. The following reasons why one does not make large errors by assuming spherical particles have been put forward by Craig Bohren:[18]

> The orientationally averaged extinction cross section of a convex particle that is large compared with the wavelength is one-half its surface area. The absorption cross section of a large, nearly transparent particle is proportional to its volume almost independent of its shape. The closer the real part of the particle's refractive index is to 1, the more irrelevant the particle shape. The asymmetry parameter of large particles is dominated by near-forward scattering, which does not depend greatly on particle shape.

In sea ice we must consider absorption by pure ice, scattering and absorption by brine pockets and the solid salts trapped within the ice, and scattering by air bubbles. Volume fractions of brine, air, and solid salts may be derived in an approximate manner from the ice temperature, density, and salinity. Assuming again that the brine pockets, air bubbles, and solid salts are spherical in shape, we may obtain the optical properties from Mie–Debye computations. Each component of the sea ice interacts differently with radiation. Pure ice acts mainly as an absorber. Although air bubbles and brine pockets both scatter strongly, air bubbles scatter more efficiently, whereas brine pockets also absorb radiation. The overall optical properties of the sea ice depend on the volume occupied by each component.

Salinity, density, and temperature may vary within the sea ice. Conditions representative for multiyear ice in the central Arctic in mid-May and September are: ice thickness 3 m, salinity 0.3%, density 0.9 mg $\cdot$ m^{-3}, and surface temperature -10°C. There is strong absorption in the infrared portion of the solar spectrum and relatively weak absorption in the visible region. Salinity is unimportant for the optical properties of seawater but plays a significant role for sea ice, because of brine pocket development associated with brine rejection when sea ice melts. Therefore, the albedo of first-year ice is only about half the value of multiyear ice due to air bubble and brine pocket formation, leading to increased scattering in the uppermost layer of multiyear ice. This is illustrated in Fig. 9.8, which shows a comparison between measured[19] and computed[20] results. No attempt was made to tune the model to get better agreement.

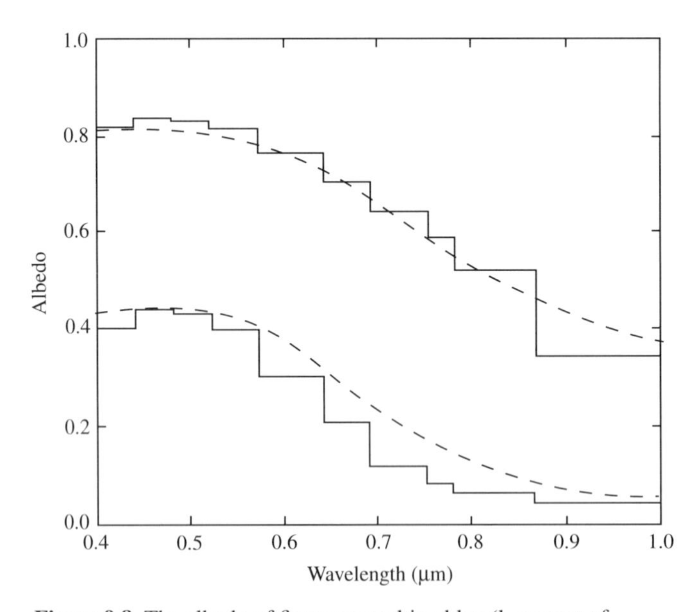

Figure 9.8 The albedo of first-year melting blue (lower set of curves) and multiyear white (upper set of curves) sea ice. The salinity and density were taken to be constant within the ice, and the temperature was assumed to vary linearly with depth from the surface value to a bottom temperature fixed at -2°C.

9.4 Modeling of Shortwave Radiative Effects in the Atmosphere

Band models will be discussed in more detail in Chapter 10. They are of great practical importance because in most atmospheric chemistry and climate models we cannot afford to compute photolysis rates and warming/cooling rates line by line throughout the spectrum. In Chapter 10 we describe how band models, which allow inclusion of multiple scattering, can be constructed to compute the radiation field in various parts of the spectrum. Here we shall consider the relatively simple case of UV/visible solar radiation. The simplicity stems from the fact that in the spectral range 200–700 nm the absorption cross sections of the main absorbers – ozone and molecular oxygen – vary quite smoothly with wavelength. This makes it straightforward to construct a band model by simply averaging over suitably chosen spectral intervals. By contrast, in the near-IR part of the solar spectrum (beyond 700 nm) and in the thermal IR the absorption cross sections vary erratically and substantially over small spectral intervals. This circumstance implies that simple averaging becomes inaccurate.

Integration over spectral ranges for which the absorption cross sections change rapidly is discussed in the next chapter. Only a brief introduction is provided here. To illustrate the problem let us consider the transmission of a normal beam of radiation through a slab of optical thickness $\Delta\tau_\nu$. We write the mean beam transmission over a wavelength interval $\Delta\nu$ as

$$\langle \mathcal{T}_\mathrm{b} \rangle = \frac{1}{\Delta\nu} \int_{\Delta\nu} d\nu e^{-\Delta\tau_\nu}, \tag{9.24}$$

where $\Delta\tau_\nu = \alpha_n(\nu)\mathcal{N}$, and $\mathcal{N}$ is the column density of the absorbing gas. If the wavelength interval is chosen to be narrow enough that the absorption cross section $\alpha_n(\nu)$ is approximately constant, then the transmission reduces to the exponential Extinction Law. Also, if $\alpha_n(\nu)$ varies smoothly and slowly across the wavelength interval $\Delta\nu$, we may approximate $\alpha_n(\nu)$ with a constant, which again leads to an exponential behavior. Finally, if $\alpha_n(\nu)$ varies rapidly and erratically across $\Delta\nu$, we may attempt to approximate the transmission function with a sum of exponential functions. This is convenient because we have the same mathematical form (i.e., exponential attenuation) as in the previous case. The *exponential sum fitting of the transmission (ESFT)* procedure has been used frequently in the past to deal with gaseous absorption/emission (see Eq. 10.37). So has a physically different, but mathematically similar, approach called *the k-distribution method*, which also reduces the transmission function to a sum of exponentials. As discussed in Chapter 10, the advantage of this reduction to a sum of exponentials is that multiple scattering by cloud and aerosol particles is easy to include in the calculation. This in turn implies that a unified treatment of radiative transfer may be achieved across the solar and terrestrial spectra.

The cross section for molecular oxygen is shown in Fig. 4.12. This figure shows the weak Herzberg continuum in the spectral range 175–200 nm, and the more significant O_2 absorption in the Schumann–Runge bands between 175 and 195 nm. At still shorter wavelengths, the structure is very complex. The very deep O_2 minimum, which coincides almost exactly with the hydrogen Lyman-α line at 121.6 nm, has important consequences in the upper atmosphere (above about 70 km). For $\lambda < 102.65$ nm, atmospheric absorption is caused by bound–free ionizing transitions. The cross section in the Schumann–Runge bands varies rapidly over small wavelength intervals. It also depends on temperature and therefore on altitude due to Doppler broadening. This implies that it is better treated by line-by-line models or by the *correlated-k distribution* and similar techniques discussed in Chapter 10.[21]

Example 9.1 Computational Strategies

To compute warming and photolysis rates we need only the irradiance or the mean intensity of the radiation field as explained in Chapter 5. We may therefore start with the azimuthally averaged version of the equation describing the transfer of diffuse monochromatic radiation at wavelength λ in a scattering and absorbing atmosphere:

$$u \frac{dI_\lambda}{n(z)dz} = -[\alpha(\lambda) + \sigma(\lambda)]I_\lambda(z, u) + \frac{\sigma(\lambda)F_\lambda^s}{4\pi} p_\lambda(-\mu_0, u)e^{-\tau_\lambda(z)/\mu_0}$$
$$+ \frac{\sigma(\lambda)}{2} \int_{-1}^{1} du' p_\lambda(u', u) I_\lambda(z, u') e^{-\tau_\lambda(z)/\mu_0}, \tag{9.25}$$

where $\tau_\lambda(z)$ is the extinction optical depth. For clarity we have written Eq. 9.25 for a single-constituent atmosphere with number density $n(z)$. Generalization to a multiconstituent medium is easily achieved by summing over species. By dropping the solar forcing term in Eq. 9.25, so that $I_\lambda(z, u)$ refers to the total diffuse plus direct intensity, and then integrating over 4π steradians, we find that the irradiance F and the mean intensity $\bar{I}$ are related by

$$\frac{dF(z)}{dz} = -4\pi n(z)\alpha(\lambda)\bar{I}, \tag{9.26}$$

where we have dropped the λ subscripts. Since the warming rate is proportional to the divergence of the irradiance, and the photolysis rate depends on the mean intensity, we see that *it is sufficient to compute the mean intensity of the radiation field to derive both warming and photolysis rates.*

From a numerical point of view it is advisable to use the mean intensity to compute the irradiance divergence because we avoid: (i) taking the difference between the upward and downward irradiance to compute the net irradiance and (ii) then taking another difference between the net irradiances at two nearby levels to compute the divergence medium, there will be little difference between the net irradiance at two nearby levels, so that we end up subtracting two numbers that are almost equal, which is numerically inaccurate. It is interesting to note that estimations of the irradiance divergence in clouds from measurements of upward and downward hemispherical irradiances have been fraught with difficulties for basically the same reason. The measurement of upward and downward irradiance must be very accurate to avoid loss of accuracy associated with the numerical differentiation used to derive the irradiance divergence from such measurements.

It is clear that for the purpose of testing models it would be better to measure the mean intensity than the irradiance divergence.

9.4.1 Spectral Averaging Procedure: The Chandrasekhar Mean

Figure 1.1 shows that the solar irradiance at the top of the atmosphere falls off rapidly with decreasing wavelength below 350 nm, whereas the ozone absorption cross section increases rapidly between 350 and 250 nm. Because of these steep gradients in UV flux and ozone cross sections, we must choose the wavelength bins carefully to avoid errors caused by a wavelength resolution that is too coarse. Rather than using an arithmetic mean of the cross section for any bin, we may define a mean absorption cross section by weighting it with the solar flux across the bin as follows:

$$\langle \alpha_n \rangle \equiv \frac{\int_{\lambda_1}^{\lambda_2} d\lambda \alpha_n(\lambda) F_\lambda^S}{\int_{\lambda_1}^{\lambda_2} d\lambda F_\lambda^S}, \tag{9.27}$$

and analogously for the mean scattering cross section, $\langle \sigma_n \rangle$. It is important to note that all cross sections used in the calculations should be weighted in this manner.[22] Thus, we must define an average value for the absorption and scattering cross sections by applying Eq. 9.27 to each of them, and in the integral defining the photolysis rate, the photoabsorption cross section, $\alpha_n^i(\lambda)$, and the quantum efficiency, $\eta^i(\lambda)$, must be similarly averaged.

9.4.2 Solar Warming Rates Due to Ozone, Aerosols, and Clouds

In the UV/visible part of the spectrum the major warming is due to absorption by ozone in the Hartley–Huggins band for wavelengths between 200 and 350 nm. Ozone also absorbs in the Chappuis band between 400 and 700 nm, but as shown in Fig. 1.1 the cross section here is much weaker than in the Hartley band. Since there are no other absorbers that provide real competition with ozone between 200 and 700 nm (except the Schumann–Runge bands below 200 nm, which we do not include here for reasons stated in the beginning of this chapter), we may ignore other absorbers and focus on ozone for demonstration purposes. In Fig. 9.9 we show warming rates computed for a solar zenith angle of 30°. We note that the warming rate has a broad maximum around 45 km altitude. It is proportional to the product of the mean intensity and the absorber density. If we were to plot the mean intensity versus decreasing altitude, we would find that intensity falls off rapidly, while the ozone density increases down to about 20 km (see solid curve in Fig. 9.2). Thus, the product will maximize at the altitude where the two curves cross each other. This explains the maximum in the warming rate, which was first pointed out in the 1930s by S. Chapman (see Problem 9.7). Therefore, the

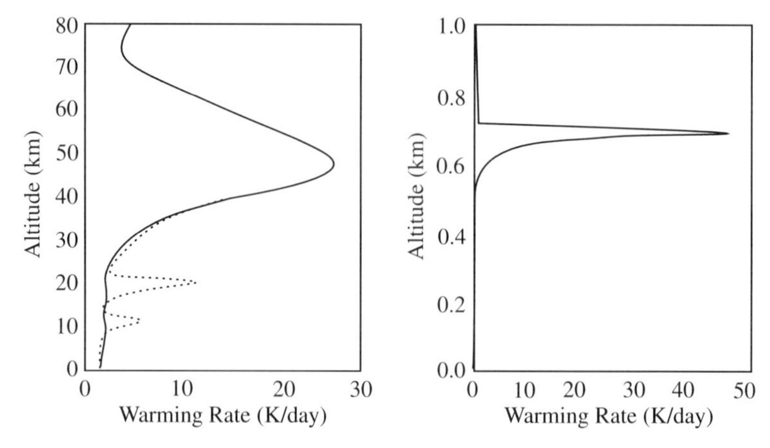

Figure 9.9 Left panel: Instantaneous warming rate due to ozone absorption between 200 and 700 nm for a solar zenith angle of 30°. For this computation optical properties derived from O_3 cross sections and the midlatitude summer atmosphere (see Appendix C) were used as input to the DISORT radiative transfer code (§8.8). Instantaneous warming rates for stratospheric aerosols and a cirrus cloud are also shown. Right panel: Warming rate for a simulated low-level warm cloud (see text) where the solar zenith angle is 60°.

shape of the curve is commonly referred to as a *Chapman profile*. Warming rates for other species exhibit similar behavior.

To illustrate the radiative effects of stratospheric aerosols we show in Fig. 9.9 the instantaneous solar warming rates for an extreme case of volcanic aerosol loading. We note that this extreme volcanic aerosol loading enhances the solar warming rate by up to 10 K/day within the peak of the aerosol layer. This is caused by the increased near-IR absorption by the aerosols. We also note that backscattering of radiation by the aerosol layer leads to enhanced warming by about 0.3 K/day throughout the stratosphere above the layer. For more moderate aerosol loadings the warming rates are similar but smaller in magnitude. In contrast, in the troposphere, the reduced solar transmission results in a cooling of the surface, although this response is delayed and modulated by the slower ocean response. Compared to the solar warming, stratospheric aerosols have a relatively modest effect on the terrestrial IR warming/cooling rate. For an extreme volcanic aerosol model, although the IR effect is significant, it is a factor of five smaller than the solar effect. Compared to the clear-sky case the extreme volcanic aerosol model gives rise to a warming (less cooling) below the aerosol layer due to upwelling IR radiation that is backscattered by the aerosol layer and absorbed by the underlying atmosphere. There is a corresponding cooling above the layer due to the attenuation of the upwelling IR radiation. In summary, the effect of a stratospheric aerosol layer on atmospheric warming/cooling rates is dominated by solar radiation, so that the net (solar plus IR) radiative impact is to induce cooling below and warming within and above the layer. This behavior is in agreement with measurements of atmospheric temperature following volcanic eruptions. The size of the tropospheric cooling depends on the magnitude of the eruption. For example, the recent Mount Pinatubo eruption had a significant cooling[23] effect at the surface, about 0.5°C, large enough to offset temporarily the warming due to increases in anthropogenic trace gases.

A cooling effect of tropospheric aerosols has been recently identified. This is due to absorption from anthropogenic sulfate aerosols, which arise through a complex chain of reactions involving gas-to-particle conversion. The globally averaged forcing is estimated to range from -0.3 up to $-2\ \mathrm{W \cdot m^{-2}}$ by various investigators.[24] Thus along with ozone depletion, the increase of aerosols may have at least partially mitigated the positive forcing from greenhouse gases. Since the source of sulfate aerosols is atmospheric pollution, the problem is complicated because the effect is strongly regional, concentrating in large population centers mainly in the northern hemisphere.

A cirrus cloud scatters solar radiation efficiently because of its high single-scattering albedo (see Fig. 9.4). The maximum instantaneous warming rate for a solar zenith angle of 30° is about 5–6 K/day for this particular cloud. Backscattering of solar radiation from the cirrus deck leads to a slight increase in the solar warming rate above the cloud and a corresponding decrease below because the cloud prevents radiation from reaching those levels.

A tropospheric water cloud strongly affects solar warming rates and net irradiances. An example of the warming rate produced by a low-level cloud is displayed in the right panel of Fig. 9.9. This particular cloud was simulated with a simple model[25] mimicking condensation of water vapor on cloud condensation nuclei. The model provides a self-consistent treatment of the interaction between radiative and cloud microphysical processes important for cloud formation in the Arctic. This model employs a convective adjustment scheme similar to that discussed in Chapter 12 in connection with simple climate models. As a consequence droplet spectra are mixed, leading to a significant broadening of the droplet size distribution at cloud top. The shape of the computed droplet spectra are in general agreement with observations. The droplet effective radius changed from about 12 μm at cloud top to about 7 μm at cloud bottom. The liquid water content for this cloud is $0.8\ \mathrm{g \cdot m^{-3}}$ at cloud top decreasing linearly toward cloud bottom. The droplet concentration is nearly constant throughout the cloud. Thus, the increase in liquid water content with height is caused by the increased droplet size. The cloud formed (from clear air) between about 200 and 800 m.

9.4.3 Computation of Photolysis Rates

To compute the photolysis rate we need to integrate the radiation field multiplied by the photoabsorption cross section and by the quantum efficiency for any particular species of interest as indicated in Eq. 5.80. There are many chemical reactions that need to be considered depending on the problem of interest. For example, one may be interested in the effects of anthropogenic release of chlorofluorocarbons on the stratospheric ozone abundance or the effects of industrial pollution on the tropospheric sulfur cycle. Caution must be exercised in choosing bin sizes for the computation of photolysis rates for all the different chemical processes in operation. The problem is that the cross sections for these various photoabsorption processes vary substantially across the UV/visible spectral range. Some of these cross sections are sharply peaked over a narrow spectral range, whereas others vary more smoothly over a wider spectral region. There is also overlap between cross sections of different species in some regions of the spectrum. The presence of these factors emphasizes the importance of choosing the bin sizes very carefully to avoid inaccuracies in the computed rates. A relatively coarse grid that may provide good accuracy for one particular species may be inadequate for another species whose cross section varies more rapidly over the same spectral range. In practice this means that we need to be somewhat conservative in the choice of grid size for the purpose of computing photolysis rates for a multitude of species.

By choice of a proper spectral grid an accurate integration over wavelength may be reduced to a sum of monochromatic problems. However, there may still be errors in the computation associated with the numerical solution of each of these monochromatic problems owing to the choice of solution method. We know, for example, that a two-stream approximation is less accurate than a high-order multistream method. However, the larger the number of streams, the more expensive the computation. It is therefore of interest to find out how well we can do in the lowest order of approximation, that is, the two-stream approximation. As an example we consider photolysis of ozone (O_3) in which the oxygen atom is left either in the ground state $O(^3P)$ ($\lambda > 310$ nm) or in the excited state $O(^1D)$ ($\lambda < 310$ nm). Numerical results for a clear Rayleigh-scattering atmosphere indicate that the error in photolysis rates computed by the two-stream approximation as compared to an accurate multistream computation depends on solar zenith angle and surface albedo and varies with altitude.[26] It is noteworthy, however, that the error is generally small compared to uncertainties in other parameters used in photochemical modeling. Thus, for this particular example, the error incurred by use of the two-stream approximation is less than 3% in the stratosphere and a maximum of about 8% in the troposphere. The errors are therefore less than the combined uncertainty in cross section and quantum yield, which is in the range 10–20%.

9.4.4 UV Transmission: Relation to Ozone Abundance

In spite of its small abundance, the large cross section of ozone causes it to absorb very effectively the UV radiation shortward of 300 nm. For this reason ozone is said to provide a protective "shield" against harmful UV radiation. To illustrate this shielding effect we have plotted in Fig. 9.10a the optical depth of ozone versus wavelength for several ozone column abundances. The Rayleigh-scattering optical depth is also shown for comparison. The corresponding flux transmittance is shown in Fig. 9.10b.

Comparison of computed results for downward irradiance and mean intensity at the surface, for dark and bright surfaces, shows that the mean intensity is much more strongly affected by changes in surface albedo than is the downward flux. Since both

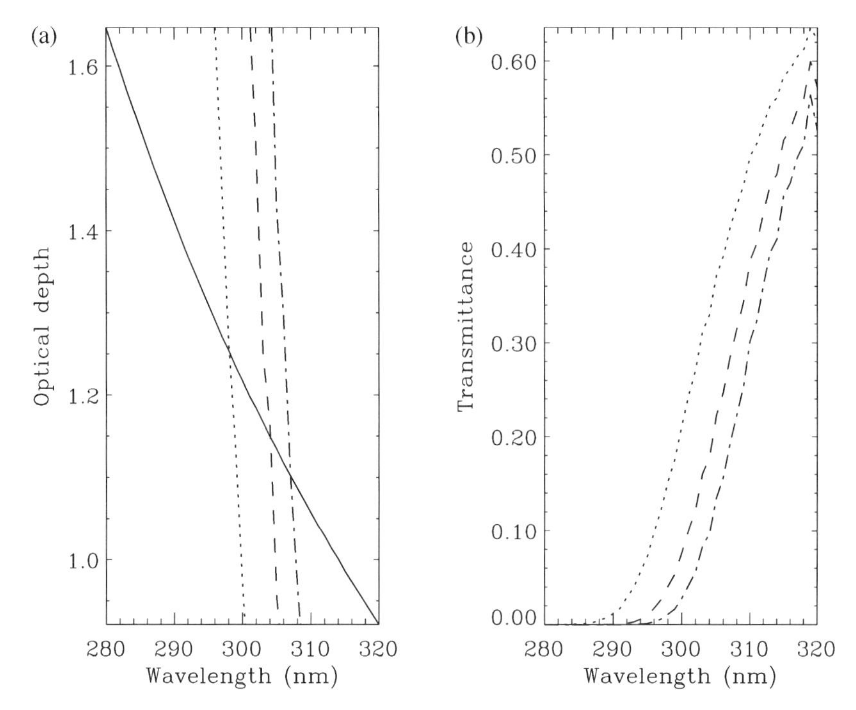

Figure 9.10 (a) Wavelength dependence of total optical depth (Rayleigh scattering plus ozone absorption) for several column amounts (100, 200, 300 DU).
(b) Corresponding flux transmittances. The solar zenith angle is 0° and the underlying surface is totally absorbing. The U.S. Standard Atmosphere was used as input and scaled to 100, 200, and 300 DU, and the DISORT code was used to do the radiative transfer computation.

photolysis rates and warming rates are proportional to the mean intensity, they will therefore be significantly affected by the value of the surface reflectance.

The strong (λ^{-4}) wavelength dependence of Rayleigh scattering is evident in Fig. 9.10. This implies that molecular scattering will become progressively more important toward shorter wavelengths. To illustrate the importance of multiple scattering we show in Fig. 9.11 the direct and diffuse irradiance at the surface for two solar zenith angles. It is clear that the smaller the solar elevation angle, the relatively more important is the diffuse component. In fact, when the Sun is close to the horizon the diffuse component dominates. This is clearly seen in the right panel of Fig. 9.11, which pertains to a solar zenith angle of 75°.

9.4.5 UV Transmission and Dose Rates at the Earth's Surface

The biological effect of UV radiation may be expressed in terms of the dose rate defined as a convolution of an action spectrum with the irradiance spectrum. Irradiance applies if the exposure applies to a flat surface (see Eq. 5.81). However, the mean radiance applies if our interest lies in the rate at which a small spherical particle receives energy

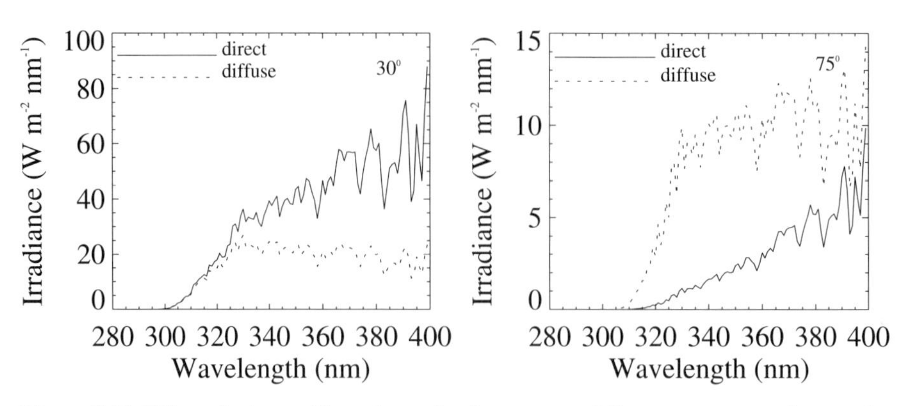

Figure 9.11 Effect of solar zenith angle on the direct versus diffuse components of spectral irradiance for solar zenith angles 30° and 75° as indicated.

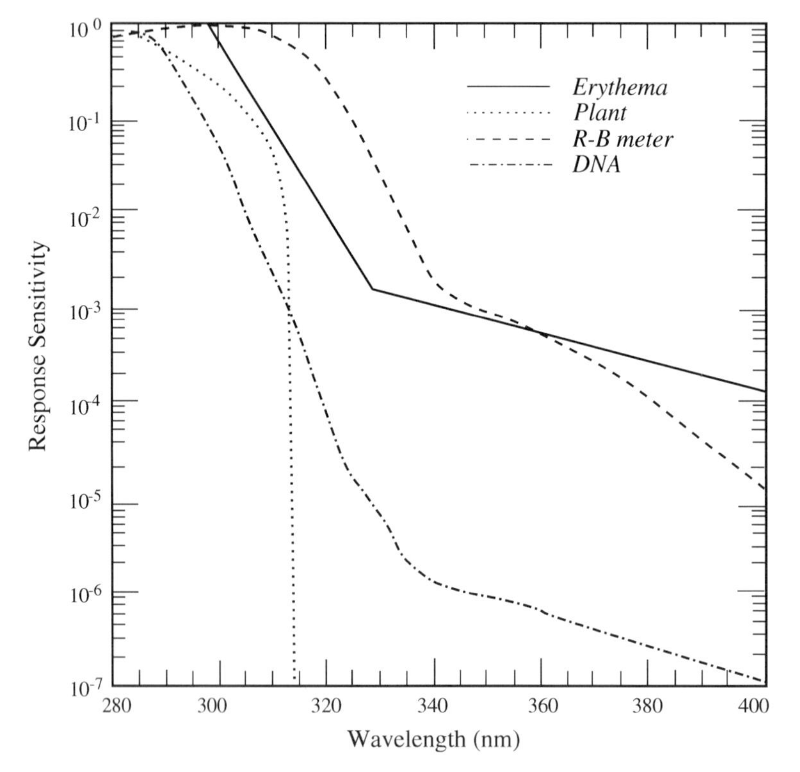

Figure 9.12 Action spectra for various biological responses. R-B meter stands for a measuring device, known as the *Robertson–Berger meter*, that was designed to mimic the sunburning response of Caucasian skin.

from the radiation field (see Eq. 5.82). Four different types of action spectra are shown in Fig. 9.12: generalized damage spectra for (1) the DNA molecule and (2) plant response, (3) a weighting spectrum for erythema (sunburning), and (4) the response spectrum for one particular measuring device – the *Robertson–Berger* (R–B) meter – designed to approximate the response of Caucasian skin.

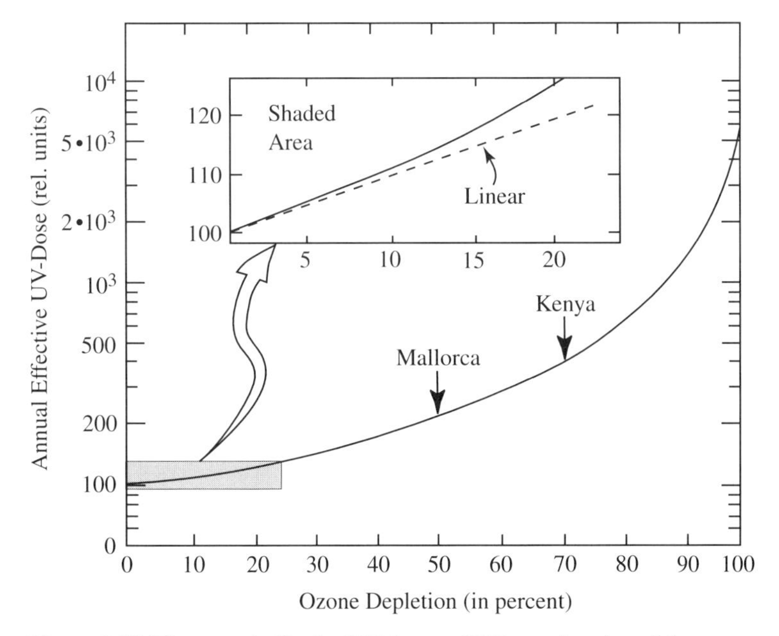

Figure 9.13 The annual effective UV dose at 60°N as a function of the ozone depletion (logarithmic scale). The annual UV dose, with normal ozone conditions throughout the year, is set to 100. The inset exhibits the dotted area with the dose axis enlarged and given a linear scale. The annual UV dose for latitude of 40°N (Mediterranean countries, California) and countries along the equator, with normal ozone conditions, are indicated by Mallorca and Kenya, respectively.

The computed annual effective UV dose for the erythema spectrum shown in Fig. 9.12 varies by about 4% per degree in latitude in the nothern hemisphere. Predictions based on ozone trend data, which show a small negative trend in ozone values from 1969 to 1986 in the northern hemisphere, indicate a change in effective annual dose that is small but positive for all latitudes. UV measurements at high elevations in the Swiss Alps from 1981 to 1989 indicate a slight increase in solar UV radiation of about 1% per year since 1981, in qualitative agreement with the reported ozone depletion of about 3% from 1969 to 1986.[27]

The annual radiation dose at 60°N is shown in Fig. 9.13, as a function of ozone depletion. The percent increase in UV dose per percent decrease in ozone abundance is commonly referred to as the *radiation amplification factor*. For small ozone depletions (less than 5%) the annual UV dose increases with an amplification factor close to 1 (see inset of Fig. 9.13). For larger ozone depletions the increase in annual UV dose is enhanced, that is, the curve becomes nonlinear. Thus, according to Fig. 9.13, a 10% depletion yields a 11.7% increase in UV dose, and a 20% depletion gives a 25.7% increase. These values are smaller than those obtained for the DNA action spectrum (Fig. 9.12), which falls off quite sharply with wavelength. This rapid falloff implies that DNA is more sensitive to UV-B radiation, and thus to changes in ozone amount, than the erythemal response, which has a significant response to UV-A radiation. Note also that the UV-A radiation is almost unaffected by ozone changes and is of much larger magnitude than the UV-B component (see Fig. 1.4). An amplification factor of 2 has often been adopted, implying that a 1% percent decrease in ozone gives rise to a 2% increase in UV dose. Such a value is obtained for the DNA action spectrum for typical midlatitude solar elevations and changes in ozone of 10% or less, but the response of humans and animals to UV radiation is perhaps more properly described by the erythema action spectrum. Hence, the amplification factor will depend on the particular type of biological response considered as well as on latitude and season.

Measurements of UV spectral irradiance during the Arctic summer have shown that the diurnal variation in the UV-B region is much stronger than in the UV-A. This demonstrates that diffuse radiation dominates the direct component in the UV part of the spectrum. The proportion of diffuse radiation is less pronounced at lower latitudes where the Sun is higher in the sky (see Fig. 9.11). In the Antarctic, the spectral distribution of UV radiation reaching the surface has been measured with scanning spectrometers deployed at four stations. The consequence of the Antarctic ozone hole[28] is evident in the data. For example, data taken at Palmer Station (65°S) during austral spring 1988 show that irradiances at wavelengths shorter than 310 nm measured in October often equaled or exceeded values measured at summer solstice.[29] Similarly, the UV radiation levels at McMurdo Station (78°S) in 1990 were substantially increased. UV irradiances in October exceeded the solstice value by a factor 1.5 for a ten-day period; a threefold increase (compared to the "normal" value) of the integrated daily dose of UV irradiance was measured at the surface.[30] The spectral measurements in both the Arctic and the Antarctic show that cloud cover provides substantial attenuation of UV radiation as well as large day-to-day variability. It would be desirable to make routine measurements with sufficient spectral information to identify ozone-related UV-B trends and to discriminate them from effects due to natural cloud variability, because spectrally integrated data will not allow for this kind of discrimination. In fact, it has recently been demonstrated that narrowband filter instruments with two or more channels in the UV are sufficient for such discrimination.[31] Spectral measurements obtained in New Zealand[32] during two separate campaigns eight years apart showed no change in UV exposure between 1980 and 1988, but more data are obviously needed to establish a climatology of UV irradiance. Finally, at other midlatitude locations in the Southern Hemisphere, increases in UV-B have been observed in connection with intrusions of ozone-depleted air from Antarctica. For example, high UV-B levels in Melbourne, Australia have been associated with air coming from the Antarctic ozone hole after the breakup of the Antarctic vortex.[33]

More recently UV measurements in Europe have been reported by several groups.[34] Similar efforts are also ongoing in the United States and Canada[35] and in the Southern Hemisphere as well.[36]

9.4.6 Comparison of Measured and Computed UV Irradiance at the Surface

Spectrally resolved measurements of UV irradiances transmitted to the surface can be used to test model computations. To compare computed and measured irradiance we should ideally use the solar irradiance measured at the top of the atmosphere with the same instrument as input to our calculations. The measured solar irradiance extrapolated to zero air mass by the Bouguer–Langley method would also be useful for this purpose. Because such solar irradiance data are usually not available, the comparison focuses on the ratio of the measured direct and diffuse irradiances to the computed ratio, which is independent of the solar irradiance at the top of the atmosphere.

Figure 9.14 shows a comparison between the measured and computed ratio of diffuse to direct irradiance.[37] This figure is for January 12, 1991 (which was prior to the eruption of Mt. Pinatubo). The deviations between the measured and computed results are also shown. Tropospheric aerosols are included in the computation. The aerosol content is based on measured values. The surface albedo was not measured,

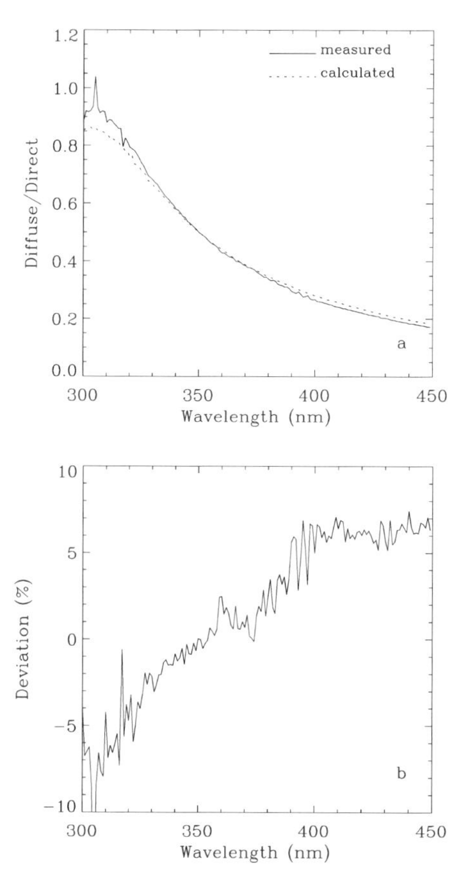

Figure 9.14 Comparison between measured and computed ratios of diffuse to direct spectral irradiance during cloud-free skies.

but it was varied within reasonable bounds in an effort to improve agreement between the measured and computed ratio. We can see from the right panel that the deviation of diffuse/direct irradiance ratio between measurements and model computation is less than 8% if the spikes in the measurement data on both days are ignored.[38] Since the ratios are formed from combinations of measurements, the difference is reasonable.

The ozone content used in the calculation was derived from the measurements by taking the ratio of the measured downward irradiance (diffuse plus direct) at two UV wavelengths (one of which is appreciably more absorbed by ozone than the other) and comparing this ratio to a computed ratio derived from a radiative transfer calculation. This in itself provides another example of the use of radiative transfer theory: To extract ozone abundance from global (diffuse plus direct) spectral irradiance measurements, the column ozone content in the model is varied until agreement between measured and computed irradiance ratio is achieved. Comparisons with other methods (TOMS and Dobson measurements) indicate that this method is reliable and accurate.[39] Ground-based retrieval of column ozone amounts under overcast conditions usually relies on measurements of the zenith sky radiance, which has to be corrected for cloud effects. This correction is usually based on empirical charts created for a particular location. Use of the global irradiance instead of the zenith-sky radiance makes the ground-based ozone retrieval insensitive to cloud conditions and obviates the need for such corrections.[40]

9.5 Modeling of Shortwave Radiation in the Ocean

9.5.1 Diffuse Radiation: Attenuation in the Ocean

The concept of exponential light penetration loses its usefulness when the multiply scattered component becomes important. One might think that scattering effects cannot be described by this simple beam-penetration notion. However, it is possible to extend its definition to describe the height at which the *total* radiation field is attenuated by the factor $(1/e)$. This concept has been used extensively in ocean optics, where measurements show that the total illumination is approximately exponential with ocean depth z. Thus it is natural to define an *apparent extinction coefficient* K, which may be defined in several different ways – in terms of the variation of the upward (upwelling) irradiance, the downward (downwelling) irradiance, the mean intensity (scalar irradiance), or the upwelling zenith intensity:

$$K_{\mathrm{u}}(\nu) \equiv -\frac{d\left(\ln\left[F_\nu^+(z)\right]\right)}{dz}, \qquad K_{\mathrm{d}}(\nu) \equiv -\frac{d\left(\ln\left[F_\nu^-(z)\right]\right)}{dz},$$
$$K_0(\nu) \equiv -\frac{d(\ln[\bar{I}_\nu(z)])}{dz}, \qquad K_{I^+}(\nu) \equiv -\frac{d\left(\ln\left[I_\nu^+(z,\mu=1)\right]\right)}{dz}. \tag{9.28}$$

These are also called the *diffuse attenuation coefficients*. These *apparent* optical properties are distinguished from the *intrinsic* properties (such as extinction, scattering, and absorption coefficients, phase functions, etc.). Clearly their values depend not only upon the ocean properties, but also upon the surface illumination, reflection from the ocean bottom, etc. Although they are not very satisfactory in this respect, they are easily measured and have practical value in conveniently describing the general falloff of radiation with depth. For example, apparent properties are useful in determining the maximum vertical extent to which photosynthesis would occur (the *euphotic zone*).

To be consistent with common usage in ocean optics we shall in the remainder of this section take z to be positive downward (from the ocean surface) and the polar angle θ to be 0° along the positive z axis. (Note that ocean optics people refer to the upwelling radiance in the direction $\theta = 180^\circ$ as the "nadir" radiance and to the downwelling radiance in the direction $\theta = 0^\circ$ as the "zenith" radiance.)

9.5.2 Two-Stream Model Appropriate for Deep Water

Consider a body of water deep enough that the radiation field in the upper layers is unaffected by radiation reflected from the bottom. This situation is typical of the open ocean. Measurements indicate that if the solar irradiance incident at the ocean surface is $F^-(0)$, the depth dependence of the irradiance can be expressed as

$$F^-(z) = F^-(0)e^{-Kz}, \tag{9.29}$$

where K is the attenuation coefficient for downward irradiance defined above (we have dropped the d subscript). For simplicity let us assume that a layer of the upper ocean near the surface has constant optical properties. Then by inserting Eq. 9.29 into Eq. 7.139, we find that (see Problem 9.2) the two-stream approximation for the attenuation coefficient is

$$K = \frac{(\alpha^+ + \alpha^-)}{2}\left[1 - \frac{4\beta^+\beta^-}{(\alpha^+ + \alpha^-)^2}\right]^{\frac{1}{2}} - \frac{(\alpha^+ - \alpha^-)}{2}, \tag{9.30}$$

where $\alpha^\pm$ and $\beta^\pm$ are defined in Eqs. 7.136. The expression for the irradiance reflectance is easily found to be

$$\rho \equiv \frac{F^+(z)}{F^-(z)} = \frac{\alpha^+ + \alpha^-}{2\beta^-}\left\{1 - \left[1 - \frac{4\beta^+\beta^-}{(\alpha^+ + \alpha^-)^2}\right]^{\frac{1}{2}}\right\}. \tag{9.31}$$

A useful relation between K and ρ that does not involve $\beta^\pm$ is (see Problem 9.3)

$$K(1 - \rho) = \alpha\left[\frac{1}{\bar{\mu}^+} + \frac{\rho}{\bar{\mu}^-}\right], \tag{9.32}$$

where $\bar{\mu}^\pm$ are defined in Eqs. 7.136 and α is the absorption coefficient (not to be confused with $\alpha^\pm$).

It can be shown (see Problem 9.4) that ρ increases monotonically when the single-scattering albedo $a = \sigma/(\sigma + \alpha)$ increases. In fact, for small values of $a = \sigma/(\sigma + \alpha)$, Eq. 9.31 reduces to

$$\rho = \frac{r^+}{1 + \bar{\mu}^+/\bar{\mu}^-} \frac{\sigma b}{\alpha}.$$

Physically this makes sense because we expect the reflectance to decrease with increasing absorption in the near-surface waters.

If other factors remain the same, the attenuation coefficient will increase if α increases (more absorption), or if σb increases, where b is the backscattering coefficient. If absorption dominates over backscattering ($\sigma b/\alpha$ is small enough) the attenuation coefficient (Eq. 9.30) can be approximated by (see Problem 9.3)

$$K \approx (\alpha + r^+ \sigma b)/\bar{\mu}^+ \approx \alpha/\bar{\mu}^+. \tag{9.33}$$

Thus, when $\sigma b/\alpha$ is sufficiently small, K depends primarily on the absorption coefficient α and the downward average cosine $\bar{\mu}^+$.

When backscattering dominates over absorption ($\sigma b/\alpha \gg 1$), K can be approximated by (see Problem 9.2)

$$K \approx \frac{\alpha}{\bar{\mu}^+} \frac{1 + r^+/r^-}{1 + (r^+ \bar{\mu}^-/r^- \bar{\mu}^+)}. \tag{9.34}$$

Thus, while K depends linearly on the backscattering coefficient σb (Eq. 9.33) when absorption dominates over backscattering, K becomes independent of σb (Eq. 9.34) when backscattering dominates over absorption.

9.5.3 Backscattering by Ocean Particles: The Role of Shape Factors

In recent years there has been significant emphasis in the ocean optics community on understanding the role of particulate scattering in determining the angular shape of the upwelling underwater radiation field. Considerable effort has been expended on devising simple means to describe the upwelling radiance in terms of inherent optical properties. $I^+(0, -\mu, \phi)$ is the upwelling radiance that leaves the water and is available for measurement by sensors deployed on a ship, an airborne platform, or a satellite.[41]

We consider the equation describing the upwelling radiance

$$-\mu \frac{dI^+(z, -\mu, \phi)}{dz} = -k(z) I^+(z, -\mu, \phi) + j^+(z, -\mu, \phi). \tag{9.35}$$

Note that, although we use $-\mu$ to denote the upward hemisphere ($\mu = |\cos\theta|$), we retain the notation I^+ for upwelling radiance. The term $j^+ = kS^+$ is the corresponding emission coefficient (see §2.8) due to multiple scattering, given by (see Eq. 6.9)

$$j^{+}(z,-\mu,\phi) = \frac{\sigma(z)}{4\pi}\int_0^{2\pi} d\phi' \left[\int_0^1 d\mu' p(z,\mu',\phi';-\mu,\phi) I^{-}(z,\mu',\phi') + \int_0^1 d\mu' p(z,-\mu',\phi',-\mu,\phi) I^{+}(z,-\mu',\phi')\right]. \tag{9.36}$$

Note that j^+ has the units of energy emitted per unit volume ([W · m^{-3}]), and differs from the source function ([W · m^{-2}]) by the factor of the extinction coefficient k. Note also that z is positive downward so that the first integral (involving integrations over $+\mu'$) represents the contribution to the emission coefficient from backscattered downwelling radiation. To isolate the influence of backscattered radiation on the upwelling radiance we introduce a dimensionless shape factor defined by

$$f_{\mathrm{b}}(z,-\mu,\phi) \equiv \frac{\int_0^{2\pi} d\phi' \int_0^1 d\mu' p(z,\mu',\phi';-\mu,\phi) I^{-}(z,\mu',\phi')}{4\pi b(z)\bar{I}^{-}(z)}. \tag{9.37}$$

Here b is the backscattering coefficient defined in Eq. 7.116, and $\bar{I}^{-}(z) = (2\pi)^{-1}\int_0^{2\pi} d\phi \int_0^1 d\mu I^{-}(z,\mu,\phi)$ is the mean downward intensity (note that $2\pi\bar{I}^{-}(z)$ is the downward scalar irradiance in ocean optics terminology).

The shape factor f_{b} is the ratio of downwelling radiation backscattered into direction $(-\mu,\phi)$ to the radiation that would be backscattered in the direction $(-\mu,\phi)$ if (1) the downwelling radiation were equal to $I^{-}(z,\mu',\phi')$ and (2) the phase function were constant and equal to $b(z)/4\pi$.

To account for the radiation scattered in the forward direction we introduce another shape factor defined by

$$f_{I^+}(z,-\mu,\phi) \equiv \frac{\int_0^{2\pi} d\phi' \int_0^1 d\mu' p(z,-\mu',\phi';-\mu,\phi) I^{+}(z,-\mu',\phi')}{4\pi[1-b(z)]\bar{I}^{+}(z,-\mu,\phi)}. \tag{9.38}$$

We see that $f_{I^+}(z,-\mu,\phi)$ is the ratio of upwelling radiation scattered into direction $(-\mu,\phi)$ to the amount of radiation that would be scattered into direction $(-\mu,\phi)$ if (1) the upwelling radiation were equal to $I^{+}(z,-\mu',\phi')$ and (2) the phase function were constant and equal to $2(1-b)$.

With these definitions the emission coefficient may be written

$$j^{+}(z,\mu,\phi) = f_{\mathrm{b}}(z,-\mu,\phi)\sigma(z)b(z)\bar{I}^{-}(z) + f_{I^+}(z,-\mu,\phi)\sigma(z)[1-b(z)]I^{+}(z,-\mu,\phi). \tag{9.39}$$

Thus, the radiative transfer equation for the upward radiance becomes

$$-\mu \frac{dI^+(z,-\mu,\phi)}{dz} = -k(z)I^+(z,-\mu,\phi) + f_b(z,-\mu,\phi)\sigma(z)b(z)\bar{I}^-(z) + f_{I^+}(z,-\mu,\phi)\sigma(z)[1-b(z)]I^+(z,-\mu,\phi). \quad (9.40)$$

We define the radiance attenuation coefficient as

$$K_I(z,\mu,\phi) \equiv -\frac{1}{I^+(z,-\mu,\phi)} \frac{dI^+(z,-\mu,\phi)}{dz}. \quad (9.41)$$

Substituting Eq. 9.41 into Eq. 9.40 and solving for $I^+(z,-\mu,\phi)/\bar{I}^-(z)$, we find

$$\rho_{\bar{I}}(z,-\mu,\phi) \equiv \frac{I^+(z,-\mu,\phi)}{\bar{I}^-(z)} = \frac{f_b(z,-\mu,\phi)\sigma(z)b(z)}{\mu K_I(z,\mu,\phi) + k(z) - f_I(z,-\mu,\phi)\sigma(z)[1-b(z)]}, \quad (9.42)$$

which is the upwelling intensity normalized to the mean downwelling intensity. This quantity, which is referred to as the *remotely sensed reflectance*, plays an important role in attempts to determine the inherent optical properties from remote sensing. For a zenith-viewing instrument measuring radiation upwelling in the nadir direction ($\mu = 1$), we have ($K_I(z,-1) = K_{I^+}$; see Eq. 9.29)

$$\rho_{\bar{I}}(z,-1) = \frac{f_b(z,-1)\sigma(z)b(z)}{K_{I^+} + k(z) - f_I(z,-1)\sigma(z)[1-b(z)]}, \quad (9.43)$$

where

$$f_b(z,-1) = \frac{\int_0^1 d\mu' \bar{p}(z,\mu',-1)I^-(z,\mu')}{2b(z)\bar{I}^-(z)} \quad (9.44)$$

and

$$f_{I^+}(z,-1) = \frac{\int_0^1 d\mu' \bar{p}(z,-\mu',-1)I^+(z,-\mu')}{2[1-b(z)]I^+(z,-1)}. \quad (9.45)$$

Here $\bar{p}(z,\mu',-1)$ is the azimuthally averaged phase function. Note that Eq. 9.43 is exact because so far we have made no approximations. We have merely rewritten Eq. 9.35 in terms of the shape factors, f_b and f_{I^+}, defined above.

According to Eqs. 9.43 and 9.44 the remotely sensed radiance is due primarily to single scattering of downwelling light at the angle where the radiance distribution maximizes. We recall from §7.2.1 that the single-scattering approximation is valid when $a\tau^* \ll 1$. Thus, when absorption is sufficiently low, $\rho_{\bar{I}}(z,-1)$ is directly proportional to the phase function, which is useful in determining the identity of the ocean particles.

9.5.4 Approximate Expressions for the Remotely Sensed Reflectance

From Eq. 9.42 the remotely sensed reflectance, $\rho_{\bar{I}}(z, -1)$, is directly proportional to $f_{\rm b}(z, -1)$. If we assume that the incident light is dominated by light from a particular direction $\mu_{\rm m}$ (which is usually the solar zenith angle), then we may approximate the expression for $f_{\rm b}(z, -1)$ as

$$f_{\rm b}(z, -1) \approx \frac{\bar{p}(z, \mu_{\rm m}, -1)\int_0^1 d\mu' I^-(z, \mu')}{2b(z)\bar{I}^-(z)} = \frac{\bar{p}(z, \mu_{\rm m}, -1)}{2b(z)},$$

where $0 \le \mu_{\rm m} \le \mu_{\rm c}$, and $\mu_{\rm c}$ is the critical angle defined by Eq. 6.45. Substituting this approximate expression for $f_{\rm b}(z, -1)$ into Eq. 9.43, we have

$$\rho_{\bar{I}}(z, \mu_{\rm m}, -1) \approx \frac{1}{2}\frac{\bar{p}(z, \mu_{\rm m}, -1)\sigma(z)}{k(z) + K_{I^+} - f_I(z, -1)\sigma(z)[1 - b(z)]}. \tag{9.46}$$

We have removed the unmeasurable parameter $f_{\rm b}(z, -1)$, but we need to find an approximation for the parameter $f_I(z, -1)$, which is not easily measured either. However, $f_{\rm L}(z, -1)$ depends on the ratio $I(z, \mu')/I^+(z, -1)$, which is greater than unity because the upwelling radiance is smallest in the nadir direction ($\mu' = 1$). Thus, we expect $f_{\rm l}(z, -1) \ge 1$. We may rewrite Eq. 9.46 as (with $k(z) = \alpha(z) + \sigma(z)$)

$$\rho_{\bar{I}}(z, \mu_{\rm m}, -1) \approx \frac{1}{2}\frac{\bar{p}(z, \mu_{\rm m}, -1)\sigma(z)}{\alpha(z) + K_{I^+} + \{1 - f_I(z, -1)[1 - b(z)]\}\,\sigma(z)}.$$

The above formula can be used for turbid waters when scattering is large compared to absorption. For many oceanic situations the term in curly brackets in the denominator is small compared to the sum of the first two terms (which are approximately equal to 2α). For such situations the above formula simplifies to

$$\rho_{\bar{I}}(z, \mu_{\rm m}, -1) \approx \frac{\bar{p}(z, \mu_{\rm m}, -1)\sigma(z)}{2[\alpha(z) + K_{I^+}]}.$$

Finally, since only the numerator in Eq. 9.46 depends on the downwelling light field, the ratio of the remotely sensed reflectance at two different values of $\mu_{\rm m}$ becomes

$$\frac{\rho_{\bar{I}}(z, \mu_{\rm m_1}, -1)}{\rho_{\bar{I}}(z, \mu_{\rm m_2}, -1)} = \frac{\bar{p}(z, \mu_{\rm m_1}, -1)}{\bar{p}(z, \mu_{\rm m_2}, -1)}.$$

Thus, the shape of the phase function can be obtained from the remotely sensed reflectance, and this information can, at least in principle, be inverted to provide insight into the particulate properties of the near-surface layer of the ocean.

9.5.5 Modeling the UV Transmission into the Ocean

Many marine organisms are sensitive to UV radiation. The increase in UV exposure as a function of geographic location and depth in the ocean due to ozone depletion remains uncertain. The extent to which these marine organisms will be able to adapt to the expected increases in UV exposure is also unknown. Recent investigations indicate that increased levels of UV-B radiation may impact phytoplankton communities by (i) initiating changes in cell size and taxonomic structure, (ii) reducing productivity, (iii) influencing protein content, dry weight, and pigment concentration, (iv) inducing chloroplast damage, and (v) directly affecting the proteins of the photosynthetic apparatus.[42]

Ultraviolet light penetration in the ocean is strongly influenced by small plankton and thus by biological productivity, which provides a close link between biological and optical oceanography. Important aspects of the ozone depletion issue include the effects of increased UV levels on algae, plankton, and fish larvae. As sources of atmospheric sulfur compounds involved in cloud formation, plankton may indirectly affect atmospheric transmission, thereby linking atmospheric radiative transfer with ocean biology.

The impact of decreased ozone levels on aquatic systems may be assessed by a radiation model that provides a solution of the coupled radiative transfer equations for the atmosphere–ocean system. We assume that this system consists of two strata with different indices of refraction separated by a plane interface, as explained previously in Chapters 5 and 8. The relative refractive index for air is taken to be unity and that of the ocean to be 1.33 (neglecting for simplicity the slight wavelength dependence). To account for the vertical inhomogeneity of the atmosphere and the water, we divide each stratum into a sufficient number of layers to resolve the changes in optical properties with height in the atmosphere and depth in the ocean. To estimate the UV penetration through this coupled atmosphere–ocean system we need the spectral distribution of the radiation incident at the top of the atmosphere (see Fig. 1.1, top panel) as well as the optical properties of the atmosphere and water media. For a clear atmosphere the optical properties are determined mainly by ozone absorption (absorption cross sections for ozone and molecular oxygen are shown in the bottom panel of Fig. 1.1) and molecular (Rayleigh) scattering.

Phytoplankton dwell in the top layers of the water column, the euphotic zone, because of their requirement for solar radiation (PAR) to drive photosynthesis. In the euphotic zone, they would be exposed to any increase in UV radiation. If all other factors remain constant, ozone depletion would lead to increased transmission of UV-B radiation through the atmosphere and into the water column. Model results indicate that UV-B radiation is significantly absorbed in the first couple of meters into the water. At high latitudes (70°) an ozone depletion of 30% (compared to normal) will increase UV-B exposure 10 meters below the surface by as much as 33% on October 1 in the Southern Ocean and by 23% at summer solstice. Thus, the relative amount of UV-B increase in the water column due to ozone depletion is most pronounced in spring, which happens to be the time when polar ozone depletion is most severe. Chlorophyll pigment in the water reduces the penetration of UV-B radiation into the water. The larger the chlorophyll concentration, the less the UV-B transmission; however the UV-B to PAR ratio is relatively unaffected by the chlorophyll concentration. Thus, the species that depend on a certain level of PAR for photosynthesis, and therefore adjust their depth in the water to optimize PAR, will be exposed to a similar level of UV-B regardless of chlorophyll content.

9.5.6 Measured and Computed UV Irradiance in the Ocean

Underwater spectral irradiance was measured in situ with a UV/visible spectrometer submersed into the ocean off Palmer Peninsula, Antarctica.[43] Figure 9.15 compares the measured and computed ratio of the irradiance integrated across the UV-B range ($F_{\text{UV-B}}$, 280–320 nm) to that integrated across the complete measured range (F_{total}, 280–700 nm). Ratios for both undepleted ozone levels (350 DU, labeled "outside hole") and depleted levels (150 DU, labeled "inside hole") are shown. The aim of this

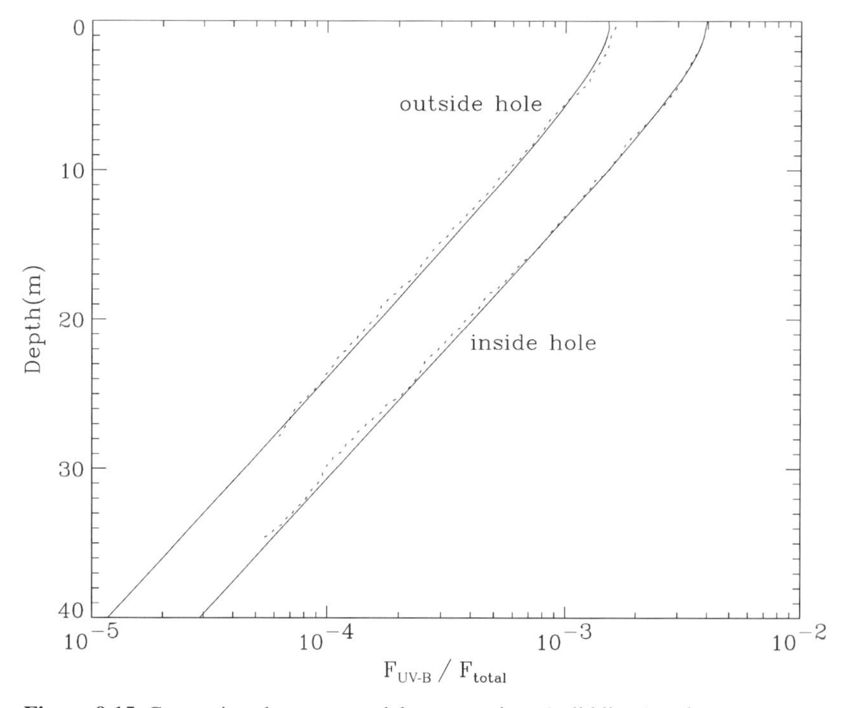

Figure 9.15 Comparison between model computations (solid lines) and measurements (dotted lines) of depth versus $F_{\text{UV-B}}/F_{\text{total}}$. Inside the ozone hole, the ozone abundance was 150 DU, the solar zenith angle was 56°, and the vertical distribution of chlorophyll concentration was 0.57 mg · m^{-3} from the surface to 20 m depth, and 0.47 mg · m^{-3} below 20 m. Outside the ozone hole, the ozone abundance was 350 DU, the solar zenith angle was 57°, and the vertical distribution of chlorophyll concentration was 1.9 mg · m^{-3} from the surface to 10 m depth, 1.6 mg · m^{-3} from 10 to 20 m, and 1.5 mg · m^{-3} below 2 m.

comparison is to examine how well the model reproduces the underwater downwelling irradiance, when it is constrained to yield the correct surface irradiance by adjusting the cloud/aerosol optical depth (which was not measured). There is good agreement between computed and measured ratios below the surface, but some curvature in the computed ratio just below the surface is absent in the ratio inferred from the measurements. A possible reason for this discrepancy is the neglect of surface waves in the model, which assumes a plane atmosphere–ocean interface.

Solar zenith angle has an important influence on UV radiation reaching the Earth's surface. It was overcast during the measurements, but the solar zenith angle was almost the same for the measurements taken inside and outside the ozone hole. Since the optical properties of clouds and aerosols depend weakly on wavelength, the impact of clouds and aerosols on the ratio ($F_{\text{UV-B}}/F_{\text{total}}$) is expected to be small. This circumstance allows us to investigate the impact of changes in ozone abundance on the surface and submarine irradiance ratio. Ozone depletion will increase the surface and underwater UV irradiance. Although the vertical distributions of chlorophyll in the water were different under and outside the ozone hole, the impact of this difference on the vertical variation in the ratio $F_{\text{UV-B}}/F_{\text{total}}$ is small. Thus, the vertical attenuation coefficients are nearly the same inside and outside the ozone hole. Therefore, if UV-B exposure is doubled at the surface, it will be doubled at all depths, and the critical depth above which UV-B damage may occur will be correspondingly deeper in the water column.

9.6 Interaction of Solar Radiation with Snow and Ice

At high latitudes snow and ice are present for long periods of the year. In the polar regions sea ice plays an important role in the radiative energy balance by reflecting solar radiation in summer and insulating the upper ocean from the atmosphere in winter. Therefore the absorption, scattering, and emission of radiation by snow, ice, and water surfaces play major roles in radiative transfer on our planet. The radiative transfer model pertinent for the coupled atmosphere–ocean system described previously in Chapter 5 may be applied to describe radiative energy transfer in a stratified system consisting of the atmosphere above a slab of snow and ice overlying the ocean.[44] The solar radiative energy distribution throughout this system as well as the radiative energy absorbed within the ice and transmitted into the ocean depend on atmospheric structure (including cloudiness), snow properties, sea ice state, and the optical properties of the seawater below the ice. Radiation absorbed within the snow or sea ice may change the internal structure of snow and ice and thereby their optical properties. These changes result in an alteration of the radiative energy transmitted into the ocean and reflected back to the atmosphere, which, in turn, affects the stratification and circulation of the atmosphere and ocean.

A study of the radiative transfer process within the coupled atmosphere–sea ice–ocean system would allow us to assess how the physical properties of each subsystem affect the partitioning of the solar radiative energy within the atmosphere, sea ice, and

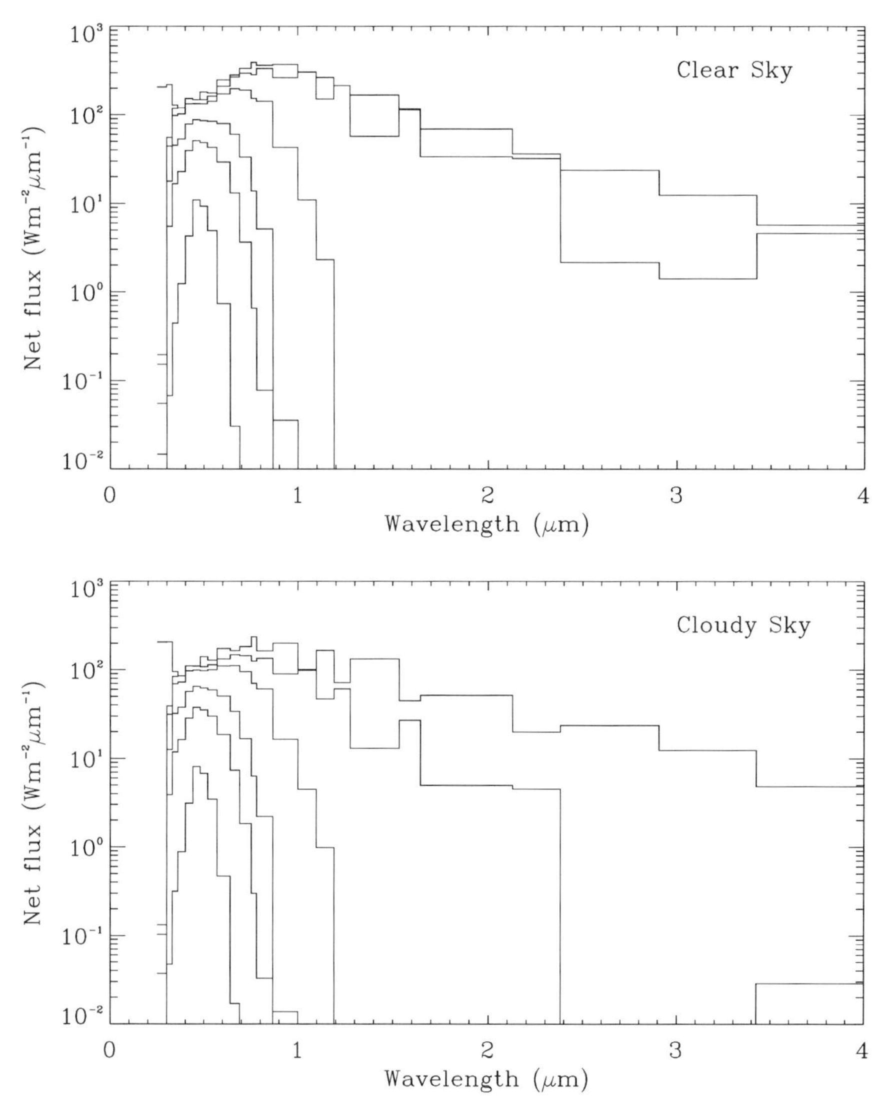

Figure 9.16 Spectral distribution of net downward irradiance at the top of the atmosphere, the surface, and several levels in the sea ice for clear- (top panel) and overcast-sky conditions (bottom panel). The computation was done with a comprehensive radiative transfer model based on the multistream discrete-ordinate approximation to the radiative transfer equations appropriate for the coupled atmosphere–sea ice–ocean system (as described in Chapters 5 and 8, and in the reference by Jin et. al. provided in Endnote 16.) The curves show from top to bottom the net flux at the following levels: top of the atmosphere, just beneath the ice surface (−0 m), 0.1 m in the ice, 0.5 m in the ice, 1.0 m in the ice, and 3.0 m in the ice.

ocean. To do this we must take into account over the entire solar spectrum the multiple scattering and absorption by atmospheric gases, clouds, snow, ice, and seawater, as well as inclusions in the ice consisting of air bubbles and brine pockets. To account for vertical inhomogeneity, each stratum of the model must be divided into a sufficient number of layers to resolve variations in optical properties.

In the following example, the solar spectrum is divided into twenty-four bands, and the ESFT method described in Chapter 10 is used to approximate the atmospheric absorption by water vapor, carbon dioxide, oxygen, and ozone within each band. A high-latitude model atmosphere divided into twenty-five layers is adopted.

The computed net downward irradiance for this spectral range is shown in the upper panel of Fig. 9.16 at various levels for solar elevation of 30° and clear-sky conditions. Since the net downward irradiance is just the difference between the downward and upward irradiances, the vertical distance between any two lines in Fig. 9.16 represents the radiative energy absorbed between the corresponding two levels. Thus, the area beneath the bottom line represents the total absorption of radiation by the ocean. Figure 9.16 indicates that much of the solar energy will be absorbed within the uppermost 0.1 m of the ice. The absorption also varies greatly with wavelength. Only visible radiation can penetrate into the deeper layers of the ice and into the ocean. The bottom panel of Fig. 9.16 is for the same conditions except that a typical arctic stratus cloud (mean droplet radius of 7 μm, and liquid water path of 60 $\text{g} \cdot \text{m}^{-2}$) was added. This leads to a substantial increase in atmospheric absorption, especially at the longer wavelengths. Therefore, absorption by the ice and the ocean will decrease. This reduction is most pronounced in the uppermost layer of the ice.

9.7 Summary

In this chapter we have given sample applications of the theory presented in the previous chapters for the interaction of solar radiation with the atmosphere and underlying ocean and surface. To provide a suitable context and focus for the subsequent discussion we introduced two problems of great current interest in which radiation plays a vital role. The first issue concerns ozone depletion and the related harmful effects of enhanced levels of solar UV radiation reaching the biosphere. It is also well established that UV/visible radiation plays an important role in the ozone chemistry through its impact on photolysis rates. Several important chemical reactions in the ozone chemistry occur at temperature-dependent rates. Therefore warming/cooling rates, which involve radiation across the solar and terrestrial regions of the spectrum, must be correctly computed in models aimed at predicting the evolution of ozone abundance. These rates in turn affect dynamical processes. It becomes clear that in the ozone-depletion problem there is a tight coupling between radiation, chemistry, and dynamics. The second issue concerns the potential for global warming and its relation to radiative effects of enhanced abundances of greenhouse gases. This issue is considered in Chapter 12.

In the thermal-IR and the near-IR part of the spectrum, rapid and erratic variations of the gaseous absorption cross sections across small spectral intervals complicate the radiative transfer treatment because the Extinction Law does not apply. An exception is the UV/visible part of the spectrum: Here ozone is the main absorber and its absorption cross section varies relatively smoothly with wavelength between 200 and 700 nm. In this chapter we focused mainly on the UV/visible region of the spectrum where the Extinction Law can be assumed to hold over broad spectral regions.

Comparisons of computed and observed spectra between 200 and 300 nm in the stratosphere illustrated that, in one case, there is significant disagreement between theoretical predictions and observations. Comparisons between computed and observed spectral UV and visible irradiances and intensities at the ground (including snow- and ice-covered surfaces) and in the ocean were provided for illustration purposes and to give a sense of the state of the art in this field. Several examples were provided to illustrate how radiation penetrates through the atmosphere and to various levels in the coupled atmosphere–ocean system. The tight coupling of the atmosphere and ocean through solar radiation should be emphasized. Solar radiation drives ocean biology, which in turn modifies the optical properties of the ocean and therefore affects radiative transfer. Thus, there are numerous atmosphere–ocean interactions and feedbacks involving radiative transfer. The study and illumination of such interactions and feedbacks will require radiative transfer models of the type described in this book that are applicable to the coupled system. Some of these were briefly demonstrated in this chapter.

Problems

9.1 This problem involves the determination of the *solar insolation* Q, which is the daily received solar energy at the top of the atmosphere at latitude, λ, and day of the year, t. The solar inclination angle δ is the angular distance of the Sun north (positive) or south (negative) of the equator. The angle δ varies with time of year according to the approximate formula $\delta = 23.5^\circ \times \sin(360t/365.25)$.

(a) Show that the angle between the local vertical and the Sun's rays (the solar zenith angle) is given by

$$\cos\theta_0 = \sin\lambda\sin\delta + \cos\lambda\cos\delta\cos(h),$$

where h is the solar hour angle – the angle through which the Earth must turn to bring the meridian directly under the Sun.

(b) The daily insolation Q is derived by integrating the instantaneous flux $F = S(r_\oplus/r)^2\cos\theta_0$ over all hour angles from sunrise to sunset:

$$Q = S\left(\frac{r_\oplus}{r}\right)^2 \int_{-H}^{H} \frac{dh}{\omega}[\sin\lambda\sin\delta + \cos\lambda\cos\delta\cos(h)].$$

Here $r_\oplus$ is 1 AU, r is the instantaneous Earth–Sun distance in AU, and $\omega = 2\pi$ rad /day is the angular velocity of Earth's rotation. H is the half-day length, given by

$$H = \frac{\tan\phi \sin\epsilon \sin\lambda}{(1 - \sin^2\epsilon \sin^2\lambda)^{1/2}},$$

where $\epsilon = 23.5°$ is the obliquity of the ecliptic. Carry out the above integration analytically, and evaluate Q in units of $\mathrm{J \cdot m^{-2}}$. Assume r varies sinusoidally with a maximum on January 3 (where $r_{\oplus}^2/r^2 = 1.0344$) and a minimum on July 5 (where $r_{\oplus}^2/r^2 = 0.9674$). Compare your contour plot of $Q(\lambda, t)$ with Fig. 2.10 on p. 47 of K-N. Liou, *An Introduction to Atmospheric Radiation*, Academic Press, New York.

9.2

(a) Derive Eq. 9.30 for the attenuation coefficient.

(b) Show that when $\sigma b/\alpha$ is sufficiently small the attenuation coefficient becomes

$$K \approx (\alpha + r^+ b)/\bar{\mu}^+ \approx \frac{\alpha}{\bar{\mu}^+}.$$

(c) If $\sigma b/\alpha \gg 1$, show that the value for K may be written

$$K \approx \frac{\alpha}{\bar{\mu}^-} \frac{1 + r^+/r^-}{1 + (r^+\bar{\mu}^-/r^-\bar{\mu}^+)}.$$

9.3 Derive Eqs. 9.32, and 9.33.

9.4

(a) Show that when the absorption dominates over backscattering ($\sigma b/\alpha \ll 1$) the irradiance reflectance given by Eq. 9.31 is written

$$\rho \approx \frac{r^+}{1 + \bar{\mu}^+/\bar{\mu}^-} \frac{\sigma b}{\alpha}.$$

(b) Use the above expression to explain how brown algae give rise to green ocean color.

(c) Show that when $\sigma b/\alpha \gg 1$ the reflectance ρ asymptotically approaches the result $r^+\bar{\mu}^-/r^-\bar{\mu}^+$.

(d) Discuss the significance of the above expressions for remote sensing of ocean properties.

9.5 Atmospheric ozone shields the Earth from harmful ultraviolet radiation by absorbing most of the radiation with wavelengths shorter than 300 nm. For longer wavelengths some radiation penetrates to the surface, but the absorption by ozone is strongly wavelength dependent, being much stronger at 310 nm than at 330 nm. By measuring direct sunlight at these two wavelengths we may determine the amount of ozone in the atmospheric column above the observation point.

(a) Show that if we know the solar irradiance incident at the top of the atmosphere F^{s}, and we measure the corresponding direct irradiance F_{dir} penetrating to the surface, then the column density, $\mathcal{N}_{\text{ozone}}$, can be determined from the formula

$$\mathcal{N}_{\text{ozone}} = \frac{\left[\log\left(F'_{\text{dir}}/F_{\text{dir}}\right) - \log(F^{s\prime}/F^{s})\right]\mu}{(\alpha - \alpha')} - \frac{(\sigma - \sigma')}{(\alpha - \alpha')},$$

where μ is the solar zenith angle. The primed quantities refer to the longer wavelength (330 nm), and the unprimed ones to the shorter one (310 nm). Here α is the absorption coefficient for ozone, and σ is the scattering coefficient for atmospheric constituents.

(b) Sketch a graph of $\log(F'_{\text{dir}}/F_{\text{dir}})$ as a function of μ, and explain how one might use observations from the ground to infer the solar irradiance at the top of the atmosphere.

The UV radiation backscattered to space is used to infer the ozone amount by instruments on satellites. Here the ratios of the backscattered radiation to the solar irradiance (both of which are measured) at two or more wavelengths are used to determine the ozone amount using the same principle as from the ground.

(c) Volcanic eruptions sometimes eject large quantities of aerosols into the ozone layer. Assume that these aerosols scatter radiation but are nonabsorbing. Explain briefly how the presence of these aerosol particles influences our ability to determine the ozone amount from space. Will ignoring the aerosols lead to an overestimation or an underestimation of ozone?

(d) Discuss how these aerosols will influence the amount of UV radiation at the surface. Show that they may lead to an increase or a decrease in UV exposure depending on solar zenith angle.

(e) Most of the ozone in the atmosphere resides in the stratosphere (10–30 km above the surface) with a much smaller tropospheric component. However, the tropospheric component is increasing due to industrial pollution. Assume that the total column amount of ozone is unchanged but that the stratospheric component is decreased and the tropospheric component increased by the same amount. Discuss how this situation might influence the UV irradiance at the surface. Do you expect an increase or a decrease, and will there be a solar elevation dependence?

9.6 Consider an isothermal, one-constituent atmosphere in hydrostatic equilibrium.

(a) Show that the warming rate is proportional to

$$\frac{\partial T}{\partial t} \propto e^{-\tau^*/\mu_0} e^{-z/H},$$

where τ^* is the optical thickness and H is the scale height.

(b) At what altitude z_{max} does the warming rate maximize?

9.7 Consider a partially reflecting surface illuminated by a collimated beam of radiation of incidence angle $\cos^{-1}\mu_0$ and flux (normal to the beam) F^{s}. Surfaces are sometimes modeled as a scattering and absorbing medium in much the same way as an atmosphere.

(a) Assume that the surface can be considered a semi-infinite medium with single-scattering albedo a and phase function $p(\cos\Theta)$, where Θ is the scattering angle. Show that the single-scattering contribution to the reflected intensity can be expressed as

$$I_{ss} = \frac{aF^s}{4\pi}\frac{\mu_0}{\mu_0+\mu}p(\cos\Theta),$$

where $\cos^{-1}\mu$ is the angle of reflection.

(b) Assume for simplicity that the multiple-scattering contribution can be derived under the assumption of isotropic scattering $[p(\cos\Theta) = 1]$, and show that then (using the expression above valid for single scattering) the total BRDF becomes

$$\rho(\mu, \mu_0, \phi) = \frac{a}{4\pi}\frac{1}{\mu_0+\mu}[p(\cos\Theta) + H(\mu_0)H(\mu) - 1],$$

where $H(\mu)$ is Chandrasekhar's H-function.

If the dimensions of the scattering elements are considerably larger than the wavelength of the incident radiation, and the elements are sufficiently close together, the scattering elements cast a shadow on each other. This leads to a phenomenon called the opposition effect, which gives enhanced scattering in the backward direction.

(c) Assume that the opposition effect can be described by simply replacing the phase function $p(\cos\Theta)$ by $p(\cos\Theta)[1 + B(\cos\Theta)]$, where $B(\cos\Theta)$ is a backscatter function. Show that then the BRDF becomes

$$\rho(\mu, \mu_0, \phi) = \frac{a}{4\pi}\frac{1}{\mu_0+\mu}[p(\cos\Theta)(1 + B(\cos\Theta)) + H(\mu_0)H(\mu) - 1].$$

(d) Under what circumstances (i.e., for what values of a and for what kind of phase functions $p(\cos\Theta)$) do you expect the above expression to be accurate?

(e) Generalize the approach in (c) above by relaxing the assumption of isotropic multiple scattering, and show that the BRDF then can be written

$$\rho(\mu, \mu_0, \phi) = \frac{a}{4\pi}\frac{1}{\mu_0+\mu}p(\cos\Theta)B(\cos\Theta) + F(\tau = 0; \mu, \mu_0, \phi)$$

and define $F(\tau = 0; \mu, \mu_0, \phi)$.

9.8 Consider light penetration in and backscattering from the ocean. To make the problem simple let us assume that the ocean is infinitely deep and that its scattering and absorption properties are independent of depth (i.e., the ocean is optically homogeneous). Further let us assume that we have scaled the phase function to become isotropic by using the δ-isotropic approximation discussed in §6.7.

(a) Show that if we ignore the coupling between the atmosphere and the ocean, then the angular distribution of direct solar radiation diffusely reflected from a calm ocean surface can be expressed in terms of Chandrasekhar's H-function as

$$I(\tau = \tau^{\mathrm{a}}, +\mu) = F^{\mathrm{s}} e^{-\tau^{\mathrm{a}}/\mu_0} \frac{H(a,\mu)}{1+\mu/\mu_{0m}} H(a,\mu_{0m}),$$

where a is the single-scattering albedo of the ocean, τ^{a} is the total optical depth of the atmosphere, $\mu_{0m} = \sqrt{1-(1-\mu_0^2)m_{\mathrm{rel}}^2}$, and m_{rel} is the real index of refraction in the ocean relative to the atmosphere.

(b) Assume that single scattering prevails in the atmosphere as well as in the oecan, and derive an expression in terms of HG-functions for the diffusely reflected radiation due to single scattering. (Assume that the atmosphere too can be represented by a homogeneous slab for this purpose.)

Assume now that the sky is overcast so that the radiation arriving at the ocean surface can be taken to be predominantly diffuse and approximately uniform. The uniform cloud field is taken to be nonabsorbing and optically thick. Assume that we measure the downwelling intensity at the ocean surface I_{dn} = constant independent of direction (by assumption).

(c) Derive an expression for the intensity distribution of the light diffusely reflected from the ocean in this case.

(d) What is the albedo of the ocean under the given assumptions?

Assume that the downwelling radiation in the ocean can be described in terms of an effective bulk attenuation coefficent, K.

(e) How does the angular distribution of the reflected light depend on K? Could measurements of the angular distribution be used to characterize the bulk attenuation coefficient K?

9.9 Consider the light reflected from the atmosphere–ocean system. We assume that the atmosphere is a homogeneous slab and that the ocean is a homogeneous semi-infinite medium.

(a) Assume that single scattering prevails in the atmosphere as well as the ocean, and derive an expression for the radiation reflected at the top of the atmosphere.

(b) Assume some properties about the atmosphere and the ocean and evaluate the angular distribution seen by a satellite.

(c) Quantify how the reflected radiation depends on the optical properties of the ocean.

9.10 Consider light penetration into the ocean. Use the solution of Problem 7.12 to study ocean heating.

(a) Derive an expression for the heating rate using the two-stream solution of Problem 7.12.

(b) At what depth does the heating rate maximize?

(c) How does the heating rate depend on the optical properties?

9.11 Consider the angular distribution of the light field in the ocean.

(a) Use the expression for the source function provided in Problem 7.12 to derive expressions for the upward and downward intensity fields.

(b) What is the asymptotic value of the intensity distribution?

(c) Explore the relationship between the light distribution described in terms of apparent optical properties versus inherent ones.

9.12 It is difficult to measure the inherent absorption in the ocean directly. Therefore, many attempts have been made to infer it from measurements of bulk quantities such as reflectances, $R^{\pm}(z)$, irradiance attenuation coefficients, $K^{\pm}(z)$, and ratios of hemispherical intensities (scalar irradiances) and irradiances, $D^{\pm}(z) = 2\pi \int_0^1 d\mu I^{\pm}(z,\mu)/2\pi \int_0^1 d\mu\mu I^{\pm}(z,\mu)$. Here the plus (+) sign refers to upwelling and the minus (−) sign to downwelling light.

One such formula is

$$x = \frac{K^-(z) - R^-(z)K^+(z)}{D^-(z) + R^-(z)D^-(z)},$$

where x is some measure of the inherent absorption. Use the two-stream solutions provided in Problem 7.12 to demonstrate that this formula gives a measure of the inherent absorption, and quantify this measure.

9.13 Consider the transfer of radiation through a medium illuminated by a collimated beam of radiation. Let the medium consist of nonabsorbing molecules and scattering and absorbing particles. The molecular optical depth is assumed to be $\tau_{\rm RAY} = 0.1$ to mimic the Earth's atmosphere at 500 nm. The medium overlies a partially reflecting surface, which we will assume acts as a Lambert reflector with albedo $\rho_{\rm L}$.

(a) Assume that the particles have total optical depth $\tau_{\rm p} = \tau_{\rm s} + \tau_{\rm a}$, where $\tau_{\rm s}$ is the scattering and $\tau_{\rm a}$ the absorption optical depth. Assume that the medium is homogeneous so that the particles and molecules have a constant mixing ratio. Show that the single-scattering albedo and the phase function for this medium can be expressed as

$$a = 1 - \tau_{\rm a}/\tau_{\rm t}, \qquad p(\cos\Theta) = b p_{\rm RAY}(\cos\Theta) + (1-b)p_{\rm p}(\cos\Theta),$$

where $\tau_{\rm t} = \tau_{\rm RAY} + \tau_{\rm p}$, $b = \tau_{\rm RAY}/(\tau_{\rm s} + \tau_{\rm RAY})$, and $p_{\rm RAY}(\cos\Theta)$ and $p_{\rm p}(\cos\Theta)$ are the phase functions for molecular (Rayleigh) and particle scattering, respectively.

(b) Assume that the particles scatter according to the Henyey–Greenstein phase function with asymmetry factor g. Show that the Legendre polynomial moments of the phase function for the medium (molecules plus particles) are

$$\chi_0 = 1, \qquad \chi_1 = (1-b)g, \qquad \chi_2 = 0.1b + (1-b)g^2,$$
$$\chi_l = (1-b)g^l \quad \text{for } l > 2.$$

(c) Use DISORT to check the duality relations pertaining to a homogeneous slab connecting the reflectance and transmittance for parallel beam incidence to the emerging intensities for uniform illumination. Do this by computing the reflectance and transmittance for a nonreflecting surface by first applying beam incidence (*IBCDN* = 0 in DISORT), and then repeating the computation for uniform incidence (*IBCDN* = 1 in DISORT). Use $a = 0.99999$ and $a = 0.5$; $g = 0.75$ and

$g = 0.5$. Compute reflectance and transmittance for $\mu = 0.1, 0.2, \ldots, 0.9, 1.0$ both ways.

Notes

1 Brasseur, G. and S. Solomon, *Aeronomy of the Middle Atmosphere*, 2nd ed., D. Reidel, Dordrecht, 1986.

2 This is another example of a term (like the greenhouse effect) that evokes an incorrect physical picture. The Sun's output is *not* constant. It is now known to vary slightly ($<0.1\%$) over time periods short compared to a rotation period (27 days as seen from Earth) up to time scales as long as an 11-year solar cycle.

3 The *Dobson unit* (DU) is named after the pioneer in ozone research, G.M.B. Dobson. 300 DU corresponds to a column having 3 milliatmospheres of constant pressure; 1 DU is 2.69×10^{20} molecules per square meter of horizontal area in a vertical column extending from the surface to the top of the atmosphere.

4 Anderson, J. G., D. W. Toohey, and W. H. Brune, "Free radicals within the Antarctic vortex: The role of CFCs in Antarctic ozone loss," *Science*, **251**, 39–46, 1991.

5 Dobson, G. M. B., "A photo electric spectrophotometer for measuring the amount of atmospheric ozone," *Proc. Phys. Soc. London*, **43**, 324–39, 1931.

6 Farman, J. C., G. Gardiner, and J. D. Shanklin, "Large losses of total ozone in Antarctica reveal seasonal ClO_x/NO_x interaction," *Nature*, **315**, 207–10, 1985.

7 Adapted from Meier, R. R., "Ultraviolet spectroscopy and remote sensing of the upper atmosphere," *Space Sci. Rev.*, **58**, 1–185, 1991.

8 Ramanathan, V. et al., "Cloud-radiative forcing and climate: Results from the radiation budget experiment," *Science*, **243**, 57–63, 1989.

9 This follows from setting the bulk density of water to $1\ g \cdot cm^{-3}$, a result clearly appropriate only in cgs units.

10 Hu, Y.-X. and K. Stamnes, "An accurate parameterization of the radiative properties of water clouds suitable for use in climate models," *J. Climate*, **6**, 728–42, 1993.

11 See Liou, K.-N. and Y. Takano, "Light scattering by nonspherical particles: Remote sensing and climatic implications," *Atmospheric Research*, **31**, 271–98, 1994.

12 The material below is taken from Fu, Q. and K. N. Liou, "Parametrization of the radiative properties of cirrus clouds," *J. Atmos. Sci.*, **21**, 115–50, 1993. A recent review of ice crystal optical properties is provided by Liou, K.-N. and Y. Takano (see Endnote 11). A recent special issue of *Journal of Quantitative Spectroscopy & Radiative Transfer* (volume 55, number 5, May, 1996) is devoted to scattering by nonspherical particles.

13 Liou and Takano, 1994; see previous endnote for full citation.

14 This ill-conditioning is quite similar to that described in Chapter 8 in connection with the scaling transformation needed to make sure the matrix that had to be inverted to obtain the constants of integration in the discrete-ordinate method did not have some very large and some very small elements (see §8.7)

15 This model, summarized by Frette, O., J. J. Stamnes, and K. Stamnes, "Optical remote sensing of marine constituents in coastal waters: A feasibility study," *Applied Optics*, **37**, 8218–8326, 1998, is based on work described in the following papers: Smith, R. C. and K. S. Baker, "Optical properties of the clearest natural waters," *Applied Optics*, **20**,

177–184, 1981; Bricaud, A., A. Morel, and L. Prieur, "Absorption by dissolved organic matter of the sea (yellow substance) in the UV and visible domains," *Limnol. Oceanogr.*, **26**, 43–53, 1981; Prieur, L. and S. Sathyendranath, "An optical classification of coastal and oceanic waters based on the specific spectral absorption of phytoplankton pigments, dissolved organic matter and particulate materials," *Limnol. Oceanogr.*, **26**, 671–89, 1981; Sathyendranath, S. and L. Prieur, "A three-component model of ocean colour and its application to remote sensing of phytoplankton pigments in coastal waters," *Int. J. Remote Sensing*, **10**, 1373–94, 1989; Morel, A. and B. Gentili, "Diffuse reflectance of natural waters: Its dependence on the sun angle as influenced by the molecular scattering contribution," *Applied Optics*, **30**, 4427–38, 1991. A discussion of the use of ocean color to characterize the marine environment is provided by Gordon, H. R., and A. Morel, *Remote Assessment of Ocean Color for Interpretation of Satellite Visible Imagery: A Review*, Springer-Verlag, Berlin, 1983.

16 See the papers by Smith and Baker and by Prieur and Sathyendranath, provided in the previous endnote.

17 The optical properties of pure seawater are taken from Smith, R. C. and K. S. Baker, "Optical properties of the clearest natural waters," *Applied Optics*, **20**, 177–84, 1981. The chlorophyll-specific coefficients are redrawn from Sathyendranath, S., L. Lazzarra, and L. Prieur, "Variations in the spectral values of specific absorption of phytoplankton," *Limnol. Oceanogr.*, **32**, 403–15, 1987. The scattering coefficients for three values of the chlorophyll concentration follow Eq. 9.20.

18 As quoted by Grenfell, T., S. G. Warren, and P. C. Mullen, "Spectral albedo of Antarctic snow," *J. Geophys. Res.*, **99**, 18,669–84, 1994.

19 Grenfell, T. C. and G. A. Maykut, "The optical properties of snow and ice in the Arctic basin," *J. Glaciol.*, **18**, 445–63, 1977.

20 Jin, Z., K. Stamnes, W. F. Weeks, and S. C. Tsay, "The effect of sea ice on the solar energy budget in the atmosphere–sea ice–ocean system: A model study," *J. Geophys. Res.*, **99**, 25,281–94, 1994.

21 Minschwaner, K., R. J. Thomas, and D. W. Rusch, "Scattered ultraviolet radiation in the upper stratosphere 1: Observations," *J. Geophys. Res.*, **100**, 11,157–63, 1995; Minschwaner, K., G. P. Anderson, L. A. Hall, J. H. Chetwynd, R. J. Thomas, D. W. Rusch, A. Berk, and J. A. Conant, "Scattered ultraviolet radiation in the upper stratosphere 2: Models and measurements," *J. Geophys. Res.*, **100**, 11,165–71, 1995; Koopers, G. A. A. and D. P. Murtagh, "Model studies of the influence of O_2 photodissociation parameterizations in the Schumann–Runge bands on ozone related photolysis in the upper atmosphere," *Ann. Geophysicae*, **14**, 68–79, 1996.

22 Two other definitions of the mean absorption cross section (the Planck and the Rosseland mean) are given in Chapter 12 (§12.2).

23 See, for example, Dutton, G. E. and J. R. Christy, "Solar radiative forcing at selected locations and evidence for lower tropospheric cooling following the eruptions of El Chichón and Pinatubo," *Geophys. Res. Lett.*, **19**, 2,313–16, 1992.

24 Kiehl, J. T. and B. P. Briegleb, "The relative roles of sulfate aerosols and greenhouse gases in climate forcing," *Science*, **260**, 311–14, 1993. Marine aerosols originating from DMS have been proposed to exert a so-called biocontrol of climate through their albedo effects by Shaw, G. E., "Planetary homeostasis through the sulfur cycle," in *Scientists on Gaia*, ed. S. Schneider and P. J. Boston, The MIT press, Cambridge, MA, 1991.

25 Lie-Svendsen, O., Q. Zhang, J. Simmons, and K. Stamnes, "The role of radiation and microphysics in the formation of Arctic Stratus clouds," submitted to Tellus, 1998.
26 See Kylling, A., K. Stamnes, and S. C. Tsay, "A reliable and efficient two-stream algorithm for spherical radiative transfer: Documentation of accuracy in realistic layered media," *J. Atmos. Chem.*, **21**, 115–50, 1995.
27 See Blumthaler, M. and W. Ambach, "Indication of increasing solar ultraviolet-B radiation flux in Alpine regions," *Science*, **248**, 206–8, 1990.
28 See Solomon, S., "Progress towards a quantitative understanding of Antarctic ozone depletion," *Nature*, **347**, 347–54, 1990, for a review of this issue.
29 Lubin, D., Frederick, J. E., Booth, C. R., Lucas, T., and Neuschuler, D., "Measurements of enhanced springtime ultraviolet radiation at Palmer Station, Antarctica," *Geophys. Res. Lett.*, **16**, 783, 1989.
30 Stamnes, K., Z. Jin, J. Slusser, C. R. Booth, and T. Lucas, "Several-fold enhancement of biologically effective ultraviolet radiation at McMurdo Station, Antarctica during the 1990 ozone hole," *Geophys. Res. Lett.*, **19**, 1,013–16, 1992.
31 Dahlback, A., "Measurements of biologically effective UV-doses, total ozone abundances, and cloud effect with multi-filter, moderate bandwith filter instruments," *Applied Optics*, **35**, 6,514–21, 1996.
32 Bittar, A. and R. L. McKenzie, "Spectral ultraviolet measurements at 45 deg S: 1980 and 1988," *J. Geophys. Res.*, **95**, 5,597, 1990.
33 Roy, C. R., Gies, H. P., and Elliot, G., "Ozone depletion," *Nature*, **347**, 235, 1990.
34 See, for example, Bordewijk, J. A., H. Slaper, H. A. J. Reinen, and E. Schlamann, "Total solar radiation and the influence of clouds and aerosols on the biologically effective UV," *Geophys. Res. Lett.*, **22**, 2,151–4, 1995; Blumthaler, M., J. Grobner, M. Hubner, and W. Ambach, "Measuring spectral and spatial variations of UV-A and UV-B sky radiance," *Geophys. Res. Lett.*, **23**, 547–50, 1996; Grobner, J., M. Blumthaler, and W. Ambach, "Experimental investigation of spectral global irradiance measurement errors due to a non ideal cosine response," *Geophys. Res. Lett.*, **23**, 2,493–6, 1996; Krzyscin, J. W., "UV controlling factors and trends derived from the ground-based measurements taken at Belsk, Poland," *J. Geophys. Res.*, **101**, 16,797–805, 1996; Mayer, B., G. Seckemeyer, and A. Kylling, "Systematic longterm comparison of spectral UV measurements and UVSPEC modeling results," *J. Geophys. Res.*, **102**, 8,755–67, 1997; Seckemeyer, G. and R. L. McKenzie, "Increased ultraviolet radiation in New Zealand (45°S) relative to Germany (48°N)," *Nature*, **359**, 135–7, 1992; Slaper, H., G. J. M. Velders, J. S. Daniel, F. R. de Gruijl, and J. C. van der Leun, "Estimates of ozone depletion and skin cancer to examine the Vienna Convention achievements," *Nature*, **384**, 256–9, 1995; Slaper, H., H. A. J, Reinen. M. Blumthaler, M. Hubner, and F. Kuik, "Comparing ground-level spectrally-resolved solar UV measurements using various instruments: A technique resolving effects of wavelength shift and slit width," *Geophys. Res. Lett.*, **22**, 2,721–4, 1995; Variotsos, C., K. Y. Kondratyev, and S. Katsikis, "On the relationship between total ozone and solar ultraviolet radiation at St. Petersburg, Russia," *Geophys. Res. Lett.*, **22**, 3,481–4, 1995.
35 Bodhaine, B. A., R. L. McKenzie, P. V. Johnston, D. J. Hofmann, E. G. Dutton, R. C. Schnell, J. E. Barnes, S. C. Ryan, and M. Kotkamp, "New ultraviolet spectroradiometer measurements at Mauna Loa Observatory," *Geophys. Res. Lett.*, **23**, 2,121–4, 1996; Herman, J. R., P. K. Barthia, J. Ziemke, Z. Ahmad, and D. Larko, "UV-B increases

(1979–1992) from decreases in total ozone," *Geophys. Res. Lett.*, **16**, 2117–20, 1996; Kerr, J. B. and C. T. McElroy, "Response to Analyzing ultraviolet-G radiation: Is there a trend?," *Science*, **264**, 1,342–3, 1994; Michaels P. J., S. F. Singer, and P. C. Knappenberger, "Analyzing ultraviolet-B radiation: Is there a trend?," *Science*, **264**, 1,341–2, 1994; Schafer, J. S., V. K. Saxena, B. N. Wenny, W. Barnard, and J. J. De Luisi, "Observed influence of clouds on ultraviolet-B radiation," *Geophys. Res. Lett.*, **23**, 2625–8, 1996.

36 S. B. Diaz, J. E. Frederick, T. Lucas, C. R. Booth, and I. Smolskaia, "Solar ultraviolet irradiance at Tierra del Fuego: Comparison of measurements and calculations over a full annual cycle," *Geophys. Res. Lett.*, **23**, 355–8, 1996; McKenzie, R. L., M. Kotkamp, and W. Ireland, "Upwelling UV spectral irradiances and surface albedo measurements at Lauder, New Zealand," *Geophys. Res. Lett.*, **23**, 1,757–60, 1996; McMinn, A., H. Heljnls, and D. Hodgson, "Minimal effects of UV-B radiation on Antarctic diatoms over the past 20 years," *Nature*, **370**, 547–9, 1994.

37 Zeng, J., R. McKenzie, K. Stamnes, M. Wineland, and J. Rosen, "Measured UV spectra compared with discrete ordinate method simulations," *J. Geophys. Res.*, **99**, 23,019–30, 1994.

38 The spike in the measured data may be due to gain switching during the measurements, which sometimes results in larger than the estimated measurement errors for the individual measurements.

39 A description of this method is provided in Stamnes, K., J. Slusser, and M. Bowen, "Derivation of total ozone abundance and cloud effects from spectral irradiance measurements," *Applied Optics*, **30**, 4,418–26, 1991.

40 See discussion in Stamnes, K., S. Pegau, and J. Frederick, "Uncertainties in total ozone amounts inferred from zenith sky observations: Implications for ozone trend analyses," *J. Geophys. Res.*, **45**, 16,523–8, 1990.

41 See, for example, Zaneveld, J. R. V., "Remotely sensed reflectance and its dependence on vertical structure: A theoretical derivation," *Applied Optics*, **21**, 4,146–50, 1982; Gordon, H. R., "Dependence of the diffuse reflectance of natural waters on sun angle," *Limnol. Oceanogr.*, **34**, 1,484–9, 1989; Stavn, R. H. and A. D. Weidemann, "Shape factors, two-flow models, and the problem of irradiance inversion in estimating optical parameters," *Limnol. Oceanogr.*, **34**, 1,426–41, 1989; Morel, A., K. J. Voss, and B. Gentili, "Bidirectional reflectance of oceanic waters: A comparison of modeled and measured upward radiance fields," *J. Geophys. Res.*, **100**,13,143–50, 1995; Zaneveld, J. R. V., "A theoretical derivation of the dependence of the remotely sensed reflectance of the ocean on the inherent optical properties," *J. Geophys. Res.*, **100**, 13,135–42, 1995; Weidemann, A. D., R. H. Stavn, J. R. V. Zaneveld, and M. R. Wilcox, "Error in predicting hydrosol backscattering from remotely sensed reflectance," *J. Geophys. Res.*, **100**, 13,163–77, 1995.

42 See, e.g., Karentz, D., "DNA repair mechanisms in Antarctic marine organisms," *Antarctic. J. of the U.S.*, **23**, 114, 1988; El-Sayed, S. Z., F. C. Stephens, R. R. Bidigard, and M. E. Ondrusek, "Effect of ultraviolet on Antarctic marine phytoplankton," in *Antarctic Ecosystems, Ecological Change and Conservation*, edited by K. R. Kerry and G. Hempel, Springer-Verlag, Berlin, 1990; Hader, D. P., R. C. Worrest, and H. D. Kumpar, *Aquatic Ecosystem, Environmental Effects of Ozone Depletion: 1991 update*, 33, 1991; Smith, R. C. and K. S. Baker, "Stratospheric ozone, middle ultraviolet radiation and phyto-

plankton productivity," *Oceanogr. Mag.*, **2**, 4, 1989; Smith, R. C., B. B. Prezelin, K. S. Baker, R. R. Bidigare, N. P. Boucher, T. Coley, D. Karentz, S. McIntyre, H. A. Matlick, D. Menzies, M. Ondrusek, Z. Wan, and K. J. Waters, "Ozone depletion: Ultraviolet radiation and phytoplankton biology in Antarctic waters," *Science*, **255**, 952–9, 1992; Worrest, R. C., "The effect of solar UV-B radiation on aquatic systems: An overview," in *Effects of Changes in Stratospheric Ozone and Global Climate, Overview*, edited by J. G. Titus, U.S. Environmental Protection Agency and United Nations Environmental Program, **1**, 175, 1986.

43 These measurements are described in Smith et al., 1992 (see previous endnote for full citation).

44 See Jin et al. for details (full citation provided in Endnote 20).

Chapter 10

Transmission in Spectrally Complex Media

10.1 Introduction

Having established some concepts of absorption and line-broadening processes in Chapters 3 and 4, we may now discuss more practical aspects of determining the transmission and radiative transfer within spectrally complex media. We first consider how radiation is transmitted through a medium characterized by absorption within a single, spectrally isolated line. This introduces the historically important quantity, the equivalent width. A generalization of this notion to include progressively more realistic absorption properties brings us to the various parameterizations of complex transmission processes, known as molecular band models. A large number of such models have been introduced over the years, all attempting to replace a very messy transmission problem with one or more analytic functions having a minimum number of parameters. These band parameters are derived from either comparisons with laboratory data or, since the advent of fast computers, with accurate line-by-line (LBL) computations. Since these classical methods are limited in accuracy, and cannot generally accommodate the simultaneous effects of scattering and absorption, it has been a goal in the field for many years to improve these descriptions. Many attempts have been made toward this end, and considerable progress has been made. Since many other references have covered band-model theory, we will provide only a few examples, before discussing the more modern methods, with emphasis on performing calculations for realistic inhomogeneous gaseous media.

There are two basic reasons why the absorption properties of a molecular gas depend very strongly upon wavenumber: (1) The line strengths can vary drastically over a given band, and (2) within a given line, the absorption coefficient changes many orders of magnitude over small wavenumber intervals. An additional complication is the fact that the radiation field itself will generally have strong spectral variations. At low spectral resolution the solar radiation field approximates a continuous spectrum. However, at higher resolution, it reveals a rich structure indicative of the physical conditions in the solar photosphere and chromosphere. The terrestrial IR radiation field has these same general characteristics, consisting of: (1) a near-blackbody component, that part emitted by the surface or the ocean, and (2) a more complicated component arising from the atmospheric emission. The latter will largely mirror the complex absorption properties of the medium through Kirchhoff's Law. Thermal emission is comparatively high at the centers of strong absorption lines and low in the transparent spectral windows. Even though photons are emitted most copiously at line center, their mean free path can be very short. However, very few photons are emitted within the spectral window regions, and yet they can be transmitted a long distance in the medium. Consequently, it is not obvious which frequencies contribute to a given quantity, for example, the flux or flux divergence. In transmission calculations care must be given not to truncate a spectral line wing too close to line center, since an important part of the energy in the wings could be missed. As usual, some insight can be gained from even a simple example. We will begin with consideration of a single spectral line.

10.2 Transmission in an Isolated Line

We define the *spectral beam transmittance* $\mathcal{T}_b(\tilde{\nu})$ (§5.2.10) and *beam absorptance* $\alpha_b(\tilde{\nu})$ as the ratios of the transmitted and absorbed intensity $[\tau(\tilde{\nu})]$ to the incident intensity I_0:

$$\mathcal{T}_b(\tilde{\nu}) = I[\tau(\tilde{\nu})]/I_0 = \exp[-\tau(\tilde{\nu})], \qquad \alpha_b(\tilde{\nu}) = 1 - \mathcal{T}_b(\tilde{\nu}). \tag{10.1}$$

The term "beam" refers to collimated radiation, to distinguish it from the flux transmittance, which applies to an integration of the angular distribution of the radiation over a hemisphere. For a single line, the optical path at wavenumber $\tilde{\nu}$ along the path $0 \to s$ is given by

$$\tau(\tilde{\nu}, s) = \int_0^s ds' \mathcal{S} n(s') \Phi(\tilde{\nu}). \tag{10.2}$$

Here $\mathcal{S}$ is the (frequency-integrated) line strength (see Eq. 4.58) in units of $\mathrm{m}^2 \cdot \mathrm{s}^{-1}$. In general $\mathcal{S}$ and the line profile function Φ remain inside the integral because their path depends on pressure and temperature. This complication will be avoided for the present, by assuming a homogeneous optical path. In practical terms, we can think of a horizontal optical path approximating this condition. Under these assumptions

$$\tau(\tilde{\nu}, s) = \mathcal{S}\mathcal{N}\Phi(\tilde{\nu}), \tag{10.3}$$

where $\mathcal{N}$ is the column number (cm^{-2}) of absorbing molecules, assumed to be a single-absorbing species. The situation of two or more molecular bands that overlap in wavenumber is discussed in §10.3.3 and 11.2.4. It is more common to describe the path in terms of the *column mass*, u. In this case the line strength is defined[1] per unit mass, and $\tau(\tilde{\nu}) = \mathcal{S}u\Phi(\tilde{\nu})$.

Suppose a homogeneous medium is illuminated by a collimated beam, which has a flat, or "white" spectrum. In practice, this need only apply to a rather narrow spectral region centered on a given spectral line, which we assume is well separated from other lines. We allow this radiation to pass through successively longer and longer path lengths. We imagine measuring the emergent light with a spectrometer of very high resolution. This situation is illustrated in Fig. 10.1. For optically thin paths ($\tau(\tilde{\nu}) \ll 1$ for all $\tilde{\nu}$) the radiation at each wavenumber will be attenuated in proportion to the respective optical path. The shape of the absorption line will be proportional to the product of the line profile $\Phi(\tilde{\nu})$ and the mass path u. For larger u, the exponential in Eq. 10.1 will begin to be important, and the absorption will no longer be proportional to the optical path.

Suppose instead we are interested in the mean absorption over the entire line. We choose an averaging wavenumber interval $\Delta\tilde{\nu}$ to encompass the line center, $\tilde{\nu}_0$, and to be large compared to the line width. The *mean beam absorptance* and *mean beam*

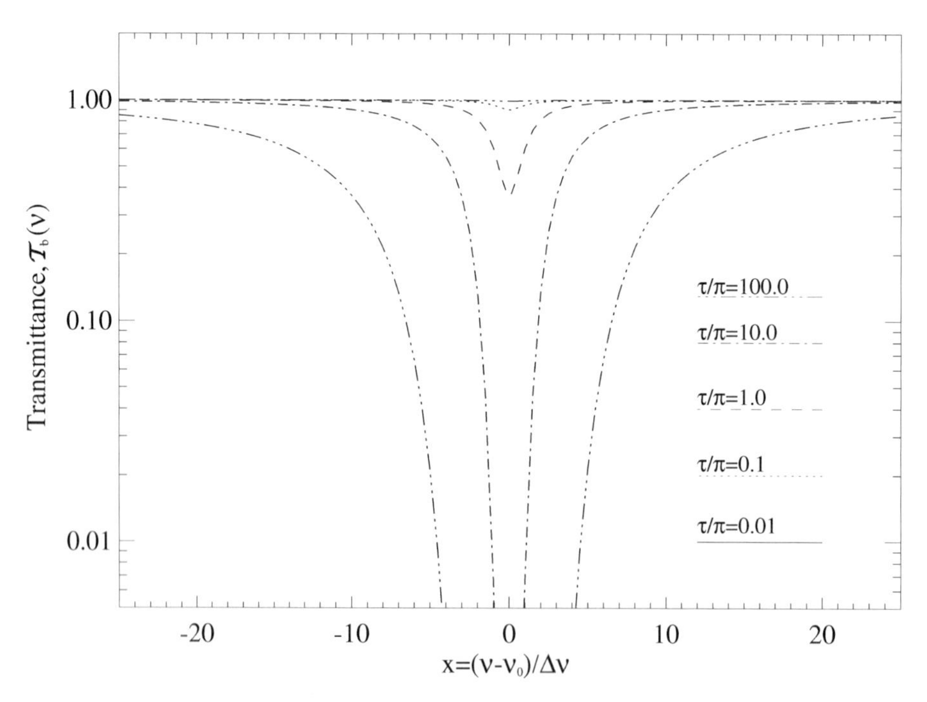

Figure 10.1 Transmittance of a homogeneous medium having a Lorentz profile, of varying line-center optical depths, τ/π (0.01, 0.1, 1.0, 10.0, and 100.0). The transmittance is almost unity for $\tau/\pi = 0.01$ and decreases rapidly with increasing τ so that the line center is opaque for $\tau/\pi > 10$. When multiplied by the line half-width α_L, the area under the curve is called the *equivalent width*.

transmittance for the spectral line are defined as

$$\langle \alpha_b \rangle \equiv 1 - \langle \mathcal{T}_b \rangle \equiv \frac{1}{\Delta\tilde{\nu}} \int\limits_{\Delta\tilde{\nu}} d\tilde{\nu} \alpha_b(\tilde{\nu}) = \frac{1}{\Delta\tilde{\nu}} \int\limits_{\Delta\tilde{\nu}} d\tilde{\nu} \left[1 - e^{-\tau(\tilde{\nu})}\right] \equiv \frac{W}{\Delta\tilde{\nu}}. \tag{10.4}$$

The product $W = \langle \alpha_b \rangle \Delta\tilde{\nu}$ (defined here in units of wavenumber) is called the *equivalent width*. This quantity is the area between the horizontal line and the lower curves in Fig. 10.1. Suppose we were to replace this area with a rectangle having the same area with complete absorption inside and zero absorption outside. The width of this equivalent rectangle is just $\langle \alpha_b \rangle \Delta\tilde{\nu}$, giving rise to the term "equivalent width." An important consideration for the energetics is that W is a measure of the radiative energy removed by the gas. The relationship of W to u is called the *curve of growth*, which generally depends upon the detailed line profile.

10.2.1 Isolated Lorentz Line

Suppose the spectral profile is given by Eq. 3.10,

$$\Phi_L(\tilde{\nu}) = \frac{\alpha_L}{\pi\left[(\tilde{\nu} - \tilde{\nu}_o)^2 + \alpha_L^2\right]}, \tag{10.5}$$

where α_L is the Lorentz half-width [cm^{-1}]. Setting $x \equiv (\tilde{\nu} - \tilde{\nu}_0)/\Delta\tilde{\nu}$ and $y \equiv \alpha_L/\Delta\tilde{\nu}$, and introducing the dimensionless mass path $\tilde{u} \equiv \mathcal{S}u/2\pi\alpha_L$, we find that the mean beam absorptance is written

$$\langle \alpha_b(u) \rangle = \int\limits_{-\infty}^{+\infty} dx \left[1 - \exp\left(-\frac{2\tilde{u}y^2}{x^2 + y^2}\right)\right]. \tag{10.6}$$

We have extended the lower wavenumber limit to $-\infty$, which should be of no consequence. This integral can be evaluated[2] in closed form to give

$$\langle \alpha_b(u) \rangle = 2\pi y \tilde{u} e^{-\tilde{u}} \left[I_0(\tilde{u}) + I_1(\tilde{u})\right] \equiv 2\pi y L(\tilde{u}). \tag{10.7}$$

$L(\tilde{u})$ is called the *Ladenburg–Reiche function*, after the authors who first described it in the literature, in 1913. It is expressed in terms of the *Bessel function of imaginary argument*,[3] I_n $(n = 0, 1, \ldots)$. We can find analytic results in two limits. First, let $\tilde{u} \ll 1$, the *weak-line limit*. Then $I_0 \approx 1$ and $I_1 \approx -\tilde{u}/2$, and

$$\langle \alpha_b(u) \rangle \approx 2\pi y \tilde{u} = \frac{\mathcal{S}u}{\Delta\tilde{\nu}} \quad (\tilde{u} \ll 1). \tag{10.8}$$

Equation 10.8 tells us that in this limit the mean beam absorptance is directly proportional to the column mass of absorbing molecules. This is also called the *linear regime* and describes an optically thin situation. This result is independent of the line-broadening mechanism. To see how this occurs, we expand the exponential inside the integral for small argument. Then we find

$$\langle \alpha_b(u) \rangle = \frac{1}{\Delta\tilde{\nu}} \int_{\Delta\tilde{\nu}} d\tilde{\nu}[1 - \exp(-\mathcal{S}u\Phi(\tilde{\nu}))] \cong \frac{\mathcal{S}u}{\Delta\tilde{\nu}} \int_{\Delta\tilde{\nu}} d\tilde{\nu}\Phi(\tilde{\nu}) = \frac{\mathcal{S}u}{\Delta\tilde{\nu}},$$

where we used the normalization property of $\Phi(\tilde{\nu})$.

Now consider the opposite case, the *strong-line limit* in which $\tilde{u} \gg 1$. We can ignore the y^2 term, in comparison to x^2, and write

$$\langle \alpha_b(u) \rangle \approx \int_{-\infty}^{+\infty} dx[1 - \exp(-2\tilde{u}y^2/x^2)]. \tag{10.9}$$

Let $\epsilon = 2\tilde{u}y^2/x^2$. Changing the variable of integration, we find

$$\langle \alpha_b(u) \rangle \approx y\sqrt{2\tilde{u}} \int_0^{\infty} \frac{d\epsilon}{\epsilon^{3/2}}(1 - e^{-\epsilon}) = 2y\sqrt{2\pi\tilde{u}} = \frac{2}{\Delta\tilde{\nu}}\sqrt{\mathcal{S}\alpha_L u}. \tag{10.10}$$

Because of the square-root dependence on the column mass, this part of the curve of growth is called the *square-root* or *saturated regime*. The square-root law has been verified by numerous spectroscopic experiments, and it is to be contrasted with the exponential Extinction Law, $1 - \langle \alpha_b \rangle = \exp(-ku)$, where $k = \mathcal{S}\Phi(\tilde{\nu})$ is a constant. The departure of absorption from an exponential behavior is the most important mathematical complication of nongrayness. It means we must abandon the convenient analytical properties of the solutions found in, for example, the prototype problems of Chapters 7 and 8. We now consider an array of overlapping lines.

10.3 Band Models

10.3.1 The Elsasser Band Model

Consider a band with equally spaced lines. Imagine that the mass path is large enough to be in the square-root regime, but the widths of the absorption lines are small compared to the line separation. As the mass path increases, the effect of line overlap becomes substantial because of absorption in the far wings. Then the mean absorptance cannot continue to grow like $\sqrt{u}$. A further increase in u can produce only a small increase in mean absorptance, and eventually it comes to a halt as the entire band becomes "blacked out." These effects are described in the highly idealized *Elsasser band model*, which approximates a band with a periodic pattern of lines of equal intensity and of equal width. At a given wavenumber $\tilde{\nu}$, it is necessary to consider that *all* lines in the band contribute to the absorption at that wavenumber. The Elsasser band assumes an *infinite* number of lines, all separated by the line spacing δ. For Lorentzian lines the mass absorption coefficient is written as

$$\alpha_m(\tilde{\nu}) = \sum_{n=-\infty}^{n=+\infty} \frac{\mathcal{S}}{\pi} \frac{\alpha_L}{\left[(\tilde{\nu} - n\delta)^2 + \alpha_L^2\right]}. \tag{10.11}$$

Elsasser showed that this function is mathematically identical to the following periodic function of wavenumber (see Problem 10.1):

$$\alpha_m(\tilde{\nu}) = \frac{\mathcal{S}}{\delta} \frac{\sinh(2\pi y)}{\cosh(2\pi y) - \cos 2\pi x}, \tag{10.12}$$

where $x \equiv \tilde{\nu}/\delta$ and $y \equiv \alpha_L/\delta$. We may then evaluate the expression for the mean transmittance by averaging with respect to $x = \tilde{\nu}/\delta$ from $x = -1/2$ to $x = +1/2$, that is, from the wavenumber of minimum absorption to maximum absorption, and back again. The expression for the beam transmittance is from Eq. 10.4,

$$\langle \mathcal{T}_b(y, u) \rangle = \int_{-1/2}^{+1/2} dx \exp\left[- \frac{2\pi \tilde{u} y \sinh(2\pi y)}{\cosh(2\pi y) - \cos 2\pi x} \right]. \tag{10.13}$$

The beam absorptance for the Elsasser model, $\langle \alpha_b(y, u) \rangle = 1 - \langle \mathcal{T}_b(y, u) \rangle$, is plotted in Fig. 10.2 versus the dimensionless mass path $\tilde{u} \equiv \mathcal{S}u/2\pi\alpha_L$ for a variety of y values. (Since we have selected the averaging interval $\Delta\tilde{\nu}$ to be the mean line spacing δ, then $y = \alpha_L/\delta$.) Here y can be thought of as a *grayness parameter*. For small y the behavior of the mean absorption departs radically from monochromatic absorption: At small $\tilde{u}$, only the line centers absorb, and most radiation passes through the medium between the lines. At larger y, the regions between the lines begin to absorb strongly as the lines become saturated, thus accounting for the more pronounced dependence on $\tilde{u}$. For large y ($y \geq 10$), the lines completely overlap and the beam transmittance may be shown (Problem 10.2) to be given by $\exp(-2\pi y\tilde{u}) = \exp(-\mathcal{S}u/\alpha_L)$, the gray limit.

Another asymptotic region[4] is the strong-line limit ($\tilde{u} \gg 1$) for which it may be shown (Problem 10.3) that $\langle \alpha_b \rangle \approx \text{erf}\,[\pi y \sqrt{2\tilde{u}}]$, where erf denotes the *error function*.

Few bands in nature resemble the regular array visualized by Elsasser. However, this model has played an important role in the historical development of the subject. Furthermore, it is useful in exploring the general nature of nongray absorption. For example, the width of the nongray *cooling-to-space* function for an Elsasser band (described in Chapter 11) is found to depend upon the single parameter y.

10.3.2 Distributed Line Intensities

A glance at any real absorption spectrum reveals that line strengths are distributed over a wide range of values and that line separations are also far from a constant, as assumed in the Elsasser model. (Variations of line widths within a given band are usually small and thus will be neglected here.) We first consider the situation of an array of nonoverlapping lines. If we use an averaging interval that contains a large

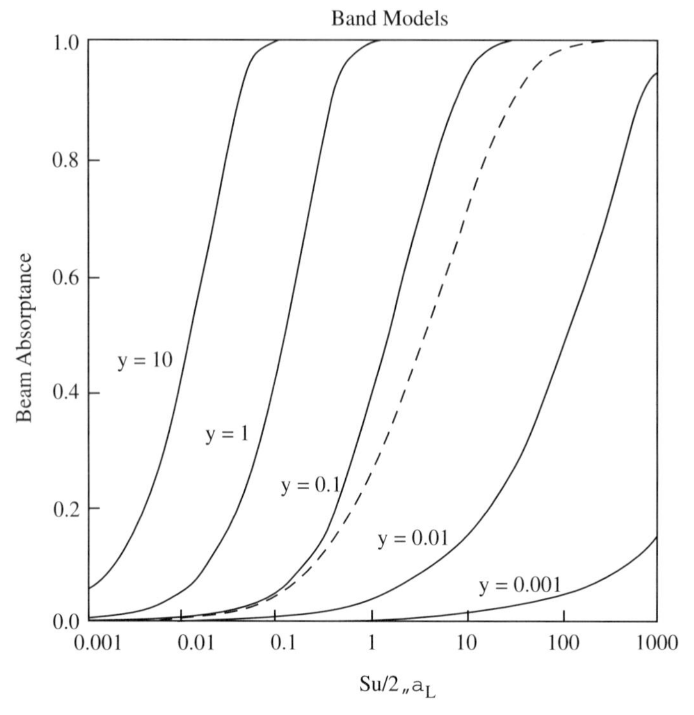

Figure 10.2 Mean beam absorptance, $\langle\alpha_b(y, u)\rangle$, versus absorber amounts $\tilde{u}$ for the Lorentz–Elsasser model (Eq. 10.13, solid lines) and the random–Lorentz–Malkmus model (shown only for $y = 0.1$, Eq. 10.25, dashed line). Each curve applies to a particular value of $y = \alpha_L/\Delta\tilde{\nu}$, the *grayness parameter*. For $y \gg 1$, the absorptance obeys the Extinction Law for a gray-absorbing medium.

number of lines of varying strength, it is possible to define a continuous distribution function of line strengths, $p(\mathcal{S})$, such that $\mathcal{S}$ and $\mathcal{S} + d\mathcal{S}$ is $p(\mathcal{S})d\mathcal{S}$. In the limit of an infinite number of lines, the mean beam absorptance is written

$$\langle\alpha_b(u)\rangle = \int_0^\infty d\mathcal{S}p(\mathcal{S}) \int_{\Delta\tilde{\nu}} \frac{d\tilde{\nu}}{\Delta\tilde{\nu}}\{1 - \exp[-\mathcal{S}u\Phi(\tilde{\nu})]\}. \tag{10.14}$$

Several analytic line-strength distributions are in use. These are chosen more for analytic convenience than for their fidelity in reproducing actual distributions, which are usually quite irregular.[5] Three of the most common are

Exponential Distribution:

$$p(\mathcal{S}) = (1/\bar{\mathcal{S}})\exp(-\mathcal{S}/\bar{\mathcal{S}}); \tag{10.15}$$

Godson Distribution:

$$p(\mathcal{S}) = \bar{\mathcal{S}}/(\mathcal{S}_{\max}\mathcal{S}) \quad (\mathcal{S} < \mathcal{S}_{\max}),$$
$$p(\mathcal{S}) = 0 \quad (\mathcal{S} > \mathcal{S}_{\max}); \tag{10.16}$$

Malkmus Distribution:

$$p(\mathcal{S}) = (1/\mathcal{S})\exp(-\mathcal{S}/\bar{\mathcal{S}}); \tag{10.17}$$

where the average line strength is defined as

$$\bar{\mathcal{S}} = \int_0^\infty d\mathcal{S}\mathcal{S}p(\mathcal{S}). \tag{10.18}$$

These distributions are normalized so that $\int_0^\infty d\mathcal{S}p(\mathcal{S}) = 1$. An additional distribution is the trivial one in which all values of $\mathcal{S}$ collapse to a single value. This is called the δ-*function distribution*, $p(\mathcal{S}) = \delta(\mathcal{S} - \bar{\mathcal{S}})$. An example of this distribution is the Elsasser band model.

Example 10.1 Absorptance in a Lorentz-Exponential Model

We illustrate the usefulness of the analytic $\mathcal{S}$-distributions to derive an analytic expression for the beam absorptance. Assuming the exponential distribution (Eq. 10.15), and interchanging orders of integration in Eq. 10.14, we can carry out the inner integration analytically:

$$\langle \alpha_b(u) \rangle = \frac{1}{\Delta\tilde{\nu}} \int_{\Delta\tilde{\nu}} d\tilde{\nu} \int_0^\infty d\mathcal{S} e^{-\mathcal{S}/\bar{\mathcal{S}}} \left[1 - e^{-\mathcal{S}u\Phi(\tilde{\nu})}\right] = \frac{1}{\Delta\tilde{\nu}} \int_{\Delta\tilde{\nu}} d\tilde{\nu} \frac{\bar{\mathcal{S}}u\Phi(\tilde{\nu})}{1 + \bar{\mathcal{S}}u\Phi(\tilde{\nu})}. \tag{10.19}$$

Adopting a Lorentz profile for $\Phi(\tilde{\nu})$, Eq. 10.5, we may integrate analytically to obtain

$$\langle \alpha_b(u) \rangle = \frac{\bar{\mathcal{S}}u/\Delta\tilde{\nu}}{\sqrt{1 + \bar{\mathcal{S}}u/\pi\alpha_L}} = \frac{2\pi y\tilde{u}}{\sqrt{1 + \tilde{u}/2}}. \tag{10.20}$$

To proceed further, we need to determine the numerical values for the parameters $\bar{\mathcal{S}}$ and $\bar{\mathcal{S}}/\alpha_L$. Both can be determined by requiring that the results match the accurate asymptotic expressions in the strong- and weak-line limits (see Problem 10.4).

Example 10.2 Absorptance in a Lorentz–Malkmus Model

Using the Malkmus distribution (Eq. 10.17), we proceed as in Example 10.1, by interchanging orders of integration. The result for the beam absorptance is

$$\langle \alpha_b(u) \rangle = \frac{1}{\Delta\tilde{\nu}} \int_{\Delta\tilde{\nu}} d\tilde{\nu} \ln[1 + \bar{\mathcal{S}}\Phi(\tilde{\nu})u]. \tag{10.21}$$

Assuming a Lorentz profile, we integrate analytically to obtain

$$\langle \alpha_b(u) \rangle = \frac{\pi\alpha_L}{2\Delta\tilde{\nu}} \left[\sqrt{1 + 4\bar{\mathcal{S}}u/\pi\alpha_L} - 1\right] = (\pi y/2) \left[\sqrt{1 + 8\tilde{u}} - 11\right]. \tag{10.22}$$

Note that Eqs. 10.20 and 10.22 are not valid for values of the parameters for which the lines overlap.

10.3.3 Random Band Model

In some irregular spectral bands (e.g., water vapor), the line positions appear to vary randomly over the spectrum. What is described below comes under various names, including the *statistical*, *random*, and *Goody–Meyer band models*. We delve only briefly into this formulation, because there are excellent references on the subject.[6]

Let the band interval of width $\Delta\tilde{\nu}$ consist of n lines of average separation δ, so that $\Delta\tilde{\nu} = n\delta$. If the line positions are uncorrelated, the transmission of the band can be written in terms of the products of the individual line transmittances $\langle \mathcal{T}_i \rangle$ and absorptances $\langle \alpha_i \rangle$:

$$\langle \mathcal{T}_b \rangle = \langle \mathcal{T}_1 \rangle \langle \mathcal{T}_2 \rangle \cdots \langle \mathcal{T}_n \rangle = \left[\frac{1}{\Delta\tilde{\nu}} \int_{\Delta\tilde{\nu}} d\tilde{\nu} \int_0^\infty d\mathcal{S} p(\mathcal{S}) e^{-\mathcal{S}u\Phi(\tilde{\nu})} \right]^n$$

$$= \left[1 - \frac{\langle \alpha_b \rangle}{n} \right]^n. \tag{10.23}$$

Taking the limit $n \to \infty$, and noting that $(1 - x/n)^n \to e^{-x}$, we find in general that

$$\langle \mathcal{T}_b(u) \rangle = e^{-\langle \alpha_b(u) \rangle}. \tag{10.24}$$

This result says that *the beam transmittance of randomly placed overlapping lines is equal to the exponential of (minus) the nonoverlapped single-line beam absorptance.*

Applying this result to the Lorentz–Malkmus band model, we find

$$\langle \mathcal{T}_b(u) \rangle = \exp\left[-\frac{\pi \alpha_L}{2\Delta\tilde{\nu}} \left\{ \sqrt{1 + \frac{4\mathcal{S}u}{\pi \alpha_L}} - 1 \right\} \right] = e^{-(\pi y/2)(\sqrt{1+8\tilde{u}}-1)}, \tag{10.25}$$

where $y \equiv \alpha_L/\delta$ and $\tilde{u} \equiv \mathcal{S}u/2\pi\alpha_L$. Equation 10.25 fits laboratory data and transmission values derived from LBL calculations with very good accuracy. In particular, it better accounts for the weak lines (for example in the H_2O 6.3 μm band) than does the random-exponential model. The curve of growth for this model is shown in Fig. 10.2 for $y = 0.1$ (dashed line) in order to compare with the Elsasser model. Note that for this value of y, the random model yields more transmission (less absorption) for all optical paths. This occurs because of the occasional wide gap that is not present in regular-array models. The random-Malkmus model (Eq. 10.25) is one of the best all-round analytic band formulations and will be discussed later in connection with the k-distribution method (§10.4.2).

10.3.4 MODTRAN: A Moderate-Resolution Band Model

This computer code[7] was devised to be efficient, user friendly, upgradeable, well maintained, and readily available to the community. It provides the user with the means to quickly evaluate atmospheric transmittance for a large variety of user-specified atmospheric conditions. Three band-model parameters are used in MODTRAN: 1. an effective absorption coefficient $\mathcal{S}/d$, 2. a line density, and 3. a line width d. For a given wavenumber interval $\Delta\tilde{\nu}_i$, the effective absorption coefficient is a measure of the total strength of lines in this interval, the line density is the average number of lines, and the line width is the average wavenumber bin width (both weighted by the line strengths):

$$\frac{\mathcal{S}_i}{d_i} \equiv \frac{1}{\Delta\tilde{\nu}} \sum_{J=1}^{N} \mathcal{S}_J^i(T), \tag{10.26}$$

$$\frac{1}{d_i} \equiv \left[\sum \mathcal{S}_J^i\right]^2 \Big/ \Delta\tilde{\nu} \sum \left(\mathcal{S}_J^i\right)^2 \tag{10.27}$$

where $\mathcal{S}_J^i$ is the line strength (see Eq. 4.64) of the Jth line of a specific molecule in wavenumber bin i at temperature T. The line strengths are scaled from the HITRAN data base using Eq. 4.64 with a reference temperature of 296 K, the standard value adopted in HITRAN. This definition of the line spacing is appropriate for the finite bin widths considered in MODTRAN. The transmittance in MODTRAN is based on the random band model (see §10.3.3) in which the parameters are calculated in 1 cm^{-1} intervals. The parameters are computed from line data in the HITRAN data base (see §4.5.3), which contains data on 709,308 molecular lines in the wavenumber range 0 to 23,000 cm^{-1}. Computation of the transmittance across a bin involves four steps:

i. integration of the Voigt line shape of an "average" line over the 1 cm^{-1} interval;
ii. if a bin contains more than one line of a given species, the lines are assumed to be randomly distributed with statistical overlap;
iii. the contribution from lines with centers in nearby bins is calculated as a molecular "continuum" component;
iv. the H–C–G approximation (see §10.5.1) is used to replace an inhomogeneous path with a homogeneous one by using average values for the band-model parameters.

The molecular transmittance is based on the random band model (Eq. 10.23) for a finite number of lines within a spectral interval $\Delta\tilde{\nu}$:

$$\langle \mathcal{T}_\text{b} \rangle = \left[1 - \frac{\langle \alpha_\text{b} \rangle}{\langle n \rangle}\right]^{\langle n \rangle}, \tag{10.28}$$

where $\langle \alpha_\text{b} \rangle \Delta\tilde{\nu}$ is the single-line equivalent width and $\langle n \rangle$ is the path-averaged effective number of lines in the bin $\langle n \rangle = \Delta\tilde{\nu}\langle 1/d \rangle$. Here $\langle 1/d \rangle$ is the path-averaged line spacing or line density. For mixed Lorentz–Doppler absorption, the single-line equivalent width $\langle \alpha_\text{b} \rangle \Delta\tilde{\nu}$ is computed from

$$\langle \alpha_\text{b} \rangle \Delta\tilde{\nu} = \Delta\tilde{\nu}\left[1 - \langle \mathcal{T}_\text{b} \rangle\right] = \int_{\Delta\tilde{\nu}} d\tilde{\nu}\left[1 - e^{-\mathcal{S}u\Phi_\text{V}(\tilde{\nu})}\right], \tag{10.29}$$

where $\Phi_\text{V}(\tilde{\nu})$ is the Voigt profile (see §3.3.3).

To compensate for the line wings outside the finite bin width, the contributions from lines located in the wing interval (outside of a given bin but within ± 25 cm^{-1}) are computed separately from line wing band-model parameters determined by integrating the Lorentz profile over the wing interval. Wing contributions beyond ± 25 cm^{-1} are ignored.

The MODTRAN code includes the previous coarser resolution (20 cm^{-1}) band model LOWTRAN[8] as an option. In fact, since spectroscopic data in the HITRAN data base (to which MODTRAN is linked) stop at 23,000 cm^{-1}, MODTRAN converts to LOWTRAN as a default for higher wavenumbers.

Unfortunately, most band models are incompatible with the requirements for "quasi-monochromatic" treatment of multiple scattering. Thus, inclusion of multiple scattering requires the use of methods like the k-distribution approach described below. In fact, the popular LOWTRAN code with 20 cm^{-1} resolution uses this approach. In contrast, the MODTRAN code, which has a "fixed-wavenumber" sampling of 1 cm^{-1} (and a nominal resolution of 2 cm^{-1}), is assumed to be quasi-monochromatic and is automatically compatible with multiple-scattering treatments. The original two-stream algorithm for multiple scattering in MODTRAN has now been replaced by the more accurate DISORT algorithm. The adequacy of this quasi-monochromatic treatment of multiple scattering must be checked against line-by-line computations. Tests indicate that MODTRAN performs well, but more extensive tests are needed across the solar and thermal infrared regions. We note, however, that the MODTRAN code is far too expensive (computationally) to use in global climate models and is most affordable for one-dimensional models.

10.4 Spectral Mapping Transformations for Homogeneous Media

Although a great deal of effort and mathematical ingenuity has been expended on band models over the past fifty years or so, they still have definite limitations, given the necessity in modern climate models to obtain radiative fluxes and heating rates to 1% accuracy. Fortunately it is now possible to calculate these quantities by brute-force line-by-line (LBL) methods to 0.1% accuracy. (This assumes perfect accuracy of the spectroscopic parameters, which of course is far from true.) Since the heating rates must be computed at nearly every grid point of a General Circulation Model (GCM), LBL calculations represent significant computational demands even on modern supercomputers. Thus, LBL results are most useful as a standard of comparison for less accurate, but much more efficient algorithms.

An algorithm that has received much attention in the recent literature is the *k-distribution* and its associated *correlated k-distribution* method. These methods provide much better accuracy than the conventional band models, and yet they require two to three orders of magnitude less computer time than LBL methods (described in §10.5.2). Furthermore, such methods can accommodate multiple scattering in a straightforward manner. A disadvantage is that an LBL calculation is needed to derive the parameters. But since the simplified calculations are then repeated many times over, the cost savings can be substantial.

10.4.1 Method of the k-Distribution

Consider a spectral interval $\Delta\tilde{\nu} = \tilde{\nu}_1 - \tilde{\nu}_2$, which is large enough to contain a significant number (say >20) of spectral lines, but small enough that the Planck function is essentially constant over $\Delta\tilde{\nu}$. The definition of the beam transmittance over a homogeneous mass path u is

$$\langle \mathcal{T}_b(u) \rangle = \frac{1}{\Delta\tilde{\nu}} \int_{\tilde{\nu}_1}^{\tilde{\nu}_2} d\tilde{\nu}\, e^{-k(\tilde{\nu})u}, \tag{10.30}$$

where $k(\tilde{\nu})$ denotes the mass extinction coefficient,[9] which is equal to the mass absorption coefficient for a purely absorbing medium.

To compute $\langle \mathcal{T}_b(u) \rangle$ accurately using the above form would require division of the spectral interval into subintervals $\delta\tilde{\nu}$ small enough so that $k(\tilde{\nu})$ is essentially constant. This requires $\delta\tilde{\nu}$ to be $\approx 10^{-3}$–10^{-5} cm^{-1}, which requires a total of $\Delta\tilde{\nu}/\delta\tilde{\nu} \approx 10^4$–$10^6$ quadrature points for a small part (say 10 cm^{-1}) of the spectrum. This must be repeated over the entire band, for all absorption bands, and over the full range of u. In addition it must be repeated for the range of pressures and temperatures encountered in the atmosphere. Clearly, this kind of "frontal attack" is computationally prohibitive.

A more efficient approach involves a transformation of Eq. 10.30 that recognizes the fact that the same value of k is encountered many times over in the spectral interval. If we were to combine, or bin, all the values of k into groups, and perform the transmittance calculation only once for a given k, then we could eliminate all the redundancy of Eq. 10.30. Furthermore, if we order the groups into monotonically increasing values of k, we will obtain a much-more "orderly" function $f(k)$, which has more desirable characteristics than the wildly varying $k(\tilde{\nu})$. Choosing a suitably small interval Δk, the *k-distribution* can be formally defined by the following grouping algorithm:

$$f(k) \equiv \frac{1}{\Delta\tilde{\nu}} \sum_{l=1}^{M} \left| \frac{d\tilde{\nu}}{dk} \right| W_l(k). \tag{10.31}$$

Here $W_l(k)$ is the "window" function, equal to unity when $k^l_{\min} \le k \le k^l_{\max}$, and zero otherwise. M is the number of monotonically varying wavenumber subintervals in which the absorption coefficient switches from increasing to decreasing values (or vice versa). In the lth subinterval, the absorption coefficient varies from $k^l_{\min}$ to $k^l_{\max}$. The absolute value of the derivative is taken, because we want to count the value whether or not k is increasing or decreasing with wavenumber. Note that k is considered to be a continuous variable. Equation 10.30 can then be rewritten as a finite sum

$$\langle \mathcal{T}_b(u) \rangle \approx \sum_{j=1}^{N} \Delta k_j f(k_j) e^{-k_j u}, \tag{10.32}$$

where N is the total number of monotonic subintervals over the entire range of k values. In the limit of $\Delta k \to 0$ (assuming that the number of lines is always large within

Δk), the above sum becomes an integral

$$\langle \mathcal{T}_b(u) \rangle = \int_{k_{min}}^{k_{max}} dk f(k) e^{-ku}, \tag{10.33}$$

where k_{min} and k_{max} are the minimum and maximum values of k over the entire spectral interval, $\Delta\tilde{\nu}$.

It is clear that if we sum $f(k)$ over all binned values of k, we should get unity, that is,

$$\sum_{j=1}^{N} f(k_j)\Delta k_j = 1, \quad \text{or for } \Delta k_j \to 0, \quad \int_{k_{min}}^{k_{max}} dk f(k) = 1. \tag{10.34}$$

If we sum the distribution up to some value of $k_n < k_{max}$, we may define the *cumulative k-distribution* as

$$g(k_n) \equiv \sum_{j=1}^{n} f(k_j)\Delta k_j, \quad \text{or for } \Delta k_j \to 0, \quad g(k) = \int_0^k dk' f(k'). \tag{10.35}$$

We can now write Eqs. 10.32 and 10.33 as

$$\langle \mathcal{T}_b(u) \rangle \approx \sum_{j=1}^{N} e^{-k_j u}\Delta g_j, \quad \text{or for } \Delta k_j \to 0, \quad \langle \mathcal{T}_b(u) \rangle = \int_0^1 dg e^{-k(g)u}. \tag{10.36}$$

Note that the upper limit of unity is consistent with g being a cumulative k-distribution, that is, the total number of k values smaller than k. Writing out Eq. 10.36 as

$$\langle \mathcal{T}_b(u) \rangle \approx \Delta g_1 e^{-k_1 u} + \Delta g_2 e^{-k_2 u} + \cdots + \Delta g_N e^{-k_N u}, \tag{10.37}$$

we see that we have an approximation to the transmittance as a *weighted sum of monochromatic transmittances*. This approximation, known as the *exponential-sum fit transmittance* or ESFT approximation, has a long history in the field. Clearly, it is desirable to reduce the nongray problem to a finite number of gray problems, since we have developed many computational tools to handle the latter problem. If we had knowledge of the transmittance measured in the laboratory under low spectral resolution, or if we had access to LBL calculations of $\langle \mathcal{T}_b(u) \rangle$, we could in principle perform a nonlinear least-squares fit of Eq. 10.37 to the "data" to any desired accuracy. This would yield the "coefficients" of the fit ($g_1, g_2, \ldots; k_1, k_2, \ldots$). Unfortunately, this problem is mathematically ill posed, and special analysis techniques must be applied for this method to be practical.[10]

Fortunately, the current availability of accurate synthetic absorption spectra (see Fig. 10.3) means that we can compute the k-distribution directly, and the coefficients of the EFST approximation may be determined without least-squares fitting, simply by numerical quadrature. It is a straightforward, but time-consuming, task for the

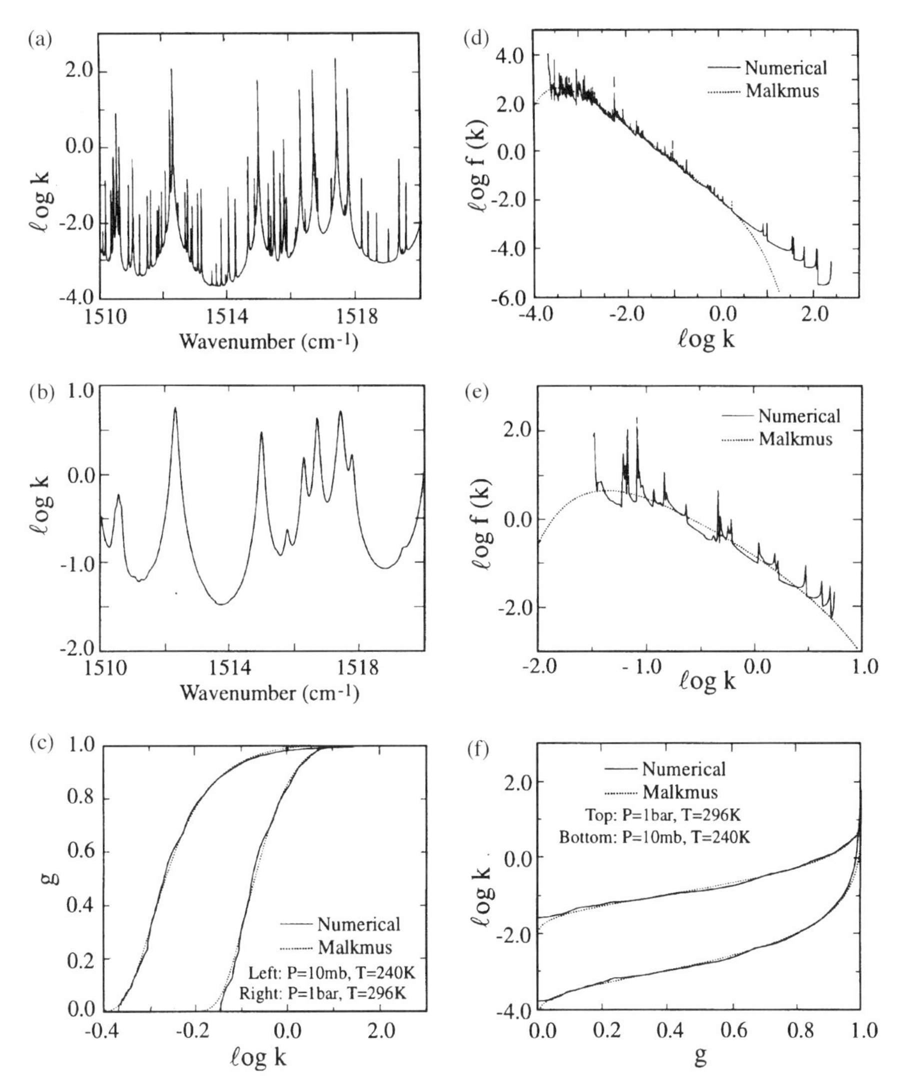

Figure 10.3 Absorption coefficient spectrum for the 1510–1520 cm^{-1} portion of the 6.3-μm water vapor band. (a) Line-by-line absorption coefficients for a pressure of 10 mbar and temperature 240 K and (b) for a pressure of 1 bar and 296 K. The spectra utilize the HITRAN 1982 atmospheric line compilation and include 141 lines with absorption contributions from outside of the interval within a 5 cm^{-1} Lorentz wing cutoff limit. (d) and (e) Absorption coefficient frequency distributions corresponding to absorption spectra in Figs. (a) and (b), respectively. (c) Cumulative frequency distributions of the absorption coefficient spectra in Figs. (a) and (b). (f) The k distributions of the absorption coefficient spectra Figs. (a) and (b). Malkmus band model equivalents, obtained to provide best fit to line-by-line transmission, are shown by dashed lines. From Lacis and Oinas, Fig. 1, see Endnote 11.

computer to construct sorted tables of absorption coefficients to derive $f(k)$ and $g(k)$ from say, a spectroscopic data base (such as HITRAN). The *inverse k-distribution* $k(g)$ is also needed in order to perform spectral mapping. An "inverse table" k versus g is easily constructed by computer methods. Examples of numerical determinations of $f(k)$, $g(k)$, and $k(g)$ are shown in Fig. 10.3.[11]

Example 10.3 k-Distribution for a Spectral Line

There is an interesting analogy between $k(g)$ and the normalized line profile $\Phi(x)$, where $x = \tilde{\nu}/\Delta\tilde{\nu}$. We can best understand this relationship by deriving $f(k)$ and $g(k)$ in the simple case of an isolated line. For a single line, Eq. 10.31 simplifies to

$$f(k) = \frac{2}{\Delta\tilde{\nu}} \frac{d\tilde{\nu}(k)}{dk} W(k), \tag{10.38}$$

where we have redefined the spectral interval to run from $-\Delta\tilde{\nu}/2$ to $+\Delta\tilde{\nu}/2$. Noting the symmetrical form of $k(\tilde{\nu})$, we have considered only half of the line. The factor of 2 comes from the fact that there are two values of k in the whole-line distribution. Note also that the absolute value has been dropped, since we have picked the small-$\tilde{\nu}$ half of the line where the derivative is positive. We now integrate the above equation to find $g(k)$, noting that $W = 1$ over the range of integration:

$$g(k) = \frac{2}{\Delta\tilde{\nu}} \int_{k_{\min}}^{k} dk' \frac{d\tilde{\nu}(k')}{dk'} W(k') = \frac{2}{\Delta\tilde{\nu}} [\tilde{\nu}(k) - \tilde{\nu}(k_{\min})]. \tag{10.39}$$

This correspondence between g and $\tilde{\nu}$ means that there is a unique *spectral mapping* from the variable g to a specific wavenumber $\tilde{\nu}$ (and vice versa). As g increases from 0 to 1, the normalized wavenumber varies from $\tilde{\nu}(k_{\min})/(\Delta\tilde{\nu}/2)$ to 0. As $\Delta\tilde{\nu} \rightarrow \infty$, the wings are truncated less and less, until the line is completely defined.

Example 10.4 k-Distribution for a Lorentz Line

For a Lorentz line, Eq. 10.5, we set the variable $\tilde{\nu}$ equal to zero at line center so that

$$k_{\mathrm{L}}(\tilde{\nu}) = \frac{S\alpha_{\mathrm{L}}}{\pi[\tilde{\nu}^2 + \alpha_{\mathrm{L}}^2]}. \tag{10.40}$$

Note that the range of $\tilde{\nu}$ is $-\tilde{\nu}(k_{\min}) < \tilde{\nu} < +\tilde{\nu}(k_{\min})$. Solving for $\tilde{\nu}$, we find

$$\tilde{\nu}_{\mathrm{L}}(k) = \pm\sqrt{(S\alpha_{\mathrm{L}}/\pi k) - \alpha_{\mathrm{L}}^2}. \tag{10.41}$$

Selecting the positive sign and carrying out the derivative we find

$$f_{\mathrm{L}}(k) = \frac{1}{(\Delta\tilde{\nu}/2)} \frac{d\tilde{\nu}}{dk} = \frac{1}{\Delta\tilde{\nu}\pi k} \frac{S\alpha_{\mathrm{L}}}{\sqrt{(S\alpha_{\mathrm{L}} k/\pi) - \alpha_{\mathrm{L}}^2 k^2}}. \tag{10.42}$$

At small k (corresponding to the line wings), $f(k) \propto k^{-3/2}$, which is characteristic of the Lorentzian wings. This behavior also shows up in the k-distribution in more complicated bands. The cumulative distribution follows from Eqs. 10.39 and 10.41. Since we are considering a finite bandwidth $\Delta\tilde{\nu}$ over which to average, we should set $\tilde{\nu}(k_{\min}) = \Delta\tilde{\nu}/2$. Thus,

$$g_{\mathrm{L}}(k) = \frac{2}{\Delta\tilde{\nu}} \sqrt{(S\alpha_{\mathrm{L}}/\pi k) - \alpha_{\mathrm{L}}^2} - 1. \tag{10.43}$$

Finally, the inverse of the cumulative distribution is obtained by solving for k:

$$k_{\mathrm{L}}(g) = g_{\mathrm{L}}^{-1}(k) = \frac{S\alpha_{\mathrm{L}}/\pi}{[(\Delta\tilde{\nu}/2)(1+g)]^2 + \alpha_{\mathrm{L}}^2}. \tag{10.44}$$

Note the resemblance between Eqs. 10.44 and 10.40.

10.4.2 k-Distribution for the Malkmus Band Model

We now ask whether we can derive analytic expressions for $f(k)$, given analytic band-model representations of $\langle \mathcal{T}_b(u) \rangle$. To illustrate how this can be done in certain cases, we examine a limiting form of Eq. 10.33, where we set the limits of integration on k equal to $(0, \infty)$. This is permissible since large and small k values contribute very little to the integral. Then

$$\langle \mathcal{T}_b(u) \rangle = \int_0^\infty dk f(k) e^{-ku} = \mathcal{L} f(k), \tag{10.45}$$

where $\mathcal{L}$ denotes the *Laplace transform*. For a given analytic expression for the transmission, we may find $f(k)$ provided the *inverse Laplace transform* of $\langle \mathcal{T}_b(u) \rangle$ can be determined, so that

$$f(k) = \mathcal{L}^{-1} \langle \mathcal{T}_b(u) \rangle. \tag{10.46}$$

It can be shown (see Problem 10.5) that the inverse transform exists in analytic form for the statistical-Malkmus model,

$$f(k) = (1/2) k^{-3/2} \sqrt{\mathcal{S} y / 2\pi} \exp\left[(y/8) \left(2 - \frac{\mathcal{S}}{k} - \frac{k}{\mathcal{S}} \right) \right]. \tag{10.47}$$

If the parameter $y = 2\pi \alpha_L / \Delta\tilde{\nu}$ becomes large, we have very broad, overlapping lines and the distribution becomes gray, that is, $k = \text{constant}$, as discussed previously for the Elsasser band. In the other limit of y small, the line overlap becomes less and less important, and smaller and smaller values of k will appear in the absorption minimum (the spectral "window") between adjacent lines. The cumulative distribution (but not the inverse cumulative distribution) can also be determined in closed form.[12] These distributions are shown in Fig 10.3.

10.5 Transmission in Nongray Inhomogeneous Media

Up to now we have dealt with homogeneous paths, that is, the pressure and temperature are assumed constant along the beam. However, unless the beam direction is horizontal, this is never valid, and we must deal with the inhomogeneous nature of the medium. In general, the mean beam transmittance in a slab medium over an inhomogeneous path of a beam making an angle θ with the vertical is given by

$$\langle \mathcal{T}_b(u, \theta) \rangle = \frac{1}{\Delta\tilde{\nu}} \int_{\Delta\tilde{\nu}} d\tilde{\nu} \left\{ \exp\left[- \int_0^u du' \mathcal{S}(u') \Phi(u', \tilde{\nu}) \sec\theta \right] \right\}. \tag{10.48}$$

To understand better the structure of Eq. 10.48, let us reconsider the simple case of an isolated Lorentz line, but now let the line of sight be taken over an inhomogeneous optical path. This leads to an analytic result, only in the special case of a well-mixed gas and a pure-Lorentzian line profile. Although it is not necessary, we will assume a vertical path ($\theta = 0$). The volume density of the absorber is given by $w^m \rho(z)$. Here w^m is its mass mixing ratio (assumed constant) and $\rho(z)$ is the total atmospheric mass density. The optical path between the heights z and z' may be transformed from the mass path variable to the geometric height variable dz'', so that $du'' = -w^m \rho(z'')dz''$. Then

$$\tau(z, z') = \int_z^{z'} dz'' w^m \rho(z'') \frac{S\alpha_L(z'')}{\pi[(\tilde{\nu} - \tilde{\nu}_0)^2 + \alpha_L(z'')^2]}$$

$$= \left[\frac{Spw^m}{2\pi\alpha_L(z)g}\right] \ln\left[\frac{(\tilde{\nu} - \tilde{\nu}_0)^2 + \alpha_L(z)^2}{(\tilde{\nu} - \tilde{\nu}_0)^2 + \alpha_L(z')^2}\right], \tag{10.49}$$

where the hydrostatic equation, $dp = -\rho(z)gdz$, was used. To obtain the second form we changed integration variables from dz'' to $d\alpha_L$ and integrated analytically. It is easily checked that the leading term in the second result is dimensionless, as it must be. It is also a constant, since $\alpha_L(z)$ is proportional to $p(z)$, the total gas pressure (see Eq. 3.11).

We will simplify the algebra to a situation in which the path is from z to $z' \to \infty$. Now the vertical mass path is simply $u = w^m p/g$ (assuming g is constant with height). Assuming $\alpha_L(z')$ approaches α_D, the Doppler width, we find for a well-mixed gas

The absorption coefficient for an inhomogeneous vertical path:

$$k_L^*(\tilde{\nu})u \equiv \frac{Su}{2\pi\alpha_L} \ln\left[\frac{(\tilde{\nu} - \tilde{\nu}_0)^2 + \alpha_L(z)^2}{(\tilde{\nu} - \tilde{\nu}_0)^2 + \alpha_D^2}\right]. \tag{10.50}$$

To be more accurate, a Voigt line shape should be used in the limit $z' \to \infty$.

Equation 10.50 describes the behavior of the absorption profile of a spectral line for very high resolution. With modern spectroscopic techniques, the resolving power is now sufficient to test such predictions. An example is shown in Fig. 4.5 of an individual CO_2 line profile measured by absorption of near-IR sunlight. The observed shape closely resembles the theoretical prediction, showing that the Lorentz profile (applicable to a homogeneous, horizontal path, Eq. 3.6) does not apply to a column-integrated inhomogeneous path.

10.5.1 The H–C–G Scaling Approximation

One might hope that the simple expressions derived for the homogeneous case, say for absorptance, $\langle \alpha_b(u) \rangle_{hom}$, could somehow be used for the inhomogeneous case, if only we could define equivalent values for the *effective pressure* $\langle p \rangle$, *effective temperature* $\langle T \rangle$, and *effective path length* $\langle u \rangle$, such that

$$\langle \alpha_b(u) \rangle_{inhom} = \langle \alpha_b(\langle u \rangle, \langle p \rangle, \langle T \rangle) \rangle_{hom}. \tag{10.51}$$

If we compare Eqs. 10.49 or 10.50 for an inhomogeneous medium with that for the absorption coefficient in a homogeneous medium, Eq. 10.6, it is clear that there are significant differences in the wavenumber dependence. This tells us that *scaling relations* can at best be approximations, except under certain circumstances. Perhaps the most accurate such scaling relation is that due to Van de Hulst, Curtis, and Godson, the *H–C–G approximation*. For a constant mixing ratio and temperature along the path, this approximation yields the following: $\langle u \rangle = u$, $\langle p \rangle = \frac{1}{2}(p + p')$, where p and p' are the pressures of the beginning and end points. In Problem 10.6 it is shown that this H–C–G scaling yields the same as the accurate value of both (1) the absorption coefficient in the wings of the line ($\tilde{\nu} - \tilde{\nu}_0 \gg \alpha_L$) and (2) the integrated line absorption (i.e., the area under the line). The H–C–G approximation is probably the best scaling approximation available, particularly for well-mixed species. However, for species such as water vapor and ozone whose mixing ratio varies appreciably with height, this scaling can produce errors in heating rate as large as 10%.

10.5.2 LBL Transmission Computation: Inhomogeneous Paths

The problem is that lines are pressure broadened at low altitudes (for which a Lorentz profile is adequate), while they become Doppler broadened at high altitudes. Spectral sampling strategies allowing for efficient yet accurate integration of the Voigt profile for a given temperature and pressure (i.e., a homogeneous path) were briefly discussed in §10.4. The extension to nonuniform atmospheric paths with varying temperature, pressure, and absorber concentration is usually done by approximating the real atmosphere by a series of homogeneous layers in which the parameters are taken to be constant in each layer but are allowed to vary from layer to layer.

Using FASCODE[13] as an example we note that it uses a sampling interval that is a suitable fraction of the average half-width of the line. Since the pressure decreases exponentially with altitude, the average half-width and therefore the sampling interval become smaller at higher rather than at lower levels in the atmosphere. The absorption coefficient for each layer may then be merged with those from neighboring layers in such a way that the absorptance for a path through two adjacent layers has the resolution of the higher layer. This is accomplished by interpolating the coarser-resolution results for the lower layer into the finer-resolution ones of the higher layer. In FASCODE this procedure is executed in a systematic manner so that the spectral absorptance for a given atmospheric slant path is obtained with the finest spectral resolution at all atmospheric levels. The transmittance between any two boundaries may now be computed and used to obtain the radiance along a given path (assuming LTE so that the emission is given by the Planck function), depending exclusively on wavenumber and temperature.

The need to speed up the computation by using bigger wavenumber steps in the line wings and a fine grid over the line center is recognized in most algorithms designed to perform line-by-line computations. For example, in another generally available LBL code, GENLN2 (A General Line-by-Line Atmospheric Transmittance and Radiance Code[14]), the user-specified spectral range is first divided into a number of wide mesh intervals, which may be of constant or variable spacing. In GENLN2, the line-by-line computation proceeds in two stages: a "wide-pass" stage followed by a "fine-pass" computation. The wide-pass stage computes absorption due to line wings of lines whose centers fall into a fixed range from the wide mesh boundaries. This step also considers absorption due to precomputed continua accounting for the absorption of the line wings at separations greater than 25 cm^{-1} from the line center. In addition, high-resolution cross-section data are used to account for absorption by molecules for which line data are lacking. The fine-pass stage uses a fine spectral grid, which is determined by the width of the narrowest line encountered over a particular path. For applications to the Earth's atmosphere, this implies that the width of the lines in the uppermost atmospheric layer of the path (where the lines are narrowest) determines the resolution. All lines within the wide mesh boundaries are included in the fine-pass stage, and line wings up to 25 cm^{-1} from the line center are taken into account.

Finally, the wide-pass absorptions, interpolated to the fine spectral grid points, are added to the fine-pass absorptions at these same points to yield the monochromatic absorption coefficient. For a vertical path the optical depth is obtained as the product of the absorption coefficient and the absorber amount. For several absorbing gases the total optical depth is obtained by adding the contributions from the individual gases. Thus, the total monochromatic absorbing optical depth may be written as

$$\tau_a(\tilde{\nu}) = \sum_j \tau(\tilde{\nu})_j^{\text{nearlines}} + \tau(\tilde{\nu})_j^{\text{linewings}} + \tau(\tilde{\nu})_j^{\text{broadband}}, \tag{10.52}$$

where the broadband absorption includes the contribution from continua and molecular cross sections. The sum extends over all absorbing species.

Comparing the line-by-line computation in FASCODE and GENLN2, we see that, despite differences in detailed approach, they both compute absorption at a fine spectral grid spacing, and both codes employ a variable computational grid to produce the final result: an overall absorption coefficient or optical path (summed over absorbing species in the path) at a fine spectral resolution that may be considered monochromatic. The main difference between the two codes lies in the subsequent computation of transmittance and radiance. Whereas GENLN2 proceeds with the transmittance and radiance computation based on the fine grid spectral resolution, which is the same in every atmospheric layer, FASCODE attempts to speed up this part of the computation by computing transmittance with a spectral resolution that varies from layer to layer, as mentioned above.

10.5.3 Inclusion of Multiple Scattering in LBL Computations

In principle, it is almost trivial to include multiple scattering in LBL computations. The reason is that LBL computations are monochromatic and therefore automatically consistent with multiple-scattering algorithms. Thus, if a line-by-line code can be used to compute gaseous optical depths layer by layer throughout the medium, this information can be combined with data on scattering and absorption coefficients (Eqs. 3.9–3.10) of other scattering and absorbing species to obtain the layer-by-layer optical depth, single-scattering albedo, and phase function. These are the data required to perform monochromatic radiative transfer computations including multiple scattering. The gaseous absorption contribution to the optical depth is given by Eq. 10.52.

In view of the above, it may appear surprising to learn that, in practice, the inclusion of multiple scattering in LBL codes has not been done in a satisfactory manner. In an effort to explain this situation we will briefly discuss below the design of two such codes. These are the FASCODE and GENLN2 codes discussed previously. The main problem is that these codes (and most available LBL codes) do not employ radiative transfer schemes that are well suited to perform multiple-scattering computations. The radiative transfer schemes were designed to work in the thermal IR, where scattering can be safely ignored unless aerosols and clouds are present. Thus, there was little incentive to employ radiative transfer methods that are suitable for multiple scattering: *The radiative transfer scheme was not designed for the computation of multiple-scattering effects.* As a consequence, the radiative transfer schemes used in most LBL codes integrate along the line of sight to obtain the intensity, assuming that the source function is known, which is the case when scattering is ignored. However, we know from Chapters 7 and 8 that the source function due to multiple scattering depends on the intensity, which is an unknown. Hence, it is a nontrivial task to compute the multiply scattered radiation field. This implies that most LBL codes will require a major overhaul to remedy this shortcoming.

The original version of FASCODE treated particle scattering as equivalent to absorption so that all scattered radiation is treated as reemitted energy that was previously absorbed.[15] An approximate treatment of multiple scattering was later introduced by using a two-stream approximation combined with an adding algorithm. This approach was chosen because it is consistent with the radiance/transmittance computation in FASCODE, which treats one layer at a time, but which employs a spectral step size that may vary from layer to layer. However, this approach is inconsistent with monochromatic multiple-scattering treatments, which require the use of a fixed wavenumber throughout the medium.

Scattering is not considered in the GENLN2 code. However, this code can be used to compute monochromatic absorption optical depth, because it utilizes the same spectral step size in all atmospheric layers. Thus, the spectral sampling in GENLN2 is compatible with monochromatic multiple-scattering algorithms. Similarly, the gaseous optical depths computed in FASCODE can be interpolated to the same spectral step size in all atmospheric layers and thereby become compatible with monochromatic multiple-scattering algorithms.

The bottom line is: Both of these codes can be used to compute the quantity required for LBL multiple-scattering computations, namely the monochromatic (fixed wavenumber) absorption optical depth; however, the radiative transfer schemes employed in most existing LBL codes, including FASCODE and GENLN2, are ill suited to perform multiple-scattering computations. Efforts to design and implement comprehensive LBL multiple-scattering codes are currently in progress.

Finally, we note that properly designed LBL codes that include multiple scattering in a rigorous manner would provide a testbed for benchmark computations against which alternative approaches aimed at enhanced efficiency may be tested. One possible way to proceed in the pursuit of an efficient yet accurate inclusion of multiple scattering in LBL codes would be to attempt to reduce the need for multiple-scattering computations by exploiting the existing redundancy in absorption coefficients across a given spectral interval over which the particle scattering and absorption coefficients do not vary appreciably. In principle, such an approach resembles the philosophy underlying the k-distribution method.

10.5.4 The Correlated-k Method

The average transmittance, Eq. 10.48, may be written in k-distribution form for an inhomogeneous path, in analogy to Eq. 10.33, as

$$\langle \mathcal{T}_b(u) \rangle = \int_{k_{min}}^{k_{max}} dk f^*(k) \exp\left[-\int_0^u du' k(u') \right] \tag{10.53}$$

or, in finite-difference form,

$$\langle \mathcal{T}_b(u_{l_1}, u_{l_2}, \theta) \rangle = \sum_{j=1}^{N} f^*(k_j) \exp\left[-\sum_{l=l_1}^{l_2} k_l \Delta u_l \right]. \tag{10.54}$$

We have assumed a vertical path, $\theta = 0$. The k_j are the absorption coefficients appropriate to the jth layer, assuming each of the layers are thin enough to be considered homogeneous. The integration extends from the center of the layer identified by u_{l_1} to the center of the layer u_{l_2}. Here f^* and g^* denote the distribution and the cumulative distribution (respectively) of the absorption coefficients along the inhomogeneous path. Clearly f^* and g^* are not equal to the distributions f and g, discussed previously, since these applied to a homogeneous path.

In terms of the cumulative distribution variable $g^*(k^*) = \int_0^{k^*} dk' f^*(k')$, we can write Eq. 10.54 as

$$\langle \mathcal{T}_b(u) \rangle = \int_0^1 dg^* \exp\left[-\int_0^u du' k(g^*, u') \right]. \tag{10.55}$$

Note carefully the difference between Eqs. 10.55 and 10.36, the latter equation applying to a homogeneous path. The function $k(g^*, u')$ refers to the distribution function appropriate to the particular level u'. However, the distribution g^* is the cumulative distribution of k values *for the inhomogeneous line of sight*. The finite-difference form of the above equation is

$$\langle \mathcal{T}_b(u) \rangle = \sum_{j=1}^{N} \Delta g_j^* \exp\left[-\sum_{l=l_1}^{l_2} k_l(g_j^*) \Delta u_l \right]. \tag{10.56}$$

The *correlated-k* (c-k) method consists of replacing Eqs. 10.55 and 10.56 with

$$\langle \mathcal{T}_b(u) \rangle = \int_0^1 dg \exp\left[- \int_0^u du' k(g, u') \right] \tag{10.57}$$

and

$$\langle \mathcal{T}_b(u) \rangle \approx \sum_{j=1}^{N} \Delta g_j \exp\left[- \sum_{l=l_1}^{l_2} k_l(g_j) \Delta u_l \right]. \tag{10.58}$$

The replacement of the variable g^* with g implies that the single variable g maps into the distribution functions at all levels u'. Since g is effectively a wavenumber variable, this says that there is a one-to-one correspondence, or mapping, of wavenumbers from one level to another. Thus the optical depth at a specific wavenumber g is given by the integral of $k(g, u')$ over the appropriate range of u', with g fixed. This applies uniquely to every value of g so that the net transmittance is the weighted average over all values of g. Equations 10.57 and 10.58 essentially describe the integration over a *pseudo-spectral line*, as described earlier.

Example 10.5 Isolated Lorentz Line

It is instructive to once again consider the simplest inhomogeneous case for which we have exact analytic solutions. We may find the k-distributions for the inhomogeneous line of sight from Eq. 10.50, where for simplicity we let $z' \to \infty$. Denoting k_L^* by simply k and solving for $\tilde{\nu}(k)$ we find

$$\tilde{\nu}_L(k) = \frac{\alpha_L}{\sqrt{e^{2\pi\alpha_L k/\mathcal{S}} - 1}}. \tag{10.59}$$

Note that this result depends upon the height z through the dependence of α_L on z. (More generally, it depends upon both z and z' and upon θ.)

The k-distribution for the inhomogeneous path is

$$f_L^*(k) = \frac{2}{\Delta\tilde{\nu}} \frac{d\tilde{\nu}_L(k)}{dk} = \frac{(\pi\alpha_L/\mathcal{S}) e^{2\pi\alpha_L k/\mathcal{S}}}{\frac{\Delta\tilde{\nu}}{2}\left(e^{2\pi\alpha_L k/\mathcal{S}} - 1\right)^{3/2}}. \tag{10.60}$$

Note also that since $k_{\max} = \mathcal{S}/2\pi\alpha_L$, the denominator $\to 0$, and $f^*(k) \to \infty$ as $k \to k_{\max}$. This presents no difficulty in practice. (Note that the same thing occurs for the homogeneous case, Eq. 10.42.) Proceeding as in the homogeneous case, we can solve analytically for the cumulative distribution $g^*(k)$ and its inverse

$$g_L^*(k) = \frac{\alpha_L}{(\Delta\tilde{\nu}/2)\sqrt{e^{2\pi\alpha_L k/\mathcal{S}} - 1}} - 1, \tag{10.61}$$

$$k_L^*(g) = \frac{\mathcal{S}}{2\pi\alpha_L} \ln\left\{ \frac{[(\Delta\tilde{\nu}/2)(1+g)]^2 + \alpha_L^2}{[(\Delta\tilde{\nu}/2)(1+g)]^2} \right\}. \tag{10.62}$$

As in the homogeneous case, the correspondence of $\tilde{\nu}$ in the band and the variable g in the pseudo-line are one to one.

For an isolated line, the monotonic ordering by strength of absorption coefficients retains the relative spectral alignment of absorption lines between different levels in the atmosphere. We can carry out a mapping from the variable $\tilde{\nu}$ to the variable g at one height, say z_1; we then go to a second height, and map this same variable g back into a wavenumber $\tilde{\nu}'$ at the height z_2. The wavenumber $\tilde{\nu}'$ will be found to be exactly the same wavenumber $\tilde{\nu}$, or, in other words, there is a perfect *spectral correlation* at different pressure levels (see Fig. 10.4). Problem 10.7 shows how this mapping works in detail for the isolated Lorentz line. Clearly, this should work for any isolated line profile, so long as the line center remains fixed in wavenumber and the broadening maintains a symmetrical line shape. It can also be shown to apply to an Elsasser band and to a band for which the lines are randomly distributed in a spectral interval $\Delta\tilde{\nu}$. In the latter case, it is valid if the averaged absorptance corresponding to a single line can be identified.

Unfortunately, this one-to-one uniqueness does not work for a general molecular band, except in the weak-line and strong-line limits.[16] It is clear that for a real molecular

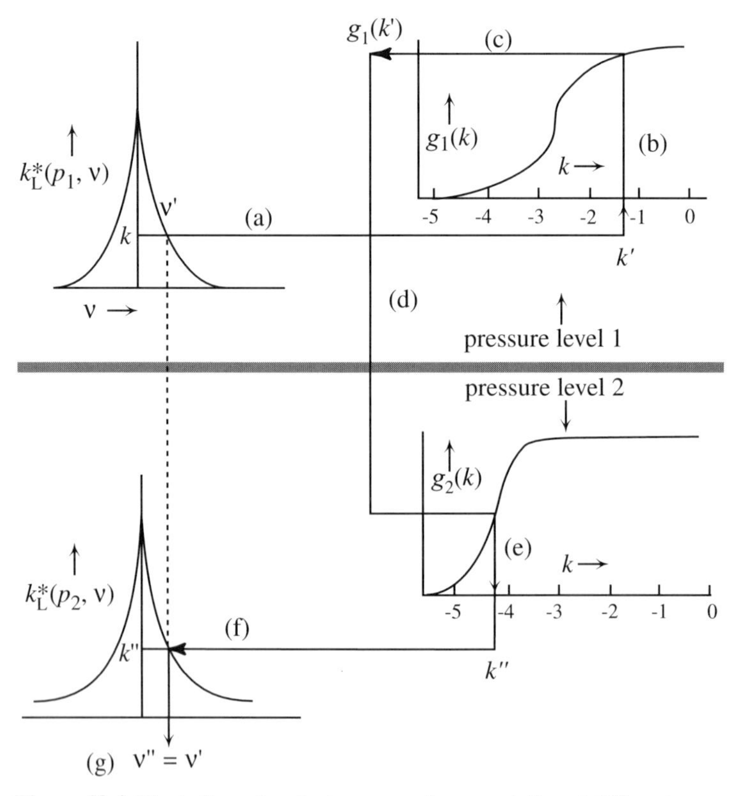

Figure 10.4 Illustration of perfect wavenumber correlation at different pressure levels for an isolated line. The mapping begins at pressure p_1 and wavenumber $\tilde{\nu}$ (upper left panel). The absorption coefficient k' corresponding to this wavenumber (actually k^*_L in Eq. 10.50) maps into the "pseudo-wavenumber" variable $g_1(k')$ (upper right panel) using Eq. 10.61. We now move to a different pressure level p_2 and map this value of g_1 using the inverse relationship to go from g to k (Eq. 10.62), but using different pressure half-width (lower right panel). This wavenumber $\tilde{\nu}''$ turns out to be the same $\tilde{\nu}'$ chosen at the initial pressure level.

band, a single value of k may correspond to a large number of wavenumber values $\tilde{\nu}_1$, $\tilde{\nu}_2$, etc. This value of k will map into a certain value of g (or more accurately a band of values Δg around g) at the height z_1. We then take this same value of g and find the corresponding value of k at a different height z_2, by using the relationship $k(z_2, g)$. A way of testing to see whether there is good correlation between levels would then be to replot the mapped spectrum at the original wavenumber values $\tilde{\nu}_1$, $\tilde{\nu}_2$, etc. If the mapped spectrum matches closely the actual spectrum calculated at z_2, then the correlation is good.

An example of this mapping is shown in Fig. 10.5, which uses two contiguous atmospheric slabs separated by a large pressure difference (0.1–1.0 bars), a more extreme example than what is normally encountered in the atmosphere.[17] The mapped spectrum (dashed lines) is seen to be very close to the actual spectrum (dark lines). Another way of assessing the error in the assumption of correlation is shown in Figs. 10.6. The left-hand panels show the CO_2 absorption spectra for two different levels in a model atmosphere. In the right-hand panels is shown a mapping from wavenumber to g such that k increases monotonically for the bottom (high-pressure) layer. The same mapping was applied to the k-spectrum of other layers in the atmosphere. The fundamental

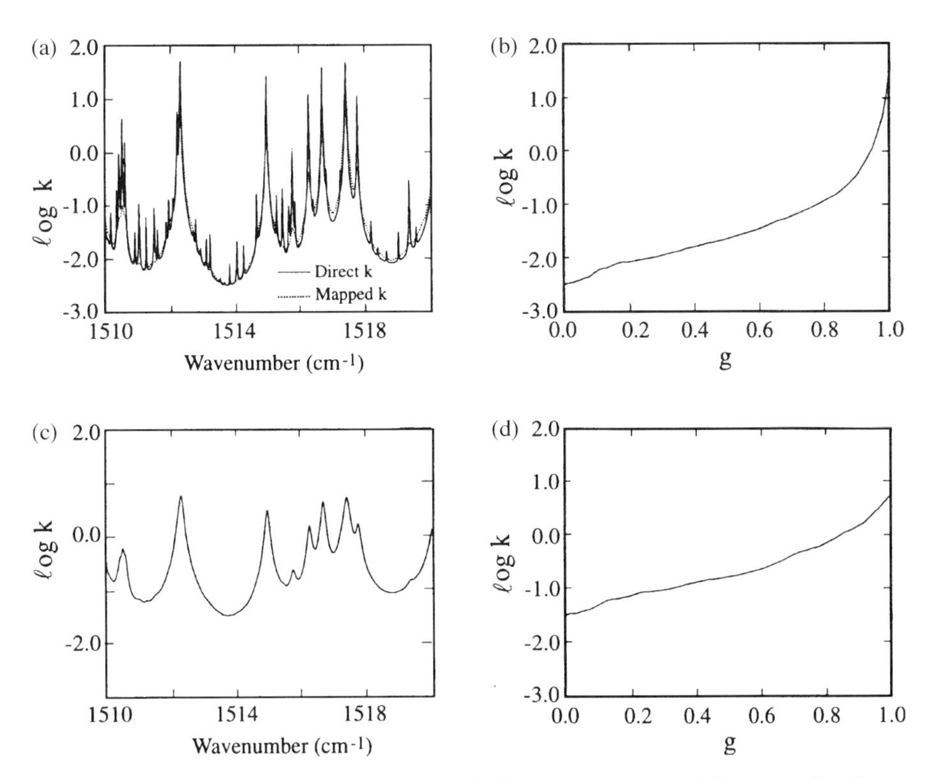

Figure 10.5 Spectral correlation for the transmission across a pressure inhomogeneity of 0.1–1.0 bars. (a) Results for the 1,510–1,520 cm^{-1} portion of the 6.3-μm water band at 0.1 bar and (c) 1.0 bar, respectively; (b) and (d) Numerical k-distributions of respective absorption spectra in (a) and (c); The dashed line in (a) is the result of mapping the absorption spectrum in (c) via the k-distribution in (b).

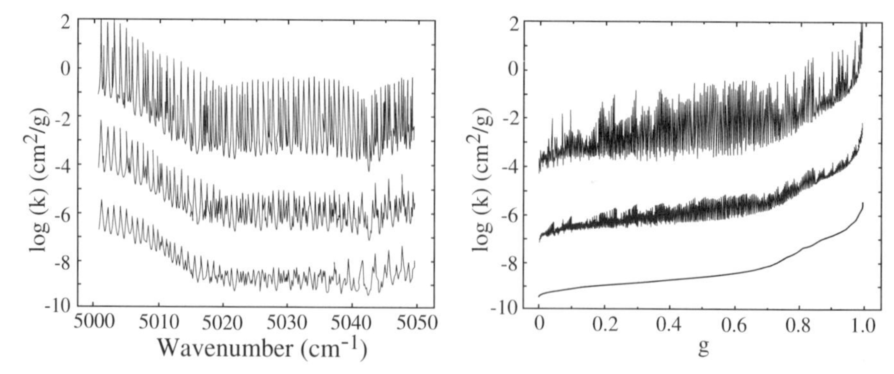

Figure 10.6 Left panel: CO_2 absorption coefficient spectra computed for three different pressure–temperature values, which (from top to bottom) are respectively (0.05 bar, 200 K), (0.25 bar, 245 K), and (0.9 bar, 295 K). The coefficients for the second layer have been multiplied by 10^{-1} and those for the third layer by 10^{-7}. Right panel: Spectra sorted by wavenumber such that k increases monotonically for the third layer (bottom curve). The wavenumber was mapped onto g according to the procedure described in the text. Note the nonmonotonic behavior, illustrating the inadequacies of this mapping procedure in this particular example.

assumption of the c-k method is that a single mapping will produce a monotonically increasing k-spectrum in every layer. On the contrary, Fig. 10.6 shows that a mapping that produces a monotonically increasing k-spectrum for one layer produces a k-spectrum for other layers in which neighboring k values fluctuate by orders of magnitude.[18] More sophisticated mapping algorithms than those described here[19] have been devised to minimize the errors in the c-k technique. In general, this approximation can produce fluxes and heating rate values with errors less than 1%. Although they are much more efficient (typically by factors of 1,000) than LBL methods, they are most suitable for one-dimensional radiative–convective models. They are generally too slow for three-dimensional GCM models, and more drastic approximations are necessary. Overlapping absorption by bands from different constituents is easily accomodated by the correlated-k technique, and these refinements are discussed elsewhere.[20]

10.5.5 Inclusion of Multiple Scattering in the Correlated-k Method

It is straightforward to include scattering particles or Rayleigh scattering from air molecules in the c-k method, since most multiple-scattering methods assume monochromatic absorption. The assumption is made that the scattering coefficient, single-scattering albedo, and phase function are constant over the spectral regions corresponding to the subintervals k_1, $k_1 + \Delta k_1$, k_2, $k_2 + \Delta k_2$, etc. The single-scattering albedo for the ith spectral interval is $a_i = \tau_{sc}(i)/(\tau_{sc}(i) + \tau_a(i) + k_i u)$, etc., where $\tau_{sc}(i)$ is the total scattering optical depth (molecules plus particles), $\tau_a(i)$ is the particle absorption optical depth, and $k_i u$ is the molecular absorption optical depth. The extinction optical depth for the ith interval is given by $\tau_i = \tau_{sc}(i) + \tau_a(i) + k_i u$. For example, the mean transmittance and mean reflectance of a nongray, scattering atmosphere are

$$\langle \mathcal{T}_b \rangle = \sum_{i=1}^{N} \mathcal{T}_b(\tau_i, a_i)\Delta g_i, \qquad \langle \rho \rangle = \sum_{i=1}^{N} \rho(\tau_i, a_i)\Delta g_i, \tag{10.63}$$

where Δg_i is the relative weighting of the ith k-interval. Note that the monochromatic transmittance, $\mathcal{T}_b(\tau_i, a_i)$, and the corresponding reflectance, $\rho(\tau_i, a_i)$, can be calculated from two-stream methods or from a doubling algorithm (described in Chapters 7 and 8) for each atmospheric layer. By combining the layers using an adding algorithm the intensity and irradiance are obtained. Alternatively, the radiative transfer equation can be solved layer by layer for each spectral subinterval, and then the layers can be combined using two-stream methods or the DISORT algorithm. Thus, the most general problem of scattering and absorption in a vertically inhomogeneous atmosphere can be solved accurately with such methods. Such techniques are the state of the art (in terms of band models) at the present time.

Example 10.6 Application of the Correlated-k Method

Suppose we wish to solve for the mean flux reflectance of an absorbing/scattering slab over the spectral interval $\Delta\tilde{\nu}$. Then

$$\langle \rho(2\pi) \rangle = \frac{1}{\Delta\tilde{\nu}} \int_{\Delta\tilde{\nu}} d\tilde{\nu}\rho(2\pi, \tilde{\nu}) = \int_{k_{\min}}^{k_{\max}} dk f(k)\rho(2\pi, k) = \int_0^1 dg\rho[2\pi, k(g)],$$

where $\rho(2\pi, k)$ is the flux reflectance calculated for a specific value of k. To be specific, let the slab be homogeneous and semi-infinite. (The same ideas apply to an inhomogeneous slab, except that analytic solutions will not exist.) We solved this problem for a specific single-scattering albedo in Chapter 7:

$$\rho(2\pi, k) = \frac{1 - \sqrt{1 - a(k)}}{1 + \sqrt{1 - a(k)}} = \frac{1 - \sqrt{1 - \sigma/(\sigma + k)}}{1 + \sqrt{1 - \sigma/(\sigma + k)}} = \frac{\sqrt{\sigma + k} - \sqrt{k}}{\sqrt{\sigma + k} + \sqrt{k}}.$$

Thus

$$\langle \rho(2\pi) \rangle = \int_0^1 dg \frac{\sqrt{\sigma + k(g)} - \sqrt{k(g)}}{\sqrt{\sigma + k(g)} + \sqrt{k(g)}}. \tag{10.64}$$

The relationship of k to g is given by the particular line or band profile – it may be an analytic result (such as in Example 10.5), or more likely it will be derived as a table of k values versus g values. For a molecular band this table will be constructed from the spectroscopic parameters of the lines within the spectral interval $\Delta\tilde{\nu}$. The mean reflectance is then evaluated by numerical quadrature of Eq. 10.64.

10.6 Summary

Radiative transfer in spectral lines and bands depends on the spectral optical path from which the spectral transmittance and absorptance are determined. An isolated Lorentz line was used (for analytical convenience) to illustrate the basic dependence of the transmission on line parameters and absorber mass path. The desire for efficient computation motivates the discussion of band models. Classical narrow-band models as well as the more sophisticated MODTRAN computational code were described and the limitations and strengths of such models were briefly discussed. A brief description of the principles behind line-by-line radiative transfer schemes was also provided and the potential for including accurate multiple scattering into such schemes was discussed. The H–C–G scaling technique for turning inhomogeneous paths into equivalent homogeneous paths was covered briefly. We discussed the concept of k-distributions by which the absorption cross section across a specified spectral region is reordered by increasing magnitude to allow for the rapidly varying magnitude with wavenumber to be replaced by a smooth, monotonically increasing function. This approach allows for rapid yet accurate computation of fluxes and heating/cooling rates. The k-distribution is valid for a homogeneous path. To generalize this approach to inhomogeneous situations the correlated k-distribution was invented to allow for accurate computations between altitude levels over which the line profile may change substantially from, say, a pressure-broadened Lorentz shape to a thermally broadened Doppler shape. This method is readily generalized to include the simultaneous effects of scattering and absorption in spectrally complex media.

Problems

10.1 Show *numerically* that the following two expressions (see Eqs. 10.11 and 10.12) are identical ($x = \tilde{\nu}/\delta$ and $y = \alpha_\mathrm{L}/\delta$):

$$\alpha_m(\tilde{\nu}) = \sum_{n=-\infty}^{n=+\infty} \frac{S}{\pi} \frac{\alpha_\mathrm{L}}{(\tilde{\nu} - n\delta)^2 + \alpha_\mathrm{L}^2} = \frac{S}{\delta} \frac{\sinh(2\pi y)}{\cosh(2\pi y) - \cos 2\pi x}.$$

10.2 Show that in the gray limit ($y \gg 1$), the Elsasser band transmittance (Eq. 10.13) reduces to $\langle \mathcal{T}_\mathrm{b} \rangle = e^{-2\pi y \tilde{u}}$.

10.3 Show that in the strong-line limit ($\tilde{u} \gg 1$), the Elsasser band transmittance is given by $\langle \alpha_\mathrm{b} \rangle \approx \mathrm{erf}\,[\pi y \sqrt{2\tilde{u}}]$, where erf denotes the *error function*, given by

$$\mathrm{erf}(x) = \frac{2}{\sqrt{\pi}} \int_0^x dt e^{-t^2}.$$

10.4 Show that when the band absorptance defined by the Lorentz-exponential model (Example 10.1) is taken in its strong- and weak-line limits, the resulting expressions can be used to empirically determine the two band parameters $\bar{S}$ and $\bar{S}/\alpha_\mathrm{L}$.

10.5 It was shown in Eq. 10.45 that the average transmittance can be written

$$\langle \mathcal{T}_\mathrm{b}(u) \rangle = \int_0^\infty dk f(k) e^{-ku} = \mathcal{L} f(k), \qquad (10.65)$$

where $\mathcal{L}$ denotes the Laplace transform. Show that for the statistical-Malkmus model, Eq. 10.25 has the following Laplace transform:

$$f(k) = \frac{1}{2} k^{-3/2} \sqrt{\alpha_L \mathcal{S}/\Delta\tilde{\nu}} \exp\left[\frac{\pi \alpha_L}{4\Delta\tilde{\nu}}\left(2 - \frac{\mathcal{S}}{k} - \frac{k}{\mathcal{S}}\right)\right]. \tag{10.66}$$

10.6 The H–C–G approximation is given by

$$\langle \alpha(u) \rangle_{\text{inhom}} = \langle \alpha(\langle u \rangle, \langle p \rangle, \langle T \rangle) \rangle_{\text{hom}}. \tag{10.67}$$

For a constant mixing ratio and temperature along the path, this approximation yields the following: $\langle u \rangle = u$, $\langle p \rangle = \frac{1}{2}(p + p')$, where p and p' are the pressures of the beginning and end points. Show that this scaling yields the exact value of the absorption coefficient in the wings of the line ($\tilde{\nu} \gg \alpha_L$) and the exact value of the integrated line absorption (i.e., the area under the line), for the optically thin limit.

10.7 Show explicitly how the spectral mapping works for the isolated Lorentz line, using Eqs. 10.59 through 10.62, and starting from Eq. 10.50,

$$k_L^*(\tilde{\nu}) = \frac{\mathcal{S}}{2\pi \alpha_L} \ln\left[\frac{(\tilde{\nu} - \tilde{\nu}_0)^2 + \alpha_L(z)^2}{(\tilde{\nu} - \tilde{\nu}_0)^2}\right].$$

Show that the wavenumber $\tilde{\nu}$ has a one-to-one correspondence with the variables k and g at the pressure p. This variable then maps into a different pressure level p', where the new absorption coefficient k' and the new wavenumber $\tilde{\nu}'$ are derived by using the inverse relationships $k(g)$ and $\tilde{\nu}(g)$. The mapping back into wavenumber space should result in the same wavenumber $\tilde{\nu}' = \tilde{\nu}$, and thus the correlated-k method is exact for this case.

Notes

1 To be consistent with our notation scheme where the subscripts n and m refer to *per molecule* and *per unit mass*, respectively, we should denote the corresponding quantities $\mathcal{S}_n$ and $\mathcal{S}_m$. However, this leads to too many confusing subscripts later on. We will hereafter refer to $\mathcal{S}_m$ as simply $\mathcal{S}$.

2 The evaluation of the integral in Eq. 10.6 in terms of Bessel functions is described by Elsasser, W. M., *Heat Transfer by Infrared Radiation in the Atmosphere*, pp. 29–30, Harvard University Blue Hill Meteorological Observatory, Boston, 1942.

3 Tabulations of the *Ladenburg–Reiche function* can be found in several references: Whittaker, E. T. and G. N. Watson, *Modern Analysis*, Cambridge Univ. Press, Cambridge, Ch. 17 and 23, 1915; Jahnke, E. and F. Emde, *Tables of Functions*, Dover, New York, 1945; Appendix 7 of Goody, R. M., and Y. L. Yung, *Atmospheric Radiation, Theoretical Basis*, 2nd ed., Oxford University Press, New York, 1989.

4 A thorough description of the various asymptotic limits is given by Goody and Yung, pp. 148–54 (see Endnote 3 for full citation).

5 For example, see Fig. 4.7 of Goody and Yung (see Endnote 3) for a plot of water vapor line strength distributions.

6 Goody and Yung (see Endnote 3 for full citation).

7 The MODTRAN model is described by Berk, A., L. S. Bernstein, and D. C. Robertson, *MODTRAN: A moderate resolution model for LOWTRAN 7*, Report GL-TR-89-0122, Geophysics Laboratory, Air Force Systems Command, United States Air Force, Hanscom, AFB, MA 01731, 1989, and Anderson, G. P. et al., "MODTRAN 2: Suitability for remote sensing," *Proceedings of the Society of Photo-Optical Instrumentation Engineers*, Vol. 1968, 514–25, 1993.

8 Progressively improved versions of the LOWTRAN code are described by Selby, J. E. A., F. X. Kneizys, E. P. Shettle, J. H. Chetwynd, and R. A. McClatchey, *Computer code LOWTRAN 4*, AFGL-TR-78-0053, Air Force Geophysics Laboratory, Hanscom, AFB, MA 01731, 1978; Kneizys, F. X., E. P. Shettle, W. O. Gallery, J. H. Chetwynd, L. W. Abreu, J. E. A. Selby, R. W. Fenn, and R. A. McClatchey, *Atmospheric transmittance/radiance: Computer code LOWTRAN 6*, AFGL-TR-83-0287, Air Force Geophysics Laboratory, Hanscom, AFB, MA 01731, 1983; Kneizys, F. X., E. P. Shettle, L. W. Abreu, J. H. Chetwynd, G. P. Anderson, W. O. Gallery, J. E. A. Selby, and S. A. Clough, *Users guide to LOWTRAN 7*, Report AFGL-TR-88-0177, Air Force Geophysics Laboratory, Hanscom, AFB, MA 01731, 1988.

9 For consistency of notation with the journal literature we have dropped the subscript m on k_m for the present purposes.

10 This problem is addressed and solved in the paper by Wiscombe, W. J. and J. W. Evans, "Exponential-sum fitting of radiative transmission functions," *J. Comput. Phys.*, **24**, 416–44, 1977.

11 This figure is based on Fig. 1 of Lacis, A. A. and V. Oinas, "A description of the correlated-k distribution method for modeling nongray gaseous absorption, thermal emission, and multiple scattering in vertically inhomogeneous atmosphere," *J. Geophys. Res.*, **96**, 9,027–63, 1991.

12 This is Eq. 21 in Lacis and Oinas (see Endnote 11 for full citation).

13 Clough, S. A., F. X. Kneizys, E. P. Shettle, and G. P. Anderson, *Atmospheric Radiance and Transmittance: FASCODE2, Proceedings of the Sixth Conference on Atmospheric Radiation*, pp. 141–144, Milliamsburg, VA, 1986.

14 The line-by-line code GENLN2 is described in Edwards, D. P., *A general line-by-line atmospheric transmittance and radiance model*, NCAR Technical note, NCAR/TN-367+STR. NCAR, Boulder, CO, 1992.

15 Inclusion of multiple scattering is described by Isaacs, R. G., W-C. Wang, R. D. Worsham, and S. Goldenberg, "Multiple scattering LOWTRAN and FASCODE models," *Applied Optics*, **26**, 1272–81, 1987.

16 This statement is proven in the reference Goody, R. M., R. West, L. Chen, and D. Crisp, "The correlated-k method for radiation calculations in nonhomogeneous atmospheres," *J. Quant. Spectrosc. Radiative Transfer*, **42**, 539–50, 1989.

17 Taken from Lacis and Oinas, Fig. 11 (see Endnote 11 for full citation).

18 Taken from the article by West, Crisp, and Chen, cited in Endnote 19.

19 West, R., D. Crisp, and L. Chen, "Mapping transformations for broadband atmospheric radiation calculations," *J. Quant. Spectrosc. Radiative Transfer*, **43**, 191–9, 1990.

20 An excellent reference for the correlated-k method is the article by Lacis and Oinas cited in Endnote 11.

Chapter 11

Radiative Transfer in Nongray Media

11.1 Introduction

In comparison with shortwave radiative transfer, where nonlocal effects of multiple scattering are important, longwave radiative transfer is simple. This simplicity stems from the fact that the LTE source function (the Planck function) is determined strictly by the local temperature and is a smooth, analytic function of wavenumber. However, it is more complicated, because of the strong dependence of the absorption coefficient on wavenumber, and if that were not enough, a dependence on the pressure and temperature along the optical path. In this chapter we concentrate on the more practical aspects of computation of atmospheric fluxes and heating rates that are important for the energy flow. Infrared remote sensing is not considered, as other fine references are available.[1] Generally speaking, IR radiative transfer in the ocean is not an important mode of energy transfer, except at the outermost "skin" (see Example 11.1), which acts like an opaque surface in transferring energy to and from the atmosphere.

We first consider the basic equations describing monochromatic transfer of radiation in a slab medium, consisting of purely absorbing (nonscattering) molecules. The necessary averaging of the flux equations over wavenumber and angle to yield quantities relevant to the energy flow pose significant computational problems. These problems are of a "mechanical" nature and have not yet been fully overcome, despite the current availability of superfast computers. Various schemes to solve this problem are discussed, including the brute-force line-by-line methods, narrow-band, and wide-band models. The IR warming (cooling) rate is derived in a wide-band context, providing a convenient framework for introducing the important cooling-to-space concept. The notion of photon escape probabilities is used to help conceptualize the cooling-to-space and radiative-exchange contributions to the heating rate. Modern correlated-k techniques are discussed briefly, and some examples of accurate cooling rates are shown.

Atmospheric particles (cloud droplets and ice particles and aerosols) add additional richness and complications to the radiative transfer problem. Here, the uncertainties become greater because of the wide ranges of cloud types and shapes, cloud particle distributions, and optical properties. Ultimately, the problem is so complicated that statistical techniques are needed, since clouds and aerosol distributions are generally not available in the detail needed for conventional radiative transfer computations. Furthermore, even if we were capable of measuring the entire microphysical structure of a particle ensemble, it would change with time and furthermore would strain even today's computational resources. Nevertheless, classical descriptions of clouds in terms of plane-parallel entities embrace a significant portion of the effects of clouds on climate. Thus we confine our attention to these nonexistent (but useful) abstractions. Weighting the clear-air and cloudy equations by the cloud "fractional coverage" is a useful device in constructing one-dimensional models of climate that are still useful today, despite its apparently crude nature. Derivation of the flux equations inside a cloud provides insight into the IR heating occurring near the cloud boundaries. Infrared cloud properties have traditionally been reduced to a few low-order moments of the particle size distribution and its vertical integral – for example, liquid (or ice) water path and the mean particle radius. In optically thin clouds,[2] scattering needs to be taken into account even in the IR. Aerosols are briefly discussed in general terms, as important components of the energy budget and as agents of climatic change.

11.2 Radiative Flux and Heating Rate: Clear-Sky Conditions

Infrared molecular absorption is due almost entirely to trace gases, of which H_2O, CO_2, and O_3 are the most abundant. Since these molecules play such a crucial role in the greenhouse warming problem, it is natural to begin with a consideration of the transfer problem under clear-sky conditions. The equations will apply to the fairly rare situation of a cloudless, particle-free atmosphere. We need to consider only thermal emission and absorption processes, since in the thermal IR ($\lambda > 3.5$ μm) molecular scattering is unimportant compared to absorption. If we assume local thermodynamic equilibrium (appropriate for heights below the mesopause), the source function is equal to the Planck function evaluated at the local value of temperature T. We also assume that T as well as the height distributions of all significant IR absorbers, are specified. To keep the equations manageable, the approximation is made that the overlying near-surface atmospheric temperature $T(\tau^*)$ is equal to the surface temperature T_s. This is not the case (see §12.2), and neglect of this difference can lead to errors in the upward flux by as much as 4 $\mathrm{W \cdot m^{-2}}$. Since it is becoming universally used in the infrared community, we will use wavenumber, $\tilde{\nu}$, in this chapter rather than frequency, ν or wavelength λ. Finally, we note that optical depth τ is due to gaseous absorption and, accordingly, is a very strong function of wavenumber.

11.2.1 Monochromatic Flux Equations

The solution of the monochromatic radiative transfer equation for the hemispherical fluxes is written (see §5.3)

$$
\begin{aligned}
F_{\tilde{\nu}}^{+}(\tau) &= \overbrace{2\pi \int_0^1 d\mu\mu \int_\tau^{\tau^*} \frac{dt}{\mu} B_{\tilde{\nu}}(t) \exp\left[-(t-\tau)/\mu\right]}^{\textit{atmospheric emission}} \\
&\quad + \overbrace{2\epsilon_{\rm F}^{\rm s}(\tilde{\nu}) \int_0^1 d\mu\mu B_{\tilde{\nu}}(T_{\rm s}) \exp\left[-(\tau^*-\tau)/\mu\right]}^{\textit{surface emission}} \\
&\quad + \overbrace{[1-\epsilon_{\rm F}^{\rm s}(\tilde{\nu})] F_{\tilde{\nu}}^{-}(\tau^*)\, 2 \int_0^1 d\mu\mu \exp[-(\tau^*-t)/\mu]}^{\textit{reflected component}}, \\
F_{\tilde{\nu}}^{-}(\tau) &= \overbrace{2\pi \int_0^1 d\mu\mu \int_0^\tau \frac{dt}{\mu} B_{\tilde{\nu}}(t) \exp\left[-(\tau-t)/\mu\right]}^{\textit{atmospheric emission}}.
\end{aligned}
\tag{11.1}
$$

The surface is assumed to have a flux emittance $\epsilon_{\rm F}^{\rm s}(\tilde{\nu}) \equiv \epsilon(\tilde{\nu}; 2\pi)$ (see Eq. 6.11). The third term in the upward flux equation is the contribution from reflection (assumed Lambertian) of the downward flux, owing to the nonblack nature of the surface.

Interchanging orders of integration, we eliminate the explicit dependence on angle by using the definition of the flux transmittance for uniform illumination (see Eq. 5.33):

$$
\mathcal{T}(\tilde{\nu}; -2\pi, -2\pi) = 2\int_0^1 d\mu\mu e^{-\tau(\tilde{\nu};z,z')/\mu} = 2E_3\left[\int_z^{z'} dz''\alpha(\tilde{\nu}, z'')\right]. \tag{11.2}
$$

Here $E_3[\tau]$ is the exponential integral (defined by Eq. 5.61) of order 3, and α is the absorption coefficient. We will subsequently use the shorthand notation, $\mathcal{T}_{\rm F}(\tilde{\nu}; z, z')$, for the diffuse flux transmittance. If we transform to height z and use $dt = -\alpha dz'$, the flux equations become

$$
\begin{aligned}
F_{\tilde{\nu}}^{+}(z) &= \int_0^z dz' \pi B_{\tilde{\nu}}[T(z')] \frac{\partial \mathcal{T}_{\rm F}(\tilde{\nu}; z', z)}{\partial z'} + \epsilon_{\rm F}^{\rm s}(\tilde{\nu}) B_{\tilde{\nu}}(T_{\rm s}) \mathcal{T}_{\rm F}(\tilde{\nu}; 0, z) \\
&\quad + \left[1-\epsilon_{\rm F}^{\rm s}(\tilde{\nu})\right] F_{\tilde{\nu}}^{-}(z=0) \mathcal{T}_{\rm F}(\tilde{\nu}; 0, z), \\
F_{\tilde{\nu}}^{-}(z) &= -\int_z^\infty dz' \pi B_{\tilde{\nu}}[T(z')] \frac{\partial \mathcal{T}_{\rm F}(\tilde{\nu}; z, z')}{\partial z'},
\end{aligned}
\tag{11.3}
$$

where we have used the relationships

$$\frac{\partial \mathcal{T}_F(\tilde{\nu}; z', z)}{\partial z'} = 2\alpha(\tilde{\nu})E_2[t - \tau(\tilde{\nu})], \qquad \frac{\partial \mathcal{T}_F(\tilde{\nu}; z, z')}{\partial z'} = -2\alpha(\tilde{\nu})E_2[\tau(\tilde{\nu}) - t].$$

Equations 11.3 express the flux in terms of height integrations over the Planck function, weighted by the monochromatic transmittance (and its derivative). Of course, to study the energy flow we require spectrally integrated quantities. Integrating over the Planck function presents no special problems, since it is a smooth analytic function of $\tilde{\nu}$ and T. However, computation of the absorption coefficient poses significant practical problems because of its nearly discontinuous variation with $\tilde{\nu}$. In the following subsections, we describe some common techniques for performing this integration in an efficient manner.

The most straightforward (but inefficient) technique to evaluate the integrated flux is to specify $\alpha(\tilde{\nu})$ at very small intervals (10^{-4} to 10^{-2} cm^{-1}). This is a tedious task, since $\alpha(\tilde{\nu})$ depends upon the locations ($\tilde{\nu}_{0i}$), strengths (S_i), and line profile functions (Φ_i) of all lines throughout the IR spectrum. As discussed in §4.5, the absorption coefficient is written as a sum over all lines (i) for which there is significant overlap at wavenumber $\tilde{\nu}$:

$$\alpha(\tilde{\nu}, T, p) = \sum_i S_i(T)\Phi(\tilde{\nu}, p, \tilde{\nu}_{0i}).$$

Computation of $\alpha(\tilde{\nu})$ and $\int d\tilde{\nu}\mathcal{T}_F(\tilde{\nu})$ is accomplished one wavenumber at a time throughout the spectrum.[3] Use of this line-by-line technique, already discussed in Chapter 10, has become widespread in recent years with the availability of high-speed computers and documented spectral line data. The main data bases of spectral parameters of atmospheric gases are HITRAN[4] and GEISA.[5] The limitation in the accuracy of LBL computations is mainly that of the line intensities (believed to be known to within 5–10% for the strongest lines), and to a lesser extent in the specification of the line profiles, $\Phi(\tilde{\nu})$. The latter are usually taken to be the Voigt profile (§3.3.3). Depending upon the particular band, some investigators have adopted sub-Lorentian line wings (which fall off faster than Voigt) or super-Lorentzian wings (which fall off more slowly) to achieve better agreement with laboratory measurements.

To compute spectrally integrated fluxes with moderate computing resources, a concept that remains quite useful is that of a band model (see Chapter 10). There are two general categories, determined by whether one selects comparatively small spectral bands over which to integrate (typically 10–100 cm^{-1} wide) or broader intervals (up to the entire IR spectrum). These are called *narrow-band* and *wide-band* models, respectively. In the following, we illustrate how the frequency-integrated flux equations can be manipulated to yield computationally useful forms.

Unfortunately, narrow-band models are computationally expensive. Therefore, the *wide-band emissivity* approach has been adopted in most climate models. We start by considering this category of band models, which has played an important part in the historical development of the subject.

11.2.2 Wide-Band Emittance Models

Instead of dividing the spectrum into small intervals, in this method one integrates over the entire IR spectrum. The spectrally integrated fluxes are written

$$F^+(z) = -\int_0^z dz' \int_0^\infty d\tilde{\nu}\pi B_{\tilde{\nu}}[T(z')] \frac{\partial \alpha_F(\tilde{\nu}; z', z)}{\partial z'}$$
$$+ \int_0^\infty d\tilde{\nu}\pi B_{\tilde{\nu}}(T_s)[1 - \alpha_F(\tilde{\nu}; 0, z)],$$
$$F^-(z) = \int_z^\infty dz' \int_0^\infty d\tilde{\nu}\pi B_{\tilde{\nu}}[T(z')] \frac{\partial \alpha_F(\tilde{\nu}; z, z')}{\partial z'}, \tag{11.4}$$

where $\alpha_F \equiv 1 - \mathcal{T}_F$ is the flux absorptance. Using the *Stefan–Boltzmann law* for the integrated Planck function, $\int_0^\infty d\tilde{\nu}\pi B_{\tilde{\nu}}(T) = \sigma_B T^4$, and defining the wide-band *Planck-weighted flux emittance* $\langle \epsilon_F \rangle$ and *Planck-weighted flux transmittance* $\langle \mathcal{T}_F \rangle$,

$$\langle \epsilon_F(z, z') \rangle \equiv 1 - \langle \mathcal{T}_F(z, z') \rangle = \frac{1}{\sigma_B T^4} \int_0^\infty d\tilde{\nu} \alpha_F(\tilde{\nu}; z, z') \pi B_{\tilde{\nu}}(T), \tag{11.5}$$

we rewrite the flux equations as

$$F^+(z) = \sigma_B T_s^4[1 - \langle \epsilon_F(0, z) \rangle] - \int_0^z dz' \sigma_B T^4(z') \frac{\partial \langle \epsilon_F(z', z) \rangle}{\partial z'},$$
$$F^-(z) = \int_z^\infty dz' \sigma_B T^4(z') \frac{\partial \langle \epsilon_F(z', z) \rangle}{\partial z'}. \tag{11.6}$$

A note about signs is in order. We have consistently defined the hemispherical fluxes to be positive. This accounts for the presence of the minus signs $(-)$ in some of the above equations. For example, $(-)\partial \langle \epsilon_F(z', z) \rangle / \partial z'$ in Eq. 11.6 is positive, since for $z > z'$, $\partial \langle \epsilon_F(z', z) \rangle / \partial z'$ is negative.

The broadband flux equations (Eq. 11.6) are written in terms of a simple integration over the entire spectrum. However, most bands occupy a restricted portion of the spectrum. For example, the 15-μm band of CO_2 and the 9.6-μm band of O_3 are confined to the intervals 540–800 cm^{-1} and 980–1,100 cm^{-1}, respectively. In practice, $\langle \epsilon_F \rangle$ is divided into a sum of specific band contributions

$$\langle \epsilon_F \rangle = \sum_i \int_{\Delta\tilde{\nu}_i} d\tilde{\nu} \frac{\pi B_{\tilde{\nu}}(T)}{\sigma_B T^4} \epsilon_F(\tilde{\nu}) \equiv \sum_i p_i(T) \langle \epsilon_F^i \rangle \Delta\tilde{\nu}_i, \tag{11.7}$$

where the sum is over all bands. Here p_i is the fraction of the Planck spectrum occupied by the ith band,

$$p_i \equiv \int_{\tilde{\nu}_i}^{\tilde{\nu}_{i+1}} d\tilde{\nu} \pi B_{\tilde{\nu}}(T)/\sigma_B T^4. \tag{11.8}$$

The flux emittance within the ith band is

$$\langle \epsilon_F^i \rangle \equiv \int_{\tilde{\nu}_i}^{\tilde{\nu}_{i+1}} d\tilde{\nu} \epsilon_F(\tilde{\nu}) \pi B_{\tilde{\nu}}(T) \Bigg/ \int_{\tilde{\nu}_i}^{\tilde{\nu}_{i+1}} d\tilde{\nu} \pi B_{\tilde{\nu}}(T). \tag{11.9}$$

The product $p_i \langle \epsilon_F^i \rangle$ is therefore the efficiency with which the band radiates like a blackbody. The quantity $\langle \epsilon_F^i \rangle \Delta\tilde{\nu}_i$ is the *band area*, $A_i(u)$, which is the molecular-band counterpart of the equivalent width (appropriate to a single line). Note that $A_i(u \to \infty) \to \Delta\tilde{\nu}_i$. As noted previously, these definitions need to be modified when bands overlap.

Figure 11.1 displays the *Planck-weighted flux absorptance* for the three major absorbers, H_2O, CO_2, and O_3, versus mass path for a homogeneous atmosphere (constant temperature and pressure).[6] The H_2O emittance is for the entire IR spectrum

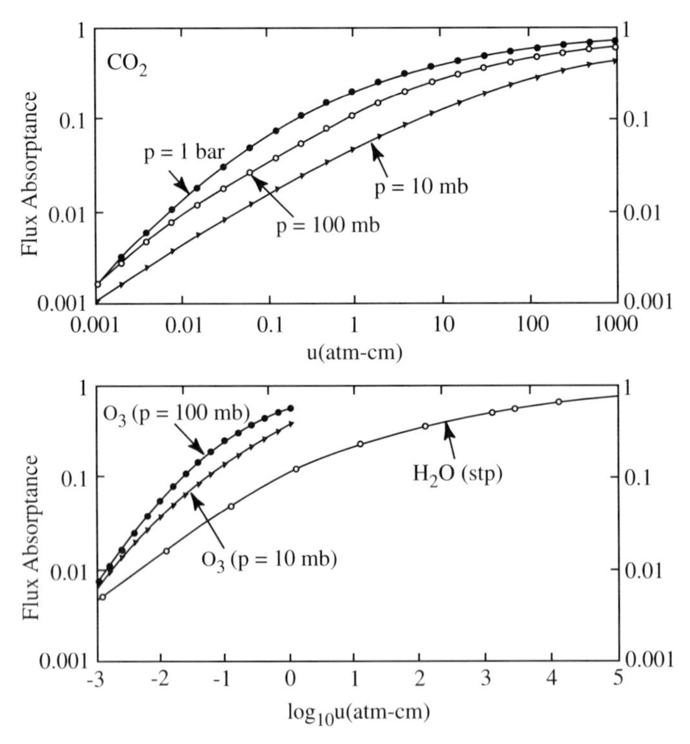

Figure 11.1 Broadband flux absorptance for CO_2 (upper panel), H_2O, and O_3 (lower panel). The CO_2 and the O_3 absorptance for the 540–800 cm^{-1} region and the 980–1,100 cm^{-1} region, respectively, are taken from LBL computations.

(0–3,000 cm^{-1}) and is computed from Eq. 11.7, where nine individual spectral intervals are used.[7] This figure shows that increasing pressure broadens the lines and increases the overall absorption at a fixed value of u. Not shown is a weaker dependence on temperature, which has a similar line-broadening effect.

A counterexample of narrow-band absorption is water vapor, which extends over most of the spectrum, particularly in the far-IR ($\tilde{\nu} < 550\,cm^{-1}$). Table 11.1 summarizes

Table 11.1. *Vibrational and rotational transitions for the important radiatively active gases in the Earth's atmosphere.*[a]

Constituent, abundance	Band, μm (cm^{-1})	Transition	Band interval (cm^{-1})	p_i ($T = 290$ K)	$\langle\epsilon_F^i\rangle(u^*)$	$P_i\langle\epsilon_F^i\rangle(u^*)$	G_i
CO_2	15 (667)	ν_2; P, Q, R, hot bands	540–800	0.268	0.761	0.204	32
356 ppmv	10.4 (961) 9.4 (1064)	overtone combination	830–1250	0.258	0.0877	2.25×10^{-2}	
H_2O	57 (175)	rotation, P, R	0–350	0.133	1	0.133	
10^{-5}–0.02	24 (425)	rot., p-type	350–500	0.147	0.988	0.145	
(trop)	15 (650)	rot., e-type	500–800	0.311	0.611	0.190	75
2–7 ppmv	8.5 (1180)	e-type, p-type	1110–1250	0.062	0.238	1.47×10^{-2}	
(strat)	7.4 (1350)	e-type, p-type	1250–1450	0.0576	0.880	5.03×10^{-2}	
	6.2 (1595)	ν_2; P, R; p-type	1450–1880	0.051	1	0.0511	
O_3 (strat) 0.2–10 ppmv	9.6 (1110)	ν_1; P, R	980–1100*	00.058	0.441	2.37×10^{-2}	10
CH_4 1.714 ppmv	7.6 (1306)	ν_4	950–1650*	0.250	0.166	0.0420	8
N_2O	7.9 (1286)	ν_1	1200–1350	0.0522	0.319	0.0170	
311 ppbv	4.5 (2224)	ν_3	2120–2270	0.003			

[a]Only those transitions are listed that are important for the radiative energy budget and global warming. Abundances are given as mixing ratio by volume. ppmv and ppbv indicate parts per million and parts per billion, respectively. In the second column the approximate location of the band center is given in both μm and wavenumber (cm^{-1}). In the third column P, Q, and R refer to branches. The H_2O continuum in the 10-μm region originates either from the distant wings of lines or the water vapor dimer $(H_2O)_2$. e-type and p-type continua refer to the self-broadened and foreign-broadened processes. The band interval $\Delta\tilde{\nu}$ is the total breadth of the completely saturated (optically thick) band. p_i is the fraction of the thermal emission within the bandwidth $\Delta\tilde{\nu}$. $\langle\epsilon_F^i\rangle$ is the broadband flux emittance for that band or collection of closely associated bands (including isotopic and hot bands). $p_i\langle\epsilon_F^i\rangle$ is the contribution of that band to the total flux emittance $\langle\epsilon_F\rangle = \sum_i p_i(T)\langle\epsilon_F^i\rangle$. The last column is the *greenhouse effect* G_i (see §12.13) attributable to the ith molecule, calculated from an accurate band model.

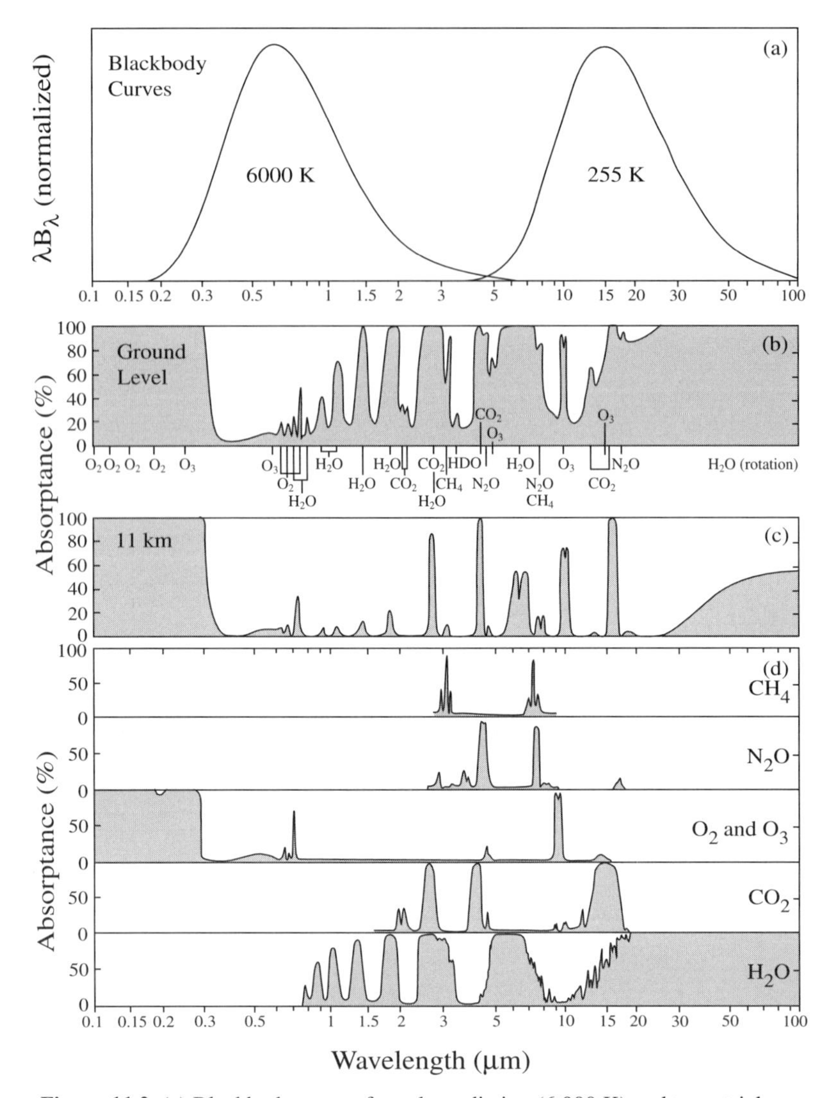

Figure 11.2 (a) Blackbody curves for solar radiation (6,000 K) and terrestrial radiation (255 K). Absorption spectra for (b) the entire vertical extent of the atmosphere, (c) the portion of the atmosphere above 11 km, and for (d) the various atmospheric gases between the top of the atmosphere and surface of the Earth.

the vibrational and rotational transitions for all atmospheric constituents with appreciable longwave absorption properties.[8]

Figure 11.2 shows the spectral position and extent of the major absorption features, along with the solar and terrestrial blackbody curves.[9]

Example 11.1 Radiative Flux from a Water Surface

To a first approximation, a liquid water medium, like most good conductors, is an opaque blackbody over the entire IR spectrum. Actually the surface emission takes place over a very thin "skin." Let the skin depth be Δz, and define depth below the surface as negative values of z. Then

$$F^+(0) \cong -\int_{-\Delta z}^{0} dz' \sigma_B T^4(z') \frac{\partial}{\partial z'} \langle \epsilon_F \rangle \cong \sigma_B \bar{T}_s^4 [\langle \epsilon_F(-\Delta z) \rangle - \langle \epsilon_F(0) \rangle] \cong \sigma_B \bar{T}_s^4 \quad (11.10)$$

since $\langle \epsilon_F(0) \rangle = 0$ and $\langle \epsilon_F(-\Delta z) \rangle = 1$. Here $\bar{T}_s$ is the average temperature over the skin layer. For a slant optical depth $2\alpha(H_2O)\Delta z \sim 1$, the skin depth is $\sim[(2\alpha(H_2O)]^{-1}$. The absorption coefficient is given by $\alpha = 4\pi m_i/\lambda$, where m_i is the imaginary index of refraction (see Appendix H). At $\lambda = 10$ μm, $m_i \sim 0.035$, and $\alpha \sim 2 \times 10^4$ m^{-1}. Thus we find $\Delta z \sim 25$ μm(!). The value of the effective skin temperature $\bar{T}_s$ is directly relevant to the heat loss from the ocean. It is usually different (up to several degrees Celsius) from sea-surface temperature values derived from microwave thermal emission, which has a thicker skin depth, or from temperature sensors on surface buoys. It is thus important to know how $\bar{T}_s$ is measured in assessing its usefulness in deriving emitted surface flux. Note that $\sigma_B \bar{T}_s^4$ does not represent the *net* loss of energy from the surface – it represents only the upward integrated flux. Not only the Sun but the atmosphere contributes an appreciable downward radiative flux at the sea surface. The atmospheric contribution to the surface irradiance (if it is considered black) is given as $F^-(0)$ in Eq. 11.6. As shown in §12.2, as temperature increases, the net flux from the sea surface can actually *decrease* as a result of the "backwarming" from increased atmospheric moisture content (and thus optical depth).

11.2.3 Narrow-Band Absorption Model

In this method, one integrates the monochromatic flux equations over spectral intervals large enough to contain a significant number of lines, but small enough for the Planck function to be considered constant. To obtain expressions that are directly comparable to those in the research literature, we hereafter approximate the surface as a blackbody. This approximation has been shown[10] to produce errors ~10% in outgoing flux values for the most highly reflecting surface (sandy desert for which $\langle \epsilon_F^s \rangle \cong 0.88$). Errors for water surfaces (0.97), ice (0.95), and nondesert land (0.95) are correspondingly smaller. If we set $\epsilon_F^s(\tilde{\nu}) = 1$, the integrated hemispherical fluxes are given by

$$\begin{aligned} F^+(i,z) &= \Delta\tilde{\nu}_i \left[\int_0^z dz' \tilde{B}_i(z') \frac{\partial \langle \mathcal{T}_F^i(z',z) \rangle}{\partial z'} + \tilde{B}_i(0) \langle \mathcal{T}_F^i(0,z) \rangle \right], \\ F^-(i,z) &= -\Delta\tilde{\nu}_i \left[\int_z^\infty dz' \tilde{B}_i(z') \frac{\partial \langle \mathcal{T}_F^i(z,z') \rangle}{\partial z'} \right], \end{aligned} \quad (11.11)$$

where $\tilde{B}_i$ denotes the value of $\pi B_{\tilde{\nu}}$ at the center of the band, $\Delta\tilde{\nu}_i$ is the spectral interval and $\langle\ \rangle$ denotes the spectral average

$$\langle \mathcal{T}_F^i(z',z) \rangle \equiv \frac{1}{\Delta\tilde{\nu}_i} \int_{\Delta\tilde{\nu}_i} d\tilde{\nu} \mathcal{T}_F(\tilde{\nu}; z', z). \quad (11.12)$$

The above equations are often the starting point in the research literature.[11] The total flux is found by summing the individual contributions, $F^{\pm}(z) = \sum_i F^{\pm}(i, z)$. The number of terms is typically $\sim$30, spanning a range from 0 to $\sim$3,000 cm^{-1}.

The key to obtaining integrated flux (and heating rate; see §10.4) is to determine the spectrally averaged transmittance $\langle \mathcal{T}_F^i \rangle$ for a homogeneous path. The most common independent variable is the *vertical path*, either the mass per unit-area column u [g · cm^{-2} or kg · m^{-2}] or the number per unit-area column when compressed to standard temperature and pressure [cm · atm]. In addition, the pressure and temperature need to be specified, since these are important for the line strengths and the line widths. After the functional dependence on these three quantities is determined, $\langle \mathcal{T}_F^i \rangle$ may be estimated by scaling the inhomogeneous path to an equivalent homogeneous one,

$$\langle \mathcal{T}_F^i \rangle (z, z') \cong \langle \mathcal{T}_F^i \rangle (\langle u \rangle, \langle p \rangle, \langle T \rangle),$$

where $\langle u \rangle$, $\langle p \rangle$, and $\langle T \rangle$ are the scaled amount of absorber mass path, pressure, and temperature for the vertical path from z to z' (§10.5.1).

11.2.4 Band Overlap

When we examine the spectral locations of the various band absorptions (see Fig. 11.2), we find that more than one molecule can absorb within the same wavenumber interval $\Delta\tilde{\nu}$. For example, H_2O and CO_2 both absorb in the spectral region 580–750 cm^{-1}. This is no problem in LBL models because the optical depth at a specific wavenumber is simply the sum of the individual optical depths. To take into account overlap in band models, reliance has been placed on the *multiplication property* of multiple bands, discussed in §10.3.3 with respect to the derivation of the random-band-model transmittance. If $u_1, u_2, \ldots, u_N$ denote the absorber amounts for the various gases, the net transmission over the ith spectral interval is written

$$\langle \mathcal{T}_F^i (u_1, u_2, \ldots, u_N) \rangle \cong \langle \mathcal{T}_F^i (u_1) \rangle \langle \mathcal{T}_F^i (u_2) \rangle \cdots \langle \mathcal{T}_F^i (u_N) \rangle. \tag{11.13}$$

Although it has been widely employed, the multiplication property actually applies only to a restricted number of transmission functions. It is valid only if the correlation between the line positions of the various gases is small. Ignoring the overlap of H_2O and CO_2 in a narrow-band model causes a significant error in the change in the downward flux when CO_2 is increased.[12]

11.2.5 The Diffusivity Approximation

The integration over angle in Eq. 11.2 is not a difficult numerical problem because of the slow variation of intensity with angle. Low-order quadrature schemes are quite accurate. The lowest-order scheme is a one-point quadrature equivalent to that used in the two-stream approximation (Chapter 7). In the band-model literature, this is called the *diffusivity approximation*. We set (see Eq. 5.61) $E_3(\tau) = \int_0^1 d\mu \mu e^{-\tau/\mu} \approx \bar{\mu} e^{-\tau/\bar{\mu}} \equiv r^{-1} e^{-r\tau}$, where $r \equiv \bar{\mu}^{-1}$ is the *diffusivity factor* and $\bar{\mu}$ is the *mean inclination*, defined in §7.3.2. This approximation was first employed by W. M. Elsasser[13] in 1942. Its use causes typical errors of order 1–2%, which is considerably less than other errors. It is customary to assume that r is constant, with 5/3 being a frequently used value. As discussed in §7.3.2, r actually varies with optical depth. For greater accuracy a convenient best-fit expression[14] is

$$r(\tau) \cong 1.5 + \frac{0.5}{1 + 4\tau + 10\tau^2}. \tag{11.14}$$

11.2.6 Equations for the Heating Rate

The atmospheric heating rate $\mathcal{H}$ was defined in §5.7 as the divergence of the radiative flux vector, or in plane-parallel geometry

$$\mathcal{H} \equiv -\int_0^\infty d\tilde{\nu} \frac{\partial}{\partial z} F_{\tilde{\nu}} = -\frac{\partial}{\partial z}(F^+ - F^-) \quad [\mathrm{W \cdot m^{-3}}]. \tag{11.15}$$

The *warming* (*cooling*) *rate* $\mathcal{W}$ was defined in §5.7 as the tendency of the atmosphere to change its temperature in response to a radiative heating, if there were no other heat transport processes,

$$\mathcal{W} = \frac{\mathcal{H}}{\rho c_p} = \frac{g}{c_p} \frac{\partial}{\partial p}(F^+ - F^-) \quad [\mathrm{K \cdot s^{-1}}], \tag{11.16}$$

where ρ is the total gas density, c_p is the specific heat of air at constant pressure, and g is the acceleration of gravity. The second form follows from the *hydrostatic equation* $dp = -\rho g dz$, where p is the total gas pressure.

In the thermal IR, $\mathcal{W}$ is generally negative. This explains why it is called a *cooling rate* in the present context. Knowledge of $\mathcal{W}$ is important for climate modeling, as it provides the basic forcing of atmospheric motions. Here, we examine how LTE cooling rates are derived from a knowledge of the temperature structure and atmospheric absorption properties. We consider the contribution from a narrow band, with the understanding that the total heating rate is the sum of the individual spectral contributions, $\mathcal{H} = \sum_i \mathcal{H}_i$.

Climate modelers usually evaluate the heating rate from Eq. 11.16 by differencing the fluxes in adjacent layers. From a computational point of view it would be preferable to compute the heating rate from the mean intensity, which is proportional to the flux divergence, because we avoid computational errors associated with numerical differencing in obtaining both the net flux and its derivative. Before considering the general case of a nonisothermal atmosphere, we first consider an idealized situation of an isothermal atmosphere in the example below.

Example 11.2 Isothermal Atmosphere: Cooling to Space

It is instructive to examine the flux and heating equations in an isothermal atmosphere. The most useful forms for this purpose are the wide-band equations. By rewriting Eqs. 11.6 in terms of the mass path ($du' = -\rho dz'$), one can show that (see Problem 11.1)

$$F^+(u) = \sigma_B T^4,$$
$$F^-(u) = \sigma_B T^4 \langle \epsilon_F(u) \rangle. \tag{11.17}$$

The upward flux is constant and equal to that which would be emitted by a blackbody at temperature T. This is because the atmosphere replaces with emission what it removes from the thermal radiation emanating upward from the surface. The downward flux is given by the blackbody flux multiplied by the emissivity $\langle \epsilon_F(u) \rangle$, which depends upon the mass path (u) above the level in question.

The heating rate for the isothermal slab also has a simple form. From Eq. 11.15 expressed in terms of u, we have

$$\mathcal{H} = \rho \frac{\partial}{\partial u}(F^+ - F^-) \equiv \mathcal{H}_{cs} = -\sigma_B T^4 \rho \frac{\partial}{\partial u} \langle \epsilon_F(u) \rangle. \tag{11.18}$$

This is the *cooling-to-space* term, which is a useful concept, even in the more general context of a nonisothermal medium (see §11.2.7 below). $\mathcal{H}_{cs}$ is always negative, that is, the medium is cooled everywhere at a rate that depends upon the product $\rho \partial \langle \epsilon_F(u) \rangle / \partial u$. Although photons will be absorbed by the surrounding medium at all levels u', these regions will also warm the region at u at a rate that exactly compensates their contribution to the cooling. The exchange of heat with its surroundings has a zero net effect only for an isothermal medium. Consequently, an isothermal medium cools as if the photons escape directly to space.

What is the behavior of $\mathcal{H}_{cs}$ with height? This is most easily understood by approximating the path dependence of $\langle \epsilon_F(u) \rangle$ with the expression $1 - \exp(-au^b)$, where a and b are coefficients determined by a least-square fit to the actual emissivity. The cooling-to-space term is easily seen to be proportional to $u^b \exp(-au^b)$, which maximizes at a level where $au^b = 1$. Thus the function is a maximum where the "optical depth" equals unity. This is in accord with the principle described in Chapter 1, that the radiation emitted by a planet, or star, at a given part of the spectrum originates at one optical depth from the surface.

Assuming that the band transmittance is given by the Elsasser model (§10.3.1), we define the dimensionless *cooling-to-space function*

$$CSF \equiv -\frac{u^* \mathcal{H}_{cs}}{\sigma_B T^4 \rho_0} = \tilde{u} \frac{\partial \langle \mathcal{T}_b(\tilde{u}) \rangle}{\partial \tilde{u}} = \tilde{u} \frac{\partial}{\partial \tilde{u}} \int_{-1/2}^{1/2} dx \exp\left[-\frac{2\pi \tilde{u} y \sinh(2\pi y)}{\cosh(2\pi y) - \cos(2\pi x)} \right]$$
$$= -\int_{-1/2}^{1/2} dx \frac{2\pi \tilde{u} y \sinh(2\pi y)}{\cosh(2\pi y) - \cos(2\pi x)} \exp\left[-\frac{2\pi \tilde{u} y \sinh(2\pi y)}{\cosh(2\pi y) - \cos(2\pi x)} \right], \tag{11.19}$$

where $\tilde{u} \equiv Su/2\pi \bar{\mu} \alpha_L$ is the (effective) dimensionless mass path for an isotropic radiation field. (see §10.2.1). Note that we have used the diffusivity approximation to transform the beam transmittance to a flux transmittance. The density ρ_0 is the mass density of the absorber at $z = 0$, which is assumed to vary exponentially with height so that $\rho \propto u/u^*$. (This is appropriate for water vapor or carbon dioxide, which may be characterized by a constant scale height H_a, so that both ρ and $u \propto \exp(-z/H_a)$, but not for ozone.). The above expression may be evaluated by numerical integration. Figure 11.3 shows how the shape of the dimensionless cooling-to-space function varies with optical path, for various values of the grayness parameter $y = \alpha_L/\Delta\tilde{\nu}$. For small values of y, the function is broad and maximizes deep within the medium. This happens as a result of photons escaping from the line centers near $u = 0$. At progressively deeper levels, the

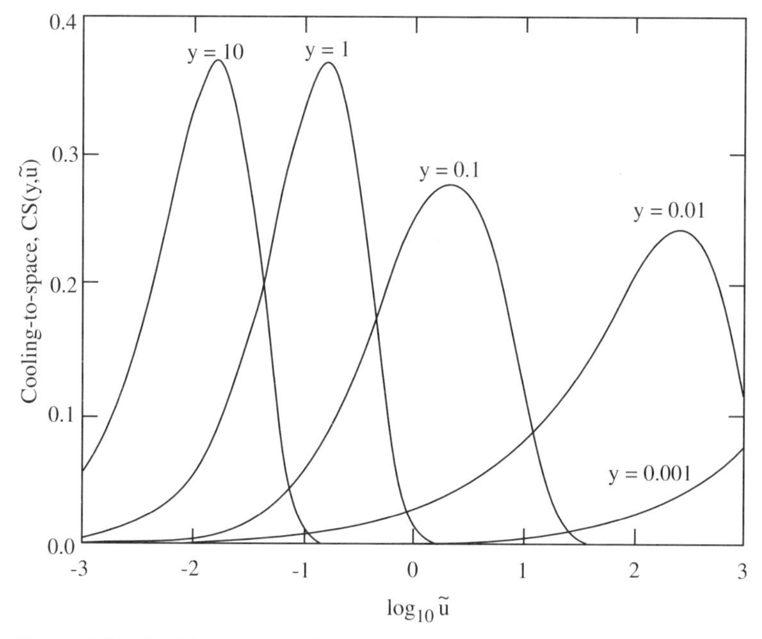

Figure 11.3 Cooling-to-space function, *CSF*, computed by numerically evaluating the integral in Eq. 11.19. *CSF* is a nondimensional cooling rate for a homogeneous atmosphere of vertical thickness $\tilde{u} = Su/2\pi\bar{\mu}\alpha_L$, which absorbs according to the Elsasser band model. $y = \alpha_L/\Delta\tilde{\nu}$ is the grayness parameter (see §10.2.1).

cooling occurs more and more from the line wings, always occurring where the monochromatic optical depth is approximately unity. For large y, the line widths exceed the line separation δ. The line profiles are blurred into a constant value of the absorption coefficient, that is, the band approximates a continuum absorber. In this gray case, the cooling occurs over a narrower layer, maximizing at an optical depth $2\pi y\tilde{u} = Su/\delta \cong 1$.

The *CSF* is related to the *Chapman function*, *CF*, defined in a different context in §9.4.2. In IR applications, the *CSF* defines the shape of the region where radiation escapes from the medium. In shortwave applications, the *CF* defines the region where collimated solar radiation is deposited. It is clear that the two concepts are closely related in an inverse sense. In fact they would be mathematically identical if IR radiation escaped in a collimated beam – or if solar radiation were isotropic. We now consider the more general situation of a variable temperature.

11.2.7 Clear-Sky Radiative Cooling: Nonisothermal Medium

By differentiating the above narrow-band expressions for flux (Eqs. 11.11), we find, after cancellation of two terms,

$$-\frac{\mathcal{H}_i}{\Delta\tilde{\nu}_i} = \tilde{B}_i(0)\frac{\partial\mathcal{T}_F^i(0,z)}{\partial z} + \int_0^z dz'\tilde{B}_i(z')\frac{\partial^2\mathcal{T}_F^i(z',z)}{\partial z\partial z'} + \int_z^\infty dz'\tilde{B}_i(z')\frac{\partial^2\mathcal{T}_F^i(z,z')}{\partial z\partial z'}. \tag{11.20}$$

This equation illustrates the fact that the heating rate depends upon the second derivative of the transmittances.

It is convenient to rewrite Eq. 11.20 to explicitly separate out terms involving exchange between layers and exchange between a given layer and the boundaries. The upper boundary is normally the half-space vacuum, which may be thought of as a boundary at zero temperature. The lower boundary is normally the surface, but it might also be a cloud deck characterized by a temperature and surface emittance. To accomplish this separation we proceed as follows.

Suppose we substituted $\tilde{B}_i(z)$ for $\tilde{B}_i(z')$ in the first integral term in Eq. 11.20. This new term simplifies as follows:

$$\int_0^z dz' \tilde{B}_i(z) \frac{\partial^2 \langle \mathcal{T}_F^i(z', z) \rangle}{\partial z \partial z'} = \tilde{B}_i(z) \int_0^z dz' \frac{\partial}{\partial z} \frac{\partial \langle \mathcal{T}_F^i(z', z) \rangle}{\partial z'}$$
$$= -\tilde{B}_i(z) \frac{\partial \langle \mathcal{T}_F^i(0, z) \rangle}{\partial z}.$$

A similar substitution in the second integral term in Eq. 11.20 yields a similar closed-form expression:

$$\int_z^\infty dz' \tilde{B}_i(z) \frac{\partial^2 \langle \mathcal{T}_F^i(z, z') \rangle}{\partial z \partial z'} = \tilde{B}_i(z) \frac{\partial \langle \mathcal{T}_F^i(z, \infty) \rangle}{\partial z}.$$

If we now add and subtract the above two terms to Eq. 11.20, we obtain the following form for the exact heating rate in the ith spectral interval:

$$\frac{\mathcal{H}_i}{\Delta \tilde{\nu}_i} = \overbrace{-[\tilde{B}_i(0) - \tilde{B}_i(z)] \frac{\partial \langle \mathcal{T}_F^i(0, z) \rangle}{\partial z}}^{\textit{exchange heating from surface}} \overbrace{- \tilde{B}_i(z) \frac{\partial \langle \mathcal{T}_F^i(z, \infty) \rangle}{\partial z}}^{\textit{cooling to space}}$$
$$\overbrace{- \int_0^z dz' [\tilde{B}_i(z') - \tilde{B}_i(z)] \frac{\partial^2 \langle \mathcal{T}_F^i(z', z) \rangle}{\partial z \partial z'}}^{\textit{exchange from below}}$$
$$\underbrace{- \int_z^\infty dz' [\tilde{B}_i(z') - \tilde{B}_i(z)] \frac{\partial^2 \langle \mathcal{T}_F^i(z, z') \rangle}{\partial z \partial z'}}_{\textit{exchange from above}}. \quad (11.21)$$

The first term represents the contribution to the heating as a result of the transfer of energy between the surface and the region at height z. Since the term $(-)\partial/\partial z \mathcal{T}_F^i \langle (0, z) \rangle$ is positive, the overall sign of this term is positive when the surface emission exceeds the local Planck function. If the surface emittance is unity, another way to say this is that a heating (or cooling) occurs when the surface temperature exceeds (or is less than) the atmospheric temperature $T(z)$.

The second term is the cooling-to-space term, appropriate to the local temperature $T(z)$. Note the resemblance of the first and second terms.

The third and fourth terms are the *exchange terms*, which describe the net effect of transfer between different layers. These cancel in an isothermal situation, as described in the previous section. It is found in practice that the fourth term is usually quite small, compared to the third term.

11.2.8 Computations of Terrestrial Cooling Rates

The approaches outlined above describe traditional ways of computing warming/cooling rates in planetary atmospheres. Unfortunately they work only under clear-sky conditions when scattering can be neglected. They are not useful in solar absorption calculations because then molecular (Rayleigh) scattering must be taken into account. In general they are not applicable in any region of the spectrum (solar or terrestrial) if clouds and/or aerosols are present. Thus, under circumstances where scattering plays a significant role, we must find ways of computing the warming/cooling rates that allow us to include the effects of multiple scattering.

In principle, we could compute the radiation field with such fine spectral resolution throughout the solar and terrestrial range that we would account for the absorption, emission, and scattering within every single spectral line, and for arbitrary distributions of particles. However, this has not been the conventional approach.

The idea of band models was introduced in Chapter 10. Band models are of great practical importance because in most atmospheric chemistry and climate models we cannot afford to compute warming/cooling rates line by line throughout the spectrum. In Chapter 10 we described how band models that allow inclusion of multiple scattering can be constructed to compute the radiation field in various parts of the spectrum. The clear-sky cooling rates due to H_2O, CO_2, and O_3 are shown in Fig. 11.4 for a standard

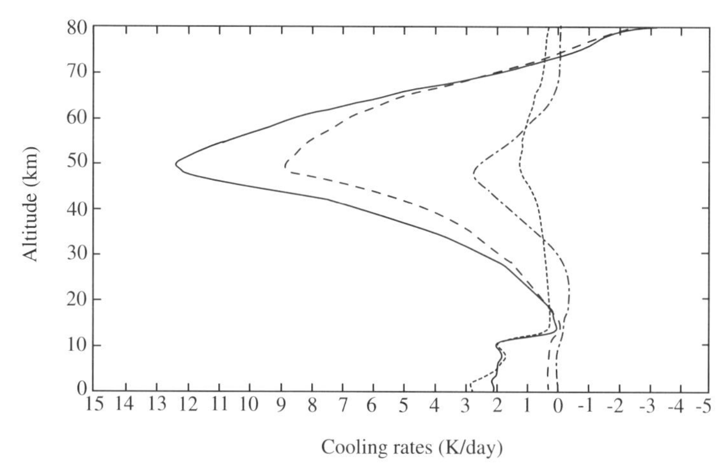

Figure 11.4 Clear-sky cooling rates based on line-by-line computations for H_2O (dotted line), CO_2 (dashed line), and O_3 (dashed-dotted line). The solid line gives the total cooling rate.

atmosphere. These are based on line-by-line computations.[15] Note that the cooling rate for each individual gas was computed in the absence of the other gases. Because of overlapping absorption the total cooling rate is less than the sum of the individual cooling rates. The *correlated-k* distribution method accurately captures the individual cooling rates as well as the overlapping absorption between different gases.[16]

11.3 The IR Radiative Impact of Clouds and Aerosols

In the presence of clouds, the flux equations need modification to account for the additional sources of opacity and emission. Suspended particles will contribute additional absorption and scattering opacity, and more importantly, destroy the plane-parallel symmetry that applies to the clear-sky case. Owing to their longer lifetimes, aerosols tend to be distributed more uniformly than clouds. In addition to geometric complexity, aerosol and cloud particles are distributed over a wide range of sizes, and in the case of ice, over a variety of shapes. Furthermore, for aerosols there may even be variation in their index of refraction from particle to particle. Consequently, it is unrealistic to expect to frame the radiative transfer equations with the same confidence, as is common in the molecular case.

11.3.1 Heating Rate in an Idealized Cloud

Let a geometrically thin cloud of thickness $\Delta z = z_t - z_b$ be imbedded in an atmosphere at a height where the temperature is $T(z_b \approx z_t) \equiv T_c$. Here z_b and z_t are the height of the cloud base and top, respectively. We first consider how the cloud interacts with the IR atmospheric radiation in physical terms; we then derive equations from a simple model to verify these expected effects. We will ignore solar heating, although it should be kept in mind that it competes with IR cooling, especially near the cloud top.

The cloud particles will absorb IR radiation from their surroundings (both other cloud particles and the atmospheric gases) in proportion to the mean intensity of the ambient radiation. The particles will also radiate, at a given wavenumber, according to the product of their absorption coefficient and the Planck function at temperature T_c. In general, the absorption and emission rates will not balance. In addition to the radiative heating, moist convection will warm the atmosphere (and the imbedded cloud particles). It is clear that the IR radiative warming is likely to be strong and positive at the cloud base, since the incident radiation is emitted by the lower, warmer atmosphere and surface. Furthermore, the tendency for the cloud base to cool is inhibited by the overlying cloud opacity. At the cloud top, the opposite occurs. The incident radiation is mainly due to the downward flux, which is small (its effective temperature is lower than T_c) because of reduced emission from the overlying cooler layers. The upward flux is attenuated by the cloud opacity. However, the cloud radiates according to the local temperature T_c. Since an optically thick cloud approximates a blackbody, its cooling to space can be quite efficient. In summary, we would expect strong warming on the bottom side of the cloud and strong cooling on the topside.

The above effects can be reproduced by making use of a simplified model of the cloud and the radiation field. We assume the cloud is gray, plane parallel, and geometrically thin. To avoid the presence of exponential integrals, we approximate the flux transmittance by the diffusivity (or two-stream) approximation (§11.2.3). The radiation field incident on the lower and upper faces of the medium are assumed to be known. We denote the incident fluxes as $F^+(z_b)$ and $F^-(z_t)$. The spectrally integrated radiation emitted by the cloud particles into a hemisphere is denoted by $\tilde{B}_c = \sigma_B T_c^4$, which is assumed constant.

The upward flux within the cloud is the sum of two contributions: (1) the transmitted atmospheric flux, $F^+(z_b)\exp[-r(\tau_c^* - \tau_c)]$, and (2) the emitted flux from the cloud particles, $B_c[1 - \exp[-r(\tau_c^* - \tau_c)]$. Here τ_c^* and τ_c are respectively, the total optical depth and the optical depth coordinate at an arbitrary height in the cloud, z. Note that we have ignored the atmospheric attenuation, which is permissible for a geometrically thin cloud. To avoid the presence of "sharp edges" (and unrealistic warming/cooling rates) we adopt a Gaussian function for the cloud–particle height distribution. Then the optical depth τ_c is given by

$$\tau_c(z) = \frac{\tau_c^*}{\Delta z\sqrt{2\pi}} \int_z^\infty dz' e^{-\frac{1}{2}[(z'-z_c)/\Delta z]^2} = \tau_c^* \{1 - C_n[(z - z_c)/\Delta z]\}, \quad (11.22)$$

where $\sqrt{2}\Delta z$ is the $1/e$ cloud thickness[17] and C_n is the *cumulative normal function*

$$C_n(x) \equiv \frac{1}{\sqrt{2\pi}} \int_{-\infty}^x dt e^{-t^2/2}.$$

The downward flux within the cloud is also the sum of the transmitted flux, $F^-(z_t)\exp(-\tau_c/\bar{\mu})$, and emitted flux, $\tilde{B}_c[(1 - \exp(-\tau_c/\bar{\mu})]$, from the cloud itself. The hemispherical fluxes interior to the cloud are therefore

$$F^+(\tau_c) = F^+(z_b)e^{-(\tau_c^*-\tau_c)/\bar{\mu}} + \tilde{B}_c\left[1 - e^{-(\tau_c^*-\tau_c)/\bar{\mu}}\right], \quad (11.23)$$

$$F^-(\tau_c) = F^+(z_t)e^{-\tau_c/\bar{\mu}} + \tilde{B}_c\left[1 - e^{-\tau_c/\bar{\mu}}\right]. \quad (11.24)$$

Note that if the cloud is optically thick ($\tau_c^* \gg 1$), the fluxes inside the cloud far away from the cloud boundaries are equal to the blackbody radiation $\tilde{B}_c$, and the radiation field is isotropic. *No heating occurs within the deep interior of an optically thick cloud.* In contrast, for a very thin cloud ($\tau_c^* \ll 1$), the fluxes are simply those of the upward and downward atmospheric fluxes.

The heating rate within the cloud is the derivative of the net flux (eqn. 5.77),

$$\mathcal{H} = -\frac{\partial}{\partial z}[F^+(z) - F^-(z)] = -\frac{\partial \tau_c}{\partial z}\frac{\partial}{\partial \tau_c}[F^+(z) - F^-(z)]. \quad (11.25)$$

From Eq. 11.22, $\partial\tau_c/\partial z = -(\tau_c^*/\Delta z)\exp[-(1/2)(z - z_b)^2/\Delta z^2]$, and we find after cancellation

$$\mathcal{H}(z) = \frac{\tau_c^*}{\bar{\mu}\Delta z} e^{-(1/2)(z-z_c)^2/\Delta z^2} \times \left\{[F^+(z_b) - \tilde{B}_c]e^{-(\tau_c^*-\tau_c)/\bar{\mu}} + [F^-(z_t) - \tilde{B}_c]e^{-\tau_c/\bar{\mu}}\right\}, \quad (11.26)$$

where τ_c is a function of z, through Eq. 11.22. The above result is easy to interpret. The first term dominates over the second term at the cloud base and is positive since $F^+(z_b) > \tilde{B}_c$. However, the second term dominates at the top, where it is negative, since $F^-(z_t) < \tilde{B}_c$. Thus we verify the expected physical processes of heating at the bottom and cooling at the top.

Warming at the cloud bottom and cooling at the top tend to destabilize the air, as described in Chapters 1 and 12. The initial effect is to cause increased convection, increased vertical motion, entrainment of dry air, and eventual dissipation of the cloud. However, clouds are often the *result* of such dynamical processes. Understanding the combined effects of latent warming/cooling, convective transport, and radiative warming/cooling on the cloud particles, and the feedback due to the change in cloud optical properties, is one of the most challenging areas of atmospheric research.[18]

11.3.2 Detailed Longwave Radiative Effects of Clouds

We now consider the more detailed interactions of cloud particles with IR radiation. We have already discussed in §9.4 the shortwave ($0.2 < \lambda < 3.5$ μm) heating due to cloud particles, and we discussed the fact that optical properties (extinction coefficient, single-scattering albedo, and the asymmetry factor) can be described in terms of two basic cloud properties, the liquid water content (*LWC*) and the effective particle radius $\langle r \rangle$. A simple parameterization of cloud optical properties that is valid across the solar and thermal IR spectral regions was also discussed. We consider first the approximation of complete absorption, that is, we ignore scattering.

It is useful to describe the cloud optical properties in the thermal IR spectrum in an approximate manner, which in turn allows us to describe in simple physical terms how clouds can affect the Earth's climate. We will focus mainly on the effects of water clouds, although we will make some remarks about the effects of high, cold, cirrus clouds on the surface temperature. If we are given the cloud *particle size distribution* $n(r, z)$, the absorption coefficient for spherical particles is given by a weighting of the absorption cross section over the particle sizes r:

$$\alpha(\nu, z) = \int_0^\infty dr\, n(r, z) Q_{\text{abs}}(\nu, r)\pi r^2, \quad (11.27)$$

where Q_{abs} is the *absorption efficiency* of a droplet of radius r, calculated from Mie–Debye theory. The above formula is of little practical use because the quantity $n(r, z)$

is rarely known. We can reformulate the problem in terms of the liquid water path within the cloud and a small number of parameters derivable from experiment. We convert $\alpha(\nu)$ to a mass absorption coefficient by dividing by the density of liquid water in the cloud, denoted previously by the *liquid water content* (*LWC*):

$$\alpha_m(\nu, z) \equiv \frac{\alpha(\nu, z)}{LWC(z)} = \frac{\int_0^\infty dr\; n(r, z) Q_{\rm abs} \pi r^2}{\int_0^\infty dr\; n(r)\rho_l (4\pi/3) r^3}. \tag{11.28}$$

Here ρ_l is the bulk density of liquid water [kg $\cdot$ m^{-3}]. Mie–Debye theory predicts that $Q_{\rm abs}$ increases linearly with particle radius when the size parameter x $(= 2\pi r/\lambda)$ is not too large ($x < 10$). For a wavelength close to the peak of the terrestrial spectrum (near 11 μm), $Q_{\rm abs}$ is in the linear regime if $r \leq 10$ μm. Since typical values for the *mean radius* $\langle r \rangle$ are in the range 5–15 μm, this is an adequate approximation. Setting $Q_{\rm abs} = c(\tilde{\nu})r$ for $r < r_m$, where $c(\tilde{\nu})$ is some wavenumber-dependent factor, the result is simply $\alpha_m(\nu) = \frac{3}{4}c(\tilde{\nu})$. For liquid water droplets, $c(\tilde{\nu})$ may be determined from Mie–Debye theory. Cloud IR optical depth is therefore written approximately as $\tau(\tilde{\nu}) = \alpha_m(\tilde{\nu})$ *LWP*, where *LWP* is the (vertical) liquid water path through the cloud usually expressed in [g $\cdot$ m^{-2}]. A typical value for α_m is 0.1 [m^2 $\cdot$ g^{-1}], which yields IR optical depths between 1 and 10 for values of *LWP* between 10 and 100 [g $\cdot$ m^{-2}].

The determination of the monochromatic IR optical depth for clouds is only the first step in computing spectrally averaged quantities, such as fluxes and heating rates. The next step is to compute absorption and scattering coefficients over suitably chosen spectral intervals as described in §9.4. Then one combines the cloud and gaseous absorption coefficients (based, for example, on the k-distribution approach; see §10.5) to obtain effective values of optical depth within the gaseous absorption bands (primarily H_2O, CO_2, and O_3) for gases and cloud droplets. To include scattering we need also effective values of the single-scattering albedo and the asymmetry factor. These are the required input parameters that must be specified spectrally and layer by layer throughout the medium to solve the radiative transfer equation. Although scattering can easily be accommodated if the k-distribution approach is adopted, it is usually ignored because it has been customary to assume that the single-scattering albedo is small in the thermal IR. As we shall see below this assumption is not always justifiable.

A more traditional method is to employ the concept of broadband emittance (see §10.3). The spectrally averaged flux emittance of the cloud is

$$\epsilon_{\rm F}(\Delta\tilde{\nu}) = \frac{\int_{\Delta\tilde{\nu}} d\tilde{\nu} B_{\tilde{\nu}}(T) \int_0^1 d\mu \mu \epsilon[\tau(\tilde{\nu})/\mu]}{\int_{\Delta\tilde{\nu}} d\tilde{\nu} B_{\tilde{\nu}}(T)},$$

where $\epsilon[\tau/\mu] = 1 - \exp[-\tau/\mu]$ is the directional emittance in the direction given by $\cos^{-1}\mu$. Note that the above relationship between the absorptance $\exp[-\tau/\mu]$

and the emittance is valid only as long as scattering can be ignored. The diffusivity approximation is usually invoked, so that

$$\int_0^1 d\mu\mu\epsilon[\tau(\tilde{\nu})/\mu] \approx 1 - \exp[-\alpha_m(\tilde{\nu})LWP/\bar{\mu}].$$

If we further assume that $\alpha_m(\tilde{\nu})$ is spectrally flat, so that $\alpha_m(\tilde{\nu}) \approx \alpha_m =$ constant, then the expression for the emittance simplifies to

$$\epsilon_F(\Delta\tilde{\nu}) \approx 1 - \exp[-\alpha_m LWP/\bar{\mu}]. \tag{11.29}$$

To determine fluxes and heating rates, one uses the above results in equations similar to those for clear-sky conditions (§11.2). In climate modeling, the difference in the outgoing overcast and clear-sky fluxes is called the *cloud forcing*.

11.3.3 Accurate Treatment Including Scattering

An accurate treatment of longwave radiative effects of water clouds must rely on Mie–Debye computations. In §9.3.3 we discussed a parameterization in terms of liquid water content and effective radius that can be used to avoid time-consuming computations. We noted that this approach yields an accurate description of shortwave net flux and warming rates. Similar results pertaining to terrestrial radiation show that the difference in the net flux computed from the parameterization and the exact computation is less than 1%, while the error in the cooling rate is within 4% of the total terrestrial cloud forcing.

Figure 11.5 displays the longwave cloud forcing at the top of the atmosphere as a function of liquid water path, *LWP*, for four different values of the effective radius, $\langle r \rangle$. The cloud forcing increases with *LWP* and reaches an asymptotic value corresponding to an optically thick, although not black, cloud (where the outgoing longwave radiation does not change with increasing *LWP*). For the smallest cloud drops ($\langle r \rangle = 3$ μm) the cloud reaches an asymptotic value (but is nonblack) when the *LWP* reaches 20 $\text{g} \cdot \text{m}^{-2}$. As the effective radius $\langle r \rangle$ increases to 50 μm, the outgoing longwave radiation increases with *LWP* until the cloud becomes very thick (*LWP* $\approx$ 150 $\text{g} \cdot \text{m}^{-2}$). This clearly demonstrates that the cloud greenhouse effect is sensitive to scattering of longwave radiation. Thus, ignoring longwave scattering effects in climate models can lead to significant errors, especially for thin clouds.

Figure 11.5 demonstrates that the longwave cloud radiative impact is determined not only by cloud-top temperature, but also by the droplet distribution. The optical properties of water clouds for longwave radiation are shown in Fig. 11.6 as a function of cloud droplet effective radius. This figure explains why there is more outgoing longwave radiation (less longwave cloud radiative forcing) for thick clouds with large

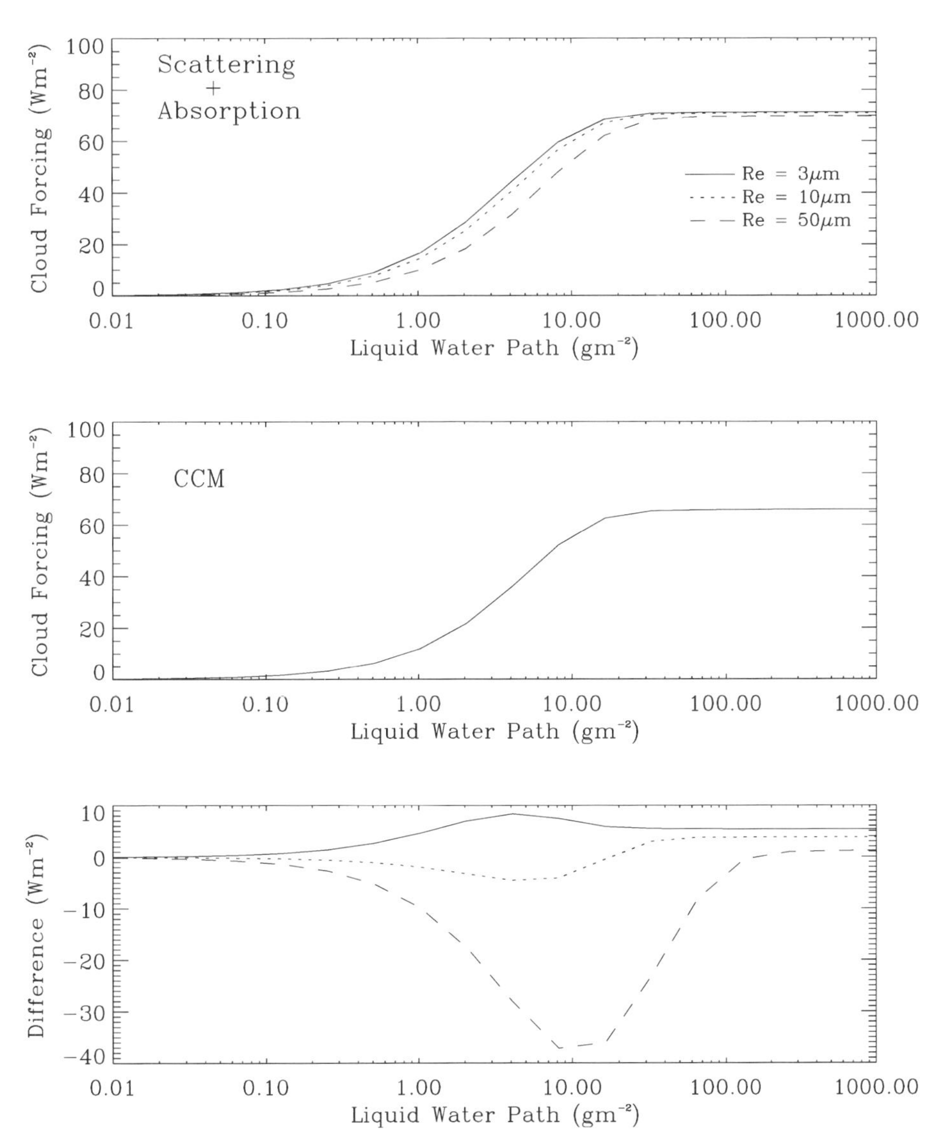

Figure 11.5 Dependence of longwave cloud forcing on liquid water path and effective droplet radius $\langle r \rangle$. The middle panel shows the result obtained by fixing the absorption coefficient at 0.1 m^{-1} as in Eq. 11.29 and ignoring scattering effects, which are approximations used in many climate models, including the NCAR Community Climate Model (CCM3). The bottom panel shows the difference between the results given in the upper and the middle panel.

equivalent radius. The difference is due to the scattering: The larger the cloud drops, the more efficient the scattering. We note that both the single-scattering albedo and the asymmetry factor increase with $\langle r \rangle$. Thus, as a result of cloud scattering the outgoing longwave radiation *increases* and the longwave cloud radiative forcing *decreases* with increasing drop size.

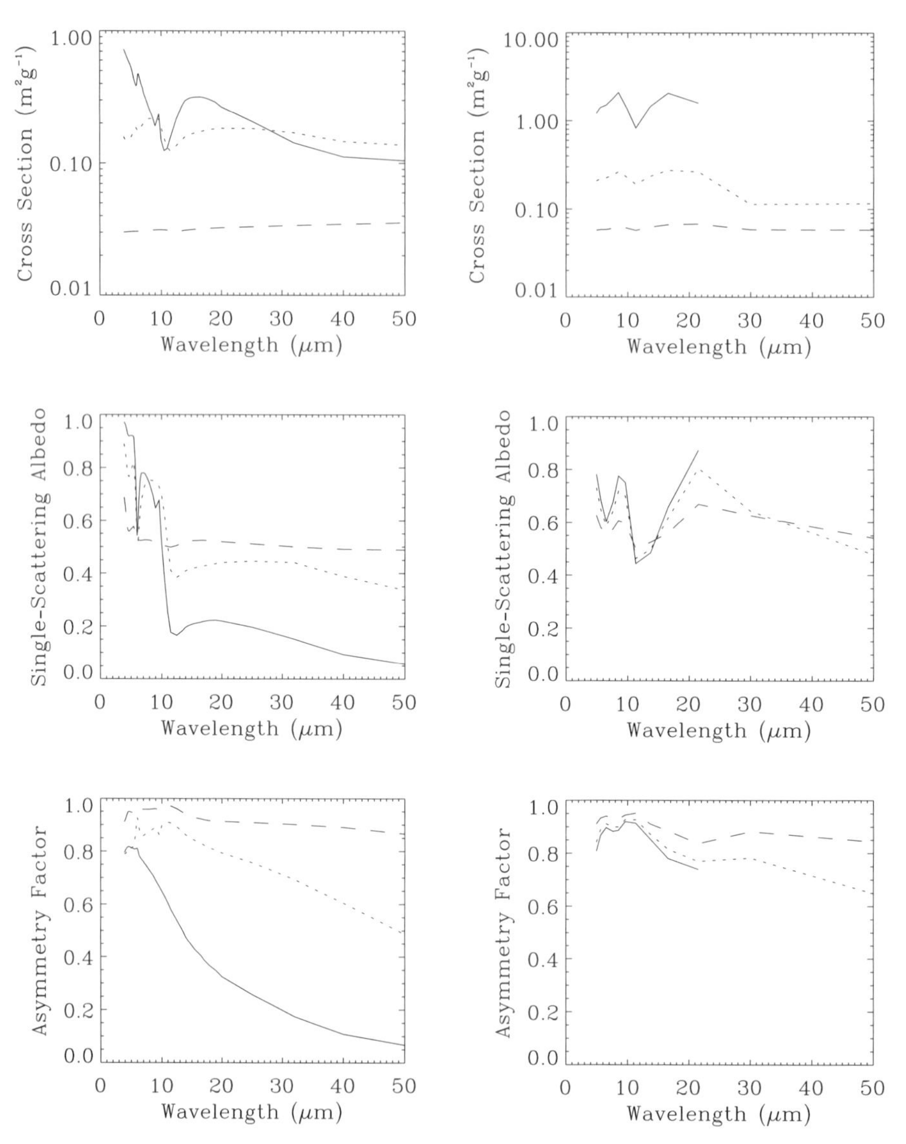

Figure 11.6 Dependence of longwave optical properties on wavelength and effective particle radius. The three left panels pertain to water clouds and the right panels to ice clouds. The curves in the left panels pertain to different effective cloud droplet radii; solid line: $\langle r \rangle = 3$ μm, dotted line: $\langle r \rangle = 10$ μm, dashed line: $\langle r \rangle = 50$ μm. Similarly, in the right panels the curves correspond to different ice particle effective diameters; solid line: $D_e = 6$ μm, dotted line: $D_e = 25$ μm, dashed line: $D_e = 100$ μm.

11.4 Summary

This chapter was devoted to radiative transfer in nongray media. We started by discussing radiative transfer under clear-sky conditions, ignoring scattering. This allowed us to focus on the physical processes responsible for warming and cooling of the atmosphere. The heating rate was explained in terms of exchange of heat between (i) the surface and the atmosphere, (ii) different layers within the atmosphere, and (iii) the atmosphere and space (cooling to space). Wide-band emittance models have played an important part in the development of fast computations of warming/cooling rates. Such models are still used in climate models to compute IR cooling rates. Although narrow-band models allow for more accurate treatment of gaseous absorption, they are computationally expensive compared to wide-band models. Such models based on the k-distribution method have the additional advantage that they reduce the integration over wavelength (or wavenumber) to a sum of "monochromatic" problems, allowing for a unified treatment of shortwave and longwave radiative transfer. This reduction also facilitates the inclusion of multiple scattering from aerosol and cloud particles, which cannot be treated accurately within the framework of wide-band models.

With increasing wavenumber interval, $\Delta\tilde{\nu}$, wide-band models tend to overestimate the true absorption.[19] Although the various LBL models are within 1% of one another, the more efficient, but less accurate, narrow-band and wide-band models show a spread of 10 to 20%. The agreement[20] is worse in calculations of changes in radiatively important variables, such as CO_2 and H_2O. Clearly, such errors could contribute substantial errors in climate-change predictions of a GCM. Thus, even in the relatively simple clear-sky case, there is still considerable room for improvement in radiative modeling. As one might expect, the situation deteriorates further when clouds and aerosols are introduced into the models.

Problems

11.1 Show that the upward and downward fluxes in an isothermal atmosphere are given by Eqs. 11.17.

11.2 This problem involves the transmittance and cooling rate in the troposphere, due to CO_2 at its present concentration (the $1 \times CO_2$ scenario) and in a future altered climate (the $2 \times CO_2$ scenario). We will use the random-Lorentz-exponential model (see Eqs. 9.20 and 9.24).

(a) Show that the 15-μm cooling-to-space approximation to the cooling rate in pressure coordinates is

$$\mathcal{W} = \frac{\mathcal{H}}{\rho c_p} = \pi B[\tilde{\nu} = 667 \text{ cm}^{-1}; T(p)]\Delta\tilde{\nu}\frac{g}{c_p}\frac{\partial \mathcal{T}_F(667 \text{ cm}^{-1}; p, 0)}{\partial p},$$

where

$$\mathcal{T}_F(667 \text{ cm}^{-1}; p, 0) = \exp\left[-\frac{\bar{S}u}{\bar{\mu}\delta}\left(1 + \frac{\bar{S}_m u/\bar{\mu}}{\pi\tilde{\alpha}_L(p, 0)}\right)^{-1/2}\right].$$

In the above $\bar{\mu}$ is the mean inclination; $\bar{S}$ is the 15-μm average band strength per unit mass; δ is the mean line spacing; $u = w_m p/g$, where w_m is the mass mixing ratio of CO_2, p is the pressure, and g is the gravitational acceleration; c_p is the specific heat per unit mass; and $\tilde{\alpha}_L(p, 0)$ is the H–C–G scaled value of the Lorentz half-width $= \alpha_L(p_0) \times (p/2p_0)$, where p_0 is 1 bar.

(b) Show that $\mathcal{T}_F$ in part (a) is of the form e^{-ap} where a is a pressure-independent quantity. Thus prove that

$$\frac{\partial \mathcal{T}_F}{\partial p} = \mathcal{T}_F \ln \mathcal{T}_F / p.$$

(c) Using the results of part (a), evaluate the flux transmittance from 30 mb to space for the $1 \times CO_2$ and $2 \times CO_2$ scenarios. Use the following values: $\bar{\mu} = 3/5$; w, the volumetric mixing ratio (see §1.6.4) for the present atmosphere, is 360 ppmv; $\bar{S}_m/\delta = 718.7$ cm^{2}g^{-1}; $\pi\alpha(p_0)/\delta = 0.448$.

(d) Using the above results find the cooling rate (in K/day) at 30 mb for the two scenarios. Let $T = 220$ K and $B[\tilde{\nu} = 667$ cm^{-1}; 220 K$]\Delta\tilde{\nu} = 24.36$ W $\cdot$ m^{-2}. Compare with the accurate results shown in Fig. 11.4.

(e) Make the argument that since a CO_2 doubling causes an increased cooling at 30 mb and throughout the stratosphere (Fig. 11.4) this would tend to cause the stratospheric temperature to be lowered, in contrast to warming in the lower troposphere. Contrast the different radiative transfer processes in the two regions.

Notes

1 Two excellent references are Stephens, G. L., *Remote Sensing of the Lower Atmosphere*, Oxford University Press, New York, 1994; and Houghton, J. T., F. T. Taylor, and C. D. Rodgers, *Remote Sounding of Atmospheres*, Cambridge University Press, Cambridge, 1984.

2 For optically thick clouds, the cloud is effectively a blackbody, and scattering becomes less important.

3 Details concerning the LBL technique are found in, for example, Scott, N. A. and A. Chedin, "A fast line-by-line method for atmospheric absorption computations: The automated atmospheric absorption atlas," *J. Appl. Meteorol.*, **20**, 801–12, 1981.

4 Rothman, L. S. et al., "The HITRAN molecular database: Editions of 1991 and 1992," *J. Quant. Spectrosc. Radiative Transfer*, **48**, 469–507, 1992.

5 Husson, N., B. Bonnet, N. A. Scott, and A. Chedin, "Management and study of spectroscopic information: The GEISA program," *J. Quant. Spectrosc. Radiative Transfer*, **48**, 509–18, 1992.

6 Chou, M.-D. and L. Kouvaris, "Calculations of the transmission functions in the infrared CO_2 and O_3 bands," *J. Geophys. Res.*, **96**, 9003–12, 1991.

7 Zhang, W. and J. D. Haigh, "Improved broadband emissivity for water vapor cooling rate calculations," *J. Atmos. Sci.*, **52**, 124–38, 1995.

8 Kiehl, J. T. and K. E. Trenberth, "Earth's annual global mean energy budget," *Bull. Am. Meteorol. Soc.*, **78**, 197–208, 1997.

9 This figure is adapted from Fig. 6.2 of Peixoto, J. P. and A. H. Oort, *Physics of Climate*, American Institute of Physics, New York, 1992.

10 Brieglieb, B. P., "Longwave band model for thermal radiation in climate studies," *J. Geophys. Res.*, **97**, 11,475–85, 1992. The merit of narrow-band models is also discussed in Tsay, S.-C., K. Stamnes, and K. Jayaweera, "Radiative transfer in stratified atmospheres: Description and validation of a unified model," *J. Quant. Spectrosc. Radiative Transfer*, **43**, 133–48, 1990.

11 For example, see Stephens, G. L., "The parameterization of radiation for numerical weather prediction and climate models," *Mon. Wea. Rev.*, **112**, 826–67, 1984.

12 Kiehl, J. T. and V. Ramanathan, "CO_2 radiative parameterization used in climate models: Comparison with narrow band models and with laboratory data," *J. Geophys. Res.*, **88**, 5,191–202, 1983.

13 Elsasser, W. M., *Heat Transfer by Infrared Radiation in the Atmosphere*, Harvard Meteorol. Stud. 6, Harvard Univ. Press, Boston, 1942.

14 Ramanathan, V., R. J. Cicerone, H. B. Singh, and J. T. Kiehl, "Trace gas trends and their potential role in climate change," *J. Geophys. Res.*, **90**, 5,547–66, 1985.

15 Adapted from A. Kylling, *Radiation Transport in Cloudy and Aerosol Loaded Atmospheres*, Ph.D. Thesis, University of Alaska, Fairbanks, 1992.

16 See Lacis, A. A. and V. Oinas, "A description of the correlated-k distribution method for modeling nongray gaseous absorption, thermal emission, and multiple scattering in vertically inhomogeneous atmosphere," *J. Geophys. Res.*, **96**, 9,027–63, 1991.

17 It is easily shown that the full width at half-maximum is $1.386\Delta z$.

18 Zender, C. S. and J. T. Kiehl, "Radiative sensitivities of tropical anvils to small ice crystals," *J. Geophys. Res.*, **99**, 25,869–80, 1994.

19 Briegleb, B. P., "Longwave band model for thermal radiation in climate studies," *J. Geophys. Res.*, **97**, 11,475–85, 1985.

20 An extensive comparison of results from 38 different research groups is described in Ellingson, R. G., J. Ellis, and S. Fels, "The intercomparison of radiation codes used in climate models: Longwave results," *J. Geophys. Res.*, **96**, 8,929–53, 1991. The issue in which this article appears contains many valuable intercomparison articles.

Chapter 12

The Role of Radiation in Climate

12.1 Introduction

As mentioned in the previous chapter, there are several radiatively active molecules – the so-called greenhouse gases – that strongly absorb and emit infrared radiation and thereby trap radiative energy that would otherwise escape to space. The global warming issue is concerned with the effects enhanced abundances of these trace gases and changes in aerosol loading (due in part to human activities) and cloud cover may have on the overall radiative energy balance of the Earth and hence on climate.

The bulk of the Earth's atmosphere (99% by mass) consists of molecular nitrogen and oxygen, in the form of radiatively inactive homonuclear, diatomic molecules. Trace amounts of polyatomic molecules are responsible for atmospheric absorption and emission of radiation in several hundred thousands of individual spectral lines arising from rotational and vibrational transitions. Water vapor, carbon dioxide, and ozone are the main absorbers (and emitters) contributing to warming and cooling of the atmosphere and underlying surface. These gases warm our planet by absorbing radiation emitted by the surface; without them the Earth would be some 33°C colder than at present and therefore uninhabitable. Hence, the so-called greenhouse effect is very important for life itself. Other trace gases make smaller contributions to warming/cooling of the atmosphere and surface. Some have natural origins, while others are partially (like methane) or wholly (like the chlorofluorocarbons) anthropogenic.

Figure 12.1 is a schematic diagram of the significant components of the Earth's energy balance. Of the incoming solar flux (342 $W \cdot m^{-2}$ averaged over the entire planet), 31% is reflected to space. The absorbed solar energy (235 $W \cdot m^{-2}$) is balanced by an equal amount radiated to space in the IR. Within the atmosphere, the land surface, and the ocean's mixed layer, the transformation of radiative energy into chemical,

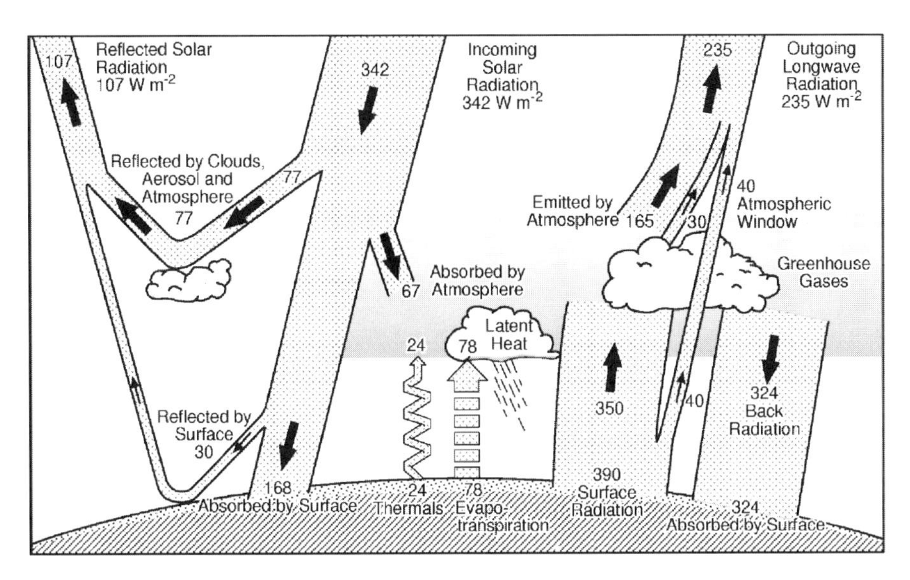

Figure 12.1 Earth's energy budget based on the study of Kiehl and Trenberth (see Endnote 16).

thermal, and kinetic energy drives the "engine" of weather and climate. Perturbations of this complex system can arise internally. Examples of internal forcing would be a change in atmospheric chemical composition or distribution of land masses. External forcing of the climate can arise from a change in the Sun's output and by changes in the orbital elements of the Earth's orbit.

The most important radiative interaction is the greenhouse effect, without which the Earth would be so cold that it would probably be in a state of permanent glaciation. This effect also explains the high surface temperature of Venus, and it may have played a key role in maintaining temperatures high enough in an early primitive atmosphere of Mars to sustain liquid water and possibly even primitive life.

The well-documented increase in CO_2 abundance, above what is believed to be the natural level existing in the preindustrial era, has been a matter of considerable concern. The reason for this concern is simply that the enhanced levels of CO_2 (and other already existing greenhouse gases or the release of new ones) absorb and trap terrestrial radiation that would otherwise escape to space. This causes an imbalance between the energy received and emitted by the planet. If the planet receives more energy from the Sun than it is able to emit to space, then by increasing its temperature it will increase the energy emitted (by the Stefan–Boltzmann law) until a new radiative equilibrium between the Sun and the Earth is established. Hence, this additional trapping of terrestrial radiation by the enhanced levels of greenhouse gases is expected to lead to a warming so as to make the net energy emitted by the planet equal to that received.

The amount of warming depends crucially on how the entire Earth climate system including the atmosphere, the land, the ocean, the cryosphere (snow and ice), and living things (the biosphere) responds to this warming. For example, could the

Earth partly compensate for this extra heat source by increasing its albedo? On the one hand, an increase in low clouds in response to warming (which is expected to enhance evaporation) may lead to increased reflection of solar energy and thus offset the warming. But more high clouds (*cirrus*) could, on the other hand, lead to additional trapping of terrestrial radiation and therefore an amplification of the warming.

In the previous two chapters we considered IR radiation and its interaction with atmospheric molecules. We now consider how both visible and IR radiative interactions affect the Earth's climate. We will not consider the manifold of climate variables (including precipitation, wind, ice and snow cover, etc.) but deal only with the temperature structure (the global warming problem). Furthermore, to isolate the basic radiative forcing, and the atmosphere's most elementary response, we employ an analytic model based on the two-stream solution to the radiative transfer equation. This model will be found to be remarkably flexible and very adaptable to problems of climate change. It will describe the static, globally averaged temperature structure of a planet's atmosphere.

The ocean serves as a heat storage medium, and in this simple picture[1] it acts only to delay the approach of the climate system to an equilibrium state. In more realistic regional models, the role of the ocean becomes one of mediating seasonal responses and of transporting heat poleward from the low-latitude regions of heating excess, to mention only a few effects. We do not mean to discount the role of the oceans in climate. Our point is simply that the ocean's true importance to climate can only be simulated using methods far beyond the level and scope of our treatment. This chapter thus offers only an introduction to a vast subject.

Rather than presenting a tutorial on the physics of climate and climate change,[2] this chapter focuses more narrowly on the radiative aspects of climate and climate change. Even in this limited context, it is necessary to simplify a complicated situation. A model that has been very useful for this purpose is the (globally averaged) one-dimensional radiative–convective approximation. Because much of the energy transfer is in the vertical direction, this model captures much of the relevant physics. It is a natural vehicle to use to study greenhouse warming in great detail. However, it cannot be used to study regional effects, land–sea differences, or transient climate change, to mention just a few limitations. We will begin this chapter with the classic radiative equilibrium problem and follow this with successive improvements to include solar absorption, convective transport, variable water content, a spectral window, clouds, aerosols, water vapor feedback, and specific greenhouse forcing from individual gas species.

12.2 Radiative Equilibrium with Zero Visible Opacity

This simplest useful solution of energy balance is based on the assumption that the atmosphere has negligible absorption for visible radiation. Although it can scatter visible light, these shortwave effects contribute only to the visible albedo. In actuality, the atmosphere absorbs about 20% of the solar energy available to the atmosphere (see Fig. 12.4). The majority of this energy is due to absorption by H_2O in the near-IR, of which ~70% is confined within 2 km of the surface. Thus our assumption that the initial solar energy deposition is at the surface is reasonable. The other neglected contribution is the small (~3%) fraction that is absorbed by stratospheric O_3 in the UV. The subsequent heating caused by this absorption is largely responsible for the stratospheric temperature inversion.

The surface is reflective in the visible and assumed to be black in the IR. Thus the surface is heated by incoming solar radiation and by downwelling IR radiation from the atmosphere. The atmosphere is heated by IR radiation, emitted by both the surface and by surrounding atmospheric layers. The absorption of outward-flowing IR energy will heat the atmosphere, setting up a diffusivelike temperature gradient throughout the optically thick region. At the upper "edge" where the slant opacity is of order unity, the atmosphere radiates to space with a globally averaged effective temperature T_e determined by the overall energy balance. For a rotating planet, we found in §1.4.5 that the effective temperature is given by

$$T_e = \left[\frac{S(1-\bar{\rho})}{4\sigma_B} \right]^{1/4}, \tag{12.1}$$

where S is the solar constant, $\bar{\rho}$ is the spherical albedo, and σ_B is the Stefan–Boltzmann constant. With no "blanketing" atmosphere to trap radiation emitted by the surface, the effective temperature is equal to the surface temperature. For the Earth, the effective temperature is $T_e = 255$ K.

The gray approximation will be used extensively in this chapter. We will solve the infrared transfer equation in a gray, perfectly absorbing atmosphere in LTE. The governing equations for this case have been previously derived in §5.3.1. As discussed above, we will impose the condition of radiative equilibrium. The boundary conditions are: (1) a black surface of temperature T_s underlying a slab atmosphere of IR optical depth τ^* and (2) zero incoming IR radiation from space. For a thermal source function the gray approximation implies

$$S(\tau) \equiv B(\tau) = \int d\nu B_\nu = \frac{\sigma_B T^4(\tau)}{\pi}. \tag{12.2}$$

If we set the single-scattering albedo $a = 0$, the radiative transfer equation for the frequency-integrated intensity becomes

$$u \frac{dI(\tau, u)}{d\tau} = I(\tau, u) - B(\tau) \tag{12.3}$$

and the corresponding half-range equations are

$$\mu \frac{dI^+(\tau, \mu)}{d\tau} = I^+(\tau, \mu) - B(\tau), \qquad -\mu \frac{dI^-(\tau, \mu)}{d\tau} = I^-(\tau, \mu) - B(\tau). \tag{12.4}$$

Adding and subtracting these equations, we obtain

$$\mu \frac{d[I^+(\tau,\mu) - I^-(\tau,\mu)]}{d\tau} = I^+(\tau,\mu) + I^-(\tau,\mu) - 2B(\tau), \tag{12.5}$$

$$\mu \frac{d[I^+(\tau,\mu) + I^-(\tau,\mu)]}{d\tau} = I^+(\tau,\mu) - I^-(\tau,\mu). \tag{12.6}$$

We now impose radiative equilibrium by setting the net flux $F(\tau)$ equal to the outgoing emitted radiation, $\sigma_B T_e^4$, which is constant for all τ. Integration of Eq. 12.3 over solid angles yields $dF/d\tau = 4\pi(\bar{I} - B) = 0$. Thus we obtain the relationship $\bar{I} = B(\tau)$, or in words, *in radiative equilibrium the source function is equal to the mean intensity,*

$$B(\tau) = \frac{1}{2}\int_{-1}^{1} du\, I(\tau,u) = \frac{1}{2}\left[\int_0^1 d\mu\, I^+(\tau,\mu) + \int_0^1 d\mu\, I^-(\tau,\mu)\right]. \tag{12.7}$$

Thus, the gray radiative transfer equation becomes

$$u\frac{dI(\tau,u)}{d\tau} = I(\tau,u) - \frac{1}{2}\left[\int_0^1 d\mu\, I^+(\tau,\mu) + \int_0^1 d\mu\, I^-(\tau,\mu)\right], \tag{12.8}$$

which in radiative equilibrium must be solved subject to the constraint

$$F = 2\pi\int_0^1 d\mu\mu[I^+(\tau,\mu) - I^-(\tau,\mu)] = 2\pi\int_0^1 d\mu\mu\frac{\sigma_B T_e^4}{\pi} = \text{constant}. \tag{12.9}$$

The two-stream approximation will play a prominent role in the results of this chapter, and so we will review and extend some of the basic ideas discussed in Chapter 7. As previously discussed, this method replaces the angular dependence of the azimuthally averaged intensity $I(\tau,\mu)$ with a constant, that is, $I(\tau,\mu = \bar{\mu})$, where $\bar{\mu}$ is the average inclination of the rays in each hemisphere. Its value will remain undetermined for now. If we use the notation $I^{\pm}(\tau) \equiv I^{\pm}(\tau,\bar{\mu})$, the two-stream equations follow immediately from the equations for the half-range intensities (Eqs. 12.5 and 12.6)

$$\bar{\mu}\frac{d(I^+ - I^-)}{d\tau} = I^+ + I^- - 2B(\tau), \tag{12.10}$$

$$\bar{\mu}\frac{d(I^+ + I^-)}{d\tau} = I^+ - I^-, \tag{12.11}$$

which must be solved subject to the constraint (Eq. 12.9)

$$F = 2\pi\bar{\mu}[I^+(\tau) - I^-(\tau)] = 2\bar{\mu}\sigma_B T_e^4. \tag{12.12}$$

Inserting $B(\tau) = (1/2)(I^+ + I^-)$ into the left-hand side of Eq. 12.11, and expressing the right-hand side in terms of the net flux $F = 2\bar{\mu}\sigma_B T_e^4 = 2\pi\bar{\mu}(I^+ - I^-)$, we obtain a differential equation for the source function,

$$\frac{dB(\tau)}{d\tau} = \frac{\sigma_B T_e^4}{2\pi\bar{\mu}} = \text{constant}. \tag{12.13}$$

We may integrate the above equation immediately, to obtain

$$B(\tau) = \frac{\sigma_B T_e^4}{2\pi\bar{\mu}}\tau + C, \tag{12.14}$$

where C is a constant of integration. This constant may be evaluated by solving for the hemispherical fluxes at the bottom of the medium ($\tau = \tau^*$). First, we note that the upward flux at the surface is simply the surface thermal emission, $F^+(\tau^*) = 2\pi\bar{\mu}I^+(\tau^*) = 2\bar{\mu}\sigma_B T_s^4$. Second, since the net flux at this point (as it is at all points) is $2\bar{\mu}\sigma_B T_e^4$, then $2\pi\bar{\mu}[I^+(\tau^*) - I^-(\tau^*)] = 2\bar{\mu}\sigma_B T_e^4$. Solving for the downward flux at the surface, we have

$$F^-(\tau^*) = 2\pi\bar{\mu}I^-(\tau^*) = 2\bar{\mu}\sigma_B\left(T_s^4 - T_e^4\right). \tag{12.15}$$

Continuing with the determination of C, we may now evaluate the source function at the bottom of the medium:

$$B(\tau^*) = \frac{\sigma_B T_e^4}{2\pi\bar{\mu}}\tau^* + C = \frac{1}{2}[I^+(\tau^*) + I^-(\tau^*)] = \frac{\left(2\sigma_B T_s^4 - \sigma_B T_e^4\right)}{2\pi}. \tag{12.16}$$

Solving for C, we find $C = (1/2\pi)[2\sigma_B T_s^4 - \sigma_B T_e^4(1 + \tau^*/\bar{\mu})]$. Thus the source function is

$$B(\tau) = \frac{1}{2\pi}\left\{2\sigma_B T_s^4 - \sigma_B T_e^4[1 + (\tau^* - \tau)/\bar{\mu}]\right\}. \tag{12.17}$$

A much simpler result is found by expressing T_s in terms of T_e and τ^*. Evaluating the source function (Eq. 12.7) at $\tau = 0$, we find $B(0) = (1/2)I^+(0)$, since $I^-(0) = 0$ from boundary condition (2). Since $F(0) = 2\pi\bar{\mu}I^+(0) = 2\bar{\mu}\sigma_B T_e^4$, we find that $B(0) = \sigma_B T_e^4/2\pi$. Setting $\tau = 0$ in Eq. 12.17, equating the two results for $B(0)$, and solving for T_s we find

$$T_s = T_e(1 + \tau^*/2\bar{\mu})^{1/4} \equiv T_e\mathcal{G}^{1/4}. \tag{12.18}$$

The quantity $\mathcal{G} = (T_s/T_e)^4$ is called the (surface) *greenhouse factor*. It may be thought of as a source function "trapping factor," which depends linearly on the infrared optical depth τ^*. An additional measure of the trapping effect is the *greenhouse effect*, defined as the difference of the surface and TOA fluxes:

$$G \equiv 2\bar{\mu}\sigma_B\left(T_s^4 - T_e^4\right) = 2\bar{\mu}\sigma_B T_e^4[\mathcal{G} - 1]. \tag{12.19}$$

From Eq. 12.15, G is also the downward flux at the surface (the "backwarming"). Eliminating T_s from the source function, Eq. 12.17; we find[3] after some algebra

$$B(\tau) = \frac{\sigma_B T_e^4}{2\pi}\left[1 + \frac{\tau}{\bar{\mu}}\right]. \tag{12.20}$$

We must now consider the relationship between the thermal source function $B(\tau)$ and the atmospheric temperature $T(\tau)$. In accordance with Eq. 12.2 we set $B(\tau) = \sigma_B T^4(\tau)/\pi$, which is consistent with the result obtained by evaluating Eq. 12.20 at $\tau = \bar{\mu}$ yielding $B(\tau = \bar{\mu}) = \sigma_B T_e^4/\pi$. Using this relationship and Eq. 12.20, we find the

Radiative equilibrium expression for the atmospheric temperature:

$$T_{re}(\tau) = T_e\left(\frac{1}{2} + \frac{\tau}{2\bar{\mu}}\right)^{1/4} \equiv T_e\mathcal{G}^{1/4}(\tau). \tag{12.21}$$

Here $\mathcal{G}(\tau)$ is the (dimensionless) greenhouse factor for an arbitrary level τ. This shows that the temperature increases monotonically downward from an outer skin temperature $T_{re}(0) = T_e/(2)^{1/4}$ to a lower boundary temperature $T_{re}(\tau^*)$ whose value is less than T_s. We now discuss this interesting discontinuity between the air immediately above the surface and the surface itself.

The relative temperature change over the interface between the air and the surface is

$$\Delta T/T_s = [T_s - T(\tau^*)]/T_s = 1 - \left[\frac{1 + \tau^*/\bar{\mu}}{2 + \tau^*/\bar{\mu}}\right]^{1/4}. \tag{12.22}$$

The value of this relative jump is $\sim$16% for optically thin media, and it decreases to zero as $\tau^* \to \infty$. This peculiarity arises from the fact that the surface is heated by both the Sun and the atmosphere, whereas the overlying atmospheric layer is heated only by the neighboring regions. Conservation of energy across this interface leads to a surface that is hotter than the immediately overlying atmosphere. This artificial condition implies ΔT varies from about 40.6 K at $\tau^*/\bar{\mu} = 0$ to 21.0 K at $\tau^*/\bar{\mu} = 2$ and 15.0 K at $\tau^*/\bar{\mu} = 4$.

Equation 12.21 ignores the possibility of dynamical heat transport across the interface. In the real world, convection tends to erase the discontinuity extremely quickly but not necessarily eliminate it.[4] Convective transport will be described in more detail in §12.4.

We shall later (§12.4) have occasion to evaluate the flux in the two-stream, gray approximation where the temperature is an arbitrary function of τ. If we approximate the exponential integrals according to

$$E_n(\tau) = \int_0^1 d\mu \mu^{n-2} e^{-\tau/\mu} \approx \bar{\mu}^{n-2} e^{-\tau/\bar{\mu}}, \tag{12.23}$$

then Eqs. 5.59 and 5.60 become

$$F^+(\tau) = 2\bar{\mu}\sigma_B T_s^4 e^{-(\tau^*-\tau)/\bar{\mu}} + 2\int_{\tau}^{\tau^*} d\tau' \sigma_B T^4(\tau')\, e^{-(\tau'-\tau)/\bar{\mu}},$$
$$F^-(\tau) = 2\int_0^{\tau} d\tau' \sigma_B T^4(\tau')\, e^{-(\tau-\tau')/\bar{\mu}}. \tag{12.24}$$

These integrals may be evaluated analytically in terms of the radiative equilibrium solutions. Thus, by inserting Eq. 12.18 for T_s^4 and Eq. 12.21 for $T^4(\tau)$ into Eqs. 12.24 and carrying out the integrations, we find

$$F_{re}^+(\tau) = 2\bar{\mu}\sigma_B T_e^4(1 + \tau/2\bar{\mu}), \qquad F_{re}^-(\tau) = \sigma_B T_e^4 \tau. \tag{12.25}$$

These are consistent with Eqs. 12.15 and 12.18 and therefore are valuable checks on the solutions.

The distribution of temperature with height z is obtained from the relationship of τ with z. For the Earth's troposphere, the dominant IR absorber is water vapor, which has a scale height $H_a \approx 2$ km (see Fig. 1.6). From Eq. 1.9 and the perfect gas law (Eq. 1.2), the optical depth at height z is given approximately by (see Eq. 1.8)

$$\tau(z) = \langle \alpha_m \rangle \int_z^{\infty} dz' \rho_0 e^{-z'/H_a} = \tau^* e^{-z/H_a}, \tag{12.26}$$

where $\tau^* = \langle \alpha_m \rangle \rho_0 H_a$ is the effective (gray) optical depth, $\langle \alpha_m \rangle$ is the spectrally averaged mass absorption coefficient, and ρ_0 is the water vapor density at the surface.

To evaluate $\langle \alpha_m \rangle$ we must average over the spectrum, weighting the absorption coefficient with the spectrally dependent radiation field. Unfortunately, there is no unique weighting scheme suitable for all situations. Two of the more commonly used schemes are the *Planck mean*,

$$\langle \alpha_m \rangle_P \equiv \int_{IR} d\nu \alpha_m(\nu) B_\nu(T)/\sigma_B T^4,$$

and the *Rosseland mean*,

$$\langle \alpha_m \rangle_R^{-1} \equiv \frac{\int_{IR} d\nu \alpha_m^{-1}(\nu)(dB_\nu/dT)}{\int_{IR} d\nu (dB_\nu/dT)}.$$

The Planck mean is applicable to optically thin problems where the transmission is governed by the strong absorption bands. The Rosseland mean is appropriate for optically thick situations where the radiative transfer takes place not in the opaque spectral regions but in the nearly transparent (window) regions. The value of $\langle \alpha_m \rangle$

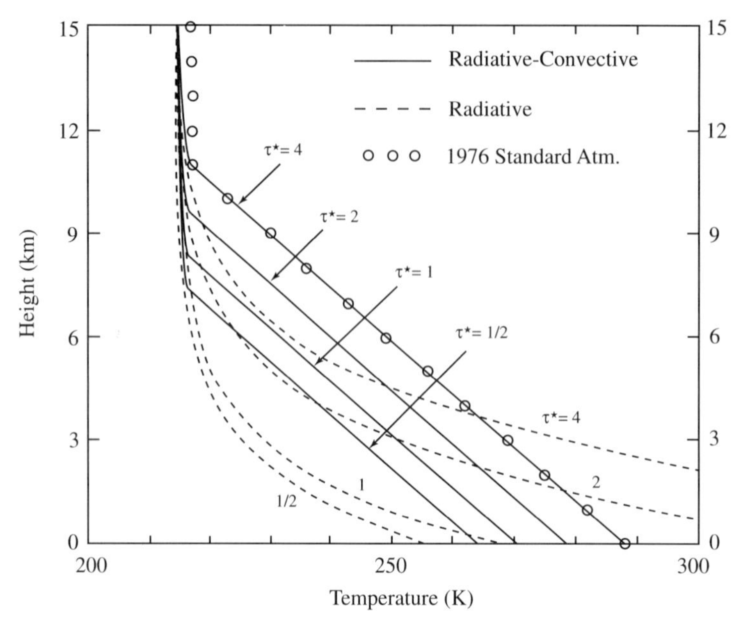

Figure 12.2 Pure-radiative (dashed lines) and radiative–convective equilibrium (solid lines) temperature profiles for four different optical depths, and for $\bar{p} = 0.30$. Open circles: The 1976 Standard Atmosphere. The radiative–convective adjustment procedure is discussed in §12.4.

produced by both these definitions depends upon height, through the dependence of the line strengths on temperature and pressure. However, the gray approximation is rather crude, and the extra labor involved in computing $\langle \alpha_m \rangle$ is seldom justified. Since we are interested in physical insight, rather than numerical accuracy, it is useful to think of τ^* or $\langle \alpha_m \rangle$ as adjustable parameters in the model, whose values are derived from fitting the model to the observed temperature profiles.

The distribution of temperature from Eq. 12.21 is shown in Fig. 12.2 (as the dashed lines) for several values of τ^*. Equation 12.21 requires that $\tau^*/2\bar{\mu} = 0.63$ to obtain a surface temperature of 288 K (assuming $T_e = 255$ K). This is the current globally averaged surface temperature. This radiative equilibrium solution was originally derived by K. Schwarszchild in 1906, and later by R. Emden in 1907, and by T. Gold and W. J. Humphreys in 1909. It became generally accepted as the explanation for the declining tropospheric temperature and the cold isothermal stratosphere.[5] We now know that the lower stratosphere is only approximately isothermal. As shown in Fig. 1.3 a temperature inversion occurs in the upper stratosphere, as a result of absorption of solar UV radiation by ozone. However, the skin temperature derived from Eq. 12.21, $T(0) = 255 \times 2^{-1/4}$ K $= 214$ K, is a good estimate for the globally averaged minimum (tropopause) temperature. In other words, the existence of a tropopause temperature minimum can be understood from purely radiative considerations. However, in the real atmosphere, the lapse rate in the troposphere is controlled largely by dynamical transport effects.

Extension of the principle of radiative balance to explain latitudinal variations yields temperatures that are too high in the tropics and too low in the polar regions. Even though the annual variation of the incoming solar flux (the *solar insolation*) is defined precisely by geometry and celestial mechanics (see Problems 1.2 and 9.1), the meridional transport of heat by air motions and ocean currents is an important heating/cooling term in the energy balance equation. Consequently, as Figure 12.8 shows, there is an (annually averaged) radiative imbalance at most latitudes, with an excess shortwave heating in the tropics and an excess longwave cooling at the higher latitudes.[6] Simple diffusive models of this horizontal transport played important roles in early studies of phenomena such as the effect of solar variations on the snow and ice cover in the polar regions.[7] This subject is discussed again in §12.7. We will return to the study of radiation by considering how shortwave absorption affects the mean radiative balance.

12.3 Radiative Equilibrium with Finite Visible Opacity

Any realistic atmosphere absorbs radiation at both IR and visible wavelengths. In addition some planets (such as Venus) have optically thick atmospheres. For our purposes such media may be regarded as semi-infinite, implying that there is no need to consider the lower boundary surface. We assume that the heating is governed by a diurnal average over all incident solar angles, so that the average cosine is $\bar{\mu}_0$. The average solar flux absorbed by a rotating planet over a diurnal period[8] is defined to be $\bar{\mu}_0 F_a^s$. Other than allowing for a spherical albedo, we ignore scattering as before. The total (visible plus IR) net radiative flux is written

$$F_{\text{tot}}(z) = -F_V(z) + F_{\text{IR}}(z) = -\bar{\mu}_0 F_a^s e^{-\tau_V/\bar{\mu}_0} + F_{\text{IR}}(z), \tag{12.27}$$

where τ_V is the frequency-averaged visible optical depth. Radiative equilibrium requires $F_{\text{tot}} = \text{constant}$, or $dF_{\text{tot}}/dz = 0$. Using the chain rule of differentiation, we have

$$\frac{dF_{\text{tot}}}{dz} = -\frac{dF_V}{dz} + \frac{dF_{\text{IR}}}{dz} = -\frac{d\left(\bar{\mu}_0 F_a^s e^{-\tau_V/\bar{\mu}_0}\right)}{d\tau_V}\frac{d\tau_V}{dz} + \frac{dF_{\text{IR}}}{dz}$$
$$= -k_V F_a^s e^{-\tau_V/\bar{\mu}_0} + \frac{dF_{\text{IR}}}{dz} = 0,$$

where k_V is the (gray) visible absorption coefficient, so that $d\tau_V = -k_V dz$. This equation provides us with the desired relationship between the infrared flux derivative and the solar heating rate $\mathcal{H}_V = -dF_V/dz = k_V F_a^s e^{-\tau_V/\bar{\mu}_0}$.

An expression relating the derivative of the infrared flux to the source function is obtained from the generalized *Gershun's Law* (see Eq. 5.74) applied to a gray, perfectly absorbing medium,

$$B(\tau) = \bar{I} - \frac{1}{4\pi}\frac{\partial F_{\text{IR}}}{\partial \tau} = \bar{I}(\tau) + \frac{F_a^s}{4\pi n} e^{-\tau/n\bar{\mu}_0},$$

where $d\tau \equiv -k_{\rm IR}dz$, and n denotes the ratio of IR and visible absorption coefficients, $n = k_{\rm IR}/k_{\rm V}$. The radiative transfer equation becomes

$$u\frac{dI(\tau,u)}{d\tau} = I(\tau,u) - B(\tau) = I(\tau,u) - \frac{1}{2}\int_{-1}^{+1} du\, I(\tau,u) - \frac{F^{\rm s}_{\rm a}}{4\pi n}e^{-\tau/n\bar{\mu}_0}, \tag{12.28}$$

which is identical to Eq. 12.8 except that it contains an "imbedded source." This equation is mathematically identical to that of *Prototype Problem 3*, provided we set $a = 1$, $p = 1$, and make the correspondences $F^{\rm s} \to F^{\rm s}_{\rm a}/n$ and $\mu_0 \to n\bar{\mu}_0$. The two-stream solution for the source function for a semi-infinite atmosphere (see Example 7.9, Eq. 7.100) was found to be

$$\begin{aligned} S(\tau) &= \frac{F^{\rm s}}{4\pi}\left[(1-m^2)e^{-\tau/\mu_0} + m(1+m)\right] \\ &= \frac{F^{\rm s}}{4\pi n}\left[(1-\gamma^2)e^{-\tau/\gamma\bar{\mu}} + \gamma(1+\gamma)\right], \end{aligned} \tag{12.29}$$

where $m \equiv \mu_0/\bar{\mu}$ and $\gamma \equiv n\bar{\mu}_0/\bar{\mu}$. Requiring that the absorbed solar energy be linked directly to (i.e., it drives) the effective temperature, we set $\mu_0 F^{\rm s} = 2\bar{\mu}\sigma_{\rm B}T_{\rm e}^4$, which yields $S(\tau) = \sigma_{\rm B}T_{\rm e}^4/\pi$ for $\gamma = 1$ (isothermal case). Thus, setting $S(\tau) = B(\tau) = \sigma_{\rm B}T^4(\tau)/\pi$ is consistent with Eq. 12.29 for $\gamma = 1$. Using the above correspondences, and $S(\tau) = \sigma_{\rm B}T^4(\tau)/\pi$, we find that the greenhouse factor may be expressed in terms of the parameter γ as

$$\mathcal{G}(\tau) \equiv T^4(\tau)/T_{\rm e}^4 = \frac{1}{2\gamma}\left[(1-\gamma^2)e^{-\tau/\gamma\bar{\mu}} + \gamma(1+\gamma)\right]. \tag{12.30}$$

In this simple model, γ is the ratio of the slant opacity in the IR ($\tau_{\rm IR}/\bar{\mu}$) to that in the visible ($\tau_{\rm V}/\bar{\mu}_0$). The temperature profiles versus optical depth for several different values of γ are shown in Fig. 12.3. There are three interesting cases:

1. $\gamma \gg 1$, or equivalently, $k_{\rm IR} \gg k_{\rm V}$ is the strong greenhouse limit where the solar radiation penetrates deeply in the atmosphere, and the resulting IR radiation is trapped. The atmosphere acts like a one-way valve allowing the incoming flux ($\bar{\mu}_0 F^{\rm s}_{\rm a}$) to enter easily but the resulting IR flux to escape with difficulty. In the deep atmosphere, the greenhouse enhancement "saturates" to the constant value $\mathcal{G}(\tau^* \to \infty) = (1+\gamma)/2$. Thus the asymptotic temperature is $T(\tau^* \to \infty) = T_{\rm e}(1/2 + n\bar{\mu}_0/2\bar{\mu})^{1/4} \approx T_{\rm e}(k_{\rm IR}\bar{\mu}_0/k_{\rm V})^{1/4}$.

   As we might expect, the two-stream approximation to the full solution of the gray radiative transfer problem is quite accurate, as one might expect for this optically thick, gray situation. The resemblance of the $\gamma \gg 1$ solutions to the temperature structure of Venus was noted in the 1960s, when observations of that planet in the microwave spectrum indicated a surface temperature near 800 K. A pure radiative-equilibrium solution is found to be a good approximation for Venus's lower atmosphere.[9] However, the modest

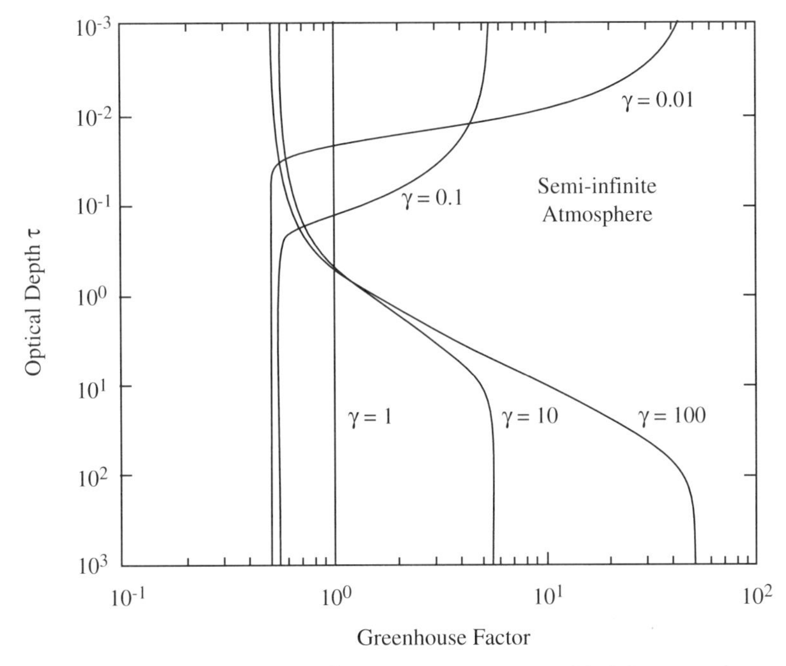

Figure 12.3 Greenhouse factor $\mathcal{G}$ for a homogeneous, semi-infinite atmosphere versus optical depth for five different values of γ, the ratio of infrared slant optical depth to the visible slant optical depth.

greenhouse trapping experienced on Earth and Mars is not well described by the above formula because of the importance of the surface in the radiative transfer, and also because of our neglect of convective heat transport (the latter process is addressed in the next section and in Problem 12.6).

2. $\gamma = 1$ represents an isothermal ($T = T_e$) situation where the solar heating exactly balances the IR escape. The $n = 1$ case also describes conservative scattering from a homogeneous, semi-infinite atmosphere for which S. Chandrasekhar[10] found an exact solution in 1949. Comparing the exact value of the ratio of source functions $S(\tau \to \infty)/S(\tau = 0) = \bar{\mu}_0\sqrt{3}$ with the two-stream result $\bar{\mu}_0/\bar{\mu}$, one immediately infers that $\bar{\mu} = 1/\sqrt{3}$ is the best value to use in optically thick situations.
3. $\gamma \ll 1$, or $k_{\rm IR} \ll k_{\rm V}$, represents the *antigreenhouse* case. This is relevant to numerous phenomena in the solar system:
   - An inverted temperature structure characterizes Earth's upper stratosphere, where high middle-UV opacity due to ozone absorption gives rise to a temperature inversion.
   - The antigreenhouse case also describes the radiative equilibrium temperature distribution for the *nuclear-winter* scenario. This (hopefully!) hypothetical situation corresponds to a reduced surface temperature caused by a perturbed stratosphere loaded with aerosols (such as soot), which absorb strongly in the visible.

- The antigreenhouse scenario may also have occurred naturally in Earth's past history. Worldwide cooling causing mass extinction is believed to have occurred 65 million years ago from the injection of massive quantities of dust, as a result of an impact of one or more 10-km-diameter meteoroids.
- Stratospheric aerosols (with optical depth estimated to range up to 10) resulting from the gigantic Mt. Toba volcanic eruption some 70,000 years ago may have been responsible for a subsequent cooling of Earth's climate for a period of 200 years.[11]
- The antigreenhouse solution is also applicable to the dusty atmospheres of Mars (compare Fig. 12.3 with Fig. 1.7) and of Titan, the largest satellite of the planet Saturn.

12.4 Radiative–Convective Equilibrium

In 1913, R. Emden[12] first pointed out that radiative-equilibrium solutions (such as those discussed in the previous section) are convectively unstable in sufficiently optically thick portions of a planetary atmosphere. When solar energy is absorbed deep within the atmosphere, the radiative-equilibrium lapse rate, $\partial T/\partial z$, can be large and negative. As discussed in Chapter 1, if this gradient exceeds a certain limit, the atmosphere becomes convectively unstable and spontaneously adjusts to a new *adiabatic lapse rate*. For a noncondensing atmosphere (no phase changes) the "dry" adiabatic lapse rate applies, $\Gamma_\mathrm{d} = -g/c_p$. For Earth's atmosphere Γ_d is about $-9.8\ \mathrm{K\cdot km^{-1}}$.

For a saturated atmosphere, condensation or deposition (evaporation or sublimation) can occur in the rising (sinking) air parcel, and latent heat exchange must be considered. The release of the latent heat of condensation or deposition partially offsets the expansional cooling in ascending air parcels. In descending air parcels, the evaporation of water droplets or sublimation of ice crystals extracts heat energy from the air, thus partially offsetting the compressional heating. The resulting *moist adiabatic lapse rate* Γ_m usually lies in the range $-9.8\ \mathrm{K\cdot km^{-1}} < \Gamma_\mathrm{m} < -3\ \mathrm{K\cdot km^{-1}}$.

The global mean tropospheric temperature gradient is about $\Gamma_\mathrm{a} = -6.5\ \mathrm{K\cdot km^{-1}}$. We will refer to this as the *environmental lapse rate*. A first-principles description of the convective transport of heat and constituent concentrations has challenged modelers for many years. A proper treatment involves a complex set of processes.[13] An artifice that yields satisfactory results is to assume that in regions of instability, resulting from imposing radiative equilibrium, the radiative-equilibrium temperature gradient is replaced by the environmental temperature gradient, Γ_a.

It is easy to identify the optical depth at which this occurs (Problem 12.4). We denote this optical depth as τ_c, and the corresponding height, z_c. Unfortunately, if we simply force the temperature gradient to be equal to the adiabatic value below the height of instability, this temperature distribution is unphysical. This is because it violates energy conservation. To see why, let us compare the radiative-equilibrium flux $F^+_\mathrm{re}(\tau_\mathrm{c}) = \sigma_\mathrm{B} T_\mathrm{e}^4(1 + \tau_\mathrm{c}/2\bar{\mu})$ immediately above z_c to the convective equilibrium flux F^+_conv

immediately below z_c. These must be equal, because at this height radiation is the only mode of heat transport. (Actually only the net fluxes need to be equal to conserve energy; however, the downward flux is unchanged, and so therefore the upward flux must also be unchanged.) The flux from the constant-gradient region below z_c is given by

$$F^+_{\text{conv}}(\tau) = 2\bar{\mu}\sigma_B T_s^4 e^{-(\tau^*-\tau)/\bar{\mu}} + 2\int_{\tau}^{\tau^*} d\tau' e^{-(\tau'-\tau)/\bar{\mu}}\sigma_B[T_s + \Gamma_a z(\tau')]^4. \quad (12.31)$$

Here $z(\tau)$ is expressed as a function of τ through the relationship $z(\tau) = H_a \ln(\tau^*/\tau)$ (see Eq. 12.26). Since the temperature $T_s + \Gamma_a z$ is less than T_{re} for all $z < z_c$, then $F^+_{\text{conv}}(\tau) < F^+_{re}$. This is only possible if there is a delta-function energy source located at z_c.

This problem is solved by recognizing that convective adjustment sets only the *gradient* of the temperature. Thus we are free to vary the actual values of the temperature. If we increase the temperature everywhere in the convective region, then clearly this will increase the flux emanating from that region. We can therefore match the radiative-equilibrium flux value, but since the intersection point of the radiative and convective curves shift upward, we need to also recompute the radiative-equilibrium flux. The procedure is straightforward but requires repeated evaluations of Eq. 12.31 by numerical integration. The net result is that the transition region z_c (the *tropopause*) is several kilometers above the "initial" point of instability. The flux across the boundary is continuous, but the slope of the temperature itself is discontinuous. Several examples of the adjusted radiative–convective equilibrium temperature distributions are shown in Fig. 12.2. Note that the surface temperature and tropopause height of the $\tau^* = 4.05$ model agree well with the 1976 global mean temperature model. The favorable comparison of the "tuned" radiative–convective temperature profile with the standard model is of course not surprising.

Net fluxes for the $\tau^* = 4.05$ case are shown in Fig. 12.4, both for incident visible radiation and for the outgoing IR radiation. These results are compared with those computed from a state-of-the-art nongray radiative–convective model.[14] This figure shows that some near-IR solar absorption by water vapor occurs in the atmosphere. The differences between the approximate and accurate IR flux computations is attributed to the neglect of this absorption. Note also that in the simple model the net IR flux becomes equal to the incoming solar flux at the tropopause, above which there is zero flux divergence (and zero solar heating). In the realistic case, the two values approach one another more gradually with height, but together they do not approach the asymptotic value (240 $\text{W}\cdot\text{m}^{-2}$) until above the stratopause, that is, above the peak of the ozone heating region.

The variation of T_s with optical depth computed from the simple two-stream radiative–convective model is shown in Fig. 12.5, along with the computed variation of tropopause height. The results of a change in albedo by $\pm 5\%$ are also shown in Fig. 12.5. Although derived from simple considerations, this figure embodies several important properties of the greenhouse effect and its possible modifications by changes

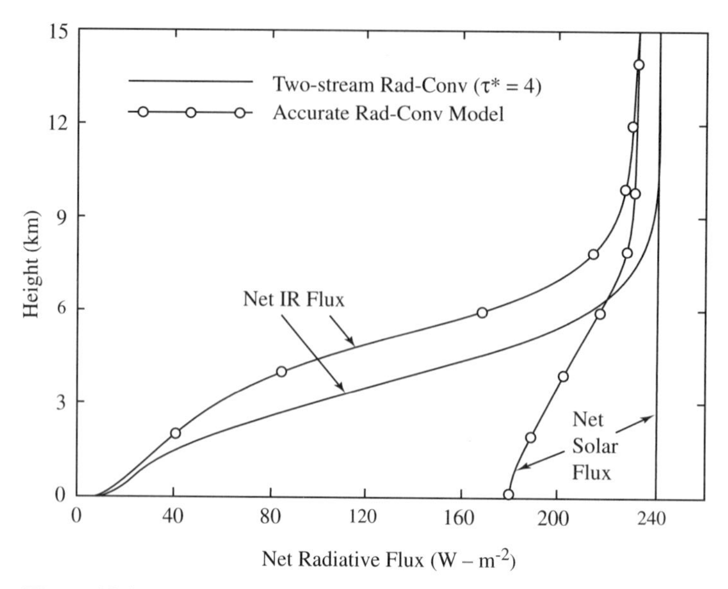

Figure 12.4 Solid curves: Net fluxes for the $\tau^* = 4.05$ case both for incident solar radiation and for the outgoing IR radiation. Open circles: Results from a nongray state-of-the-art radiative–convective model (reference given in Endnote 15). The curve for the net solar flux in the accurate model shows that ~20% of the solar flux is absorbed in the troposphere.

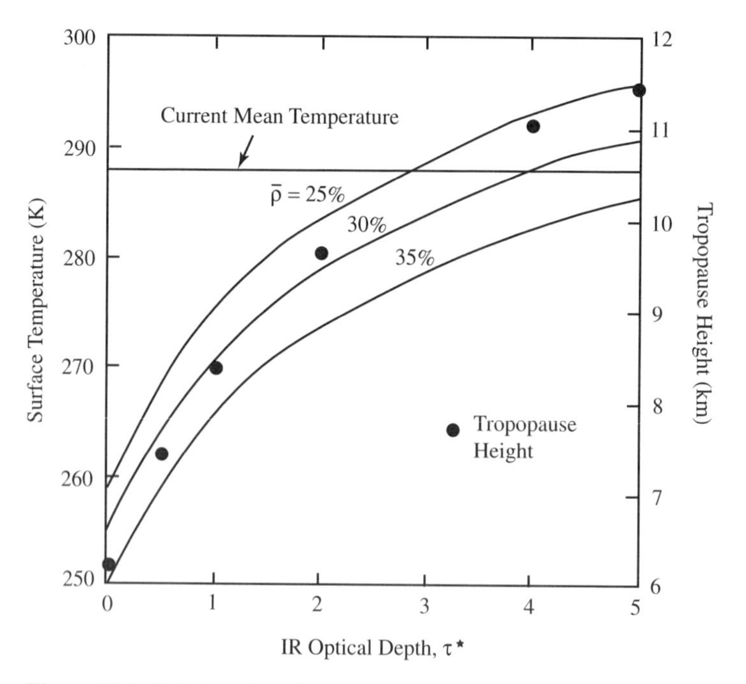

Figure 12.5 Variation of surface temperature (solid curves) and tropopause height (filled circles) with optical depth predicted by the two-stream radiative–convective model. The three curves are for three different values of the spherical albedo.

in IR optical depth (through changes in infrared active species, such as CO_2 and CH_4) and albedo (through changes in the clouds, aerosols, or snow/ice cover).

12.5 The Concept of the Emission Height

Our model requires a numerical integration to match fluxes at the tropopause. This limits its conceptual usefulness. We will adopt a simple analytic approximation for the surface temperature that fits the above results quite accurately. We introduce the concept of the *emission height* as the level where the slab medium experiences its maximum radiative cooling. The cooling has its maximum at an optical depth τ_e, where the temperature is approximately T_e. Its value in the radiative-equilibrium problem previously discussed is easily found from Eq. 12.21 by setting $T_{re}(\tau_e) = T_e$. The result is $\tau_e = \bar{\mu}$, which is physically reasonable, since a medium effectively cools where its mean slant optical depth $\tau/\bar{\mu}$ is unity. The optical depth τ_e will take on different values for other temperature distributions. The geometric height is easily found from Eq. 12.26 to be $z_e = H_a \ln(\tau^*/\tau_e)$. We set the greenhouse increase in temperature (over the effective temperature) equal to the product of the lapse rate Γ_a and z_e, or

$$T_s \cong T_e + |\Gamma_a| z_e = T_e + |\Gamma_a| H_a \ln(\tau^*/\tau_e). \tag{12.32}$$

If $\bar{\mu} = 0.5$, then the numerical results for $T_s(\tau^*)$ shown in Figs. 12.2 and 12.5 are in excellent agreement with Eq. 12.32 provided that $\tau^* > 0.4$ and τ_e depends upon τ^* according to the following numerical fit of the curve of T_s versus τ^*:

$$\tau_e^{-1} = 3.125 + 0.235/\tau^{*2}. \tag{12.33}$$

The asymptotic value for τ_e as $\tau^* \to \infty$ is 0.32. As previously mentioned, this value is slightly different from that derived from the radiative-equilibrium case ($\tau_e = \bar{\mu} = 0.5$). For a convective atmosphere, our clear-air model has the parameters $\tau^* = 4$, $H_a = 2$ km, and $z_e = 5$ km.

Using Eq. 12.32, we can tune the optical depth τ^* to simulate the greenhouse effect in a realistic atmosphere. We write τ^* as a sum of the opacities of the nonwater greenhouse gases (τ_n^*) and a linear function of the liquid water path, bw, where b is an empirical constant and $w = \rho_0 H_a$ is the *precipitable water* [$g \cdot cm^{-2}$]. To determine the numerical values of τ_n^* and b in our model, we impose two constraints: (1) The clear-sky greenhouse effect $G \equiv \sigma_B T_s^4 - \sigma_B T_e^4$ (see Eq. 12.19) for $w = 0$ should be equal to 50 $W \cdot m^{-2}$, and (2) the clear-sky greenhouse factor $\mathcal{G} = T_s^4/T_e^4$ defined by Eq. 12.32 is consistent with observed values. Constraint (1) comes from detailed modeling studies[15] and constraint (2) comes from data from the Earth Radiation Budget Experiment (ERBE) on the *Nimbus 7* spacecraft.[16] These two constraints yield the following result for the clear-sky IR optical depth:

$$\tau^* = \tau_n^* + bw, \tag{12.34}$$

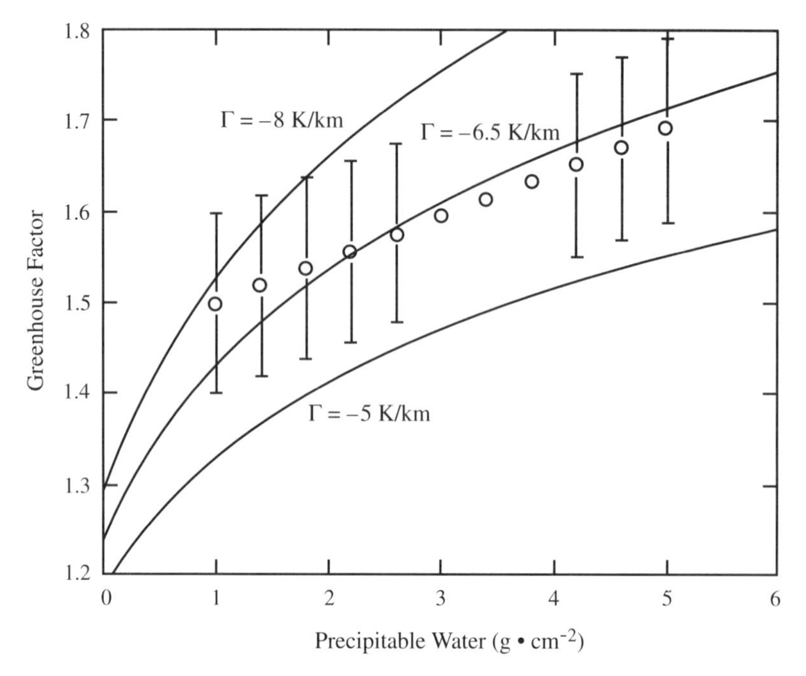

Figure 12.6 Solid curves: Greenhouse factor $\mathcal{G}$ as a function of precipitable water for three different values of the environmental lapse rate Γ_a. Open circles: $\mathcal{G}$ values inferred from satellite data are also shown for comparison. The error bars (omitted on several of the points for clarity of presentation) indicate the spread of the data points. To compare with data it was necessary to account for the observed change of F_{TOA} (or equivalently T_e) with sea-surface temperature. The empirical relationship (which applies to this figure only) was found to be $T_e \approx 153 + 70 \times \mathcal{G}$.

where $\tau_n^* = 0.788$ and $b = 1.1\ \text{cm}^2 \cdot \text{g}^{-1}$. For a global average value of $\bar{w} = 1.32\ \text{g} \cdot \text{cm}^{-2}$, the combination of CO_2, CH_4, N_2O, O_3, and the various CFCs contribute ~35% of the total clear-sky optical depth. We find $G = \sigma_B T_s^4(\tau_n^*) - \sigma_B T_e^4 = 53\ \text{W} \cdot \text{m}^{-2}$ by using $\tau_n^* = 0.788$, $\Gamma_a = -6.5\ \text{K} \cdot \text{km}^{-1}$, and $H_a = 2$ km in Eq. 12.32. The greenhouse factor, $\mathcal{G} = (T_s/T_e)^4 = (1 + |\Gamma_a| z_e/T_e)^4$, agrees reasonably well with ERBE results as shown in Fig. 12.6, provided the lapse rate is about $-6.5\ \text{K} \cdot \text{km}^{-1}$. These results will be used later in assessing the water-vapor feedback effect.

12.6 Effects of a Spectral Window

A gray model may be made more realistic by allowing for transparency in the 8–12 μm spectral window. We define the temperature-dependent *transparency factor* as the fraction of the blackbody radiation available across the window

$$\beta(T) = \int_{\tilde{\nu}_1}^{\tilde{\nu}_2} d\tilde{\nu} B_{\tilde{\nu}}(T)/\sigma_B T^4. \tag{12.35}$$

Here, $\tilde{\nu}_1$ and $\tilde{\nu}_2$ are the lower and upper limits for the spectral window. We assume that the fraction β $(0 \le \beta \le 1)$ of the blackbody emission from the surface is transmitted freely to space and that the remaining fraction $(1 - \beta)$ is trapped. Then Eq. 12.31 for the upward flux $F^+_{\rm conv}(\beta = 0; \tau)$ is modified to the following:

$$F^+_{\rm conv}(\beta; \tau) \to (1 - \beta) F^+_{\rm conv}(\beta = 0; \tau) + \beta \sigma_{\rm B} T_{\rm s}^4. \tag{12.36}$$

The expression for the downward flux is the same as for the zero transparency case with the correction factor $(1 - \beta)$,

$$F^-_{\rm conv}(\beta; z) = 2(1 - \beta) \int_0^{\tau(z)} d\tau' \sigma_{\rm B} T^4(\tau') e^{-[\tau(z) - \tau']/\bar{\mu}}. \tag{12.37}$$

Because we will henceforth refer only to a radiative–convective model we will drop the "conv" subscript. It can be shown (see Problem 12.3) that the greenhouse equation $T_{\rm s} \equiv \mathcal{G}^{1/4} T_{\rm e}$ (Eq. 12.18) is modified in the presence of a spectral window to

$$T_{\rm s} = \left[\frac{\mathcal{G}(\beta = 0)}{1 - \beta + \beta \mathcal{G}(\beta = 0)} \right]^{1/4} T_{\rm e}. \tag{12.38}$$

Note that when the atmosphere is completely transparent $(\beta = 1)$, $T_{\rm s} = T_{\rm e}$, as it must. It is also possible to show (see Problem 12.3) that in terms of the emission height, the modified surface temperature is given by

$$T_{\rm s} = \frac{T_{\rm s}(\beta = 0)}{\left[1 - \beta + \beta T_{\rm s}^4(\beta = 0)/T_{\rm e}^4\right]^{1/4}} = \frac{T_{\rm e} + |\Gamma_{\rm a}| z_{\rm e}}{\left[1 - \beta + \beta(1 + |\Gamma_{\rm a}| z_{\rm e}/T_{\rm e})^4\right]^{1/4}}. \tag{12.39}$$

It is clear from the above results that for a fixed optical depth, the effect of the window is to decrease the surface temperature because of the higher rate of surface cooling and decreased IR trapping.

The heating rate $\mathcal{H}$ is obtained by differentiating the net flux:

$$\begin{aligned} \mathcal{H}(\beta, z) &= -\frac{\partial}{\partial z}[F^+(\beta; z) - F^-(\beta; z)] \\ &= -\frac{(1 - \beta)\tau(z)}{H_{\rm a}} \frac{\partial}{\partial \tau}[F^+(\beta = 0; \tau) - F^-(\beta = 0; \tau)]. \end{aligned} \tag{12.40}$$

We find that the heating rate including a transparent window is given by $(1 - \beta)$ times the value with no window $(\beta = 0)$. Carrying out the differentiations, we find

$$\begin{aligned} \mathcal{H}(\beta; z) = \frac{(1 - \beta)\tau(z)}{\bar{\mu} H_{\rm a}} \Bigg[& -2\sigma_{\rm B} T_{\rm s}^4 e^{-[\tau^* - \tau(z)]/\bar{\mu}} \\ & + 2 \int_0^{\tau^*} \frac{d\tau'}{\bar{\mu}} \sigma_{\rm B} T^4(\tau') e^{-|\tau(z) - \tau'|/\bar{\mu}} \Bigg]. \end{aligned} \tag{12.41}$$

By setting T = constant in the above equation, we find that the cooling-to-space term (§11.2.6) is given by

$$\mathcal{H}_{\rm cs} = -\frac{2(1-\beta)\tau(z)}{\bar{\mu} H_{\rm a}} \sigma_{\rm B} T^4(\tau) e^{-\tau/\bar{\mu}}. \tag{12.42}$$

In Problem 12.3, it is shown that $\mathcal{H}_{\rm cs}$ reaches its maximum cooling rate where $\tau(z_{\rm e})/\bar{\mu} \approx 1$. In this particular case, cooling to space is a reasonable approximation to the total cooling rate (including exchange terms).

12.7 Radiative Forcing

We previously defined the greenhouse effect G to be the difference between the planetary mean flux emitted by the surface and the planetary mean outgoing flux at the top of the atmosphere. It may be defined on a local basis, and at a specific wavenumber, as $G_{\tilde{\nu}} \equiv F_{\tilde{\nu}}^{+}(\tau^*) - F_{\tilde{\nu}}^{+}(0)$. The spectrally integrated greenhouse effect is thus

$$G = \int d\tilde{\nu} G_{\tilde{\nu}} = \sigma_{\rm B} T_{\rm s}^4 - F_{\rm TOA}. \tag{12.43}$$

On a global average, $\bar{F}_{\rm TOA} = \sigma_{\rm B} T_{\rm e}^4$. Figure 12.7 shows the spectral variation of these two fluxes, illustrating how their difference maximizes in the optically thick bands of the major greenhouse gases. A change in a climate variable will cause a perturbation in G, ΔG, which gives rise to a change in the equilibrium surface temperature ($\Delta T_{\rm s}$). There are several drawbacks to the definition of ΔG that limit its usefulness as a forcing function:

1. Since G and ΔG involve surface temperature, they include the feedback *response* of the climate system.
2. ΔG is a measure of the change in the atmospheric backwarming of the surface; a more satisfactory definition would include the entire atmospheric column.
3. Surface temperature is not easily measured from satellites, especially over land surfaces.

What is needed is a strictly radiative definition that defines the *initial* forcing.[17] A quantity that fulfills these requirements is the difference between the incoming and outgoing fluxes at the top of the atmosphere,

$$N \equiv N_{\rm sw} - N_{\rm lw} \equiv (1-\rho) F^{\rm s} - F_{\rm TOA}. \tag{12.44}$$

N is equal to the instantaneous column-integrated radiative heating resulting from an imbalance between the shortwave heating and the longwave cooling. Note that this definition involves local (or regional) quantities. Thus, N can vary with latitude and even time of day. Note also that the plane albedo ρ, the solar flux $F^{\rm s}$, and the outgoing IR flux $F_{\rm TOA}$ are readily measured from space.

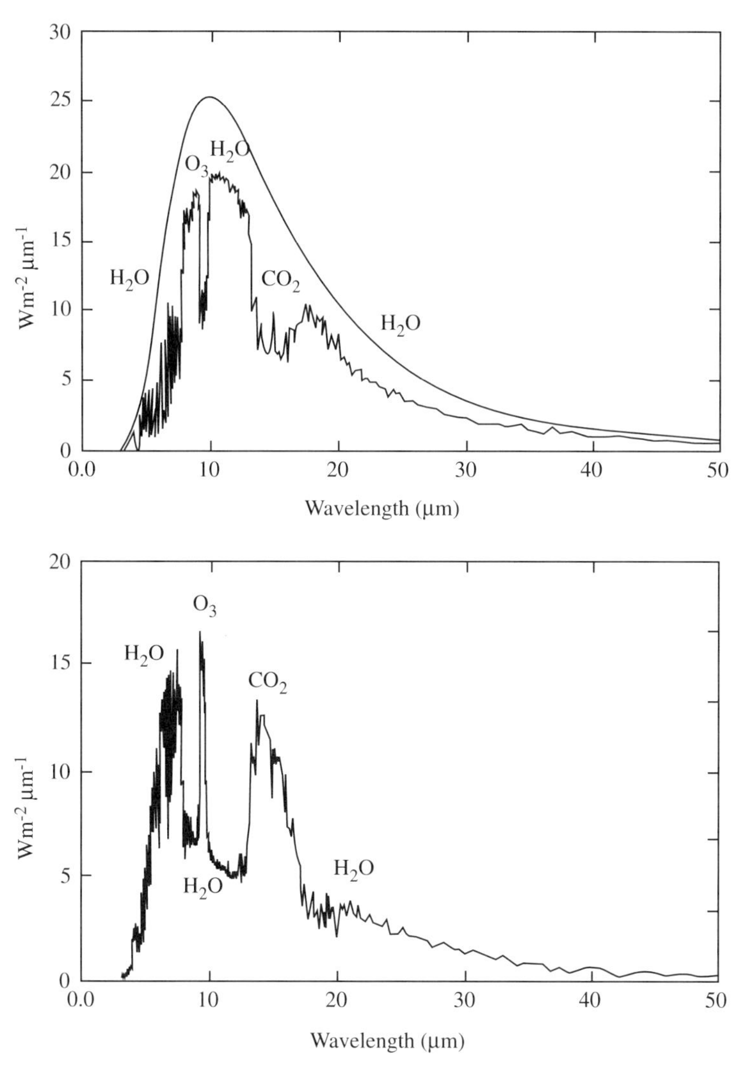

Figure 12.7 Upper panel: The upper smooth curve is the emission from the surface (assumed black) in the model of J. Kiehl and K. Trenberth (their Fig. 1; see reference in Endnote 15). The lower curve is the flux at the top of the atmosphere, computed from their narrow-band model (their Fig. 2; see reference in Endnote 15). Lower panel: greenhouse effect $G_\lambda = F_\lambda^+(z = 0) - F_\lambda^+(z \to \infty)$ for the Kiehl–Trenberth model.

To illustrate the role of N as a radiative forcing, we consider the more general energy balance equation[18] applying to a particular region,

$$\frac{\partial E_{\text{atm}}}{\partial t} = N - \int_0^\infty dz \, \text{div} \, F_{\text{h}} \equiv N - \Phi. \qquad (12.45)$$

Here E_{atm} is the zonally and column-averaged atmospheric energy (see Eq. 1.11 for a more general definition involving both the atmospheric and oceanic columns) and Φ

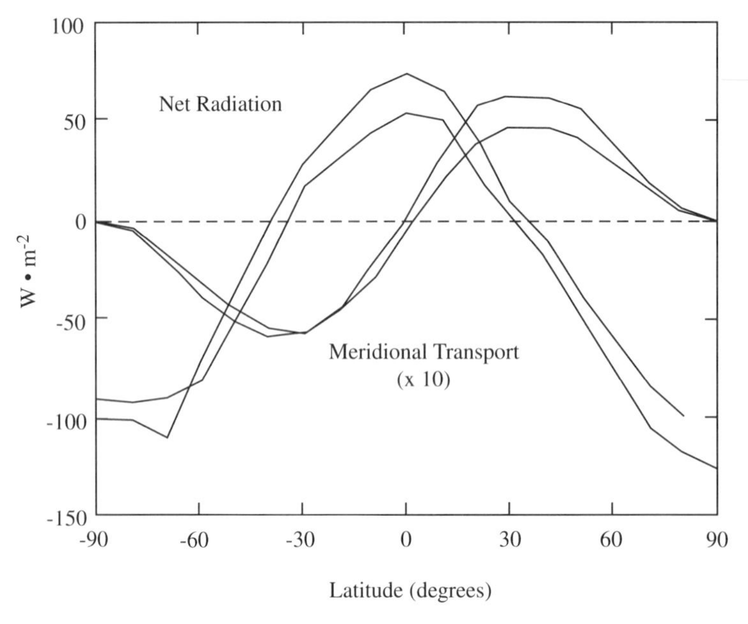

Figure 12.8 Curves labeled "Net Radiation": zonally and annually averaged net radiative flux $\langle N \rangle$ from satellite observations. Curves labeled "meridional transport" is the north-south transport $\int dz F_{\mathrm{h}}$ implied by the above variation of $\langle N \rangle$ with latitude, and assuming a balance between meridional transport and radiative heating, $\partial \langle \bar{E}_{\mathrm{atm}} \rangle / \partial t = 0$ (see Eq. 12.45). The two pairs of curves are from two different data sets. Note that the meridional transport term does not have the same units as N. Adapted from Fig. 3 of Hartmann et al. (Endnote 18).

is the zonally and column-averaged flux divergence of energy leaving the atmospheric column horizontally as a result of meridional transport by large-scale atmospheric circulation and ocean currents. If we average the above equation over the Earth's surface, the transport terms will drop out, so that $\partial \bar{E}_{\mathrm{atm}} / \partial t = \bar{N}$. Further averaging over one year (or a period of several years) yields a global balance,

$$\frac{\partial \langle \bar{E}_{\mathrm{atm}} \rangle}{\partial t} = \langle \bar{N} \rangle = 0.$$

The requirement that $\langle \bar{N} \rangle = 0$ is a useful check on the accuracy of radiation budget measurements. Figure 12.8 shows the measured variation of $\langle N \rangle$ with latitude and the inferred meridional transport of energy $\langle \int dz F_{\mathrm{h}} \rangle$, assuming $\langle \bar{N} \rangle = 0$. More detailed geographical distributions of N are found in the research literature.[19]

12.8 Climate Impact of Clouds

We now consider the combined shortwave and longwave radiative effects of clouds on the atmospheric heat balance. The climatic effects of increasing greenhouse gases can be amplified, or dampened, depending upon whether the consequent changes in clouds lead to a warming (due to an enhanced greenhouse effect) or to a cooling (due to an increased shortwave albedo). Unfortunately many uncertainties still exist in present-day climate models. Variations in cloud forcing between different climate models[20] can be as large as the entire radiative forcing due to a doubling of CO_2! A detailed discussion of such problems is beyond our scope. Fortunately, some of the basic radiative transfer aspects may be illustrated by a simple extension of the gray radiative–convective model, described in the previous section.

In §9.3.3, §9.3.4, and §11.3 we described the optical properties of *low clouds* consisting of spherical water droplets, and *high clouds*, containing irregular ice crystals. More complicated situations (middle clouds and spatially extended *cumulo-nimbus* clouds) consist of mixed water phases. We ignored their complicated three-dimensional structures even though, as will be discussed shortly, this is certainly not correct even for large-scale energy budget studies. However, some cloud types (*marine stratus* for example) can be approximated by the plane-parallel assumption. We showed earlier in Chapters 9 and 11 that this allows us to capture their salient properties by two quantities, their liquid (*LWC*) or ice water (*IWC*) content and their mean particle size $\langle r \rangle$. The absorption and scattering coefficients and the asymmetry factor may be related to the above quantities by simple algebraic relationships, determined by fitting them to accurate scattering computations. From these specifications, and by knowing the temperature of the cloud (determined by its height), we can use the computational tools described in this book to find the cloud albedo, its emittance and absorptance, its optical depth, etc.

In the following, we will illustrate how these cloud properties enter into the mean radiative energy balance. For this purpose, we once again turn to simple models.

12.8.1 Longwave Effects of Water Clouds

The IR opacity of stratus-type water-droplet clouds adds to that of the gaseous opacity, further blanketing the surface and inducing additional warming. This greenhouse warming effect is particularly apparent during a cloudy night, following a clear day of intense solar heating. The combination of a realistic scattering–absorbing cloud imbedded in an absorbing atmosphere introduces a host of complications to the radiative transfer problem, many of which have not yet been overcome. In order to examine some of the first-order effects of clouds, we will ignore most of these complications. For example, we ignore shortwave absorption within the cloud on the grounds that it is small compared with the IR absorption, and we ignore IR scattering. Also, we assume a homogeneous cloud with plane-parallel geometry, and in the spirit of the earlier sections, we adopt the gray, two-stream approximation for describing the radiative-equilibrium radiation field. As much of the radiative transfer occurs in the window region, we adopt values for the gray absorption coefficient at 11 μm. It is sufficient to approximate the cloud optical depth τ_c^* as the product of the mass absorption coefficient $\alpha_{\rm m}^{\rm c}$ in $[{\rm m}^2 \cdot {\rm g}^{-1}]$ and the liquid water path *LWP* in $[{\rm g} \cdot {\rm m}^{-2}]$. Note that a convenient unit of water content in clouds is $[{\rm g} \cdot {\rm m}^{-2}]$ (rather than $[{\rm g} \cdot {\rm cm}^{-2}]$) because clouds generally contain much less water than air.

In our first attempt to estimate the IR effect of clouds, we recognize that the cloud particles introduce additional opacity $\tau_{\rm c}^*$ in a convective atmosphere. If the cloud is introduced above the clear-air emission height, it will raise the effective emitting level to a cooler region. The lower cooling rate implies a higher surface temperature, as discussed in §12.4. In our first approximation, we therefore correct the effective cooling height $z_{\rm e}$ in Eq. 12.32 to include the additional cloud opacity. The surface temperature in a cloudy atmosphere is thus written as

$$\begin{aligned} T_{\rm s} &= T_{\rm e} + |\Gamma_{\rm a}| z_{\rm e} = T_{\rm e} + |\Gamma_{\rm a}| H \ln\left[\left\{\tau_n^* + \tau^*({\rm H_2O}) + \tau_{\rm c}^*\right\}/\tau_{\rm e}\right] \\ &= T_{\rm e} + |\Gamma_{\rm a}| H \ln\left[\left(\tau_n^* + bw + \alpha_{\rm m}^{\rm c} LWP\right)/\tau_{\rm e}\right]. \end{aligned} \tag{12.46}$$

To compare the above result with observations, we again use data for the greenhouse factor $\mathcal{G} = (T_{\rm s}/T_{\rm e})^4$, derived from satellite observations in the presence of clouds of known liquid water path. Figure 12.9 shows $\mathcal{G} = \sigma_{\rm B} T_{\rm ss}^4 / F_{\rm TOA}$ derived from satellite data[21] for: (1) the sea-surface temperature $T_{\rm ss}$, (2) the outgoing flux $F_{\rm TOA}$, and for (3) the liquid water path *LWP*.[22] We also show for comparison the prediction of the analytic formula (Eq. 12.46) for two different values of $\alpha_{\rm m}^{\rm c}$. It is encouraging that the best-fit values of $\alpha_{\rm m}^{\rm c}$ (0.10–0.14) fall within the range of theory (§10.3.3) and

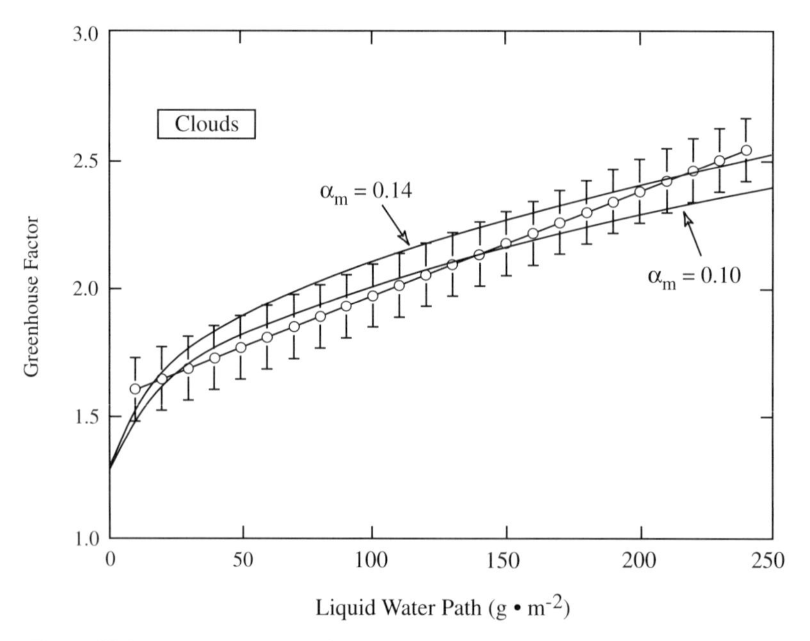

Figure 12.9 Greenhouse factor $\mathcal{G}$ for a cloudy atmosphere versus liquid water path of the cloud droplets. Data are shown as open circles with error bars, indicating the dispersion around the mean. The data are taken from Figure 9b of Stephens and Greenwald (Endnote 21). The solid curves are for the analytic expression, Eq. 12.46, for two different values of the cloud droplet absorption coefficient.

observation[23] ($0.06 < \alpha_{\rm m}^{\rm c} < 0.16$). This successful application of the theory suggests that the principal IR effect of clouds (introduced above the effective clear-air emission height) is simply to raise the effective radiating height. The importance of the cloud *height* itself will be addressed in a later section.

12.8.2 Shortwave Effects of Water Clouds

The importance of shortwave absorption by cloud particles is a question of current research interest, which is somewhat controversial.[24] The controversy stems from the interpretation of aircraft measurements that suggest a surprising amount of absorption by cloud particles, which are expected to absorb only modestly at near-IR wavelengths. The two possibilities are that either the radiation is absorbed within the cloud or it escapes from the sides of the cloud and thus avoids detection by the aircraft instruments. To explore this further would take us into areas not suitable for a textbook. The only other major effect to consider for the energy budget is the change in albedo due to the presence of clouds. Increased reflection (larger $\bar{\rho}$) causes the amount of absorbed energy $(1-\bar{\rho})F^{\rm s}$ to be lowered and thus reduces the mean surface temperature. Generally speaking, shortwave effects of water clouds tend to cool the Earth, in contrast to IR effects, which generally warm the Earth. As in the previous section, we choose a simple model to illustrate how cloud opacity and cloud fractional coverage influence the surface temperature. We will continue to rely upon our analytic approximation for the surface temperature in a radiative–convective atmosphere. Thus we need to consider only the change in the effective temperature, due to addition of a cloud.

For this purpose, we will adopt the two-stream solutions derived in Chapter 7 for *Prototype Problem 3*. We found previously that the solution for the plane albedo $\rho_{\rm c}$ of a conservative ($a = 1$) and isotropically scattering ($p = 1$), plane-parallel medium is (Eq. 7.98)

$$\rho_{\rm c} = \frac{2b_{\rm c}\tau_{\rm c}^* + (\bar{\mu} - \mu_0)\left(1 - e^{-2b_{\rm c}\tau_{\rm c}^*/\mu_0}\right)}{2b_{\rm c}\tau_{\rm c}^* + 2\bar{\mu}}, \tag{12.47}$$

where $\tau_{\rm c}^*$ is the (gray) visible optical depth of the cloud, $\bar{\mu}$ is the mean cosine of the multiply scattered photons, μ_0 is the cosine of the solar zenith angle, $b_{\rm c} = (1 - g_{\rm c})/2$ is the cloud backscattering coefficient, and $g_{\rm c}$ is the asymmetry factor.

In order to evaluate the albedo in terms of the liquid water path, we will use the parameterizations of the optical properties derived from accurate Mie-scattering results, as discussed in §9.4. The optical depth and asymmetry factor are written (see Eqs. 9.5 and 9.6)

$$\begin{aligned} \tau_{\rm c}^* &= k_{\rm m}^{\rm c} LWP \approx (a_1\langle r\rangle^{b_1} + c_1)LWP, \\ g_{\rm c} &= 1 - 2b_{\rm c} \approx a_3\langle r\rangle^{b_3} + c_3, \end{aligned} \tag{12.48}$$

where the coefficients a_i, b_i, and c_i are best-fit values, dependent upon wavelength and upon the particle size regime,[25] and $\langle r\rangle$ is the mean particle radius.

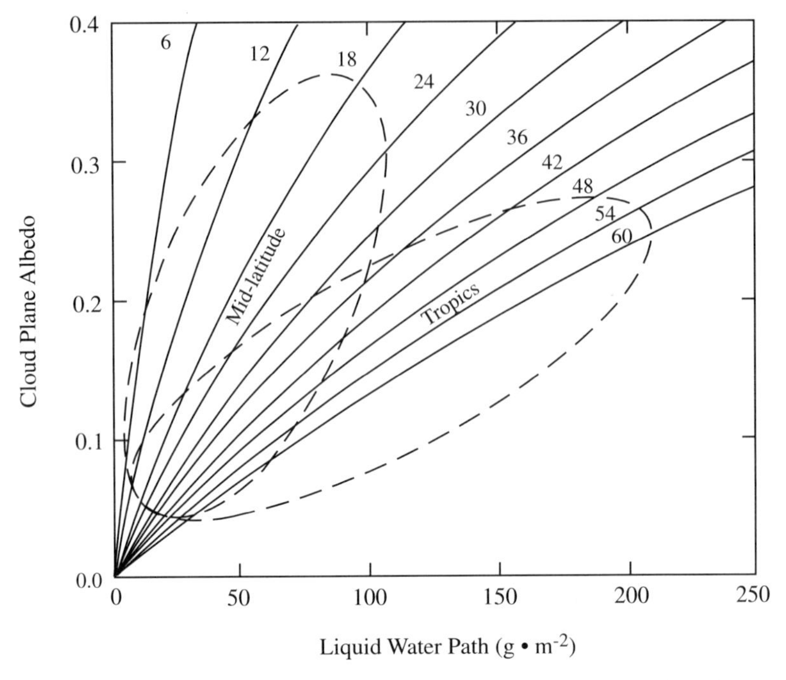

Figure 12.10 Plane albedo of a cloud versus liquid water path, calculated from the two-stream solution, Eq. 12.47. Values of the asymmetry factor and absorption coefficient are taken from Eq. 12.48. Different curves apply to different values of the mean particle radius $\langle r \rangle$ in μm. The dashed curves, taken from Stephens and Greenwald (Endnote 21) enclose 97% of all observations for the tropics and midlatitudes.

As before, Eq. 12.47 should be tested by comparison with data. Figure 12.10 shows the 97% occurrence zones (divided into latitude zones) of cloud albedo values extracted from *Nimbus 7* and ERBE satellite data.[26] For comparison, predictions are given from the two-stream formula (Eq. 12.47) using the parameterization formulas, Eq. 12.48, for a wavelength of 0.459 μm. (The scattering coefficient is spectrally flat in the range 0.3–0.7 μm; a single value characterizes the entire visible range.)

As discussed by Stephens and Greenwald[27] this figure reveals a serious discrepancy, since in situ measurements show that cloud particle sizes $\langle r \rangle$ are in the range 5–10 μm. The errors in the two-stream results are too small to explain the discrepancy. The fact that the *LWP* measurements are not sensitive to ice particles above the main clouds is also inadequate to explain the discrepancy. *The theory for plane-parallel clouds appears to predict albedo values higher than observed.* This suggested to Stephens and Greenwald that the neglected macrophysical effects of finite cloud size are probably larger than the errors introduced from the uncertainities in the knowledge of the cloud microphysics. This unsolved problem poses a significant barrier to understanding the effects of cloud feedback on climate.

We now consider how the albedo of the surface–atmosphere–cloud system enters into the overall energy balance of the planet. First we should recognize the fact that the

Earth's surface reflects visible light, with an average (spherical) albedo[28] $\bar{\rho}_s = 0.11$. This value is low, because of the very low (5–10%) reflectance of ocean water, which covers nearly three-fourths of the planet's surface. We must also include the small but still significant effects of the clear-sky (Rayleigh-scattering) contribution, as well as the shortwave solar absorption by the molecular atmosphere. The spherical albedo of the atmosphere (assuming an underlying dark surface) is $\bar{\rho}_a = 0.07$.[29] To take into account multiple reflections between the surface and the atmosphere, we apply the formula derived in §6.11 (Eq. 6.76) to obtain the effective spherical albedo of the combined atmosphere and surface:

$$\bar{\rho}_{as} = \bar{\rho}_a + \frac{\bar{\rho}_s(1-\bar{\rho}_a)^2}{1-\bar{\rho}_s\bar{\rho}_a}. \tag{12.49}$$

Substituting in the above numerical values for $\bar{\rho}_a$ and $\bar{\rho}_s$, we find $\bar{\rho}_{as} = 0.166$, in good agreement with the observed range 0.14–0.18 found by satellite measurements.[30]

Assuming that clouds of spherical albedo $\bar{\rho}_c$ (again assuming a black underlying surface) overlie the atmosphere–surface system, we can apply the same formula to estimate the spherical albedo of a cloudy atmosphere,

$$\bar{\rho}_{cas} = \bar{\rho}_c + \frac{\bar{\rho}_{as}(1-\bar{\rho}_c)^2}{1-\bar{\rho}_{as}\bar{\rho}_c}. \tag{12.50}$$

We must also include the fact that the Earth is partially covered with clouds with a fractional coverage A_c ($0 \le A_c \le 1$). Thus, we weight the clear-sky albedo ρ_{as} with the fraction of clear sky $(1 - A_c)$ so that the total spherical albedo of the Earth system is approximated by

$$\bar{\rho}_{tot} = A_c\bar{\rho}_{cas} + (1 - A_c)\bar{\rho}_{as}. \tag{12.51}$$

For the determination of cloud albedo, it is desirable to relate $\bar{\rho}_c$ to more basic cloud properties, such as liquid water path (*LWP*) or ice water path (*IWP*). Two approaches are possible: analytic and empirical. To pursue the first course, we again use the two-stream approximation. The spherical albedo is given by the average of Eq. 12.47 over all solar zenith angles. However, it is much simpler to invoke the *Duality Principle* to equate the spherical albedo for *Prototype Problem 3*, $\bar{\rho}_3$, to the plane albedo for *Prototype Problem 1*, ρ_1. The latter is given by (see Eq. 7.52) $\bar{\rho}_c \equiv \rho_1 = \bar{\rho}_3 = \tau_c^*/(\tau_c^* + 2\bar{\mu})$, or, when properly scaled to account for anisotropic scattering,

$$\bar{\rho}_c = \frac{(1-g_c)\tau_c^*}{(1-g_c)\tau_c^* + 2\bar{\mu}}. \tag{12.52}$$

(Note that the averaging procedure to derive the spherical albedo from the plane albedo amounts to simply setting $\mu_0 = \bar{\mu}$ in Eq. 12.47.) An alternative to the above formula, discussed in the next section, is to use an empirical relationship between $\bar{\rho}_c$ and *LWP*.

12.8.3 Combined Shortwave and Longwave Effects of Clouds

It should be apparent that if the total albedo $\bar{\rho}_{tot}$ is higher than some critical value $\bar{\rho}_{crit}$, clouds will result in a net cooling (referenced to a clear atmosphere). However, if $\bar{\rho}_{tot} < \bar{\rho}_{crit}$, a net warming will occur. We can estimate this critical value as a function of cloud thickness by expressing the surface temperature in terms of the products of the clear-air ($\mathcal{G}_{clr}$) and cloudy ($\mathcal{G}_{cld}$) greenhouse factors:

$$T_s^4 = (1 - \bar{\rho}_{tot})\frac{S}{4\sigma_B}\mathcal{G}_{cld} = T_s^4(LWP = 0)\frac{(1 - \bar{\rho}_{tot})}{(1 - \bar{\rho}_{as})}\left[1 + \frac{d_1}{c_1}LWP\right]. \quad (12.53)$$

The clear-air surface flux is $\sigma_B T_s^4(LWP = 0) = (1 - \bar{\rho}_{as})(S/4)\mathcal{G}_{clr}(\tau^*)$, in accordance with the results of §12.5 and Fig. 12.6. τ^* is the clear-air optical depth. We have also approximated $\mathcal{G}_{cld}$ as a linear function of liquid water path (shown in Fig. 12.9). The cloudy greenhouse factor has been written as a linear fit to data, $\mathcal{G}_{cld} = c_1 + d_1 LWP$, where $c_1 = 1.56$ and $d_1 = 4.09 \times 10^{-3}$. Now for no cloud cover ($LWP = 0$) the greenhouse factor is $\mathcal{G}_{clr}(\tau^*) = c_1$. Defining the critical albedo as that value which yields $T_s(LWP) = T_s(LWP = 0) =$ constant as liquid water path varies in Eq. 12.53, then we can solve for the critical albedo,

$$\bar{\rho}_{crit} = \frac{\bar{\rho}_{as} + (d_1/c_1)LWP}{1 + (d_1/c_1)LWP}. \quad (12.54)$$

This result is fairly general since it is based on an empirical correlation of surface temperature with cloud thickness. It results from an assumption of a steady state, but it does not require radiative equilibrium. It thus is valid at any latitude (for which the net radiative flux N may be nonzero). However, as the cloud properties vary, it does require a fixed proportion of incoming radiation, outgoing radiation, and meridional transport.

We have defined $\bar{\rho}_{crit}$ to be the albedo value for which there is no effect on surface temperature. Thus, we expect that $\bar{\rho}_{tot} > \bar{\rho}_{crit}$ will correspond to a cooling and $\bar{\rho}_{tot} < \bar{\rho}_{crit}$ to a warming. This is illustrated in Fig. 12.11, which shows that the critical albedo separates domains of cooling and warming. Clouds in midlatitude regions are seen to have a net cooling effect, whereas those in tropical regions are mainly in the warming domain. This behavior is immediately apparent from an inspection of Fig. 12.10, where it is seen that tropical clouds are much weaker functions of cloud water content than those at midlatitude. This result, although based on simple ideas, is consistent with conclusions in the research literature.[31]

To obtain an alternate model for the surface temperature we use the concept of effective radiating height in a formulation involving partial cloud coverage. Both the albedo and the effective radiating height depend upon the liquid water path of the clouds, and upon the cloud fractional coverage, A_c. The effective temperature and radiating height contributions are weighted by the contribution from clouds (Eq. 12.46)

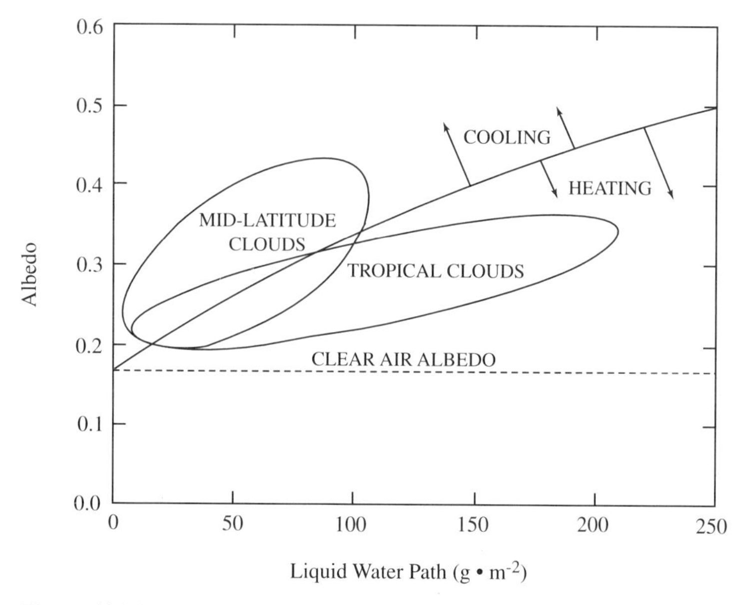

Figure 12.11 Albedo versus liquid water path of clouds. Heavy solid line: Critical planetary albedo $\bar{\rho}_{\text{crit}}$ dividing regions of net warming and net cooling, given by Eq. 12.54. The albedo of the clear air plus surface $\bar{\rho}_{\text{as}}$ is set equal to 0.166, as inferred from Eq. 12.49. $d_1/c_1 = 2.622 \times 10^{-3}$. The ellipses define the regions containing 97% of all observations for the tropics and midlatitudes. They were computed from Eq. 12.50 where $\bar{\rho}_{\text{c}}$ is given by the ellipses in Fig. 12.10. If $\bar{\rho}_{\text{tot}}$ is in the upper domain ($\bar{\rho}_{\text{tot}} > \bar{\rho}_{\text{crit}}$), net cooling occurs. If it is in the lower domain,($\bar{\rho}_{\text{tot}} < \bar{\rho}_{\text{crit}}$), net warming occurs.

and from clear air (Eq. 12.32). The result is

$$T_{\text{s}} = \left[\frac{F^{\text{s}}(1-\bar{\rho}_{\text{tot}})}{\sigma_{\text{B}}}\right]^{1/4} + A_{\text{c}} H_{\text{cld}} |\Gamma_{\text{cld}}| \ln\left[(\tau_n^* + bw + \alpha_{\text{m}}^{\text{c}} LWP)/\tau_{\text{e}}\right]$$
$$+ (1-A_{\text{c}}) H_{\text{clr}} |\Gamma_{\text{clr}}| \ln\left[\left(\tau_n^* + bw\right)/\tau_{\text{e}}\right], \qquad (12.55)$$

where $F^{\text{s}} \equiv S/4$ is the incident solar flux averaged over the planet and over twenty-four hours (see Eq. 12.1), and S is the solar constant. Note that both the scale height (either H_{clr} or H_{cld}) and the lapse rate (either Γ_{clr} or Γ_{cld}) are allowed to vary between clear and cloudy regions. The above equation may be used to study the effects on the surface temperature of a planet of variable cloud cover A_{c}, moisture content of the air w, and cloud water path LWP.

In the next section we consider cloud height, another important influence on greenhouse warming.

12.9 Climate Impact of Cloud Height

One expects that low clouds, being relatively warm, would have different effects on the energy balance than high clouds, which are relatively cold. In order to include the cloud height as an additional climate variable, we adopt a more general formulation of radiative equilibrium, while still retaining the simplicity of the gray approximation. This method employs the outgoing IR flux F_{TOA} at the top of the atmosphere and assumes that the surface temperature T_s is an adjustable variable. By adjusting T_s, we effectively adjust the temperature in the entire convective region where $T(\tau) = T_s + \Gamma z(\tau)$. For simplicity, we assume a geometrically thin cloud ($z_b \approx z_t \equiv z_c$) and a constant transparency factor β. The cloud is characterized by the flux emittance and absorptance, equal to ϵ_c and $(1 - \epsilon_c)$, respectively. In a two-stream gray approximation ϵ_c is simply $1 - e^{-\tau_c^*/\bar{\mu}}$, where τ_c^* is the cloud optical depth. The outgoing flux for both clear and cloudy regions may be considered to be a function of T_s, with specified cloud parameters ϵ_c and z_c:

$$
\begin{aligned}
F_{TOA}(T_s; \epsilon_c, z_c) = {} & 2\,\bar{\mu}(1-\beta)\epsilon_c\sigma_B T^4(z_c)e^{-\tau_c^*/\bar{\mu}} + 2\bar{\mu}\beta\epsilon_c\sigma_B T^4(z_c) \\
& + 2(1-\beta)\int_0^{\tau_c^*} d\tau'\sigma_B T^4(\tau')e^{-\tau'/\bar{\mu}} \\
& + 2\bar{\mu}(1-\beta)(1-\epsilon_c)\sigma_B T_s^4 e^{-\tau_c^*/\bar{\mu}} + 2\bar{\mu}\beta\sigma_B T_s^4(1-\epsilon_c) \\
& + 2(1-\beta)\int_{\tau_c^*}^{\tau^*} d\tau'\sigma_B T^4(\tau')(1-\epsilon_c)e^{-\tau'/\bar{\mu}}.
\end{aligned} \tag{12.56}
$$

Here τ^* is the clear-sky total optical depth. Radiative equilibrium means that the absorbed solar flux ($\sigma_B T_e^4$) and the outgoing IR flux are equal, that is,

$$
\sigma_B T_e^4 = A_c F_{TOA}(T_s; \epsilon_c, z_c) + (1 - A_c)F_{TOA}(T_s; \epsilon_c = 0). \tag{12.57}
$$

The clear and cloudy parts are weighted by the clear-air and cloudy fractions, $(1 - A_c)$ and A_c, respectively. Since T_s is a variable parameter in the above equation, it can be changed until the equality is satisfied. The results of this numerical exercise are shown in Fig. 12.12 as a function of the cloud fraction A_c. Several cloud heights and opacities have been selected to illustrate how high cold clouds warm the surface and low warm clouds cool the surface. The former effect arises because of the reduced emission to space, and the latter effect is due to an increased emission to space. The optically thick middle cloud ($z_c = 4.1$ km) model was tuned (by varying τ^*) to yield the globally average mean surface temperature of 288 K for a cloud fraction of 60%. This diagram suggests how Earth's mean temperature might be affected as a result of changing cloud coverage, cloud height, or cloud opacity.

Other factors complicate this simple picture. For example, the effects of low clouds over highly reflecting surfaces can result in an overall warming, since longwave greenhouse warming can offset the effects of the small variation in total albedo from clear to overcast conditions. More accurate radiative transfer computations show that the warming from high clouds is due, not only to suppressed cooling to space, but also to ice crystals having effectively larger values of $n = k_{IR}/k_V$ than water droplets. This is the "one-way valve" effect of radiative trapping, discussed in §12.3. As discussed previously, the latitude differences of cloud albedo properties make the surface

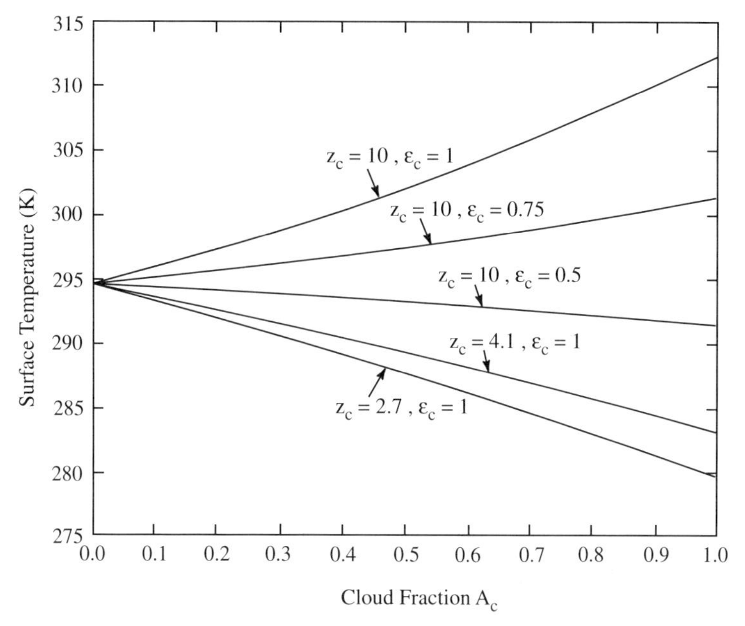

Figure 12.12 Radiative–convective equilibrium surface temperature as a function of cloud areal coverage for low (2.7 km) and middle (4.1 km) clouds of emittance unity, and a high (10 km) cloud with three different values of cloud emittance. These results were obtained from a numerical integration of Eq. 12.56 in which T_s is adjusted until Eq. 12.57 is satisfied. The clear-air albedo is assumed to be given by 0.166, and the completely overcast albedo is 0.389. The clear-air optical depth (τ^*) is assumed to be 6. Combined with the value of the window transparency factor ($\beta = 0.22$), these parameters yield $T_s = 288$ K for the middle cloud model when $A_c = 0.6$.

temperature relationship more involved than discussed above. Actually, the latitude differences of cloud heights is only one contributing factor in explaining why mid-latitude clouds cool the surface and tropical clouds warm the surface. In the following section, we discuss a concept that describes climate change in terms of a flux change, rather than as a change in the (fictitious) radiative-equilibrium temperature.

12.10 Cloud and Aerosol Forcing

To focus on the radiative effects of clouds (an identical formulation applies to aerosols), it is common to use the concept of *cloud radiative forcing*[32]:

$$CF \equiv N_{\mathrm{cld}} - N_{\mathrm{clr}} \equiv SWCF + LWCF. \tag{12.58}$$

Here N_{cld} and N_{clr} are the cloudy and clear fluxes measured at the top of the atmosphere. If CF is determined empirically, use is made of different scenes (cloudy or clear) at the same latitude and above the same type of surface. By referencing the clear atmosphere, we isolate the effects of clouds from other radiative effects, for example, changes of water vapor or surface albedo.

Cloud forcing CF (which should not be confused with the Chapman function having the same symbol) is conveniently separated into shortwave ($SWCF$) and longwave ($LWCF$) contributions. Clearly

$$\begin{aligned} SWCF &\equiv (1-\rho_{\rm cas})F^{\rm s} - (1-\rho_{\rm as})F^{\rm s} = (\rho_{\rm as} - \rho_{\rm cas})F^{\rm s}, \\ LWCF &\equiv F_{\rm TOA}({\rm clr}) - F_{\rm TOA}({\rm cld}). \end{aligned} \tag{12.59}$$

$\rho_{\rm cas}$ is the plane albedo of the cloudy atmosphere (clouds plus atmosphere plus surface) and $\rho_{\rm as}$ is that for the clear atmosphere (atmosphere plus surface). Note that since clouds usually increase the overall albedo, $SWCF$ is generally negative. Since clouds generally reduce the outgoing flux, $LWCF$ is generally positive.

Example 12.1 Longwave Cloud Forcing in a Model Atmosphere: Cloud Overlap

We will compute the longwave forcing from a model atmosphere (the 1976 U.S. Standard Atmosphere[33]) for the following distribution of clouds: (1) a high cloud with a cloud-top height $z_{\rm t}^{(1)} = 11$ km, emittance $\epsilon_{\rm c}^{(1)} = 0.06$, and fractional area coverage $A_{\rm c}^{(1)} = 0.20$; (2) a midlevel cloud with $z_{\rm t}^{(2)} = 6$ km, $\epsilon_{\rm c}^{(2)} = 0.1$, and $A_{\rm c}^{(2)} = 0.06$; and a low cloud with $z_{\rm t}^{(3)} = 2$ km, $\epsilon_{\rm c}^{(3)} = 0.1$, and $A_{\rm c}^{(3)} = 0.49$. We require the outgoing IR flux for both the clear and cloudy cases to find the longwave cloud forcing. The clear case has been computed from an accurate narrow-band spectral model by Kiehl and Trenberth.[34] Their value is $F_{\rm TOA}({\rm clr}) = 262\ {\rm W\cdot m^{-2}}$, in excellent agreement with clear-air ERBE data ($265\ {\rm W\cdot m^{-2}}$).

The outgoing flux in the cloudy case will be estimated from the gray, two-stream model with no spectral window ($\beta = 0$). We use Eq. 12.56 to include the presence of a single cloud with cloud-top height $z_{\rm t}$. It is easy to show that the flux at the top of the atmosphere is

$$\begin{aligned} F_{\rm TOA}({\rm cld}) = 2\bar{\mu}\sigma_{\rm B}T_{\rm s}^4(1-\epsilon_{\rm c})e^{-\tau^*/\bar{\mu}} + \epsilon_{\rm c}\sigma_{\rm B}T^4(z_{\rm t}) + 2\int_0^{\tau} d\tau' e^{-(\tau-t)/\bar{\mu}}\sigma_{\rm B}T^4[z(d\tau')] \\ + 2(1-\epsilon_{\rm c})\int_{\tau}^{\tau^*} d\tau' e^{-(\tau'-\tau)/\bar{\mu}}\sigma_{\rm B}T^4[z]. \end{aligned} \tag{12.60}$$

Here $\epsilon_{\rm c}$ is the cloud flux emittance, $(1-\epsilon_{\rm c})$ is the cloud flux transmittance, and τ is shorthand for $\tau(z_{\rm t})$. The cloud has been assumed to be geometrically thin.

We simplify the above expression for each of the three clouds by ignoring the atmospheric contributions (the integrals) for all except the thin upper cloud. In that case, we approximate the atmospheric contribution as $(1-\epsilon_{\rm c})F_{\rm TOA}({\rm clr})$. Then we can write down the individual contributions from each cloud, assuming that they are independent (a false assumption that we will correct shortly):

$$\begin{aligned} F_{\rm TOA}^{(1)}({\rm cld}) &\cong \epsilon_{\rm c}^{(1)}\sigma_{\rm B}T^4\left(z_{\rm t}^{(1)}\right) + \left(1-\epsilon_{\rm c}^{(1)}\right)F_{\rm TOA}({\rm clr}) = 184.5\ {\rm W\cdot m^{-2}}, \\ F_{\rm TOA}^{(2)}({\rm cld}) &\cong \sigma_{\rm B}T^4\left(z_{\rm t}^{(2)}\right)e^{-\tau(z_{\rm c}^{(2)})/\bar{\mu}} = 146.8\ {\rm W\cdot m^{-2}}, \\ F_{\rm TOA}^{(3)}({\rm cld}) &\cong \sigma_{\rm B}T^4\left(z_{\rm t}^{(3)}\right)e^{-\tau(z_{\rm c}^{(3)})/\bar{\mu}} = 262\ {\rm W\cdot m^{-2}}, \end{aligned} \tag{12.61}$$

where we have set $\epsilon_{\rm c}^{(2)} = \epsilon_{\rm c}^{(3)} = 1$ and $\bar{\mu} = 0.5$. Using the values of temperature from the 1976 Standard Atmosphere we find the numerical results given above.

To find the net flux, it is necessary to consider *cloud overlap*. The net flux is a weighted sum of fluxes,

$$F_{\mathrm{TOA}} = \sum_i p_c^i F_{\mathrm{TOA}}^{(i)} + p_{\mathrm{nc}} F_{\mathrm{TOA}}(\mathrm{clr}). \tag{12.62}$$

Here p_c^i is the probability that a given vertical line of sight intersects the ith cloud, without intersecting any other cloud, and p_{nc} is the probability of no clouds (i.e., clear air). Since A_c^i is the probability of cloud occurrence, and $(1 - A_c^i)$ is the probability of "no-cloud" occurrence, then it is clear that

$$\begin{aligned} p_c^{(1)} &= A_c^{(1)} = 0.2, \qquad p_c^{(2)} = A_c^{(2)}\left(1 - A_c^{(1)}\right) = 0.048, \\ p_c^{(3)} &= A_c^{(3)}\left(1 - A_c^{(1)}\right)\left(1 - A_c^{(2)}\right) = 0.368, \\ p_{\mathrm{nc}} &= \left(1 - A_c^{(1)}\right)\left(1 - A_c^{(2)}\right)\left(1 - A_c^{(3)}\right)\left(1 - A_c^{(4)}\right) = 0.384. \end{aligned} \tag{12.63}$$

It is easily checked that $\sum_i p_c^i + p_{\mathrm{nc}} = 1$, as it must. The effective cloud area is $\sum_i p_c^i A_c^i = 0.62$, which is less than $\sum_i A_c^i = 0.75$ because of random overlapping.

Using Eq. 12.26 for $\tau(z)$ and setting $\tau^* = 4$ from the results of §12.2, we find from Eq. 12.62 that $F_{\mathrm{TOA}}(\mathrm{cld}) = 241\ \mathrm{W \cdot m^{-2}}$, in excellent agreement with the accurate results ($235\ \mathrm{W \cdot m^{-2}}$) of Kiehl and Trenberth (who also considered cloud overlap effects). The longwave cloud forcing is thus $LWCF = F_{\mathrm{TOA}}(\mathrm{cld}) - F_{\mathrm{TOA}}(\mathrm{clr}) = 265 - 241 = 24\ \mathrm{W \cdot m^{-2}}$ (two-stream), or $262 - 235 = 30\ \mathrm{W \cdot m^{-2}}$ (accurate results).

The spectral variation of the longwave cloud forcing is shown in Fig. 12.13, calculated with an accurate narrow-band radiative transfer model. This figure is contrasted

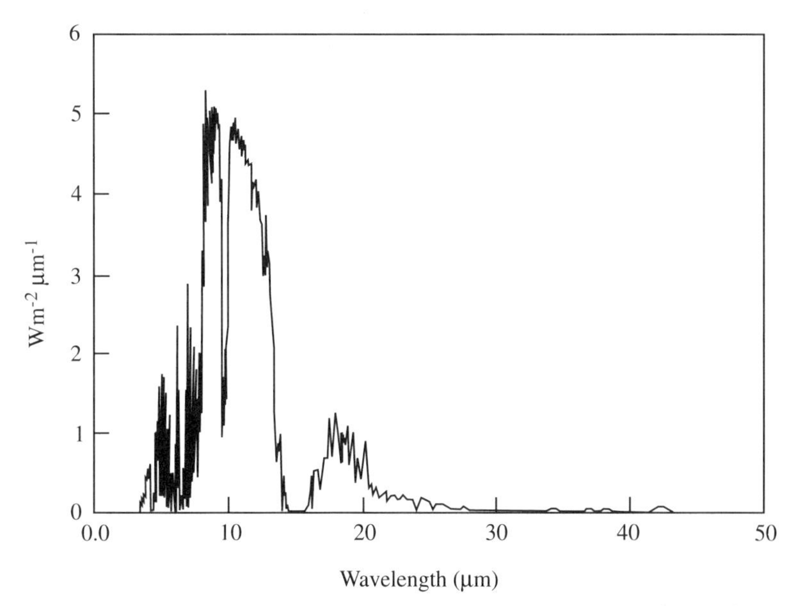

Figure 12.13 Spectral variation of longwave cloud forcing ($\mathrm{W \cdot m^{-2} \cdot \mu m^{-1}}$), the difference between the clear-air flux at the top of the atmosphere, and the cloudy outgoing longwave flux. From Kiehl and Trenberth (see Endnote 15).

with the greenhouse effect (Fig. 12.7) where the optically thick bands dominate. In the cloud case, the transparent window at 8–12 μm dominates the longwave cloud forcing.

12.10.1 Aerosol Forcing

Volcanic aerosols have long been suspected to be important agents of global climate change. Of greatest importance for climate are those particles that originate in the stratosphere as a result of an explosive injection of sulfates (mainly sulfur dioxide) from so-called Plinian volcanic eruptions. Growth of sulfuric acid occurs as a result of both heterogeneous nucleation on existing aerosols (Junge layer particles) and homogeneous nucleation in supercooled regions. These liquid particles persist in the stratosphere for several years following major eruptions, such as El Chichon in 1982 and Mt. Pinatubo in 1991. The optical properties are similar to those of small cloud particles in that they tend to cool the atmosphere in the shortwave and tend to warm it in the longwave spectral range. It turns out that the sign of the overall forcing depends upon mean particle size. If the effective radius exceeds $\sim$2 μm, the tendency is to produce warming of the surface (see Fig. 12.14). The effects of aerosol composition,

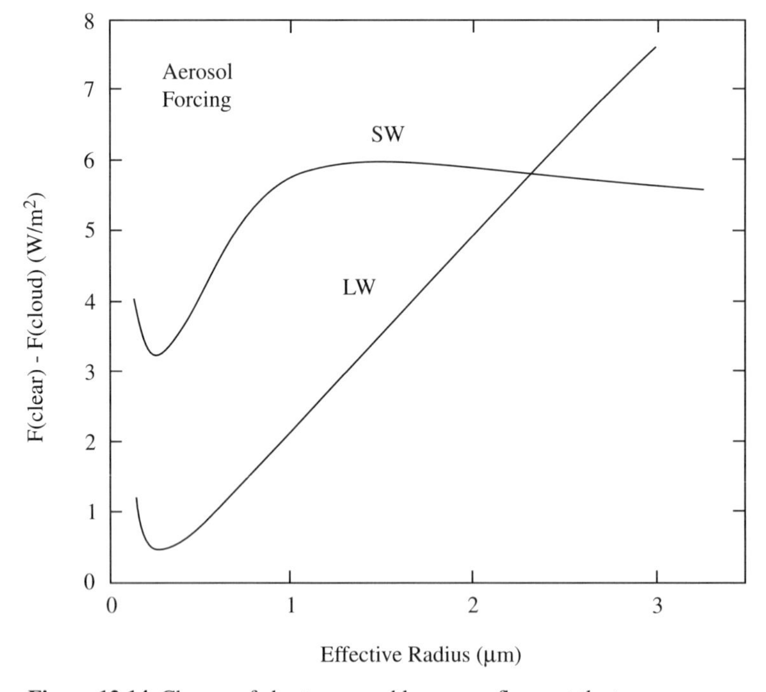

Figure 12.14 Change of shortwave and longwave fluxes at the tropopause (longwave and shortwave forcing) caused by adding a stratospheric (20–25 km) aerosol layer with $\tau(0.55\ \mu\text{m}) = 0.1$ in a one-dimensional radiative–convective model with fixed surface temperature. From A. Lacis, J. Hansen, and M. Sato, "Climate forcing by stratospheric aerosols," *Geophys. Res. Lett.*, **19**, 1607–10, 1992.

size distribution, and altitude are less important than the mean size. Since stratospheric aerosols of volcanic origin are in the size range 0.5–1.5 μm (depending upon time after the eruption), the overall effect is to cool the Earth. In fact, the largest eruption of this century, Mt. Pinatubo, caused a peak radiative forcing[35] of about 4–5 $\mathrm{W \cdot m^{-2}}$. This caused an apparent cooling in the northern hemisphere lower troposphere of 0.5–0.7°C, compared with pre-Pinatubo levels.[36]

12.11 Water-Vapor Feedback

In Chapter 1 and §12.7 we discussed the concept of *radiative forcing* $N_{\mathrm{sw}} - N_{\mathrm{lw}}$ as a quantitative measure of greenhouse warming. Here the subscripts refer to the separate shortwave (sw) and longwave (lw) forcing contributions. In our simple model, $N_{\mathrm{sw}} = (1 - \bar{\rho})F^{\mathrm{s}}$ and $N_{\mathrm{lw}} = F_{\mathrm{TOA}}$. N is the net radiative flux at the top of the atmosphere and is zero in radiative equilibrium. In Chapter 1 we showed that in a "zero-D" model, the surface temperature response to a forcing is written as the product of a direct response $\Delta T_{\mathrm{s}}^{\mathrm{d}}$ and the gain f of the system,

$$\Delta T_{\mathrm{s}} = f \Delta T_{\mathrm{s}}^{\mathrm{d}} = \frac{1}{\left(1 - \sum_i \lambda_i\right)} \frac{N}{(4F_{\mathrm{TOA}}/T_{\mathrm{s}})}. \tag{12.64}$$

The various feedback parameters, λ_i, are measures of individual feedback processes. The parameter λ_i is written in terms of a general temperature-dependent climate variable Q_i as

$$\lambda_i = [T_{\mathrm{s}}/4F_{\mathrm{TOA}}] \frac{\partial N}{\partial Q_i} \frac{\partial Q_i}{\partial T_{\mathrm{s}}}. \tag{12.65}$$

Here we consider the most important terrestrial feedback influence due to the change in IR opacity resulting from the temperature dependence of the water-vapor content. The atmosphere becomes more (less) moist in a warmer (colder) climate resulting in more (less) greenhouse warming. The underlying principle of the temperature–moisture content relationship is described by the *Clausius–Clapeyron equation*, expressing the dependence of the saturation vapor pressure on temperature. The following empirical relationship between the precipitable water w and the sea-surface temperature T_{s} has been determined from an analysis of satellite measurements[37]:

$$w = a_w e^{\mathrm{b}_w (T_{\mathrm{s}} - 288)}, \tag{12.66}$$

where T_{s} is in K, $a_{\mathrm{w}} = 1.753\ \mathrm{g \cdot cm^{-2}}$, and $b_{\mathrm{w}} = 0.0686\ \mathrm{K^{-1}}$. This result is consistent with an average surface relative humidity (RH) of 85% over the world's oceans, provided the mean scale height for water vapor is 2 km. The relative humidity tends to remain fixed as the sea-surface temperature changes, whereas the absolute water content w is exponentially dependent on temperature.

We now have the necessary equations to estimate water-vapor feedback in the clear-sky atmosphere. We previously specified a relationship (see Eq. 12.34) between the atmospheric optical depth and the column water amount: $\tau^* = \tau_n^* + bw$, where τ_n^* denotes the clear-sky opacity due to nonwater-vapor greenhouse gases and b is a constant determined from fitting the model greenhouse factor to satellite data. We note that the temperature-dependent variable in Eq. 12.65 is $Q = \tau(\mathrm{H_2O}) = bw$. Since $N_{\mathrm{lw}} = \sigma_\mathrm{B} T_\mathrm{e}^4$, which can be expressed in terms of optical depth τ^* from Eq. 12.32, $N_{\mathrm{lw}} = \sigma_\mathrm{B}[T_\mathrm{s} - |\Gamma_\mathrm{a}| H_\mathrm{a} \ln(\tau^*/\tau_\mathrm{e})]^4$, we may evaluate the feedback parameter

$$\lambda = \left(\frac{T_\mathrm{s}}{4F_{\mathrm{TOA}}}\right)\left(\frac{\partial N}{\partial \tau^*}\right)\left(\frac{\partial \tau^*}{\partial w}\right)\left(\frac{\partial w}{\partial T_\mathrm{s}}\right) = \frac{|\Gamma_\mathrm{a}| H_\mathrm{a}(b)(b_w w)}{\tau_n^* + bw}. \tag{12.67}$$

Note that $\lambda > 0$, indicating a positive feedback. Substituting in the value for the average water-vapor column amount $\bar{w} = 1.32\ \mathrm{g \cdot cm^{-2}}$ and the values for b, b_w, etc., we find $\lambda = 0.66$. This result is in excellent agreement with numerical climate models and observed variations of clear-sky temperature and water vapor.[38] More elaborate methods yield values of λ in the range 0.59 to 0.77.

If $\lambda \to 1$, $f \to \infty$, which refers to a situation called the *runaway greenhouse*, in which higher temperatures cause more water-vapor IR opacity and still higher temperatures, causing more evaporation, etc. The end result is that all the oceans are evaporated, a fate that may have been experienced by the early atmosphere of Venus. Problem 12.8 addresses the likelihood of this happening on Earth if it experiences a significantly warmer climate in the future.

12.12 Effects of Carbon Dioxide Changes

The observed increase in the mean temperature of the Earth by $\sim 0.5°\mathrm{C}$ since 1900 suggests that the rise of $\mathrm{CO_2}$ from 200 to 240 parts per million may be partly responsible. Figure 1.8 shows the radiative forcings from $\mathrm{CO_2}$, as well as other greenhouse gases, and the changes in these forcings that have occurred since the industrial era.

A formula[39] that approximates accurate climate model results for the $\mathrm{CO_2}$ radiative forcing is

$$N_{\mathrm{lw}}(\chi) = 32 + 6.3 \ln(\chi/\chi_0) \quad [\mathrm{W \cdot m^{-2}}], \tag{12.68}$$

where χ_0 is the present-day concentration of $\mathrm{CO_2}$. A standard scenario for climate change is that of $\mathrm{CO_2}$ doubling – an eventuality that is predicted to occur late in the twenty-first century at the current rate of world industrial output. The formula above predicts a forcing $N_{\mathrm{lw}} = 6.3 \ln 2 = 4.37\ \mathrm{W \cdot m^{-2}}$. The predicted climate response is, from Eq. 12.64, $\Delta T_\mathrm{s} = 1.2 f\,°\mathrm{C}$. Model analyses using empirical data suggest that the gain falls in the range $2 < f < 4$. Thus the rise in mean surface temperature for a doubling of $\mathrm{CO_2}$ is expected to be in the range 2.4–4.8 K, in agreement with accurate model[40] predictions (1.5–4.5 K). However, it should be emphasized that this is an equilibrium response. As will be seen below, the response time can be very long.

Let us use Eq. 12.64 to predict the temperature change over the period 1900 to 1990 due to the known greenhouse forcing (due to all greenhouse gas increases) that has occurred over this period (1.92 $W \cdot m^{-2}$). Equation 12.64 predicts a direct temperature change of 0.36 K. With a gain of 2.9 from the water-vapor feedback effect, the predicted change of 1 K exceeds the observed value (0.5 K). To explain this discrepancy, a number of factors may be responsible. At least part of the problem stems from the neglect of a time lag in the system response. This time lag due to the slow overturning of the mixed layer of the ocean is estimated to be over 100 years.[41] Thus ultimately, time-dependent ocean–atmosphere interactions must be included in any realistic climate model.

12.13 Greenhouse Effect from Individual Gas Species

It is convenient to isolate the radiative forcings from individual greenhouse gases, in order to assess their relative importance. For this purpose we generalize the definition of the greenhouse effect G (Eq. 12.19) to apply to individual spectral bands. The *specific greenhouse effect* is defined as the difference between the surface and TOA fluxes in the spectral interval $\Delta\tilde{\nu}_i$:

$$G_i \equiv F_i^+(z=0) - F_i^+(\text{TOA}). \tag{12.69}$$

Here F_i^+ is defined by Eq. 11.6, but for the spectral interval $\Delta\tilde{\nu}_i$

$$F_i^+(z) = p_i(T_s)\sigma_B T_s^4\left[1 - \langle\epsilon_F^i(0,z)\rangle\right] - \int_0^z dz'\sigma_B T^4(z')p_i[T(z')]\frac{\partial\langle\epsilon_F^i(z',z)\rangle}{\partial z'}. \tag{12.70}$$

Here $p_i(T) \equiv \pi B_{\tilde{\nu}}(T)/\sigma_B T^4$ is the fraction of the Planck spectrum occupied by the band if it were completely opaque over $\Delta\tilde{\nu}_i$, and $\langle\epsilon_F^i\rangle$ is the Planck-weighted (wide-band) flux emittance for the interval i. From Eqs. 11.2 and 11.5, the flux emittance is defined for the mass path u and temperature T as

$$\langle\epsilon_F^i(\tilde{\nu},u)\rangle \equiv \int_{\tilde{\nu}_i} d\tilde{\nu}\left\{1 - 2E_3\left[\int_0^u du'\alpha(\tilde{\nu},u')\right]\right\}, \tag{12.71}$$

where $\alpha(\tilde{\nu},u')$ is the absorption coefficient at wavenumber $\tilde{\nu}$ and at the height corresponding to the vertical mass path u'. E_3 is the exponential integral of order 3 (see Eq. 12.23). We now approximate the surface flux by ignoring the atmospheric contribution (second term in Eq. 12.70). Setting the upward surface flux $F_i^+(z=0)$ equal to $p_i(T_s)\sigma_B T_s^4$, substituting this into Eq. 12.70, and summing over the individual spectral bands for a given species j, we find

$$G_j(u_j) \approx \sigma_B T_s^4 \sum_i p_i(T_s)\langle\epsilon_F^i(u_j)\rangle. \tag{12.72}$$

We find that *the (clear-sky) specific greenhouse effect is approximately the fraction of the Planck flux emitted by the surface that is occupied by all the bands of a specific greenhouse gas*. Clearly, if we ignore band overlap between various spectral bands, $G = \sum_j G_j$.

In Table 11.1, the quantities $p_j \langle \epsilon_F^i \rangle$ are tabulated for current-day values of the vertical mass paths u_j for the various greenhouse gases. The last column in the table shows the specific greenhouse effects calculated from an accurate model. As can be easily verified, a comparison with results from an accurate narrow-band model shows good agreement with our simple model. The accurate results show that H_2O is the dominant contributor to the greenhouse effect (75 $W \cdot m^{-2}$), followed by CO_2 (32 $W \cdot m^{-2}$), O_3 (10 $W \cdot m^{-2}$), and CH_4 (8 $W \cdot m^{-2}$). This important result tells us that *in the atmospheric greenhouse effect the atmospheric blocking effect on surface emission is more important for surface warming than the reduced atmospheric emission from cooler regions*.

More realistic results are obtained by calculating the spectral fluxes with models containing today's globally averaged cloudiness. The model of Kiehl and Trenberth shows that the above results are modified to yield the following values of G_j: CO_2 (24 $W \cdot m^{-2}$), H_2O (51 $W \cdot m^{-2}$). O_3 (7 $W \cdot m^{-2}$), and CH_4 (4 $W \cdot m^{-2}$). Thus clouds alter the numerical values but do not change the relative importance of the various forcings.

12.14 Summary

This chapter was devoted to the fundamentals of the role of radiation in climate. The greenhouse effect was discussed in terms of simple two-stream models that nevertheless include the basic physical principles. We first discussed radiative-equilibrium situations for a clear (transparent in the visible) atmosphere whose underlying surface is heated by the Sun and by downwelling IR radiation. We then considered a more realistic case of solar absorption combined with greenhouse trapping by an optically thick atmosphere. This latter problem illustrates the stratospheric temperature inversion and also the antigreenhouse effect typified by the nuclear-winter scenario. Because most planetary atmospheres, constrained to be in radiative equilibrium, are convectively unstable, we then introduced a convective adjustment to create a simple radiative–convective model based on the notion of a constant lapse rate with an upper effective radiating region. A simple generalization of the gray model allows for transparency through a spectral window, described by the transparency factor β. The notion of radiative forcing is defined. The effects of clouds were included by adding the reflecting and IR-trapping properties of cloud particles. Low clouds were found to generally cool the Earth, and high clouds were found to produce a warming effect. Cloud radiative forcing is defined as the difference between the fluxes at the top of the atmosphere evaluated for a cloudy and clear atmosphere. Aerosol forcing was defined in an analogous manner. An example was given of a positive climate feedback, due to water-vapor increase from enhanced evaporation in a warmer climate. The standard scenario of a carbon-dioxide doubling was shown to lead to an enhanced surface temperature, whose value depends upon the feedback parameter, f, which contains all (positive and negative) feedbacks of the climate system. The time-dependent response must also include the time lag due to the slow overturning of ocean water through the mixed layer and the still slower response of the deep ocean. Finally, the specific greenhouse effect G_i was defined to describe individual clear-air radiative forcings from the principal greenhouse gases. Comparison of approximate values of G_i with accurate results show that greenhouse forcing is predominantly due to the effects of atmospheric blocking of surface emission.

Problems

12.1 Consider the problem addressed in §12.2 from a different viewpoint. Let the model atmosphere consist of a stack of N plane-parallel isothermal layers, each about one optical depth. Consider each layer to be approximately opaque, so that the flux absorptance and emittance are both unity. The IR energy will flow through the stack by a succession of emissions and absorptions. The energy emitted both upward and downward by the nth layer (with temperature T_n) is $\sigma_B T_n^4$. Because this layer is opaque in the IR, it absorbs energy from the two adjoining layers in the amounts $\sigma_B T_{n-1}^4$ and $\sigma_B T_{n+1}^4$. The uppermost layer $n = 1$ radiates to space and therefore receives no energy from above. Its temperature is T_e.

(a) Equate the energy absorbed and emitted by each layer and show that the temperature of the nth layer is

$$T_n = n^{1/4} T_e.$$

(b) In addition to the overlying IR emission, the surface also receives (unattenuated) solar energy in the amount $S(1 - \bar{\rho})/4$. Show that the surface temperature is determined from the balance of absorption and emission,

$$S(1 - \bar{\rho})/4 + \sigma_B T_N^4 = \sigma_B T_s^4,$$

and thus the temperature jump $(T_s - T_N)$ follows naturally from the energy balance at the air–surface interface.

(c) Now consider an additional optically thin layer at the top, having a temperature T_0. Identify the temperature of this layer as the skin temperature. Assume that it has flux emittance and flux absorptance $\epsilon = \alpha \ll 1$. As before, the layer loses energy in both directions but gains it from only one direction. Show that the skin temperature for this outermost layer is

$$T_0 = T_1/(2)^{1/4} = T_e/(2)^{1/4}.$$

12.2 Prove Eqs. 12.38 and 12.39 by using Eq. 12.37 for the TOA flux and requiring that $F_{\mathrm{TOA}} = \sigma_B T_e^4$ for both the window ($\beta \neq 0$) and the nonwindow ($\beta = 0$) cases.

These two equations can be related by requiring that $T(\beta; z)$ bears a fixed ratio to $T(\beta = 0; z)$ at all heights. Show that this ratio is given by $[1 + \beta(\mathcal{G} - 1)]^{-1/4}$.

12.3 By setting $\partial\mathcal{H}_{cs}/\partial z = 0$ (in Eq. 12.42) and by solving for $\tau = \tau_e$ for a radiative–convective atmosphere show that

$$\tau_e/\bar{\mu} \equiv \tau(z_e)/\bar{\mu} = 1 + \frac{4H_a|\Gamma_a|}{T_s - |\Gamma_a|z_e}.$$

12.4 For the problem of §12.2, assume that the absorbing gas is well mixed in the atmosphere and obeys the hydrostatic equation $\partial p_a/\partial z = -\rho_a g$, where p_a is the partial pressure of the absorbing gas and ρ_a is the absorber mass density.

(a) Define the optical depth as $\tau = \langle\alpha_m\rangle \int_z^\infty dz'\rho_a(z')$, where $\langle\alpha_m\rangle$ is the (gray) absorption coefficient per unit mass. Show that

$$T(p_a) = T(\tau = 0)\left[1 + \frac{3}{2}\langle\alpha_m\rangle p_a/g\right]^{1/4}.$$

(b) Show that the radiative equilibrium temperature lapse rate is given by

$$\partial T/\partial z = \frac{-g}{R_a}\frac{\frac{3}{8}\tau}{\left(1 + \frac{3}{2}\tau\right)}.$$

From this result show that the atmosphere becomes convectively unstable only if $c_p > 4R_a$. Here c_p and R_a are the specific heat and gas constant per unit mass, respectively.

(c) From the above results show that for the Earth, where the scale height of the chief absorber (H_2O) is 2 km, the height at which the temperature gradient equals $-6.5\text{K} \cdot \text{km}^{-1}$ is

$$z_{6.5} = 2\log(\tau^*/0.205).$$

Show that this equation yields numerical values in agreement with Table 9.1 of Goody and Yung, *Atmospheric Radiation, Theoretical Basis*, vol. 2, Oxford Univ. Press, 1989, p. 406.

12.5 Estimate the temperature of the stratosphere by approximating it as a single optically thin ozone layer overlying the "stacked-layer" model atmosphere of Problem 12.1. Assume the outer layer ($n = 0$) absorbs solar UV according to the absorptance $\alpha_{uv} \sim \tau_{uv}$. Show that the skin temperature, or stratopause temperature, is approximately given by

$$T_0^4 \approx (1/2)T_1^4 + (\tau_{uv}/\tau_{IR})S_{uv}/4\sigma_B.$$

Here we have approximated $\alpha_{IR} = \epsilon_{IR} = \tau_{IR}$, S_{uv} is the solar constant evaluated for wavelengths $\lambda < 290$ nm, and T_1 is the tropopause temperature, approximately equal to T_e. Find the ratio (τ_{uv}/τ_{IR}) necessary to obtain a mean stratopause temperature of 250 K, where S_{uv} is 0.1% of the solar constant S.

12.6

(a) Use the greenhouse solution $T(\gamma; \tau)$, where $\gamma > 1$ (Eq. 12.30) to show that an atmosphere becomes convectively unstable if γ exceeds a critical value γ_c, given by

$$\frac{(\gamma_c^2 - 1)}{2\gamma_c e}[T(\gamma_c; \tau = \gamma_c\bar{\mu})/T_e]^{-3} = 2/N.$$

Here N is the effective number of degrees of freedom (including translational) of the radiatively active molecular constituent (assumed to also be the only constituent), such that $c_p = NR_a/2$. (For example, $N = 5$ for CO_2.) e is the base of the natural logarithm, and the scale height H_a is assumed to be equal to R_aT_e/g. R_a is the specific gas constant (see Chapter 1).

Hint: First show that the maximum lapse rate, for a fixed γ, occurs where $\tau/\gamma\bar{\mu} = 1$.

(b) This result shows that convection can occur even in a very cold planetary atmosphere, since γ_c is independent of temperature. Estimate the range of corresponding critical values of N (and the corresponding number of atoms in the molecule). Make the argument that convective heat transport should be a common feature of planetary atmospheres, regardless of their solar distances. How would clouds or an upper haze layer influence your conclusion?

12.7 Suppose that the environmental lapse rate Γ_a were to change as a result of altered dynamical heat exchange processes.

(a) Show from Eq. 12.32 that the radiative forcing and the temperature response due to a change $\Delta\Gamma_a$ are given by

$$\Delta G = 4\sigma_B T_s^3(T_s - T_e)(\Delta\Gamma_a/\Gamma_a),$$
$$\Delta T_s = f(T_s - T_e)(\Delta\Gamma_a/\Gamma_a).$$

(b) Find the relative change in lapse rate $(\Delta\Gamma_a/\Gamma_a)$ leading to a global *cooling* that exactly cancels the warming due to a CO_2 doubling (see §12.12). This result should be independent of f.

12.8

(a) Show that the runaway greenhouse effect, $\lambda = 1$, occurs if $|\Gamma_a|H_a b_w \geq 1$ (see Eq. 12.67). Evaluate this factor for current terrestrial conditions, and comment on the likely occurrence of this phenomenon in the future. How would other feedback factors aid or abet this effect?

(b) Generalize Eq. 12.67 to include a spectral window, described by the quantity β. Repeat part (a) above.

Notes

1 In the reference by J. A. Dutton, "An analytical model of atmospheric feedback and global temperature change," *J. of Climate*, **8**, 1122–39, 1995, simple time-dependent climate models are considered. Dutton defines radiative forcing in terms of a small change in the greenhouse effect G, rather than the conventional net flux at the top of the atmosphere, N.

2 Several good introductory treatments are available: Ackerman, T. P., "The physics of the greenhouse effect," pp. 63–80, in *Global Climate Change: Implications, Challenges and Mitigation Measures*, ed. S. K. Majumdar et al., The Pennsylvania Academy of Science, Philadelphia, 1992; Covey, C., "Mechanisms of climatic change," pp. 11–33, in *Global Climate Change: Human and Natural Influences,* Paragon House Publishers, New York, ed. S. F. Singer, 1989.

3 It may be shown that this result also follows from solving *Prototype Problem 1* in the limit $a \to 1$. From Example 5.3, it is necessary to "invert" the vertical coordinate so that the incident radiation is from below. This means setting $F^{\pm} \to F^{\mp}$ and $I \to \sigma_{\mathrm{B}} T_{\mathrm{s}}^4/2\pi\bar{\mu}$ and letting $\tau^* - \tau \to \tau$.

4 The diurnal variation of solar heating will cause a significant daytime gradient across the atmosphere–surface interface.

5 For a thorough description of the state of knowledge early in this century, see Humphreys, W. J., *Physics of the Air*, 2nd ed., McGraw-Hill, New York, 1929. The isothermal stratosphere was discovered earlier by Teisserenc le Bort late in the nineteenth century.

6 For example, see Harrison, E. F., P. Minnis, B. R. Barkstrom, V. Ramanathan, R. D. Cess, and G. G. Gibson, "Seasonal variations of cloud radiative forcing derived from the Earth Radiation Budget Experiment," *J. Geophys. Res.*, **95**, 18,687–703, 1990; and Ramanathan, V., R. D. Cess, E. F. Harrison, P. Minnis, B. R. Barkstrom, E. Ahmad, and D. Hartmann, "Radiative cloud forcing and climate: Results from the Earth Radiation Budget Experiment," *Science*, **243**, 57–63, 1989.

7 Bodyko, M. I., "The effect of solar radiation variations on the climate of the Earth," *Tellus*, **21**, 611–19, 1969; and Bodyko, M. I., *Climate and Life*, Academic Press, New York, 1975.

8 $\bar{\mu}_0$ is the average cosine over the day–night period and over the globe. We are not interested in its exact evaluation, but for those who are, see Cogley, A. C. and W. J. Borucki, "Exponential approximation for daily average solar heating or photolysis," *J. Atmos. Sci.*, **33**, 1347–56, 1976.

9 Because of the height distribution of shortwave heating in Venus's lower atmosphere, the calculated temperature profile actually is quite close to adiabatic equilibrium. See Pollack, J. B., O. B. Toon, and R. Boese, "Greenhouse models of Venus' high surface temperature as constrained by *Pioneer Venus* measurements," *J. Geophys. Res.*, **85**, 8,223–31, 1980.

10 Chandrasekhar, S., *Radiative Transfer*, Dover, New York, pp. 86–87.

11 Rampino, M. R. and S. Self, "Climate–volcanism feedback and the Toba eruption of $\sim$ 74,000 years ago," *Quaternary Res.*, **40**, 269–80, 1993.

12 Early work in this area was reviewed by Pekeris, C. L., "The development and the present status of the heat balance of the atmosphere," MIT Professional Note 5, MIT, Cambridge, MA, 1932. A more modern historical account is provided by F. Möller, "Radiation in the

atmosphere," in *Meteorological Challenges: A History*, D. P. McIntyre, ed., Information Canada, Ottawa, 1972.

13 The theoretical determination of lapse rate involves transport of water in all phases, latent heat exchange by both moist and dry convection, and "wave breaking" due to atmospheric absorption of large-amplitude baroclinic waves.

14 See Fig. 3.4 of the Ph.D. thesis of Y. X. Hu, *A Study of the Link between Cloud Microphysics and Climate Change*, Geophysical Institute Report UAG-R-323, Univ. of Alaska, Fairbanks, AK, 121 pp., 1994.

15 Kiehl, J. T. and K. E. Trenberth, "Earth's annual global mean energy budget," *Bull. Am. Meteorol. Soc.*, **78**, 197–208, 1997.

16 Stephens, G. L. and T. J. Greenwald, "The Earth's radiation budget and its relation to atmospheric hydrology 1. Observations of the clear sky greenhouse effect," *J. Geophys. Res.*, **96**, 15, 311–24, 1991.

17 Caution is called for when comparing radiative forcings in the older literature with modern values. One early definition was the change in the downward surface flux. The modern definition is the change in the flux at the *tropopause*, which characterizes the heating change in the atmospheric column. From the modeling point of view, the most convenient definition includes the response of the stratosphere. Including this (comparitively fast) response into the definition of cloud forcing has the following beneficial effect: Except for ozone forcing, it causes the predicted changes in surface temperature to be dependent only upon the value of the forcing, and independent (or nearly so) of the particular mechanism, whether it be CO_2 doubling, cloud-albedo changes, etc. For more details on this somewhat confusing subject, see Ramanathan, V., "Commentary" in the book *Carbon Dioxide Review*, ed. W. C. Clark, pp. 278–83, Clarendon Press, Oxford, 1982; and *Climate Change 1995: The Science of Climate Change*, Contribution of Working Group I to the Second Assessment Report of the Intergovernmental Panel on Climate Change, ed. J. T. Houghton, L. G. Meiro Filho, B. A. Callander, N. Harris, A. Kattenburg, and K. Maskell, Cambridge Univ. Press, Cambridge, 1996.

18 Hartmann, D. L., V. Ramanathan, A. Berroir, and G. E. Hunt, "Earth radiation budget data and climate research," *Rev. Geophys.*, **24**, 439–68, 1986.

19 The reference by D. L. Hartmann et al. (see previous endnote) provided one of the first comprehensive data sets from the *ERBE* spacecraft mission. A recent study compares the distributions of N simulated by a GCM with *ERBE* data (Chen, C.-T. and E. Roeckner, "Validation of the earth radiation budget as simulated by the Max Planck Institute for Meteorology general circulation model ECHAM4 using satellite observations of the Earth Radiation Budget Experiment," *J. Geophys. Res.*, **101**, 4,269–87, 1996).

20 Cess, R. D. et al., "Intercomparison and interpretation of climate feedback processes in 19 atmospheric GCM's," *J. Geophys. Res.*, **95**, 16,601–15, 1990.

21 Stephens, G. L. and T. J. Greenwald, "The earth's radiation budget and its relation to atmospheric hydrology 2. Observations of cloud effects," *J. Geophys. Res.*, **96**, 15,325–40, 1991.

22 The study by Stephens and Greenwald (see Endnote 21) uses the following data sets: (1) *Nimbus 7* narrow field of view, broadband scanning radiometer data; (2) *ERBE* broadband scanning data; (3) sea-surface temperature data provided by the National Meteorological Center; and (4) the *Nimbus 7* SMMR microwave measurements of liquid water path.

23 See §5.3.2 of K. N. Liou's, *Radiation and Cloud Processes in the Atmosphere*, Oxford Univ. Press, New York, 1992.

24 This subject is reviewed by B. G. Levi, "Clouds cast a shadow of doubt on models of earth climate," *Physics Today*, **48**, pp. 21–23, May, 1995.

25 Hu, Y. X. and K. Stamnes, "An accurate parameterization of the radiative properties of water clouds suitable for use in climate models," *J. Climate*, **6**, 728–42, 1993.

26 See Endnote 21.

27 See Endnote 21. Note that their Fig. 11b, which presents accurate albedo results, has a strong resemblance to our two-stream results.

28 This value is derived from Fig. 12.1 by dividing the average reflected flux (30 $W \cdot m^{-2}$) by the average incident flux $(168 + 30 + 67)$.

29 Manabe, S. and R. F. Strickler, "Thermal equilibrium of the atmosphere with a convective adjustment," *J. Atmos. Sci.*, **21**, 361–85, 1964.

30 See Fig. 3 of Chen and Roeckner (see Endnote 19).

31 See Endnote 21.

32 The term *cloud forcing* was first introduced by T. P. Charlock and V. Ramanathan, "The albedo field and cloud radiative forcing produced by a general circulation model with internally generated cloud optics," *J. Atmos. Sci.*, **42**, 1,408–29, 1985.

33 See Appendix C.

34 See Endnote 16.

35 Hansen, J., A. Lacis, R. Ruedy, and M. Sato, "Potential climate impact of Mount Pinatubo eruption," *Geophys. Res. Lett.*, **19**, 215–18, 1992.

36 Dutton, E. G. and J. R. Christy, "Solar radiative forcing at selected locations and evidence for global lower tropospheric cooling following the eruptions of El Chichon and Pinatubo," *Geophys. Res. Lett.* **19**, 2,313–16, 1992.

37 Stephens, G. L., "On the relationship between water vapor over the oceans and sea surface temperature," *J. Climate*, **3**, 634–45, 1990.

38 The results from three such determinations are listed in Table 2 of the article by J. A. Dutton (see Endnote 1). Note that our definition for λ differs by the factor T_s/T_e from Dutton's definition, since he uses the change in the greenhouse effect ΔG, instead of the radiative forcing N.

39 Houghton, J. T., G. J. Jenkins, and J. J. Ephraums, eds., *IPCC, 1990: Climate Change: The Scientific Assessment*, Cambridge Univ. Press, Cambridge, 364 pp., 1990.

40 Houghton, J. T., L. G. Meiro Filho, B. A. Callander, N. Harris, A. Kattenburg, and K. Maskell, eds., *Climate Change 1995: The Science of Climate Change*, Contribution of Working Group I to the Second Assessment Report of the Intergovernmental Panel on Climate Change, Cambridge Univ. Press, 1996. See also *Solar Influences in Global Change*, National Research Council Press, Washington, DC, 1994, pp. 36–40.

41 Hansen, J., A. Lacis, D. Rind, G. Russell, P. Stone, I. Fung, R. Ruedy, and J. Lerner, "Climate response times: Dependence on climate sensitivity and ocean mixing," *Science*, **229**, 857–9, 1985.

Appendix A

Nomenclature: Glossary of Symbols

So diverse are the number of applications of radiative transfer techniques that over the years a large variety of terms, symbols, and units have been used for the same set of quantities. This has caused much confusion and wasted effort. To rectify this situation many fields of physical science have established standards, sometimes by design and sometimes by de facto recognition of the usage of a single author. Although the issue of a standard set of units is by now settled, at least in the geophysical sciences, it is another matter when it comes to nomenclature and symbols.

An attempt to set forth a uniform set of nomenclature and symbols was made in 1978 by the Radiation Commission of the International Association of Meteorology and Atmospheric Physics. Although the SI system recommended by them has been adopted by most atmospheric and oceanic journals, there is not yet any degree of acceptance of nomenclature. For example, even in texts published in the 1990s, the basic quantity of radiation, the *radiance* L, is still referred to by its traditional name (intensity) and symbol I. Also, the *irradiance* E is frequently referred to by its familiar name *flux* and symbol F. Some of the factors explaining this lack of compliance are: (1) traditional inertia of workers trained in an earlier era; (2) the strong influence of a few classic references, such as that of S. Chandrasekhar; and (3) the uneven acceptance, or lack thereof, of the agreements adopted by the international scientific bodies, such as the International Commission on Illumination, the Commission for Symbols, Units and Nomenclature of the International Union of Pure and Applied Physics.[1]

The subject that is in almost total disarray is that of *symbols*. For example, in 1978 the Radiation Commission advocated abandonment of the traditional symbol for optical depth (τ), substituting the decidedly untraditional δ. The motivation for this is to have

[1] See Wolfe, H. C., "Symbols, units and nomenclature in physics," *Phys. Today*, **15**, 19–30, 1962.

a uniform set of Greek symbols for transmittance (τ), reflectance (ρ), absorptance (α), and emittance (ϵ). We believe that this "flying in the face of tradition" is carrying consistency too far. If nothing else, τ has been the de facto symbol for optical depth for too long. Astrophysicists began this tradition, and they still cling to it today. Our opinion is that it is too firmly ingrained in the literature to change at this late date. We have chosen to retain τ as optical depth and to depart from consistency in the symbols for boundary properties, letting $\mathcal{T}$ represent the transmittance.

Another example of variety in the literature is the choice for the interaction properties – the scattering, absorption, and extinction coefficients. After some thought, we chose the symbols α for absorption coefficient in [m^{-1}], σ for the scattering coefficient, and the traditional k for extinction coefficient ($k = \alpha + \sigma$). Similarly, for the cross sections, we attach the subscript n to denote cross section per molecule, or particle. Thus cross sections [m^2] are $k_n = k/n$, $\sigma_n = \sigma/n$, and $\alpha_n = \alpha/n$. Similarly, the subscript m denotes cross section per unit mass [$m^2 \cdot kg^{-1}$], $\alpha_m = \alpha/\rho$, etc., where ρ is mass density.

With most of the symbols adopted in this book, we attempted to use a "rational" system, or perhaps it could be called more aptly a "user-friendly" system. Very simply, we use the first letter of the name of the quantity as the symbol, either as a capital Arabic letter or a Greek letter. Such schemes are useful in computer system languages, playing the same role as mnemonics. All other things being equal, it is preferable for a computer command to be given as an abbreviation of the function itself: "dir" should be a command for consulting a directory file, "del" should be a command to delete a file, etc.

We have accepted the suggestion by H. C. van de Hulst and have decided to throw on the scrap pile one traditional symbol: $\varpi = \sigma/k$ will be replaced by the symbol a for single-scattering albedo.

This fuss over symbols and nomenclature may be considered to be much ado about nothing. We would agree that there are many more serious issues. What is important is to define the terms and to be at least consistent within the same article (or series of articles by the same authors). We can only promise that our *own* book will be internally consistent (and if there is ever another book, to continue that same consistency).

In this appendix we display our system of symbols, along with their units that are in common use. Note that some symbols are unavoidably shared by more than one term. However, because the quantities represented are sufficiently different, and appear in different contexts, confusion should rarely occur. We have not attempted to list all the various subscripts and superscripts attached to the basic quantities. Hopefully these will be obvious from the context. A few examples follow: "s" refers to "specular" as in specular reflection (but when attached to a temperature refers to the surface value); "d" refers to "diffuse"; "L" refers to "Lambertian" reflection; "i" refers to the "ith" molecular species, etc. We have attempted to avoid double subscripts whenever possible.

It should also be noted that frequency ν, wavelength λ, or wavenumber $\tilde{\nu}$ are used as subscripts when the quantity is per unit frequency, or per unit wavelength, for example, I_ν or F_ν. However, when it is only a *function* of frequency, we do not use the subscript, but denote it as a functional dependence, for example $\tau(\nu)$. We use two notations for averaging: the overbar symbol ($^{-}$) refers to angular (for example, $\bar{I}_\nu$) or global averages (for example $\bar{\rho}$), whereas angle brackets ($\langle\rangle$) denote frequency or ensemble averaging. It should be noted that *esu* refers to the electrostatic unit of charge. Finally, Table A.1 contains all the symbols used more than once in the text.

Table A.1. *Glossary of symbols.*

Symbol	Name	Units
I_ν	intensity, radiance	$W \cdot m^{-2} \cdot Hz^{-1}sr^{-1}$
$\bar{I}_\nu$	average intensity, radiance	$W \cdot m^{-2} \cdot Hz^{-1}sr^{-1}$
I_ν^+, I_ν^-	half-range intensity, radiance	$W \cdot m^{-2} \cdot Hz^{-1}sr^{-1}$
I_n	nth Legendre polynomial of intensity	$W \cdot m^{-2} \cdot Hz^{-1}sr^{-1}$
S_ν	source function	$W \cdot m^{-2} \cdot Hz^{-1}sr^{-1}$
S^*	first-order source function	$W \cdot m^{-2} \cdot Hz^{-1}sr^{-1}$
λ, λ_o	wavelength, wavelength in a vacuum	μm or nm
λ	feedback parameter	
$\mathbf{F}_\nu$	vector flux, irradiance	$W \cdot m^{-2} \cdot Hz^{-1}$
F_ν	flux, irradiance	$W \cdot m^{-2} \cdot Hz^{-1}$
F_ν^+, F_ν^-	half-range flux, irradiance	$W \cdot m^{-2} \cdot Hz^{-1}$
F	frequency-integrated flux, irradiance	$W \cdot m^{-2}$
F^s	extraterrestrial solar flux, irradiance	$W \cdot m^{-2}$
$\mathcal{I}$	incident isotropic intensity	$W \cdot m^{-2} \cdot Hz^{-1}sr^{-1}$
$\mathcal{U}_\nu$	spectral energy density	$J \cdot m^{-3} \cdot Hz^{-1}$
$\mathcal{H}$	heating rate	$W \cdot m^{-3}$
$\mathcal{W}$	warming rate	$K \cdot day^{-1}$
A	area	m^2
A_{ij}	spontaneous emission coefficient ($j < i$)	s^{-1}
B_{ij}	induced absorption ($j > 1$) or emission coefficient ($j < i$)	$m^2 \cdot W^{-1}$
ω	solid angle	sr
ω	angular frequency	s^{-1}
ν	frequency	cycles $\cdot s^{-1}$ or Hz^{-1}
$\hat{\Omega}', \hat{\Omega}$	propagation unit vector in (out)	
$\Omega_x, \Omega_y, \Omega_z$	rectangular components of $\hat{\Omega}$	
$\Omega_o(\theta_o, \phi_o)$	solar beam direction	
Θ	scattering angle	
ϕ, ϕ_0	azimuthal angle, solar azimuthal angle	
θ, θ_0	polar angle, solar polar angle	
μ, μ_0	$\lvert\cos\theta\rvert$, $\lvert\cos\theta_0\rvert$	
u	$\cos\theta$	
$\bar{\mu}$	average cosine over a hemisphere	
B_ν	Planck function	$W \cdot m^{-2} \cdot Hz^{-1}sr^{-1}$

Table A.1 (*cont.*)

Symbol	Name	Units
σ_B	Stefan–Boltzmann constant	$W \cdot m^{-2} \cdot K^{-4}$
T	temperature	K, °C
ρ	mass density	$kg \cdot m^{-3}$
n	concentration or number density	m^{-3}
$k(\nu)$	extinction coefficient	m^{-1}
$\sigma(\nu)$	scattering coefficient	m^{-1}
$\alpha(\nu)$	absorption coefficient	m^{-1}
α	climate sensitivity	$K \cdot W^{-1} \cdot m^2 \cdot s$
k_m, σ_m, α_m	mass extinction, scattering absorption coefficients	$m^2 \cdot kg^{-1}$
k_n, σ_n, α_n	extinction, scattering, absorption cross sections	m^2
s	distance along a ray	
$\tau(\nu)$	extinction optical depth	
$\tau_a(\nu), \tau_s(\nu)$	absorption, scattering optical depths	
$\tau_s(\nu)$	optical path (line of sight)	
j_ν	emission coefficient	$W \cdot m^{-2} \cdot Hz^{-1} \cdot sr^{-1}$
p	scattering phase function	sr^{-1}
a	single-scattering albedo	
χ_l	Legendre polynomial moment of p	
H	scale height	km
H	Chandrasekhar's H-function	
E_{col}	column-averaged thermal energy	J
k_B	Boltzmann's constant	$J \cdot K^{-1}$
R	molar gas constant	$J \cdot K^{-1} \cdot kmol^{-1}$
R_a	specific gas constant	$J \cdot K^{-1} \cdot kg^{-1}$
$\mathcal{N}$	column number, column density	m^{-2}
$\mathcal{M}, u$	column mass, column mass density	$kg \cdot m^{-2}$
M	molecular weight	$kg \cdot kmol^{-1}$
g	acceleration of gravity	$m \cdot s^{-2}$
p, p_i	gas pressure, partial pressure	$N \cdot m^{-2}$
c_p	specific heat at constant pressure	$J \cdot kg^{-1} \cdot K^{-1}$
$\rho(-\hat{\Omega}', \hat{\Omega})$	bidirectional reflectance function (BDRF)	
$\rho(-2\pi, \hat{\Omega})$	hemispherical-directional reflectance	
$\rho(-\hat{\Omega}, 2\pi)$	directional-hemispherical, flux reflectance, or plane albedo	
$\mathcal{T}(-\hat{\Omega}', -\hat{\Omega})$	bidirectional transmittance	
$\mathcal{T}(-\hat{\Omega}', -2\pi)$	directional-hemispherical or flux transmittance	
$\mathcal{T}(-2\pi, -\hat{\Omega})$	hemispherical-directional transmittance	
ρ_∞	flux reflectance of a semi-infinite atmosphere	
$\mathcal{T}_b(-\hat{\Omega}, -\hat{\Omega})$	beam transmittance	
$\mathcal{T}_d(-\hat{\Omega}', -\hat{\Omega})$	diffuse transmittance	

Table A.1 (*cont.*)

Symbol	Name	Units
$\mathcal{T}_F$	flux transmittance	
$\rho_s(\theta)$	specular reflection function	
$\rho_d(-\hat{\Omega}', \hat{\Omega})$	diffuse reflection function	
ρ_L	Lambert reflectance	
$\bar{\rho}$	spherical or Bond albedo	
$\alpha(-\hat{\Omega})$	directional absorptance	
$\epsilon(\hat{\Omega})$	directional emittance	
$\epsilon_F, \epsilon(2\pi)$	flux emittance	
ϵ_v	volume emittance	
$\alpha_F, \alpha(-2\pi)$	flux absorptance	
$\epsilon_F, \epsilon(2\pi)$	flux emittance	
α	backscattering or phase angle	
$A(\nu)$	action spectrum	
$\eta(\nu)$	quantum yield	
$\delta(x)$	Dirac δ-function	x^{-1}
δ_{ij}	Kronecker delta	
$r, r_\oplus$	planetary, (Earth) solar distance	km or AU
E_n	exponential integral of order n	
Λ	inverse thermalization length	
Λ_l^m	normalized associated Legendre polynomial	
P_l^m	associated Legendre polynomial	
P_l	Legendre polynomial	
r	diffusivity factor $(= \bar{\mu}^{-1})$	
q	Hopf function	
G	greenhouse effect	$W \cdot m^{-2}$
$\mathcal{G}$	greenhouse factor	
I^m	azimuthal intensity component in a Legendre polynomial expansion	$W \cdot m^{-2} \cdot sr^{-1}$
$\bar{p}$	azimuthally averaged phase function	
b	backscattering fraction	
g, χ_1	asymmetry factor	
V	volume	m^3
z, z_0	height, height of reference level	
N_a	Avogadro's number	m^{-3}
$R, R_\oplus$	radius of planet, radius of Earth	km
N	radiative forcing	$W \cdot m^{-2}$
w	volume mixing ratio or molar fraction	
w^m	mass mixing ratio	$gm \cdot gm^{-1}$
h	ocean depth	m
σ	conductivity	
$\nabla \cdot$	gradient operator	m^{-1}
$\tilde{\nu}$	wavenumber	cm^{-1}

Table A.1 (*cont.*)

Symbol	Name	Units
c, c_o	speed of light, in vacuum	$m \cdot s^{-1}$
i	$\sqrt{-1}$	
m	index of refraction ($m = m_r + i \cdot m_i$)	
m_r, m_i	real, imaginary parts of index of refraction	
$I_\parallel$, $I_\perp$	intensity components of plane wave	$W \cdot m^{-2}$
$\mathcal{E}$	radiative energy	J
h	Planck's constant	$J \cdot s$
α_p	molecular polarizability	$esu \cdot m$
mfp	photon mean free path	m
f_ν	photon distribution function	$phot \cdot m^{-3} \cdot s^{-1} \cdot Hz^{-1} \cdot sr^{-1}$
$p(E_i)$	Gibbs distribution of energy E_i	J^{-1}
Q	partition function	
$\Phi(k)$	density of states with wavenumber k	$m^{-3} \cdot s^2$
$f(\hat{\Omega}_0, \hat{\Omega})$	term in Lommel–Seeliger model	
σ_{el}	inelastic collisional cross section	
k_{el}, k_{in}	elastic, inelastic excitation coefficients	$cm^3 \cdot s^{-1}$
$\cos \xi_i$	direction cosine in ith direction	
J	photolysis rate	$m^{-3} \cdot s^{-1}$
D	UV dose rate	$W \cdot m^{-3} \cdot s^{-1}$
$\bar{\rho}$	spherical (Bond) albedo	
$\bar{\mathcal{T}}$	spherical transmittance	
$\bar{\alpha}$	spherical absorptance	
$\mathcal{P}(\tau, \mu)$	directional escape probability	
$\mathcal{P}(\tau)$	hemispherical escape probability	
P_n	escape probability after n scatterings	
Ch	Chapman function	
$\Gamma(n)$	Lamda iteration operator for nth-order scattering	
ρ_L, $\rho_L(2\pi)$	Lambert reflectance, (flux) reflectance	
$\mathcal{D}$	denominator in two-stream solution	
$q(\tau)$	Hopf function	
$\mathcal{G}$	point-direction gain	sr^{-1}
f	climate gain	
Q	temperature-dependent climate parameter	
Q	extinction efficiency	
Q	solar insolation	$J \cdot m^{-2}$
D, L	ice crystal width, length	m
$X(\mu)$, $Y(\mu)$	Chandrasekhar X- and Y-functions	
ω_j	quadrature weight	
λ	eigenvalue of discrete-ordinate solution	
E_i	energy of the ith quantum state	J

Table A.1 (*cont.*)

Symbol	Name	Units
$E(\nu)$	instantaneous spectral radiative energy	
AB, AB^*	ground-state diatomic, excited molecule	
σ_c, σ_{el}	collision, elastic cross section	m^2
k_{in}	collision excitation cross section	m^2
k'_{in}	deexcitation or quenching cross section	m^2
$f_{MB}(\mathbf{v})$	Maxwell–Boltzmann distribution	$s \cdot m^{-1}$
n_j	concentration, or volume density of jth quantum state	m^{-3}
g_j	degeneracy, or statistical weight of jth state	
t_{coll}	collisional lifetime of quantum state	s
$\Phi(\nu)$	normalized absorption line profile	Hz^{-1}
$k^*(\nu)$	absorption coefficient in LTE	m^{-1}
C_{ij}	collisional excitation coefficient per molecule	s^{-1}
J	absorption profile-weighted mean intensity	$W \cdot m^{-2}$
ϵ_v	volume emittance	
$g(k)$	cumulative k-distribution	
$k(g)$	inverse k-distribution	
$f(k)$	k-distribution function	cm
$\mathcal{L}$, $\mathcal{L}^{-1}$	Laplace, inverse Laplace transform	
$\tilde{u}$	scaled mass path	$kg \cdot m^{-2}$
$\tilde{p}$	scaled pressure along path	$N \cdot m^{-2}$
G, F, T	vibrational, rotational, electronic term values	cm^{-1}
ω_e, $\omega_e x_e$	vibrational coefficients for electronic term values	cm^{-1}
K_u	diffuse extinction coefficient for upward flux	m^{-1}
K_d	diffuse extinction coefficient for downward flux	m^{-1}
K_0	diffuse extinction coefficient for mean intensity	m^{-1}
$\rho_{\bar{I}}$	remotely sensed reflectance	
ρ_l	specific gravity of liquid	
S	solar constant	$W \cdot m^{-2}$
$\mathcal{L}$	molecular angular momentum	$kg \cdot m^2 \cdot s^{-1}$
I	molecular moment of inertia	$kg \cdot m^2$
B_v, B_r	vibrational, rotational constant	cm^{-1}
D_v	anharmonic rotational constant	cm^{-1}
$\tilde{\nu}$	wavenumber	cm^{-1}
$\mathcal{S}_n$, $\mathcal{S}$	line strength per molecule, per unit mass	$m^2 \cdot kg^{-1}$
γ	natural damping coefficient	s^{-1}
m_e	electron mass	kg
e	electron charge	esu
$\mathcal{P}$	bulk polarization	$esu \cdot m^{-2}$
α_L, α_D, α_V	Lorentz, Doppler, Voigt half-width	Hz
$\langle \alpha_b \rangle$	mean beam absorptance	
$\langle \mathcal{T}_b \rangle$	mean beam transmittance	

Table A.1 (*cont.*)

Symbol	Name	Units
$\mathcal{T}_F$	flux transmittance	
W	equivalent width (wavenumbers)	cm^{-1}
$p(\mathcal{S})$	line-strength distribution of $\mathcal{S}$	$kg \cdot m^2$
$p_i(T)$	fraction of Planck flux within the ith band	
d	average line width over a spectral interval	cm^{-1}
$f(k)$	k-distribution	$kg \cdot m^2$
$g(k)$	cumulative k-distribution	
CSF	cooling-to-space function	
CF	Chapman function	
CF	Cloud forcing	$W \cdot m^{-2}$
$\Gamma_d, \Gamma_m, \Gamma_a$	lapse rate (dry, moist, environmental)	$K \cdot km^{-1}$

Appendix B

Physical Constants

Table B.1. *Physical constants.*[a]

Physical constant	Value
Stefan–Boltzmann constant, σ_B	5.67032×10^{-8} W · m^{-2} · K^{-4}
speed of light in a vacuum, c_o	2.997925×10^{8} m · s^{-1}
Boltzmann's constant, k_B	1.380622×10^{-23} J · K^{-1}
Planck's constant, h	6.626196×10^{-34} J · s
Loschmidt's number, n_L	2.686754×10^{25} mol · m^{-3}
Avogadro's number, N_a	6.022169×10^{26} mol · kmol^{-1}
molar gas constant, R	8.31434×10^{3} J · K^{-1} · kmol^{-1}
specific gas constant, R_a	2.8700×10^{2} J · K^{-1} · kg^{-1}
mean molecular mass, $\bar{M}$	28.964 kg · kmol^{-1}
standard surface pressure (1 bar)	1.01325×10^{5} Pa
standard temperature	0°C, 273.16 K
standard density	1.2925 kg · m^{-3}
atomic mass unit (amu)	1.660531×10^{-27} kg
specific heat of dry air (STP), c_p	1.006×10^{3} J · kg^{-1} · K^{-1}
mass of an electron, m_e	9.109558×10^{-31} kg
charge of an electron, e	1.602192×10^{-19} coulombs (C)
permittivity of a vacuum, ϵ_o	8.854199×10^{-12} C^{2} · N^{-1} · m^{-2}
astronomical unit, $r_\oplus$ or AU	1.4960×10^{11} m
acceleration of gravity (sea level, 45° lat.), g	9.80616 m · s^{-2}
solar constant, S	1,368 ± 5 W · m^{-2}
mean Earth radius, $R_\oplus$	6,371 km

[a] *Source:* B. N. Taylor, W. H. Parker, and D. N. Langenberg, "Determination of *e/h*, using macroscopic quantum phase coherence in superconductors: Implications for quantum electrodynamics and the fundamental physical constants," *Reviews of Modern Physics*, **41**, 375–496, 1969.

Appendix C

Model Atmospheres

Reference atmospheric models have long been used for a variety of purposes: validation of theoretical atmospheric models, intercomparison of radiation codes, and design and mission planning for aerospace systems, to mention a few applications. Standard atmospheres typically provide numerical values for the thermodynamic variables (pressure, temperature, density, etc.) and concentrations of minor species.

Atmospheric models may be considered to be the result of an ingestion of the best data sets available at the time; consequently, the older models (such as the 1976 U. S. Standard Atmosphere) do not represent the current state-of-the-art knowledge. For our purposes they are most useful for comparing the output of radiation codes.[1]

In Tables C.1 to C.6 we present six model atmospheres, which contain, among other variables, the temperature and constituent concentrations for H_2O, CO_2, O_3, and NO_2. With the exception of CH_4 (not listed) these five species are the most important infrared-active gases in Earth's atmosphere. It should be mentioned that the tabulated values of H_2O and NO_2 in the upper stratosphere and mesosphere are not to be trusted as representative of the Earth's upper regions, since these models were created more than twenty-five years ago.[2] Modern remote sensing measurements of these constituents provide much more accurate values of H_2O, O_3, and NO_2,[3] as well as other species, such as NO and O.

[1] For example, see R. G. Ellingson and Y. Fouquart, "The intercomparison of radiation codes in climate models: An overview," *J. Geophys. Res.*, **96**, 8925–27, 1991.

[2] McClatchey, R. A., R. W. Fenn, J. E. A. Selby, F. E. Volz, and J. S. Garing, "Optical properties of the atmosphere," Rep. AFCRL-71-0279, 85 pp., Air Force Cambridge Res. Lab., Bedford, MA, 1973.

[3] Rees, D., J. J. Barnett and K. Labitzke (eds.), COSPAR International Reference Atmosphere, Part II: Middle Atmosphere Models in Advances in Space Research, **10**, No. 12, 328–34, Permagon Press, Elsevier Science Inc., Tarrytown, NY, 1990.

Table C.1. *AFGL atmospheric constituent profiles, U.S. standard atmosphere 1976 (AFGL-TR-86-0110).*

z (km)	p (mb)	T (K)	air (cm^{-3})	O_3 (cm^{-3})	O_2 (cm^{-3})	H_2O (cm^{-3})	CO_2 (cm^{-3})	NO_2 (cm^{-3})
100.000	0.00032	195.100	1.187967E+13	4.756001E+06	1.902400E+12	4.756001E+06	2.318550E+09	2.021300E+03
95.000	0.00076	188.400	2.921760E+13	2.046800E+07	5.263200E+12	1.578960E+07	7.894800E+09	5.146240E+03
90.000	0.00184	186.900	7.130506E+13	4.995200E+07	1.355840E+13	6.065600E+07	2.212160E+10	1.305888E+04
85.000	0.00446	188.900	1.710073E+14	8.555000E+07	3.422000E+13	2.275630E+08	5.475200E+10	3.285120E+04
80.000	0.01050	198.600	3.829322E+14	1.149600E+08	8.008879E+13	7.855600E+08	1.256896E+11	7.740640E+04
75.000	0.02400	208.400	8.341139E+14	2.086750E+08	1.744523E+14	2.358028E+09	2.754510E+11	1.794605E+05
70.000	0.05220	219.600	1.721670E+15	5.169000E+08	3.601070E+14	6.030500E+09	5.685899E+11	3.980130E+05
65.000	0.10900	233.300	3.383947E+15	2.370200E+09	7.076740E+14	1.422120E+10	1.117380E+12	8.566580E+05
60.000	0.21900	247.000	6.421832E+15	7.068600E+09	1.343034E+15	3.052350E+10	2.120580E+12	1.831410E+06
55.000	0.42500	260.800	1.180302E+16	2.125800E+10	2.468290E+15	6.023100E+10	3.897300E+12	4.003590E+06
50.000	0.79780	270.700	2.134605E+16	6.621600E+10	4.464240E+15	1.116060E+11	7.048799E+12	9.462480E+06
47.500	1.09000	270.600	2.917498E+16	1.197200E+11	6.102800E+15	1.533000E+11	9.636000E+12	1.944720E+07
45.000	1.49100	264.200	4.087489E+16	2.147250E+11	8.548099E+15	2.137025E+11	1.349700E+13	4.703500E+07
42.500	2.06000	257.300	5.798815E+16	3.597860E+11	1.212827E+16	2.988545E+11	1.914990E+13	1.259251E+08
40.000	2.87100	250.400	8.304447E+16	6.066300E+11	1.736790E+16	4.175775E+11	2.742300E+13	3.348930E+08
37.500	4.15000	242.900	1.237464E+17	9.656401E+11	2.587420E+16	6.128100E+11	4.085400E+13	7.749880E+08
35.000	5.74600	236.500	1.759731E+17	1.380096E+12	3.680490E+16	8.628900E+11	5.811300E+13	1.282008E+09
32.500	8.01000	230.000	2.522415E+17	1.860945E+12	5.275160E+16	1.217830E+12	8.329200E+13	1.819804E+09
30.000	11.97000	226.500	3.827699E+17	2.509799E+12	8.004700E+16	1.809675E+12	1.263900E+14	2.359280E+09
27.500	17.43000	224.000	5.635873E+17	3.272892E+12	1.178760E+17	2.580300E+12	1.861200E+14	2.712840E+09
25.000	25.49000	221.600	8.331283E+17	4.266877E+12	1.742433E+17	3.689123E+12	2.751210E+14	3.118038E+09
24.000	29.72000	220.600	9.757872E+17	4.518265E+12	2.040885E+17	4.198950E+12	3.222450E+14	2.988090E+09

Table C.1. *(cont.)*

z (km)	p (mb)	T (K)	air (cm^{-3})	O_3 (cm^{-3})	O_2 (cm^{-3})	H_2O (cm^{-3})	CO_2 (cm^{-3})	NO_2 (cm^{-3})
23.000	34.67000	219.600	1.143492E+18	4.768192E+12	2.390960E+17	4.804800E+12	3.775200E+14	2.951520E+09
22.000	40.47000	218.600	1.340895E+18	4.894274E+12	2.804780E+17	5.455230E+12	4.428600E+14	2.898720E+09
21.000	47.29000	217.600	1.574064E+18	4.769100E+12	3.291750E+17	6.260625E+12	5.197500E+14	2.772000E+09
20.000	55.29000	216.700	1.847990E+18	4.768571E+12	3.864410E+17	7.211100E+12	6.101700E+14	2.570110E+09
19.000	64.67000	216.700	2.161503E+18	4.390890E+12	4.520670E+17	8.327550E+12	7.137900E+14	2.292780E+09
18.000	75.65000	216.700	2.528494E+18	4.015110E+12	5.287700E+17	9.677249E+12	8.349000E+14	1.950630E+09
17.000	88.50000	216.700	2.957987E+18	3.513520E+12	6.186399E+17	1.139600E+13	9.768000E+14	1.536240E+09
16.000	103.50000	216.700	3.459340E+18	3.012286E+12	7.235580E+17	1.367490E+13	1.142460E+15	1.104378E+09
15.000	121.10000	216.700	4.047595E+18	2.634525E+12	8.464500E+17	2.025000E+13	1.336500E+15	6.925500E+08
14.000	141.70000	216.700	4.736121E+18	2.383717E+12	9.904510E+17	2.808805E+13	1.563870E+15	3.544772E+08
13.000	165.80000	216.700	5.541629E+18	2.132992E+12	1.159114E+18	6.017410E+13	1.830180E+15	2.467970E+08
12.000	194.00000	216.700	6.484174E+18	2.008345E+12	1.356201E+18	1.236803E+14	2.141370E+15	2.044035E+08
11.000	227.00000	216.800	7.583652E+18	1.630876E+12	1.586101E+18	2.741906E+14	2.504370E+15	1.988318E+08
10.000	265.00000	223.300	8.595457E+18	1.129443E+12	1.797818E+18	6.017959E+14	2.838660E+15	2.047276E+08
9.000	308.00000	229.700	9.711841E+18	8.910379E+11	2.031271E+18	1.538518E+15	3.207270E+15	2.254808E+08
8.000	356.50000	236.200	1.093179E+19	6.526804E+11	2.286460E+18	4.011698E+15	3.610200E+15	2.516200E+08
7.000	411.10001	242.700	1.226845E+19	6.151052E+11	2.566520E+18	7.024160E+15	4.052400E+15	2.824400E+08
6.000	472.20001	249.200	1.372429E+19	5.645776E+11	2.869570E+18	1.270574E+16	4.530900E+15	3.157900E+08
5.000	540.50000	255.700	1.531006E+19	5.772576E+11	3.201880E+18	2.140204E+16	5.055600E+15	3.523600E+08
4.000	616.59998	262.200	1.703267E+19	5.771448E+11	3.561360E+18	3.677232E+16	5.623200E+15	3.919200E+08
3.000	701.20001	268.700	1.890105E+19	6.274337E+11	3.952190E+18	6.017162E+16	6.240300E+15	4.349300E+08
2.000	795.00000	275.200	2.092331E+19	6.778279E+11	4.376460E+18	9.697315E+16	6.910200E+15	4.816201E+08
1.000	898.79999	281.700	2.310936E+19	6.779402E+11	4.834170E+18	1.404222E+17	7.632900E+15	5.319900E+08
0.000	1013.00000	288.200	2.545818E+19	6.777680E+11	5.325320E+18	1.973426E+17	8.408400E+15	5.860400E+08

Table C.2. *AFGL atmospheric constituent profiles, tropical (AFGL-TR-86-0110).*

z (km)	p (mb)	T (K)	air (cm^{-3})	O_3 (cm^{-3})	O_2 (cm^{-3})	H_2O (cm^{-3})	CO_2 (cm^{-3})	NO_2 (cm^{-3})
100.000	0.00029	190.700	1.097638E+13	4.392000E+06	1.756800E+12	4.392000E+06	2.141100E+09	1.866600E+03
95.000	0.00069	184.300	2.703802E+13	1.353000E+07	4.870800E+12	1.461240E+07	7.306201E+09	4.762560E+03
90.000	0.00172	177.000	7.038287E+13	3.662360E+07	1.338170E+13	5.986550E+07	2.183330E+10	1.288869E+04
85.000	0.00440	177.100	1.799476E+14	9.005000E+07	3.602000E+13	2.341300E+08	5.763200E+10	3.457920E+04
80.000	0.01100	184.800	4.311244E+14	1.423620E+08	9.016260E+13	9.059399E+08	1.414992E+11	8.714280E+04
75.000	0.02600	201.800	9.331769E+14	1.680840E+08	1.951642E+14	3.081540E+09	3.081540E+11	2.007670E+05
70.000	0.05800	218.900	1.919084E+15	5.760000E+08	4.012800E+14	8.640000E+09	6.336000E+11	4.435200E+05
65.000	0.12100	236.000	3.713515E+15	2.415400E+09	7.766440E+14	2.006640E+10	1.226280E+12	9.401481E+05
60.000	0.23900	253.100	6.839393E+15	7.528400E+09	1.430396E+15	4.106400E+10	2.258520E+12	1.950540E+06
55.000	0.45600	263.400	1.253894E+16	2.259000E+10	2.622950E+15	7.530000E+10	4.141500E+12	4.254450E+06
50.000	0.85400	270.200	2.289203E+16	6.414800E+10	4.788190E+15	1.374600E+11	7.560300E+12	1.014913E+07
47.500	1.16000	269.600	3.116377E+16	1.076055E+11	6.518710E+15	1.840210E+11	1.029270E+13	2.077254E+07
45.000	1.59000	264.800	4.349016E+16	1.958400E+11	9.095680E+15	2.480640E+11	1.436160E+13	5.004800E+07
42.500	2.20000	259.400	6.142774E+16	3.626730E+11	1.284723E+16	3.380850E+11	2.028510E+13	1.333899E+08
40.000	3.05000	254.000	8.697170E+16	6.527249E+11	1.818927E+16	4.525560E+11	2.871990E+13	3.507309E+08
37.500	4.26000	248.500	1.241638E+17	1.093840E+12	2.597870E+16	6.090700E+11	4.101900E+13	7.781180E+08
35.000	6.00000	243.100	1.787632E+17	1.735330E+12	3.739010E+16	8.229400E+11	5.903700E+13	1.302392E+09
32.500	8.52000	237.700	2.596105E+17	2.559030E+12	5.429820E+16	1.117140E+12	8.573400E+13	1.873158E+09
30.000	12.20000	232.300	3.803842E+17	3.540510E+12	7.956630E+16	1.522800E+12	1.256310E+14	2.345112E+09
27.500	17.63000	227.000	5.625204E+17	4.390620E+12	1.176461E+17	2.026440E+12	1.857570E+14	2.707549E+09
25.000	25.70000	221.400	8.407510E+17	4.543020E+12	1.758317E+17	2.734225E+12	2.776290E+14	3.146462E+09
24.000	30.00000	219.200	9.912712E+17	4.265600E+12	2.073280E+17	3.174400E+12	3.273600E+14	3.035520E+09

Table C.2. *(cont.)*

z (km)	p (mb)	T (K)	air (cm^{-3})	O_3 (cm^{-3})	O_2 (cm^{-3})	H_2O (cm^{-3})	CO_2 (cm^{-3})	NO_2 (cm^{-3})
23.000	35.00000	217.000	1.168208E+18	3.974600E+12	2.443210E+17	3.390100E+12	3.857700E+14	3.016020E+09
22.000	40.90000	214.600	1.380402E+18	3.314400E+12	2.886290E+17	3.866800E+12	4.557300E+14	2.982960E+09
21.000	48.00000	210.700	1.650017E+18	2.971800E+12	3.450590E+17	4.375150E+12	5.448300E+14	2.905760E+09
20.000	56.50000	206.700	1.979793E+18	2.773400E+12	4.140290E+17	5.150600E+12	6.537300E+14	2.753590E+09
19.000	66.60000	202.700	2.379755E+18	2.261950E+12	4.976290E+17	6.190600E+12	7.857300E+14	2.523860E+09
18.000	78.90000	198.800	2.874567E+18	1.438500E+12	6.012930E+17	7.911750E+12	9.494100E+14	2.218167E+09
17.000	93.70000	194.800	3.483874E+18	8.715000E+11	7.285740E+17	1.010940E+13	1.150380E+15	1.809234E+09
16.000	111.00000	197.000	4.081019E+18	5.897296E+11	8.535560E+17	1.225200E+13	1.347720E+15	1.302796E+09
15.000	132.00000	203.700	4.693478E+18	5.899431E+11	9.816730E+17	1.878800E+13	1.550010E+15	8.031870E+08
14.000	156.00000	210.300	5.372756E+18	5.645850E+11	1.123793E+18	3.344494E+13	1.774410E+15	4.021996E+08
13.000	182.00000	217.000	6.074680E+18	5.646784E+11	1.270511E+18	6.018210E+13	2.006070E+15	2.705155E+08
12.000	213.00000	223.600	6.899532E+18	5.395476E+11	1.442936E+18	2.005612E+14	2.278320E+15	2.174760E+08
11.000	247.00000	230.100	7.774853E+18	5.144914E+11	1.626020E+18	5.684068E+14	2.567400E+15	2.038360E+08
10.000	286.00000	237.000	8.740364E+18	4.893947E+11	1.828123E+18	1.672426E+15	2.886510E+15	2.081786E+08
9.000	329.00000	243.600	9.782062E+18	4.894500E+11	2.045901E+18	4.011532E+15	3.230370E+15	2.271048E+08
8.000	378.00000	250.300	1.093812E+19	4.895745E+11	2.288550E+18	8.362515E+15	3.613500E+15	2.518500E+08
7.000	432.00000	257.000	1.217482E+19	5.143614E+11	2.545620E+18	1.570002E+16	4.019400E+15	2.801400E+08
6.000	492.00000	263.600	1.351859E+19	5.397117E+11	2.827770E+18	2.842653E+16	4.464900E+15	3.111900E+08
5.000	559.00000	270.300	1.497882E+19	5.646733E+11	3.132910E+18	5.015654E+16	4.946700E+15	3.447700E+08
4.000	633.00000	277.000	1.655144E+19	5.897016E+11	3.461040E+18	7.354296E+16	5.464800E+15	3.808800E+08
3.000	715.00000	283.700	1.825402E+19	6.401807E+11	3.818430E+18	1.571220E+17	6.029100E+15	4.202100E+08
2.000	805.00000	287.700	2.026599E+19	6.777576E+11	4.238520E+18	3.110952E+17	6.692400E+15	4.664400E+08
1.000	904.00000	293.700	2.229340E+19	7.027649E+11	4.662790E+18	4.348219E+17	7.362299E+15	5.131300E+08
0.000	1013.00000	299.700	2.448130E+19	7.029050E+11	5.120500E+18	6.352850E+17	8.085000E+15	5.635000E+08

Table C.3. *AFGL atmospheric constituent profiles, midlatitude summer (AFGL-TR-86-0110).*

z (km)	p (mb)	T (K)	air (cm^{-3})	O_3 (cm^{-3})	O_2 (cm^{-3})	H_2O (cm^{-3})	CO_2 (cm^{-3})	NO_2 (cm^{-3})
100.000	0.00026	190.500	9.809267E+12	3.926400E+06	1.570560E+12	3.926400E+06	1.914120E+09	1.668720E+03
95.000	0.00062	178.300	2.538870E+13	1.778700E+07	4.573800E+12	1.372140E+07	6.860700E+09	4.472160E+03
90.000	0.00164	165.000	7.198993E+13	5.403000E+07	1.368760E+13	6.123400E+07	2.233240E+10	1.318332E+04
85.000	0.00448	165.100	1.965363E+14	1.121190E+08	3.934000E+13	2.616110E+08	6.294400E+10	3.776640E+04
80.000	0.01200	174.100	4.992227E+14	9.992000E+07	1.044164E+14	1.049160E+09	1.638688E+11	1.009192E+05
75.000	0.03000	196.100	1.108040E+15	2.107100E+08	2.317810E+14	3.271550E+09	3.659700E+11	2.384350E+05
70.000	0.06700	218.100	2.225005E+15	8.908001E+08	4.654430E+14	8.239900E+09	7.349100E+11	5.144370E+05
65.000	0.13900	240.100	4.193093E+15	3.356800E+09	8.769640E+14	1.846240E+10	1.384680E+12	1.061588E+06
60.000	0.27200	257.100	7.662644E+15	9.968400E+09	1.602612E+15	3.834000E+10	2.530440E+12	2.185380E+06
55.000	0.51500	269.300	1.385105E+16	2.494800E+10	2.896740E+15	7.415099E+10	4.573800E+12	4.698540E+06
50.000	0.95100	275.700	2.498363E+16	7.000000E+10	5.225000E+15	1.375000E+11	8.250000E+12	1.107500E+07
47.500	1.29000	275.200	3.395104E+16	1.189300E+11	7.101820E+15	1.868900E+11	1.121340E+13	2.263068E+07
45.000	1.76000	269.900	4.723040E+16	2.126700E+11	9.877339E+15	2.575670E+11	1.559580E+13	5.434900E+07
42.500	2.41000	263.700	6.619401E+16	3.908160E+11	1.384416E+16	3.510720E+11	2.185920E+13	1.437408E+08
40.000	3.33000	257.500	9.366532E+16	7.076615E+11	1.958957E+16	4.780230E+11	3.093090E+13	3.777319E+08
37.500	4.64000	251.300	1.337326E+17	1.164060E+12	2.796420E+16	6.690000E+11	4.415400E+13	8.375880E+08
35.000	6.52000	245.200	1.925923E+17	1.715030E+12	4.027430E+16	9.538649E+11	6.359100E+13	1.402856E+09
32.500	9.30000	239.000	2.818363E+17	2.284200E+12	5.893800E+16	1.367700E+12	9.306000E+13	2.033220E+09
30.000	13.20000	233.700	4.090977E+17	2.865800E+12	8.556460E+16	1.924180E+12	1.351020E+14	2.521904E+09
27.500	19.07000	228.450	6.046045E+17	3.630000E+12	1.264450E+17	2.692250E+12	1.996500E+14	2.910050E+09
25.000	27.70000	225.100	8.912839E+17	4.281120E+12	1.864071E+17	3.745980E+12	2.943270E+14	3.335706E+09
24.000	32.20000	223.900	1.041630E+18	4.168000E+12	2.177780E+17	4.168000E+12	3.438600E+14	3.188520E+09

Table C.3. *(cont.)*

z (km)	p (mb)	T (K)	air (cm^{-3})	O_3 (cm^{-3})	O_2 (cm^{-3})	H_2O (cm^{-3})	CO_2 (cm^{-3})	NO_2 (cm^{-3})
23.000	37.60000	222.800	1.222319E+18	4.158200E+12	2.556070E+17	4.708550E+12	4.035900E+14	3.155340E+09
22.000	43.70000	221.600	1.428313E+18	4.144100E+12	2.986610E+17	5.144399E+12	4.715700E+14	3.086640E+09
21.000	51.00000	220.400	1.675986E+18	4.024800E+12	3.504930E+17	5.785650E+12	5.534100E+14	2.951520E+09
20.000	59.50000	219.200	1.966021E+18	3.934000E+12	4.111030E+17	6.491100E+12	6.491100E+14	2.734130E+09
19.000	69.50000	217.900	2.310146E+18	3.468000E+12	4.832080E+17	7.398400E+12	7.629600E+14	2.450720E+09
18.000	81.20000	216.800	2.712742E+18	2.715000E+12	5.674350E+17	8.552250E+12	8.959500E+14	2.093265E+09
17.000	95.00000	215.700	3.189961E+18	2.234400E+12	6.671280E+17	1.021440E+13	1.053360E+15	1.656648E+09
16.000	111.00000	215.700	3.727217E+18	2.238000E+12	7.795700E+17	1.230900E+13	1.230900E+15	1.189870E+09
15.000	130.00000	215.700	4.365209E+18	2.184000E+12	9.129120E+17	1.485120E+13	1.441440E+15	7.469280E+08
14.000	153.00000	215.700	5.137515E+18	2.262040E+12	1.074469E+18	2.570500E+13	1.696530E+15	3.845468E+08
13.000	179.00000	215.800	6.007772E+18	1.803600E+12	1.256508E+18	4.809600E+13	1.983960E+15	2.675340E+08
12.000	209.00000	222.300	6.809554E+18	1.519522E+12	1.424126E+18	2.006042E+14	2.248620E+15	2.146410E+08
11.000	243.00000	228.800	7.692404E+18	1.380251E+12	1.608882E+18	7.356209E+14	2.540340E+15	2.016876E+08
10.000	281.00000	235.300	8.649604E+18	1.128742E+12	1.809104E+18	2.139763E+15	2.856480E+15	2.060128E+08
9.000	324.00000	241.700	9.709126E+18	1.079448E+12	2.030644E+18	4.011736E+15	3.206280E+15	2.254112E+08
8.000	372.00000	248.200	1.085558E+19	9.910836E+11	2.269740E+18	7.019904E+15	3.583800E+15	2.497800E+08
7.000	426.00000	254.700	1.211414E+19	9.409968E+11	2.533080E+18	1.236240E+16	3.999600E+15	2.787600E+08
6.000	487.00000	261.200	1.350416E+19	8.657207E+11	2.823590E+18	2.040010E+16	4.458300E+15	3.107300E+08
5.000	554.00000	267.200	1.501707E+19	8.284536E+11	3.141270E+18	3.344175E+16	4.959900E+15	3.456900E+08
4.000	628.00000	273.200	1.664910E+19	8.031786E+11	3.481940E+18	6.352458E+16	5.497800E+15	3.831800E+08
3.000	710.00000	279.200	1.841852E+19	7.781146E+11	3.851870E+18	1.102851E+17	6.081900E+15	4.238900E+08
2.000	802.00000	285.200	2.036745E+19	7.528372E+11	4.259420E+18	1.972784E+17	6.725400E+15	4.687400E+08
1.000	902.00000	289.700	2.255121E+19	7.531609E+11	4.717130E+18	3.110146E+17	7.448100E+15	5.191100E+08
0.000	1013.00000	294.200	2.493898E+19	7.530432E+11	5.216640E+18	4.682496E+17	8.236800E+15	5.740800E+08

Table C.4. *AFGL atmospheric constituent profiles, midlatitude winter (AFGL-TR-86-0110).*

z (km)	p (mb)	T (K)	air (cm^{-3})	O_3 (cm^{-3})	O_2 (cm^{-3})	H_2O (cm^{-3})	CO_2 (cm^{-3})	NO_2 (cm^{-3})
100.000	0.00041	218.600	1.349841E+13	5.404000E+06	2.161600E+12	5.404000E+06	2.634450E+09	2.296700E+03
95.000	0.00088	208.300	3.049454E+13	2.441600E+07	5.493600E+12	1.648080E+07	8.240400E+09	5.371520E+03
90.000	0.00198	199.500	7.188431E+13	5.755200E+07	1.366860E+13	6.114900E+07	2.230140E+10	1.316502E+04
85.000	0.00456	199.800	1.653032E+14	9.097000E+07	3.308000E+13	2.199820E+08	5.292800E+10	3.175680E+04
80.000	0.01030	210.100	3.550774E+14	8.171900E+07	7.425770E+13	7.106000E+08	1.165384E+11	7.177059E+04
75.000	0.02220	220.400	7.295469E+14	1.825250E+08	1.525909E+14	1.971270E+09	2.409330E+11	1.569715E+05
70.000	0.04700	230.700	1.475577E+15	4.726400E+08	3.086930E+14	4.874100E+09	4.874100E+11	3.411870E+05
65.000	0.09500	240.900	2.856266E+15	1.571900E+09	5.973220E+14	1.143200E+10	9.431400E+11	7.230740E+05
60.000	0.18800	250.800	5.429278E+15	5.433000E+09	1.135497E+15	2.444850E+10	1.792890E+12	1.548405E+06
55.000	0.36200	260.600	1.006111E+16	1.711900E+10	2.104630E+15	4.883950E+10	3.323100E+12	3.413730E+06
50.000	0.68300	265.700	1.861834E+16	5.123250E+10	3.893670E+15	9.221850E+10	6.147900E+12	8.253090E+06
47.500	0.94000	265.100	2.568206E+16	9.509000E+10	5.371300E+15	1.285000E+11	8.481000E+12	1.711620E+07
45.000	1.29000	258.500	3.614440E+16	1.663820E+11	7.559530E+15	1.808500E+11	1.193610E+13	4.159550E+07
42.500	1.80000	250.800	5.198245E+16	3.069180E+11	1.087218E+16	2.601000E+11	1.716660E+13	1.128834E+08
40.000	2.53000	243.200	7.534749E+16	5.202600E+11	1.575860E+16	3.732300E+11	2.488200E+13	3.038620E+08
37.500	3.60000	235.500	1.107193E+17	7.977600E+11	2.315720E+16	5.429200E+11	3.656400E+13	6.936080E+08
35.000	5.18000	227.900	1.646256E+17	1.169370E+12	3.442230E+16	7.987950E+11	5.435100E+13	1.199016E+09
32.500	7.56000	220.400	2.484403E+17	1.690480E+12	5.195740E+16	1.193280E+12	8.203800E+13	1.792406E+09
30.000	11.10000	217.400	3.698071E+17	2.257610E+12	7.735090E+16	1.757975E+12	1.221330E+14	2.279816E+09
27.500	16.46000	215.500	5.532155E+17	3.100160E+12	1.157024E+17	2.601920E+12	1.826880E+14	2.662816E+09
25.000	24.40000	215.200	8.212197E+17	4.191180E+12	1.717562E+17	3.821370E+12	2.711940E+14	3.073532E+09
24.000	28.60000	215.200	9.625772E+17	4.527510E+12	2.013297E+17	4.431180E+12	3.178890E+14	2.947698E+09

Table C.4. *(cont.)*

z (km)	p (mb)	T (K)	air (cm^{-3})	O_3 (cm^{-3})	O_2 (cm^{-3})	H_2O (cm^{-3})	CO_2 (cm^{-3})	NO_2 (cm^{-3})
23.000	33.40000	215.200	1.124129E+18	4.837501E+12	2.351250E+17	5.118751E+12	3.712500E+14	2.902500E+09
22.000	39.10000	215.200	1.315971E+18	5.136300E+12	2.752530E+17	5.966010E+12	4.346100E+14	2.844720E+09
21.000	45.80000	215.200	1.541470E+18	5.400500E+12	3.224870E+17	6.943500E+12	5.091900E+14	2.715680E+09
20.000	53.70000	215.200	1.807356E+18	5.246100E+12	3.780810E+17	8.140500E+12	5.969700E+14	2.514510E+09
19.000	62.80000	215.200	2.113631E+18	4.864500E+12	4.420350E+17	9.517500E+12	6.979500E+14	2.241900E+09
18.000	73.60000	215.700	2.471380E+18	4.451400E+12	5.168570E+17	1.112850E+13	8.160900E+14	1.906683E+09
17.000	86.10000	216.200	2.884426E+18	4.040400E+12	6.031740E+17	1.298700E+13	9.523800E+14	1.497834E+09
16.000	100.70000	216.700	3.365754E+18	3.704800E+12	7.039120E+17	1.549280E+13	1.111440E+15	1.074392E+09
15.000	117.80000	217.200	3.928234E+18	3.537900E+12	8.215790E+17	1.847570E+13	1.297230E+15	6.722010E+08
14.000	137.80000	217.700	4.584612E+18	3.670400E+12	9.588920E+17	2.202240E+13	1.514040E+15	3.431824E+08
13.000	161.10001	218.200	5.347523E+18	3.764963E+12	1.118359E+18	2.675500E+13	1.765830E+15	2.381195E+08
12.000	188.20000	218.700	6.232793E+18	3.263198E+12	1.303533E+18	3.742200E+13	2.058210E+15	1.964655E+08
11.000	219.89999	219.200	7.266018E+18	2.635010E+12	1.519639E+18	7.271000E+13	2.399430E+15	1.905002E+08
10.000	256.79999	219.700	8.465970E+18	2.007864E+12	1.770648E+18	2.507712E+14	2.795760E+15	2.016336E+08
9.000	299.29999	225.700	9.604770E+18	1.506200E+12	2.008908E+18	5.350039E+14	3.171960E+15	2.229984E+08
8.000	347.29999	231.700	1.085652E+19	1.128354E+12	2.269740E+18	1.169622E+15	3.583800E+15	2.497800E+08
7.000	401.60001	237.700	1.223704E+19	9.666475E+11	2.560250E+18	2.843225E+15	4.042500E+15	2.817500E+08
6.000	462.70001	243.700	1.375168E+19	8.031712E+11	2.875840E+18	7.021728E+15	4.540800E+15	3.164800E+08
5.000	531.29999	249.700	1.541108E+19	7.278240E+11	3.222780E+18	1.270762E+16	5.088600E+15	3.546600E+08
4.000	608.09998	255.700	1.722488E+19	6.149508E+11	3.603160E+18	2.206720E+16	5.689200E+15	3.965200E+08
3.000	693.79999	261.700	1.920182E+19	6.150400E+11	4.016980E+18	4.013136E+16	6.342600E+15	4.420600E+08
2.000	789.70001	265.200	2.156753E+19	6.148142E+11	4.510220E+18	6.016504E+16	7.121400E+15	4.963400E+08
1.000	897.29999	268.700	2.418699E+19	6.776000E+11	5.057800E+18	8.358680E+16	7.986000E+15	5.566000E+08
0.000	1018.00000	272.200	2.708766E+19	7.531159E+11	5.665990E+18	1.170068E+17	8.946300E+15	6.235300E+08

Table C.5. *AFGL atmospheric constituent profiles, subarctic summer (AFGL-TR-86-0110).*

z (km)	p (mb)	T (K)	air (cm^{-3})	O_3 (cm^{-3})	O_2 (cm^{-3})	H_2O (cm^{-3})	CO_2 (cm^{-3})	NO_2 (cm^{-3})
100.000	0.00025	190.400	9.434015E+12	3.776400E+06	1.510560E+12	3.776400E+06	1.840995E+09	1.604970E+03
95.000	0.00061	176.800	2.482574E+13	1.987200E+07	4.471200E+12	1.341360E+07	6.706800E+09	4.371840E+03
90.000	0.00161	161.600	7.215997E+13	6.498900E+07	1.371990E+13	6.137850E+07	2.238510E+10	1.321443E+04
85.000	0.00451	161.700	2.020125E+14	1.314300E+08	4.044000E+13	2.689260E+08	6.470400E+10	3.882240E+04
80.000	0.01250	170.600	5.306923E+14	9.559800E+07	1.109999E+14	1.062200E+09	1.742008E+11	1.072822E+05
75.000	0.03200	193.600	1.197172E+15	2.396000E+08	2.503820E+14	3.234600E+09	3.953400E+11	2.575700E+05
70.000	0.07100	216.600	2.374169E+15	9.504000E+08	4.965840E+14	7.840800E+09	7.840800E+11	5.488560E+05
65.000	0.14700	239.700	4.441822E+15	3.556000E+09	9.290050E+14	1.778000E+10	1.466850E+12	1.124585E+06
60.000	0.28800	262.700	7.940434E+15	9.535200E+09	1.660714E+15	3.575700E+10	2.622180E+12	2.264610E+06
55.000	0.53700	274.000	1.419501E+16	2.415700E+10	2.969890E+15	6.891850E+10	4.689300E+12	4.817190E+06
50.000	0.98700	277.200	2.578907E+16	6.452500E+10	5.394290E+15	1.277595E+11	8.517300E+12	1.143383E+07
47.500	1.34000	276.200	3.513929E+16	1.125120E+11	7.348440E+15	1.758000E+11	1.160280E+13	2.341656E+07
45.000	1.82000	273.600	4.818004E+16	2.024820E+11	1.007589E+16	2.410500E+11	1.590930E+13	5.544150E+07
42.500	2.48000	269.500	6.665071E+16	3.601800E+11	1.394030E+16	3.335000E+11	2.201100E+13	1.447390E+08
40.000	3.40000	262.100	9.395583E+16	6.581400E+11	1.965018E+16	4.701000E+11	3.102660E+13	3.789006E+08
37.500	4.72000	254.600	1.342751E+17	1.048320E+12	2.808960E+16	6.720000E+11	4.435200E+13	8.413440E+08
35.000	6.61000	247.200	1.936711E+17	1.492260E+12	4.050420E+16	9.690000E+11	6.395400E+13	1.410864E+09
32.500	9.40000	240.000	2.836798E+17	1.958910E+12	5.933510E+16	1.419500E+12	9.368700E+13	2.046919E+09
30.000	13.40000	235.100	4.128230E+17	2.354670E+12	8.633790E+16	2.065500E+12	1.363230E+14	2.544696E+09
27.500	19.23000	231.000	6.029470E+17	3.198020E+12	1.261106E+17	2.986830E+12	1.991220E+14	2.902354E+09
25.000	27.80000	228.100	8.827370E+17	3.975300E+12	1.846306E+17	4.328660E+12	2.915220E+14	3.303916E+09
24.000	32.28000	226.600	1.031776E+18	4.338600E+12	2.158970E+17	5.010049E+12	3.408900E+14	3.160980E+09

Table C.5. *(cont.)*

z (km)	p (mb)	T (K)	air (cm^{-3})	O_3 (cm^{-3})	O_2 (cm^{-3})	H_2O (cm^{-3})	CO_2 (cm^{-3})	NO_2 (cm^{-3})
23.000	37.50000	225.200	1.206076E+18	4.465900E+12	2.522630E+17	5.829810E+12	3.983100E+14	3.114060E+09
22.000	43.60000	225.200	1.402264E+18	4.629900E+12	2.932270E+17	6.734400E+12	4.629900E+14	3.030480E+09
21.000	50.70000	225.200	1.630615E+18	4.406400E+12	3.410880E+17	7.670400E+12	5.385600E+14	2.872320E+09
20.000	59.00000	225.200	1.897560E+18	3.987900E+12	3.968910E+17	8.735400E+12	6.266700E+14	2.639610E+09
19.000	68.60000	225.200	2.206315E+18	3.753600E+12	4.614720E+17	9.936000E+12	7.286400E+14	2.340480E+09
18.000	79.80000	225.200	2.566530E+18	3.338400E+12	5.367120E+17	1.104240E+13	8.474400E+14	1.979928E+09
17.000	92.80000	225.200	2.984637E+18	2.987000E+12	6.242830E+17	1.209735E+13	9.857100E+14	1.550253E+09
16.000	108.00000	225.200	3.473499E+18	2.954600E+12	7.264840E+17	1.390400E+13	1.147080E+15	1.108844E+09
15.000	126.00000	225.200	4.052416E+18	2.838500E+12	8.474950E+17	1.622000E+13	1.338150E+15	6.934050E+08
14.000	146.00000	225.200	4.695657E+18	2.819400E+12	9.820909E+17	1.879600E+13	1.550670E+15	3.514852E+08
13.000	170.00000	225.200	5.467545E+18	2.735500E+12	1.143439E+18	2.434595E+13	1.805430E+15	2.434595E+08
12.000	197.70000	225.200	6.358433E+18	2.608830E+12	1.329867E+18	3.817800E+13	2.099790E+15	2.004345E+08
11.000	230.00000	225.200	7.397267E+18	2.257610E+12	1.547018E+18	9.844661E+13	2.442660E+15	1.939324E+08
10.000	267.70001	225.200	8.609776E+18	1.628424E+12	1.800744E+18	3.653184E+14	2.843280E+15	2.050608E+08
9.000	310.79999	232.200	9.694616E+18	1.377542E+12	2.027509E+18	1.261130E+15	3.201330E+15	2.250632E+08
8.000	359.00000	239.200	1.087039E+19	9.916032E+11	2.273920E+18	4.347648E+15	3.590400E+15	2.502400E+08
7.000	413.00000	246.100	1.215487E+19	9.409409E+11	2.541440E+18	9.692736E+15	4.012800E+15	2.796800E+08
6.000	474.00000	253.100	1.356432E+19	8.907348E+11	2.836130E+18	1.804810E+16	4.478100E+15	3.121100E+08
5.000	541.00000	260.100	1.506498E+19	8.034624E+11	3.151720E+18	3.344744E+16	4.976400E+15	3.468400E+08
4.000	616.00000	265.500	1.680459E+19	7.531996E+11	3.515380E+18	5.685160E+16	5.550600E+15	3.868600E+08
3.000	700.00000	270.900	1.871548E+19	7.280351E+11	3.914570E+18	9.027860E+16	6.180900E+15	4.307900E+08
2.000	792.90002	276.300	2.078497E+19	7.028321E+11	4.347200E+18	1.404000E+17	6.864000E+15	4.784000E+08
1.000	896.00000	281.700	2.303737E+19	6.776700E+11	4.817450E+18	2.005581E+17	7.606500E+15	5.301500E+08
0.000	1010.00000	287.200	2.547116E+19	6.148188E+11	5.327410E+18	3.043506E+17	8.411700E+15	5.862700E+08

Table C.6. *AFGL atmospheric constituent profiles, subarctic winter (AFGL-TR-86-0110).*

z (km)	p (mb)	T (K)	air (cm^{-3})	O_3 (cm^{-3})	O_2 (cm^{-3})	H_2O (cm^{-3})	CO_2 (cm^{-3})	NO_2 (cm^{-3})
100.000	0.00042	218.500	1.402170E+13	5.612000E+06	2.244800E+12	5.612000E+06	2.735850E+09	2.385100E+03
95.000	0.00091	211.000	3.113413E+13	2.492800E+07	5.608800E+12	1.682640E+07	8.413200E+09	5.484160E+03
90.000	0.00202	202.300	7.239309E+13	5.795200E+07	1.376360E+13	6.157400E+07	2.245640E+10	1.325652E+04
85.000	0.00450	213.100	1.529470E+14	1.148250E+08	3.062000E+13	2.036230E+08	4.899200E+10	2.939520E+04
80.000	0.00966	223.900	3.124891E+14	4.065100E+07	6.535430E+13	6.254000E+08	1.025656E+11	6.316540E+04
75.000	0.02000	234.700	6.172040E+14	2.038080E+08	1.290784E+14	1.667520E+09	2.038080E+11	1.327840E+05
70.000	0.04000	245.400	1.180585E+15	5.905000E+08	2.468290E+14	3.897300E+09	3.897300E+11	2.728110E+05
65.000	0.07900	248.400	2.303495E+15	1.498250E+09	4.817450E+14	9.220000E+09	7.606500E+11	5.831650E+05
60.000	0.15500	250.900	4.474483E+15	4.254100E+09	9.359020E+14	2.015100E+10	1.477740E+12	1.276230E+06
55.000	0.29900	259.100	8.358255E+15	1.338240E+10	1.748076E+15	4.056540E+10	2.760120E+12	2.835396E+06
50.000	0.57190	259.300	1.597458E+16	4.157400E+10	3.341910E+15	7.915050E+10	5.276700E+12	7.083570E+06
47.500	0.79000	253.200	2.259827E+16	6.783000E+10	4.725491E+15	1.130500E+11	7.461300E+12	1.505826E+07
45.000	1.11300	247.000	3.263698E+16	1.339060E+11	6.825940E+15	1.633000E+11	1.077780E+13	3.755900E+07
42.500	1.57000	240.800	4.722316E+16	2.410260E+11	9.877339E+15	2.363000E+11	1.559580E+13	1.025542E+08
40.000	2.24300	234.700	6.921943E+16	4.086930E+11	1.447743E+16	3.463500E+11	2.285910E+13	2.791581E+08
37.500	3.23000	228.500	1.023831E+17	6.406250E+11	2.142250E+16	5.125000E+11	3.382500E+13	6.416500E+08
35.000	4.70100	222.300	1.531661E+17	9.504599E+11	3.203970E+16	7.664999E+11	5.058900E+13	1.116024E+09
32.500	6.91000	218.500	2.290543E+17	1.352280E+12	4.790280E+16	1.146000E+12	7.563600E+13	1.652532E+09
30.000	10.20000	216.000	3.420253E+17	1.848420E+12	7.154070E+16	1.711500E+12	1.129590E+14	2.108568E+09
27.500	15.13000	213.600	5.130379E+17	2.515660E+12	1.073006E+17	2.567000E+12	1.694220E+14	2.469454E+09
25.000	22.56000	211.200	7.736723E+17	3.638740E+12	1.618078E+17	3.871000E+12	2.554860E+14	2.895508E+09
24.000	26.49000	211.800	9.058741E+17	4.169900E+12	1.894585E+17	4.532500E+12	2.991450E+14	2.773890E+09
23.000	31.09000	212.400	1.060176E+18	4.774500E+12	2.217490E+17	5.251950E+12	3.501300E+14	2.737380E+09
22.000	36.47000	213.000	1.240132E+18	5.212200E+12	2.593690E+17	6.080900E+12	4.095300E+14	2.680560E+09

Table C.6. *(cont.)*

z (km)	p (mb)	T (K)	air (cm^{-3})	O_3 (cm^{-3})	O_2 (cm^{-3})	H_2O (cm^{-3})	CO_2 (cm^{-3})	NO_2 (cm^{-3})
21.000	42.77000	213.600	1.450273E+18	5.804000E+12	3.032590E+17	7.037350E+12	4.788300E+14	2.553760E+09
20.000	50.14000	214.200	1.695418E+18	6.278900E+12	3.546730E+17	8.145601E+12	5.600100E+14	2.358830E+09
19.000	58.75000	214.800	1.981004E+18	6.144200E+12	4.142380E+17	9.414500E+12	6.540600E+14	2.100920E+09
18.000	68.82000	215.400	2.314093E+18	5.674200E+12	4.840440E+17	1.088520E+13	7.642800E+14	1.785636E+09
17.000	80.58000	216.000	2.702000E+18	5.137600E+12	5.651360E+17	1.257360E+13	8.923200E+14	1.403376E+09
16.000	94.31000	216.600	3.153632E+18	4.734000E+12	6.596040E+17	1.451760E+13	1.041480E+15	1.006764E+09
15.000	110.30000	217.200	3.678134E+18	4.417200E+12	7.693290E+17	1.674855E+13	1.214730E+15	6.294510E+08
14.000	129.10001	217.200	4.305051E+18	3.877200E+12	9.003720E+17	1.938600E+13	1.421640E+15	3.222384E+08
13.000	151.00000	217.200	5.035342E+18	3.275350E+12	1.053151E+18	2.242355E+13	1.662870E+15	2.242355E+08
12.000	176.60001	217.200	5.889016E+18	2.357200E+12	1.231637E+18	3.535800E+13	1.944690E+15	1.856295E+08
11.000	206.70000	217.200	6.892749E+18	2.414300E+12	1.441682E+18	6.898000E+13	2.276340E+15	1.807276E+08
10.000	241.80000	217.200	8.063216E+18	2.420700E+12	1.686421E+18	1.613800E+14	2.662770E+15	1.920422E+08
9.000	282.89999	217.200	9.433762E+18	1.982400E+12	1.972960E+18	2.809344E+14	3.115200E+15	2.190080E+08
8.000	330.79999	220.600	1.086105E+19	1.130480E+12	2.271830E+18	3.678408E+14	3.587100E+15	2.500100E+08
7.000	385.29999	227.300	1.227754E+19	8.912709E+11	2.568610E+18	1.806630E+15	4.055700E+15	2.826700E+08
6.000	446.70001	234.100	1.382058E+19	6.148818E+11	2.890470E+18	3.276327E+15	4.563900E+15	3.180900E+08
5.000	515.79999	240.900	1.550802E+19	5.899152E+11	3.243680E+18	6.687568E+15	5.121600E+15	3.569600E+08
4.000	593.20001	247.700	1.734551E+19	5.647208E+11	3.628240E+18	1.371093E+16	5.728800E+15	3.992800E+08
3.000	679.79999	252.700	1.948443E+19	5.395650E+11	4.075500E+18	2.273700E+16	6.435000E+15	4.485000E+08
2.000	777.50000	255.900	2.200604E+19	5.143872E+11	4.602180E+18	3.142254E+16	7.266600E+15	5.064600E+08
1.000	887.79999	259.100	2.481759E+19	5.146847E+11	5.191560E+18	4.011660E+16	8.197200E+15	5.713200E+08
0.000	1013.00000	257.200	2.852662E+19	5.144710E+11	5.966950E+18	4.011275E+16	9.421500E+15	6.566500E+08

Appendix D

Ocean Optics Nomenclature

Radiative transfer in the ocean occupies a central role in physical oceanography: Light provides the energy that powers primary productivity in the ocean; light diffusely reflected by the ocean provides the signal for the remote sensing of subsurface constituent concentrations; light absorbed by the water heats the ocean's surface layer; light absorbed by chemical species, particularly dissolved organics, provides energy for their breakup; and the rate at which light extinction varies with depth provides an estimate of planktonic activity.

From the above, it is clear that ocean optics is concerned with the distribution of *visible* light with angle, depth, and wavelength. Infrared radiation is important only for surface cooling, since the mean free path is very small (the opacity of water in the infrared is very high). Unfortunately the absorption and scattering properties of water are usually poorly known, because of the great variety of suspended particulates that occur in nature. In addition, the measurement capabilities are not yet sufficiently accurate to test the predictions of radiative transfer theory. Consequently, mathematically accurate solutions to the radiative transfer equation are not usually needed, although recently Monte Carlo techniques have become popular in the ocean optics literature. Approximate solutions are usually adequate, particularly when comparing with measurements made at great depth, where asymptotic solutions become valid.

Ocean optics developed only in the past fifty years, although some pioneering work was done by Petersson and colleagues in the 1930s. Although it was influenced by early workers in atmospheric radiation, such as Chandrasekhar and van de Hulst, its close connections with experiment and emphasis on empirical aspects has distinguished it from the atmospheric radiation community, which in the 1960s and 1970s was, in general, more detached from experiment. This detachment resulted from the fact that atmospheric absorption and scattering could be successfully reproduced by theoretical modeling to a much greater extent than its aqueous counterpart. Careers could be built by becoming an expert in atmospheric radiative transfer calculations. In contrast, ocean optics remained largely an empirical science, evolving separately with its own nomenclature and jargon. One important difference in approach that emerged was that of *apparent* and *inherent* properties of ocean water. Due to the behavior of the marine radiation field varying approximately exponentially with depth, the decay could be described by Beer's Law (our *Extinction law*) coefficients of absorption and scattering. Generally speaking, these coefficients included the effects of multiple scattering, and therefore they were apparent properties, rather than inherent properties of the medium (§9.4.7). However, they are easily measured and may be interpreted in terms of the actual absorption and scattering properties, for example in certain asymptotic limits (see §9.4.8). In these cases, measurements may be "inverted" to yield physical properties of the liquid medium.

Table D.1. *Ocean optics nomenclature.*

Ocean optics nomenclature	Definition and remarks	Nomenclature (this book)
n	index of refraction	m
W, Q	quantity of energy	E
Φ	radiant flux ($\Phi \equiv dW/dt$)	dE/dt
I	intensity, the radiant flux emitted by a point source, $I \equiv d\Phi/d\omega$ (ω = solid angle)	not used
L	radiance, intensity $L = d^2\Phi/dA\cos\theta d\omega$	I
E_d, E_u	downward, upward irradiance	F^-, F^+
$\vec{E}$	vector irradiance, (+ downward)	$-\mathbf{F}$
E_0	scalar irradiance	$4\pi\bar{I}$
E_{0d}, E_{0u}	downward, upward scalar irradiance	$2\pi I^-, 2\pi I^+$
E_s	spherical irradiance	$\pi\bar{I}$
$\bar{\mu}, \bar{\mu}_d, \bar{\mu}_u$	average cosine in a hemisphere	$\bar{\mu}, \bar{\mu}^-, \bar{\mu}^+$
R	irradiance ratio, $R \equiv E_u/E_d$	F^+/F^-
A	flux absorptance	$\alpha_F, \alpha(-2\pi)$
T	flux transmittance	$\mathcal{T}$
a, b	absorption, scattering coefficient	α, σ
c	extinction coefficient, $c \equiv a + b$	k
$\beta(\theta' \to \theta)$	volume scattering coefficient	$\sigma p(\theta', \theta)$
a^*, b^*, c^*	specific absorption, scattering, extinction $a_i^* \equiv a_i/n_i$ (n_i is particle concentration)	α_n, σ_n, k_n
ω_0, ϖ	single-scattering albedo, (b/c)	a
b_b, b_f	back-, forward-scattering coefficient	$b, 1 - b$
$\bar{a}$	coalbedo, (a/c)	$1 - a$

In recent decades with the advent of coupled ocean–atmosphere models, and with increasing interest in atmospheric radiation effects on the biosphere, there has grown a need for the two areas to combine forces. This effort is complicated by the usual interdisciplinary obstacles. Since it is clear that the two fields really describe the same basic physical phenomena, we believe that more communication is vital to an understanding of many problems dealing with this important interface. In this spirit, we present in this appendix a "translation table" of terms, which hopefully will make it easier for students and specialists alike to read the research papers and books in both fields. Although ocean scientists have not yet reached complete consensus on their own system, they have nevertheless reached a degree of standardization that is at least as good, if not superior to, that in the field of atmospheric radiation. In particular, the community has adopted most of the 1985 recommendations of the IAPSO Working Group on Symbols, Units and Nomenclature in Physical Oceanography.[1] In Table D.1 we compare the material from Table II.4 of the IAPSO report, prepared by A. Morel, with the quantities defined in this book. We include only those quantities that differ from our notation and only those basic quantities that are of central importance to a description of the radiation field and the optical properties of the medium. As usual we consider only those aspects dealing with unpolarized radiation.

[1] UNESCO Technical Papers in Marine Science, IAPSO Scientific Publication No. 32, *The International System of Units (SI) in Oceanography*, Report of the IAPSO Working Group on Symbols, Units and Nomenclature in Physical Oceanography, UNESCO, 1985.

Appendix E

Reflectance and Transmittance at an Interface

When a light wave encounters the interface between two media with different indices of refraction, the ray splits into two components, a transmitted (refracted) ray and a reflected ray. We will assume that in both media there is neither scattering nor absorption. With reference to Fig. E.1, it is shown in Problem 1 that the angle of reflection is given by $\theta_\mathrm{r} = \pi - \theta_i$ and that the angle of refraction is given by *Snell's law*:

$$\sin\theta_i / \sin\theta_\mathrm{t} = c_i / c_\mathrm{t} = m_\mathrm{t} / m_i. \tag{E.1}$$

Here c_i and c_t are the propagation speeds in the first and second media, respectively, and m_i and m_r denote the respective indices of refraction. No loss of generality results from assuming $m_\mathrm{t} > m_i$, so that $\sin\theta_\mathrm{t} = (m_i/m_\mathrm{t})\sin\theta_i < \sin\theta_i$. This means there is a real angle θ_t for every angle of incidence. (In the case $m_\mathrm{t} < m_i$ total reflection can occur for a range of incidence angles.) We first derive the expressions for the transmitted and reflected electric and magnetic field components in terms of the respective incident quantities. This is achieved by relating the two fields through the plane-wave relationships, and imposing continuity of the tangential components of the fields across the interface. The key observation is that the time-dependent part of the plane wave is the same for all components. This part is denoted by $\phi \equiv \omega(t - \frac{\mathbf{r}\cdot\hat{\Omega}}{c})$, where ω is the angular frequency, $\mathbf{r}$ the position vector of a point in space, $\hat{\Omega}$ the propagation vector of the wave, and c the propagation speed in the medium. Carrying out the dot product, we write for the incident wave

$$\phi_i = \omega\left(t - \frac{\mathbf{r}\cdot\hat{\Omega}^i}{c_i}\right) = \omega\left(t - \frac{x\sin\theta_i + z\cos\theta_i}{c_i}\right). \tag{E.2}$$

We now consider the amplitudes of the electric field components. We write these

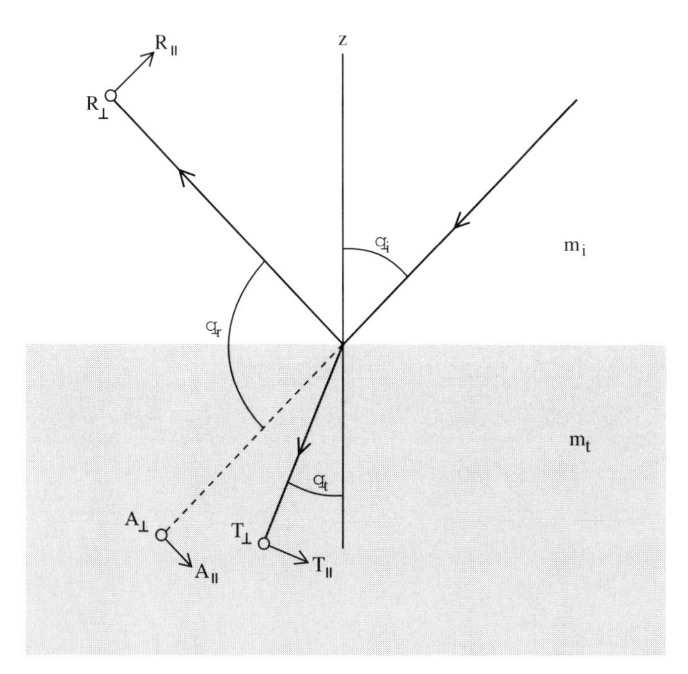

Figure E.1 Vectors are in the plane of incidence. Those components marked with "⊥" indicate the component of the electric or magnetic field perpendicular to the plane of incidence, and those marked with "∥" denote the component parallel to the plane of incidence.

components as $A_\parallel$ and $A_\perp$, parallel to and perpendicular to the plane of incidence. From the figure, it can be verified the rectangular components of the electric field are given by

$$E_x^i = -A_\parallel \cos\theta_i e^{-i\phi_i}, \qquad E_y^i = A_\perp e^{-i\phi_i}, \qquad E_z^i = A_\parallel \sin\theta_i e^{-i\phi_i}. \tag{E.3}$$

Here we have ignored the constant part of the phase angle. The components of the magnetic field vector are obtained from the plane-wave relationship, $\mathbf{H} = \sqrt{\epsilon/\mu}\hat{\Omega} \times \mathbf{E} = (m/\mu c_o)\hat{\Omega} \times \mathbf{E}$. From the definition of the cross product, we have

$$\mu c_o \mathbf{H}^i = m \begin{pmatrix} \hat{i} & \hat{j} & \hat{k} \\ \hat{\Omega}_x^i & \hat{\Omega}_y^i & \hat{\Omega}_z^i \\ E_x^i & E_y^i & E_z^i \end{pmatrix} = m e^{-i\phi_i} \begin{pmatrix} \hat{i} & \hat{j} & \hat{k} \\ \sin\theta_i & 0 & \cos\theta_i \\ -A_\parallel \cos\theta_i & A_\perp & A_\parallel \sin\theta_i \end{pmatrix}. \tag{E.4}$$

Using the rules of matrix expansion, we find

$$\mu c_o H_x^i = -A_\perp m_i \cos\theta_i e^{-i\phi_i}, \qquad \mu c_o H_y^i = -A_\parallel m_i e^{-i\phi_i},$$
$$\mu c_o H_z^i = A_\perp m_i \sin\theta_i e^{-i\phi_i}. \tag{E.5}$$

If T and R represent the complex amplitudes of the transmitted and reflected waves, respectively, we can show in an analogous fashion that the corresponding components of the transmitted field vectors are

$$E_x^{\rm t} = -T_\parallel \cos\theta_{\rm t} e^{-i\phi_{\rm t}}, \qquad E_y^{\rm t} = T_\perp e^{-i\phi_{\rm t}}, \qquad E_z^{\rm t} = T_\parallel \sin\theta_{\rm t} e^{-i\phi_{\rm t}}, \tag{E.6}$$

$$\mu c_{\rm o} H_x^{\rm t} = -T_\perp \cos\theta_{\rm t} m_{\rm t} e^{-i\phi_{\rm t}}, \qquad \mu c_{\rm o} H_y^{\rm t} = -T_\parallel m_{\rm t} e^{-i\phi_{\rm t}},$$
$$\mu c_{\rm o} H_z^{\rm t} = T_\perp \sin\theta_{\rm t} m_{\rm t} e^{-i\phi_{\rm t}}, \tag{E.7}$$

where

$$\phi_{\rm t} = \omega\left(t - \frac{\mathbf{r}\cdot\hat{\Omega}^{\rm t}}{c_{\rm t}}\right) = \omega\left(t - \frac{x\sin\theta_{\rm t} + z\cos\theta_{\rm t}}{c_{\rm t}}\right). \tag{E.8}$$

The procedure is identical for the reflected component:

$$E_x^{\rm r} = -T_\parallel \cos\theta_{\rm t} e^{-i\phi_{\rm r}}, \qquad E_y^{\rm r} = T_\perp e^{-i\phi_{\rm r}}, \qquad E_z^{\rm r} = T_\parallel \sin\theta_{\rm r} e^{-i\phi_{\rm r}}, \tag{E.9}$$

$$\mu c_{\rm o} H_x^{\rm r} = -T_\perp \cos\theta_{\rm r} m_i e^{-i\phi_{\rm r}}, \qquad \mu c_{\rm o} H_y^{\rm r} = -T_\parallel m_i e^{-i\phi_{\rm r}},$$
$$\mu c_{\rm o} H_z^{\rm r} = T_\perp \sin\theta_{\rm r} m_{\rm r} e^{-i\phi_{\rm r}}, \tag{E.10}$$

where

$$\phi_{\rm r} = \omega\left(t - \frac{\mathbf{r}\cdot\hat{\Omega}^{r}}{c_i}\right) = \omega\left(t - \frac{x\sin\theta_{\rm r} + z\cos\theta_{\rm r}}{c_i}\right). \tag{E.11}$$

The boundary conditions for the electric and magnetic fields require that the *tangential* (x- and y-components) of **E** and **H** be continuous across the interface, that is,

$$E_x^i + E_x^{\rm r} = E_x^{\rm t}, \qquad E_y^i + E_y^{\rm r} = E_y^{\rm t}, \tag{E.12}$$

$$H_x^i + H_x^{\rm r} = H_x^{\rm t}, \qquad H_y^i + H_y^{\rm r} = H_y^{\rm t}. \tag{E.13}$$

By substituting the expressions above into these two equations, one can verify that the following four relationships are obtained:

$$\cos\theta_i \left(A_\parallel - R_\parallel\right) = \cos\theta_{\rm t} T_\parallel, \tag{E.14}$$

$$A_\perp + R_\perp = T_\perp, \tag{E.15}$$

$$m_i \cos\theta_i \left(A_\perp - R_\perp\right) = m_{\rm t} \cos\theta_{\rm t} T_\perp, \tag{E.16}$$

$$m_i \left(A_\parallel + R_\parallel\right) = m_{\rm t} T_\parallel. \tag{E.17}$$

Here we have ignored the small change in the magnetic susceptibility μ between the two media.[1] We note that there are two pairs of equations dealing *separately* with the parallel and perpendicular components. This shows that the two waves are *independent*. It can be verified that, by solving for the transmitted and reflected components, the results are

$$R_\perp = \frac{m_i\cos\theta_i - m_{\rm t}\cos\theta_{\rm t}}{m_i\cos\theta_i + m_{\rm t}\cos\theta_{\rm t}} A_\perp, \qquad R_\parallel = \frac{m_{\rm t}\cos\theta_i - m_i\cos\theta_{\rm t}}{m_i\cos\theta_{\rm t} + m_{\rm t}\cos\theta_i} A_\parallel,$$

$$T_\perp = \frac{2m_i\cos\theta_i}{m_i\cos\theta_i + m_{\rm t}\cos\theta_{\rm t}} A_\perp, \qquad T_\parallel = \frac{2m_i\cos\theta_i}{m_i\cos\theta_{\rm t} + m_{\rm t}\cos\theta_i} A_\parallel.$$

[1] We assume that the entire change in index of refraction is due to the differences between the permittivity ϵ.

The above four equations are called the *Fresnel formulae*, and were derived by Fresnel in 1823 on the basis of his elastic theory of light (but in a somewhat less general form). We found in Appendix H that the time-averaged *Poynting vector* for the plane wave describes the energy flow normal to the beam,

$$\langle S \rangle = \frac{m}{2\mu c_o}\left\{\frac{\epsilon}{2}\left[\mathcal{E}_\parallel \cdot \mathcal{E}_\parallel^* + \mathcal{E}_\perp \cdot \mathcal{E}_\perp^*\right]\right\}\delta(\hat{\Omega} - \hat{\Omega}_0) \equiv \frac{m}{2\mu c_o} A^2 \delta(\hat{\Omega} - \hat{\Omega}_0),$$

where the second equation defines A^2. When the wave is incident on an interface at the angle θ_i, the incident energy per unit area per unit time is

$$F_i^- = \langle S \rangle_i \cos\theta_i = \frac{m_i}{2\mu c_o}|A|^2 \cos\theta_i.$$

In a similar way, we write the fluxes of the reflected and transmitted waves as

$$F_\mathrm{r}^+ = \frac{m_i}{2\mu c_o}|R|^2 \cos\theta_\mathrm{r}, \qquad F_\mathrm{t}^- = \frac{m_\mathrm{t}}{2\mu c_o}|T|^2 \cos\theta_\mathrm{t}.$$

The ratios of the fluxes for the flux albedo and flux transmittance for an incoming collimated plane wave are therefore

$$\rho_\mathrm{F} \equiv F_\mathrm{r}^+/F_i^- = \frac{|R|^2}{|A|^2}, \qquad \mathcal{T}_\mathrm{F} \equiv F_\mathrm{t}^-/F_i^- = \frac{m_\mathrm{t}\cos\theta_\mathrm{t}|T|^2}{m_i \cos\theta_i |A|^2}.$$

Now let us assume that the incident light is unpolarized. Then $|A|^2 = A_\parallel^2 + A_\perp^2$, and $A_\parallel^2 = A_\perp^2$. The flux albedo and transmittance are thus

$$\rho_\mathrm{F} = \frac{1}{2}\left[\left(\frac{\mu_i - m_\mathrm{r}\mu_\mathrm{t}}{\mu_i + m_\mathrm{r}\mu_\mathrm{t}}\right)^2 + \left(\frac{\mu_\mathrm{t} - m_\mathrm{r}\mu_i}{\mu_\mathrm{t} + m_\mathrm{r}\mu_i}\right)^2\right], \tag{E.18}$$

$$\mathcal{T} = 2m_\mathrm{t}\mu_i\mu_\mathrm{t}\left[\left(\frac{1}{\mu_i + m_\mathrm{r}\mu_\mathrm{t}}\right)^2 + \left(\frac{1}{\mu_\mathrm{t} + m_\mathrm{r}\mu_i}\right)^2\right]. \tag{E.19}$$

Here we used Snell's law (Eq. E.1 and denoted $\mu_i = \cos\theta_i$ and $\mu_\mathrm{t} = \cos\theta_\mathrm{t}$). We recall that the bidirectional reflectance function (BRDF) $\rho(-\hat{\Omega}', \hat{\Omega})$ (see §5.2.4.) was defined as the sum of a specular reflectance ρ_s and a diffuse reflectance ρ_d:

$$\rho(-\hat{\Omega}', \hat{\Omega}) \equiv \rho_\mathrm{d}(-\hat{\Omega}', \hat{\Omega}) + \rho_\mathrm{s}\delta(\cos\theta_i + \cos\theta_\mathrm{r})\delta(\pi + \phi_i - \phi_\mathrm{r}). \tag{E.20}$$

The Dirac δ-functions contribute only when the reflection angle is equal to the incident angle, and the reflected ray is in the plane of incidence. In the same way, we may define the bidirectional transmittance function (see §5.2.9) as

$$\mathcal{T}(-\hat{\Omega}', \hat{\Omega}) \equiv \mathcal{T}_\mathrm{d} + \mathcal{T}_\mathrm{s}\delta(m_i \sin\theta_i - m_\mathrm{t}\sin\theta_\mathrm{t})\delta(\phi_i - \phi_\mathrm{t}), \tag{E.21}$$

where $\mathcal{T}_\mathrm{s}$ is the specular transmittance. We may now define these quantities in terms of the incident, reflected, and transmitted intensities. We consider a *cone* of incident light waves instead of a collimated beam. I_i, I_r and I_t are the incident, reflected and transmitted intensities, respectively. A cone of incident light within an infinitesimal solid angle $d\omega_i$ will shrink or expand (depending the value of the relative index of refraction of the two media) to the solid angle $d\omega_t$ after passing the interface, and their relationship can be derived by making use of Snell's law,

$$\frac{d\omega_t}{d\omega_i} = \frac{2\pi \sin\theta_t d\theta_t}{2\pi \sin\theta_i d\theta_i} = \frac{\mu_i}{m_r^2 \mu_t}. \tag{E.22}$$

Applying the definition of ρ_s and $\mathcal{T}_s$ for the irradiance and making use of Eq. E.22, we find

$$\mathcal{T}_s = \frac{I_t \cos\theta_t d\omega_t}{I_i \cos\theta_i d\omega_i} = \left(\frac{I_t}{m_t^2}\right) \Big/ \left(\frac{I_i}{m_i^2}\right), \tag{E.23}$$

$$\rho_s = \frac{I_r \cos\theta_i d\omega_i}{I_i \cos\theta_i d\omega_i} = \left(\frac{I_r}{m_i^2}\right) \Big/ \left(\frac{I_i}{m_i^2}\right). \tag{E.24}$$

Equations E.23 and E.24 show that if we define the transmittance and reflectance with respect to the invariant intensity, I/m^2, instead of the actual intensity, then the expressions for the reflectance and transmittance will have the same forms as for the parallel light beam. Such a definition has the additional advantage of satisfying reciprocity for the lightwave propagating in the opposite direction. If we denote the reflectance and transmittance specified by Eqs. E.18 and E.19 by $\rho_s(-\mu_i, m_t/m_i)$ and $\mathcal{T}_s(-\mu_i, m_t/m_i)$ for downwelling incidence, and by $\rho_s(\mu_t, m_i/m_t)$ and $\mathcal{T}_s(\mu_t, m_i/m_t)$ for upwelling incidence, then the reciprocity relationship can be expressed as

$$\rho_s(\mu_t, m_i/m_t) = \rho_s(-\mu_i, m_t/m_i), \qquad \mathcal{T}_s(\mu_t, m_i/m_t) = \mathcal{T}_s(-\mu_i, m_t/m_i). \tag{E.25}$$

Finally, it is easy to show that $\rho_s + \mathcal{T}_s = 1$, which of course follows from our assumption of a medium that neither scatters nor absorbs ($\rho_d = 0$ and $\mathcal{T}_d = 0$).

Problem

E.1

(a) Show that if we require the time dependence of all three waves to be the same, that is, $\phi_i(t) = \phi_t(t) = \phi_r(t)$, we obtain

$$\frac{\hat{\Omega}_x^i}{c_i} = \frac{\hat{\Omega}_x^r}{c_i} = \frac{\hat{\Omega}_x^t}{c_t}, \qquad \frac{\hat{\Omega}_y^i}{c_i} = \frac{\hat{\Omega}_y^r}{c_i} = \frac{\hat{\Omega}_y^t}{c_t}.$$

(b) Argue that the above equations prove that $\hat{\Omega}^t$ and $\hat{\Omega}^r$ lie in the plane of incidence.

(c) Using $\hat{\Omega}_z^i = \cos\theta_i > 0$, $\hat{\Omega}_z^r = \cos\theta_r < 0$, and $\hat{\Omega}_z^t = \cos\theta_t > 0$, show that these equations yield

$$\frac{\sin\theta_i}{c_i} = \frac{\sin\theta_r}{c_i} = \frac{\sin\theta_t}{c_t}$$

and thus argue that the above equations prove both the law of reflection and *Snell's Law*, Eq. E.1.

Index